D1618451

The Handbook of Homogeneous Hydrogenation

Edited by
Johannes G. de Vries
and Cornelis J. Elsevier

1807–2007 Knowledge for Generations

Each generation has its unique needs and aspirations. When Charles Wiley first opened his small printing shop in lower Manhattan in 1807, it was a generation of boundless potential searching for an identity. And we were there, helping to define a new American literary tradition. Over half a century later, in the midst of the Second Industrial Revolution, it was a generation focused on building the future. Once again, we were there, supplying the critical scientific, technical, and engineering knowledge that helped frame the world. Throughout the 20th Century, and into the new millennium, nations began to reach out beyond their own borders and a new international community was born. Wiley was there, expanding its operations around the world to enable a global exchange of ideas, opinions, and know-how.

For 200 years, Wiley has been an integral part of each generation's journey, enabling the flow of information and understanding necessary to meet their needs and fulfill their aspirations. Today, bold new technologies are changing the way we live and learn. Wiley will be there, providing you the must-have knowledge you need to imagine new worlds, new possibilities, and new opportunities.

Generations come and go, but you can always count on Wiley to provide you the knowledge you need, when and where you need it!

William J. Pesce
President and Chief Executive Officer

Peter Booth Wiley
Chairman of the Board

The Handbook of Homogeneous Hydrogenation

Edited by
Johannes G. de Vries and Cornelis J. Elsevier

Volume 2

WILEY-VCH Verlag GmbH & Co. KGaA

The Editors

Prof. Dr. Johannes G. De Vries
DSM Pharmaceutical Products
Advanced Synthesis, Catalysis, and Development
P.O. Box 18
6160 MD Geleen
The Netherlands

Prof. Dr. Cornelis J. Elsevier
Universiteit van Amsterdam
HIMS
Nieuwe Achtergracht 166
1018 WV Amsterdam
The Netherlands

All books published by Wiley-VCH are carefully produced. Nevertheless, authors, editors, and publisher do not warrant the information contained in these books, including this book, to be free of errors. Readers are advised to keep in mind that statements, data, illustrations, procedural details or other items may inadvertently be inaccurate.

Library of Congress Card No.: applied for

British Library Cataloguing-in-Publication Data
A catalogue record for this book is available from the British Library.

Bibliographic information published by the Deutsche Nationalbibliothek
The Deutsche Nationalbibliothek lists this publication in the Deutsche Nationalbibliografie; detailed bibliographic data are available in the Internet at http://dnb.d-nb.de.

© 2007 WILEY-VCH Verlag GmbH & Co. KGaA, Weinheim, Germany

All rights reserved (including those of translation into other languages). No part of this book may be reproduced in any form – by photoprinting, microfilm, or any other means – nor transmitted or translated into a machine language without written permission from the publishers. Registered names, trademarks, etc. used in this book, even when not specifically marked as such, are not to be considered unprotected by law.

Printed in the Federal Republic of Germany
Printed on acid-free paper

Composition K+V Fotosatz GmbH, Beerfelden
Printing Betz-Druck GmbH, Darmstadt
Bookbinding Litges & Dopf Buchbinderei GmbH, Heppenheim

ISBN: 978-3-527-31161-3

Contents

The Handbook of Homogeneous Hydrogenation.
Edited by J.G. de Vries and C.J. Elsevier
Copyright © 2007 WILEY-VCH Verlag GmbH & Co. KGaA, Weinheim
ISBN: 978-3-527-31161-3

18
Dehalogenation Reactions

Attila Sisak and Ottó Balázs Simon

18.1
Introduction

Reductive dehalogenation – one of the earliest reactions described in the organic chemical literature – has achieved special significance since the 1980s when the harmful properties of numerous halogenated (chiefly chlorinated) hydrocarbons became clear. The identification of methods capable of neutralizing or, at least, diminishing these dangers, remains a major challenge for chemistry. It is reasonable to convert stocks of the prohibited chemicals (e.g., polychlorobiphenyls, PCBs; chlorofluorocarbons, CFCs) to valuable products as far as possible. At the same time, the halogen-containing wastes should be detoxified by degradation. During the past two decades, the mainly heterogeneous (but also homogeneous) catalytic dehalogenation provided a major share towards solving these problems. Within this period, substantial progress was also made in the application of these reactions in organic syntheses.

Hydrodehalogenation – that is, hydrogenolysis of the carbon–halogen bond – involves the displacement of a halogen bound to carbon by a hydrogen atom. This chapter is devoted to dehalogenations mediated by transition-metal complexes (Eq. (1)):

$$\begin{array}{c} | \\ -\mathrm{C}-\mathrm{X} \\ | \end{array} \xrightarrow[{[\mathrm{L}_n\mathrm{M}]}]{\text{reducing agent}} \begin{array}{c} | \\ -\mathrm{C}-\mathrm{H}, \\ | \end{array} \qquad (1)$$

where X=F, Cl, Br, I, and $[L_nM]$=a transition metal complex. The use of a wide variety of reducing agents (H_2, hydrides of metals and metalloids, organic reductants, etc.) under the most diverse reaction conditions (e.g., one- or two-phase systems) has been reported. Organic halides have also been reduced by electrochemical and photochemical methods in the presence of compounds of the type $[L_nM]$. Reductive transformations of organic halides not relevant to Eq. (1) (e.g., coupling reactions) and dehalogenation of acyl halides will not be included here.

The Handbook of Homogeneous Hydrogenation.
Edited by J. G. de Vries and C. J. Elsevier
Copyright © 2007 WILEY-VCH Verlag GmbH & Co. KGaA, Weinheim
ISBN: 978-3-527-31161-3

The reactivity of the carbon–halogen bond in Eq. (1) depends on several factors:

- the nature of the halogen atom;
- the environment of the halogen atom in the molecule; and
- the reagents and conditions used in Eq. (1) [1].

The order of reactivity of the C–X bond (generally: I > Br > Cl > F) is consistent with its strength. For instance, the experimentally found dissociation energies for phenyl halides (D_{Ph-X}) are 528, 402, 339, and 272 kJ mol^{-1} at 298 K for X = F, Cl, Br, and I, respectively [2]. Consequently, catalytic defluorination in the literature is comparatively rare. The different reactivity of the C–X bonds renders possible the selective dehalogenation of compounds containing two dissimilar halides, leaving intact the stronger C–X bond.

Table 18.1 Catalytic hydrodechlorination of chlorobenzene. [a]

Catalyst/Reductant(s)	Solvent	Temperature [K]	TON [b]	TOF [c] [h^{-1}]	Reference
$[py_3RhCl_3]/NaBH_4/H_2$	DMF	298	9.9 [d]	0.76 [d]	11
$[(1,5\text{-hexadiene})RhCl]_2/H_2/Et_2NH$	*p*-Xylene/H_2O	323	>64 [d]	–	12
$[(Cy_3P)_2Rh(H)Cl_2]/H_2/PhCH_2NEt_3^+Cl^-/NaOH$	Toluene	298	19.4 [d]	0.81 [d]	13
$[Cp^*RhCl_2]_2/H_2/NEt_3$	*i*-PrOH	348	≥16.6 [d, e]	5.55 [d]	16
$[(COD)RuCl_2]/2PCy_3/H_2/NaOH$	2-BuOH	353	39.6 [d]	39.6 [d]	20
$RhCl_3$/Aliquat 336/H_2	CH_2Cl_2	353	≥48.8 [d, e]	2.71 [d]	21
$[(PPh_3)_4Ni]/MgH_2$	THF	298	13.3 [d]	0.74 [d]	33
$NiCl_2/MgH_2$	THF	298	≥50 [d]	2.5 [d]	34
$[Cp_3Ln]/NaH$	THF	319	5.6 [d]	0.12 [d]	35
$Ni(OAc)_2/[Cp_2TiCl_2]/NaH$	THF	339	≥16.7 [d]	8.33 [d]	37
$[(PPh_3)_3Ni]/NaBH_4$	DMF	343	5.40 [d]	0.23 [d]	43
$[(PhCH_2CN)PdCl_2]/NaAlH_2(OCH_2CH_2OMe)_3$	Benzene	343	43.3 [d]	9.63 [d]	53
$Pd(OAc)_2$/Polymethylhydrosiloxane/KF	THF/H_2O	298	19.0 [d]	66.7 [d]	77
$PdCl_2$/indoline	MeOH	413	6.79 [d]	1.70 [d]	92
$Pd(OAc)_2$/2dippp [f]/MeOH/NaOH	MeOH	373	90 [d]	4.5 [d]	100
$Pd(OAc)_2$/2dippp [f]/HCO_2Na	MeOH	373	≥100 [d]	≈5 [d]	100
$[Pd(dba)_2]$/3L/*i*-PrOH/K_2CO_3 [g]	*i*-PrOH	353	1019	46.3 [d]	103
$[Cp^*Rh(OAc)_2]$ H_2O/2-BuOH/KOH	2-BuOH	371	38.4 [d, i]	2.26 [d]	106
$[(IPr)Pd(allyl)Cl]$ [h]/*t*-BuONa	*i*-PrOH	333	≥380 [d, i]	109 [d]	108
Ni(0)/IMes HCl [f]/*i*-PrONa	THF	338	3.2 [d]	3.2 [d]	109
$[Cp_2TiCl_2]$/BuMgCl	THF	298	7.4 [d]	0.15 [d]	116

a) Benzene product unless otherwise stated.
b) Mol converted substrate/mol catalyst.
c) Mol converted substrate/(mol catalyst × h).
d) Calculated from the experimental data of the author(s).
e) Cyclohexane product.
f) See text.
g) L = Dicyclohexyl-{(2,4,6-triisopropyl)phenyl}phosphine; 4-chloro-*α*-methylstyrene substrate, methylstyrene product.
h) IPr: *N*,*N*′-bis(2,6-diisopropyl-phenyl)imidazol-2-ylidene.
i) 4-Chlorotoluene substrate, toluene product.

Table 18.2 Hydrodechlorination of PCBs and related compounds.[a)]

Catalysts/Reductant(s)	Substrate	Solvent	Temperature [K]	Dechlorination [%][b)]	TOF[c)] $[h^{-1}]$	Reference
$[(Et_3P)_2NiCl_2]/NaBH_2(OCH_2CH_2OMe)_2$	3,3′,4,4′-tetrachlorobiphenyl	THF	341	>99[d)]	–	31
$[(Et_3P)_2NiCl_2]/NaBH_2(OCH_2CH_2OMe)_2$	Arochlor 1232[e)]	THF	341	75[d, f)]	–	31
$[Cp_2TiCl_2]/NaBH_4/py$	Octachlorodibenzofuran	Bis(2-methoxyethyl)ether	398	100[g)]	1.4×10^{-2}[h)]	48
$Ni(acac)_2/NaAlH_2(OCH_2CH_2OMe)_2$	Delor 103[i)]	Toluene	383	>99[j)]	19.7[h)]	56
$Co(acac)_2/NaAlH_2(OCH_2CH_2OMe)_2$	Delor 103[i)]	Toluene	383	>99[j)]	56.9[h)]	56
$[(dppf)PdCl_2]$[k)]$/NaBH_4$	4,4′-dibromobiphenyl	THF/TMEDA[i)]	298	100[l)]	22.8[h)]	67
$Pd(OAc)_2/2PPh_3/i\text{-}PrOH/NaOH$	Kanechlor 600[m)]	*i*-PrOH	355	50[d, f)]	0.24[h)]	94
[ZnPc][j, n)]	Arochlor 1232[e)]	DDAB[o)]/dodecane/H_2O (microemulsion)	298	>99.8	–	167
[ZnPc][j, n)]	Arochlor 1260[m)]	DDAB[o)]/dodecane/H_2O (microemulsion)	298	94[f)]	–	167
$CeCl_3/LiAlH_4$	4,4′-dichlorobiphenyl	DME	358	>99	0.22[h)]	214
$NiCl_2/NaBH_2(OCH_2CH_2OMe)_2$	Arochlor 1016[p)]	THF	341	90[f, r)]	3.3×10^{-2}[h)]	215

a) Biphenyl product unless otherwise stated.
b) Conversion of PCB in % (w/w) unless otherwise stated.
c) mol converted substrate/(mol catalyst × h).
d) Determined by analysis of the Cl-content of the product.
e) Refined compound of mono- to tetrachlorobiphenyls with a majority of mono-, di-, and trichlorobiphenyls.
f) The product is a mixture.
g) Dibenzofuran product.
h) Calculated from the experimental data of the author(s).
i) Refined compound of di- to pentachlorobiphenyls with a majority of trichlorobiphenyls.
j) Determined by GC analysis.
k) See text.
l) Debromination % (w/w).
m) Refined compound of penta- to enneachlorobiphenyls with a majority of hexachlorobiphenyls.
n) Electrolysis: 1.07 mA cm^{-2} on lead cathode.
o) Didodecyldimethylammonium bromide.
p) Refined compound of mono- to pentachlorobiphenyls with a majority of trichlorobiphenyls.
r) mol.%.

Table 18.3 Hydrodechlorination of carbon tetrachloride. [a)]

Catalyst/Reductant(s)	Solvent	Temperature [K]	TON [b)]	TOF [c)] $[h^{-1}]$	Reference
$[(PPh)_3RuCl_2]/H_2/(CH_2OH)_2$	Xylene	298	81	–	24
$[(PPh)_3RuCl_2]H_2/Et_3N$	Xylene	298	69	–	24
$[(PPh)_3RuCl_2]/Et_3SiH$	–	353	778 [d)]	156 [d)]	74
$[(PN)PtMeCl]/HSiMe_2Ph$	Benzene	333	≥50 [d)]	11.1 [d)]	82
$[(PTA)_3Ru(H_2O)_3](tosylate)_2$ [e)]/ HCOONa	H_2O	353	478	–	111
$[RuCl_2(TPPMS)_2]$ [f)]/HCOONa	H_2O	353	≈480 [d, g)]	≈160 [d)]	111
$[RuCl_2(TPPMS)_2]$ [f)]/$HCOONH_4$	H_2O	353	416 [d)]	139 [d)]	111
CoTMPyP-SG [g)]	Phosphate buffer (pH=7.5)	298	–	37.3	121
CoPcTs [h)]	Phosphate buffer (pH=7.5)	298	–	35.8	121
Vitamin B_{12}/Ti(III)-citrate	MeOH	298	>48 [d, g)]	>2.66 [d)]	130
Cobinamide-dicyanide/Ti(III)-citrate	MeOH	298	>48 [d, g)]	>2.66 [d)]	130
Fe-porphyrin/cysteine	Phosphate buffer (pH=7.0)	298	≈0.8 [d, j)]	–	136
$Na_4[W_{10}O_{32}]$ [k)]	DMF/*i*-PrOH [f)]	295	≈49 [d)]	–	173
$[Bu_4N][W_{10}O_{32}]$ [k, l)]	DMF	295	40 [d, g)]	5 [d)]	173

a) Chloroform product unless otherwise stated.
b) Mol converted substrate/mol catalyst.
c) Mol converted substrate/(mol catalyst×h).
d) Calculated from the experimental data of the author(s).
e) PTA: 1,3,5-triaza-7-phosphaadamantane.
f) See text.
g) Mixture of products.
h) 5,10,15,20-tetrakis(1-methyl-4-pyridino)porphinecobalt(II) tetrachloride supported on silica gel.
i) Cobalt(II) phthalocyanine-tetrasulfonic acid, tetrasodium salt.
j) Complete degradation.
k) Irradiation with 1000-W high-pressure Hg lamp.
l) Under pure O_2.

The diversity of the substrates, catalysts, and reducing methods made it difficult to organize the material of this chapter. Thus, we have chosen an arrangement related to that used by Kaesz and Saillant [3] in their review on transition-metal hydrides – that is, we have classified the subject according to the applied reducing agents. Additional sections were devoted to the newer biomimetic and electrochemical reductions. Special attention was paid mainly to those methods which are of preparative value. Stoichiometric hydrogenations and model reactions will be discussed only in connection with the mechanisms.

In order to facilitate the comparison of the effectiveness of the very diverse methods, turnover numbers (TON), and/or turnover frequencies (TOF) (if they were given by the author or could be calculated based on their data) are summarized in the tables. The limited space available for this chapter allowed results to be compiled only for the dehalogenation of chlorobenzene (Table 18.1), for polychlorinated biphenyls and related compounds (Table 18.2), and for carbon tetrachloride (Table 18.3). To emphasize the environmental concerns, two of the three chosen compounds are common pollutants.

Despite catalytic hydrodehalogenation having been reviewed on several occasions, even when such reports contained data for homogeneous catalysis it was the information for heterogeneous which generally dominated [1, 2, 4–6]. The only comprehensive volume discussing only homogeneous hydrogenolysis, the classical handbook of James [7], was published more than 30 years ago.

18.2 Catalytic Dehalogenation with Various Reducing Agents

18.2.1 Molecular Hydrogen

This type of hydrodehalogenation has been performed generally in the presence of organic or inorganic bases to neutralize the hydrogen halides formed. Among published results, the use of rhodium complexes as catalysts dominates, but palladium and ruthenium complexes have also been applied on a frequent basis.

Although relatively few investigations have been published on the use of homogeneous hydrogenolysis of organic halides with molecular hydrogen, the first examples are rather old. During the 1960s, Kwiatek and coworkers [8] reduced a number of halogenated alkanes, alkenes, and alkynes with alkaline solutions of pentacyanocobaltate(II) under atmospheric hydrogen at 25 °C. The catalytic results, as well as their mechanistic interpretation, have been exhaustively reviewed [8, 9].

Roček and coworkers [10] attempted to compare the activity of various transition-metal complexes in the hydrodehalogenation of 5-iodouracil with atmospheric H_2 at 70–80 °C. The majority of the examined complexes decomposed to metal, but $[(PPh_3)_2CoCl_2]$, and $[(PPh_3)_3RuCl_2]$ in DMA (*N,N*-dimethyl-acetamide) remained homogeneous. The more active Ru complex catalyzed also the preparation of [5-2H]uracil using D_2 instead of H_2.

Love and McQuillin [11] have shown that the catalyst system $[py_3RhCl_3]/NaBH_4$ (1 : 1) in DMF (py = pyridine, DMF = *N,N*-dimethylformamide) is active in halogen/hydrogen exchange concerning several alkyl and aryl halides with atmospheric H_2. Interestingly, the reactivity orders $ClC_6H_5 > BrC_6H_5$, and $PhCH_2Cl > PhCH_2Br$ have been found. (–)-2-Chloro-2-phenylpropanoic acid and its methyl ester were dechlorinated to produce almost totally racemic products.

Markó and coworkers [12] dehalogenated several alkyl and benzyl halides as well as halobenzenes (X = Cl, Br, I) with $[(1,5\text{-hexadiene})RhCl]_2/PPh_3$ in the presence of Et_2NH using a medium of *p*-xylene/water at 50 °C and atmospheric H_2 pressure. Grushin and Alper [13] elaborated a similar (but more effective) method for the hydrogenolysis of chloroarenes using $[L_2Rh(H)Cl_2]$ (**1**) (L=PCy_3, $Pi\text{-}Pr_3$) as catalysts in a toluene/40% NaOH solvent system with benzyltriethylammonium chloride as a phase-transfer agent. The reaction occurred under mild conditions (25–100 °C, 0.1 MPa), and many functional groups (e.g., R, OR, CF_3, COAr, COOH, NH_2) were compatible with the C–Cl bond cleavage. Some chloro-substituted heterocycles were also readily dehalogenated. Hydrogenolysis of the C–F bond of 1-fluoronaphthalene with a similar catalytic system (**1**, L=PCy_3) could be performed at 95 °C and 0.5 MPa [14]. The use of the two-phase system improved the chance for the recovery of the catalysts [13, 14].

Setti and Mascaretti [15] realized the highly chemoselective and stereocontrolled hydrodehalogenation of the carbon-6-halogen bond of (pivaloyloxy)-methyl-6,6-dihalopenicillanate by $[(PPh_3)_3RhCl]$ in EtOAc and/or MeOH solvent systems with atmospheric H_2. For the diiodo derivative (Eq. (2)):

$[(PPh_3)_3RhCl]$, 0,1MPa H_2; EtOAc and/or MeOH, $CaCO_3$, 25°C

$CO_2CH_2OC(O)C(CH_3)_3$ → major + minor (2)

Ferrughelli and Horváth [16] converted chloroaromatics into the corresponding saturated hydrocarbons by a system generated from $[Cp_2RhCl_2]_2$ (Cp=η^5-cyclopentadienyl) in the presence of Et_3N and 2-propanol under 4.1 MPa H_2 at 75 °C. For example, 1 mmol 1,2,4-trichlorobenzene was successively reduced to cyclohexane within 6 h using 0.03 mmol of catalyst. Aizenberg and Milstein [17] have reported on the effective and selective hydrogenolysis of some polyfluorinated arenes. Heating $[L_3RhY]$ (L=PMe_3, Y=C_6F_5 or H) at 95–100 °C in C_6F_6 or C_6F_5H in the presence of a base under 0.6 MPa of H_2 led to the displacement of one of the F atoms by H with selectivities >92%. Jones' group realized a quasicatalytic hydrogenolysis of hexafluorobenzene: heating the solution of $[Cp^*Rh(H)_2PMe_3]$ (Cp^*=1,2,3,4,5-pentamethylcyclopentadienyl) and C_6F_6 at 135 °C for 25 days under 0.1 MPa of H_2, they produced 1.4 equiv. of C_6F_5H based on the starting rhodium complex [18].

Angeloff, Brunet and colleagues [19] have found $[(PPh_3)_4Pd]$ to be an efficient catalyst for the selective conversion of 2,3-dichloronitrobenzene into 3-chloronitrobenzene at 120 °C under atmospheric H_2. The reaction could be stopped at the maximum selectivity (>90%). Nolan, Grubbs, and associates [20] have described that $[(PCy_3)_2RuH_2(H_2)_2]$ and *in-situ*-generated $[L_2Ru(H)Cl(H_2)_2]$ (from $[(COD)\text{-}RuCl_2]_x$ (COD=1,5-cyclooctadiene) and 2L, L=PCy_3 or $Pi\text{-}Pr_3$) hydrogenolyze aryl chlorides completely in alcohols as solvents at 80 °C under 0.3 MPa of H_2 within 1 h. $[(PCy_3)_2Ru(H)Cl(H_2)_2]$ was very active also in the transfer hydrogenolysis of chloroarenes in *sec*-butyl alcohol. The systems exhibited significant functional group tolerance.

The heterogenization as a method for recycling of organotransition metal catalysts has also been studied in the case of haloarene hydrogenolysis. Blum's group [21] reported that the SiO_2 sol-gel entrapped ion pair $[(C_8H_{17})_3NMe]^+[RhCl_4 \cdot nH_2O]^-$, generated from $RCl_3 \cdot 3H_2O$, Aliquat 336, and $Si(OMe)_4$ catalyzes the hydrogenation of aryl fluorides and chlorides to give cyclohexane derivatives at 80 °C and 1.6 MPa of H_2. Fluorobenzenes, however, were proved to lose HF in a noncatalytic process. The catalyst was leach-proof and recyclable, with little or no loss in activity. Angelici and coworkers [22] have developed a process using the complexes $[(COD)Rh(\mathbf{2})]BF_4$ and $[(COD)Rh(\mathbf{3})]BF_4$ tethered on a heterogeneous catalyst, Pd/SiO_2. Fluorobenzene and 1,2-difluorobenzene were reduced under very mild conditions (0.4 MPa H_2, 70 °C) in the presence of NaOAc. The supported catalyst was more active than Pd/SiO_2 or the starting rhodium complexes alone.

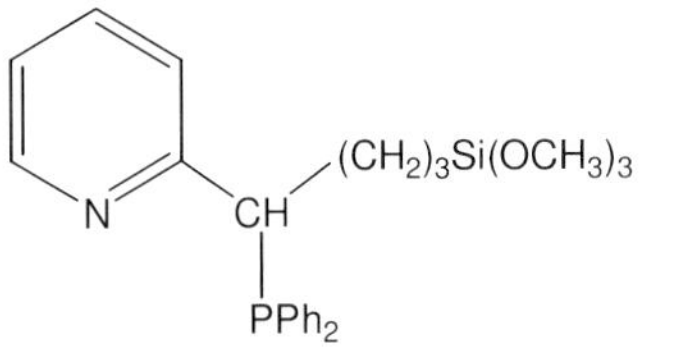

N
CH
$(CH_2)_3Si(OCH_3)_3$
PPh_2

(2)

N
$C(O)NH(CH_2)_3Si(OC_2H_5)_3$
N
$C(O)NH(CH_2)_3Si(OC_2H_5)_3$

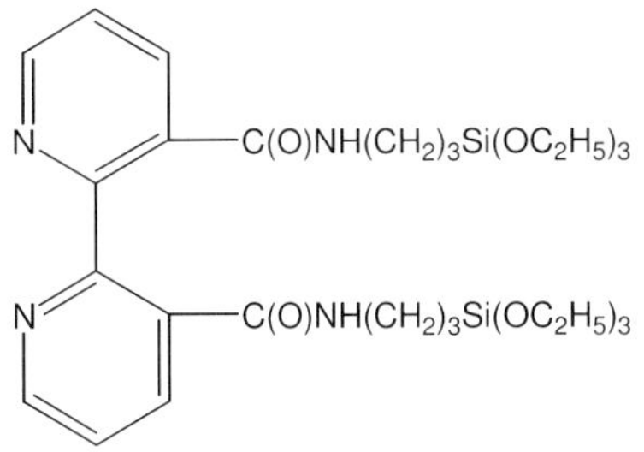

(3)

It has not been shown unambiguously, however, whether the rhodium remains coordinated to the ligands **2** and **3** during the catalysis. Kantam and associates [23] have used a Pd(II) catalyst anchored onto MCM-41/silylamine (MCM-41: a molecular-sieve) for hydrodehalogenation of aryl halides by atmospheric hydrogen. The catalyst was reused for several cycles with constant activity.

During the past ten years, several research groups have attempted to catalyze homogeneously the dechlorination of chlorofluorocarbon (CFC) and hydrochlorofluorocarbon (HCFC) compounds. Roundhill and associates [24] have studied the transformation of CCl_4 to $CHCl_3$, and that of $CFCl_3$ to $CHFCl_2$ in the presence of $[(PPh_3)_3RuCl_2]$ and $[(dppe)_2RuCl_2]$ (dppe = 1,2-bis(diphenylphosphino) ethane) at 25 °C and 0.11 MPa. The first transformation occurred more easily than the second (TONs up to 90 could be reached in 5–14 days in the former case; see Table 18.3). Lee and colleagues [25, 26] achieved efficient and selective hydrodechlorination of CF_3CCl_3 to CF_3CHCl_2, as well as $CF_2ClCFCl_2$ to $CF_2ClCHClF$ and $CFCl{=}CCl_2$, by the use of Group VIII transition metal complexes at 80–150 °C and about 1 MPa H_2 pressure. In the case of the most effective catalyst, $[RhCl(PPh_3)_3]$, a slightly polar solvent such as tetrahydrofuran (THF) was found to be appropriate for selective hydrogenolysis of CFCs. Mingos and Vilar [27, 28] synthesized the novel methylidyne cluster compounds $[(Pt\text{-}Bu_3)_4Pd_4(\mu_3\text{-}CY)(\mu\text{-}Cl)_3]$ (**4**) from $[Pd_2(dba)_3]$ (dba = dibenzylideneacetone) and $Pt\text{-}Bu_3$ in the presence of the halides $CYCl_3$ (Y = H, F). In toluene/Et_3N, **4** catalyzed the transformation of $CFCl_3$ into CH_3F and $CHFCl_2$ with atmospheric H_2. Under similar conditions with $[(Pt\text{-}Bu_3)_2Pd]$, $CFCl_2H$ was the only hydrogenation product. Recently, Sisak and coworkers [29] have found that $[py_3RhCl_3]$ – in the presence of pyridine as a base – and *in-situ*-generated $[(Pi\text{-}Pr_3)_3Pd]$, surpassed Pd/Al_2O_3, the most active heterogeneous catalyst tested in the selective conversion of CF_3CHFCl into CF_3CH_2F at 120 °C and 8–10 MPa H_2 pressure. The dechlorination of $CHFCl_2$ and CF_2Cl_2 with the same catalysts, however, could be performed only at higher temperatures, and hydrodefluorination of the substrates also occurred.

18.2.2
Simple and Complex Metal Hydrides

Simple and complex metal hydrides are capable of reducing organic halides due to their nucleophilic character. The efficiency of these reducing agents can be increased considerably by adding stoichiometric or catalytic amounts of various transition-metal compounds. Initially, hydrides and simple metal salts (without any stabilizing ligands) were combined, and although these catalytic systems were qualified only rarely as definitely *heterogeneous* (for example, see [30]), it is questionable whether they are in fact *homogeneous*, or not (see [31, 32]). Similar problems arose in the case of the application of simple metal salts together with other reducing agents (vide infra).

Carfagna and coworkers [33] found that Ni(0) and Pd(0) tertiary phosphine complexes activate MgH_2 and MgD_2 in the reduction of organic halides. The more active $[(PPh_3)_4Ni]$ promoted the reduction of chlorobenzene, whereas $[(PPh_3)_4Pd]$ promoted only that of bromo- and iodobenzene to benzene (deuterobenzene) at room temperature or at 67 °C, respectively, in THF. The same authors [34] also tested several transition-metal salts as promoters in the hydrodehalogenation of halobenzenes with MgH_2. $NiCl_2$ proved to be the most active, hydrogenolyzing even the C–F bond at 67 °C. A $[Cp_3Ln]/NaH/THF$ system has been reported to dehalogenate aryl halides with moderate activity at 45–65 °C [35]. Bromochlorobenzenes gave chlorobenzene selectively. The dimeric organolanthanide hydrides, $[O(CH_2CH_2C_5H_4)_2LaH]_2$ hydrogenolyzed *p*-bromoanisole and 1-bromohexadecane catalytically under similar conditions [36].

Aromatic halides such as chlorobenzene and *p*-fluorotoluene were rapidly hydrogenolyzed in 100% conversion by NaH of nanometric size in the presence of homogeneous catalysts. One- or two-component (e.g., $Ni(OAc)_2/TiCl_4$) systems were effective. The combination of ytterbium chloride and a transition-metal chloride showed a remarkable synergistic effect [37, 38].

Dzhemilev and associates [39–41] prepared monohalocyclopropanes and cyclopropanes by the reductive dehalogenation of alkyl- and aryl-substituted *gem*-dihalocyclopropanes with *i*-Bu_2AlH in the presence of catalytic amounts of titanium and zirconium complexes. The stereochemistry of the reductions has been also studied. Very recently, Knight and coworkers [42] hydrogenolyzed racemic bromoalkanes using *i*-Bu_2AlH or *i*-PrMgCl in stoichiometric, and chiral, single-enantiomer titanium complexes in catalytic amounts at 0 to 110 °C. There was no detectable difference in the reduction rate between the two enantiomers of the alkyl halide. An enantiomerically pure secondary bromide was reduced under the same conditions without racemization during the course of the reaction.

As early as 1979, Roth's group [43] reported that aryl bromides could be reduced with a homogeneous system generated from $[(PPh_3)_2NiCl_2]$ and $NaBH_4$ in DMF at 70 °C. The same group [44] later studied the reduction of a series of pure PCB congeners with $NaBH_4$. Extensive hydrogenolysis occurred in the presence of a Ni(0) triphenylphosphine complex at ambient temperatures in DMF. High selectivity for 2-, 3-, and 4-chloro displacement from di- and trichlorobiphenyls has been found. The solvent properties determined the catalytic efficiency in the transformation of 1,2,3-trichlorobenzene with a related system at 70 °C: using ethanol/pyridine solvent mixtures, benzene was produced a high rate [45].

The $[Cp_2TiCl_2]/NaBH_4$ system has been studied by several groups. Meunier [46] observed that the dehalogenation of iodobenzenes in DMF at 70 °C requires the presence of molecular oxygen. In a later and more exhaustive study, Schwartz and coworkers [47, 48] found the reaction scope and mechanism of the haloarene reduction by this system to be solvent-dependent. At 85–93 °C in DMF, halogen/hydrogen exchange and the formation of dimethylamino-substituted byproducts were observed. In DMA or in ethers, however, only dechlorinated products resulted. The $[Cp_2TiCl_2]/NaBH_4$/amine systems reduced catalytically pollutants such as PCBs and DDT (1,1-bis(4-chlorophenyl)-2,2,2-trichloroethane) [49]. Kim and coworkers [50, 51] reported that $[Cp'_2MCl_2]$ (Cp′ = Cp or Cp*; M = Ti, Zr, Hf) in the presence of several metal hydrides or alkyls catalyzes the conversion of monohalopyridines to give pyridine at room temperature. The effectiveness decreased as follows: Ti > Zr > Hf. The order of the aryl halide reactivity was surprising: C–F > C–Cl > C–Br. The reaction rate was boosted by adding 4 Å molecular sieves. Very recently, monofluoroaromatics have been reduced to the corresponding aromatic hydrocarbons when treated with the $NbCl_5/LiAlH_4$ system [52].

Several Czech research groups have studied the dehalogenation with $NaAlH_2(OCH_2CH_2OCH_3)_2$ (**5**). Hydrogenolysis of haloarenes with this hydride was accelerated by transition-metal species formed *in situ* from $[(RCN)_2PdCl_2]$ (R = Ph, $PhCH_2$) [53] or from 2,4-pentanedionates [54–56]. In the latter case, the efficiency decreased in the order Co ≈ Ni ≈ Pd > Cu ≫ Mn > Fe [54]. Complete conversion of the PCB liquid Delor 103 (42.6% Cl) to biphenyl has been effected in toluene with **5** and catalytic amounts of Ni(II) and Co(II) 2,4-pentanedionates at elevated temperatures [56].

Sharf's group [57–59] has investigated Rh and Ru complexes with various ligands immobilized on the surface of silica gels modified with, for example, γ-aminopropyl groups (γ-AMPS), or polymers containing 3(5)-methyl-pyrazole and imidazole groups. $[Rh_2(OAc)_4]$ immobilized on γ-AMPS dehalogenated *p*-bromotoluene by transfer of hydrogen from $NaBH_4$ and 2-propanol [57]. The immobilized binuclear Ru(II)-Ru(III) tetraacetate exhibited higher catalytic activity in the hydrogenolysis of *p*-bromotoluene than the heterogenized mononuclear systems [58]. The same authors [59] hydrogenolyzed *gem*-dihalocyclopropanes partially in the presence of the system Rh(I)-γ-AMPS/$NaBH_4$/CaO/2-propanol.

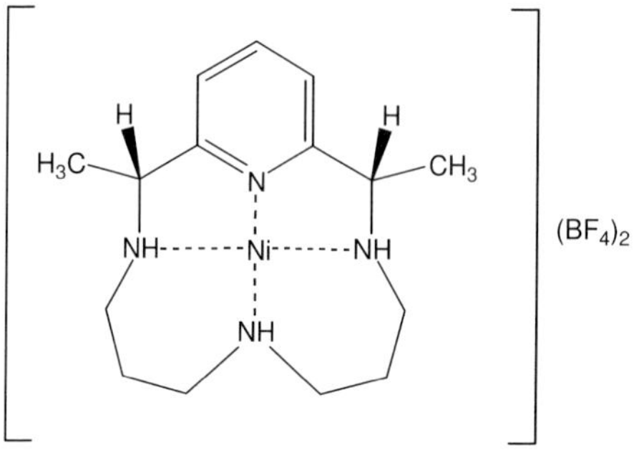

$(BF_4)_2$

(6)

$(BF_4)_2$

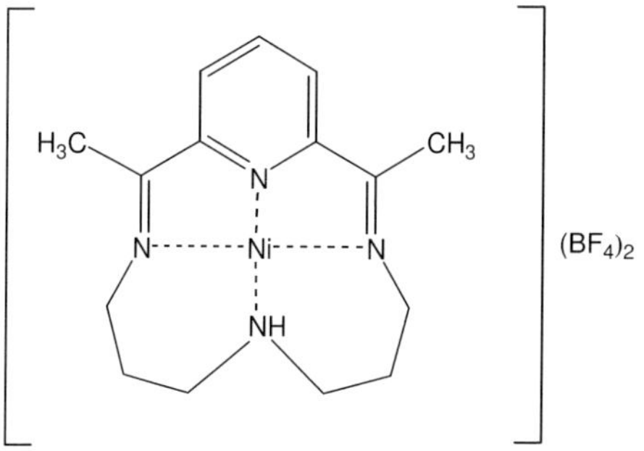

(7)

The Ni(II) complexes **6** and **7** have been found by Stiles [60] to be soluble catalysts for reductive dehalogenation when combined with $NaBH_4$ or hydrazine at 25–45 °C in protic solvents. Reactivity toward the reducing system increased with the halogen content of the substrate. Aryl bromides were converted much faster than chlorides, polychlorobenzenes, however, reacted readily with stepwise loss of chlorine.

The dehalogenation of several types of polyhalogenated arenes has been studied by Hor's group [61–67]. Brominated thiophenes and bithiophenes were reduced regioselectively using $NaBH_4$ in the presence of numerous Ni and Pd complexes as catalysts (e.g., 2,3,5-tribromothiophene afforded isomerically pure 2,3- or 2,4-dibromothiophene depending on the catalyst). The extent of debromination was controlled by the stoichiometry and substrate quantity. Pd-containing catalysts produced greater regioselectivity [61–63]. A variety of catalysts hydrogenolyzed 4,4′-dibromobiphenyl [64] as well as polybromobenzenes [65, 66] at room temperature, and complete conversions to biphenyl and benzene, respectively, has been achieved. The most satisfactory catalysts were generated *in situ* from the complexes [(dppf)$PdCl_2$] or [$(dppf)_2Pd$] (dppf=1,1′-bis(diphenylphosphino)ferrocene), $NaBH_4$ as a hydrogen source, and TMEDA (*N,N,N′,N′*-tetramethylethylenediamine) as a supporting base and THF as a solvent (the addition sequence of the reactants was critical) [64, 65] (Eq. (3)):

$$\mathrm{ArX} + \mathrm{NaBH_4} + {}^{1}/_{2}\,\mathrm{TMEDA} \xrightarrow[\mathrm{THF}]{\text{“Pd(dppf)”}} \mathrm{ArH} + \mathrm{NaX} + {}^{1}/_{2}\,\mathrm{TMEDA} \cdot 2\mathrm{BH_3} \tag{3}$$

A convenient one-pot system was developed also for the conversion of highly chlorinated benzenes to less chlorinated ones at room temperature, with reasonable conversion rates using the system [(dppf)$PdCl_2$]/$NaBH_4$/TMEDA/THF [67]. Degradation to benzene could not be achieved. Removal of chlorines in *meta*-position was preferred over those in *ortho*- or *para*-positions. The effectiveness of the method has been tested on the PCB mixtures Aroclor 1242, 1248, and 1254 at 67 °C.

King and coworkers [31] have used homogeneous organophosphorus–nickel complexes to detoxify PCBs by catalyzed hydrodechlorination using $NaBH_2(OCH_2CH_2OCH_3)_2$ in boiling THF. In model experiments with decachlorobiphenyl, the cone angle of the organophosphorus ligand was shown to be a key factor controlling the magnitude and position of chlorine displacement. Significantly, the highly toxic, coplanar dioxin precursor 3,3′,4,4′-tetrachlorobiphenyl, a *meta-para* chlorine-substituted congener, was dechlorinated quantitatively with a PEt_3-containing catalyst system.

Organic analogues of metal hydrides have been applied recently as reductants in homogeneous hydrodehalogenations. Treatment of aryl halides containing other functional groups with catalytic amounts of [$(PPh_3)_2NiCl_2$]/PPh_3 in the presence of 1 equiv. $Me_2NH \cdot BH_3$ and a base under mild conditions resulted in halogen/hydrogen exchange products. Noteworthy was the clean dehalogenation of densely functionalized haloperidol benzoate (Eq. (4)), attesting to the functional group compatibility offered by this combination of reagents [68].

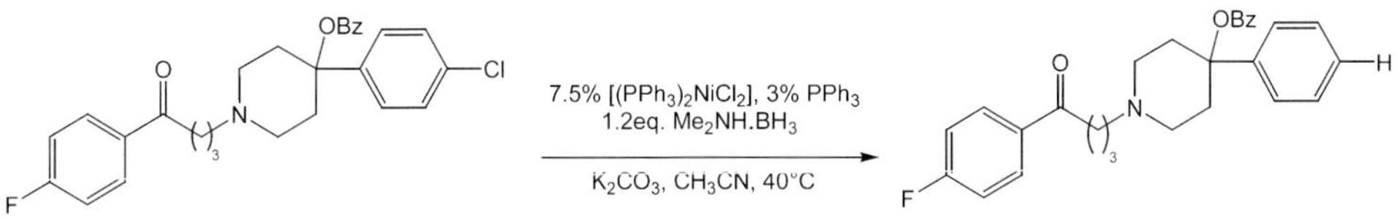

(4)

The novel, low-melting-point salt [*N*-pentylpyridinium][*closo*-$CB_{11}H_{12}$] has been used as solvent in several dehalogenations of mono- and polychlorides and -bromides, catalyzed by several Pd phosphine complexes [69]. The debromination of hexabromo- and 1,2,4,5-tetrabromobenzene was accomplished quite rapidly, whereas the dechlorination of 1,2,4-trichlorobenzene proceeded more slowly, but with excellent selectivity to 1,2-dichlorobenzene. The system could be recycled at least seven times without noticeable decrease of activity.

18.2.3 Hydrosilanes and Hydrostannanes

Hydrosilanes have a low ability to donate hydride or a hydrogen atom (cf. Section 18.2.2), but they are also useful reducing agents in combination with transition-metal complexes (Eq. (5)) [70]:

$$R'X + R_3SiH \xrightarrow[{[L_nM]}]{} R'H + R_3SiX \tag{5}$$

The problem of *homogeneity-heterogeneity* mentioned in Section 18.2.2 arises also here in the cases of using a simple metal salt instead of a ligand-stabilized salt (see [70, 71]).

Freidlina and coworkers [72, 73] were the first to use hydrosilanes for reducing organochlorine derivatives catalyzed by transition-metal complexes. For example, 1,1,1,5-tetrachloropentane was transformed into 1,1,5-trichloropentane by [$Fe_2(CO)_9$], and 1,3,3,5-tetrachlorodecane into 1,3,5-trichlorodecane by [$Mn_2(CO)_{10}$], respectively. Kono's group [74] has described the dechlorination of polychloroalkanes by silicon hydrides and in the presence of Ru(II) phosphine complexes as catalysts. CCl_4, CH_3CCl_3 and 1,1,1,3-tetrachloroalkanes have been reduced selectively to give $CHCl_3$, CH_3CHCl_2 and 1,1,3-trichloroalkanes with high turnovers at 80–100 °C (see Table 18.3). Pri-Bar and Buchman [75] hydrodehalogenated aryl iodides and bromides by polymethylhydrosiloxane (PHMS) using [$(PPh_3)_4Pd$] as a catalyst and Bz_3N as a base in Me_2SO/MeCN solvent mixtures at 60–110 °C. The debromination took place with a notable functional group tolerance. Recently, Maleczka and coworkers [76] developed a similar, but more powerful, method by fluoride activation of PMHS. In this case, the hydrodehalogenations of bromo- and iodoarenes were carried out in an amine-free system using THF as a solvent with relatively low loads of [$(PPh_3)_2PdCl_2$] catalyst. The same authors [77] reduced chloroarenes using catalytic amounts of $Pd(OAc)_2$ in combination with PMHS and aqueous KF at room temperature. The mildness of these methods was demonstrated by its functional group tolerance.

Aizenberg and Milstein [78] have found rhodium complex-catalyzed reactions between polyfluorobenzenes and hydrosilanes which resulted in the substitution of fluorine atoms by hydrogen and were both chemoselective and regioselective (Eq. (6):

$$1,2,4,5\text{-}C_6F_4H_2 + (EtO)_3SiH \xrightarrow[90\ °C]{[(PMe_3)_3RhC_6F_5]} 1,2,4,5\text{-}C_6F_4H_2 + (EtO)_3SiF \quad (6)$$

Esteruelas, Herrero and associates [79–81] have studied the simultaneous dechlorination of polychlorinated substrates and chlorination of Et_3SiH catalyzed by several complexes of Group VIII. The Os and Ir derivatives were less effective catalysts than those of Ru and Rh, and the former ones underwent deactivation during polychloroarene hydrogenolysis [79, 80]. Very recently, the same group showed that hexachlorocyclohexanes and Et_3SiH in the presence of various Rh and Ru complexes (e.g., $[(PPh_3)_3RhCl]$, $[(P\mathit{i}\text{-}Pr_3)_2RhH_2Cl]$, $[(PPh_3)_3Ru(H)Cl]$, and $[(P\mathit{i}\text{-}Pr_3)_2Ru(H)Cl(H_2)]$), afford cyclohexane/cyclohexene/benzene mixtures and Et_3SiCl. The reactivities of the substrates decreased in the order γ->α->δ-hexachlorocyclohexane [81].

Schubert's group [82, 83] has discovered that the reactivity of Pt(II) complexes is enhanced towards organosilanes by employing hemilabile chelating ligands such as $Me_2NCH_2CH_2PPh_2$ ($P\cap N$). Thus, the complex $[(P\cap N)PtMeCl]$ catalyzed the hydrogenolysis of R–Cl compounds by Me_2PhSiH to give R–H and $Me_2PhSiCl$ at 60 °C in benzene [82].

The dehalogenation of organic halides by organotin hydrides takes place in most cases with a free-radical mechanism [1, 84, 85]. The stereospecific reduction of 1,1-dibromo-1-alkenes with Bu_3SnH discovered by Uenishi and coworkers [86–89], however, did not occur in the absence of palladium complexes and did not involve radicals. For the synthesis of (*Z*)-1-bromo-1-alkenes, $[(PPh_3)_4Pd]$ proved to be the most effective catalyst which could also be generated *in situ*. The reaction in Eq. (7) proceeded at room temperature and a wide range of solvents could be used.

$$RCH{=}CBr_2 \xrightarrow[\text{toluene, } 25\ °C]{4\ mol\%\ [(PPh_3)_4Pd],\ Bu_3SnH} (Z)\text{-}RCH{=}CHBr \quad (7)$$

The authors determined the optimal reaction conditions and illustrated the scope of the method with 32 different starting compounds including alkenyl-, alkynyl-conjugated and 2,2-disubstituted 1,1-dibromo-1-alkenes.

18.2.4
Hydrogen Donors other than Hydrides

Various reagents have been found to be capable of donating hydrogen to transition metals in low valent state (e.g., formates, cyclic amines, alcohols in the presence of ligands and/or in basic medium). The reduction of transition-metal salts with metal alkyls (e.g., Grignard reagents with alkyl groups having β-hydrogen) may be regarded in close relation to their reactions with alcohol/base systems. The hydrogen transfer takes place from a metal-coordinated group both for alcoholates and for metal alkyls [3].

The dehalogenation of activated organic halides (e.g., benzyl halides and α-haloketones) was first published with various non-hydridic hydrogen donors. Grigg and coworkers [90] hydrogenolyzed activated organic halides with some noble metal chlorides (e.g., $RhCl_3 \cdot 3H_2O$) as catalysts in the presence of an alcohol and an excess of PPh_3. A very intriguing reducing agent, *N*-benzyl-1,4-dihydronicotinamide (BNAH) as a NAD(P)H model has been applied by Yasui and associates [91]. Aryl iodides and activated alkyl halides were reduced with the system $[(PPh_3)_3RhCl]$/BNAH in good yields at 70 °C in MeCN. Secondary cyclic amines proved to be useful reducing agents for the transfer hydrogenolysis of aryl halides with $PdCl_2$ as a catalyst in methanol at 140 °C. Indoline had the highest hydrogen-donating ability, and various bases (e.g., KOH) promoted the reaction [92].

Cortese and Heck [93] were first to report the application of palladium(0) triaryl phosphine complexes for homogeneous catalytic halogen/hydrogen exchange. The $Pd(OAc)_2/PAr_3/HCOOH/Et_3N$ (Ar = Ph, *o*-tolyl, 2,5-*i*-$Pr_2C_6H_3$) systems reduced bromoarenes at 50 °C, and the P(*o*-tolyl)$_3$-containing catalyst converted *m*-bromonitrobenzene selectively to nitrobenzene. Okamoto and Oka [94] dehalogenated aryl bromides by heating them with NaOH in alcoholic solvents in the presence of $RhCl_3$ or $Pd(OAc)_2$ and PPh_3. The method was suitable for destroying aromatic polyhalides. For example, $(C_6Br_5)_2O$ was transformed into $(C_6H_5)_2O$ in 2-propanol with more than 80 turnovers when treated for 5 h at 82 °C. Helquist has applied sodium methoxide [95] and formate [96] as reducing agents for the hydrogenolysis of haloarenes with 5 mol.% of $[(PPh_3)_4Pd]$ in DMF at 100 °C. In the case of formate, the debromination was really compatible with various functional groups.

Sasson and Rempel [97] showed that the system $[(PPh_3)_3RuCl_2]$/secondary alcohol is suitable for the selective transformation of 1,1,1,3-tetrachloro into 1,1,3-trichloro compounds. Similarly, Blum and coworkers [98, 99] employed $[(PPh_3)_3RuCl_2]$ as well as polystyrene-anchored Rh, Ru and Ir complexes for the hydrogen transfer from alcohols to trihalomethyl compounds, leading to dihalomethyl derivatives. For example, one of the Cl atoms of 2,2,2-trichloro-1-phenylethanol was displaced by H at 140–160 °C in 2-propanol. The polymer-anchored catalysts proved to be resistant to leaching [99].

Milstein and colleagues [100] have developed very efficient methods using basic, chelating phosphine ligands. Even aryl chlorides underwent reductive dechlorination to the corresponding arenes with $[(dippp)_2Pd]$ as catalyst (dippp, dippe, dippb = 1,3-bis(diisopropylphosphino)propane, -ethane, -butane) with high yields. The systems exhibited high functional group tolerance. Base-sensitive groups (e.g., CHO, CN) did not survive the conditions of the NaOH/MeOH reducing system, but remained unaffected when treated with HCO_2Na in MeOH or DMF. Dippp homologues were also effective ligands (reactivity order: dippp > dippb > dippe). Beletskaya and associates [101] have reported that using the system $[Pd(dba)_2]$/dppf/NaO*t*-Bu/dioxane the formation of dialkyl anilines from *m*- and *p*-dibromobenzenes and secondary amines – that is, the exchange of one of the Br atoms to H became the main process. $[(PPh_3)_2PdCl_2]$ or $[(PPh_3)_4Pd]$ catalyzed the deiodination of 5-iodopyrrole-2-carboxylates with HCOONa to give the corresponding 5-unsubstituted pyrrole-2-carboxylates in good yields [102]. Very recently, the C–Cl bond of various chloroarenes was reduced by $[Pd(dba)_2]$/phosphine/K_2CO_3/*i*-PrOH systems at 80 °C with very remarkable turnovers (see Table 18.1) [103].

Dicyclohexyl-{(2,4,6-triisopropyl)phenyl}-phosphine proved to be the most suitable ligand for the functional group-friendly hydrogenolysis. Rapid and specific deuterium labeling has been achieved through the microwave-enhanced dehalogenation of a number of *N*-4-picolyl-4-halogenobenzamides using deuterated formate as solid deuterium donor and $Pd(OAc)_2$, $RhCl_3$ or $[(PPh_3)_3RhCl]$ as catalysts in dimethyl sulfoxide (DMSO) [104]. Polychloroarenes could be dehalogenated with HCOONa in 2-propanol using $[(PPh_3)_3RhCl]$ catalyst. For example, the hydrogenolysis of 1,2,4-trichlorobenzene proved selective for 1,2-dichlorobenzene [105]. Fujita and colleagues [106] have achieved effective transfer hydrodechlorination of aryl chlorides catalyzed by $[Cp^*RhCl_2]_2$ and related complexes using refluxing 2-butanol as solvent and hydrogen source in the presence of bases. The system showed high compatibility with functional groups.

Intriguing transition-metal complex/NHC/alkoxide systems (NHC = *N*-heterocyclic carbene) have been applied recently for haloarene reduction. Nolan and coworkers [107] have found that SIMes·HCl ((2,4,6-trimethylphenyl)dihydroimidazolium chloride) is the most effective of the NHC precursors when combined with $[Pd(dba)_2]$ for the dehalogenation of mono- and polyhalogenated arenes at 100 °C in dioxane. Strong bases having β-hydrogens both performed deprotonation of the imidazolium salt and behaved as hydrogen sources. Very recently, these authors have used a series of air- and moisture-stable [(NHC)Pd(allyl)Cl] complexes for dehalogenation processes with microwave assistance (120 °C) or with conventional heating (60 °C) affording very high yields in both cases [108]. In related studies, Fort and associates [109] have applied $[Ni(acac)_2]$ as a catalyst. The complex associated to IMes·HCl (1,3-bis(2,4,6-trimethylphenyl)imidazolium chloride) and *in-situ*-generated *i*-PrONa in refluxing THF proved to have high efficiency and functional group tolerance in the reduction of various mono- and polyhalogenated arenes.

Water-soluble transition-metal complexes have been used recently for transfer hydrogenolysis of halocarbons. Paetzold and Oehme [110] have realized the reductive dehalogenation of allyl or benzyl halides in the presence of $[(phosphine)_2PdCl_2]$ complexes with sulfonated phosphines as ligands (e.g., $Ph_2P(CH_2)_3SO_3K$) by means of formates in a biphasic water/heptane system at 90 °C. This reaction could be promoted by addition of polyethers of different types as phase-transfer agents, the best results being obtained with triethylene glycol. Joó and associates [111] catalyzed the transformation of C–X bonds into C–H bonds by water-soluble Ru(II) phosphine complexes. CCl_4 was converted into $CHCl_3$ and CH_2Cl_2. Excellent TOFs up to 1000 h^{-1} could be achieved at 80 °C when, for example, an aqueous solution of HCOONa was the hydrogen donor and $[(TPPMS)_2RuCl_2]$ the catalyst (TPPMS = *m*-sulfophenyldiphenylphosphine Na salt) (see Table 18.3). Ogo, Watanabe and colleagues [112] have reported a pH-dependent transfer dehalogenation of water-soluble substrates with organoiridium(III) aqua complexes such as $[Cp^*Ir(bpy)(H_2O)]^{2+}$ (**8**) as catalyst precursors and formates as hydrogen donors. For example, **8** reacted with the formate ions at pH 5.0 to form the hydride $[Cp^*Ir(bpy)(H)]^+$, which acted as active catalysts in the hydrogenolysis of 2- and 3-halopropanoic acids.

Perfluoroaromatic Grignard reagents (R_FMgBr) were formed from alkylmagnesium bromides and perfluoroarenes in the presence of transition-metal halides (mainly $CoCl_2$) as catalysts in THF at about 0 °C, as had been reported during the late 1960s. Hydrolysis of the reaction mixtures resulted in the formation of R_FH [113]. Colomer and Corriu have applied the system $[Cp_2TiCl_2]$/*i*-PrMgBr in Et_2O for debromination, and *p*-BrC_6H_4Cl gave chlorobenzene selectively [114]. Deiodination and debromination of aromatic halides proceeded smoothly with high yields when treated with RMgX (where R groups have *β*-hydrogens) and a catalytic amount of $[Cp_2ZrCl_2]$ [115]. From $[Cp_2TiCl_2]$ and the above type of Grignard reagents, an effective catalyst system was obtained even for dechlorination [116]. Addition of $[(dppf)PdCl_2]$ or [(dppf)Pd] markedly accelerated the reduction of alkyl halides with Grignard reagents in Et_2O, but in THF the reduction was independent of palladium [117]. Benzyl halides underwent halogen/hydrogen exchange with equimolar amounts of Et_2Zn in DMF at room temperature using $[(PPh_3)_4Pd]$ as a catalyst. The method tolerated a variety of functional groups [118]. Very recently, Kotora and coworkers [119] have developed iron- and ruthenium-containing catalytic systems capable of reductive dehalogenation of 2-chloro-*α*,*ω*-dienes when combined with trialkylaluminum reagents.

18.2.5
Biomimetic Dehalogenations

Halocarbons are common soil, sediment, and groundwater pollutants, many of them being toxic, mutagenic materials [6, 120]. Although certain anaerobic microorganisms are capable of reductively degrading halocarbons, these processes are often slow, and high pollutant concentrations may be toxic, even for the bacteria. Thus, the problem of toxicity may be eliminated rather by biomimetic catalysis than by biocatalysis [121]. Reduction of organic halides has been investigated in the presence of transition metal-containing coenzymes vitamin B_{12} (Co, **9**), F_{430} (Ni, **10**), hematin (Fe, **11**), and related complexes. In the most cases, corrinoids have been applied (for earlier results, see [122]).

CONH2
CH3 H3C
CONH2
H2NOC
H3C
N
CN
N
CONH2
H3C
Co+
H
N
N
CH3
H2NOC
CH3
CH3
CH3
O
CONH2
HN
O
H3C
O
O
P
O−
N
CH3
H
O
OH
N
CH3
O
CH2OH

(9)

CO2H
O
H
HN
H3C
CH3
H2NOC
N
N
CO2H
Ni+
X−
H
N
N
HO2C
CO2H
H
O
CO2H

(10)

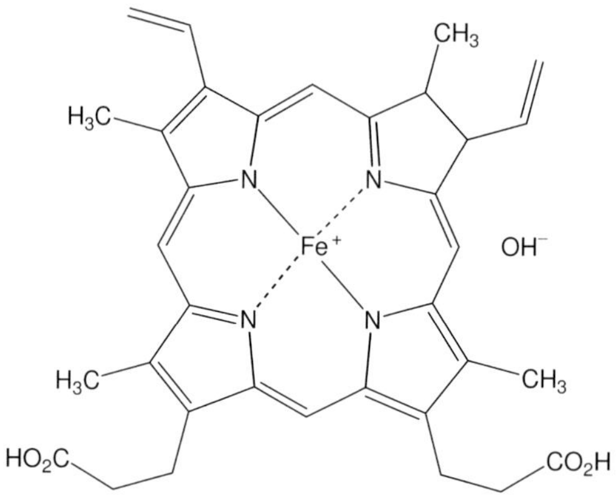

(11)

It should be emphasized that the aim of the great part of these biomimetic studies was *degradation* of the substrate and not the halogen/hydrogen exchange resulting in one or more product(s). This concerns above all the detoxification of polyhalogenated methanes and related pollutants (cf. [123–127]). The product composition depends strongly on the reducing agents: Ti(III) citrate [121, 128–131], dithiothreitol [121, 130], cysteine, and sulfides [130, 132, 133] have been used most frequently. The results of Morra and colleagues [130] may serve as an example. They catalyzed the dehalogenation of CCl_4 with corrins (vitamin B_{12}, cobinamide dicyanide, or aquocobalamin). Products in the presence of Ti(III) citrate were mostly hydrogenolytic and included predominantly CH_3Cl and CH_4, whereas they were CH_2Cl_2, CO and formate in the case of dithiothreitol. Sulfide/cysteine reductants were least reactive against CCl_4, giving the major products $CHCl_3$, CS_2, 2-oxothiazolidine carboxylic acid, and 2-thioxo-4-thiazolidinecarboxylic acid. In addition to cobalt complexes, a pyridine-2,6-bis(thiocarboxylic acid)-Cu(II) complex [134], and an iron porphyrin in the presence of cysteine [135, 136] have been found to be active in the degradation of CCl_4 and related compounds.

Vitamin B_{12} catalyzed also the dechlorination of tetrachloroethene (PCE) to trichloroethene (TCE) and 1,2-dichloroethene (DCE) in the presence of dithiothreitol or Ti(III) citrate [137–141], but zero-valent metals have also been used as bulk electron donors [142, 143]. With vitamin B_{12}, carbon mass recoveries were 81–84% for PCE reduction and 89% for TCE reduction; *cis*-1,2-DCE, ethene, and ethyne were the main products [138, 139]. Using Ni(II) humic acid complexes, TCE reduction was more rapid, leading to ethane and ethene as the primary products [144, 145]. Angst, Schwarzenbach and colleagues [140, 141] have shown that the corrinoid-catalyzed dechlorinations of the DCE isomers and vinyl chloride (VC) to ethene and ethyne were pH-dependent, and showed the reactivity order: 1,1-DCE > VC > *trans*-DCE > *cis*-DCE. Similar results have been obtained by Lesage and colleagues [146]. Dror and Schlautmann [147, 148] have demonstrated the importance of specific core metals and their solubility for the reactivity of a porphyrin complex.

Several research groups have investigated catalytic systems related to F_{430}. Gantzer and Wackett [149] have found different reactivity orders for the substrates examined: in the case of vitamin B_{12} (**9**) and coenzyme F_{430} (**10**), $CCl_4 > C_2Cl_4 > C_6Cl_6$, for hematin (**11**), $CCl_4 > C_6Cl_6 > C_2Cl_4$. TCE was dechlorinated stereoselectively to *cis*-1,2-dichloroethene with **9–11**. Arai and colleagues [150] catalyzed efficiently the hydrodehalogenation of cycloalkyl halides yielding cycloalkanes by [Ni(tmtaa)] (**12**; tmtaa = dianion of 6,8,15,17-tetramethyl-5,14-dihydro-dibenzo[b,i][1,4,8,11]tetraaza-cyclotetradecine) in combination with $NaBH_4$ or $NaBH(OMe)_3$ under mild conditions.

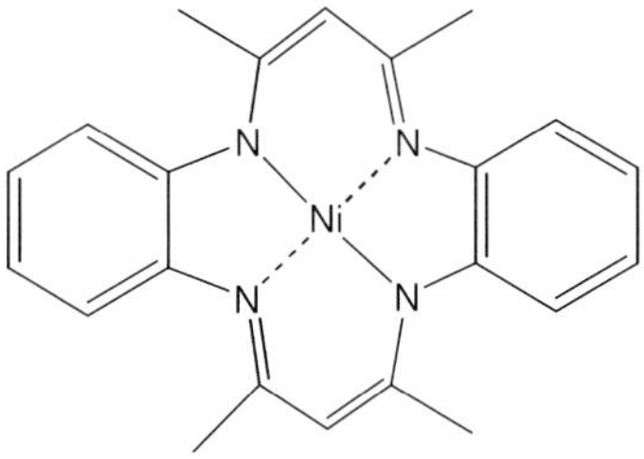

(**12**)

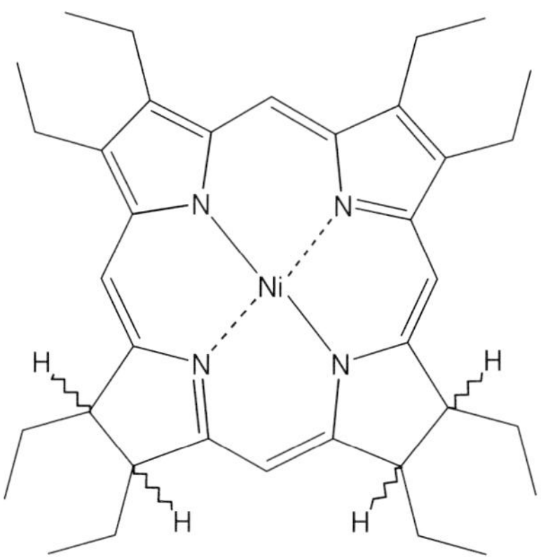

(**13**)

Stolzenberg's group [151–153] has studied the ability of various nickel(II) macrocycle and coordination complexes (e.g., **7**, **10**, and [Ni(OEiBC)] (**13**; OEiBC = octaethylisobacteriochlorin, mixture of isomers)). The facility of catalytic reduction of cyclohexyl bromide by $NaBH_4$ varied markedly with the structure of the ligands and the solvent composition. The highest TOFs (up to 70 h^{-1}) were obtained by the complex **7** (see Section 18.2.2) in diglyme/ethanol. Morra and coworkers [154] have shown that a combination of aquocobalamin or **10** and Fe(0) may effectively promote dehalogenation.

18.2.6
Electrochemical Reductions

Electrochemical methods are available for the direct dehalogenation of organic halides to a limited extent: fluorides and monochlorides are generally not reducible [1]. In the presence of transition-metal complexes as mediators (Med), however, the electrolysis of halocarbons (RX) can be performed more effectively and selectively under various conditions [155–158]. Mediated electroreduction is most efficient when the electron transfer step E° (Med/Med$^{\bullet-}$) is more negative than E° (RX/RX$^{\bullet-}$) [157] (cf. Section 18.4.1).

Pletcher and associates [155, 159, 160] have studied the electrochemical reduction of alkyl bromides in the presence of a wide variety of macrocyclic Ni(II) complexes. Depending on the substrate, the mediator, and the reaction conditions, mixtures of the dimer and the disproportionation products of the alkyl radical intermediate were formed (cf. Section 18.4.1). The same group [161] reported that traces of metal ions (e.g., Cu^{2+}) in the catholyte improved the current density and selectivity for several cathodic processes, and thus the conversion of trichloroacetic acid to chloroacetic acid. Electrochemical reductive coupling of organic halides was accompanied several times by hydrodehalogenation, especially when Ni complexes were used as mediators. In many of the reactions examined, dehalogenation of the substrate predominated over coupling [162–165].

The use of electrochemical methods for the destruction of aromatic organochlorine wastes has been reviewed [157]. Rusling, Zhang and associates [166, 167] have examined a stable, conductive, bicontinuous surfactant/soil/water microemulsion as a medium for the catalytic reduction of different pollutants. In soils contaminated with Arochlor 1260, 94% dechlorination was achieved by [Zn(pc)] (H_2pc = phthalocyanine) as a mediator with a current efficiency of 50% during a 12-h electrolysis. Conductive microemulsions have also been employed for the destruction of aliphatic halides and DDT in the presence of $[Co(bpy)_3]^{2+}$ (bpy = 2,2′-bipyridine) [168] or metal phthalocyanine tetrasulfonates [169].

Nünnecke and Voss [158] reduced aryl chlorides electrochemically in methanol using $[(bpy)NiCl_2]$ and $[(cyclam)NiCl_2]$ (cyclam = 1,4,8,11-tetraazacyclotetradecane) as mediators. More highly chlorinated benzenes were converted to chlorobenzene, whereas chlorodibenzofurans gave unsubstituted dibenzofuran as the major product. Due to the mediators, higher selectivities could be achieved, and the formation of hydrogenated products was completely suppressed. Peters and associates [170–174] have applied electrogenerated Co(I) salen (H_2salen = bis(salicylidene)ethylenediamine) or Co(I) salophen (H_2salophen = bis(salicylidene)-1,2-phenylenediamine) for the catalytic reduction of various halogenated substrates. In the case of 1,1,2-trichloro-1,2,2-trifluoroethane, cyclic voltammetry and controlled-potential electrolysis resulted in the formation of $CFCl{=}CF_2$ and CF_3CH_2F as main products, respectively, using DMF as a solvent and Bu_4NBF_4 as a supporting electrolyte [174]. Mugnier, Harwey and coworkers [175, 176] activated the C–Br and the C–I bonds by the cluster $[(dppm)_3Pd_3(CO)]^{2+}$ (dppm = bis(diphenylphosphino)-

methane). Catalytic dehalogenation of 2,3,4-tri-*O*-acetyl-5-thioxylopyranosyl bromide (Xyl-Br, both α- and β-isomers) provided Xyl-H as the major organic product at the potential of –0.9 V (versus SCE).

18.2.7
Miscellaneous Reducing Methods

Martin and associates [124, 125] have studied the dehalogenation of $CHCl_3$ in boiling methanol by Schiff-base complexes of some transition metals in the presence of TMEDA. The kinetics of chloride ion formation has been measured without characterizing the organic products. Nahar and Mukhedkar [126, 127] found that the reactivities of related Schiff-base complexes in the above reaction decreased in the order Pd > Pt > Ni > Cu > Zn.

Several researchers hydrogenolyzed – mostly activated – alkyl halides under carbonylation conditions and/or in the presence of metal carbonyls. Alper and coworkers debrominated bromomethyl ketones with $[Co_2(CO)_8]$ as a catalyst under phase-transfer conditions [177]. Brunet and Taillefer [178, 179] catalyzed the reduction of aryl iodides by *in-situ*-generated $K[HFe(CO)_5]$ in methanol (up to 18 cycles) under 0.1 MPa CO pressure at 60 °C; several functional groups were tolerated. An intriguing method has been developed by Cavinato and Toniolo [180] for the synthesis of γ-keto acids of the type $ArC(O)CH_2CH_2COOH$ via carbonylation-decarboxylation of $ArC(O)CH_2CHClCOOH$. The reactions were carried out in the presence of Pd(II) phosphine complexes, typically at 2–3 MPa CO and 100–120 °C in acetone/H_2O. When the same authors [181] attempted the carbonylation of 2-chlorocyclohexanone with the system $[(PPh_3)_2PdCl_2]/PPh_3/EtOH/H_2O$ (at 100 °C and 10 MPa CO, P/Pd = 2.5), a hydrogen transfer occurred leading to halogen/hydrogen exchange. Trabuco and Ford [182] have shown that homogeneous catalysts prepared from $RhCl_3$ in aqueous aromatic amines reduce C–Cl bonds under mild water gas shift conditions (100 °C and 0.1 MPa CO). In a 4-picoline/H_2O solvent mixture, 1,2-dichloroethane was transformed to ethene and ethane. An ambient temperature liquid carbonylmetallate, $[bmim][Co(CO)_4]$ ($[bmim]^+$ = 1-butyl-3-methylimidazolium cation), has been prepared by Dyson and coworkers [183]. The mixture of the ionic liquid and NaOH catalyzed the debromination of 2-bromoketones.

An interesting new homogeneous catalytic process was developed by Buijs [184] for the reductive dehalogenation of polychlorinated and -brominated aromatic hydrocarbons and ethers. Cu(I) benzoate catalyzed the reaction under Dow-Phenol conditions in the absence of air at 235 °C (Eq. (8)):

$$Ar'COOH + ArX + H_2O \xrightarrow[Cu(I)]{} Ar'OH + ArH + HX + CO_2 \,. \tag{8}$$

Starichenko and colleagues [185–187] have studied the hydrogenolysis of polychloro- and polyfluoroaromatic compounds with the $[(N\cap N)NiCl_2]/Zn$ reducing system ($N\cap N$ = bpy or phen (phen = 1,10-phenanthroline)). Using DMF or DMA solvents in the presence of water or NH_4Cl, the displacement of the Cl or F atoms by H took place at 50–70 °C. Interestingly, the systems catalyzed the regioselective *ortho*-hydrodefluorination of pentafluorobenzoic acid to 2,3,4,5-tetrafluorobenzoic and 3,4,5-trifluorobenzoic acids in high yields [187].

The applications of polyoxometalates in catalytic dehalogenation of halocarbons have been succinctly reviewed by Hill and coworkers [188]. This reaction involves the photocatalytic transformation of organic halides coupled with the oxidation of sacrificial organic reductants (secondary alcohols or tertiary amides) (Eq. (9)) [189, 190]:

$$3CCl_4 + 3\,(CH_3)_2CHOH + W_{10}O_{32}^{4-} \xrightarrow[h\nu]{} CHCl_3 + C_2Cl_6 + 3(CH_3)_2CO + H_2W_{10}O_{32}^{4-} \tag{9}$$

Very recently, Gkika and colleagues [191] realized the degradation of diversified pesticides (e.g., lindane) to CO_2, H_2O and the corresponding inorganic anions by photolysis in the presence of polyoxotungstates. A stable "hydrophobic vitamin B_{12}", heptamethyl cobyrinate perchlorate catalyzed efficiently the reduction of DDT using a visible light irradiation system containing a $[Ru(bpy)_3]Cl_2$ photosensitizer [192].

18.3 Mechanistic Considerations

The mechanisms of the homogeneous catalytic hydrodehalogenation have been examined by the following methods:

- spectroscopic investigations to detect intermediates in the reaction mixtures;
- isolation and characterization of possible intermediates or related model compounds;
- qualitative and quantitative analysis of the organic products;
- kinetic measurements; and
- systematic variation of the structure of the substrate and/or the catalyst. Theoretical investigations (e.g., MO calculations) have also been made.

In spite of the wide variety of substrates, hydrogen sources, and catalysts applied in the hydrodehalogenations of organic halides, some general statements can be made on the reaction pathways. We shall examine the following crucial steps of the dehalogenations: activation of the C–X bond, steps involving the reducing agents, and formation of the products.

18.3.1 Activation of the C–X Bond

18.3.1.1 Oxidative Addition

This type of activation may proceed by various mechanisms, which have been discussed exhaustively [2, 193, 194].

S_N2 and S_NAr Reactions In these reactions the metal atom attacks aliphatic or aromatic carbon bonded to X, respectively. A stronger nucleophilic metal as well as a better leaving group X (I > Br > Cl > F) facilitates, whereas steric hindrance in R slows these types of oxidative addition [193, 194]. S_NAr reactions are favored by electron-withdrawing substituents Y in the case of the substrates 4-YC_6H_4X [2]. S_N2 [27, 29, 89, 117, 180, 181] and S_NAr [31, 33, 62–67, 95, 100, 107–109] mechanisms have been suggested frequently for zerovalent d^{10} complexes such as $[L_nM]$ (M = Ni, Pd, Pt; L = tertiary phosphine; n = 2,3,4). For example:

$$[(PPh_3)_3Pd] + Ar - X \longrightarrow [(PPh_3)_2Pd(X)Ar] + PPh_3 \tag{10}$$

Products of S_NAr-type oxidative additions in some active Pd- or Ni-containing hydrodehalogenating systems have been isolated and characterized structurally (e.g., $[(PEt_3)_2NiCl(p\text{-}C_6Cl_5C_6Cl_4)]$ [31], $[(PPh_3)_2PdBr$(3,4,5-tribromo-2-thienyl)] [62]). The reactivity order Ni > Pd > Pt has been found for the oxidative addition of aryl halides. Steric and electronic properties, and the numbers of L as well as chelate effects, play an important role [65, 194–196]. For example, Pd(0) complexes of basic chelating phosphines react substantially more easily with chlorobenzenes than their nonchelating analogues (see Section 18.2.4) [2, 100, 196]. In contrast to $[L_nPd]$, oxidative addition of aryl halides on $[L_nNi]$ often proceed by single electron transfer mechanism [2, 197]. S_N2 and S_NAr types of oxidative addition as a step of the catalytic dehalogenation have also been proposed in the literature for low-valent Ti [40, 114], Zr(II) [115], Ru(II) [20, 74, 81, 98], Rh(I) [18, 29, 68, 81, 91] and Pt(II) [82, 83] complexes.

Atom Transfer Atom Transfer (AT) takes place typically in the case of d^7 complexes, which abstract the halogen atom from RX. The radical formed combines then with a second metal [193, 194]. A "classical" example of this mechanism is the hydrodehalogenation with cyanocobaltates(II) (see Section 18.2.1) [8, 9], but an analogous pathway was suggested recently for the Co(II) corrin-catalyzed dechlorination of CCl_4 in the presence of S^{2-}/cysteine as reductant (Eqs. (11)–(12)) [130]:

$$BCo^{II} + CCl_4 \rightarrow BCo^{III}Cl + CCl_3^{\bullet} \tag{11}$$

$$BCo^{II} + CCl_3^{\bullet} \rightleftarrows BCo^{III}CCl_3 \ldots \tag{12}$$

Single Electron Transfer A single electron transfer (SET) mechanism is often difficult to distinguish from an S_N2 reaction because the principal product of these two pathways is the same, apart from the stereochemistry at carbon (racemization instead of inversion). The radicals formed can recombine rapidly in a solvent cage (*inner-sphere ET*) [2, 193, 194]. The $[HFe(CO)_5]^-$-catalyzed deiodination of iodobenzene may serve as an example [179] (Eq. (13)).

$$[HFe(CO)_5]^- + PhI \rightarrow [HFe(CO)_5^{\bullet}, PhI^{\bullet -}] \rightarrow [HFe(CO]_5^{\bullet}, Ph^{\bullet}, I^-] \rightarrow [HFePh(CO)_5, I^-] \qquad (13)$$

Coordinatively unsaturated complexes and those giving easily such species by ligand dissociation favor pathways related to that described in Eqs. (10) and (13). Coordinatively saturated complexes reduce halocarbons via *outer-sphere ET* [193, 194]. In cases of electrochemical dehalogenations, the species formed by one-electron reduction of the mediators on the cathode often react in this way [156, 157, 198]. For example (Eq. (14)) [157, 166]:

$$[Zn(pc)]^{\bullet -} + ArX \rightarrow [Zn(pc)] + [ArX]^{\bullet -} \rightarrow [Zn(pc)] + Ar^{\bullet} + X^- \qquad (14)$$

A SET process has been postulated between Rh(III) oxidative adducts and an NAD(P)H model compound (cf. Section 18.2.4) [91]. Oxidative adducts formed by S_N2, S_NAr, or inner-sphere SET pathways may produce radicals by homolytic M–C bond cleavage [130, 155, 176, 199].

The transformation of the radical ($R^{\bullet}$) (which may escape also from the solvent cage) affords several products. Usually, RH is formed by hydrogen abstraction from the reducing agent or the solvent [36, 91, 150, 157, 169, 173, 179, 198], but dimerization [173, 194, 198], disproportionation (formation of RH and R(–H) simultaneously) [155, 158–160, 170], or rearrangement [43, 49, 55, 165, 194] can also take place. For example, the formation of the cyclized product in the reaction of Eq. (15) requires the intermediacy of an aryl radical [49, 55]:

Br

(15)

The C–F bond activations in C_6F_6 and related compounds with ruthenium [200, 201] and rhodium [17, 78, 201] complexes, for which an S_NAr mechanism is energetically unfavorable, have been explained by SET pathways. Both S_N2 [128, 129, 131, 170–174, 199, 202] and SET [130, 132, 199] mechanisms have been proposed for the reaction of Co(I) complexes with alkyl and vinyl halides.

Carbanions may be formed in the electrochemical reductions of aryl halides [157, 158] (Eq. (16)):

$$Ar^{\bullet} + Med^{\bullet -} \rightarrow Ar^{-} + Med\,, \tag{16}$$

and of *gem*-di- or trihaloaliphatics [174, 198], as well as in the hydrogenolysis of the latter type of substrates with corrinoids and related complexes [130, 199, 203]. The cleavage of a halide ion from a polyhalocarbanion (or from its complex) affords (di)halocarbene (or its complex) which will be transformed depending on the reaction conditions [130, 203, 204]. Equation (17) shows some of the possible transformations of CCl_4 in the presence of corrins [130]:

$$CCl_4 \xrightarrow[\text{-Cl}^-]{2e^-} :CCl_3^- \xrightarrow[\text{-Cl}^-]{} :CCl_2 \xrightarrow{H_2O} CO + 2HCl\ldots \tag{17}$$

Radical Chain Mechanism This mechanism also requires a coordinatively unsaturated metal and the presence of a radical initiator $Q^{\bullet}$ (trace of O_2, $h\nu$, etc.). Such a pathway has been proposed for a Ni(II) complex-catalyzed dehalogenation of polyhaloarenes [60], and it occurs frequently in the stoichiometric C–X activations with early transition-metal complexes (see [205–207]).

$$Q^{\bullet} + [Cp_2^*ZrH_2] \rightarrow [Cp_2^*Zr^{\bullet}H_2] + QH \tag{18}$$

$$[Cp_2^*Zr^{\bullet}H_2] + RF \rightarrow [Cp_2^*ZrHF] + R^{\bullet} \tag{19}$$

$$[Cp_2^*ZrH_2] + R^{\bullet} \rightarrow [Cp_2^*Zr^{\bullet}H_2] + RH \tag{20}$$

In addition to initiation (Eq. (18)) and propagation (Eqs. (19) and (20)), termination steps are also possible (resulting in, for example, dimers of $R^{\bullet}$). Radical traps inhibit this type of oxidative addition [194].

18.3.1.2 σ-Bond Metathesis

This is the simultaneous breaking and formation of bonds to the metal with a four-membered ring transition state (Eq. (21)):

$$\begin{matrix} M-H \\ X-R \end{matrix} \longrightarrow \left(\begin{matrix} M\cdots H \\ \vdots \quad \vdots \\ X\cdots R \end{matrix}\right)^{\ddagger} \longrightarrow \begin{matrix} M & H \\ | & | \\ X & R \end{matrix} \tag{21}$$

This concerted process may operate in the case of d^0 early metal complexes where the oxidative addition is forbidden [194]. Nevertheless, it was postulated also in the interaction of a dihalo-ruthenium(II) intermediate and a hydrosilane [74].

18.3.1.3 S_N2 Attack of the Hydride Ligand

Another route not involving an M–C bond formation has been suggested recently for some water-soluble Ir(III) complexes [112]. The hydride pushes the halide from the carbon atom directly by an S_N2 way with the **14** transition state.

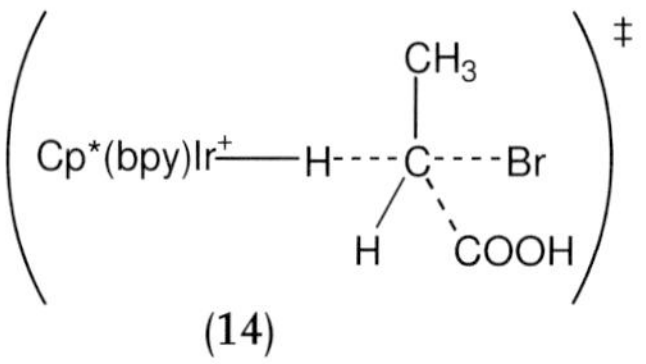

$$\left(\mathrm{Cp^*(bpy)Ir^{\pm}{-}H \cdots \underset{H\ \ \ COOH}{\overset{CH_3}{C}} \cdots Br} \right)^{\ddagger}$$

(14)

18.3.1.4 1,2-Insertion

1,2-Insertion of the C=C bond into an M–H bond precedes frequently the C–X bond activation in halogenated alkenes. Such a pathway has been suggested for the cobalamin-mediated dechlorinations of *cis-* and *trans-*DCE, as well as VC with Ti(III) citrate as a reducing agent [141].

18.3.2
Reaction Steps Involving the Reducing Agents

The double role of the reducing agents in homogeneous hydrodehalogenations is:

(i) transformation of the transition-metal complex into a state capable of the activation of the substrate;
(ii) formation of the M–H bond to cleave – directly or indirectly – the C–X bond of the substrate [3, 208].

Processes corresponding to both of these roles are not involved by all means in each catalytic cycle. They may also take place in one step [24, 74]. Process (i) is needed generally to transfer – mostly to reduce – the precursor complex into a catalytically active form. Reactions related to that of Eq. (22) (L=PCy_3, P*i*-Pr_3 [13]) are promoted by the addition of bases [2, 20, 24, 28, 29, 45, 47, 65, 106, 108]:

$$[L_2Rh(H)Cl_2] + H_2 + NaOH \rightarrow [L_2Rh(H)_2Cl] + NaCl + H_2O \qquad (22)$$

In outer-sphere SET reductions (e.g., in electrochemical dehalogenations), hydrogen abstraction by $R^\bullet$ leads to the product RH (i.e., no step related to (ii) is required to occur). Process (ii) follows generally the activation of the substrate in the proposed hydrodehalogenation cycles, but we know also of opposite examples [77, 82, 106, 112].

The nature of the M–H bond-forming step, (ii), in a given catalytic cycle depends strongly on the reducing agent used. Dihydrogen [13, 14, 17, 20, 24, 29] and hydrosilane [78, 81, 82] react mostly by *oxidative addition* [193, 209, 210]. For example, the product of the reaction in Eq. (23) – which is involved in an actual "working" catalytic cycle – has been isolated and characterized by its X-ray structure ($L=PMe_3$) [78]:

$$[L_3RhC_6F_5] + (EtO)_3SiH \rightarrow [L_3RhH(C_6F_5)Si(OEt)_3] \quad (23)$$

The *heterolytic activation* of H_2 has been considered in systems containing basic ligands such as pyridine [18, 29, 209]. *Transmetalation* proposed for Bu_3SnH as a reductant resulted in the transformation of the Pd–Br bond into Pd–H [89].

Simple and complex metal hydrides as strong nucleophiles easily displace the halide bound to the central atom of the catalyst by *hydride transfer* [40, 45, 62–67]. Alcoholates [97, 98, 100, 106, 108], formates [93, 96, 100, 112], and metal alkyls [114, 115, 117] also substitute the halogen atom on the transition metal, after which the hydrogen transfer takes place from the alkoxy, formato and alkyl ligands formed [3, 211]. A good model for the halobenzene reduction catalyzed by the $[L_nPd]/HCOO^-$ system [98, 100, 112] has been found by Alper and co-workers [211]. Decomposition of the organopalladium formato complex, $[(PPh_3)_2PdPh(HCOO)]$, gave benzene, indicating the intermediacy of a Pd–H species (Eq. (24), $L=PPh_3$) [211]:

$$[L_2PdPh(HCOO)] \xrightarrow[-CO_2]{\Delta} [L_2PdPh(H)] \xrightarrow[-RH]{} [L_2Pd] \quad (24)$$

18.3.3
Formation of the Product

The product-forming steps of dehalogenations by free radical pathways were discussed earlier (see Section 18.3.1.1). In non-radical mechanisms, the dehalogenated products (RH) will be formed mostly by reductive elimination [193, 194]; however, concerted processes lead directly from RX to RH (see Sections 18.3.1.2 and 18.3.1.3).

Catalytic dehalogenation cycles with a binuclear reductive elimination step have not yet been reported, but many examples are known with a single metal. The RH product eliminates easily from d^8 metals (Ni(II), Pd(II)) or d^6 metals (Ru(II), Rh(III)) (Eq. (25)) [196]. This reaction is believed to go by a three-center transition state [193, 194, 212]:

$$L_nM(H)R \longrightarrow \left(L_nM \begin{smallmatrix} H \\ \vdots \\ R \end{smallmatrix} \right)^{\ddagger} \longrightarrow L_nM + RH \quad (25)$$

In the case of R=aryl with electron-withdrawing substituents, however, kinetic measurements indicated a rapid, reversible η^2-arene complex formation followed by the rate-determining loss of the arene [213].

Since the reaction in Eq. (26) is generally the last step of the catalytic cycle, the $[L_nM]$ fragment should survive long enough to react with the substrate again. The presence of an excess of phosphine ligand can facilitate the reductive elimination and can also stabilize the $[L_nM]$ species [3, 194].

18.4 Concluding Remarks

Our knowledge regarding transition metal-mediated (catalytic and stoichiometric) hydrodehalogenation has advanced significantly during the past two decades. One favorable aspect of this progress is that such progress has been achieved mainly in the activation of the more stable C–Cl and C–F bonds. In the former case, environmental concerns have dominated, including the detoxification of chlorocarbon pollutants by biomimetic, electrochemical and other methods in solving problems caused by chemical industrial processes. From the synthetic viewpoint, the most intriguing results have been achieved in fluorine/hydrogen exchange, as in addition to stoichiometric transformations a number of catalytic processes have also been developed. In the near future, a step-up in the efficiency of the homogeneous hydrodehalogenation – that is, increasing the turnovers under the mildest possible reaction conditions – appears to be a real possibility, both in environmental and synthetic applications.

Acknowledgments

The authors gratefully acknowledge the Hungarian Academy of Sciences (Grant nos. OTKA T 031934/2000, T 037817/2002) for financial support of these studies. O. B. S. also thanks the Ministry of Education (Hungary) for a PhD fellowship.

Abbreviations

γ-AMP	γ-aminopropyl
BNAH	*N*-benzyl-1,4-dihydronicotinamide
CFC	chlorofluorocarbon
DCE	1,2-dichloroethene
DMF	dimethylformamide
DMSO	dimethyl sulfoxide
HCFC	hydrochlorofluorocarbon
NHC	*N*-heterocyclic carbene
OEiBC	octaethylisobacteriochlorin
PCB	polychlorinated biphenyl
PCE	tetrachloroethene

PHMS polymethylhydrosiloxane
SCE standard calomel electrode
SET single electron transfer
TCE trichloroethene
TMEDA *N,N,N′,N′*-tetramethylethylenediamine
TOF turnover frequency
TON turnover number
TPPMS *m*-sulfophenyldiphenylphosphine
VC vinyl chloride

References

1 A. R. Pinder, *Synthesis* **1980**, *6*, 425.
2 V. V. Grushin, H. Alper, *Chem. Rev.* **1994**, *94*, 1047.
3 H. D. Kaesz, R. B. Saillant, *Chem. Rev.* **1972**, *72*, 231.
4 B. R. James, *Homogeneous Hydrogenation*, John Wiley & Sons, New York, **1973**.
5 M. Hudlicky, In: B. M. Trost, I. E. Fleming (Eds.), *Comprehensive Organic Synthesis, Volume 8*, Pergamon, Oxford, **1991**, p. 895.
6 O. Hutzinger, S. Safe, V. Zitko, *The Chemistry of PCBs*, CRC Press, Cleveland, OH, **1974**.
7 Z. Ainbinder, L. E. Manzer, M. J. Nappa, in: G. Ertl, H. Knözinger (Eds.) *Handbook of Heterogeneous Catalysis, Vol. 4*, VCH, Weinheim, **1997**, p. 1677.
8 J. Kwiatek, J. K. Seyler, *Adv. Chem.* **1968**, *70*, 207.
9 B. R. James, *Homogeneous Hydrogenation*, John Wiley & Sons, New York, **1973**, p. 139.
10 J. Rocek, V. Svata, L. Leseticky, *Collection Czechoslov. Chem. Commun.* **1985**, *50*, 1244.
11 C. J. Love, F. J. McQuillin, *J. Chem. Soc., Perkin I Trans.* **1973**, 2509.
12 P. Kvintovics, B. Heil, J. Palágyi, L. Markó, *J. Organomet. Chem.* **1978**, *148*, 311.
13 V. V. Grushin, H. Alper, *Organometallics* **1991**, *10*, 1620.
14 R. J. Young, V. V. Grushin, *Organometallics* **1999**, *18*, 294.
15 E. L. Setti, O. A. Mascaretti, *J. Org. Chem.* **1989**, *54*, 2233.
16 D. T. Ferrughelli, I. T. Horváth, *J. Chem. Soc. Chem. Commun.* **1992**, 806.
17 M. Aizenberg, D. Milstein, *J. Am. Chem. Soc.* **1995**, *117*, 8674.
18 B. L. Edelbach, W. D. Jones, *J. Am. Chem. Soc.* **1997**, *119*, 7734.
19 A. Angeloff, J. J. Brunet, P. Legars, D. Neibecker, D. Souyri, *Tetrahedron Lett.* **2001**, *42*, 2301.
20 M. E. Cucullu, S. P. Nolan, T. R. Belderrain, R. H. Grubbs, *Organometallics* **1999**, *18*, 1299.
21 J. Blum, A. Rosenfeld, F. Gelman, H. Schumann, D. Avnir, *J. Mol. Catal. A-Chem.* **1999**, *146*, 117.
22 H. Yang, H. R. Gao, R. J. Angelici, *Organometallics* **1999**, *18*, 2285.
23 M. L. Kantam, A. Rahman, T. Bandyopadhyay, Y. Haritha, *Synth. Commun.* **1999**, *29*, 691.
24 S. Xie, E. M. Georgiev, D. M. Roundhill, *J. Organomet. Chem.* **1994**, *482*, 39.
25 O. J. Cho, I. M. Lee, K. Y. Park, H. S. Kim, *J. Fluorine Chem.* **1995**, *71*, 107.
26 H. S. Kim, O. J. Cho, I.M. Lee, S. P. Hong, C. Y. Kwag, B. S. Ahn, *J. Mol. Catal. A-Chem.* **1996**, *111*, 49.
27 R. Vilar, S. E. Lawrence, D. M. P. Mingos, D. J. Williams, *J. Chem. Soc. Chem. Commun.* **1997**, 285.
28 R. Vilar, D. M. P. Mingos, *J. Organomet. Chem.* **1998**, *557*, 131.
29 A. Sisak, O. B. Simon, K. Nyíri, *J. Mol. Catal. A-Chem.* **2004**, *213*, 163.
30 R. A. Egli, *Helv. Chim. Acta* **1968**, *51*, 2090.
31 C. M. King, R. B. King, N. K. Bhattacharyya, M. G. Newton, *J. Organomet. Chem.* **2000**, *600*, 63, and references therein.

32 J. P. Collman, L. S. Hegedus, J. R. Norton, R. G. Finke, *Principles and Applications of Organotransition Metal Chemistry*, University Science Books, Mill Valley, CA, **1987**, p. 673, and references therein.
33 C. Carfagna, A. Musco, R. Pontellini, *J. Mol. Catal.* **1989**, *54*, L23.
34 C. Carfagna, A. Musco, R. Pontellini, *J. Mol. Catal.* **1989**, *57*, 23.
35 C. Qian, D. Zhu, Y. Gu, *J. Mol. Catal.* **1990**, *63*, L1.
36 Z. W. Xie, C. T. Qian, Y. Z. Huang, *J. Organomet. Chem.* **1991**, *412*, 61.
37 H. Q. Li, S. J. Liao, Y. Xu, *Chem. Lett.* **1996**, 1059.
38 Y. K. Zhang, S. J. Liao, Y. Xu, D. R. Yu, Q. Shen, *Synth. Commun.* **1997**, *27*, 4327.
39 U. M. Dzhemilev, R. L. Gaisin, *Bull. Acad. Sci. USSR Div. Chem. Sci.* **1988**, *37*, 2332.
40 U. M. Dzhemilev, R. L. Gaisin, A. A. Turchin, N. R. Khalikova, I. P. Baikova, G. A. Tolstikov, *Bull. Acad. Sci. USSR Div. Chem. Sci.* **1990**, *39*, 967.
41 U. M. Dzhemilev, R. L. Gaisin, A. A. Turchin, G. A. Tolstikov, *Bull. Acad. Sci. USSR Div. Chem. Sci.* **1991**, *40*, 2084.
42 A. R. Abbott, J. Thompson, L. C. Thompson, K. S. Knight, *Transition Met. Chem.* **2003**, *28*, 305.
43 S. T. Lin, J. A. Roth, *J. Org. Chem.* **1979**, *44*, 309.
44 J. A. Roth, S. R. Dakoji, R. C. Hughes, R. E. Carmody, *Environ. Sci. Technol.* **1994**, *28*, 80.
45 A. Scrivanti, B. Vicentini, V. Beghetto, G. Chessa, U. Matteoli, *Inorg. Chem. Commun.* **1998**, *1*, 246.
46 B. Meunier, *J. Organomet. Chem.* **1981**, *204*, 345.
47 Y. M. Liu, J. Schwartz, *J. Org. Chem.* **1994**, *59*, 940.
48 C. L. Cavallaro, Y. M. Liu, J. Schwartz, P. Smith, *New J. Chem.* **1996**, *20*, 253–257.
49 Y. M. Liu, J. Schwartz, *Tetrahedron* **1995**, *51*, 4471.
50 H. G. Woo, B. H. Kim, S. J. Song, *Bull. Korean Chem. Soc.* **1999**, *20*, 865.
51 B. H. Kim, H. G. Woo, W. G. Kim, S. S. Yun, T. S. Hwang, *Bull. Korean Chem. Soc.* **2000**, *21*, 211.
52 K. Fuchibe, T. Akiyama, *Synlett* **2004**, 1282.
53 I. Simunek, M. Kraus, *Collection Czechoslov. Chem. Commun.* **1973**, *38*, 1786.
54 J. Vcelák, J. Hetflejs, *Collection Czechoslov. Chem. Commun.* **1994**, *59*, 1645.
55 M. Czakóová, J. Hetflejs, J. Vcelák, *React. Kinet. Catal. Lett.* **2001**, *72*, 277.
56 J. Hetflejs, M. Czakoova, R. Rericha, J. Vcelák, *Chemosphere* **2001**, *44*, 1521.
57 V. I. Isaeva, Z. L. Dykh, L. I. Lafer, V. I. Yakerson, V. Z. Sharf, *Bull. Russ. Acad. Sci. Div. Chem. Sci.* **1992**, *41*, 49.
58 V. Z. Sharf, V. I. Isaeva, Y. V. Smirnova, Z. L. Dykh, G. N. Baeva, A. N. Zhilyaev, T. A. Fomina, I. B. Baranovskii, *Russ. Chem. Bull.* **1995**, *44*, 64.
59 V. F. Dovganyuk, V. Z. Sharf, L. G. Saginova, I. I. Antokolskaya, L. I. Bolshakova, *Bull. Acad. Sci. USSR Div. Chem. Sci.* **1989**, *38*, 777.
60 M. Stiles, *J. Org. Chem.* **1994**, *59*, 5381.
61 Y. Xie, S. C. Ng, T. S. A. Hor, H. S. O. Chan, *J. Chem. Res. (S)* **1996**, *3*, 150.
62 Y. Xie, S. C. Ng, B. M. Wu, F. Xue, T. C. W. Mak, T. S. A. Hor, *J. Organomet. Chem.* **1997**, *531*, 175.
63 Y. Xie, B.M. Wu, F. Xue, S. C. Ng, T. C. W. Mak, T. S. A. Hor, *Organometallics* **1998**, *17*, 3988.
64 B. Wei, S. H. Li, H. K. Lee, T. S. A. Hor, *J. Mol. Catal. A-Chem.* **1997**, *126*, L83.
65 B. Wei, T. S. A. Hor, *J. Mol. Catal. A-Chem.* **1998**, *132*, 223.
66 S. H. Li, H. S. Ngew, S. O. H. Chan, S. C. Ng, H. K. Lee, T. S. A. Hor, *Environ. Monitor. Assess.* **1997**, *44*, 481.
67 L. Lassova, H. K. Lee, T. S. A. Hor, *J. Org. Chem.* **1998**, *63*, 3538.
68 B. H. Lipshutz, T. Tomioka, S. S. Pfeiffer, *Tetrahedron Lett.* **2001**, *42*, 7737.
69 Y. H. Zhu, C. B. Ching, K. Carpenter, R. Xu, S. Selvaratnam, N. S. Hosmane, J. A. Maguire, *Appl. Organomet. Chem.* **2003**, *17*, 346.
70 R. Boukherroub, C. Chatgilialoglu, G. Manuel, *Organometallics* **1996**, *15*, 1508.
71 D. Villemin, B. Nechab, *J. Chem. Res. (S)* **2000**, *9*, 432.
72 E. C. Chukovskaia, N. A. Kuzmina, R. K. Freidlina, *Bull. Acad. Sci. USSR Div. Chem. Sci.* **1967**, *16*, 1031.
73 L. N. Kiseleva, N. A. Rybakova, R. K. Freidlina, *Bull. Acad. Sci. USSR Div. Chem. Sci.* **1986**, *35*, 10302.

74 H. Kono, H. Matsumoto, Y. Nagai, *J. Organomet. Chem.* **1978**, *148*, 267.

75 I. Pri-Bar, O. Buchman, *J. Org. Chem.* **1986**, *51*, 734.

76 R. E. Maleczka, R. J. Rahaim, R. R. Teixeira, *Tetrahedron Lett.* **2002**, *43*, 7087.

77 R. J. Rahaim, R. E. Maleczka, *Tetrahedron Lett.* **2002**, *43*, 8823.

78 M. Aizenberg, D. Milstein, *Science* **1994**, *265*, 359.

79 M. A. Esteruelas, J. Herrero, F. M. Lopez, M. Martin, L. A. Oro, *Organometallics* **1999**, *18*, 1110.

80 J. Diaz, M. A. Esteruelas, J. Herrero, L. Moralejo, M. Olivan, *J. Catal.* **2000**, *195*, 187.

81 M. A. Esteruelas, J. Herrero, M. Olivan, *Organometallics* **2004**, *23*, 3891.

82 F. Stöhr, D. Sturmayr, U. Schubert, *Chem. Commun.* **2002**, 2222.

83 U. Schubert, J. Pfeiffer, F. Stöhr, D. Sturmayr, S. Thompson, *J. Organomet. Chem.* **2002**, *646*, 53.

84 I. Terstiege, R. E. Maleczka, *J. Org. Chem.* **1999**, *64*, 342.

85 D. P. Curran, *Synthesis* **1988**, 417.

86 J. Uenishi, R. Kawahama, O. Yonemitsu, *J. Org. Chem.* **1996**, *61*, 5716.

87 J. Uenishi, R. Kawahama, Y. Shiga, O. Yonemitsu, J. Tsuji, *Tetrahedron Lett.* **1996**, *37*, 6759.

88 J. Uenishi, R. Kawahama, A. Tanio, S. Wakabayashi, O. Yonemitsu, *Tetrahedron* **1997**, *53*, 2439.

89 J. Uenishi, R. Kawahama, O. Yonemitsu, *J. Org. Chem.* **1998**, *63*, 8965.

90 R. Grigg, T. R. B. Mitchell, S. Sutthivaiyakit, *Tetrahedron Lett.* **1979**, *12*, 1067.

91 S. Yasui, K. Nakamura, M. Fujii, A. Ohno, *J. Org. Chem.* **1985**, *50*, 3283.

92 H. Imai, T. Nishiguchi, M. Tanaka, K. Fukuzumi, *J. Org. Chem.* **1977**, *42*, 2309.

93 N. A. Cortese, R. F. Heck, *J. Org. Chem.* **1977**, *42*, 3491.

94 T. Okamoto, S. Oka, *Bull. Chem. Soc. Jpn.* **1981**, *54*, 1265.

95 A. Zask, P. Helquist, *J. Org. Chem.* **1978**, *43*, 1619.

96 P. Helquist, *Tetrahedron* **1978**, *22*, 1913.

97 Y. Sasson, G. L. Rempel, *Synthesis* **1975**, 448.

98 J. Blum, S. Shtelzer, P. Albin, *J. Mol. Catal.* **1982**, *16*, 167.

99 Y. Migron, J. Blum, *J. Mol. Catal.* **1983**, *22*, 187.

100 Y. Ben-David, M. Gozin, M. Portnoy, D. Milstein, *J. Mol. Catal.* **1992**, *73*, 173.

101 I. P. Beletskaya, A. G. Bessmertnykh, R. Guilard, *Tetrahedron Lett.* **1999**, *40*, 6393.

102 S. H. Leung, D. G. Edington, T. E. Griffith, J. J. James, *Tetrahedron Lett.* **1999**, *40*, 7189.

103 X. Bei, A. Hagemayer, A. Volpe, R. Saxton, H. Turner, A.S. Guram, *J. Org. Chem.* **2004**, *69*, 8626.

104 J. R. Jones, W. J. S. Lockley, S. Y. Lu, S. P. Thompson, *Tetrahedron Lett.* **2001**, *42*, 331.

105 M. A. Atienza, M. A. Esteruelas, M. Fernandez, J. Herrero, M. Olivan, *New J. Chem.* **2001**, *25*, 775.

106 K. Fujita, M. Owaki, R. Yamaguchi, *J. Chem. Soc. Chem. Commun.* **2002**, 2964.

107 M. S. Viciu, G. A. Grasa, S. P. Nolan, *Organometallics* **2001**, *20*, 3607.

108 O. Navarro, H. Kaur, P. Mahjoor, S. P. Nolan, *J. Org. Chem.* **2004**, *69*, 3173.

109 C. Desmarets, S. Kuhl, R. Schneider, Y. Fort, *Organometallics* **2002**, *21*, 1554.

110 E. Paetzold, G. Oehme, *J. Prakt. Chem.* **1993**, *335*, 181.

111 A.C. Bényei, S. Lehel, F. Joó, *J. Mol. Catal. A-Chem.* **1997**, *116*, 349 and references therein.

112 S. Ogo, N. Makihara, Y. Kaneko, Y. Watanabe, *Organometallics* **2001**, *20*, 4903.

113 W.L. Respess, C. Tamborski, *J. Organomet. Chem.* **1969**, *18*, 263.

114 E. Colomer, R. Corriu, *J. Organomet. Chem.* **1974**, *82*, 367.

115 R. Hara, W. H. Sun, Y. Nishihara, T. Takahashi, *Chem. Lett.* **1997**, 1251.

116 R. Hara, K. Sato, W. H. Sun, T. Takahashi, *J. Chem. Soc. Chem. Commun.* **1999**, 845.

117 K. Yuan, W.J. Scott, *J. Org. Chem.* **1990**, *55*, 6188.

118 K. A. Agrios, M. Srebnik, *J. Org. Chem.* **1993**, *58*, 6908.

119 D. Necas, M. Kotora, I. Cisarova, *Eur. J. Org. Chem.* **2004**, 1280.

120 L. N. Zanaveskin, V. A. Averyanov, Y. A. Treger, *Uspekhi Khimii* **1996**, *65*, 667.

121 L. Ukrainczyk, M. Chibwe, T. J. Pinnavaia, S. A. Boyd, *Environ. Sci. Technol.* **1995**, *29*, 439.

122 B. R. James, *Homogeneous Hydrogenation*, John Wiley & Sons, New York, **1973**, p. 193, and references therein.

123 L. Wilputte-Steinert, *Transition Met. Chem.* **1978**, *3*, 172.

124 D. F. Martin, *J. Inorg. Nucl. Chem.* **1975**, *37*, 1941.

125 D. F. Martin, K. A. Hewes, S. G. Maybury, B. B. Martin, *Inorg. Chim. Acta* **1986**, *111*, 5.

126 C. T. Nahar, A. J. Mukhedkar, *J. Indian Chem. Soc.* **1980**, *57*, 961.

127 C. T. Nahar, A. J. Mukhedkar, *J. Indian Chem. Soc.* **1981**, *58*, 343.

128 U. E. Krone, R. K. Thauer, H. P. C. Hogenkamp, *Biochemistry* **1989**, *28*, 4908.

129 U. E. Krone, R. K. Thauer, H. P. C. Hogenkamp, K. Steinbach, *Biochemistry* **1991**, *30*, 2713.

130 T. A. Lewis, M. J. Morra, P. D. Brown, *Environ. Sci. Technol.* **1996**, *30*, 292.

131 P. C. Chiu, M. Reinhard, *Environ. Sci. Technol.* **1995**, *29*, 595.

132 N. Assaf-Anid, K. Y. Lin, *J. Environ. Engineering-ASCE* **2002**, *128*, 94.

133 P. C. Chiu, M. Reinhard, *Environ. Sci. Technol.* **1996**, 30, 1882.

134 T. A. Lewis, A. Paszczynski, S. W. Gordon-Wylie, S. Jeedigunta, C. H. Lee, R. L. Crawford, *Environ. Sci. Technol.* **2001**, *35*, 552.

135 J. A. Perlinger, J. Bushmann, W. Angst, R. P. Schwarzenbach, *Environ. Sci. Technol.* **1998**, *32*, 2431.

136 J. Buschmann, W. Angst, R. P. Schwarzenbach, *Environ. Sci. Technol.* **1999**, *33*, 1015.

137 B. D. Habeck, K. L. Sublette, *Appl. Biochem. Biotechnol.* **1995**, *51/52*, 747.

138 D. R. Burris, C. A. Delcomyn, M. H. Smith, A. L. Roberts, *Environ. Sci. Technol.* **1996**, *30*, 3047.

139 D. R. Burris, C. A. Delcomyn, B. L. Deng, L. E. Buck, K. Hatfield, *Environ. Toxicol. Chem.* **1998**, *17*, 1681.

140 G. Glod, W. Angst, C. Holliger, R. P. Schwarzenbach, *Environ. Sci. Technol.* **1997**, *31*, 253.

141 G. Glod, U. Brodmann, W. Angst, C. Holliger, R. P. Schwarzenbach, *Environ. Sci. Technol.* **1997**, *31*, 3154.

142 Y. H. Kim, E. R. Carraway, *Environ. Technol.* **2002**, *23*, 1135.

143 J. K. Gotpagar, E. A. Grulke, D. Bhattacharyya, *J. Hazardous Materials* **1998**, *62*, 243.

144 E. J. O'Loughlin, D. R. Burris, C. A. Delcomyn, *Environ. Sci. Technol.* **1999**, *33*, 1145.

145 H. Ma, E. J. O'Loughlin, D. R. Burris, *Environ. Sci. Technol.* **2001**, *35*, 717.

146 S. Lesage, S. Brown, K. Millar, *Environ. Sci. Technol.* **1998**, *32*, 2264.

147 I. Dror, M. A. Schlautman, *Environ. Toxicol. Chem.* **2003**, *22*, 525.

148 I. Dror, M. A. Schlautman, *Environ. Toxicol. Chem.* **2004**, *23*, 252.

149 C. J. Gantzer, L. P. Wackett, *Environ. Sci. Technol.* **1991**, *25*, 715.

150 T. Arai, K. Kashitani, H. Kondo, S. Sakari, *Bull. Chem. Soc. Jpn.* **1994**, *67*, 705.

151 G. K. Lahiri, L. J. Schussel, A. M. Stolzenberg, *Inorg. Chem.* **1992**, *31*, 4991.

152 G. K. Lahiri, A. M. Stolzenberg, *Inorg. Chem.* **1993**, *32*, 4409.

153 M. Stolzenberg, Z. Zhang, *Inorg. Chem.* **1997**, *36*, 593.

154 M. J. Morra, V. Borek, J. Koolpe, *J. Environ. Quality* **2000**, *29*, 706.

155 J. Y. Becker, J. B. Kerr, D. Pletcher, R. Rosas, *J. Electroanal. Chem.* **1981**, *117*, 87.

156 C. P. Andrieux, A. Merz, J. M. Saveant, R. Tomabogh, *J. Am. Chem. Soc.* **1984**, *106*, 1957.

157 N. J. Bunce, S. G. Serica, J. Lipkowski, *Chemosphere* **1997**, *35*, 2719.

158 D. Nünnecke, J. Voss, *Acta Chem. Scand.* **1999**, *53*, 824.

159 C. Gosden, D. Pletcher, *J. Organomet. Chem.* **1980**, *186*, 401.

160 C. Gosden, J. B. Kerr, D. Pletcher, R. Rosas, *J. Electroanal. Chem.* **1981**, *117*, 101.

161 D. Pletcher, A. J. Sheridan, *Electrochim. Acta* **1998**, *43*, 3105.

162 M. Troupel, Y. Rollin, S. Sibille, J. Perichon, *J. Organomet. Chem.* **1980**, *202*, 435.

163 Y. Rollin, M. Troupel, D.G. Tuck, J. Perichon, *J. Organomet. Chem.* **1986**, *303*, 131.

164 A. Bakac, J.H. Espenson, *J. Am. Chem. Soc.* **1986**, *108*, 719.

165 M.A. Fox, D.A. Chandler, C. Lee, *J. Org. Chem.* **1991**, *56*, 3246.

166 S.P. Zhang, J.F. Rusling, *Environ. Sci. Technol.* **1993**, *27*, 1375.

167 S.P. Zhang, J.F. Rusling, *Environ. Sci. Technol.* **1995**, *29*, 1195.

168 J.F. Rusling, G.N. Kamau, *J. Electroanal. Chem.* **1985**, *187*, 355.

169 J.F. Rusling, S. Schweizer, S. Zhang, G.N. Kamau, *Colloids and Surfaces A: Physicochemical and Engineering Aspects* **1994**, *88*, 41.

170 K.S. Alleman, D.G. Peters, *J. Electroanal. Chem.* **1998**, *451*, 121.

171 K.S. Alleman, D.G. Peters, *J. Electroanal. Chem.* **1999**, *460*, 207.

172 C. Ji, D.G. Peters, J.A. Karty, J.P. Reilly, M.S. Mubarak, *J. Electroanal. Chem.* **2001**, *516*, 50.

173 A.J. Moad, L.J. Klein, D.G. Peters, J.A. Karty, J.P. Reilly, *J. Electroanal. Chem.* **2002**, *531*, 163.

174 J.D. Persinger, J.L. Hayes, L.J. Klein, D.G. Peters, J.A. Karty, J.P. Reilly, *J. Electroanal. Chem.* **2004**, *568*, 157.

175 D. Brevet, D. Lucas, H. Cattey, F. Lemaitre, Y. Mugnier, P.D. Harvey, *J. Am. Chem. Soc.* **2001**, *123*, 4340.

176 D. Brevet, Y. Mugnier, F. Lemaitre, D. Lucas, S. Samreth, P.D. Harvey, *Inorg. Chem.* **2003**, *42*, 4909.

177 H. Alper, K.D. Logbo, H. des Abbayes, *Tetrahedron Lett.* **1977**, *33*, 2861.

178 J.J. Brunet, M. Taillefer, *J. Organomet. Chem.* **1988**, *348*, C5.

179 J.J. Brunet, D. Demontauzon, M. Taillefer, *Organometallics* **1991**, *10*, 341.

180 G. Cavinato, L. Toniolo, *J. Mol. Catal.* **1993**, *78*, 121.

181 G. Cavinato, L. Toniolo, *J. Mol. Catal. A-Chem.* **1999**, *143*, 325.

182 E. Trabuco, P.C. Ford, *J. Mol. Catal. A-Chem.* **1999**, *148*, 1.

183 R.J.C. Brown, P.J. Dyson, D.J. Ellis, T. Welton, *J. Chem. Soc. Chem. Commun.* **2001**, 1862.

184 W. Buijs, *Catal. Today* **1996**, *27*, 159.

185 N.Y. Adonin, V.F. Starichenko, *Mendeleev Commun.* **2000**, *2*, 60.

186 D.V. Trukhin, N.Y. Adonin, V.F. Starichenko, *Russ. J. Org. Chem.* **2000**, *36*, 1227.

187 N.Y. Adonin, V.F. Starichenko, *J. Fluor. Chem.* **2000**, *101*, 65.

188 C.L. Hill, M. Kozik, J. Winkler, Y.Q. Hou, C.M. Prosser-McCartha, *Adv. Chem. Ser.* **1993**, *238*, 243.

189 D. Sattari, C.L. Hill, *J. Chem. Soc. Chem. Commun.* **1990**, 634.

190 D. Sattari, C.L. Hill, *J. Am. Chem. Soc.* **1993**, *115*, 4649.

191 E. Gkika, P. Kormali, S. Antonaraki, D. Dimoticali, E. Papaconstantinou, A. Hiskia, *Int. J. Photoenergy* **2004**, *6*, 227.

192 H. Shimakoshi, M. Tokunaga, T. Baba, Y. Hisaeda, *J. Chem. Soc. Chem. Commun.* **2004**, *16*, 1806.

193 J.P. Collman, L.S. Hegedus, J.R. Norton, R.G. Finke, *Principles and Applications of Organotransition Metal Chemistry*, University Science Books, Mill Valley, CA, **1987**, p. 278.

194 R.H. Crabtree, *The Organometallic Chemistry of the Transition Metals*, 2nd edn. John Wiley & Sons, New York, **1994**, p. 140.

195 C. Amatore, E. Carré, A. Jutand, M.A. M'Barki, *Organometallics* **1995**, *14*, 1818.

196 M. Portnoy, D. Milstein, *Organometallics* **1993**, *12*, 1665.

197 T.T. Tsou, J.K. Kochi, *J. Am. Chem. Soc.* **1979**, *101*, 6319.

198 L. Eberson, M. Ekström, *Acta Chem. Scand. B* **1987**, *41*, 41.

199 K.M. McCauley, S.R. Wilson, W.A. van der Donk, *Inorg. Chem.* **2002**, *41*, 393.

200 M.K. Whittlesey, R.N. Perutz, M.H. Moore, *J. Chem. Soc. Chem. Commun.* **1996**, 787.

201 J. Burdeniuc, B. Jedlicka, R.H. Crabtree, *Chem. Ber./Recueil* **1997**, *130*, 145.

202 G.N. Schrauzer, E. Deutsch, *J. Am. Chem. Soc.* **1969**, *91*, 3341.

203 K.L. Brown, X. Zou, M. Richardson, W.P. Henry, *Inorg. Chem.* **1991**, *30*, 4834.

204 K.L. Brown, G.Z. Wu, *Organometallics* **1993**, *12*, 496.

205 R. J. Kinney, W. D. Jones, R. G. Bergman, *J. Am. Chem. Soc.* **1978**, *100*, 7902.

206 S. C. Kao, M. Y. Darensbourg, *Organometallics* **1984**, *3*, 646.

207 W. D. Jones, *J. Chem. Soc. Dalton Trans.* **2003**, *21*, 3991 and references therein.

208 J. P. Collman, L. S. Hegedus, J. R. Norton, R. G. Finke, *Principles and Applications of Organotransition Metal Chemistry*, University Science Books, Mill Valley, CA, **1987**, p. 89.

209 J. P. Collman, L. S. Hegedus, J. R. Norton, R. G. Finke, *Principles and Applications of Organotransition Metal Chemistry*, University Science Books, Mill Valley, CA, **1987**, p. 286.

210 R. H. Crabtree, *The Organometallic Chemistry of the Transition Metals*, 2nd edn. John Wiley & Sons, New York, **1994**, p. 229.

211 V. V. Grushin, C. Bensimon, H. Alper, *Organometallics* **1995**, *14*, 3259.

212 R. A. Michelin, S. Faglia, P. Uguagliati, *Inorg. Chem.* **1983**, *22*, 1831.

213 A. D. Selmeczy, W. D. Jones, R. Osman, R. N. Perutz, *Organometallics* **1995**, *14*, 5677

214 T. Imamoto, T. Takeyama, T. Kusumoto, *Chem. Lett.* **1985**, 1491.

215 S. M. H. Tabaei, C. U. Pittman, Jr., K. T. Mead, *J. Org. Chem.* **1992**, *57*, 6669.

19
Homogeneous Catalytic Hydrogenation of Polymers

Garry L. Rempel, Qinmin Pan, and Jialong Wu

19.1
General Introduction

Chemical modification of polymers via catalysis is of great importance as it provides an efficient synthetic route for the production of novel polymers with desirable physical properties. It also allows the introduction of functional groups that are often inaccessible by conventional polymerization techniques. One of the most important chemical modifications is the hydrogenation of unsaturated carbon-carbon double bonds in polymers. Basically, there are two technical routes for the hydrogenation of polymers: homogeneous and heterogeneous. This chapter reviews research and process development with respect to homogeneous catalytic hydrogenation of diene-based polymers.

19.1.1
Diene-Based Polymers

Diene polymers refer to polymers synthesized from monomers that contain two carbon-carbon double bonds (i.e., diene monomers). Butadiene and isoprene are typical diene monomers (see Scheme 19.1). Butadiene monomers can link to each other in three ways to produce *cis*-1,4-polybutadiene, *trans*-1,4-polybutadiene and 1,2-polybutadiene, while isoprene monomers can link to each other in four ways. These dienes are the fundamental monomers which are used to synthesize most synthetic rubbers. Typical diene polymers include polyisoprene, polybutadiene and polychloroprene. Diene-based polymers usually refer to diene polymers as well as to those copolymers of which at least one monomer is a diene. They include various copolymers of diene monomers with other monomers, such as poly(butadiene-styrene) and nitrile butadiene rubbers. Except for natural polyisoprene, which is derived from the sap of the rubber tree, *Hevea brasiliensis*, all other diene-based polymers are prepared synthetically by polymerization methods.

The Handbook of Homogeneous Hydrogenation.
Edited by J. G. de Vries and C. J. Elsevier
Copyright © 2007 WILEY-VCH Verlag GmbH & Co. KGaA, Weinheim
ISBN: 978-3-527-31161-3

$H_2C{=}CH{-}CH{=}CH_2$ $\qquad$ $CH_2{=}C(CH_3){-}CH{=}CH_2$

Scheme 19.1 The monomers butadiene and isoprene.

19.1.2 Hydrogenation of Diene-Based Polymers

Polymers obtained by polymerizing or copolymerizing conjugated dienes are widely utilized for commercial purposes. These polymers have residual double bonds in their polymer chains. A limited amount of these residual double bonds are advantageously utilized for vulcanization, yet the residual double bonds have a disadvantage in that they lack the stability to resist weather conditions, oxidation, and ozone. Such disadvantages are more severe for block copolymers of conjugated dienes and vinyl aromatic hydrocarbons when used as thermoplastic elastomers. Moreover, the disadvantages are even more severe when the polymers are used as modifiers and transparent impact-resistant materials for styrenic resins and olefinic resins, or when they are used to make parts of equipment for utilization in solvent/oily/high-temperature environments. This deficiency in stability can be notably improved by hydrogenating the conjugated diene polymers in order to eliminate the residual double bonds that persist within the polymer chains.

With the availability of a large number of unsaturated polymers of differing microstructures, the selective reduction of carbon-carbon double bonds offers a means of producing a wide variety of specialty polymers. By reducing the unsaturation level of the polymers, their physical properties – such as tensile strength, elongation, thermal stability, light stability and solvent resistance – may be optimized. For example, the removal of the C=C unsaturation in polybutadiene (PB) provides a tough semi-crystalline polymer similar to linear polyethylene or an elastomer such as poly(ethylene-*co*-butylene), depending on the relative levels of units with 1,2 or 1,4 structure (Scheme 19.2 a). Hydrogenation of a styrene-butadiene-styrene triblock copolymer (SBS) with a moderate amount of 1,2-addition units in the center block yields a copolymer with a poly(ethylene-*co*-butylene) center segment (Scheme 19.2 b). This modified polymer has greatly increased thermal and oxidative stability, together with processability and serviceability at higher temperature, by virtue of its poly(ethylene-*co*-butylene) center block. The catalytic hydrogenation of acrylonitrile-butadiene rubber (NBR) (Scheme 19.2 c) is an especially important commercial example, resulting in its tougher and more stable derivative, hydrogenated nitrile butadiene rubber (HNBR), which has been widely used in the automotive industry.

Numerous methods have been employed for hydrogenating conjugated diene-based polymers in the presence of suitable and effective hydrogenation catalysts. Typical hydrogenation catalysts can be classified into two types: heterogeneous and homogeneous catalysts.

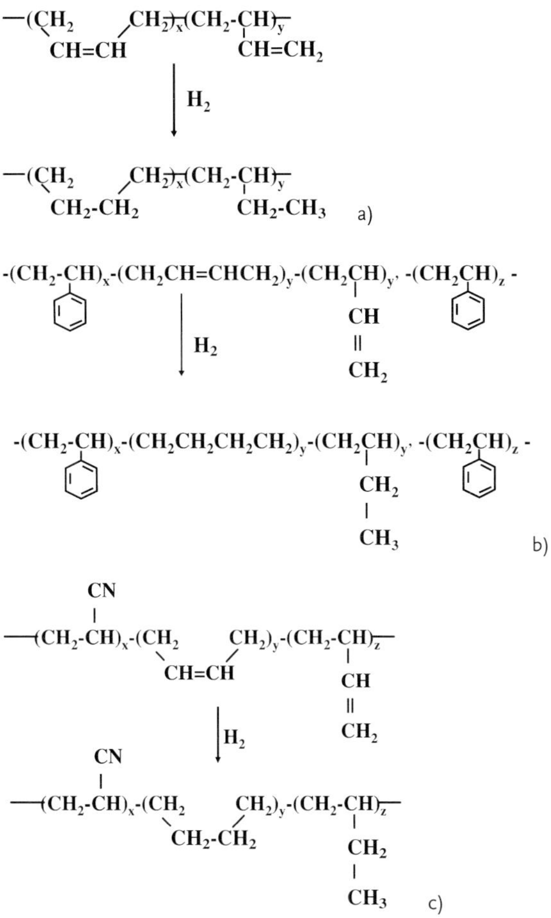

$$-(CH_2\text{–}CH{=}CH\text{–}CH_2)_x\text{–}(CH_2\text{–}CH(CH{=}CH_2))_y-$$

$$\downarrow H_2$$

$$-(CH_2\text{–}CH_2\text{–}CH_2\text{–}CH_2)_x\text{–}(CH_2\text{–}CH(CH_2\text{–}CH_3))_y- \quad \text{a)}$$

$$-(CH_2\text{-}CH(C_6H_5))_x\text{-}(CH_2CH{=}CHCH_2)_y\text{-}(CH_2CH(CH{=}CH_2))_{y'}\text{-}(CH_2CH(C_6H_5))_z-$$

$$\downarrow H_2$$

$$-(CH_2\text{-}CH(C_6H_5))_x\text{-}(CH_2CH_2CH_2CH_2)_y\text{-}(CH_2CH(CH_2\text{-}CH_3))_{y'}\text{-}(CH_2CH(C_6H_5))_z- \quad \text{b)}$$

$$-(CH_2\text{-}CH(CN))_x\text{-}(CH_2\text{-}CH{=}CH\text{-}CH_2)_y\text{-}(CH_2\text{-}CH(CH{=}CH_2))_z-$$

$$\downarrow H_2$$

$$-(CH_2\text{-}CH(CN))_x\text{-}(CH_2\text{-}CH_2\text{-}CH_2\text{-}CH_2)_y\text{-}(CH_2\text{-}CH(CH_2\text{-}CH_3))_z- \quad \text{c)}$$

Scheme 19.2 (a) Hydrogenation of polybutadiene.
(b) Hydrogenation of the copolymer of butadiene and styrene.
(c) Hydrogenation of acrylonitrile butadiene rubber.

19.1.2.1 **Heterogeneous Catalysts**

Heterogeneous catalysts, which are not soluble in the diene-based polymers to be hydrogenated, have involved metals such as nickel, platinum, or palladium deposited onto supports such as activated carbon, silica, alumina, or calcium carbonate. Hydrogenation using a heterogeneous catalyst is described as follows. First, the polymer to be hydrogenated is dissolved in a suitable solvent, after which the polymer is brought into contact with hydrogen in the presence of a heterogeneous catalyst. During hydrogenation of the polymer, contact between the polymer and catalyst is difficult due to the influence of the high viscosity of the reaction system and the influence of steric hindrance of the polymer chain. In addition the polymer, once hydrogenated, tends to remain on the surface of the catalyst and to interfere with subsequent access to active centers of the catalyst with the non-hydrogenated polymer.

The hydrogenation also requires higher temperature and pressure, and hence decomposition of the polymer and gelation of the reaction system often tends to occur. For the hydrogenation of a copolymer of a conjugated diene with a vinyl aromatic hydrocarbon, even hydrogenation of the aromatic ring portion may take place to some extent due to the high temperature and hydrogen pressure. Thus, it is often difficult to selectively hydrogenate only the double bonds in the conjugated diene portion of the polymer. Although widely used in industry, heterogeneous catalysts generally have lower activity than homogeneous catalysts, and usually a larger amount of heterogeneous catalyst is required. Furthermore, since the polymer may be strongly adsorbed onto the heterogeneous catalyst, it is impossible to completely remove the catalyst from the hydrogenated polymer solution, although the separation of catalyst from heterogeneous catalytic systems is usually easier than from homogeneous ones.

19.1.2.2 **Homogeneous Catalysts**

Homogeneous catalysts, which are soluble in the solutions of the diene-based polymers to be hydrogenated, include:

- Ziegler-type catalysts obtained from an organic acid salt or acetylacetone salt of nickel, cobalt, iron, or chromium which reacts with a reducing agent such as an organic aluminum compound.
- Metallocene complexes.
- Organometallic catalysts containing Ru-, Rh-, Os-, Ir-, Pd-complexes, etc.

The reaction operation using a homogeneous catalyst is similar to that using heterogeneous catalysts, except for the nature of the contact between the catalyst and the polymer. First, the polymer to be hydrogenated is dissolved in a suitable solvent. The catalyst can be prepared in a solution or solid form. Then, the catalyst is mixed with the polymer solution in the presence of hydrogen. During hydrogenation of the polymer, contact between the polymer and catalyst is much more effective compared to a heterogeneous catalyst system. Therefore, compared to the heterogeneous catalyst, the homogeneous catalyst generally shows higher activity, and only a small amount of catalyst enables hydrogenation to be effected at mild temperature and low pressure. In addition, homogeneous catalysts usually have high selectivity. When appropriate hydrogenation conditions are selected, it is possible to preferentially hydrogenate the conjugated diene portion of a copolymer of a conjugated diene with a vinyl aromatic hydrocarbon, while not hydrogenating the aromatic ring portion. Compared to heterogeneous catalysis, the disadvantage of homogeneous catalysis is that the catalyst recovery from a homogeneous system is often much more difficult than from a heterogeneous system. Therefore, in recent years, a new class of catalysts termed "heterogenized homogeneous catalysts" have received significant attention and have become an area of extensive research interest.

This chapter provides a review of the progress in reaction art, reactor techniques and process technology with respect to homogeneous catalytic hydrogenation of diene-based polymers, in accordance with the homogeneous hydrogenation theme of this handbook.

19.2 Reaction Art

19.2.1 Catalyst Techniques

Although the overall polymer hydrogenation reaction is exothermic, a high activation energy prevents it from taking place under normal conditions in the absence of a catalyst. An efficient catalyst is needed to circumvent this restriction and it is in fact the key in realizing a successful homogeneous hydrogenation. Metals from Group VIIIB (e.g., platinum, palladium, nickel, rhodium, ruthenium, iridium, osmium) and from Group IV (e.g., titanium) are among the most widely used hydrogenation catalysts, and they usually can be activated for hydrogenation at a temperature of less than 200 °C.

An early US Patent [1] disclosed that the catalyst $RhHCO(PPh_3)_3$ was effective for a limited degree of hydrogenation of PB. However, because of the very low reaction temperature and hydrogen pressure employed (25 °C, 2 atm, 3 h), the concentration of catalyst used was about 1100 μM, which was very high compared to the level now used in industrial polymer hydrogenation processes. Moreover, the hydrogenation degree was very low where mainly some of the vinyl C=Cs were hydrogenated. However, it indicated there was some degree of efficacy of the rhodium catalyst in hydrogenating PB. The low hydrogenation efficiency was successfully overcome for the hydrogenation of diene-based polymers by increasing the hydrogenation temperature to 110 °C or higher [2], in which a much lower catalyst concentration was used and a hydrogenation degree of 99.95% was achieved.

Research in this area over the past 30 years has focused on the design of efficient catalytic systems and on improving the hydrogenation operations. Besides rhodium-based species, a variety of catalysts has been investigated for the homogeneous hydrogenation of diene-based polymers, including Os-, Ru-, Ir-, Ni- and Ti-based catalysts. The patents most relevant to the hydrogenation technology of diene-based polymers are summarized in Table 19.1.

Two critical reviews with respect to polymer hydrogenation have been published [63, 64]. Among the investigated catalysts, Wilkinson's catalyst ($RhCl(PPh_3)_3$) is still considered as the most preferable for the hydrogenation of diene-based polymers as it provides high selectivity towards the olefin double bonds with minimized crosslinking problems. The only concern for using the rhodium catalyst is its relatively high cost.

In order to overcome the catalyst recovery problem in homogeneous hydrogenation operations, the significant enhancement of catalyst activity has been pur-

Table 19.1 Outline of major patents with respect to catalysts for diene-based polymer hydrogenation.

Catalyst	Polymer	Representative example [a)]	Assignee	Reference (Year)
$RhHCO(PPh_3)_3$; $RuHCl(PPh_3)_3$	PB	PB in toluene (2.3 wt.%); $RhHCO(PPh_3)_3$: 5 g per 100 g PB; P_{H2}: 0.2 MPa; *T*: 25 °C; *t*: 3 h; Conversion: 10.4%	The Firestone Tire & Rubber Company (Akron, USA)	1 (1976)
$RhH(PPh_3)_4$; $RhCl(PPh_3)_3$/epoxidized soybean oil ($R^1OCH_2CH(OR^2)CH_2)R^3$)	NBR, SBS, methacrylic acid-acrylonitrile – butadiene copolymer	NBR in chlorobenzene (2.5 wt.%); $RhH(PPh_3)_4$: 1 g per 100 g NBR; PPh_3: 10 g per 100 g NBR; P_{H2}: 2.8 MPa; *T*: 160 °C; *t*: 1 h; Conversion: >99%	Polysar Limited/Bayer Inc. (Sarnia, Canada)	2 (1984) 3 (1985) 4 (2004)
$OsHX(CO)(L)(PR_3)_2$, where X: Cl, BH_4 or CH_3COO; L: O_2 or no ligand; R: cyclohexyl or isopropyl	NBR, styrene butadiene random copolymer (SBR), methacrylic acid-acrylonitrile – butadiene copolymer	NBR in chlorobenzene (2.4 wt.%); $OsHCl(CO)(P(cyclohexyl)_3)_2$: 0.25 g per 100 g NBR; P_{H2}: 2.07 MPa; *T*: 130 °C; *t*: 1 h; Conversion: 99.3%	University of Waterloo (Waterloo, Canada)	5 (1996)
$RuX_m(L^1)(L^2)_n$, where X: H, Halogen or $SnCl_3$, etc.; L^1: H, Halogen or substituted indenyl, etc.; L^2: phosphane or arsane, etc.	NBR	NBR in butanone (9.7 wt.%); $RuCl_2(PPh_3)_3$: 0.16 g per 100 g NBR; P_{H2}: 14 MPa; *T*: 130 °C; *t*: 4 h; Conversion: 99.7%	Bayer Aktiengesellschaft (Leverkusen, Germany)	6 (1986) 7 (1989) 8 (1991)

Table 19.1 (continued)

Catalyst	Polymer	Representative example[a)]	Assignee	Reference (Year)
$HRuCl(CO)(PPh_3)_3$; $RuCl_2(PPh_3)_3$; $RuH_2(CO)(PPh_3)_3$, etc.	NBR, SBS, styrene butadiene diblock copolymer (SB), acrylonitrile-isoprene copolymer	NBR in chlorobenzene (1.7 wt.%); $HRuCl(CO)(PPh_3)_3$: 0.8 g per 100 g NBR; C_2H_5COOH: 0.2 g per 100 g NBR; P_{H2}: 4.1 MPa; T: 140 °C; t: 1.8 h; Conversion: >99%	University of Waterloo (Waterloo, Canada)	9 (1989) 10 (1989) 11 (1991)
$HRuCl(CO)(PCy_3)_3$, etc./ Amine(RNH_2)	NBR	NBR in chlorobenzene (1.7 wt.%); $HRuCl(CO)(PCy_3)_3$: 0.05 g per 100 g NBR; octylamine: 0.39 g per 100 g NBR; P_{H2}: 5.5 MPa; T: 145 °C; t: 5 h; Conversion: >99%	University of Waterloo (Waterloo, Canada)	12 (1991)
$HRuCl(CO)(PPh_3)_3$, etc.	Ring-opening polymers	Poly(8-methyl-8-methoxy-carbonyltetracyclo-3-dodecene) in toluene (20 wt.%); $HRuCl(CO)(PPh_3)_3$: 0.01 g per 100 g polymer; P_{H2}: 4.0 MPa; T: 160 °C; t: 4 h; Conversion: 99.7%	Japan Synthetic Rubber Co., Ltd. (Tokyo, Japan)	13 (1993)
$RuCl_2(PPh_3)_4$, etc.	Ring-opening metathesis polymers	Polymer in tetrahydrofuran (11.4 g L^{-1}); $RuCl_2(PPh_3)_4$: 0.05 g per 100 g polymer; Et_3N: 0.021 g per 100 g polymer; P_{H2}: 8.3 MPa; T: 165 °C; t: 5 h; Conversion: 100%	Mitsui Chemicals Inc. (Tokyo, Japan)	14 (1999)

Table 19.1 (continued)

Catalyst	Polymer	Representative example [a)]	Assignee	Reference (Year)
Bimetallic complexes, Rh/Ru ($RhCl(PPh_3)_3/RuCl_2(PPh_3)_3$), etc.	NBR	NBR in xylene (7 g per 100 mL); $RhCl(PPh_3)_3/RuCl_2(PPh_3)_3$ (Rh : Ru = 3 : 1, mole ratio): 0.4 g per 100 g NBR, PPh_3: 2 g per g catalyst; P_{H2}: 0.8 MPa; *T*: 145 °C; *t*: 4 h; Conversion: 98.5%	Nantex Industry Co., Ltd., (Taiwan)	15 (2000)
Nickel acetylacetonate ($(CH_3COCH{=}C(O\text{-})CH_3)_2Ni$)/ *p*-nonylphenol ($C_9H_{19}C_6H_4OH$)/ *n*-butyllithium ($CH_3(CH_2)_3Li$)	Polyisoprene, EPDM	Polyisoprene in hexane (2.5 wt.%); Catalyst (Nickel acetylacetonate: *p*-nonylphenol: *n*-butyllithium = 8 : 8 : 25, mol ratio): 4 mmol nickel per 100 mL solution; P_{H2}: 0.34 MPa; *T*: room temperature; *t*: 0.8 h; Conversion: 78%	Uniroyal Inc. (New York, USA)	16 (1976)
Cobalt or nickelbenzohydroxamic acid ($(C_6H_5CONH\text{-}O\text{-})_2M$, M = Co, Ni)/ organoaluminum (R_3Al, R = alkyl)	PB, SBR	SBR in hexane (10 wt.%); Catalyst (nickel benzohydroxamic acid : triisobutyl aluminum = 1 : 3, mol ratio): 0.01 mmol Ni per g SBR; P_{H2}: 1.4 MPa; *T*: 180 °C; *t*: 2 h; Conversion: 99%	The Firestone Tire & Rubber Company (Akron, USA)	17 (1976)

Table 19.1 (continued)

Catalyst	Polymer	Representative example a)	Assignee	Reference (Year)
Nickel-2-ethylhexanoate (($CH_3(CH_2)_3CH(C_2H_5)CO_2)_2Ni$)/alkylaluminum ($R_3Al$, R = alkyl) or alkylalumoxane ($(\text{-Al(R)O-})_n$) or hydrocarbyl-substituted silicon alumoxane ($(Et_2AlO)_2SiO_2$, etc.)	SBS, Styrene-butadiene-isoprene copolymers	SBS in cyclohexane (18 wt.%); Catalyst (nickel-2-ethylhexanoate/triethyl aluminum, Al : Ni = 2.2 : 1, atomic ratio): 0.05 g Ni per 90 g SBS; P_{H2}: 6.2 MPa; *T*: 90 °C; *t*: 3 h; Conversion: 93.4%	Shell Oil Company (Houston, USA)	18 (1991) 19 (1991) 20 (1991) 21 (1989) 22 (1991)
Nickel acetylacetonate (($CH_3COCH{=}C(O\text{-})CH_3)_2Ni$)/alkylaluminum ($R_3Al$, R = alkyl)	SBR	SBR (0.9 L BD + 0.5 L ST anionic polymerization) in 4.8 L cyclohexane; Catalyst: 0.3 g nickel acetylacetonate (in 9 mL toluene) + 11 mL aluminum triisobutyl solution (10% in hexane); H_2O: 14 mol per mol Ni; P_{H2}: 1 MPa; *T*: 25–30 °C; *t*: 1 h; Conversion: 93.4%	BASF Aktiengesellschaft (Germany)	23 (1980)
Nickel bis(acetylacetonate) or 2-ethylhexanoic acid nickel (($CH_3COCH{=}C(O\text{-})CH_3)_2Ni$ or $(CH_3(CH_2)_3CH(C_2H_5)CO_2)_2Ni$)/alkylaluminum ($R_3Al$, R = alkyl)	Styrene-isoprene-styrene block copolymer (SIS), polyisoprene, PB	PB in cyclohexane (20 wt.%); Catalyst (2-ethylhexanoic acid nickel/triisobutyl aluminum, Ni : Al = 1 : 3): 0.74 mol Ni per 100 g PB; P_{H2}: 0.98 MPa; *T*: 70–80 °C; *t*: 6 h; Conversion: 97%	Kuraray Co., Ltd. (Tokyo, Japan)	24 (2002) 25 (2002) 26 (2003)

Table 19.1 (continued)

Catalyst	Polymer	Representative example [a)]	Assignee	Reference (Year)
Nickel or Cobalt acetylacetonate ($(CH_3COCH{=}C(O\text{-})CH_3)_nM$, n = 2,3; M = Ni, Co)/alkylaluminum (R_3Al, R = alkyl); titanium compounds (Cp_2TiCl_2, etc.)/organoaluminum (R_3Al, etc., R = alkyl) or organolithium (RLi, R = alkyl)	Metathesis polymers	Poly(dicyclopentadiene) in cyclohexane (7.6 wt.%); Catalyst (cobalt(III) acetylacetonate: triisobutyl-aluminum = 1.9 : 4.2, weight ratio): 0.2 g per g polymer; P_{H2}: 0.98 MPa; *T*: 60 °C; *t*: 0.5 h; Conversion: 99.9%	Nippon Zeon Co., Ltd (Tokyo, Japan)	27 (1996)
Cobalt-2-ethylhexanoate ($(CH_3(CH_2)_3CH(C_2H_5)CO_2)_2Co$)/ alkylaluminum ($R_3Al$, R = alkyl)	Polymers containing ketone groups	SBS (include ketone groups) in cyclohexane (~5 wt.%); Catalyst (cobalt-2-ethylhexanoate: triethyl aluminum = 1 : 3.9, weight ratio): [Co] = 880 ppm (added in increments); P_{H2}: 3.45 MPa; *T*: 70–90 °C; *t*: 6 h; Conversion: 78%	Shell Oil Company (Houston, USA)	28 (1996) 29 (1997)
Palladium-2-ethylhexanoate ($(CH_3(CH_2)_3CH(C_2H_5)CO_2)_2Pd$)/ organoaluminum ($(\text{-}Al(CH_3)O\text{-})_n$)	NBR	NBR in methylethyl ketone (5 wt.%); Catalyst (palladium-2-ethylhexanoate/methyl alumoxane, Al : Pd = 0.7 : 1, atomic ratio): 8.5 mmol Pd per lb NBR; P_{H2}: 6.2 MPa; *T*: 60 °C; *t*: 2 h; Conversion: 90%	Shell Oil Company (Houston, USA)	30 (1989) 31 (1989)

Table 19.1 (continued)

Catalyst	Polymer	Representative example [a)]	Assignee	Reference (Year)
Titanocene ($Cp_2Ti(C_6H_2(R^1)(R^2)(R^3))C_6H_2(R^4)(R^5)(R^6)$, $R^1 \sim R^6$: alkyls or hydrogen atoms)	PB, polyisoprene, styrene/isoprene or butadiene copolymers	SBS in cyclohexane (5 wt.%); Catalyst (di-*p*-tolyl-bis(η-cyclopentadienyl) titanium): 4 μM; P_{H2}: 0.49 MPa; *T*: 90 °C; *t*: 2 h; Conversion: 99%	Asahi Kase Kogyo Kabushiki Kaisha (Osaka, Japan)	32 (1987)
Titanocene ($Cp_2Ti(R^1)R^2$, R^1 and R^2: halogen atoms, aryloxy groups, etc.)/alkyloxy lithium (R^3OLi, R^3: a hydrocarbon group, etc.); Titanocene or zirconocene $Cp_2M^1(R^1)R^2$, M^1: Ti or Zr; R^1 and R^2: halogen atoms, aryloxy groups, etc.)/organoalkali (M^2OR^4, M^2: alkali metal, R^4: alkyl group, etc.)/organoaluminum or organomagnesium ($M^2M^3R^5{}_n$, M^2: alkali metal; M^3: Al or Mg; n: 3, 4; R^5: halogen, alkyl, aryl or alkoxy group, etc.)	SB, SBS, SIS	SB in cyclohexane (20 wt.%); Catalyst (bis(cyclopentadienyl) titanium dichloride: 2,6-di-butyl-4-methylphenoxylithium = 1 : 6, mole ratio): 25 μmol Ti per 100 g SB; P_{H2}: 0.98 MPa; *T*: 70 °C; *t*: 1 h; Conversion: 99%	Japan Synthetic Rubber Co. Ltd. (Tokyo, Japan)	33 (1990) 34 (1992)
Titanocene ($Cp_2Ti(R^1)R^2$, etc., R^1 and R^2: halogen atoms, aryloxy groups, etc.) or Tebbe's reagent ($Cp_2Ti(\mu\text{-}Cl)(\mu\text{-}CH_2)Al(CH_3)_2$)/co-catalyst and/or promoter (LiR, R: alkyl, etc.)	Styrene, butadiene and/or isoprene living copolymers	SBS in cyclohexane (18 wt.%); Catalyst (2,5-diphenylphospholyl (cyclopentadienyl) titanium dichloride): 0.5 mg per g SBS; P_{H2}: 1 MPa; *T*: 70 °C; *t*: 3 h; Conversion: 100%	Shell Oil Company (Houston, USA)	35 (1992) 36 (1992) 37 (1992) 38 (1992) 39 (1993) 40 (1994) 41 (1998) 42 (1999) 43 (1999) 44 (2002)

Table 19.1 (continued)

Catalyst	Polymer	Representative example[a]	Assignee	Reference (Year)
Titanocene or Zirconocene $((A)(E)M(R^4)^+$, A and E represent: $(R^1)_m$–cyclopentadienyl or $(R^3)_p$–indenyl–$(R^2)_q$; R^1–R^4: halogens, hydrocarbonyl groups, etc.)/non-coordinating stable anion ($B(Ar)_4^-$, Ar: C_6F_5, etc.)	Polyisoprene	Polyisoprene in bromobenzene (8 wt.%); Catalyst ($Cp^*Cp(tBu)ZrMe_2$: $(Ph_3C)(B(C_6F_5)_4$=1.1:1, mole ratio): 60 μmol per g polymer; P_{H2}: 0.1 MPa; *T*: not available; *t*: 0.75 h; Conversion: 100%	Shell Oil Company (Houston, USA)	45 (1999)
Rare earth (Sm) metallocene (Cp_2SmR)/organolithium (LiR, etc.)	SBS	SBS in cyclohexane (5 wt.%); Catalyst (*sec*-butyllithium: $(Cp^*_2SmH)_2$=7.5:1, mole ratio): 2.6 mmol per 100 g SBS; P_{H2}: 3.45 MPa; *T*: 40–66 °C; *t*: 3 h; Conversion: 81%	Shell Oil Company (Houston, USA)	46 (1992)
Rare earth (Sm) metallocene ($(C_5R_5)_2SmX$, X is an inert substituent capable of replacement by hydrogen)	PB	PB in cyclohexane (1.4 wt.%); Catalyst (bis(pentamethylcyclopentadienyl) (bis(trimethylsilyl)methyl) samarium: 0.018 g per g PB; P_{H2}: 2.8 MPa; *T*: 90 °C; *t*: 3 h; Conversion: 99.5%	The Dow Chemical Company (Midland, USA)	47 (1990)

Table 19.1 (continued)

Catalyst	Polymer	Representative example[a)]	Assignee	Reference (Year)
Monocyclopentadienyl titanium compound (R_nCpSmL_m, R: anion or dianion non-Cp group; L: neutral ligand)	SBR, PB	PB in cyclohexane (~10 wt.%); Catalyst (pentamethylcyclopenta-dienyl tribenzyl titanium): 0.66 mmol per 100 g PB; P_{H2}: 2.1 MPa; *T*: 55 °C; *t*: 24.6 h; Conversion: 71%	The Dow Chemical Company (Midland, USA)	48 (1998)
Metallocene (CpCp′MD, M: metal; D: conjugated, neutral diene)	Unsaturated polymers		The Dow Chemical Company (Midland, USA)	49 (2002)
Monocyclopentadienyl titanium compound ($CpTi(R^1)(R^2)R^3$, $R^1 \sim R^3$: non-Cp groups)	SBS, SBR, PB, SIS	SBS in cyclohexane (12.5 wt.%); Catalyst ($CpTiCl_2(\text{-}OC_5NH_4)$): 0.4 mmol per 100 g SBS; P_{H2}: 0.98 MPa; *T*: 80 °C; *t*: 3 h; Conversion: 99.4%	Korea Kumho Petrochemical Co., Ltd. (Seoul, Rep. of Korea)	50 (1999) 51 (1999) 52 (2000) 53 (2002)
Titanocene ($Cp_2Ti(PhOR)_2$ or Cp_2TiR_2)	PB, SBS, SB	SBS in cyclohexane (2 wt.%); Catalyst ($Cp_2Ti(4\text{-}CH_3OPh)_2$): 4 mmol per 100 g SBS; P_{H2}: 0.59 MPa; *T*: 85 °C; *t*: 2 h; Conversion: 91%	Repsol Quimica S.A. (Madrid, Spain)	54 (1996)
Zirconocene (Cp_2ZrR_2, R: halogen or alkyl group, etc.)/ alumoxane ($(\text{-}Al(CH_3)O\text{-})_n$, etc.)	SBS (star-form)	SBS in toluene (5.5 wt%); Catalyst (Cp_2ZrCl_2/methylalumoxane, Zr:Al = 1:118, atomic ratio): 3.08 mmol Zr per 100 g SBS; P_{H2}: 2 MPa; *T*: 90 °C; *t*: 0.65 h; Conversion: 80%	Neste Oy (Provoo, Finland)	55 (1998)

Table 19.1 (continued)

Catalyst	Polymer	Representative example [a]	Assignee	Reference (Year)
Cobaltocene or Nickellocene (Cp_2MR_2, M: metal)/Organic lithium (LiR, etc.)	SBS	SBS in cyclohexane (5.8 wt.%); Catalyst (bis(cyclopentadienyl) cobalt(II)/butyllithium): 4.1 mmol Co per 100 g SBS; P_{H2}: 2.45 MPa; *T*: 80 °C; *t*: 2 h; Conversion: 99.2%	Taiwan Synthetic Rubber Corporation (Taiwan)	56 (1998)
Titanocene (Cp_2TiR_2)/silyl hydride ($RSiO\text{-}(SiH(R)O\text{-})_nSiR_3$, etc.)	SBS	SBS in cyclohexane (13 wt.%); Catalyst (Cp_2TiMe_2 : ethylhydrocyclosiloxane = 1 : 1.5, mole ratio): 0.08 mmol per 100 g SBS; P_{H2}: 1.4 MPa; *T*: 60 °C; *t*: 1 h; Conversion: 99%	Industrial Technology Research Institute (Taiwan), Chi Mei Co. (Taiwan)	57 (2001)
Titanocene (Cp_2TiR_2)/organoderivate (MR_2, M: Mg or Zn)/modifier (ROR, etc.) or another organoderivate (AlR_3)	Styrene, butadiene and/or isoprene copolymers	SBS in cyclohexane (12.5 wt.%); Catalyst (bis-cyclopentadienyltitanium dichloride: 1,2-di-*n*-butoxy-ethane: diisobutyl magnesium = 1 : 0.1 : 0.25, mole ratio): 0.19 mmol per 100 g SBS; P_{H2}: 0.49 MPa; *T*: 70 °C; *t*: 1 h; Conversion: > 98%	Enichem S.p.A. (Milan, Italy)	58 (1999) 59 (2001)
Titanocene (Cp_2TiR_2)	SBS, SIS	SBS in cyclohexane (12.5 wt.%); Catalyst ($Cp_2Ti((C_5H_8)_2C_4H_9)$): 0.2 mmol Ti per 100 g SBS; P_{H2}: 0.8 MPa; *T*: 100 °C; *t*: 0.5 h; Conversion: > 98%	Enichem S.p.A. (Rome, Italy); Polimeri Europa S.p.A. (Brindisi, Italy)	60 (2004)

Table 19.1 (continued)

Catalyst	Polymer	Representative example[a)]	Assignee	Reference (Year)
Titanocene (Cp_2TiR_2)/titanium alkoxide (R_3TiOR)/trialkylaluminum (AlR_3)	SBS	SBS in cyclohexane (9.7 wt.%); Catalyst (titanium(IV) isopropoxide: bis(cyclopentadienyl) titanium dichloride: triisobutyl aluminum = 1 : 0.5 : 3, mole ratio): 0.17 mmol Ti per 100 g SBS; P_{H2}: 2.5 MPa; *T*: 80 °C; *t*: 1 h; Conversion: 97%	TSRC Corporation (Taiwan)	61 (2005)
Titanocene (Cp_2TiR_2)/alkyllithium (LiR)	Styrene, butadiene or isoprene copolymers	PB in cyclohexane and toluene (5 wt.%); Catalyst (bis(cyclopentadienyl) titanium dichloride): 0.4 mmol per 100 g PB; P_{H2}: 0.49 MPa; *T*: 40 °C; *t*: 2 h; Conversion: 97%	Asahi Kasei Kogyo Kabushiki Kaisha (Osaka, Japan)	62 (1985)

P_{H2}: pressure of hydrogen; *T*: temperature; *t*: time.

sued such that the amount of catalyst required may be significantly reduced. Parent et al. [65] described the selective hydrogenation of C=C within NBR using the homogeneous catalyst precursor, $OsHCl(CO)(O_2)(PCy_3)_2$ in solution, which has a high hydrogenation activity for hydrogenation of NBR as well as poly-isoprene [66]. However, this catalyst is not as effective as Wilkinson's catalyst in suppressing the polymer crosslinking problem, which tends to occur during the later stages of hydrogenation. Pan and Rempel [67] described an efficient catalytic system for the hydrogenation of styrene-butadiene rubber (SBR) in solution using a ruthenium complex ($Ru(CH{=}CHPh)Cl(CO)(PCy_3)_2$, where Ph = phenyl, Cy = cyclohexyl). Because of the high efficiency of the ruthenium catalyst, the catalyst required to realize the hydrogenation is used at a very low level. Indeed, even if all of the catalyst were to be retained in the final products the metal residue would still be less than 7 ppm. Ruthenium-based catalysts have also been used for the hydrogenation of NBR [6–12, 15, 68], polyisoprene [68, 69] and PB [68, 70]. As with the osmium-based catalyst, the ruthenium-based catalysts are efficient but more susceptible with respect to causing crosslinking in NBR during hydrogenation. The mechanism of crosslinking is not well understood. Although a Michael-type addition mechanism (see Scheme 19.3) was speculated to account for this problem, and there were some signs to support this mechanism (e.g., the hydrogenation of NBR catalyzed by ruthenium-based catalysts in the presence of an amine helped to suppress crosslinking [12]), this mechanism has not been substantiated by definitive experimental results [71].

Hsu et al. [15] applied a bimetallic catalyst comprising rhodium and ruthenium for the hydrogenation to combine the high selectivity of the rhodium complex with the lower cost of the ruthenium complex. When the amount of each metal is identical, the catalytic activity of the bimetallic complex catalyst system was similar to that of the single rhodium-complex catalyst, containing

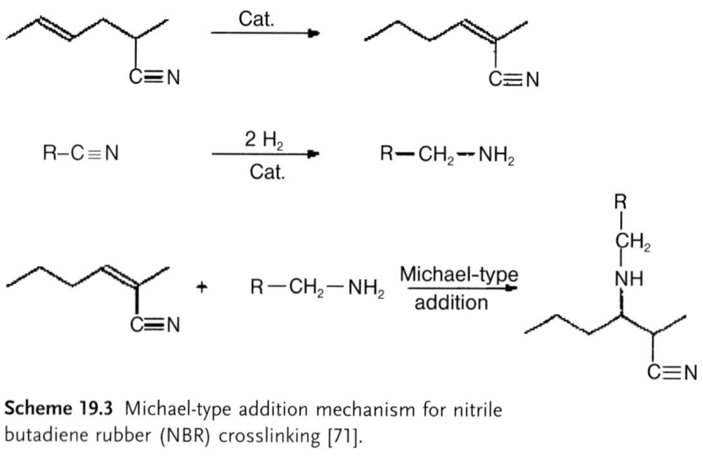

$$R{-}C{\equiv}N \xrightarrow[\text{Cat.}]{2\,H_2} R{-}CH_2{-}NH_2$$

Scheme 19.3 Michael-type addition mechanism for nitrile butadiene rubber (NBR) crosslinking [71].

the same total amount of metal. However, since part of the rhodium is substituted by ruthenium, the bimetallic catalyst becomes less expensive.

Few reports have been published on the hydrogenation of diene polymers using iridium complexes. Gilliom [72] and Gilliom and Honnell [73] described the use of $[Ir(COD)L_2]PF_6$, where COD = cyclooctadiene and L = a phosphine, for the hydrogenation of bulk PB. It was found that $[Ir(COD)(PMePh_2)_2]PF_6$, where Me = methyl and Ph = phenyl, when compared with Wilkinson's catalyst, resulted in a faster hydrogenation than the rhodium complex in the early stages of the reaction; however, the degree of hydrogenation achieved was less than that with the Wilkinson's catalyst. This study presented a rare example of polymer hydrogenation occurring in a pure polymer matrix. In this case an organic solvent was used initially to disperse the catalyst into the polymer matrix, with subsequent removal of the solvent before the hydrogenation operation.

Some examples of the hydrogenation of unsaturated rubber catalyzed by palladium complexes have been reported by Bhattacharjee et al. [74–78]. When palladium acetate ($(CH_3CO_2)_2Pd$) was used as a catalyst for the hydrogenation of NBR [74] and styrene-isoprene-styrene triblock copolymer (SIS) [75], the catalyst showed good selectivity. However the observed maximum conversion was 96% for NBR and 90% for SIS. This catalyst also showed activity for the hydrogenation of natural rubber and epoxidized natural rubber [76]. The results showed that the catalyst was highly selective in reducing olefinic unsaturation in the presence of epoxy groups, and an increase in the epoxy content of the rubber resulted in a decrease in hydrogenation rate and a decrease in the maximum attainable hydrogenation level. Another palladium complex, namely, a six-membered cyclopalladate complex of 2-benzoyl pyridine ($(Pd(CH_3COO)(C_6H_5COC_5H_4N))_2$), was used as catalyst for the selective hydrogenation of NBR, carboxylated nitrile rubber (XNBR) and PB [77, 78]. The reported maximum conversion of C=C using this complex was lower than that obtained with palladium acetate. The main drawback of the palladium complex catalysts is that the degree of hydrogenation achieved cannot satisfy the commercial requirement.

One type of Ziegler-catalyst system, used for polymer hydrogenation, consisted of an organic acid salt or acetylacetone salt of Ni, Co, Pd, etc. and a reducing agent such as an organoaluminum compound. Nickel-based catalysts, such as nickel 2-ethyl hexanoate ($(CH_3(CH_2)_3CH(C_2H_5)CO_2)_2Ni$)/triisobutyl aluminum ($((CH_3)_3CH)_3Al$), nickel acetylacetonate ($(CH_3COCH{=}C(O\text{-})CH_3)_2Ni$)/triisobutyl aluminum, and nickel benzohydroxamic acid ($(C_6H_5CONH\text{-}O\text{-})_2M$, M = Co, Ni)/triisobutyl aluminum, demonstrated activities for the hydrogenation of PB, SBR, and polyisoprene [16, 17, 79]. The Shell Oil Company (Houston, Texas) has filed a series of patents for the application of a Ni/Al catalyst system [18–22]. Besides nickel, cobalt [27–29] and palladium [30, 31] salts together with alkylaluminum compounds or alkylalumoxane have also been used as catalyst systems for polymer hydrogenation. The chief benefit of these systems is that they are relatively inexpensive in terms of the cost of metals used. However, these types of catalysts and co-catalysts need to be used at relatively high concentration in order to achieve favorable reaction rates at the chosen reaction conditions due to their relative lower activities compared with organometallic catalysts containing Pt-group metals. The high metal concentration in the solution is also a major obstacle when removal of the metal is necessary, as the residual catalyst adversely affects the stability of the hydrogenated product and is detrimental to various applications. Another drawback of these types of catalysts is their poor selectivity as hydrogenation of both ethylenic and aromatic unsaturation may occur, though under certain conditions they show some degree of selectivity for the hydrogenation of ethylenic unsaturations as a result of steric hindrance [21, 22].

Another type of Ziegler-catalyst, which was investigated for polymer hydrogenation, is that of metallocene catalysts. These catalysts consist mainly of halides or aryls of cyclopentadienyl Group III or IV metals (e.g., $RMX^1X^2X^3$, where R is an unsubstituted or substituted cycopentadienyl group; M is metal; X^1, X^2 and X^3 may be either the same or different selected from halogen atoms, aryl groups, aryloxy groups or carbonyl groups, etc., and one of them may be an unsubstituted or substituted cyclopentadienyl group) which are often treated with organolithium reducing agents, and have been applied for homogeneous polymer hydrogenation [32–62, 80]. The most widely used of these catalysts are substituted or unsubstituted bis(cyclopentadienyl)-titanium compounds; however, zirconium and hafnium complexes have also been reported for the purpose of polymer hydrogenation [34, 45, 55]. When these titanocene-based catalysts are used for the hydrogenation of unsaturated living polymers, the hydrocarbon lithium or lithium hydride compounds which are produced from the termination reactions of the hydrocarbon or hydrogen with the alkyllithium initiator serve as a reducing agent for the hydrogenation catalyst. In this case, the addition of an organolithium compound which is necessary for reduction of the complex during the hydrogenation process can be omitted [35–38, 41–44]. In this catalyst system, when the alkyllithium complex (lithium hydride generated *in situ*) or the alkoxy lithium compound is used to activate the bis(cyclopentadienyl) titanium catalyst for effective hydrogenation of polymers, an excess amount of the lithium species may also induce reduction of the titanium compounds, resulting in decomposition of the catalyst component as well as a reduction in catalyst activity. In this case some reagents such as ethanol or difluorodiphenyl silane can be added to the system to adjust the ratio of lithium hydride to titanium [37, 44]. Some of these catalysts may be used without the addition of Group I, II, IIIA alkyl compounds such as substituted titanocene biaryl compounds, etc. [32, 54].

Related to these catalysts are the systems based on lanthanide metal systems or rare earth metal complexes [46, 47]. The main problem with these catalyst systems is their instability. When the catalyst solution is prepared by reacting a metallocene with an organolithium compound in a polar solvent, the prepared catalyst solution is unstable and decomposes quickly, even under a nitrogen atmosphere. The activity of these catalysts can be high only if the catalyst is added to the polymer solution immediately after preparation. Attempts have been made to overcome the stability problem by using an additive in the system to improve the stability and the activity of the catalyst [33–35, 41, 57, 58, 61]. Recently, it was also found that monocyclopentadienyl titanium compounds are more stable and less sensitive to the extra lithium hydride [50–53] than bis(cyclopentadienyl) titanium compounds. Another disadvantage of the above catalysts is that the Group III and IV metal halides (as well as the lithium halides) formed from the catalyst system tend to corrode the metal reactors used in the hydrogenation process. This results in increased investment costs to provide expensive, corrosion-resistant metal alloy reactor systems. It may be possible to overcome such a problem by replacing the titanium with a Group VIIIB metal such as cobalt or nickel [56], as no corrosion of the reactor system was observed when a bis(cyclopentadienyl) cobalt (or nickel) complex and an organolithium compound were used as a catalyst system for the hydrogenation of polymers. Furthermore, this cobalt or nickel catalyst system showed much higher stability than the titanocene catalysts. In fact, the bis(cyclopentadienyl) cobalt(II)/*n*-butyllithium catalyst still had high activity for the hydrogenation of the polymer two weeks after it was prepared and stored under nitrogen.

The metallocene catalysts have good selectivity and high activity under mild reaction conditions. Given these advantages, it is to be expected that these systems would show promise for the homogeneous hydrogenation of polymers on a commercial scale. However, there is no report on the application of this catalyst for the hydrogenation of NBR. One possible reason for this is that the polar (CN) groups in NBR may bond to the active metal center and deactivate the catalyst.

19.2.2 Hydrogenation Kinetic Mechanism

An understanding of the kinetics and catalytic mechanism of polymer hydrogenation is essential in order to optimize the reaction conditions, to control the reaction systems, and to design commercial production processes. Catalytic kinetic mechanisms for Rh-, Os- and Ru-complex polymer hydrogenation systems have been extensively investigated, and are summarized in the following sections.

19.2.2.1 Rhodium-Based Catalysts

$RhCl(PPh_3)_3$ has been used for the homogeneous hydrogenation of various diene-based polymers, and its catalytic mechanism is understood to a considerable extent. Parent et al. [81] proposed a mechanism which has been found to be consistent with the kinetic data for various diene-based polymer hydrogenation systems and an understanding of the coordination chemistry of $RhCl(PPh_3)_3$ in solution. The main points comprising the mechanism are outlined as follows:

- $RhCl(PPh_3)_3$ oxidatively adds molecular hydrogen to form a five-coordinate dihydride complex which is consistent with the previous understanding [82] for olefin hydrogenation. This was also confirmed by experiments conducted by Mohammadi and Rempel [83] for NBR hydrogenation, where at 65 °C under 1 bar H_2 the reaction is quantitative towards formation of the dihydride:

$$RhCl(PPh_3)_3 + H_2 \leftrightarrows RhClH_2(PPh_3)_3 \quad (1)$$

- The dissociation of phosphine from $RhClH_2(PPh_3)_3$, whilst limited at room temperature, increases at higher hydrogenation temperatures:

$$RhClH_2(PPh_3)_3 \leftrightarrows RhClH_2(PPh_3)_2 + PPh_3 \quad (2)$$

- The initial coordination of the substrate is rate-limiting; however, the reductive elimination of the alkane is considered to be rapid:

$$RhClH_2(PPh_3)_2 + C{=}C \rightarrow RhClH_2(C{=}C)(PPh_3)_2 \quad (3)$$

$$RhClH_2(C{=}C)(PPh_3)_2 \rightarrow RhCl(PPh_3)_2 + C{-}C \quad (4)$$

- A potential ligand such as the nitrile present in a nitrile-butadiene copolymer may inhibit the catalytic hydrogenation cycle:

$$RhClH_2(PPh_3)_2 + \sim\!\sim CNR \leftrightarrows RhClH_2(CNR)(PPh_3)_2 \quad (5)$$

$$RhCl(PPh_3)_2 + \sim\!\sim CNR \leftrightarrows RhCl(CNR)(PPh_3)_2 \quad (6)$$

$[Rh(CO)(MeCN)(PPh_3)_2]ClO_4$ has been prepared [84] and $[Rh(PPh_3)_3$-$(MeCN)][BF_4]$ analyzed crystallographically by Pimblett et al. [85]. Schrock and Osborn [86] report that the use of acetonitrile as a solvent has a deleterious effect on the hydrogenation activity of $[Rh(diene)(PPh_3)_2]A$ ($A = ClO_4$, BF_4 or PF_6). To date, a detailed study of the propensity of nitrile to associate with complexes derived from $RhCl(PPh_3)_3$ is lacking, although Ohtani, Yamagishi and Fujimoto [87] have presented some spectrophotometric data on the system. As nitrile likely coordinates by donation of its lone pair of electrons, it experiences little of the steric hindrance which affects coordination of the olefin. It may therefore compete effectively with olefin for coordination to coordinatively unsaturated metal complexes.

Based on the above reactions, an overall mechanism for the hydrogenation of NBR catalyzed by Wilkinson's catalyst was proposed (see Scheme 19.4), which is also applicable to the kinetic performance of the homogeneous hydrogenation of PB [88] and styrene-butadiene *copolymers* [89], where K_2 and K_5 vanish.

Based on Scheme 19.4, the following mathematical equation can be derived for calculating the hydrogenation rate:

$$R_H = \frac{k' K' K K_1 [H_2][Rh][C{=}C]}{K K_1 + K'[PPh_3] + K K'[H_2][PPh_3] + K K_1 K'[H_2] + K K_1 K_5 [CN] + K K_1 K_2 K'[H_2][CN]} \quad (7)$$

The investigated experimental ranges and experimental estimations of the kinetic parameters for various hydrogenation systems are listed in Table 19.2,

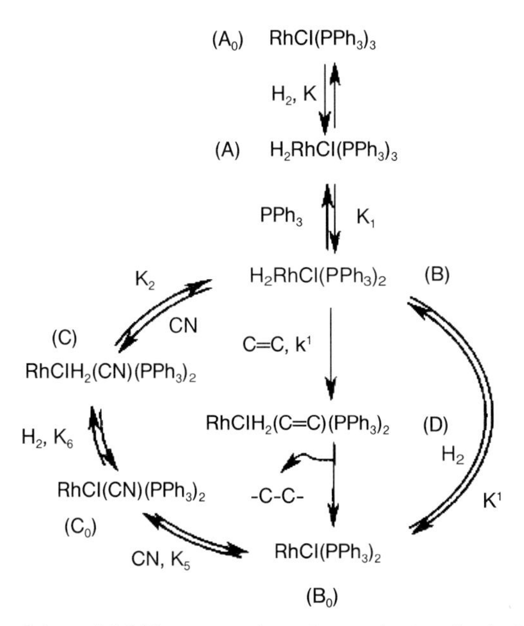

Scheme 19.4 The proposed reaction mechanism for the $RhCl(PPh_3)_3$/NBR system [90].

Table 19.2 Kinetic parameters for homogeneous hydrogenation of diene-based polymers.

Polymer (solvent)	K [mM^{-1}]	K_1 [mM]	K' [mM^{-1}]	K_2 [mM^{-1}]	K_5 [mM^{-1}]	k' [mMs^{-1}]	E [kJ mol^{-1}]	Reference
NBR (chlorobenzene)	∞	1.44	3.41×10^{-3}	3.98×10^{-2}	2.71×10^{-2}	1.19	73.5	81
NBR (butanone)	∞	0.198	0.276	6.5×10^{-2}	0	4.23×10^{-4}	87.3	83
1,4-PB (*o*-di-cholorobenzene)	0.60	4.45	0.59	0	0	1.28×10^{-3}	98.5	88
SBS (*o*-di-cholorobenzene)	0.31	3.13	0.63	0	0	3.26×10^{-4}	78.8	89
SB (toluene)	1.23	4.70	0.72	0	0	4.77×10^{-4}	60.8	89

with typical reaction conditions provided in Table 19.3. These parameters are consistent with the observed experimental results. For the hydrogenation of NBR in chlorobenzene using $RhCl(PPh_3)_3$, Bhattacharjee et al. [91] reported a value of 22 kJ mol^{-1} for the apparent activation energy, though this was not consistent with the normal range of reaction activation energy and also quite different from the value shown in Table 19.2. However, these authors did not provide sufficient data to account for the discrepancy. It would appear from this value of activation energy that Bhattacharjee et al.'s catalytic hydrogenation study was possibly mass transfer-controlled.

Table 19.3 Typical reaction conditions for the hydrogenation of polybutadiene (PB), styrene-butadiene diblock copolymer (SB), styrene-butadiene-styrene triblock copolymer (SBS) and nitrile butadiene rubber (NBR).

Polymer	[Rh] [mM]	$[PPh_3]$ [mM]	Temperature [K]	H_2 [mM] (P_{H2} [MPa])	Solvent	Reference
NBR	0.080	4.0	418.2	101 (2.37)	Chlorobenzene	81
NBR	1.958	0	313.2	3.142 (0.10)	Butanone	83
1,4-PB	1.99	7.40	338.2	3.90 (0.10)	*o*-Dichlorobenzene	88
SBS	1.99	7.40	338.2	3.90 (0.10)	*o*-Dichlorobenzene	89
SB	2.04	0	324.2	3.17 (0.10)	Toluene	89

19.2.2.2 Ruthenium-Based Catalysts

Ruthenium catalysts, such as $Ru(CHCH(Ph))Cl(CO)(PCy_3)_2$, have been found to be active for catalyzing the hydrogenation of various diene-based polymers. The catalytic mechanism for the hydrogenation of NBR, SBR and PB has been investigated [68].

Kinetic results show that the hydrogenation reaction rate exhibits a first-order dependence on both hydrogen concentration, $[H_2]$, and the total ruthenium concentration, $[Ru]_T$ and an inverse dependence on the nitrile concentration, [CN]. The catalytic mechanism proposed for polymer hydrogenation is illustrated in Scheme 19.5 and the main points of the mechanism are outlined below:

- Rapid hydrogenation of the styryl group in $Ru(CH{=}CH(Ph))Cl(CO)(PCy_3)_2$, to give the active species $RuHCl(CO)(PCy_3)_2$, and styrene, which is subsequently rapidly hydrogenated to ethyl benzene.

$$Ru(CH{=}CH(Ph))Cl(CO)(PCy_3)_2 + H_2 \rightarrow RuHCl(CO)(PCy_3)_2 + \text{styrene} \tag{8}$$

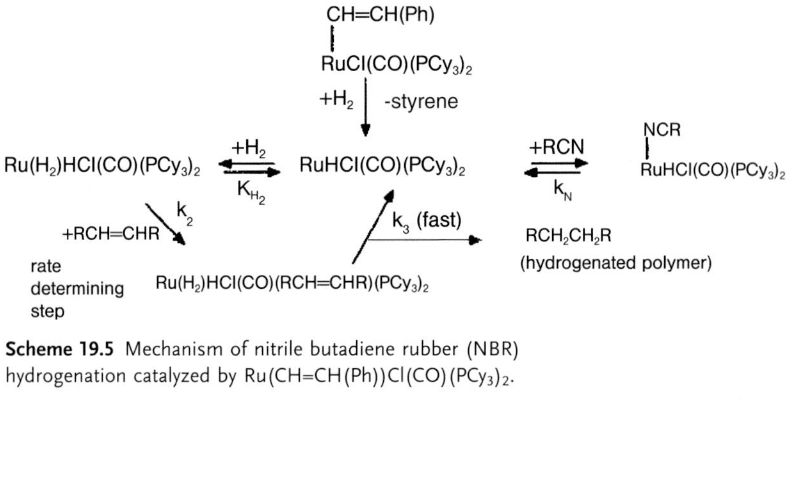

CH=CH(Ph)
$RuCl(CO)(PCy_3)_2$
$+H_2$ -styrene
$Ru(H_2)HCl(CO)(PCy_3)_2$ $+H_2$ K_{H_2} $RuHCl(CO)(PCy_3)_2$ +RCN K_N NCR $RuHCl(CO)(PCy_3)_2$
+RCH=CHR k_2 rate determining step
k_3 (fast)
RCH_2CH_2R (hydrogenated polymer)
$Ru(H_2)HCl(CO)(RCH{=}CHR)(PCy_3)_2$

Scheme 19.5 Mechanism of nitrile butadiene rubber (NBR) hydrogenation catalyzed by $Ru(CH{=}CH(Ph))Cl(CO)(PCy_3)_2$.

- The coordination of H_2 to $RuHCl(CO)(PCy_3)_2$ is the initial step in the catalytic cycle.

$$RuHCl(CO)(PCy_3)_2 + H_2 \leftrightarrows Ru(H_2)HCl(CO)(PCy_3)_2 \tag{9}$$

- The coordination of olefin before the final rapid elimination of products and regeneration of $RuHCl(CO)(PCy_3)_2$, which is assumed to be the rate-determining step:

$$Ru(H_2)HCl(CO)(PCy_3)_2 + C{=}C \rightarrow Ru(H_2)HCl(CO)(C{=}C)(PCy_3)_2 \tag{10}$$

$$Ru(H_2)HCl(CO)(C{=}C)(PCy_3)_2 \rightarrow RuHCl(CO)(PCy_3)_2 + \text{-C}-\text{C-} \tag{11}$$

- A potential ligand such as a nitrile may inhibit the hydrogenation cycle:

$$RuHCl(CO)(PCy_3)_2 + CNR \leftrightarrows Ru(CNR)HCl(CO)(PCy_3)_2 \tag{12}$$

Based on the above reaction mechanism, and with some reasonable simplification, the hydrogenation rate can be expressed as:

$$R_H = \frac{k_1[H_2][Ru]_T[C{=}C]}{1 + K_N[RCN]} \tag{13}$$

The mechanism for this catalyst for the hydrogenation of *cis*-1,4-polyisoprene (CPIP) is slightly different [69]. As there was no clear evidence that coordination of hydrogen occurs prior to the coordination of C=C to the $RuHCl(CO)(PCy_3)_2$, there may be two possible pathways for the hydrogenation of CPIP in the presence of $Ru(CH{=}CH(Ph))Cl(CO)(PCy_3)_2$, namely an unsaturated path and a hydride path. The catalytic mechanism for these two pathways is represented in Scheme 19.6.

Based on this mechanism, if the unsaturated pathway ($A_0 \rightarrow B \rightarrow D$) is undertaken, with the reaction of the alkyl complex with hydrogen serving as the rate-limiting step, the hydrogenation rate can be expressed as:

$$R_H = \frac{k_3 K_{C=C}[H_2][Ru]_T[C{=}C]}{1 + K_{C=C}[C{=}C]} \tag{14}$$

If the hydride pathway ($A_0 \rightarrow B' \rightarrow C' \rightarrow D$) is undertaken, the hydrogenation rate can be expressed as:

$$R_H = \frac{K_2 K_{H_2}[H_2][Ru]_T[C{=}C]}{1 + K_{H_2}[H_2]} \tag{15}$$

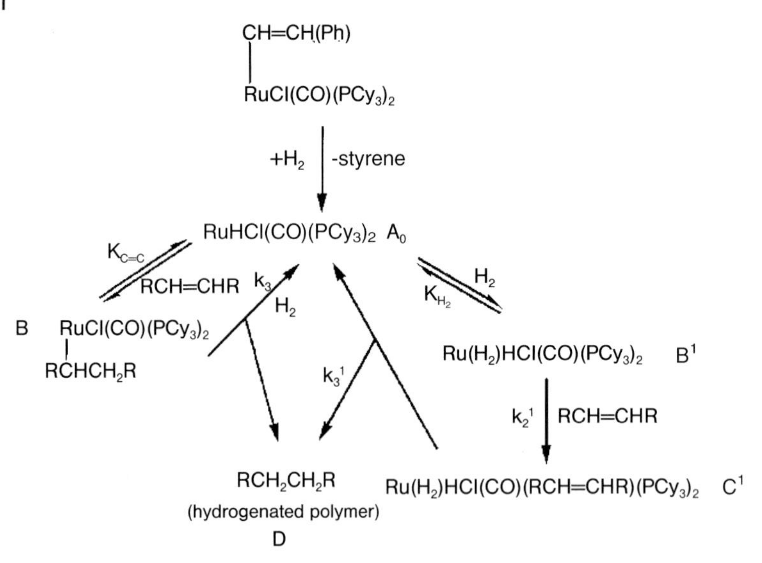

Scheme 19.6 Proposed mechanism of *cis*-1,4-polyisoprene (CPIP) hydrogenation using $Ru(CH{=}CH(Ph))Cl(CO)(PCy_3)_2$.

If both K_{H2} and $K_{C=C}$ are assumed to be very small, depending on which pathway is considered, then both equations will reduce to

$$R_H = k[H_2][Ru]_T[C{=}C] \tag{16}$$

which is consistent with the observed experimental kinetics.

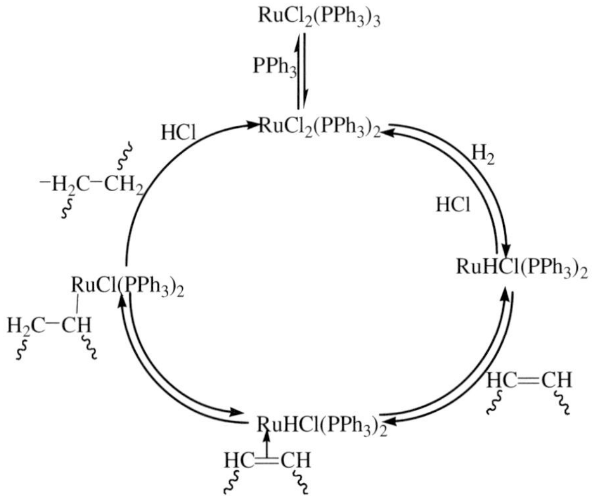

$RuCl_2(PPh_3)_3$

PPh3

HCl $RuCl_2(PPh_3)_2$ H2

$-H_2C-CH_2$ HCl

$RuHCl(PPh_3)_2$

$RuCl(PPh_3)_2$

H_2C-CH

HC=CH

$RuHCl(PPh_3)_2$

HC=CH

Scheme 19.7 Proposed mechanism for polybutadiene (PB) hydrogenation using $RuCl_2(PPh_3)_3$.

Rao et al. [70] investigated the hydrogenation of PB catalyzed by $RuCl_2(PPh_3)_3$, and showed a degree of hydrogenation >99% to be obtained within 6 h using 0.3 mol.% catalyst at 100 °C and under 50 bar H_2 pressure. These authors proposed the possible mechanism shown in Scheme 19.7, but provided no detailed kinetic data to check the reliability of the above mechanism.

19.2.2.3 **Osmium-Based Catalysts**

For the hydrogenation of NBR, osmium complexes such as $OsHCl(CO)(O_2)(PCy_3)_2$ have shown distinctive performance which permits their discrimination from Rh- and Ru-based catalysts [65]. A somewhat difficult element of the kinetic NBR hydrogenation data is the apparent second-to-zero-order dependence of the reaction rate with respect to $[H_2]$. A second-order behavior requires 2 molecules of H_2 either to produce an active complex or to participate in the rate-determining step. Bakhmutov et al. [92] identified an exchange between the apical hydride and the *trans*-coordinated dihydrogen ligand, but concluded that the trihydride could at most be a reactive intermediate. The addition of a second molecule of H_2 to the catalyst intermediate had not been observed. Another distinguishing characteristic in the Os catalytic system is that of severe inhibition of hydrogenation activity when additional PCy_3 is present. Based on these phenomena, two possible hydrogenation mechanisms have been proposed (Schemes 19.8 a and b).

A rate expression, as shown in Eq. (17), may be obtained from Scheme 19.8 a:

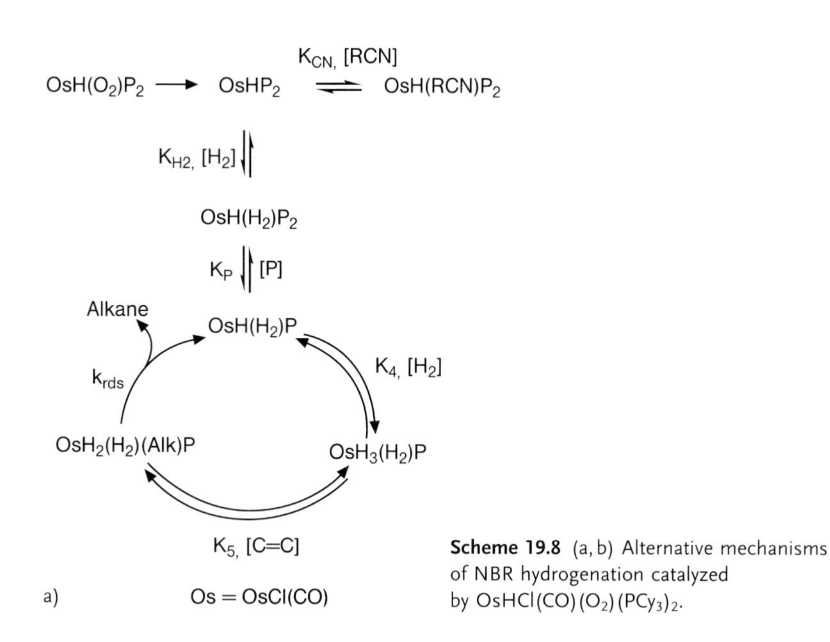

Scheme 19.8 (a, b) Alternative mechanisms of NBR hydrogenation catalyzed by $OsHCl(CO)(O_2)(PCy_3)_2$.

K_{CN}, [RCN]

$OsH(O_2)P_2 \longrightarrow OsHP_2 \rightleftharpoons OsH(RCN)P_2$

K_{H2}, $[H_2]$

$OsH(H_2)P_2$

K_P [P]

Alkane

$OsH(H_2)P$

k_{rds}

K_4, [C=C]

$OsH_2(H_2)(Alk)P$

$OsH_2(Alk)P$

K_5, $[H_2]$

Os = OsCl(CO)

Scheme 19.8 b

$$R_H = \frac{k_{rds} K_{H_2} K_p K_4 K_5 [Os]_T [H_2]^2 [C{=}C]}{[P](1 + K_{CN}[CN] + K_{H_2}[H_2]) + K_{H_2} K_p [H_2] + K_{H_2} K_p K_4 [H_2]^2 (1 + K_5 [C{=}C]))} \tag{17}$$

The rate expression is consistent with the observed kinetic data. A similar mechanism was also proposed for the hydrogenation of polyisoprene catalyzed by $OsHCl(CO)(O_2)(PCy_3)_2$ [93].

Scheme 19.8 b illustrates a more conventional mechanism, which results in a rate expression as shown in Eq. (18):

$$R_H = \frac{k_{rds} A [Os]_T [H_2]^2 [C{=}C]}{[P](1 + K_{CN}[CN] + K_{H_2}[H_2]) + K_{H_2} K_p [H_2] + B[C{=}C][H_2](1 + C[H_2]))} \tag{18}$$

While this mechanism involves lower formal coordination numbers than that of Scheme 19.8 a, its shortcoming is a zero-order dependence on $[H_2]$ that must be accompanied by a zero-order reaction dependence on [C=C].

19.2.2.4 **Palladium Complexes**

Palladium acetate has an unusual structure that comprises three palladium atoms in a triangular arrangement, held together by six bridging acetate groups. Details of the structure of the complex have been reported [75]. The catalyst has been used for the hydrogenation of both nitrile rubber and natural rubber [74–76].

Another palladium complex, namely, a six-membered cyclopalladate complex of 2-benzoyl pyridine, has also been used for the hydrogenation of polymers [77, 78]. Possible catalytic mechanisms for the hydrogenation of natural rubber [76] and NBR [77] catalyzed by these two complexes were proposed, but unfortunately the authors did not provide sufficient evidence to support their proposed mechanisms.

Although Ziegler-type catalysts have been widely investigated for the homogeneous hydrogenation of polymers, their catalytic mechanism remains unknown. One possible reason for this may be the complexity of the coordination catalysis and the instability of the catalysts. Metallocene catalysts are highly sensitive to impurities, and consequently it is very difficult to obtain reproducible experimental data providing reliable kinetic and mechanistic information.

19.2.3
Kinetic Mechanism Discrimination

Modeling and analysis of the kinetics represent powerful tools by which a better understanding of the catalysis involved in polymer hydrogenation may be obtained. These approaches also form the basis for a more thorough design of chemical reactors, and for a better insight into the behavior of existing reactors. In homogeneous catalytic systems, in the presence of organometallic catalysts, complicated equilibrium reaction cycles often exist and in general the networks composed of the elementary reactions are highly complex. Thus, model discrimination, selection on the basis of experiment of the best rate equation among a set of rivals, is very important. Fortunately, on the basis that overall catalytic intermediates can be considered to be reasonably constant, model discrimination may be significantly simplified, and a generalized method has quite recently been proposed for the hydrogenation of diene-based polymers [94].

19.3
Engineering Art

Homogeneous polymer hydrogenation operation involves a number of unit operations, and some typical procedures are shown in Figure 19.1 with respect to NBR hydrogenation. The major unit operations include homogeneous catalytic hydrogenation, catalyst recovery and solvent recycling, in addition to emulsion polymerization before the hydrogenation stage. Because of the high exothermicity in the initial stage of the hydrogenation reaction, and because gaseous hydrogen is used, heat transfer and mass transfer are major concerns for hydrogenation reactors. As a high degree of hydrogenation is usually required for high-performance hydrogenated elastomers, a suitable flow pattern in the reactor is also required for a continuous operation, besides a need for superior mass transfer and heat transfer. Because of the high price of the catalyst used, and also because of potential catalyst toxicity, catalyst recovery is a critical stage for

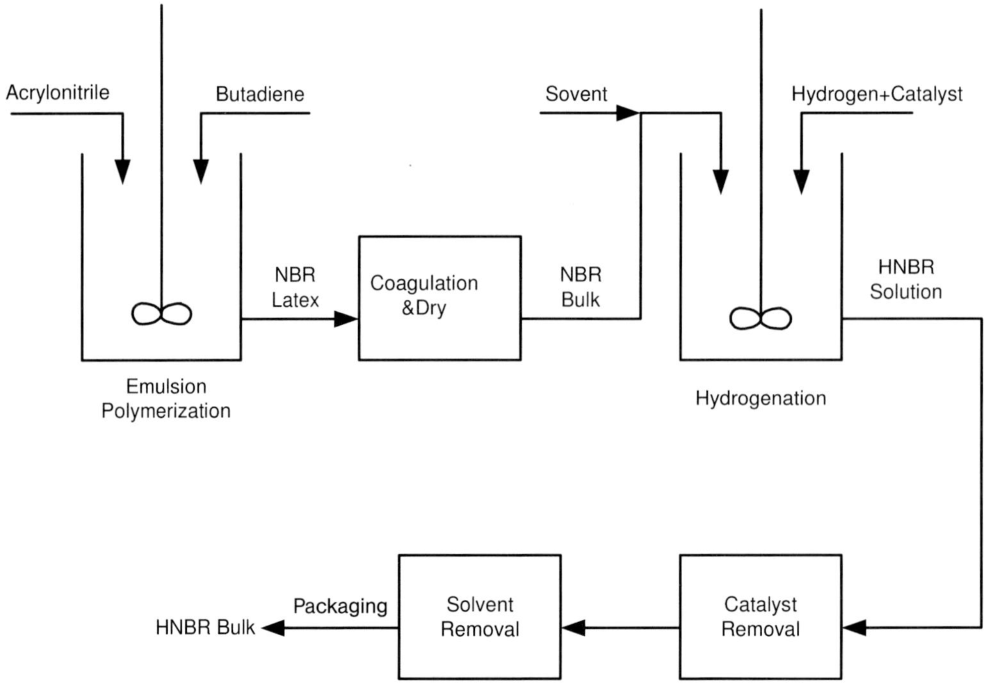

Fig. 19.1 Schematic process for the production of hydrogenated nitrile butadiene rubber (HNBR).

both batch and continuous operations. Likewise, as large amounts of organic solvent are required to realize homogeneous hydrogenation, an efficient design for solvent removal and recycle is necessary. Together, these three aspects determine the commercial feasibility of the hydrogenation process.

19.3.1
Catalyst Recovery

Catalyst recovery is a universal problem in the field of homogeneous catalysis. For the homogeneous hydrogenation of unsaturated polymers, the incentives of catalyst removal dwell on both improving the quality of the polymer product and on reducing the high costs of the process that relate mainly to the expensive noble metals often used as catalysts. Catalyst removal is especially difficult from polymer solutions because of the high viscosity and the good compatibility between the catalysts and the functionalized polymers. In general, two main methods are used for the removal of metal catalysts. The first method is to precipitate the catalyst from the polymer solution and then to separate it by filtration or gravity settling. The second method is to adsorb the catalyst from the polymer solution by ion-exchange resins or other absorbents which have an affinity for the spent catalyst.

19.3.1.1 **Precipitation**

Precipitation of the catalyst can be effected by treating the polymer solution with acid/base and/or oxidants. Poloso and Murray [95] proposed a method to recycle the nickel octanoate ($(CH_3(CH_2)_6CO_2)_2Ni$)/triethylaluminum($(C_2H_5)_3Al$) catalyst from a styrene-butadiene polymer solution. The polymer solution containing the catalysts was refluxed with 4 wt.% glacial acetic acid (relative to polymer) for 4 h, followed by treatment with 1.4 wt.% anhydrous ammonia. The solution was then filtered through a diatomaceous earth. The nickel content in the polymer was decreased from 310 ppm to 5.6 ppm.

In a method developed by Kang [96], 2 equiv. dimethylglyoxime (based on nickel content) was used to treat a butadiene-styrene polymer solution. The reddish bis(dimethylglyoximato) nickel complex was precipitated and then removed by filtration. By using this method, the nickel content in the polymer was reduced to less than 1 ppm. Alternatively, Hoxmeier [97] mixed the polymer solution with azelaic acid or adipic acid (50 wt.% excess based on the metals) in organic solvents, after which a phase separation was realized by gravity settling. Subsequently, it was claimed that nickel levels within the polymers were reduced to 2 ppm. The process was later improved by using hydrogen peroxide to oxidize the nickel catalyst in the polymer solution initially, and then by adding azelaic acid or adipic acid as an aqueous solution to precipitate the catalyst [98]. Nickel-based catalysts have also been removed by contacting the polymer solution with molecular oxygen with subsequent treatment with activated carbon [99]. Other similar methods to precipitate catalysts from the polymer solutions include contacting the polymer solution with a trialkylaluminum compound in the presence of water [100], treating the polymer solution with a chelating resin which is comprised of iminodiacetate ions [101], bubbling oxygen/nitrogen in the presence of an aqueous solution of an acid [102], reacting with an aqueous solution of ammonia and carbon dioxide [103], and contacting with an aqueous solution of a weak acid followed by contacting with an aqueous solution of a weak base [104]. In selecting a suitable agent to remove metallic catalyst residues from hydrogenated polymer solutions, the following criteria should be considered: (i) the agent should be substantially inert toward the polymer and the polymer solvent; and (ii) it should be capable of turning the catalyst into an insoluble compound, efficiently. Contamination of the polymer by the catalyst recovery procedure must also be considered.

19.3.1.2 **Adsorption**

Adsorption is commonly used for catalyst removal/recovery. The process involves treating the polymer solution with suitable materials which adsorb the catalyst residue and are then removed by filtration. Panster et al. [105] proposed a method involving adsorbers made from organosiloxane copolycondensates to recover rhodium and ruthenium catalysts from solutions of HNBR. These authors claimed that the residual rhodium could be reduced to less than 5 ppm, based on the HNBR content which had a hydrogenation conversion of over

99%. Silicates, including calcium silicate, magnesium silicate and diatomaceous earth, have also been used as adsorbers [106]. When these were applied to a hydrogenated SBS solution containing a nickel-based catalyst, the nickel content of the polymer was reduced to less than 1 ppm. Rhodium-based catalyst residues from HNBR solution can also be removed by using an ion-exchange resin with thiourea functional groups present on a resin which is both macroporous and monodispersed [107]. The criterion for selecting adsorbers is similar to that for selecting a precipitating agent: an efficient catalyst adsorber should be inert to both the polymer and the solvent.

19.3.2 Solvent Recycling

Solvent recycling is an important post-treatment stage of the homogeneous catalytic hydrogenation of polymers, for both economical and environmental reasons. This topic has not been widely investigated for homogeneous polymer hydrogenation operations, but in present-day processes the organic solvents used are stripped by steam, which is itself an energy-intensive operation.

In other related areas, such as solution polymerization and bulk polymerization, the removal/recycling of solvents or unreacted monomer has been extensively investigated [108–112]. The methods used are based on lateral heat-dependent operations such as evaporation and steam-stripping, or non-lateral heat-dependent operations that include a variety of extraction procedures.

Many types of equipment have been developed to improve the evaporation of solvent in order to provide energy savings. The most widely used techniques for devolatilization are the falling strand devolatilizer (FSD), the thin-film evaporator and the vented extruder [113].

The FSD is a flash evaporator, whereby the preheated polymer solution/melt falls within the vessel primarily by gravity, while the volatiles evaporate during falling. This method is normally used with process streams that are not exceptionally temperature-sensitive and where the concentration of volatiles is relatively high.

Thin-film and surface renewal evaporators are mostly applied to materials with medium to high viscosity, or to high-boiling contaminated mixtures. A typical thin-film evaporator employs a unique rotor with an array of discrete, plow-like blades attached to the rotor core. The blades transport the viscous concentrate or melt through the evaporator while simultaneously forming films to facilitate heat and mass transfer.

Single-screw and double-screw extruders are normally used for polymer melts to accomplish the deaeration or devolatilization of residual volatiles. Devolatilization in an extruder is effected through formation of the venting zone inside the chamber by carefully designed upstream and downstream screw sections.

Many techniques have also been developed to improve devolatilization efficiency, including steam stripping [114], second fluid-assisted devolatilization [115], supercritical fluid devolatilization [116], and a variety of specially designed devolatilizers. Some of the above methods can also be used for solvent recycling in polymer hydrogenation processes.

19.3.3 Reactor Technology and Catalytic Engineering Aspects

There are many commercial advantages to hydrogenating relatively inexpensive polymers in order to create new, more valuable materials. Polymer hydrogenation is not widely developed today on a commercial scale, however, mainly because of the high production costs associated with catalyst and reactor technologies. The development of an efficient reactor for homogeneous hydrogenation of diene-based polymers remains a major challenge for several reasons:

- In order to reduce production costs, high polymer concentrations are preferred in hydrogenation operations. However, the viscosity of polymer solutions rises rapidly as the polymer concentration increases. In present-day commercial processes, polymer concentrations do not normally exceed 15 wt.%.
- The hydrogenation operation involves hydrogen transfer from the gas phase into the viscous polymer solution phase. If the mass transfer capacity is not sufficiently superior, the hydrogenation could be significantly retarded and product quality adversely affected.
- The hydrogenation reaction is highly exothermic. Thus, heat release during the initial hydrogenation may be a serious problem, and a reactor with a superior heat transfer capacity tunability will be required.
- Due to the viscous and highly exothermic characteristics of the hydrogenation reactions, the reactor should have superior mixing capabilities in order to avoid possible crosslinking induced by hot spots in the reactor.
- As high hydrogenation conversion is usually desired, and crosslinking of the polymer should be avoided, correct control of the reaction conditions is very important.

Among present-day commercial processes for the homogeneous hydrogenation of polymers, a semi-batch operation system prevails where the hydrogen gas supply is provided continuously and the liquid phase is operated batchwise. A semi-batch operation is suitable for relatively small volume production, but it is not economical for high-production yields. Such semi-batch reactor systems cannot meet the needs of the growing demand for hydrogenated polymers, and the development of more efficient reactor systems is needed. A Japanese Patent [117] disclosed a production process for hydrogenated polymer wherein a polymer solution containing olefinic unsaturated groups, hydrogen gas, and a hydrogenation catalyst are continuously supplied to a stirred-tank reactor and the reaction product is continuously removed. Another Japanese Patent [118] disclosed a continuous production process of a hydrogenated polymer containing olefinic unsaturated groups wherein plural reactors are connected in series and hydrogen is supplied to at least one of the reactors from the lower portion thereof. An improved method was proposed [119] to obtain a polymer having a desirable degree of hydrogenation steadily for a long period in the above continuous reactors by recycling one part of the hydrogenated polymer solution. Pan and Rempel et al. [120–122] investigated the effect of various reactor performances on the hydrogenation of diene-based polymers via modeling and simulation, and proposed that an optimal reactor for diene-based polymer hydrogenation would be a plug flow reactor with an instantaneous mixing component in the inlet zone. These studies provide useful information for commercial continuous hydrogenation process development.

19.4
A Commercial Example: Production of HNBR via a Homogeneous Hydrogenation Route

HNBR has an intriguing combination of properties [123], including high tensile strength, low permanent set (especially at high temperatures), very good abrasion resistance and high elasticity. HNBR also shows excellent stability towards heat, being able to resist temperatures of up to ca. 150 °C (NBR under the same conditions is stable up to only ca. 120 °C); it also demonstrates better properties at low temperatures (lower brittle point) than other heat- and oil-resistant elastomers. This combination of properties is opening up a broad range of applications for these materials, particularly in the automotive industry. HNBR is now widely used for timing belts in cars, due to its good static as well as dynamic properties at under-the-hood operation temperatures, and it also exhibits good retention of properties under continuous heat exposure. In addition, new grades of the material with improved low-temperature flexibility are extending the HNBR service temperature range, allowing new applications in seals and mounts. For example, HNBR is also proving useful for seals and moldings of motor car engines that run on new fuels such as rapeseed oil methyl ester. Seal applications also include air conditioner O-rings, shock absorbers, power steering systems and water pumps. HNBR has also been widely used in industrial seals for oil field exploration and processing, as well as rolls for steel and paper mills.

There are two major commercial producers of HNBR worldwide. Nippon Zeon Corporation manufactures HNBR (heterogeneous hydrogenation) under the tradename Zetpol®, while Lanxess Inc. produces HNBR (homogeneous hydrogenation) in Orange, Texas, and in Leverkusen, Germany under the tradename Therban®. The manufacturing process of HNBR is shown schematically in Figure 19.1. The process begins with the production of an emulsion-polymerized NBR which is then dissolved in an appropriate solvent (chlorobenzene). When dissolution is complete, the addition of hydrogen gas, in conjunction with a precious-metal catalyst at a designated temperature and pressure, brings about a selective hydrogenation to produce the hydrogenated polymer. The solvent and catalyst are then recovered and the remaining polymer crumb is dried. After vulcanization, the HNBR is ready for industrial use.

19.5
Future Outlook and Perspectives

Hydrogenated polymers have many desirable properties over their parent polymers, although the high cost of hydrogenated products still restricts their widespread application. The following aspects should be considered for the sustainable development of the hydrogenated polymer industry:

- The development of highly efficient and easily recoverable catalyst systems. Today, the high cost of hydrogenated polymers is due mainly to the cost of the metal catalyst and its recovery operation.
- A reduction in the amounts of organic solvent used in the hydrogenation process. Considerable energy costs result from solvent recycling. Investigations into the hydrogenation of NBR in supercritical fluid media have been carried out, and positive results obtained [124, 125]. Considerable cost savings could be realized if an efficient catalyst system were to be developed for the hydrogenation of polymers in aqueous latex form [126, 127].
- The development of high-efficiency reactors (e.g., continuous/flexible systems) for the hydrogenation process [117–122].
- The extension of new applications for hydrogenated polymers.
- An overall improvement on the hydrogenation process to reduce production costs is required.

Abbreviations

CPIP *cis*-1,4-polyisoprene
FSD falling strand devolatilizer
HNBR hydrogenated nitrile butadiene rubber
NBR acrylonitrile-butadiene rubber
PB polybutadiene
SBR styrene-butadiene rubber
SBS styrene-butadiene-styrene triblock copolymer
SIS styrene-isoprene-styrene triblock copolymer
XNBR carboxylated nitrile rubber

References

1 Kang, J. US Patent 3,993,855, **1976** (to The Firestone Tire & Rubber Company, Akron, OH).
2 Rempel, G. L., Azizian, H., US Patent 4,464,515, **1984** (to Polysar Limited, Sarnia, CA).
3 Rempel, G. L., Azizian, H., US Patent 4,503,196, **1985** (to Polysar Limited, Sarnia, CA).
4 Guo, S., Nguyen, P., US Patent 6,683,136, **2004** (to Bayer Inc., Sarnia, CA)
5 Rempel, G. L., McManus, N. T., Parent, J. S., US Patent 5,561,197,**1996** (to University of Waterloo, CA),

6 Buding, H., Fiedler, P., Konigshofen, H., Thormer, J., US Patent 4,631,315, **1986** (to Bayer Aktiengesellschaft, Leverkusen, DE).
7 Himmler, T., Fiedler, P., Braden, R., Buding, H., US Patent 4,795,788, **1989** (to Bayer Aktiengesellschaft, Leverkusen, DE).
8 Buding, H., Thormer, J., Nolte, W., Hohn, J., Fiedler, P., Himmler, T., US Patent 5,034,469, **1991** (to Bayer Aktiengesellschaft, Leverkusen, DE).
9 Rempel, G. L., Mohammadi, N. A., Farwaha, R., US Patent 4,812,528, **1989** (to University of Waterloo, CA).
10 Rempel, G. L., Mohammadi, N. A., Farwaha, R., US Patent 4,816,525, **1989** (to University of Waterloo, CA).
11 Rempel, G. L., McManus, N. T., Mohammadi, N. A., US Patent 5,057,581, **1991** (to University of Waterloo, CA).
12 Rempel, G. L., McManus, N. T., US Patent 5,075,388, **1991** (to University of Waterloo, CA).
13 Iio, A., Oshima, N., Ohira, Y., Sakamoto, M., Oka, H., US Patent 5,202,388, **1993** (to Japan Synthetic Rubber Co. Ltd., Tokyo, Japan).
14 Tadahiro, S., Masutada, O., Tadashi, A., Japanese Patent 11-193,323, **1999** (to Mitsui Chemicals Inc., Japan).
15 Hsu, K., Wu, G., Xu, R., Yue, D., Zhou, S., US Patent 6,084,033, **2000** (to Nantex Industry Co., Ltd., Taiwan).
16 Loveless, F. C., Miller, D. H., US Patent 3,932,308, **1976** (to Uniroyal Inc., New York, NY).
17 Halasa, A. F., US Patent 3,988,504, **1976** (to The Firestone Tire & Rubber Company, Akron, OH).
18 Hoxmeier, R. J., Slaugh, L. H., US Patent 5,013,798, **1991** (to Shell Oil Company, Houston, TX).
19 Hoxmeier, R. J., Slaugh, L. H., US Patent 5,030,799, **1991** (to Shell Oil Company, Houston, TX).
20 Hoxmeier, R. J., Slaugh, L. H., US Patent 5,061,668, **1991** (to Shell Oil Company, Houston, TX).
21 Hoxmeier, R. J., US Patent 4,879,349, **1989** (to Shell Oil Company, Houston, TX).
22 Hoxmeier, R. J., US Patent 5,001,199, **1991** (to Shell Oil Company, Houston, TX).
23 Ladenberger, V., Bronstert, K., Fahrbach, G., Grog, W., US Patent 4,207,409, **1980** (to BASF Aktiengesellschaft, DE).
24 Mizuho, M., Kenichi, T., Japanese Patent 14-308,905, **2002** (to Kuraray Co Ltd, Japan).
25 Mizuho, M., Kenichi, T., Japanese Patent 14-317,008, **2002** (to Kuraray Co Ltd, Japan).
26 Kazuya, O., Mizuho, M., Kenichi, T., Japanese Patent 15-002,917, **2003** (to Kuraray Co Ltd, Japan).
27 Tsungae, Y., Mizuno, H., Kohara, T., Natsuume, T., US Patent 5,539,060, **1996** (to Nippon Zeon Co., Ltd., Tokyo, Japan).
28 Willis, C., US Patent 5,521,254, **1996** (to Shell Oil Company, Houston, TX).
29 Willis, C., US Patent 5,597,872, **1997** (to Shell Oil Company, Houston, TX).
30 Hoxmeier, R. J., Slaugh, L. H., US Patent 4,876,314, **1989** (to Shell Oil Company, Houston, TX).
31 Hoxmeier, R. J., US Patent 4,892,928, **1989** (to Shell Oil Company, Houston, TX).
32 Kishimoto, Y., Masubuchi, T., US Patent 4,673,714, **1987** (to Asahi Kase Kogyo Kabushiki Kaisha, Osaka, Japan).
33 Teramoto, T., Goshima, K., Takeuchi, M., US Patent 4,980,421, **1990** (to Japan Synthetic Rubber Co., Ltd., Tokyo, Japan).
34 Hashiguchi, Y., Katsumata, H., Goshima, K., Termoto, T., Takemura, Y., US Patent 5,169,905 **1992** (to Japan Synthetic Rubber Co., Ltd., Tokyo, Japan).
35 Chamberlain, L. R., Gibler, C. J., US Patent 5,132,372, **1992** (to Shell Oil Company, Houston, TX).
36 Chamberlain, L. R., Gibler, C. J., Kemp, R. A., Wilson, S. E., US Patent 5,141,997, **1992** (to Shell Oil Company, Houston, TX).
37 Chamberlain, L. R., Gibler, C. J., US Patent 5,173,537, **1992** (to Shell Oil Company, Houston, TX).
38 Gibler, C. J., Chamberlain, L. R., Hoxmeier, R. J., US Patent 5,242,986, **1992** (to Shell Oil Company, Houston, TX).

39 Gibler, C. J., Wilson, S. E., US Patent 5,244,980, **1993** (to Shell Oil Company, Houston, TX).
40 Gibler, C. J., Wilson, S. E., US Patent 5,334,566, **1994** (to Shell Oil Company, Houston, TX).
41 De Boer, E. J. M., Hessen, B., Van Der Huizen, A. A., De Jong, W., Van Der Linden, A. J., Ruisch, B. J., Schoon L., De Smet, H. J. A., Van Der Steen, F. H., Van Strien, H. C. T. L., Villena, A., Walhof, J. J. B., US Patent 5,814,709, **1998** (to Shell Oil Company, Houston, TX).
42 De Boer, E. J. M., Hessen, B., Van Der Huizen, A. A., De Jong, W., Van Der Linden, A. J., Ruisch, B. J., Schoon L., De Smet, H. J. A., Van Der Steen, F. H., Van Strien, H. C. T. L., Villena, A., Walhof, J. J. B., US Patent 5,886,107, **1999** (to Shell Oil Company, Houston, TX).
43 De Boer, E. J. M., Hessen, B., Van Der Huizen, A. A., De Jong, W., Van Der Linden, A. J., Ruisch, B. J., Schoon L., De Smet, H. J. A., Van Der Steen, F. H., Van Strien, H. C. T. L., Villena, A., Walhof, J. J. B., US Patent 5,925,717, **1999** (to Shell Oil Company, Houston, TX).
44 Van Der Heijden, H., Van De Weg, H., US Patent 6,461,993, **2002** (to Shell Oil Company, Houston, TX).
45 Van Der Heijden, H., Van De Weg, H., US Patent 5,592,430, **1999** (to Shell Oil Company, Houston, TX).
46 Chamberlain, L. R., Gibler, C. J., Kemp, R. A., Wilson, S. E., Brownscombe, T. F., US Patent 5,177,155, **1992** (to Shell Oil Company, Houston, TX).
47 Wilson, D. R., Stevens, J. C., US Patent 4,929,699, **1990** (to The Dow Chemical Company, Midland, MI).
48 Hahn, S. F., Wilson, D. R., US Patent 5,789,638, **1998** (to The Dow Chemical Company, Midland, MI).
49 Devore, D. D., Stevens, J. C., Hahn, S. F., Timmers, F. J., Wilson, D. R., US Patent 6,476,283, **2002** (The Dow Chemical Company, Midland, MI).
50 Ko, Y. H., Kim, J. Y., Hwang, J. M., US Patent 5,910,566, **1999** (to Korea Kumho Petrochemical Co., Ltd., Seoul, Rep. of Korea).
51 Ko, Y. H., Kim, H. C., US Patent 5,994,477, **1999** (to Korea Kumho Petrochemical Co., Ltd., Seoul, Rep. of Korea).
52 Ko, Y. H., Kim, H. C., US Patent 6,020,439, **2000** (to Korea Kumho Petrochemical Co., Ltd., Seoul, Rep. of Korea).
53 Ko, Y. H., Kim, H. C., Kim, J. Y., Hwang, J. M., US Patent 6,410,657, **2002** (to Korea Kumho Petrochemical Co., Ltd., Seoul, Rep. of Korea).
54 Parellada Ferrer, M. D., Barrio Calle, J. A., US Patent 5,583,185, **1996** (to Repsol Quimica S. A., Madrid, Spain).
55 Rekonen, P., Kopola, N., Koskimies, S., Andell, O., Oksman, M., US Patent 5,814,710, **1998** (to Neste Oy, Provoo, FI).
56 Tsiang, R. C., Hsien, H. C., Yang, W., US Patent 5,705,571, **1998** (to Taiwan Synthetic Rubber Corporation, Taipei, Taiwan).
57 Tsai, J., Chang, W., Chao, Y., Chu, C., Huang, C., Hsiao, H., US Patent 6,313,230, **2001** (to Industrial Technology Research, Hsinchu, Taiwan; Chi Mei Co., Tainan, Taiwan).
58 Vallieri, A., Cavallo, C., Viola, G. T., US Patent 5,948,869, **1999** (to Enichem S.p.A., Milan, Italy).
59 Viola, G. T., Vallieri, A., Cavallo, C., US Patent 6,228,952, **2001** (to Enichem S.p.A., Milan, Italy).
60 Masi, F., Sommazzi, A., Santi, R., US Patent 6,831,135, **2004** (to Enichem S.p.A., Roma, IT; Polimeri Europa S.p.A., Brindisi, Italy).
61 Lin, F., Tsai, C., Liu, S., US Patent 6,881,797, **2005** (to TSRC Corporation, Kaohsiung, Taiwan).
62 Kishimoto, Y., Morita, H., US Patent 4,501,857, **1985** (to Asai Kasei Kogyo Kabushiki Kaisha, Osaka, Japan).
63 McManus, N. T., Rempel, G. L., *J. Macromol. Chem. Physics* **1995**, *35*(2), 239–285.
64 Singha, N. K., Bhattacharjee, S., Sivaram, S., *Rubber Chem. Technol.* **1997**, *70*(3), 309.
65 Parent, J. S., McManus, N. T., Rempel, G. L., *Ind. Eng. Chem. Res.* **1998**, *37*, 4253.
66 Charmondusit, K., Prasassarakich, P., McManus, N. T., Rempel, G. L., *J. Appl. Polym. Sci.* **2003**, *89*, 142.

67 Pan, Q., Rempel, G. L., *Macromolec. Rapid Commun.* **2004**, *25*, 843.
68 Martin, P., McManus, N. T., Rempel, G. L., *J. Mol. Catal. A: Chemical* **1997**, *126*, 115.
69 Tangthongkul, R., Prasassarakich, P., McManus, N. T., Rempel, G. L., *J. Appl. Polym. Sci.* **2004**, *91*(5), 3259.
70 Rao, P. V. C., Upadhyay, V. K., Pillai, S. M., *Eur. Polym. J.* **2001**, *37*, 1159.
71 Parent, J. S., McManus, N. T., Rempel, G. L., *J. Appl. Polym. Sci.* **2001**, *79*, 1618.
72 Gilliom, L. R., *Macromolecules* **1989**, *22*, 662.
73 Gilliom, L. R., Honnell, K. G., *Macromolecules*, **1992**, *25*, 6066.
74 Bhattacharjee, S., Bhowmick, A. K., Avasthi, B. N., *J. Polym. Sci., Polym. Chem.* **1992**, *30*, 471.
75 Bhattacharjee, S., Rajagopalan, P., Bhowmick, A. K., Avasthi, B. N., *J. Appl. Polym. Sci.* **1993**, *49*, 19717.
76 Bhattacharjee, S., Bhowmick, A. K., Avasthi, B. N., *Polymer* **1993**, *34*, 5168.
77 Bhattacharjee, S., Bhowmick, A. K., Avasthi, B. N., *J. Appl. Polym. Sci.* **1990**, *41*, 1357.
78 Bhattacharjee, S., Bhowmick, A. K., Avasthi, B. N., *J. Polym. Sci., Polym. Chem.* **1992**, *30*, 1961.
79 Velichkova, R., Toncheva, V., Antonov, V., Alexandrov, V., Pavlova, S., Dubrovina, L., Cadkova, E., *J. Appl. Polym. Sci.* **1991**, *42*, 3083.
80 Yang, W., Hsieh, H. C., Tsiang, R. C., *J. Appl. Polym. Sci.* **1999**, *72*, 1807.
81 Parent, J. S., McManus, N. T., Rempel, G. L., *Ind. Eng. Chem. Res.* **1996**, *35*, 4417.
82 Halpern, J., *J. Chem. Soc. Chem. Commun.* **1973**, 629.
83 Mohammadi, N. A., Rempel, G. L., *Macromolecules* **1987**, *20*, 2362.
84 Booth, B. L., Haszeldine, R. N., Holmes, R. G.G., *J. Chem. Soc. Chem. Commun.* **1976**, *13*, 489.
85 Pimblett, B., Garner, C. D., Clegg, W., *J. Chem. Soc., Dalton Trans.* **1985**, 1977.
86 Schrock, R. R., Osborn, J. A., *J. Am. Chem. Soc.* **1976**, *98*, 2134.
87 Ohtani, Y., Yamagishi, A., Fujimoto, M., *Bull. Chem. Soc. Jpn.* **1979**, *52*, 2149.
88 Guo, X., Rempel, G. L., *J. Mol. Catal.* **1990**, *63*, 279.
89 Guo, X., Parent, J. S., Rempel, G. L., *J. Mol. Catal.* **1992**, *72*, 193.
90 Pan, Q., Rempel, G. L., *Ind. Eng. Chem. Res.* **2000**, *39*, 277.
91 Bhattacharjee, S., Bhowmick, A. K., Avasthi, B. N., *Ind. Eng. Chem. Res.* **1991**, *30*, 1086.
92 Bakhmutov, V. I., Bertran, J., Esteruelas, M. A., Lledos, A., Maseras, F., Modrego, J., Oro, L. A., Sola, E., *Chem. Eur. J.* **1996**, *2(7)*, 815.
93 Charmondusit, K., Prasassarakich, P., McManus, N. T., Rempel, G. L., *J. Appl. Polym. Sci.* **2003**, *89*, 142.
94 Pan, Q., Rempel, G. L. *Computer-Aided Modeling and Analysis of Complex Catalysis Kinetics*, October, 49th CSChE, Saskatoon, October **1999**.
95 Poloso, A., Murray, J. G., US Patent 4,028,485, **1977** (to Mobil Oil Corporation, New York, NY).
96 Kang, J., US Patent 4,098,991, **1978** (to The Firestone Tire & Rubber Company, Akron, OH).
97 Hoxmeier, R. J., US Patent 4,595,749, **1986** (to Shell Oil Company, Houston, TX).
98 Tsiang, R. C., US Patent 5,073,621, **1991** (to Shell Oil Company, Houston, TX).
99 Madgavkar, A. M., Daum, D. W., Gibler, C. J., US Patent 5,089,541, **1992** (to Shell Oil Company, Houston, TX).
100 Gibler, C. J., US Patent 5,821,696, **1994** (to Shell Oil Company, Houston, TX).
101 Diaz, Z., Gibler, C. J., US Patent 5,212,285, **1993** (to Shell Oil Company, Houston, TX).
102 Gibler, C. J., Austgen, Jr., D. M., Parker, R. A., US Patent 6,177,521, **2001** (to Shell Oil Company, Houston, TX).
103 Wilkey, J. D., US Patent 6,207,795, **2001** (to Shell Oil Company, Houston, TX).
104 Hofman, A. H., De Smet, H. J. A., Villena, A., Wirts, A. G. C., US Patent 6,800,725, **2004** (to KRATON Polymers U. S. LLC, Houston, TX).
105 Panster, P., Wieland, S., Buding, H., Obrecht, W., US Patent 5,403,566, **1995** (to Bayer AG, Leverkusen, DE).
106 Madgavkar, A. M., Gibler, C. J., Daum, D. W., US Patent 5,104,972, **1992** (to Shell Oil Company, Houston, TX).

107 Nguyen, P., Bender, H., Arsenault, G., Spadola, I., Mersmann, F.-J., US Patent 6,646,059, **2003** (to Bayer Inc., Sarnia, CA).

108 Biesenberger, J.A., Sebastian, D.H., *Principles of Polymerization Engineering*, Chapter 6, John Wiley Interscience, New York, NY, **1983**.

109 Pan, Q., Liu, Q., Sun, J., Li, C., Liang, A., Xie, F. Zhang, T., Chinese Patent ZL 97110481 (**1997**).

110 Pan, Q., Liu, Q., Sun, J., Li, C., Xie, F., Zhang, S., Chinese Patent, ZL 97113303 (**1997**).

111 Pan, Q., Zhou, X., Song, H., Feng, L., Xie, F., Jin, F., Chinese Patent, ZL 97219386 (**1997**).

112 Nauman, E.B., Flash devolatilization, in: *Encyclopedia of Polymer Science and Engineering*, Supplement Volume, Wiley, New York, **1989**, p. 317.

113 Biesenberger, J.A., *Devolatilization of Polymers: Fundamentals, Equipment, Applications*. Hanser Publishers, Munich, Vienna, New York, **1983**.

114 Flock, J.W., Matson, S.L., US Patent 4,408,040, **1983** (to General Electric Company, Schenectady, NY).

115 Houslay, R.J.G., US Patent 4,049,897, **1977** (to Imperial Chemical Industries Limited, London, UK).

116 Alsoy, S., Duda, J.L., *Aiche J.*, **1998**, *44(3)*, 582.

117 Hiroshi, Y., Shigeru, S., Japanese Patent 08-109,219, **1996** (to Asahi Chem. Ind. Co. Ltd, Japan).

118 Taizo, K., Tomohiro, Y., Yoshihiro, M., Kazumi, N., Japanese Patent 11-286,513, **1999** (to JSR CORP, Japan).

119 Miyamot, K., Yamakoshi, Y., Shiraki, T., US Patent 6,815,509, **2004** (to Asahi Kasei Kabushiki Kaisha, Osaka, Japan).

120 Pan, Q., Rempel, G.L., Ng, F.T.T., *Polymer Eng. Sci.* **2002**, *42*(5), 899.

121 Pan, Q., Kehl, A., Rempel, G.L., *Ind. Eng. Chem. Res.* **2002**, *41*(15), 3505.

122 Pan, Q., Rempel, G.L., *Int. J. Chemical Reactor Eng.* **2003**, *1*, A59.

123 Kempermann, T., Koch, S., Sumner, J., *Manual for the Rubber Industry*. Bayer AG, **1993**.

124 Rempel, G.L., Li, G.H., Pan, Q., Ng, F.T.T., *Macromol. Symp.* **2002**, *186*, 23.

125 Li, G., Pan, Q., Rempel, G.L., Ng, F.T.T., *Macromol. Symp.* **2003**, *204*, 141.

126 Rempel, G.L., Guo, X., US Patent 5,208,296, **1993** (to Polysar Rubber Corporation, Sarnia, CA).

127 Rempel, G.L., Guo, X., US Patent 5,210,151, **1993** (to Polysar Rubber Corporation, Sarnia, CA).

20
Transfer Hydrogenation Including the Meerwein-Ponndorf-Verley Reduction

Dirk Klomp, Ulf Hanefeld, and Joop A. Peters

20.1
Introduction

The first homogeneous transfer hydrogenation was reported in 1925 when Meerwein and Schmidt described the reduction of ketones and aldehydes using alcohols as reductants and aluminum alkoxides as the catalysts (Scheme 20.1) [1]. The major difference from previous studies was the hydrogen source; instead of molecular hydrogen, a small organic molecule was utilized to provide the hydrogen necessary to reduce the carbonyl compound. The scope of the reaction was independently investigated by Verley [2], Ponndorf [3], and Lund [4]. Some 12 years later, Oppenauer recognized the possibility of reversing the reaction into an oxidation procedure [5]. Ever since that time, the Meerwein-Ponndorf-Verley (MPV) reduction and the Oppenauer oxidation have been taken as textbook examples of highly selective and efficient reactions under mild conditions.

More recently, Ln^{III} alkoxides were shown to have much higher catalytic activity in this reaction, which allowed their use in only catalytic amounts [6, 7]. Later, however, much higher reactivities for Al^{III}-catalyzed Meerwein-Ponndorf-Verley and Oppenauer (MPVO) reactions have also been achieved with dinuclear Al^{III} complexes [8,9] and with Al^{III} alkoxides generated *in situ* [10]. Several reviews on the MPVO reactions have been published [11–14].

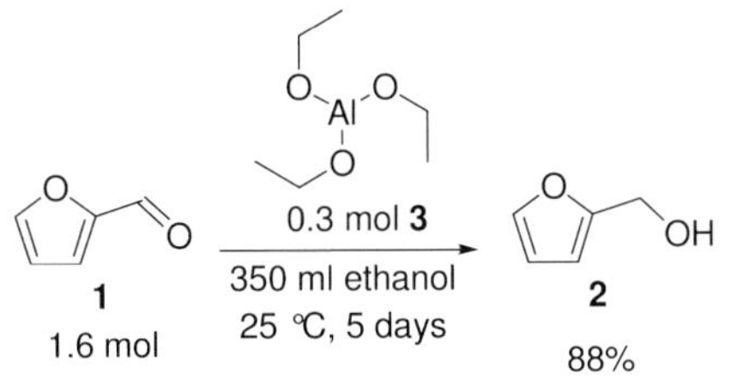

1 (1.6 mol) → [0.3 mol 3, 350 ml ethanol, 25 °C, 5 days] → 2 (88%)

Scheme 20.1 The first homogeneous transfer hydrogenation reduction of furfural (**1**) to furfurylalcohol (**2**), described by Meerwein and Schmidt, using 0.18 equiv. aluminum ethoxide (**3**) [1].

The Handbook of Homogeneous Hydrogenation.
Edited by J. G. de Vries and C. J. Elsevier
Copyright © 2007 WILEY-VCH Verlag GmbH & Co. KGaA, Weinheim
ISBN: 978-3-527-31161-3

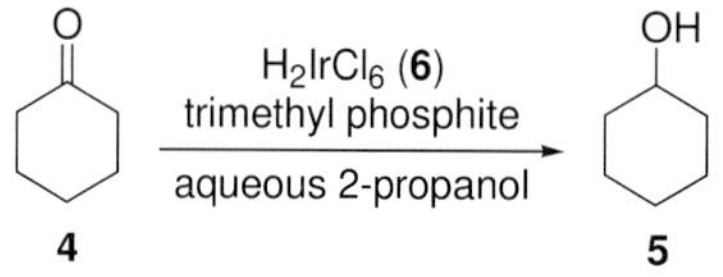

H_2IrCl_6 (**6**)
trimethyl phosphite
aqueous 2-propanol

4 → **5**

Scheme 20.2 A new approach towards the transfer hydrogenation of ketones, performed in 1964.

Pioneering studies on a different class of transfer hydrogenation catalysts were carried out by Henbest et al. in 1964 [15]. These authors reported the reduction of cyclohexanone (**4**) to cyclohexanol (**5**) in aqueous 2-propanol using chloroiridic acid (H_2IrCl_6) (**6**) as catalyst (Scheme 20.2). In the initial experiments, turnover frequencies (TOF) of 200 h^{-1} were reported.

A major step forward at this time was the introduction of the Wilkinson catalyst ($RhCl(PPh_3)_3$) (**7**) for hydrogen transfer reactions [16]. Although actually designed for hydrogenation with molecular hydrogen, this catalyst has been intensively used in transfer hydrogenation catalysis. Ever since, iridium, rhodium and also ruthenium complexes have been widely used in reductive transfer hydrogenations. The main advantage of these catalysts over the MPVO catalysts known until then was their comparatively higher catalytic activity. The TOFs of the transition-metal catalysts could be improved even further by the use of a base as additive, which deprotonates the substrate, facilitating complexation of the substrate to the metal ion in the intermediate complex [17–21]. Numerous reviews have been published on the topic of transition metal-catalyzed transfer hydrogenations [22–28].

The scope of hydrogen transfer reactions is not limited to ketones. Imines, carbon–carbon double and triple bonds have also been reduced in this way, although homogeneous and heterogeneous catalyzed reductions using molecular hydrogen are generally preferred for the latter compounds.

The advantages of hydrogen transfer over other methods of hydrogenation comprise the use of readily available hydrogen donors such as 2-propanol, the very mild reaction conditions, and the high selectivity. High concentrations of the reductant can be applied and the hydrogen donor is often used as the solvent, which means that mass transfer limitations cannot occur in these reactions. The uncatalyzed reduction of ketones requires temperatures of 300 °C [29].

Hydrogen transfer reactions are reversible, and recently this has been exploited extensively in racemization reactions in combination with kinetic resolutions of racemic alcohols. This resulted in dynamic kinetic resolutions, kinetic resolutions of 100% yield of the desired enantiopure compound [30]. The kinetic resolution is typically performed with an enzyme that converts one of the enantiomers of the racemic substrate and a hydrogen transfer catalyst that racemizes the remaining substrate (see also Scheme 20.31). Some 80 years after the first reports on transfer hydrogenations, these processes are well established in synthesis and are employed in ever-new fields of chemistry.

20.2 Reaction Mechanisms

Since the first use of catalyzed hydrogen transfer, speculations about, and studies on, the mechanism(s) involved have been extensively published. Especially in recent years, several investigations have been conducted to elucidate the reaction pathways, and with better analytical methods and computational chemistry the catalytic cycles of many systems have now been clarified. The mechanism of transfer hydrogenations depends on the metal used and on the substrate. Here, attention is focused on the mechanisms of hydrogen transfer reactions with the most frequently used catalysts. Two main mechanisms can be distinguished: (i) a direct transfer mechanism by which a hydride is transferred directly from the donor to the acceptor molecule; and (ii) an indirect mechanism by which the hydride is transferred from the donor to the acceptor molecule via a metal hydride intermediate (Scheme 20.3).

In the direct transfer mechanism, the metal ion coordinates both reactants enabling an intramolecular reaction, and activates them via polarization. Consequently, strong Lewis acids including Al^{III} and the Ln^{III} ions are the most suitable catalysts in this type of reactions. In the hydride mechanism, a hydride is transferred from a donor molecule to the metal of the catalyst, hence forming a metal hydride. Subsequently, the hydride is transferred from the metal to the acceptor molecule. Metals that have a high affinity for hydrides, such as Ru, Rh and Ir, are therefore the catalysts of choice. The Lewis acidity of these metals is too weak to catalyze a direct hydride transfer and, vice versa, the affinity of Al^{III} and Ln^{III} to hydride-ions is too low to catalyze the indirect hydrogen transfer. Two distinct pathways are possible for the hydride mechanism: one in which the catalyst takes up two hydrides from the donor molecule; and another in which the catalyst facilitates the transfer of a single hydride.

All hydrogen transfer reactions are equilibrium reactions. Consequently, both a reduction and an oxidation can be catalyzed under similar conditions. The balance of the reaction is determined by the thermodynamic stabilities of the spe-

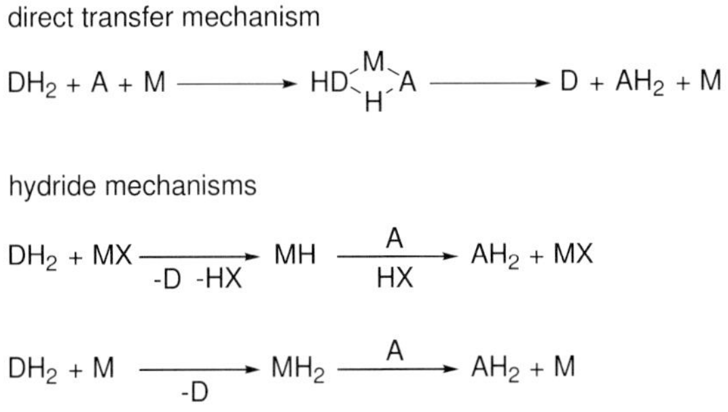

direct transfer mechanism

$$DH_2 + A + M \longrightarrow HD\overset{M}{\underset{H}{\diamond}}A \longrightarrow D + AH_2 + M$$

hydride mechanisms

$$DH_2 + MX \xrightarrow[-D\ -HX]{} MH \xrightarrow[HX]{A} AH_2 + MX$$

$$DH_2 + M \xrightarrow[-D]{} MH_2 \xrightarrow{A} AH_2 + M$$

Scheme 20.3 Schematic representation of the two different hydrogen transfer mechanisms (D = donor molecule; A = acceptor molecule; M = metal).

cies in the redox equilibrium involved and by the concentrations of the hydride donors and acceptors.

20.2.1
Hydrogen Transfer Reduction of Carbonyl Compounds

Transfer hydrogenations of carbonyl compounds are often conducted using 2-propanol as the hydrogen donor. One advantage of this compound is that it can be used simultaneously as a solvent. A large excess of the hydrogen donor shifts the redox equilibrium towards the desired product (see also Section 20.3.1).

Studies aimed at the elucidation of reaction mechanisms have been performed by many groups, notably by those of Bäckvall [28]. In test reactions, typically enantiopure 1-phenylethanol labeled with deuterium at the 1-position (**8**) is used. The compound is racemized with acetophenone (**9**) under the influence of the catalyst and after complete racemization of the alcohol, the deuterium content of the racemic alcohol is determined. If deuterium transfer proceeds from the α-carbon atom of the donor to the carbonyl carbon atom of the acceptor the deuterium is retained, but if it is transferred to the oxygen atom of the acceptor it is lost due to subsequent exchange with alcohols in the reaction mixture (Scheme 20.4).

20.2.1.1 Meerwein-Ponndorf-Verley Reduction and Oppenauer Oxidation

The most common catalysts for the Meerwein-Ponndorf-Verley reduction and Oppenauer oxidation are Al^{III} and Ln^{III} isopropoxides, often in combination with 2-propanol as hydride donor and solvent. These alkoxide ligands are readily exchanged under formation of 2-propanol and the metal complexes of the substrate (Scheme 20.5). Therefore, the catalytic species is in fact a mixture of metal alkoxides.

The catalytic cycle of the reaction is depicted in Scheme 20.6 [31]. After the initial ligand exchange, the ketone (**10**) is coordinated to the metal ion of **11** (a), yielding complex **12**. A direct hydride transfer from the alkoxide to the ketone takes place via a six-membered transition state (b) in which one alkoxy group is oxidized (**13**). The acetone (**14**) and the newly formed alcohol (**15**) are released

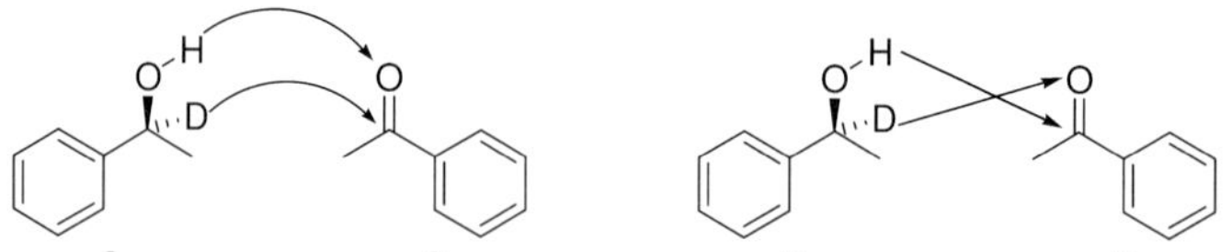

O-H D **8** O **9**
oxygen-to-oxygen/carbon-to-carbon

O-H D **8** O **9**
oxygen-to-carbon/carbon-to-oxygen

Scheme 20.4 Possible pathways of hydrogen transfer during the racemizations of alcohols using the corresponding carbonyl compound and a hydrogen transfer catalyst.

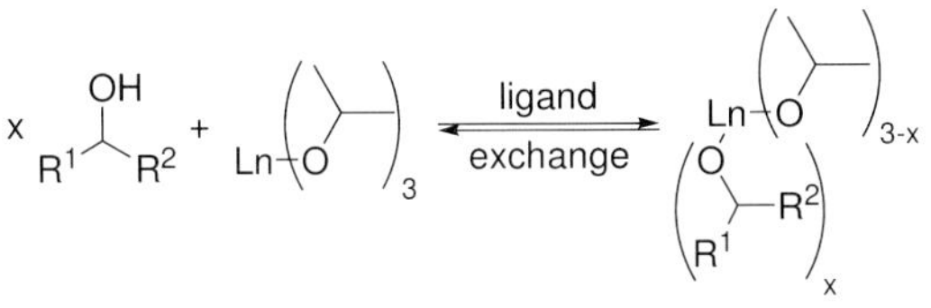
ligand
exchange

Scheme 20.5 Ligand exchange in MPVO reactions.

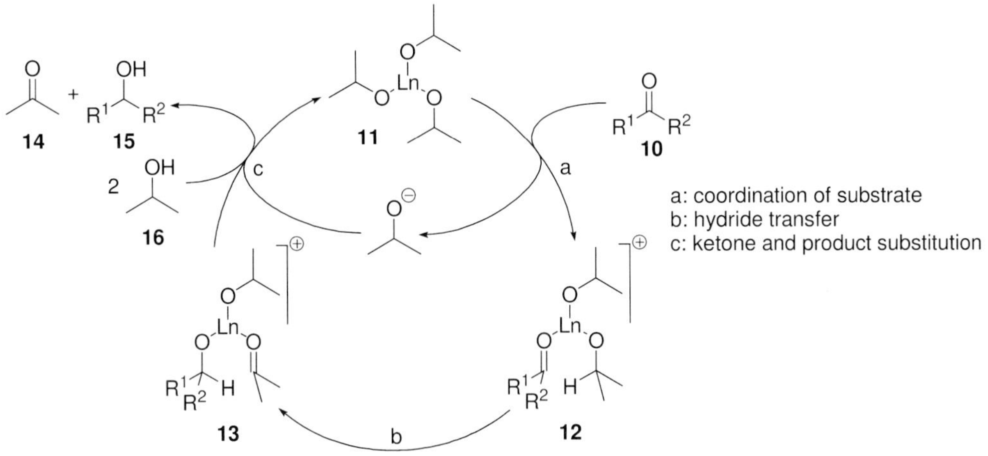
a: coordination of substrate
b: hydride transfer
c: ketone and product substitution

Scheme 20.6 Mechanism of the Meerwein-Ponndorf-Verley-Oppenauer reaction.

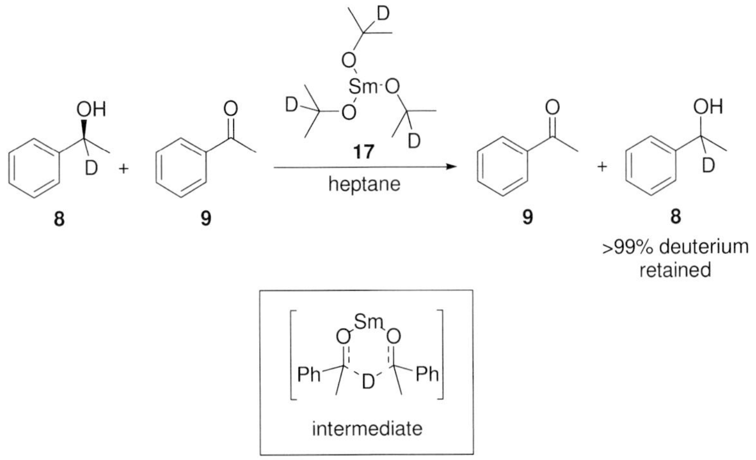
heptane
>99% deuterium retained
intermediate

Scheme 20.7 Racemization of (*S*)-1-deutero-1-phenylethanol (**9**) with deuterated samarium(III) isopropoxide (**17**).

from the metal center by substitution for new donor molecules (**16**) (c) completing the cycle.

The mechanism of the MPVO reactions has been investigated and questioned on several occasions, and a variety of direct hydrogen-transfer pathways have been suggested (see Scheme 20.4) [31–35]. Recently, racemization of D-labeled 1-phenylethanol with deuterated samarium(III) isopropoxide (**17**) proved that the MPVO reaction occurs via a direct hydrogen transfer from the α-position of the isopropoxide to the carbonyl carbon of the substrate (Scheme 20.7) [31].

The selectivity of the hydrogen transfer is excellent. When employing a catalyst with deuterium at the α-positions of the isopropoxide ligands (**17**), complete retention of the deuterium was observed. A computational study using the density functional theory comparing the six-membered transition state (as in Scheme 20.3, the direct transfer mechanism) with the hydride mechanism (Scheme 20.3, the hydride mechanism) supported the experimental results obtained [36]. A similar mechanism has been proposed for the MPV alkynylations [37] and cyanations [38].

20.2.1.2 **Transition Metal-Catalyzed Reductions**

The Wilkinson catalyst, ($RhCl(PPh_3)_3$) (**7**), is not only an excellent hydrogenation catalyst when using molecular hydrogen as hydrogen donor, but can also be employed as a hydrogen transfer catalyst. It is a square-planar, 16-electron complex, which catalyzes these reactions via different pathways depending on the hydrogen donor. The intermediate rhodium complexes tend to retain a four-coordinated square-planar configuration, whereas the molecular hydrogen pathway proceeds through an octahedral state [35, 39–42] (Scheme 20.8).

In transfer hydrogenation with 2-propanol, the chloride ion in a Wilkinson-type catalyst (**18**) is rapidly replaced by an alkoxide (Scheme 20.9). β-Elimination then yields the reactive 16-electron metal monohydride species (**20**). The ketone substrate (**10**) substitutes one of the ligands and coordinates to the catalytic center to give complex **21** upon which an insertion into the metal hydride bond takes place. The formed metal alkoxide (**22**) can undergo a ligand exchange with the hydride donor present in the reaction mixture, liberating the product (**15**).

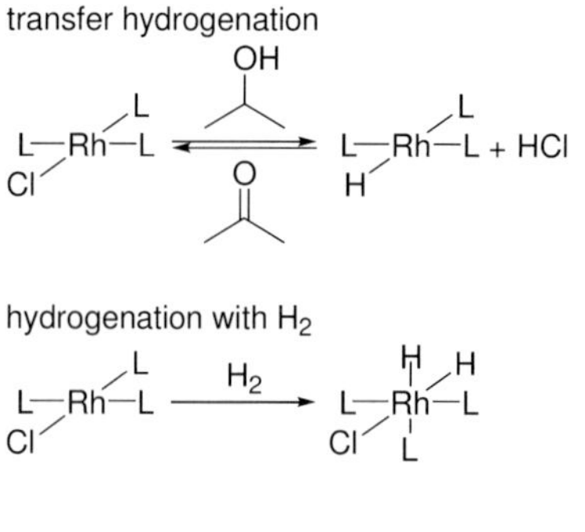

transfer hydrogenation

hydrogenation with H_2

Scheme 20.8 Different behavior of the Wilkinson catalyst (**7**) for transfer hydrogenation and hydrogenation using molecular hydrogen.

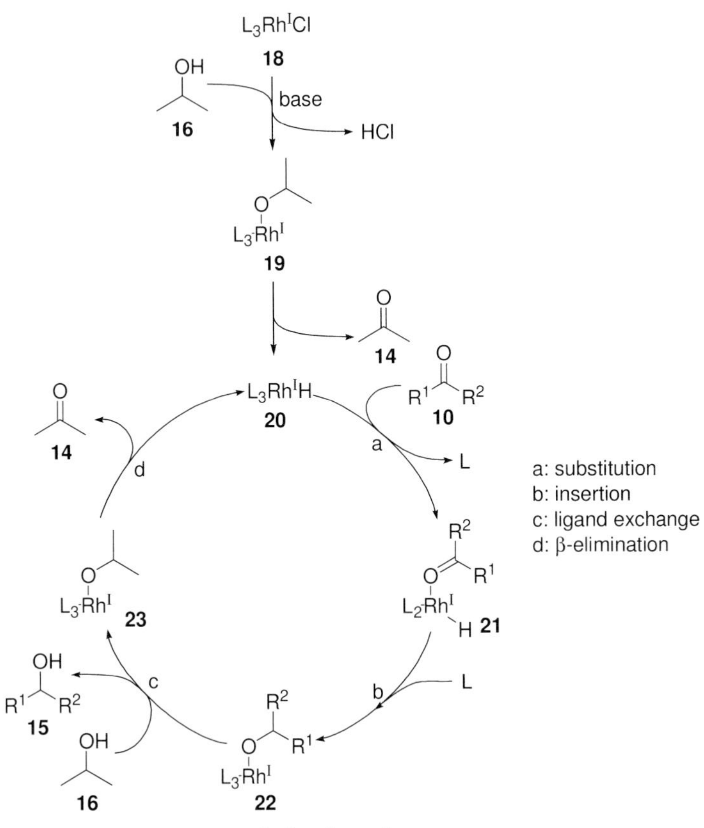

Scheme 20.9 Transition metal alkoxide mechanism.

After β-elimination, acetone is released and the metal monohydride (**20**) is obtained again from **23**, closing the catalytic cycle.

Mechanistic studies show that the extent of deuterium-labeling at the α-position of (*S*)-1-deutero-1-phenylethanol remains almost unchanged during a racemization reaction with this system (Scheme 20.10) [35]. This indicates that a single hydride is transferred from the α-position of the donor to the α-position of the acceptor. Only a slight decrease in deuterium content occurs (5%), which may be attributed to exchange with traces of water. In catalysts bearing phenyl phosphine ligands, the loss of deuterium can also be explained by orthometalation [43], leading to H/D exchange. Several other catalysts have been shown to operate via the same mechanism as the Wilkinson catalyst (Fig. 20.1).

A different mechanism is operative with the 16-electron complex $RuCl_2(PPh_3)_3$ (**24**) (Scheme 20.11). Here, the dichloride complex (**25**) is rapidly converted into a dihydride species (**26**) by substitution of both chloride ligands with alkoxides and subsequent eliminations similar to the conversion of **18** to **20** described above [46, 47]. Subsequently, the ruthenium dihydride species **26**

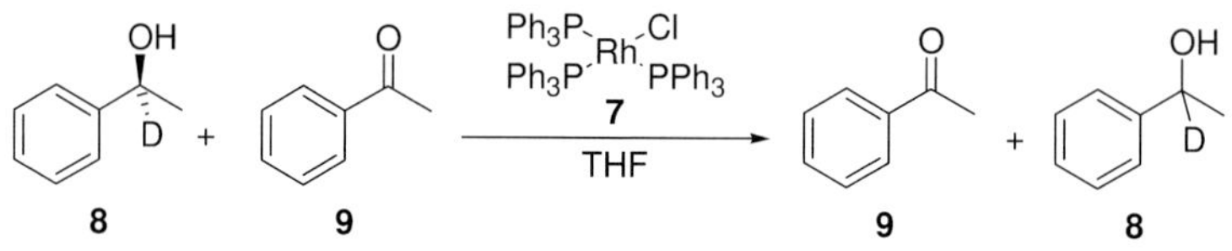

Scheme 20.10 Racemization of (*S*)-1-deutero-1-phenylethanol (**8**) with the Wilkinson catalyst (**7**).

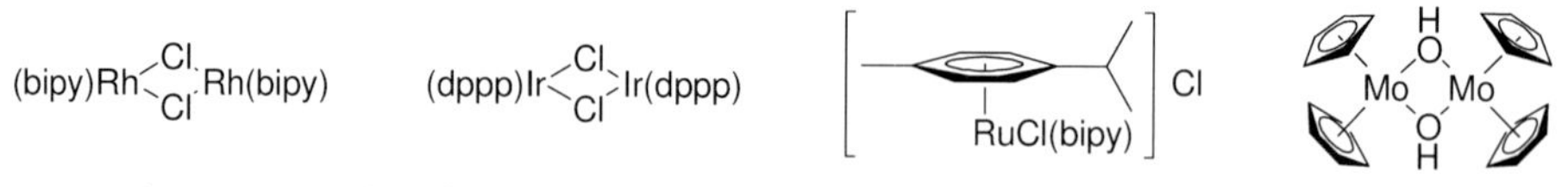

Fig. 20.1 Examples of catalysts operating via the same mechanism as the Wilkinson catalyst (bipy = bipyridine; dppp = 1,3-bis(diphenylphosphinopropane) [35, 44, 45].

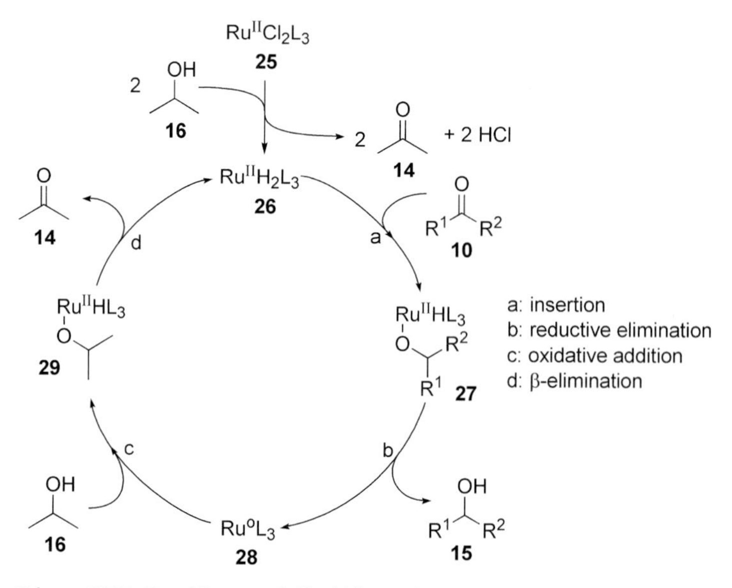

Scheme 20.11 Transition metal dihydride mechanism.

reacts (a) with a substrate molecule (**10**) to give the monohydride alkoxide complex (**27**). Reductive elimination (b) liberates the product (**15**) and a Ru^0 species (**28**). Oxidative addition (c) of an alcohol (**16**) yields a new monohydride alkoxide complex (**29**). After a β-elimination step (d), Ru^{II} dihydride (**26**) is formed again. This mechanism is supported by the fact that the racemization of (*S*)-1-deutero-1-phenylethanol (**8**), catalyzed by this and similar catalysts, decreased to about 40% [35]. In theory, the mechanism depicted in Scheme 20.11 leads to an equal distribution of the deuterium-label over the α-position of the alcohol and its hy-

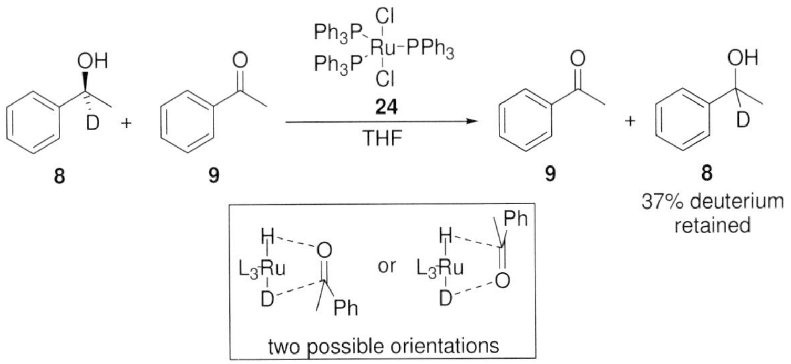

Scheme 20.12 Racemization of (*S*)-1-deutero-1-phenylethanol (**8**) with $RuCl_2(PPh_3)_3$ (**24**).

droxyl function (Scheme 20.12). The deuterium content of the product is probably somewhat lower due to H/D exchange between the alcohol function and traces of water and other alcohols in the reaction mixture.

In the transition metal-catalyzed reactions described above, the addition of a small quantity of base dramatically increases the reaction rate [17–21]. A more elegant approach is to include a basic site into the catalysts, as is depicted in Scheme 20.13. Noyori and others proposed a mechanism for reactions catalyzed with these 16-electron ruthenium complexes (**30**) that involves a six-membered transition state (**31**) [48–50]. The basic nitrogen atom of the ligand abstracts the hydroxyl proton from the hydrogen donor (**16**) and, in a concerted manner, a hydride shift takes place from the *α*-position of the alcohol to ruthenium (a), re-

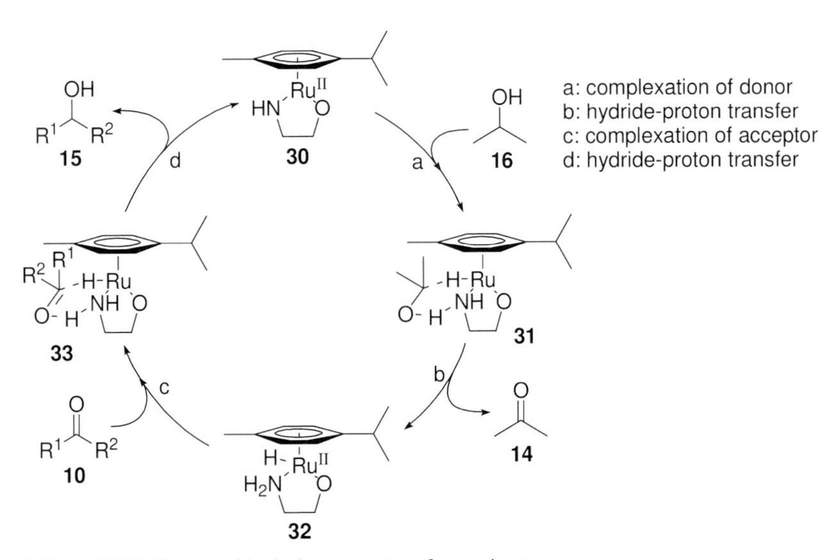

Scheme 20.13 Concerted hydride-proton transfer mechanism.

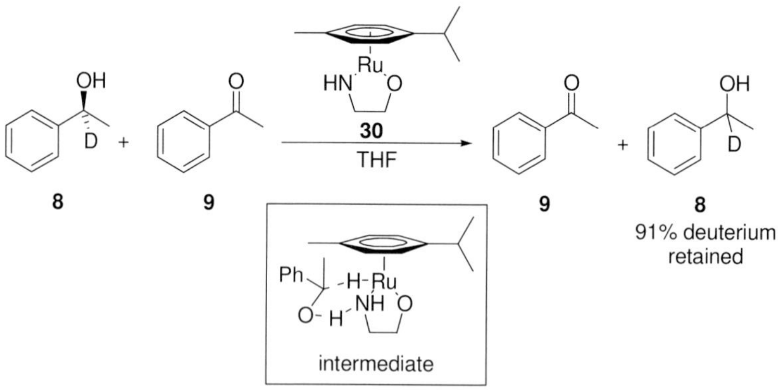

Scheme 20.14 Racemization of (*S*)-1-deutero-1-phenylethanol (**8**) with $Ru(OC_2H_4NH)$(cymene) (**30**).

leasing a ketone (**14**) (b). The Ru^{II} monohydride (**32**) formed is now able to bind to the substrate ketone (**10**) (c) and, in another concerted reaction, the alcohol (**15**) is formed together with the 16-electron ruthenium complex (**30**) (d).

Formation of the Ru^{II} hydride species is supported by the findings of deuterium-labeling studies (Scheme 20.14) [35]. The deuterium label remains at the α-position of the alcohol during racemizations; this is due to the orientation of the alcohol when it coordinates to the Ru-complex. Once again, some deuterium

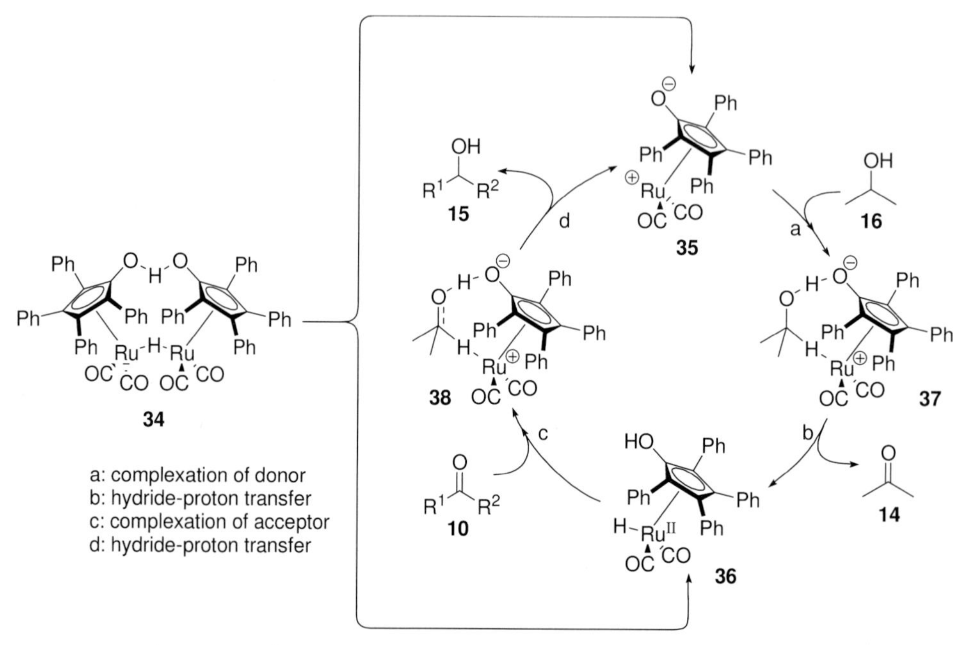

Scheme 20.15 The catalytic cycle of the Shvo catalyst (**34**).

is probably lost due to H/D exchange with traces of water or alcohol in the reaction mixture.

The most reactive transition metal transfer hydrogenation catalysts identified to date have bidentate ligands. Studies towards active catalysts are mainly directed towards the size and nature of the bridge in the ligand [51] and towards the nature of coordinating atoms to the metal [52–54]. It seems that ligands containing both a phosphorus and a nitrogen atom possess the best properties for these types of reactions (see also Section 20.3.3).

Of particular interest is the dinuclear Ru complex **34**, the so-called Shvo catalyst [55, 56]. It has been established that, under the reaction conditions, this complex is in equilibrium with two monometal complexes (**35** and **36**) [57–59]. Both of these resemble catalytic intermediates in the concerted proton-hydride transfer pathway (Scheme 20.13), and will react in a similar way (Scheme 20.15) involving the six-membered transition state **37** and the reduction of the substrate via **38**.

20.2.2
Transfer Hydrogenation Catalysts for Reduction of C–C Double and Triple Bonds

The reduction of C–C double and triple bonds using molecular hydrogen is generally preferred over transfer hydrogenation. However, some interesting examples of transfer hydrogenations of alkenes and alkynes are known. As an illustration of the mechanism of a typical transfer hydrogenation, the reduction of an alkene with dioxane as the hydride donor and the Wilkinson catalyst (**7**) is discussed. The reduction does not necessarily have to be performed with dioxane, but this hydride donor is rather common in these reductions. The use of hydrogen donors and their distinct advantages and disadvantages are discussed in Section 20.3.1.

The first step consists of the substitution of one of the ligands (L) of **18** by dioxane (**39**) in an oxidative addition (a) (Scheme 20.16). β-Elimination of **40** releases 2,3-dihydro-dioxine (**41**) and the 16-electron dihydrogen rhodium complex (**42**) (b). Alkene **43** coordinates to the vacant site of **42** (c) to give complex **44**. A hydride insertion then takes place (d), affording complex **45**. After a reductive elimination (e) of the product **46**, the coordination of a ligand reconstitutes the Wilkinson-type catalyst (**18**).

The coordination of dioxane and subsequent oxidative addition to the catalytic species (step (a) in Scheme 20.16) probably proceeds after the oxygen atom coordinates to the rhodium (**47**), followed by abstraction of a hydrogen atom. The cationic species (**48**) then rearranges to a complex in which the dioxane is bound to the rhodium via the carbon atom (**40**) (Scheme 20.17) [60].

Transfer hydrogenations are typically equilibrium reactions; however, when formic acid (**49**) is utilized as the hydrogen donor, carbon dioxide (**50**) is formed which escapes from the reaction mixture [61–64].

Here, an example is given for the reduction of itaconic acid (**51**) with a rhodium catalyst precursor (**52**) and a phosphine ligand (**53**) (Scheme 20.18). The

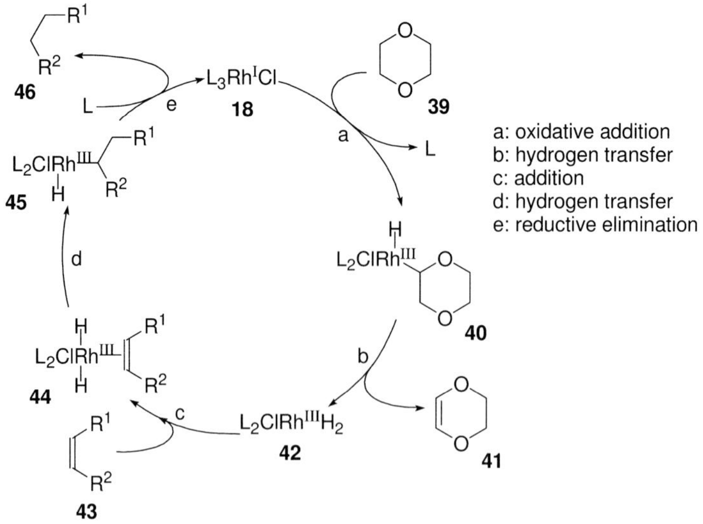

a: oxidative addition
b: hydrogen transfer
c: addition
d: hydrogen transfer
e: reductive elimination

Scheme 20.16 Alkene reduction with dioxane (**39**) as hydride donor and a Wilkinson-type catalyst (**18**).

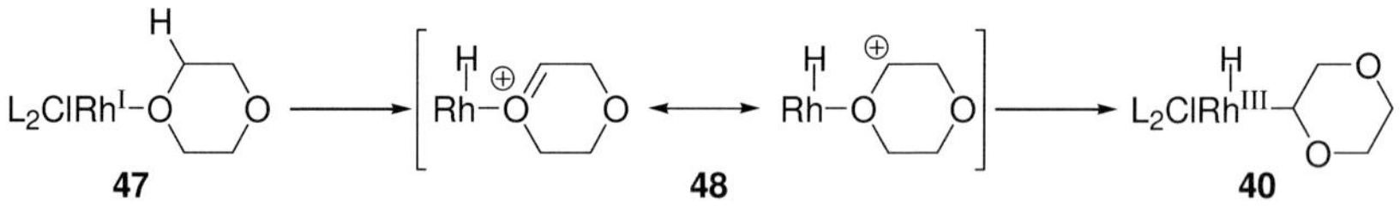

Scheme 20.17 Step (a) of Scheme 20.16: the coordination and oxidative addition of dioxane.

itaconic acid (**51**) is a good chelating ligand for the catalyst, and when the 16-electron Rh^{I} active species **54** is formed, an oxidative addition of formic acid (**49**) takes place (a). Decarboxylation (b) of **55** liberates CO_2 (**50**), forming a Rh^{III}-dihydride (**56**). A hydride transfer (c) leading to a pentacoordinated metal (**57**) and subsequent reductive elimination (d), in which the product (**58**) is liberated and a new substrate (**51**) is coordinated, closes the cycle.

This system is very selective towards the reduction of C–C double bonds, and the oxygen of the acid group that coordinates to the metal is important for good catalytic properties. In the reaction mixture, triethylamine is added in a ratio of formic acid : triethylamine of 5 : 2, which is the commercially available azeotropic mixture of these compounds.

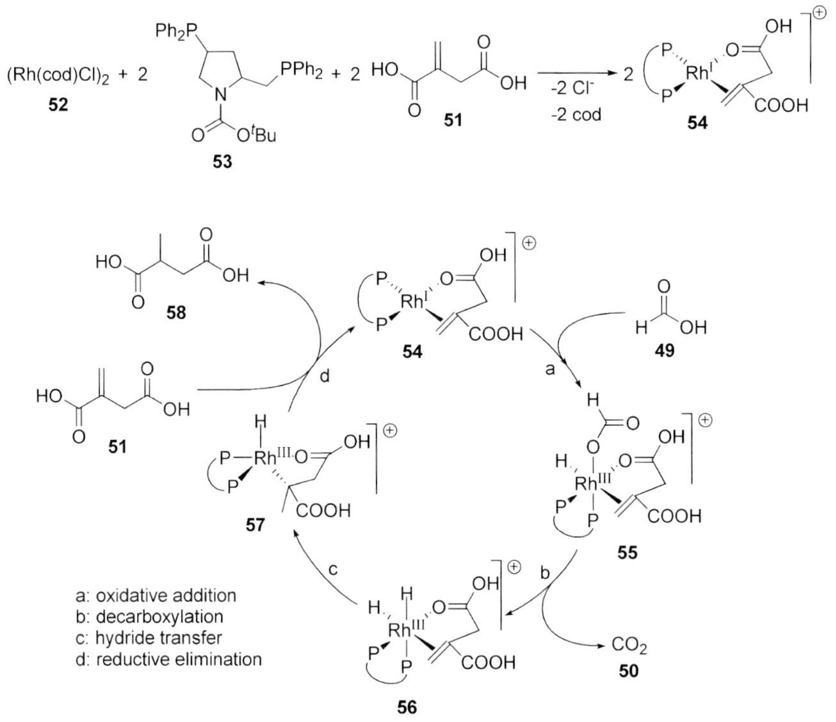

Scheme 20.18 Reduction of the C–C double bond of itaconic acid (**51**) utilizing a rhodium catalyst (**54**) and formic acid (**49**) as hydrogen donor.

20.3 Reaction Conditions

20.3.1 Hydrogen Donors

By definition, hydrogen transfer is a reaction during which hydrogen is transferred from a source other than molecular hydrogen. In theory, the donor can be any compound that has an oxidation potential which is low enough to allow hydrogen abstraction under influence of a catalyst under mild conditions. Another requirement is that the donor is able to coordinate to the catalytic center and does not bind tightly after donation of the hydrogen.

The hydrogen donors vary widely from heteroatom-containing compounds such as alcohols, amines, acids and cyclic ethers to hydrocarbons such as alkanes (Table 20.1). The choice of donor is largely dependent on several issues:

- the *type* of reaction: MPVO or transition metal-catalyzed;
- the *affinity* of the substrate for the metal concerned;

- the *exchange rates* of the substrate between the metal-bound and the free form;
- its *solubility* in the reaction medium or its ability to *dissolve* all other reaction ingredients;
- its influence on the *equilibrium* of the reaction;
- the *temperature* at which the reaction is taking place;
- its ability to avoid harmful *side products*; and
- the *nature* of the functional group to be reduced.

Alcohols have always been the major group of hydrogen donors. Indeed, they are the only hydrogen donors that can be used in Meerwein-Ponndorf-Verley (MPV) reductions. 2-Propanol (**16**) is most commonly used both in MPV reductions and in transition metal-catalyzed transfer hydrogenations. It is generally available and cheap, and its oxidation product, acetone (**14**), is nontoxic and can usually be removed readily from the reaction mixture by distillation. This may have the additional advantage that the redox equilibrium is shifted even more into the direction of the alcohol. As a result of sigma inductive electronic ef-

Table 20.1 Hydrogen donors and their oxidized products.

Entry	Donor	Acceptor
1 [a)]	$R^1CH(OH)R^2$	$R^1C(=O)R^2$
2 [b)]	$R^1OCH_2CH_2R^2$	$R^1OCH=CHR^2$
3	$R^1CH(NH_2)R^2$	$R^1C(=NH)R^2$
4	$R^1NHCH_2CH_2R^2$	$R^1NHCH=CHR^2$
5 [c)]	cyclopentane	cyclopentene or cyclopentadienyl anion
6 [d)]	cyclohexane	cyclohexene or benzene
7	$HCOOH$	CO_2

a) Both primary and secondary alcohols.
b) Typically cyclic ethers as dioxane and THF; only one pair of hydrogens is abstracted.
c) The cyclopentadienyl ring coordinates to the catalyst.
d) The reaction preferably stops at cyclohexene.

fects, secondary alcohols are generally better hydrogen donors than primary ones. However, many examples of the use of primary alcohols have been reported. Ethanol, as already pointed out by Meerwein and Schmidt [1], yields acetaldehyde which, even at room temperature, leaves the reaction mixture and results in irreversible reductions. Unfortunately, the aldehydes resulting from primary alcohols as donors are known to act as catalyst poisons. Furthermore, they may decarbonylate, forming CO, which may modify the catalysts and consequently change their activity [65, 66].

Other alcohols, such as diols [67–69], polyols such as furanoses, pyranoses [70, 71] and polyvinyl alcohol [72] have been reported to enable the reduction of ketones to alcohols.

Heterocyclic compounds are frequently used as hydrogen donors in the reduction of C–C double and triple bonds catalyzed by complexes of transition metals. Cyclic ethers such as [1,4]dioxane (**39**) and 2,3-dihydrofuran are known to donate a pair of hydrogen atoms to this type of compound. 2,3-Dihydro-[1,4]dioxine (**41**), the product of dioxane (**39**), is not able to donate another pair of hydrogen atoms [46, 60, 73, 74]. These heterocyclic compounds are in general also very good solvents for both the catalyst and the substrates.

Nitrogen-containing heterocyclic compounds, including 1,2,3,4-tetrahydroquinoline, piperidine, pyrrolidine and indoline, are also popular hydrogen donors for the reduction of aldehydes, alkenes, and alkynes [75, 76]. With piperidine as hydrogen donor, the highly reactive 1-piperidene intermediate undergoes trimerization or, in the presence of amines, an addition reaction [77]. Pyridine was not observed as a reaction product.

Hydrocarbons are also able to donate hydrogen atoms. In particular, indan and tetralin, which are able to form conjugated double bonds or a fully aromatic system, are used [74].

Once again, use of these donors as solvent may shift the reaction equilibrium towards the desired product. Since the reactivity of olefins is lower than that of carbonyl compounds, higher reaction temperatures are usually required to achieve acceptable TOFs, and then the relatively higher boiling hydrogen donating solvents mentioned above may be the best choice.

Henbest and Mitchell [78] have shown that water can be used as hydrogen source with chloroiridic acid (**6**) as the catalyst through oxidation of phosphorous acid (**59**) to phosphoric acid (**60**) in aqueous 2-propanol. Under these conditions, no hydrogen transfer occurs from 2-propanol. However, iridium complexes with sulfoxide or phosphine ligands show the usual transfer from 2-propanol [79–81].

$$R^1C(O)R^2\ (\mathbf{10}) + HO\text{-}P(OH)_2\ (\mathbf{59}) + H_2O \xrightarrow{H_2IrCl_6} R^1CH(OH)R^2\ (\mathbf{15}) + HO\text{-}P(O)(OH)\text{-}OH\ (\mathbf{60})$$

Scheme 20.19 Transfer hydrogenation with the Henbest system.

Hydrogen transfer reactions are highly selective and usually no side products are formed. However, a major problem is that such reactions are in redox equilibrium and high TOFs can often only be reached when the equilibria involved are shifted towards the product side. As stated above, this can be achieved by adding an excess of the hydrogen donor. (For a comparison, see Table 20.2, entry 8 and Table 20.7, entry 3, in which a 10-fold increase in TOF, from 6 to 60, can be observed for the reaction catalyzed by neodymium isopropoxide upon changing the amount of hydrogen donor from an equimolar amount to a solvent. Removal of the oxidation product by distillation also increases the reaction rate. When formic acid (**49**) is employed, the reduction is a truly irreversible reaction [82]. This acid is mainly used for the reduction of C–C double bonds. As the proton and the hydride are removed from the acid, carbon dioxide is formed, which leaves the reaction mixture. Typically, the reaction is performed in an azeotropic mixture of formic acid and triethylamine in the molar ratio 5:2 [83].

In summary, the most popular hydrogen donors for the reduction of ketones, aldehydes and imines are alcohols and amines, while cyclic ethers or hydroaromatic compounds are the best choice for the reduction of alkenes and alkynes.

20.3.2 Solvents

As mentioned above, the hydrogen donor is the solvent of choice in hydrogen-transfer reactions. However, if for any reason another solvent is needed, it is important to select one that does not compete with the substrate or the ligands of

Table 20.2 Racemization of (*S*)-1-phenylethanol (**61**) in different solvents (Scheme 20.20).[a]

Entry	Solvent	Time [h]	ee [%][b]	TOF [h^{-1}][c]
1	Acetonitrile	>48	>99	0
2	Dioxane	5	28	2
3	THF	3.5	0	4
4	Diisopropyl ether	3.5	0	4
5	MTBE	3.5	0	4
6	Toluene	3	0	5
7	Hexane	2	0	6
8	Heptane	2	0	6

a) Solvent (12 mL), zeolite NaA (30 mg, dried at 400 °C), (*S*)-1-phenylethanol (**61**) (0.24 mL, 2 mmol), acetone (**14**) (0.15 mL, 2 mmol, 1 equiv.), 1,3,5-triisopropylbenzene (int. std.) (0.2 mL) and neodymium(III) isopropoxide (120 mg, 0.37 mmol, 0.185 equiv.) were stirred at 50 °C.
b) ee (starting material) >99%.
c) As determined in the first 15 min of the reaction; in this period predominantly oxidation takes place.

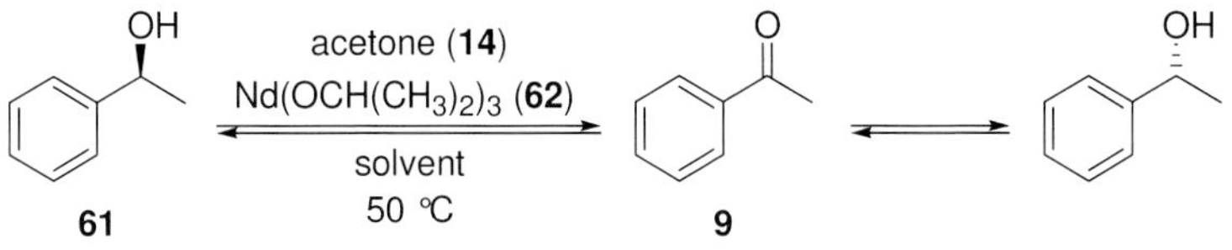

OH
acetone (**14**)
$Nd(OCH(CH_3)_2)_3$ (**62**)
solvent
50 °C
61 **9** O OH

Scheme 20.20 Racemization of **61** with 19% neodymium(III) isopropoxide (**62**) and 1 equiv. acetone (**14**).

the catalyst. By replacing ligands of the catalyst, the electron density of the metal changes, and this may have a detrimental effect on the activity of the catalyst.

As an example, Table 20.2 lists the rate of the racemization of **61** via an MPVO procedure utilizing the catalyst neodymium(III) isopropoxide (**62**) as a function of the solvent. In this case, an equimolar amount of acetone was applied as the oxidant. The best results were obtained with hydrocarbons such as hexane (entry 7) and heptane (entry 8) as solvents, while the reaction rates in dioxane (entry 2) and acetonitrile (entry 1) were much lower due to inactivation of the catalyst by coordination of the solvent to the metallic center (Table 20.2) [84].

20.3.3 Catalysts and Substrates

Meerwein-Ponndorf-Verley-Oppenauer catalysts typically are aluminum alkoxides or lanthanide alkoxides (see above). The application of catalysts based on metals such as ytterbium (see Table 20.7, entries 6 and 20) and zirconium [85, 86] has been reported.

Lanthanide(III) isopropoxides show higher activities in MPV reductions than $Al(OiPr)_3$, enabling their use in truly catalytic quantities (see Table 20.7; compare entry 2 with entries 3 to 6). Aluminum-catalyzed MPVO reactions can be enhanced by the use of TFA as additive (Table 20.7, entry 11) [87, 88], by utilizing bidentate ligands (Table 20.7, entry 14) [89] or by using binuclear catalysts (Table 20.7, entries 15 and 16) [8, 9]. With bidentate ligands, the aluminum catalyst does not form large clusters as it does in aluminum(III) isopropoxide. This increase in availability per aluminum ion increases the catalytic activity. Lanthanide-catalyzed reactions have been improved by the *in-situ* preparation of the catalyst; the metal is treated with iodide in 2-propanol as the solvent (Table 20.7, entries 17–20) [90]. Lanthanide triflates have also been reported to possess excellent catalytic properties [91].

One drawback of all these catalysts is their extreme sensitivity to water. To avoid this problem, reactions should be carried out under an inert atmosphere and, if possible, in the presence of molecular sieves [92]. The molecular sieves also suppress aldol reactions, as will be discussed in Section 20.4.

For the reduction of carbonyl groups or the oxidation of alcohols in the presence of C–C double and triple bonds, MPVO catalysts seem to be the best choice with respect to selectivity for the carbonyl group, as reductions with complexes of transition metals are less selective (see Section 20.3.4). In the vast majority of syntheses, aluminum(III) isopropoxide is used as the catalyst. From a catalytic point of view, this is not the best choice, since it typically must be added in equimolar amounts. Probably due to its availability in the laboratory and ease of handling, it is the most frequently used MPVO-catalyst, despite the development of the more convenient lanthanide(III)isopropoxides. An advantage of the aluminum catalyst in industrial processes is the possibility to distil off the products while the catalyst remains active in the production vessel.

In recent years, many active transition-metal catalysts have been developed (see Table 20.7, entries 21–53). Careful design of the ligands of the transition-metal complexes has led to the development of catalysts with high activities. Mixed chelate ligands containing both a phosphorus- and a nitrogen-binding site were employed to prepare catalysts with unusual electronic properties (Table 20.7, entries 24–29, 40–42, 44). In particular, the catalyst in entry 44 shows a very high TOF for the reduction of acetophenone (10^6 h^{-1}). Other very good catalysts have bidentate phosphine ligands and TOFs of up to 2300 h^{-1} (entries 34 and 35), contain both nitrogen and phosphorus ligands and TOFs of up to 900 h^{-1} (entries 45–47), or have different bidentate moieties (entries 48 and 50–52) and TOFs of up to 14700 h^{-1}.

Neutral mixed chelate ligands containing both phosphorus- and nitrogen-binding sites often show a hemilabile character (they are able to bind via one or two atoms to the metal; Fig. 20.2), which allows for the temporary protection and easy generation of reactive sites in the complexes.

Furthermore, the acidity of PCH_2 protons in oxazoline ligands (**63**) enables easy deprotonation of the chelate, giving rise to a static (non-dissociating) anionic four-electron-donating ligand (**64**). These properties give rise to a high activity (Fig. 20.2) [52].

Transition-metal catalysts are, in general, more active than the MPVO catalysts in the reduction of ketones via hydrogen transfer. Especially, upon the introduction of a small amount of base into the reaction mixture, TOFs of transition-metal catalysts are typically five- to 10-fold higher than those of MPVO catalysts (see Table 20.7, MPVO catalysts: entries 1–20, transition-metal catalysts: entries 21–53). The transition-metal catalysts are less sensitive to moisture than MPVO catalysts. Transition metal-catalyzed reactions are frequently carried out in 2-propanol/water mixtures. Successful transition-metal catalysts for transfer hydrogenations are based not only on iridium, rhodium or ruthenium ions but also on nickel [93], rhenium [94] and osmium [95]. It has been reported that

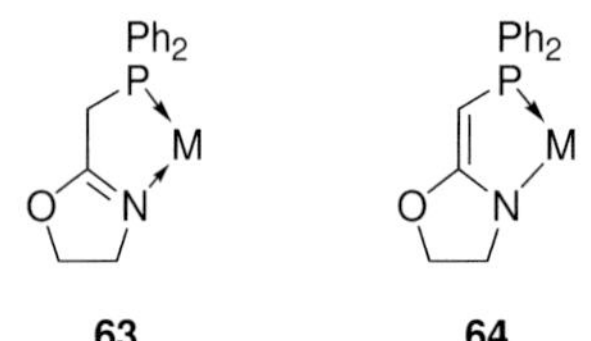

Fig. 20.2 The neutral PCH_2-oxazoline ligand (**63**) and the anionic PCH-oxazoline ligand (**64**).

MPV reductions with aluminum(III) isopropoxide as the catalyst can be hugely enhanced by microwave irradiation [96].

In summary, the reduction of ketones and aldehydes can both be performed with MPV and transition-metal complexes as catalysts. Reductions of alkenes, alkynes, and imines require transition-metal catalysts; MPV reductions with these substrates are not possible.

Hydrogen transfer towards imines is in general slower than towards the corresponding carbonyls. Nonetheless, the reduction can be performed using the same catalysts, although harsher reaction conditions may have to be applied [97]. This is probably a result of the relative stability of imines with respect to carbonyls. In general, the hydrogen transfer of imines proceeds faster with aldimines than with ketimines. The Shvo catalyst (**34**), however, is slightly more reactive towards the latter [56].

In general, the activity of transition-metal catalysts is higher in hydrogenation reactions than in hydrogen transfer reactions. In the few cases where both hydrogenation methods were performed with the same catalyst, it has been shown that reaction rates are lower for transfer hydrogenations. Some examples are known in which transfer hydrogenation is faster than hydrogenation with H_2 [98–100]. The simplicity of the transfer hydrogenation protocol and the abundance of selective and active catalysts make this method very competitive with hydrogenations utilizing H_2, and it is often the preferred reaction.

20.3.4
Selectivity

As mentioned above, MPVO catalysts are very selective towards carbonyl compounds. Alkenes, alkynes or other heteroatom-containing double bonds are not affected by these catalysts, while they can be reduced by transition-metal catalysts. Examples of the reduction of *α,β*-unsaturated ketones and other multifunctional group compounds are compiled in Table 20.3.

Transition metals can display selectivities for either carbonyls or olefins (Table 20.3). $RuCl_2(PPh_3)_3$ (**24**) catalyzes reduction of the C–C double bond function in the presence of a ketone function (Table 20.3, entries 1–3). With this catalyst, reaction rates of the reduction of alkenes are usually higher than for ketones. This is also the case with various iridium catalysts (entries 6–14) and a ruthenium catalyst (entry 15). One of the few transition-metal catalysts that shows good selectivity towards the ketone or aldehyde function is the nickel catalyst (entries 4 and 5). Many other catalysts have never been tested for their selectivity for one particular functional group.

In total syntheses where a homogeneously catalyzed transfer hydrogenation is applied, almost exclusively aluminum(III) isopropoxide is utilized as the catalyst. At an early stage in the total synthesis of (–)-reserpine (**65**) by Woodward [106], an intermediate with two ketone groups and two C–C double bonds is formed (**66**) by a Diels-Alder reaction of *para*-benzoquinone (**67**) and vinylacrylate (**68**). The two ketone groups were reduced with aluminum(III) isopropoxide

Table 20.3 Selectivity towards functional groups.

Entry	Catalyst	Substrate ([S]/[C]) [a]	Product	Reductant	Temperature (time) [°C, h] [b]	Conversion ratio [%]	TOF [h^{-1}]	Reference
1	$RuCl_2(PPh_3)_3$	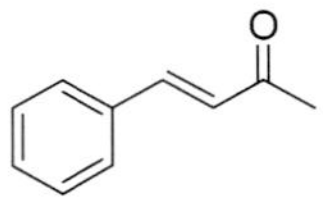 (400)	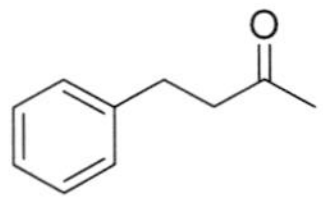	1-phenylethanol	180 (1)	95	380	101
2	$RuCl_2(PPh_3)_3$	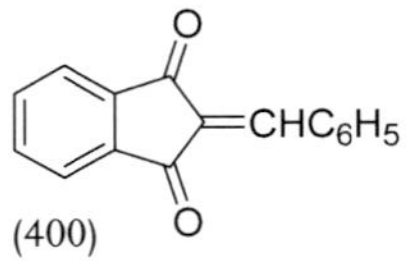 (400)	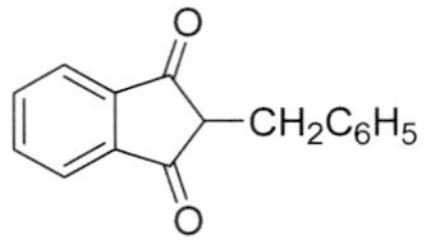	1-phenylethanol	180 (1)	45	180	101
3	$RuCl_2(PPh_3)_3$	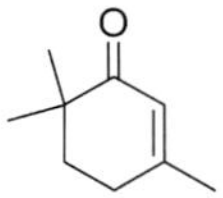 (400)	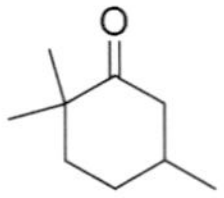	1-phenylethanol	180 (1)	54	216	101
4	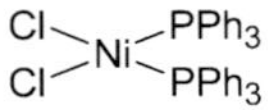 $Cl_2Ni(PPh_3)_2$	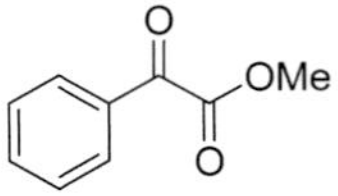 (7)	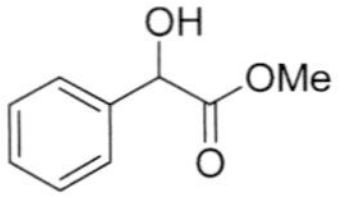	2-propanol	82 (36)	65	0.12	102
5	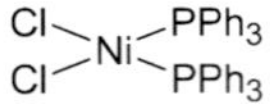 $Cl_2Ni(PPh_3)_2$	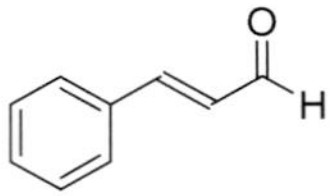 (7)	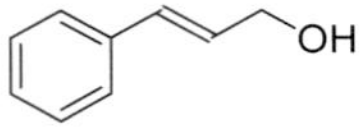	2-propanol	82 (30)	51	0.11	102

Table 20.3 (continued)

Entry	Catalyst	Substrate ([S]/[C]) [a]	Product	Reductant	Temperature (time) [°C, h] [b]	Conversion ratio [%]	TOF [h^{-1}]	Reference
6	$[Ir(cod)Cl]_2$	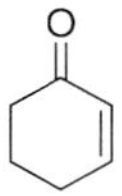 (500)	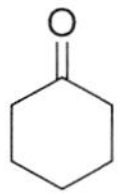	2-propanol	80 (4)	99	12.0	103
7	$[Ir(cod)Cl]_2$	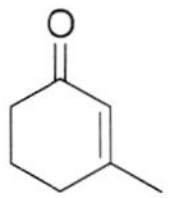 (500)	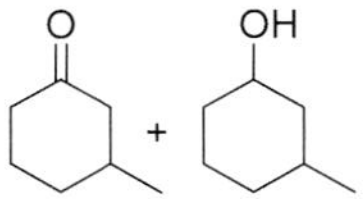 +	2-propanol	80 (4)	91 (3:1)	23.0	103
8	$[Ir(cod)Cl]_2$	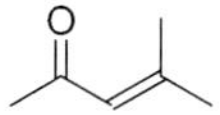 (500)	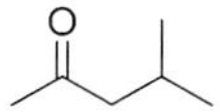	2-propanol	80 (4)	99	12.0	103
9	$[Ir(cod)Cl]_2$	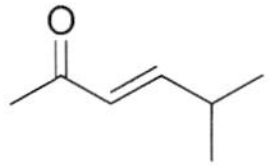 (500)	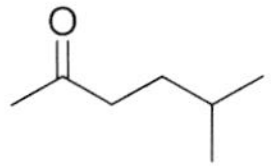	2-propanol	80 (4)	90	6.0	103
10	$[Ir(cod)Cl]_2$	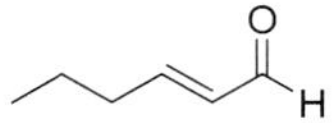 (500)	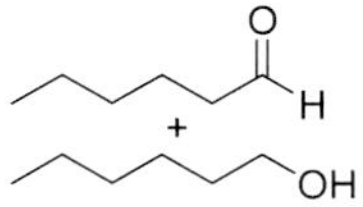 +	2-propanol	80 (4)	75 (10:1)	9.0	103

Table 20.3 (continued)

Entry	Catalyst	Substrate ([S]/[C]) [a]	Product	Reductant	Temperature (time) [°C, h] [b]	Conversion ratio [%]	TOF [h^{-1}]	Reference
11	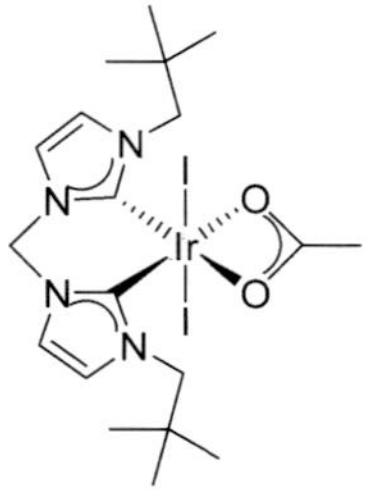 N N Ir O O N N I I	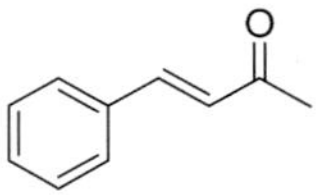 O (1000)	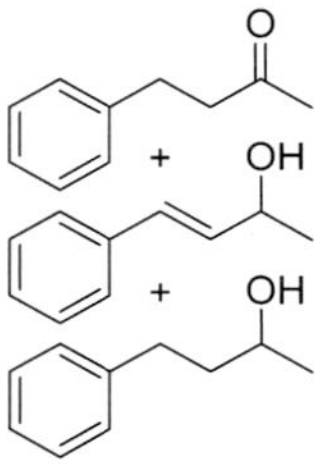 O + OH + OH	2-propanol	82 (20)	92 (9:3:2)	1.0	104
12	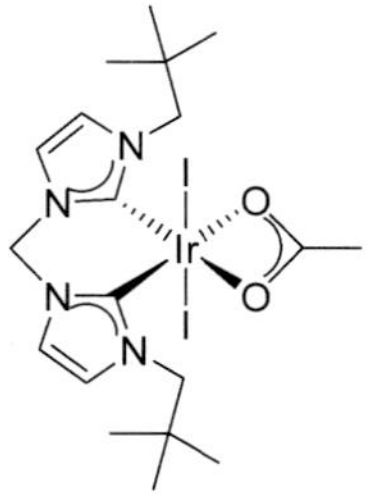 N N Ir O O N N I I	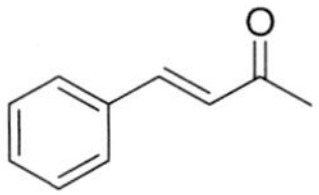 O (200)	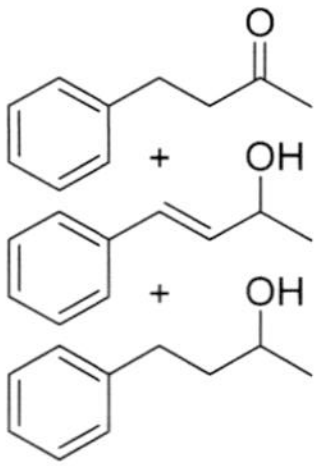 O + OH + OH	2-propanol	82 (4)	>98 (0:0:1)	1.0	104
13	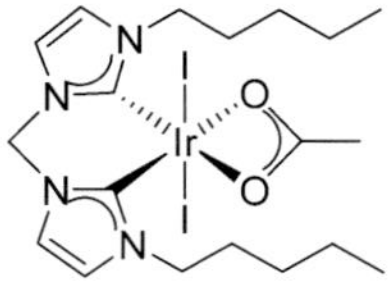 N N Ir O O N N I I	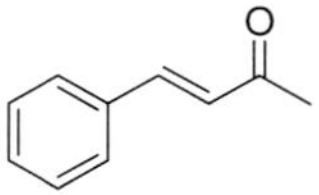 O (1000)	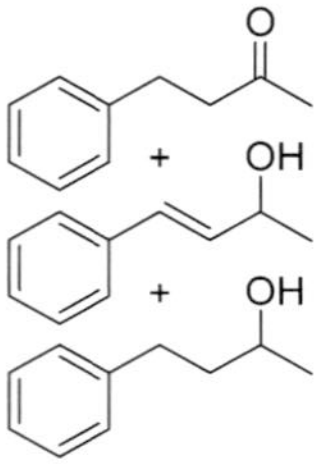 O + OH + OH	2-propanol	82 (20)	66 (7:1:0)	0.7	104

Table 20.3 (continued)

Entry	Catalyst	Substrate ([S]/[C])[a]	Product	Reductant	Temperature (time) [°C, h][b]	Conversion ratio [%]	TOF $[h^{-1}]$	Reference
14	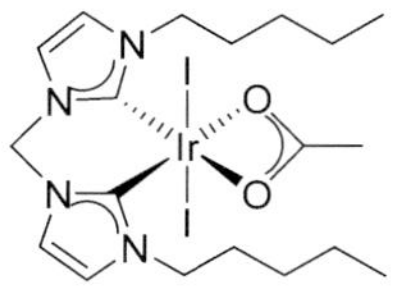 N N Ir O O N N	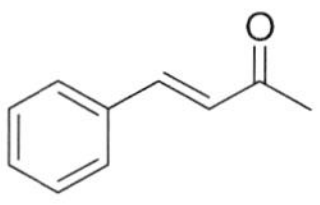 O (200)	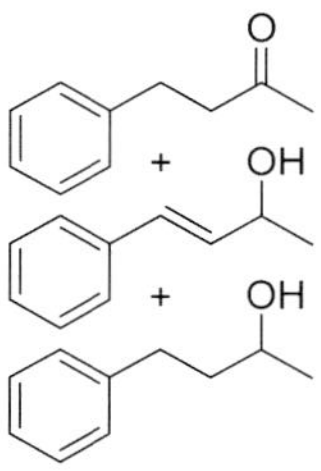 O + OH + OH	2-propanol	82 (7)	100 (1:0:5)	0.6	104
15	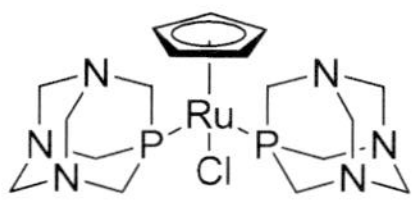 N N N P Ru Cl P N N N	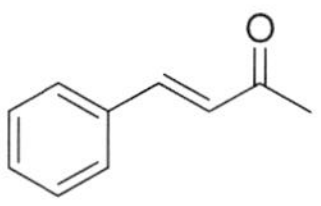 O (100)	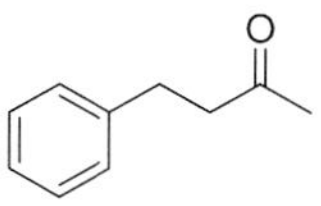 O	HCO_2Na/MeOH/ H_2O	90 (6)	97	16.0	105

a) Substrate : catalyst ratio shown in parentheses.
b) Reaction temperature (reaction time in parentheses).

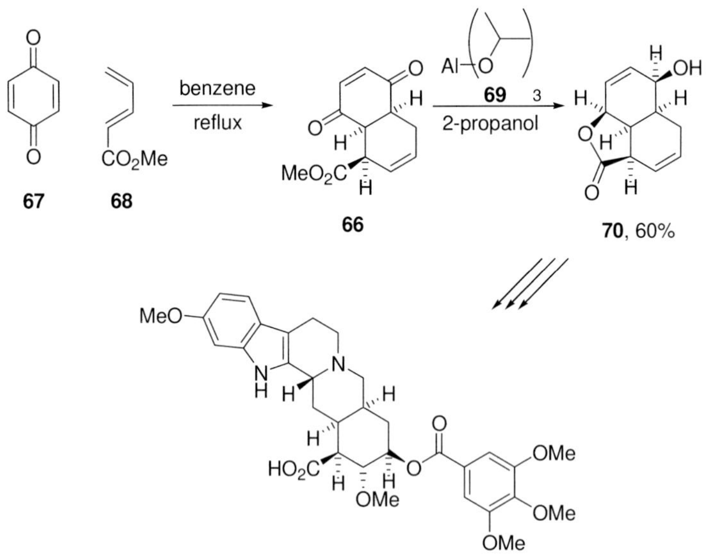

benzene reflux → 69, 2-propanol

67 68 66 70, 60%

(-)-reserpine (65)

Scheme 20.21 Woodward's total synthesis of (–)-reserpine (**65**).

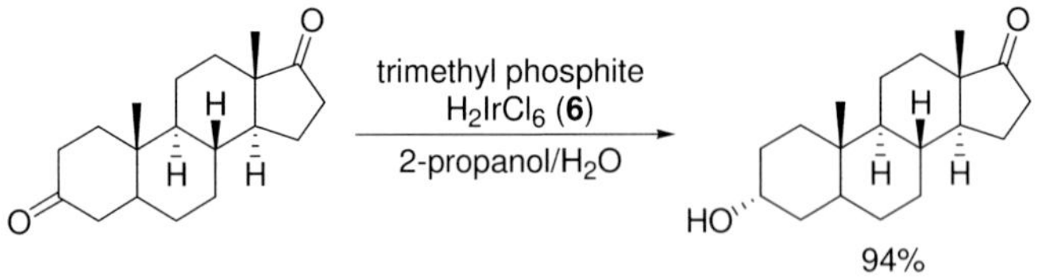

trimethyl phosphite H_2IrCl_6 (**6**) 2-propanol/H_2O

androstan-3,17-dione (**71**) androsterone (**72**) 94%

Scheme 20.22 The reduction of androstan-3,17-dione (**71**) using an iridium catalyst.

(**69**) while leaving the remainder of the molecule unaltered. The resulting di-alcohol is immediately lactonized to the tricyclic compound **70** (Scheme 20.21).

One of the very few examples of a practical application of a transition-metal catalyst in total synthesis is shown in Scheme 20.22 [107]. The chloroiridic acid catalyst (H_2IrCl_6) (**6**) reduces **71** to androsterone (**72**) by selective attack of the sterically less hindered ketone in the 3-position of **71**.

20.4
Related Reactions and Side-Reactions

20.4.1
Aldol Reaction

In the MPVO reaction, several side-reactions can occur (Scheme 20.23). For example, an aldol reaction can occur between two molecules of acetone, which then leads to the formation of diacetone alcohol. The latter acts as a good ligand for the metal of the MPVO catalyst, rendering it inactive. Moreover, the aldol product may subsequently eliminate water, which hydrolyzes the catalyst. The aldol reaction can be suppressed by adding zeolite NaA [84, 92].

20.4.2
Tishchenko Reaction

When aldehydes are reduced, the Tishchenko reaction may be a side-reaction. It is the result of an attack of the oxygen atom of the alkoxide on the carbonyl function of the aldehyde. In particular, aldehydes lacking an α-hydrogen atom such as benzaldehyde are prone to form esters (Scheme 20.24) [108]. It has been reported that many aldehydes can be converted into Tishchenko esters at room temperature, almost quantitatively and with high turnovers, using SmI_2 catalysts [109] or a bi-aluminum catalyst [8].

20.4.3
Cannizzaro Reaction

In the Cannizzaro reaction [110, 111] two aldehyde functionalities disproportionate into the corresponding hydroxyl and carboxyl functions, either as separate compounds or as an ester (Scheme 20.25). The reaction conditions needed are rather harsh, except when R^1 or R^2 is a phenyl group. Typically, an excess of so-

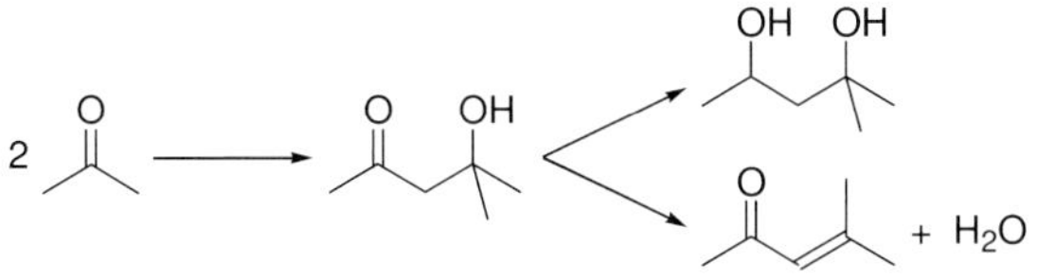

2 O O OH OH OH O + H_2O

Scheme 20.23 The aldol reaction.

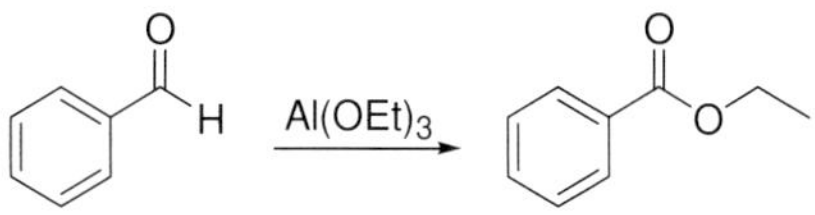

O H $Al(OEt)_3$ O O

Scheme 20.24 The original Tishchenko reaction.

Scheme 20.25 The Cannizzaro reaction.

dium or potassium hydroxide is needed. Therefore, in general, during MPVO reactions only traces of Cannizzaro products are formed.

20.4.4 Decarbonylation

Aldehydes may sometimes pose a problem in transfer hydrogenations catalyzed by transition metals. They can poison the catalyst or decarbonylate, forming CO, which may coordinate to the metal complex and result in a change in activity (Scheme 20.26) [65, 66].

20.4.5 Leuckart-Wallach and Eschweiler-Clarke Reactions

The reductive alkylation of amines is called the Leuckart-Wallach reaction [112–115]. The primary or secondary amine reacts with the ketone or aldehyde. The formed imine is then reduced with formic acid as hydrogen donor (Scheme 20.27). When amines are reductively methylated with formaldehyde and formic acid, the process is termed the Eschweiler-Clarke procedure [116, 117].

20.4.6 Reductive Acetylation of Ketones

In the presence of an active acyl donor such as isopropenyl acetate, a reductive acetylation of a ketone can be performed in the presence of MPVO catalysts

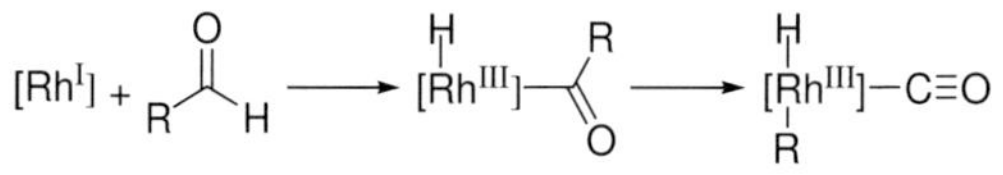

Scheme 20.26 Decarbonylation of an aldehyde under influence of a transition-metal catalyst.

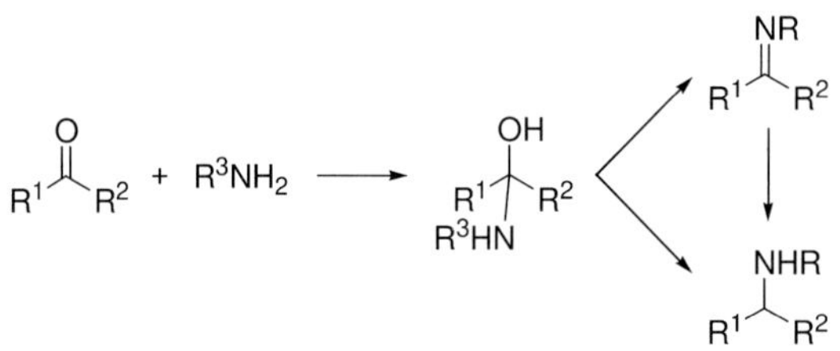

Scheme 20.27 The Leuckart-Wallach reaction.

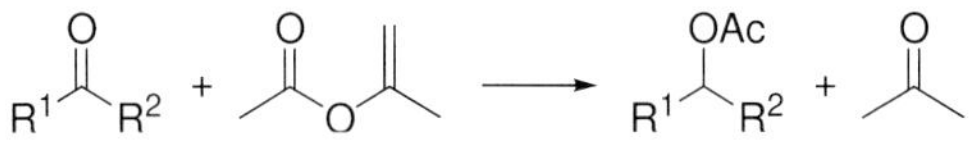

Scheme 20.28 Reductive acetylation of ketones.

(Scheme 20.28) [84, 118]. The first step in this procedure is reduction of the ketone, followed by the acetylation of the formed alkoxide. It may be noted that aluminum(III) isopropoxide and zirconium(IV) isopropoxide do not catalyze the acetylation. With these catalysts, the alcohol is obtained.

20.4.7 Other Hydrogen Transfer Reactions

A few remarkable, but rather uncommon, transfer hydrogenations also deserve mention within the context of this chapter: namely, the reduction of alkynes to alkenes using a chromium catalyst, and the reduction of double bonds using diimines.

In the reduction of C–C triple bonds with chromous sulfate in water, the key intermediate consists of a dichromium complex with the alkyne (Scheme 20.29) [119]. This configuration assures the selective formation of *trans* double bonds. Various substrates have been reduced in excellent yields without the occurrence of isomerizations or byproduct formation (Table 20.4).

One very fast and reliable method for the reduction of double bonds is that of transfer hydrogenation with diimine (Scheme 20.30). Under the influence of traces of copper ion and oxygen from air, hydrazine is rapidly transformed into diimine. This compound is able to hydrogenate double bonds with great success under the formation of nitrogen [120].

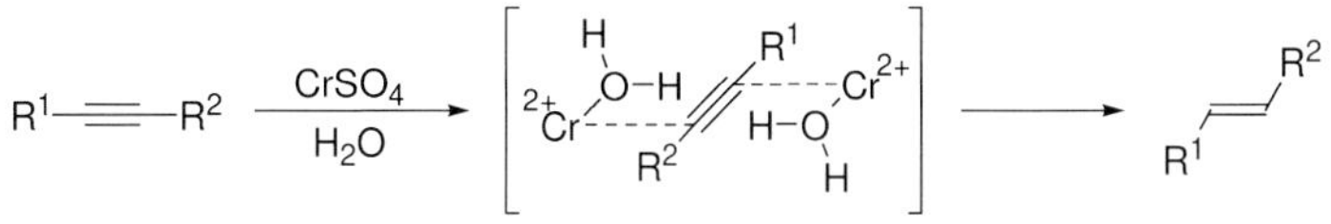

Scheme 20.29 Reduction of alkynes to *trans*-alkenes by chromous sulfate.

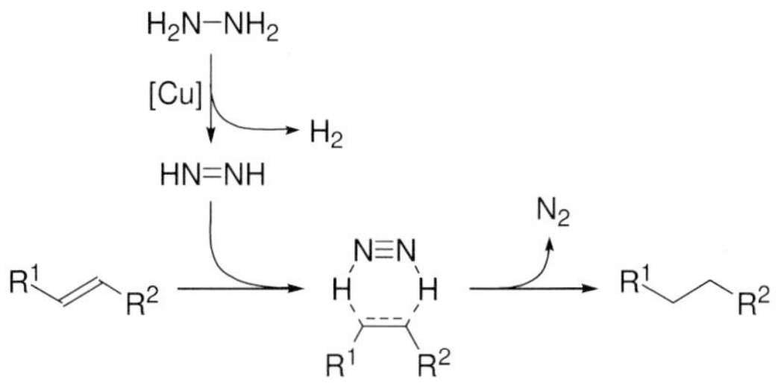

Scheme 20.30 Reduction of alkenes with hydrazine.

Table 20.4 Reduction of alkynes to *trans*-alkenes by chromous sulfate.

Entry	Substrate	Product	Reaction time [h]	Yield [%]	TOF [h^{-1}]
1	$HC{\equiv}C{-}CH_2OH$	$H_2C{=}CH{-}CH_2OH$	0.08	89	5.0
2	$Ph{-}C{\equiv}C{-}CO_2H$	$Ph{-}CH{=}CH{-}CO_2H$	0.25	91	2.5
3	$CH_3{-}C{\equiv}C{-}CH_2OH$	$CH_3{-}CH{=}CH{-}CH_2OH$	2	84	0.4
4	2-($PhC{\equiv}C$)$C_6H_4CO_2H$	2-($PhCH{=}CH$)$C_6H_4CO_2H$	24	85	0.02

20.5 Racemizations

Since transfer hydrogenation reactions of carbonyls are always equilibrium reactions, it is possible to perform both a reduction and an oxidation of a substrate simultaneously. In this way, these reactions can be utilized for both racemizations and epimerizations.

In the contemporary production of enantiopure compounds this feature is highly appreciated. Currently, kinetic resolution of racemates is the most important method for the industrial production of enantiomerically pure compounds. This procedure is based on chiral catalysts or enzymes, which catalyze conversion of the enantiomers at different rates. The theoretical yield of this type of reaction is only 50%, because the unwanted enantiomer is discarded. This generates a huge waste stream, and is an undesirable situation from both environmental and economic points of view. Efficient racemization catalysts that enable recycling of the undesired enantiomer are, therefore, of great importance.

In order to accomplish a racemization rather than an oxidation of an alcohol, the hydrogen acceptor should be added in an equimolar or lower concentration. This is illustrated in Table 20.5 [84]. In order to achieve acceptable reaction times and yields, the amount of hydrogen donor must be adjusted to meet every single reactant.

Racemizations are not limited to alcohols; indeed, some racemizations of amines have also been reported [121].

The next step in the use of transfer hydrogenation catalysts for recycling of the unwanted enantiomer is the dynamic kinetic resolution. This is a combination of two reaction systems: (i) the continuous racemization of the alcohol via hydrogen transfer; and (ii) the enantioselective protection of the alcohol using a

Table 20.5 Racemization of a conjugated and a non-conjugated ketone.

Entry	Substrate	Acetone [equiv.]	Time [h] [a]	Ketone formed [%]
1	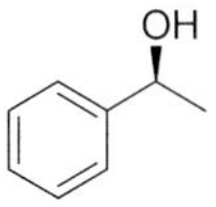 1-phenylethanol	1	1.5	50
2	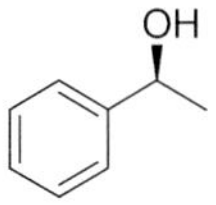 1-phenylethanol	0.1	3	10
3	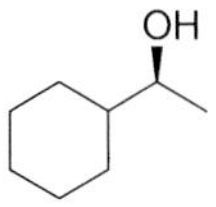 1-cyclohexylethanol	1	2.5	25
4	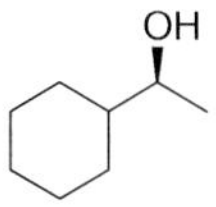 1-cyclohexylethanol	0.1	>7	9

a) Time needed for complete racemization.

stereoselective catalyst, typically an enzyme [122–127]. As was first demonstrated by Williams and colleagues [30], a dynamic kinetic resolution of racemic alcohols by the combination of two catalysts provides a mild and effective means of obtaining enantiomerically pure alcohols in high yields and selectivities (Scheme 20.31).

It is important that the catalysts are stable in each other's presence. Typically, kinetic resolution of the reaction is performed with an enzyme, which always will contain traces of water. Hence, MPVO catalysts and water-sensitive transition-metal catalysts cannot be used in these systems. The influence of the amount of the hydrogen acceptor in the reaction mixture during a dynamic kinetic resolution is less pronounced than in a racemization, since the equilibrium of the reaction is shifted towards the alcohol side.

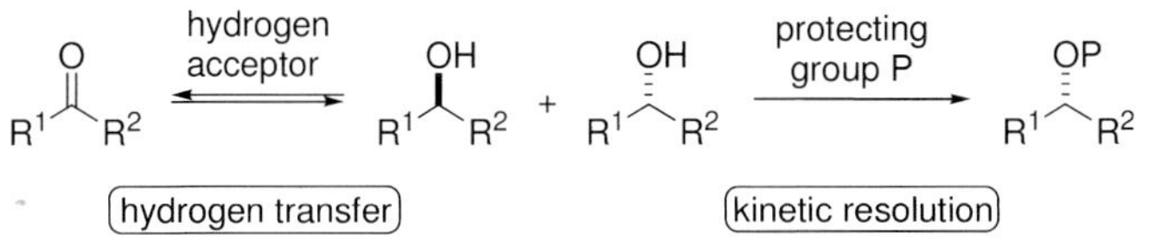

hydrogen acceptor; protecting group P; hydrogen transfer; kinetic resolution

Scheme 20.31 The dynamic kinetic resolution of a racemic alcohol.

Table 20.6 Recent examples of successful dynamic kinetic resolutions.

Entry	Substrate	Product	Racemization catalyst [a]	SCR [b]	Time [h]	Tempera-ture [°C]	Yield [%]	ee [%]	Reference
1	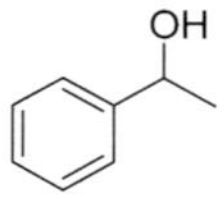	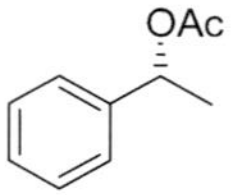	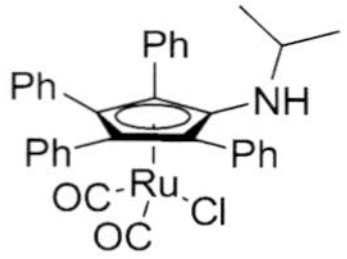	25	31	25	95	>99	128
2	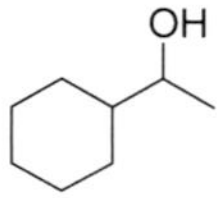	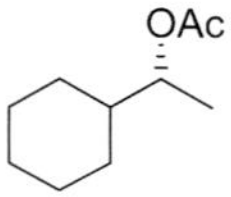	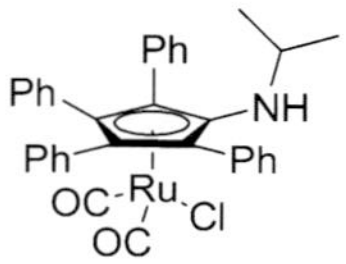	25	48	25	86	>99	128
3	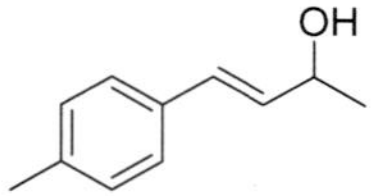	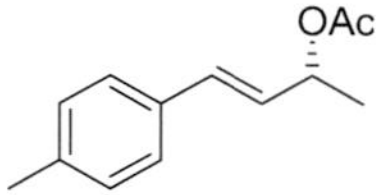	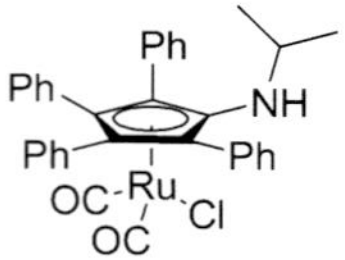	25	96	25	92	>99	128
4	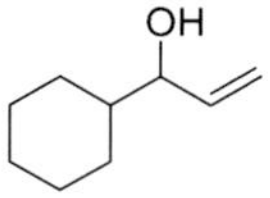	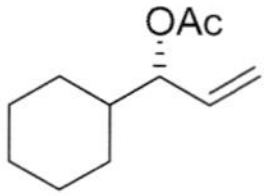	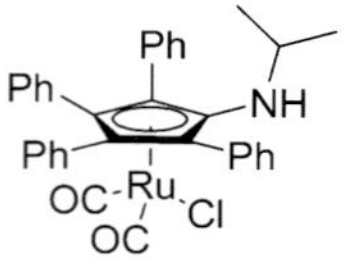	25	96	25	90	>99	128
5	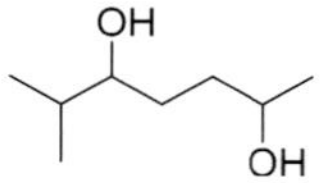	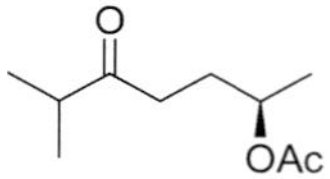	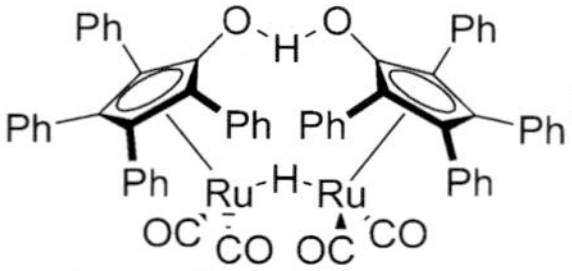	20	18	70	77	89	129

Table 20.6 (continued)

Entry	Substrate	Product	Racemization catalyst[a)]	SCR[b)]	Time [h]	Tempera-ture [°C]	Yield [%]	ee [%]	Reference
6	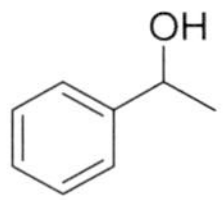 OH	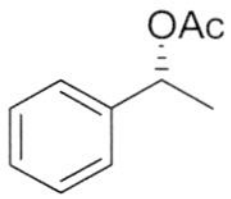 OAc	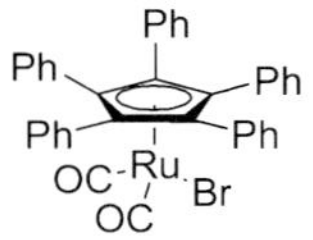 Ph, Ph, Ph, Ph, Ph; OC, OC, Ru, Br	25	3	25	98	>99	130
7	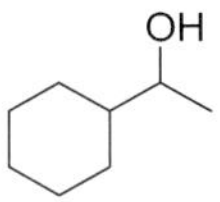 OH	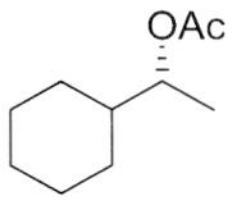 OAc	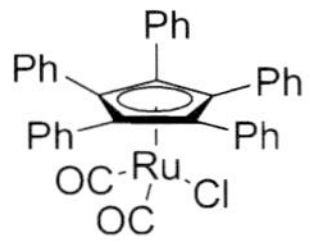 Ph, Ph, Ph, Ph, Ph; OC, OC, Ru, Cl	20	17	25	98	>99	130
8	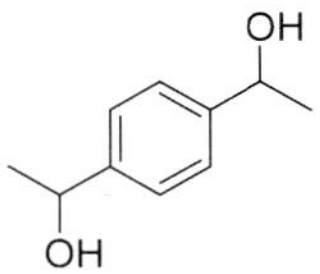 OH, OH	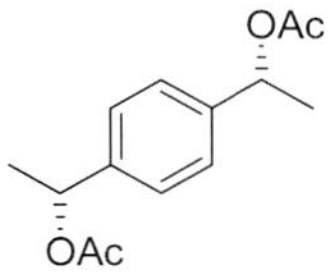 OAc, OAc	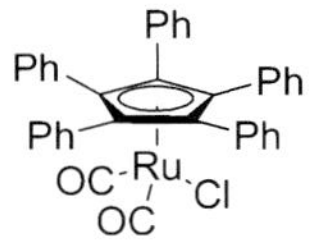 Ph, Ph, Ph, Ph, Ph; OC, OC, Ru, Cl	20	10	25	94	>99 (99:1 d.r.)	130

a) The catalyst for the kinetic resolution is in all cases *Candida antarctica* Lipase B.
b) SCR = substrate:catalyst ratio between the racemic alcohol and the hydrogen transfer catalyst.

Table 20.7 A short overview of the reduction of acetophenone with MPVO catalysts and transition metal catalysts developed during the past five years.

Entry	Catalyst	Reducing agent [a]	Additive/Solvent [a]	Temperature [°C]	SCR	Conversion [%]	Reaction time [h]	TOF [h^{-1}]	Reference
1	tBuOSmI_2	O [c]	THF	65	10	98	24	0.4	131
2	$Al(OCH(CH_3)_2)_3$	OH		50	10	1	1	0.1	7
3	$Nd(OCH(CH_3)_2)_3$	OH		50	100	57	1	60	7
4	$Gd(OCH(CH_3)_2)_3$	OH		50	100	58	1	60	7
5	$Er(OCH(CH_3)_2)_3$	OH		50	100	22	1	20	7
6	$Yb(OCH(CH_3)_2)_3$	OH		50	100	5	1	5	7
7	$La(OCH(CH_3)_2)_3$	O [c]		80	20	75	60	0.3	132
8	$Ce(OCH(CH_3)_2)_3$	O [c]		80	20	15	48	0.1	132

Table 20.7 (continued)

Entry	Catalyst	Reducing agent[a)]	Additive/Solvent[a)]	Temperature [°C]	SCR	Conversion [%]	Reaction time [h]	TOF [h^{-1}]	Reference
9	$Sm(OCH(CH_3)_2)_3$	O [c]		80	20	70	24	0.6	132
10	$Yb(OCH(CH_3)_2)_3$	O [c]		80	20	98	24	0.8	132
11	$Al(OCH(CH_3)_2)_3$	OH	TFA (1)	25	12	44	22	0.3	88
12	$Pu(N(Si(CH_3)_3)_2)_3$	OH		25	20	91	24	0.8	133
13	$Al(CH_3)_3$	OH		65	10	80	12	0.7	10
14	O, Al, O, $NSO_2C_8F_{17}$	OH (10)	CH_2Cl_2	25	10	85	5	1.5	84

Table 20.7 (continued)

Entry	Catalyst	Reducing agent a)	Additive/Solvent a)	Temperature [°C]	SCR	Conversion [%]	Reaction time [h]	TOF [h^{-1}]	Reference
15	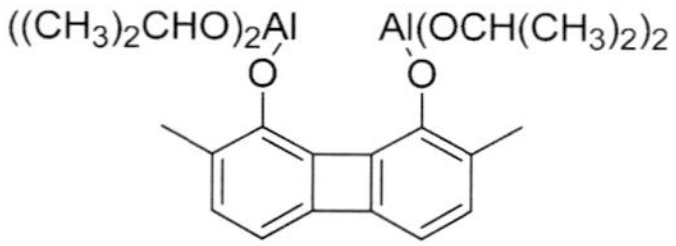 $((CH_3)_2CHO)_2Al$ … $Al(OCH(CH_3)_2)_2$	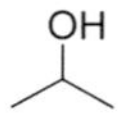 OH (3)	CH_2Cl_2	25	20	96	1	20	8
16	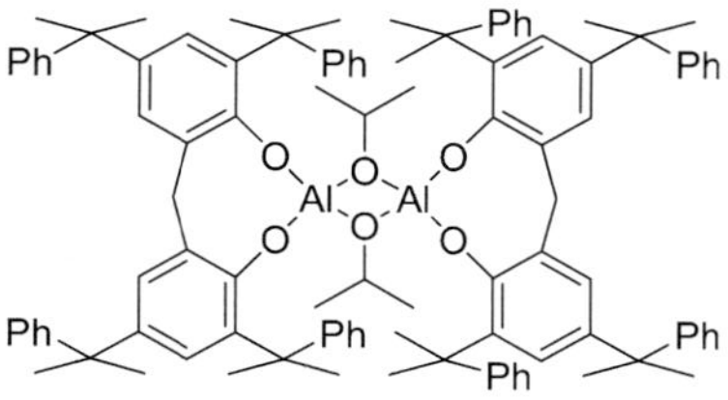 Ph, Ph, Ph, Ph, O, O, Al, O, Al, O, O, O, Ph, Ph, Ph, Ph	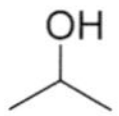 OH (100)	Toluene	111	40	93	1	40	9
17	La + 5% I_2	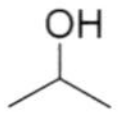 OH		25	1	48	20	0.048	90
18	Ce + 5% I_2	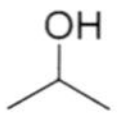 OH		25	1	34	20	0.039	90
19	Sm + 5% I_2	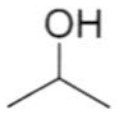 OH		25	1	96	20	0.05	90
20	Yb + 5% I_2	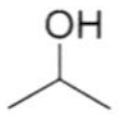 OH		25	1	24	20	0.036	90

Table 20.7 (continued)

Entry	Catalyst	Reducing agent a)	Additive/Solvent a)	Temperature [°C]	SCR	Conversion [%]	Reaction time [h]	TOF [h^{-1}]	Reference
21	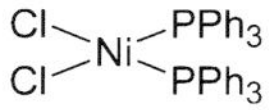 $Cl_2Ni(PPh_3)_2$	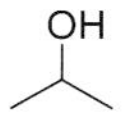 OH	NaOH (0.33)	82	7	82	30	0.2	102
22	$NiBr_2$	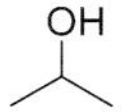 OH	NaOH (85)	95	250	60	4	10	134
23	$SnTf_3$ 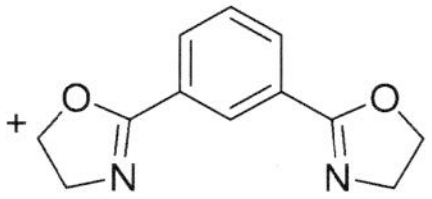 +	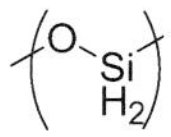 O–Si H2	MeOH	25	10	98	12	0.8	135
24	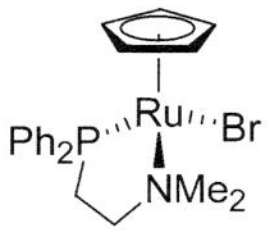 Ph_2P Ru Br NMe_2	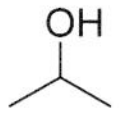 OH	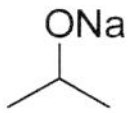 ONa (0.02)	82	200	>99	48	5	136
25	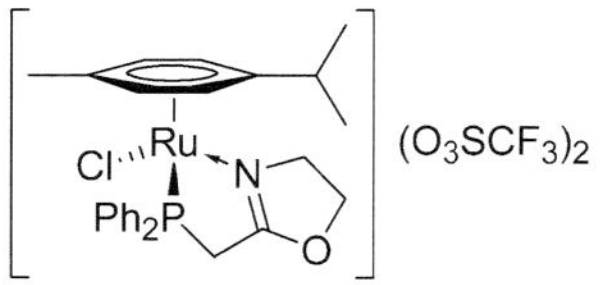 [Cl Ru N Ph_2P O] $(O_3SCF_3)_2$	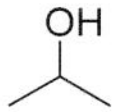 OH	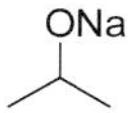 ONa (0.025)	82	200	61	1	120	52

Table 20.7 (continued)

Entry	Catalyst	Reducing agent [a)]	Additive/Solvent [a)]	Temperature [°C]	SCR	Conversion [%]	Reaction time [h]	TOF [h^{-1}]	Reference
26	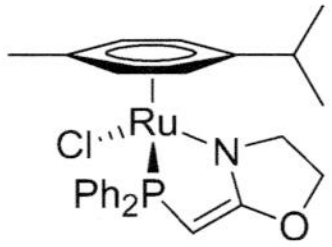 Cl, Ru, N, Ph_2P, O	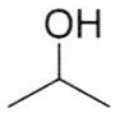 OH	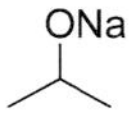 ONa (0.025)	82	200	25	0.25	200	52
27	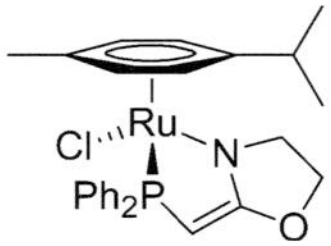 Cl, Ru, N, Ph_2P, O	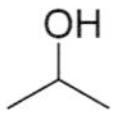 OH	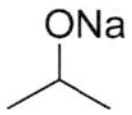 ONa (0.025)	82	200	94	6	30	52
28	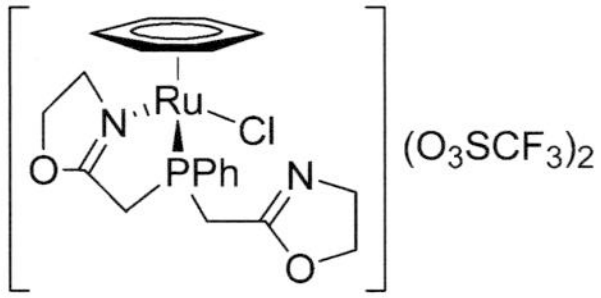 [N, Ru, Cl, O, PPh, N, O] $(O_3SCF_3)_2$	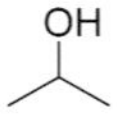 OH	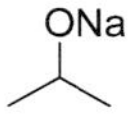 ONa (0.12)	82	200	54	1	110	52
29	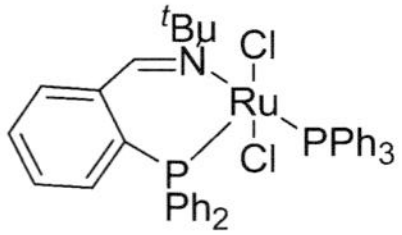 ^{t}Bu, N, Cl, Ru, PPh_3, Cl, P, Ph_2	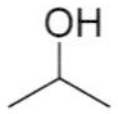 OH	NaOH (0.5)	90	5	91	0.5	910	137
30	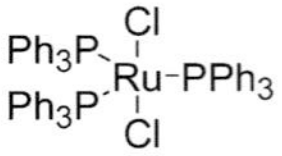 Ph_3P, Cl, Ru–PPh_3, Ph_3P, Cl	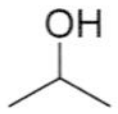 OH	NaOH (0.024)	82	500	50	2	2.0	138

Table 20.7 (continued)

Entry	Catalyst	Reducing agent[a)]	Additive/Solvent[a)]	Temperature [°C]	SCR	Conversion [%]	Reaction time [h]	TOF [h^{-1}]	Reference
31	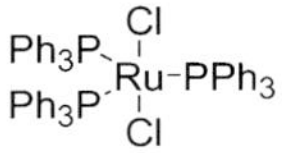 Cl, Ph_3P, Ru–PPh_3, Ph_3P, Cl	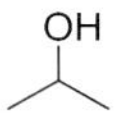 OH	NaOH (0.024) $Yb(OTf)_3$ (0.004)	82	500	86	2	3.5	138
32	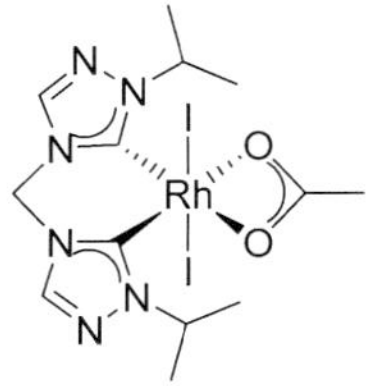 N, N, I, N, O, Rh, N, O, I, N, N	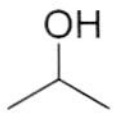 OH	KOH (0.5)	82	1000	>98	10	100	139
33	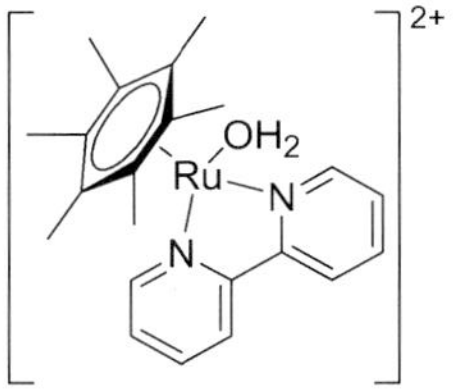 [OH_2, Ru, N, N]$^{2+}$	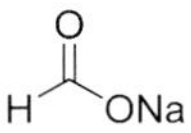 O, H, ONa (30)	H_2O	70	200	98	4	50	140
34	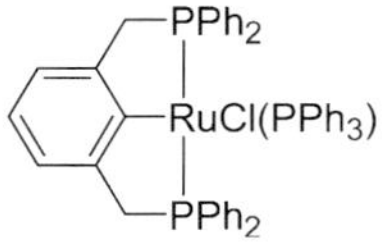 PPh_2, RuCl(PPh_3), PPh_2	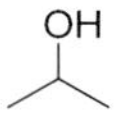 OH		82	1000	50	0.22	2300	141

Table 20.7 (continued)

Entry	Catalyst	Reducing agent [a]	Additive/Solvent [a]	Temperature [°C]	SCR	Conversion [%]	Reaction time [h]	TOF [h^{-1}]	Reference
35	(PPh$_2$, RuCl(PPh$_3$), PPh$_2$)$_6$	OH		82	1000	50	0.23	2200	141
36	$[Cp^*Ir_{III}(bpy)(H_2O)]^{2+}$	H, O, OH (5)	H_2O	70	200	97	1	194	142
37	MeC_6H_4, Cl–Ru-PPh$_2$, Cl, O, H	OH	NaOH (0.048)	82	500	98	10	50	143
38	MeC_6H_4, Cl–Ru-PPh$_2$, Cl, O, H	OH	NaOH (0.048)	82	500	97	9	55	143

Table 20.7 (continued)

Entry	Catalyst	Reducing agent[a)]	Additive/Solvent[a)]	Temperature [°C]	SCR	Conversion [%]	Reaction time [h]	TOF [h^{-1}]	Reference
39	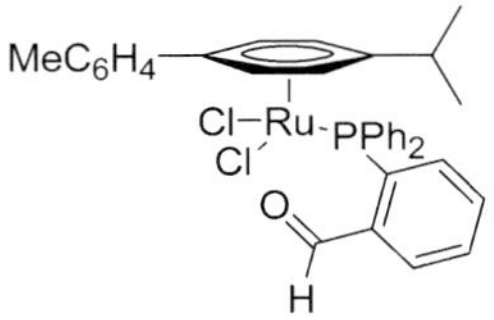 MeC_6H_4, Cl–Ru-PPh_2, Cl, O, H	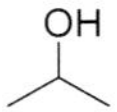 OH	NaOH (0.048)	82	500	96	6	80	143
40	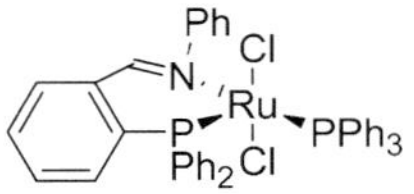 Ph, Cl, N, Ru, PPh_3, P, Ph_2Cl	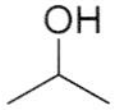 OH	NaOH (0.048)	82	500	98	1	485	144
41	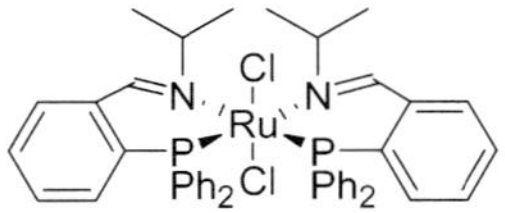 Cl, N, Ru, N, P, Ph_2Cl, P, Ph_2	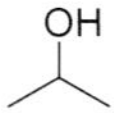 OH	NaOH (0.048)	82	500	97	1	240	144
42	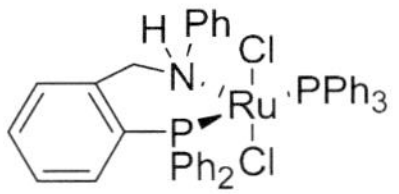 H, Ph, Cl, N, Ru, PPh_3, P, Ph_2Cl	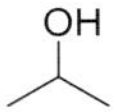 OH	NaOH (0.048)	82	500	96	2	240	144
43	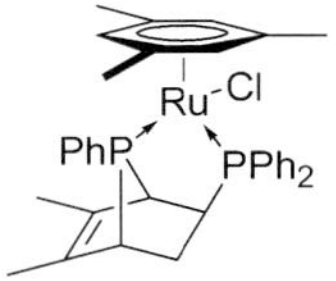 Ru, Cl, PhP, PPh_2	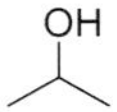 OH	KOH (0.025)	82	200	>99	0.5	400	145

Table 20.7 (continued)

Entry	Catalyst	Reducing agent a)	Additive/Solvent a)	Temperature [°C]	SCR	Conversion [%]	Reaction time [h]	TOF [h^{-1}]	Reference
44	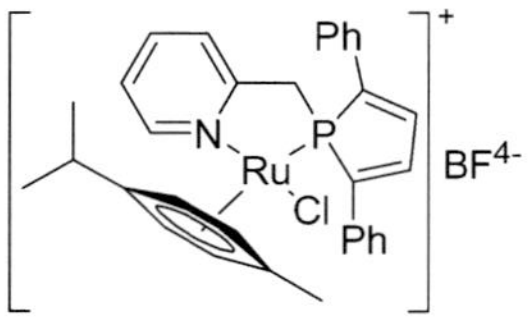 Ph, N, P, Ru, Cl, Ph, BF^{4-}, +	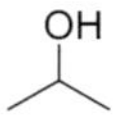 OH	KOH (0.5)	90	$60 \cdot 10^6$	>99	60	$1 \cdot 10^6$	53
45	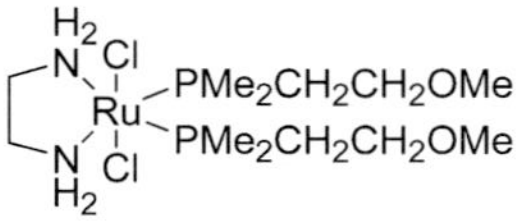 H_2, N, Cl, Ru, $PMe_2CH_2CH_2OMe$, $PMe_2CH_2CH_2OMe$, N, Cl, H_2	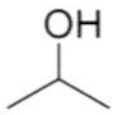 OH	KOH (0.1)	82	500	93	0.5	935	146
46	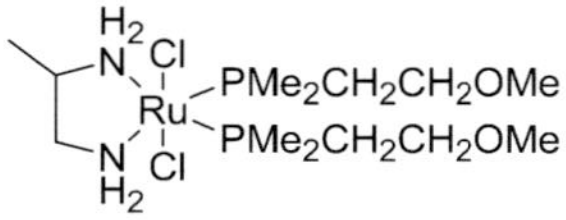 H_2, N, Cl, Ru, $PMe_2CH_2CH_2OMe$, $PMe_2CH_2CH_2OMe$, N, Cl, H_2	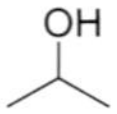 OH	KOH (0.1)	82	500	82	0.5	815	146
47	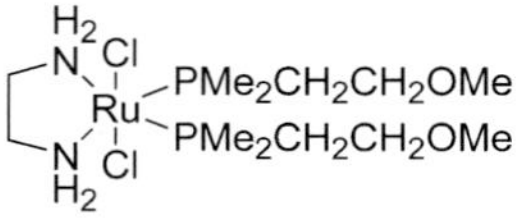 H_2, N, Cl, Ru, $PMe_2CH_2CH_2OMe$, $PMe_2CH_2CH_2OMe$, N, Cl, H_2	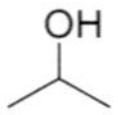 OH	KOH (0.1)	82	500	90	0.5	900	146

Table 20.7 (continued)

Entry	Catalyst	Reducing agent [a]	Additive/Solvent [a]	Temperature [°C]	SCR	Conversion [%]	Reaction time [h]	TOF [h^{-1}]	Reference
48	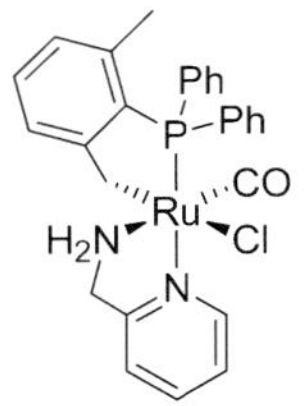	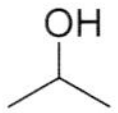	NaOH (0.02)	82	400	98	0.08	4700	54
49	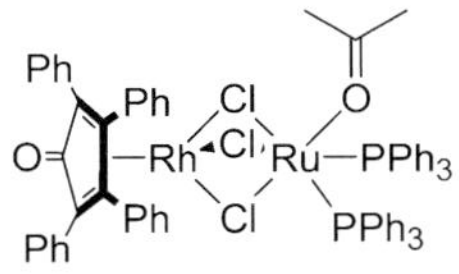	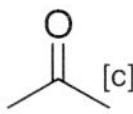 [c]	K_2CO_3 (1)	RT	1000	70	6	60	147
50	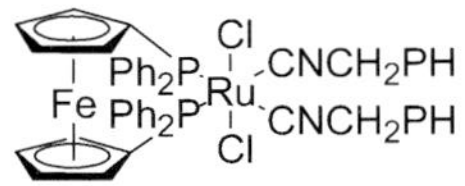	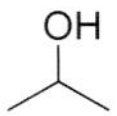	NaOH (0.096)	250	250	84	0.17	1260	148
51	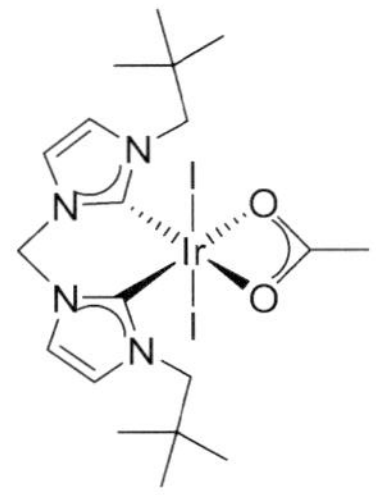	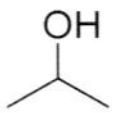	KOH (0.005)	82	1000	98	0.07	14700	51

Table 20.7 (continued)

Entry	Catalyst	Reducing agent [a]	Additive/Solvent [a]	Temperature [°C]	SCR	Conversion [%]	Reaction time [h]	TOF [h^{-1}]	Reference
52		OH	K_2CO_3 (0.5)	82	1000	97	0.67	1455	51
53		OH	KOH (0.005)	82	1000	80	2	400	51

a) The number in parentheses denotes the number of equivalents used. If no number is given, the donor/additive is used as solvent.
b) TOF = [substrate]/[catalyst] h^{-1}, calculated from the yield and stated reaction time.
c) Oxidation of 1-phenylethanol.
SCR = substrate : catalyst ratio.

Several groups have been active in the field of dynamic kinetic resolution since its introduction, and products have been obtained at almost quantitative yields and with excellent enantiomeric excesses (Table 20.6) [128–130].

Abbreviations

MPV	Meerwein-Ponndorf-Verley
MPVO	Meerwein-Ponndorf-Verley and Oppenauer
TFA	trifluoroacetic acid
TOF	turnover frequency
TON	turnover number

References

1 H. Meerwein, R. Schmidt, *Justus Liebigs Ann. Chem.* **1925**, *444*, 221.
2 A. Verley, *Bull. Soc. Chim. Fr.* **1925**, *37*, 537.
3 W. Ponndorf, *Angew. Chem.* **1926**, *29*, 138.
4 H. Lund, *Ber. Dtsch. Chem. Ges.* **1937**, *70*, 1520.
5 R.V. Oppenauer, *Recl. Trav. Chim. Pays-Bas* **1937**, *56*, 137.
6 J.L. Namy, J. Souppe, J. Collin, H.B. Kagan, *J. Org. Chem.* **1984**, *49*, 2045.
7 T. Okano, M. Matsuoka, H. Konishi, J. Kiji, *Chem. Lett.* **1987**, 181.
8 T. Ooi, T. Miura, Y. Itagaki, H. Ichikawa, K. Maruoka, *Synthesis* **2002**, 279.
9 Y.-C. Liu, B.-T. Ko, B.-H. Huang, C.-C. Lin, *Organometallics* **2002**, *21*, 2066.
10 E.J. Campbell, H. Zhou, S.T. Nguyen, *Org. Lett.* **2001**, *3*, 2391.
11 A.L. Wilds, *Org. React.* **1944**, *2*, 178.
12 C. Djerassi, *Org. React.* **1953**, *6*, 207.
13 C.F. de Graauw, J.A. Peters, H. van Bekkum, J. Huskens, *Synthesis* **1994**, 1007.
14 K. Nishide, M. Node, *Chirality* **2002**, *14*, 759.
15 Y.M.Y. Haddad, H.B. Henbest, J. Husbands, T.R.B. Mitchell, *Proc. Chem. Soc.* **1964**, 361.
16 J.A. Osborn, F.H. Jardine, J.F. Young, G. Wilkinson, *J. Chem. Soc. A* **1966**, 1711.
17 R. Uson, L.A. Oro, R. Sariego, M.A. Esteruelas, *J. Organomet. Chem.* **1981**, *214*, 399.
18 P. Kvintovics, B.R. James, B. Heil, *Chem. Commun.* **1986**, 1810.
19 S. Gladiali, G. Chelucci, G. Chessa, G. Delogu, F. Soccolini, *J. Organomet. Chem.* **1987**, *327*, C15.
20 S. Gladiali, L. Pinna, G. Delogu, S. De Martin, G. Zassinovich, G. Mestroni, *Tetrahedron. Asymm.* **1990**, *1*, 635.
21 R.L. Chowdhury, J.-E. Bäckvall, *Chem. Commun.* **1991**, 1063.
22 N.C. Deno, H.J. Peterson, G.S. Saines, *Chem. Rev.* **1960**, *60*, 7.
23 G. Brieger, T.J. Nestrick, *Chem. Rev.* **1974**, *74*, 567.
24 R.A.W. Johnstone, A.H. Wilby, I.D. Entwistle, *Chem. Rev.* **1985**, *85*, 129.
25 P.A. Chaloner, M.A. Esteruelas, F. Joó, L.A. Oro, *Homogeneous hydrogenation.* Kluwer Academic Publishers, Dordrecht, Boston, London, **1994**, Chapter 3, p. 87.
26 P.A. Chaloner, M.A. Esteruelas, F. Joó, L.A. Oro, *Homogeneous hydrogenation.* Kluwer Academic Publishers, Dordrecht, Boston, London, **1994**, Chapter 5, p. 183.
27 S. Gladiali, G. Mestroni, Transfer hydrogenations. In: M. Beller, C. Bolm (Eds.), *Transition Metals for Organic Synthesis.* Wiley-VCH, Weinheim, New York, Chichester, Brisbane, Singapore, Toronto, **1998**, Chapter 3, p. 97.

28 J.-E. Bäckvall, *J. Organomet. Chem.* **2002**, *652*, 105.
29 L. Sominsky, E. Rozental, H. Gottlieb, A. Gedanken, S. Hoz, *J. Org. Chem.* **2004**, *69*, 1492.
30 P.M. Dinh, J.A. Howarth, A.R. Hudnott, J.M.J. Williams, W. Harris, *Tetrahedron Lett.* **1996**, *37*, 7623.
31 D. Klomp, T. Maschmeyer, U. Hanefeld, J.A. Peters, *Chem. Eur. J.* **2004**, *10*, 2088.
32 E.D. Williams, K.A. Krieger, A.R. Day, *J. Am. Chem. Soc.* **1953**, *75*, 2404.
33 W.N. Moulton, R.E. van Atta, R.R. Ruch, *J. Org. Chem.* **1961**, *26*, 290.
34 E.C. Ashby, J.N. Argyropoulos, *J. Org. Chem.* **1986**, *51*, 3593.
35 O. Pàmies, J.-E. Bäckvall, *Chem. Eur. J.* **2001**, *7*, 5052.
36 R. Cohen, C.R. Graves, S.T. Nguyen, J.M.L. Martin, M.A. Ratner, *J. Am. Chem. Soc.* **2004**, *126*, 14796.
37 T. Ooi, T. Miura, K. Maruoka, *J. Am. Chem. Soc.* **1998**, *120*, 10790.
38 T. Ooi, T. Miura, K. Takaya, H. Ichikawa, K. Maruoka, *Tetrahedron* **2001**, *57*, 867.
39 H. Imai, T. Nishiguchi, K. Fukuzumi, *J. Org. Chem.* **1974**, *39*, 1622.
40 T. Nishiguchi, K. Tachi, K. Fukuzumi, *J. Org. Chem.* **1975**, *40*, 240.
41 J. Halpern, *Inorg. Chim. Acta* **1981**, *50*, 11.
42 H.A. Brune, J. Unsin, R. Hemmer, M. Reichhardt, *J. Organomet. Chem.* **1989**, *369*, 3352.
43 M.A. Bennett, S.K. Bhargava, M. Ke, A.C. Willis, *J. Chem. Soc., Dalton Trans.* **2000**, 3537.
44 L.Y. Kuo, D.M. Finigan, N.N. Tadros, *Organometallics* **2003**, *22*, 2422.
45 G. Mestroni, G. Zassinovich, A. Camus, F. Martinelli, *J. Organomet. Chem.* **1980**, *198*, 87.
46 H. Imai, T. Nishiguchi, K. Fukuzumi, *J. Org. Chem.* **1976**, *41*, 2688.
47 A. Aranyos, G. Csjernyik, K. J. Szabó, J.-E. Bäckvall, *Chem. Commun.* **1999**, 351.
48 K.-J. Haack, S. Hashiguchi, A. Fujii, T. Ikariya, R. Noyori, *Angew. Chem. Int. Ed. Engl.* **1997**, *36*, 285.
49 M. Yamakawa, H. Ito, R. Noyori, *J. Am. Chem. Soc.* **2000**, *122*, 1466.
50 J.-W. Handgraaf, J.N.H. Reek, E.J. Meijer, *Organometallics* **2003**, *22*, 3150.
51 J.R. Miecznikowski, R.H. Crabtree, *Polyhedron* **2004**, *23*, 2857.
52 P. Braunstein, F. Naud, S.J. Rettig, *New J. Chem.* **2001**, *25*, 32.
53 C. Thoumazet, M. Melaimi, L. Ricard, F. Mathey, P. le Floch, *Organometallics* **2003**, *22*, 1580.
54 W. Baratta, P. Da Ros, A. Del Zotto, A. Sechi, E. Zangrando, P. Rigo, *Angew. Chem. Int. Ed. Engl.* **2004**, *43*, 3584.
55 Y. Shvo, D. Czarkie, Y. Rahamim, D.F. Chodosh, *J. Am. Chem. Soc.* **1986**, *108*, 7400.
56 J.S.M. Samec, J.-E. Bäckvall, *Chem. Eur. J.* **2002**, *8*, 2955.
57 H.M. Jung, S.T. Shin, Y.H. Kim, M.-J. Kim, J. Park, *Organometallics* **2001**, *20*, 3370.
58 C.P. Casey, S.W. Singer, D.R. Powell, R.K. Hayashi, M. Kavana, *J. Am. Chem. Soc.* **2001**, *123*, 1090.
59 J.B. Johnson, J.-E. Bäckvall, *J. Org. Chem.* **2003**, *68*, 7681.
60 T. Nishiguchi, K. Fukuzumi, *J. Am. Chem. Soc.* **1974**, *96*, 1893.
61 W. Leitner, J. Brown, H. Brunner, *J. Am. Chem. Soc.* **1993**, *115*, 152.
62 A. Bucsai, J. Bakos, M. Laghmari, D. Sinou, *J. Mol. Catal. A: Chemical* **1997**, *116*, 335.
63 D. Heller, R. Kadyrov, M. Michalik, T. Freier, U. Schmidt, H.W. Krause, *Tetrahedron Asymmetry* **1996**, *7*, 3025.
64 A. Harthun, R. Kadyrov, R. Selke, J. Bargon, *Angew. Chem. Int. Ed. Engl.* **1997**, *36*, 1103.
65 M.C. Baird, C.J. Nyman, G. Wilkinson, *J. Chem. Soc. A* **1968**, 348.
66 K. Ohno, J. Tsuji, *J. Am. Chem. Soc.* **1968**, *90*, 99.
67 Y. Sasson, M. Cohen, J. Blum, *Synthesis* **1973**, 359.
68 Y. Sasson, J. Blum, E. Dunkelblum, *Tetrahedron Lett.* **1973**, *14*, 3199.
69 S. Mukhopadhyay, A. Yaghmur, A. Benichou, Y. Sasson, *Org. Proc. Res. Dev.* **2000**, *4*, 571.
70 G. Descotes, D. Sinou, J.-P. Praly, *Carbohydr. Res.* **1980**, *78*, 25.
71 G. Descotes, D. Sinou, *Tetrahedron Lett.* **1976**, *17*, 4083.
72 G. Descotes, J. Sabadie, *Bull. Soc. Chim. Fr.* **1978**, *II*, 158.

73 T. Nishiguchi, K. Tachi, K. Fukuzumi, *J. Am. Chem. Soc.* **1972**, *94*, 8916.
74 H. Imai, T. Nishiguchi, K. Fukuzumi, *J. Org. Chem.* **1976**, *41*, 665.
75 H. Imai, T. Nishiguchi, K. Fukuzumi, *J. Org. Chem.* **1977**, *42*, 431.
76 H. Imai, T. Nishiguchi, M. Tanaka, K. Fukuzumi, *J. Org. Chem.* **1977**, *42*, 2309.
77 T. Nishiguchi, K. Tachi, K. Fukuzumi, *J. Org. Chem.* **1975**, *40*, 237.
78 H.B. Henbest, T.R.B. Mitchell, *J. Chem. Soc. C* **1970**, 785.
79 M. Gullotti, R. Ugo, S. Colonna, *J. Chem. Soc. C* **1971**, 2652.
80 Y.M.Y. Haddad, H.B. Henbest, J. Trocha-Grimshaw, *J. Chem. Soc. Perkin Trans. I* **1974**, 592.
81 Y.M.Y. Haddad, H.B. Henbest, J. Husbands, T.R.B. Mitchell, J. Trocha-Grimshaw, *J. Chem. Soc. Perkin Trans. I* **1974**, 596.
82 M.E. Vol'pin, V.P. Kukolev, V.O. Chernyshev, I.S. Kolomnikov, *Tetrahedron Lett.* **1971**, *12*, 4435.
83 H. Brunner, E. Graf, W. Leitner, K. Wutz, *Synthesis* **1989**, 743.
84 D. Klomp, K. Djanashvili, N. Cianfanelli Svennum, N. Chantapariyavat, C.-S. Wong, F. Vilela, T. Maschmeyer, J.A. Peters, U. Hanefeld, *Org. Biomol. Chem.* **2005**, *3*, 483.
85 Y. Ishii, T. Nakano, A. Inada, Y. Kishigami, K. Sakurai, M. Ogawa, *J. Org. Chem.* **1986**, *51*, 240.
86 B. Knaver, K. Krohn, *Liebigs Ann.* **1995**, 677.
87 K.G. Akamanchi, N.R. Varalakshmy, *Tetrahedron Lett.* **1995**, *36*, 3571.
88 K.G. Akamanchi, N.R. Varalakshmy, *Tetrahedron Lett.* **1995**, *36*, 5085.
89 T. Ooi, H. Ichikawa, K. Maruoka, *Angew. Chem. Int. Ed. Engl.* **2001**, *40*, 3610.
90 S. Fukuzawa, N. Nakano, T. Saitoh, Eur. *J. Org. Chem.* **2004**, 2863.
91 C. Bisi Castellani, O. Carugo, A. Perotti, D. Sacchi, *J. Mol Catal.* **1993**, *85*, 65.
92 A. Lebrun, J.L. Namy, H.B. Kagan, *Tetrahedron Lett.* **1991**, *32*, 2355.
93 M.D. le Plage, D. Poon, B.R. James, *Catalysis of Organic Reactions.* Marcel Dekker, Inc., New York, Basel, Chapter 6, p. 61.
94 D. Baudry, M. Ephritikhine, H. Felkin, R. Holmes-Smith, *J. Chem. Soc. Chem. Commun.* **1983**, 788.
95 M. Aracama, M.A. Esteruelas, F.J. Lahoz, J.A. Lopez, U. Meyer, L.A. Oro, H. Werner, *Inorg. Chem.* **1991**, *30*, 288.
96 D. Barbry, S. Torchy, *Tetrahedron Lett.* **1997**, *38*, 2959.
97 G.-Z. Wang, J.-E. Bäckvall, *Chem. Commun.* **1992**, 980.
98 T. Ohkuma, H. Ooka, S. Hashiguchi, T. Ikariya, R. Noyori, *J. Am. Chem. Soc.* **1995**, *117*, 2675.
99 M. Ito, M. Hirakawa, K. Murata, T. Ikariya, *Organometallics* **2001**, *20*, 379.
100 V. Rautenstrauch, X. Hoang-Cong, R. Churlaud, K. Abdur-Rashid, R.H. Morris, *Chem. Eur. J.* **2003**, *9*, 4954.
101 Y. Sasson, J. Blum, *J. Org. Chem.* **1975**, *40*, 1887.
102 S. Iyer, J. P. Varghese, *J. Chem. Soc. Chem. Commun.* **1995**, 465.
103 S. Sakaguchi, T. Yamaga, Y. Ishii, *J. Org. Chem.* **2001**, *66*, 4710.
104 M. Albrecht, J. R. Miecznikowski, A. Samuel, J.W. Faller, R.H. Crabtree, *Organometallics* **2002**, *21*, 35964.
105 S. Bolaño, L. Gonsalvi, F. Zanobini, F. Vizza, V. Bertolasi, A. Romerosa, M. Peruzzini, *J. Mol. Catal. A* **2004**, *224*, 61.
106 R.B. Woodward, F.E. Bader, H. Bickel, A.J. Frey, R.W. Kierstead, *Tetrahedron* **1958**, *2*, 1.
107 P.A. Browne, D.N. Kirk, *J. Chem. Soc. C* **1969**, 1653.
108 W. Tischtschenko, *Chem. Zentralbl.* **1906**, *77*, 1309.
109 J. Collin, J.L. Namy, H.B. Kagan, *Nouv. J. Chim.* **1986**, *10*, 229.
110 S. Cannizzaro, *Liebigs Ann.* **1853**, *88*, 129.
111 T.A. Geissman, *Org. React.* **1944**, *2*, 94.
112 R. Leuckart, *Ber. Dtsch. Chem. Ges.* **1885**, *18*, 2341.
113 O. Wallach, *Ber. Dtsch. Chem. Ges.* **1891**, *24*, 3992.
114 M.L. Moore, *Org. React.* **1949**, *5*, 301.
115 M. Kitamura, D. Lee, S. Hayashi, S. Tanaka, M. Yoshimura, *J. Org. Chem.* **2002**, *67*, 8685.
116 W. Eschweiler, *Ber. Dtsch. Chem. Ges.* **1905**, *38*, 880.

117 H.T. Clarke, H.B. Gillespie, S.Z. Weisshaus, *J. Am. Chem. Soc.* **1933**, *55*, 4571.
118 Y. Nakano, S. Sakaguchi, Y. Ishii, *Tetrahedron Lett.* **2000**, *41*, 1565.
119 C.E. Castro, R.D. Stephens, *J. Am. Chem. Soc.* **1964**, *86*, 4358.
120 D.J. Pasto, D.M. Chipman, *J. Am. Chem. Soc.* **1979**, *101*, 2290.
121 O. Pàmies, A.H. Éll, J.S.M. Samec, N. Hermanns, J.-E. Bäckvall, *Tetrahedron Lett.* **2002**, *43*, 4699.
122 K. Faber, *Chem. Eur. J.* **2001**, *7*, 5005.
123 F.F. Huerta, A.B.E. Minidis, J.-E. Bäckvall, *Chem. Soc. Rev.* **2001**, *30*, 321.
124 M.-J. Kim, Y. Ahn, J. Park, *Curr. Opin. Biotechnol.* **2002**, *13*, 578.
125 M.T. El Gihani, J.M.J. Williams, *Curr. Opin. Biotechnol.* **1999**, *3*, 11.
126 O. Pàmies, J.-E. Bäckvall, *Chem. Rev.* **2003**, *103*, 3247.
127 O. Pàmies, J.-E. Bäckvall, *Trends Biotechnol.* **2004**, *22*, 130.
128 J.H. Choi, Y.K. Choi, Y.H. Kim, E.S. Park, E.J. Kim, M.-J. Kim, J. Park, *J. Org. Chem.* **2004**, *69*, 1972.
129 B. Mártin-Matute, J.-E. Bäckvall, *J. Org. Chem.* **2004**, *69*, 9191.
130 B. Mártin-Matute, M. Edin, K. Bogár, J.-E. Bäckvall, *Angew. Chem. Int. Ed. Engl.* **2004**, *43*, 6535.
131 J.L. Namy, J. Souppe, J. Collin, H.B. Kagan, *J. Org. Chem.* **1984**, *49*, 2045.
132 A. Lebrun, J.-L. Namy, H.B. Kagan, *Tetrahedron Lett.* **1991**, *32*, 2355.
133 B.P. Warner, J.A. D'Alessio, A.N. Morgan, III, C.J. Burns, A.R. Schake, J.G. Watkin, *Inorg. Chim. Acta* **2000**, *309*, 45.
134 M.D. Le Page, B.R. James, *Chem. Commun.* **2000**, 1647.
135 N.J. Lawrence, S.M. Bushell, *Tetrahedron Lett.* **2000**, *41*, 4507.
136 C. Standfest-Hauser, C. Slugovc, K. Mereiter, R. Schmid, K. Kirchner, L. Xiao, W. Weissensteiner, *J. Chem. Soc., Dalton Trans.* **2001**, 2989.
137 P. Crochet, J. Gimeno, S. García-Granda, J. Borge, *Organometallics* **2001**, *20*, 4369.
138 H. Matsunaga, N. Yoshioka, T. Kunieda, *Tetrahedron Lett.* **2001**, *42*, 8857.
139 M. Albrecht, R.H. Crabtree, J. Mata, E. Peris, *Chem. Commun.* **2002**, 32.
140 S. Ogo, T. Abura, Y. Watanabe, *Organometallics* **2002**, *21*, 2964.
141 H.P. Dijkstra, M. Albrecht, S. Medici, G.P.M. van Klink, G. van Koten, *Adv. Synth. Catal.* **2002**, *344*, 1135.
142 T. Abura, S. Ogo, Y. Watanabe, S. Fukuzumi, *J. Am. Chem. Soc.* **2003**, *125*, 4149.
143 P. Crochet, M.A. Fernández-Zumel, C. Beauquis, J. Gimeno, *Inorg. Chim. Acta* **2003**, *356*, 114.
144 P. Crochet, J. Gimeno, J. Borge, S. García-Granda, *New. J. Chem.* **2003**, *27*, 414.
145 K.Y. Ghebreyessus, J.H. Nelson, *J. Organomet. Chem.* **2003**, *669*, 48.
146 Z.-L. Lu, K. Eichele, I. Warad, H.A. Mayer, E. Lindner, Z.-J. Jiang, V. Schurig, *Z. Anorg. Allg. Chem.* **2003**, *629*, 1308.
147 S. Gauthier, R. Scopelliti, K. Severin, *Organometallics* **2004**, *23*, 3769.
148 V. Cadierno, P. Crochet, J. Díez, S.E. García-Garrido, J. Gimeno, *Organometallics* **2004**, *23*, 4836.

21
Diastereoselective Hydrogenation

Takamichi Yamagishi

21.1
Introduction

The stereochemical control of a reaction is a continuing challenge in the synthesis of complex organic compounds, especially those with stereogenic center(s) or molecular chirality. Homogeneous catalytic hydrogenation is a simple and widely applicable method for the construction of stereogenic centers and enantioselective hydrogenation using chiral transition metal complexes. It has undergone striking development during the past two decades [1]. In enantioselective hydrogenation, prochiral substrates with unsaturated linkages are converted to chiral compounds by using chirally modified transition metal complexes based on rhodium, iridium, ruthenium, cobalt, and lanthanide metals, etc. In enantioselective hydrogenation, stereochemical control is performed through the selection of one diastereomeric intermediate composed of a prochiral (achiral) substrate and a chiral metal complex. For this purpose, many chiral ligands (representatives of which include chiral diphosphine ligands) have been developed to realize the production of almost homochiral products with stereogenic center(s). In the hydrogenation of compounds with a stereogenic center, an achiral metal complex can induce a new stereogenic center selectively, by utilizing the steric factor of the stereogenic center in the substrate to afford compounds with two or more stereogenic centers. These diastereoselective hydrogenations also go through the selection of diastereomeric intermediates by differentiating the reaction face of unsaturated bonds. This is termed "intramolecular asymmetric induction".

21.2
Hydrogenation of Alkenes, Ketones, and Imines

In the hydrogenation of alkenes, rhodium–, ruthenium– and iridium–phosphine catalysts are typically used [2–4]. Rhodium–phosphine complexes, such as Wilkinson's catalyst, are effective for obtaining alkanes under atmospheric pres-

The Handbook of Homogeneous Hydrogenation.
Edited by J. G. de Vries and C. J. Elsevier
Copyright © 2007 WILEY-VCH Verlag GmbH & Co. KGaA, Weinheim
ISBN: 978-3-527-31161-3

Table 21.1 Diastereoselective hydrogenation of olefinic bonds.

	Substrate	Major diastereomer	Catalyst	mol%	P_{H_2}	Solvent	Temp. [°C]	Diastereo-meric ratio	TOF	Reference
1		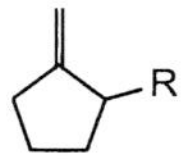 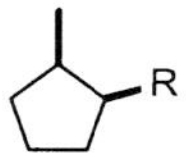 R = *i*-Bu	Sm	10	1 atm	c-C_5H_{10}	r.t.	100:0	1.7	6
2		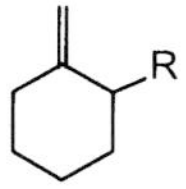 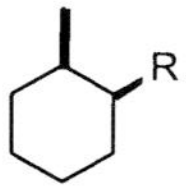 R = Me	Sm	3	1 atm	c-C_5H_{10}	−20	93:7	5.1	6
		R = *t*-Bu	Sm	5	1 atm	c-C_5H_{10}	r.t.	100:0	6.3	6
		R = Ph	Sm	5	1 atm	c-C_5H_{10}	50	100:0	6.4	6
		R = $(CH_2)_3NMe_2$	Sm	3	1 atm	c-C_5H_{10}	50	91:9	6.4	6
3		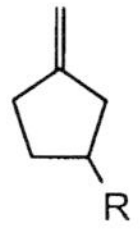 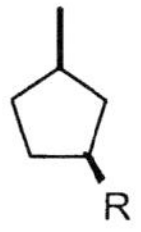 R = *n*-Bu	Sm	3	1 atm	c-C_5H_{10}	0	60:40	5.6	6
4		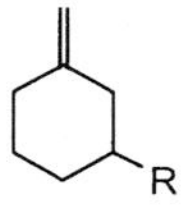 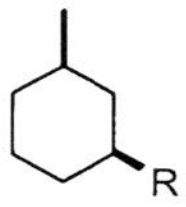 R = Me	Yb	3	1 atm	c-C_5H_{10}	−20	61:39	4.9	6
		R = Et	Yb	3	1 atm	c-C_5H_{10}	−20	73:27	5.3	6
5		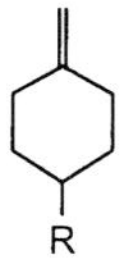  R = *t*-Bu	Yb	3	1 atm	c-C_5H_{10}	−20	77:23	4.9	6

Cat: $Cp^*_2LnCH(SiMe_3)_2$ (Ln = Sm, Yb)

Table 21.1 (continued)

	Substrate	Major diastereomer	Catalyst	mol%	P_{H_2}	Solvent	Temp. [°C]	Diastereomeric ratio	TOF	Reference
6	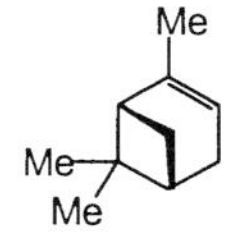 (1*S*,5*S*)-α-pinene	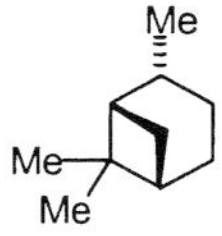 (1*S*,2*R*,5*S*)-*cis*-pinane	$Ru_2(\text{-}CO)_4(OAc)_2(PPh_3)_2$	0.9	50 bar	THF	90	98.3:1.7	2.7	7
7	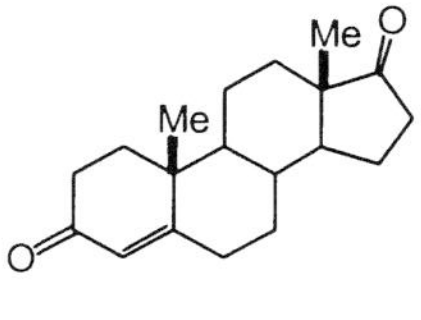	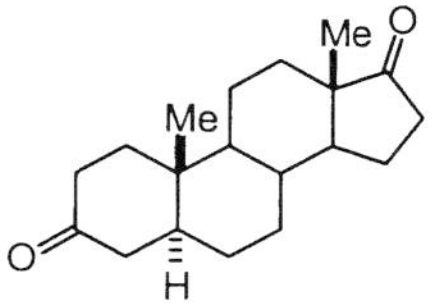	$[Ir(PCy_3)(py)(nbd)]^+$	20	1 atm	CH_2Cl_2	r.t.	100:0	0.87	8
8	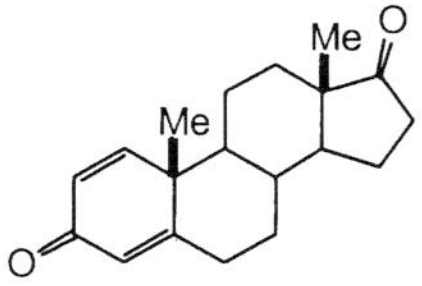	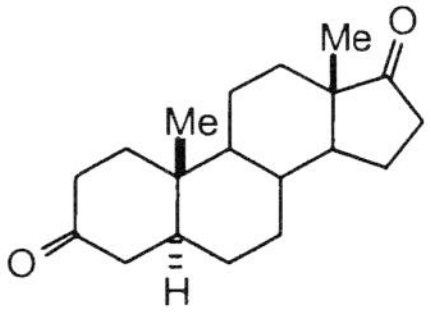	$[Ir(PCy_3)(py)(nbd)]^+$	15	1 atm	CH_2Cl_2	r.t.	100:0	0.27	8
9	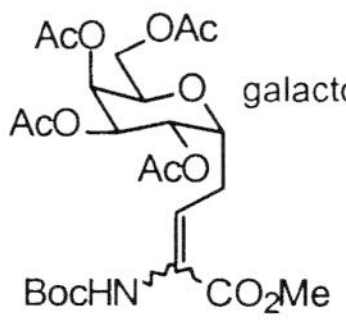 galactose	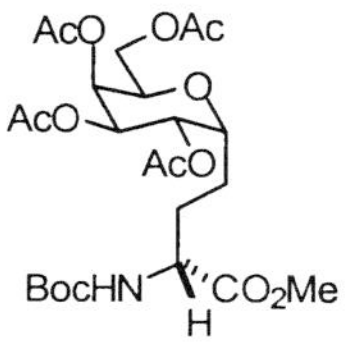	$[Rh(diphos\text{-}4)]^+$ $[Rh((R,R)\text{-}Me\text{-}Duphos)]^+$ $[Rh((R,R)\text{-}Et\text{-}Duphos)]^+$ $[Rh((R,R)\text{-}Pr\text{-}Duphos)]^+$		90 psi 90 psi 90 psi 90 psi	MeOH MeOH MeOH MeOH		50:50 79:21 91:9 88.5:11.5		9

Table 21.1 (continued)

	Substrate	Major diastereomer	Catalyst	mol%	P_{H_2}	Solvent	Temp. [°C]	Diastereo-meric ratio	TOF	Reference
10	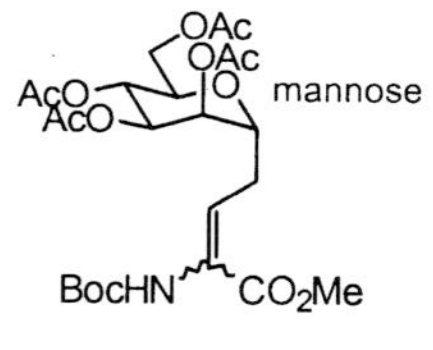 mannose	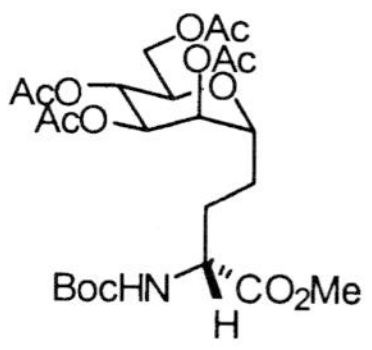	$[Rh(diphos\text{-}4)]^+$		90 psi	MeOH		50:50		9
			$[Rh((R,R)\text{-Me-Duphos})]^+$		90 psi	MeOH		87:13		
			$[Rh((R,R)\text{-Et-Duphos})]^+$		90 psi	MeOH		>97.5:2.5		
			$[Rh((R,R)\text{-Pr-Duphos})]^+$		90 psi	MeOH		>97.5:2.5		
11	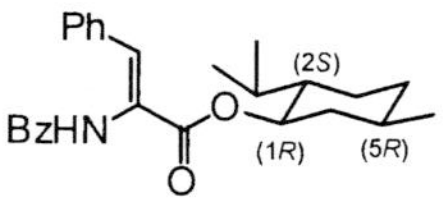 (2S) (1R) (5R)	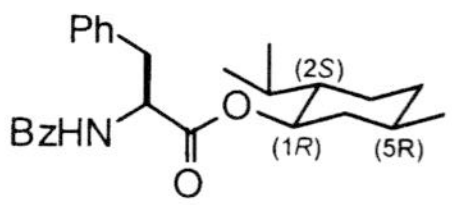 (2S) (1R) (5R)	$[Rh(Ph\text{-}\beta\text{-glup-OH})]^+$	1	0.1 MPa	MeOH		96.3:3.7		10
			$[Rh(Ph\text{-}\beta\text{-glup-OH})]^+$	1	0.1 MPa	benzene		91:9		
			$[Rh(Me\text{-}\alpha\text{-glu})]^+$	1	0.1 MPa	acetone		86:14		
			$[Rh(Me\text{-}\alpha\text{-glu})]^+$	1	0.1 MPa	benzene		75:25		
12	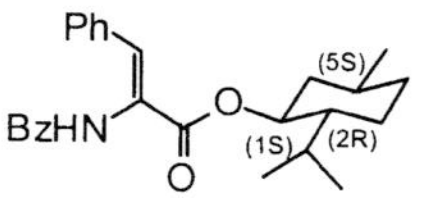 (5S) (1S) (2R)	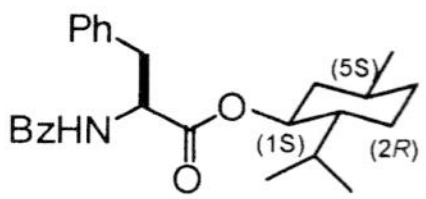 (5S) (1S) (2R)	$[Rh(Ph\text{-}\beta\text{-glup-OH})]^+$	1	0.1 MPa	MeOH		97.4:2.6		10
			$[Rh(Ph\text{-}\beta\text{-glup-OH})]^+$	1	0.1 MPa	benzene		79:21		
			$[Rh(Me\text{-}\alpha\text{-glu})]^+$	1	0.1 MPa	MeOH		66:34		
			$[Rh(Me\text{-}\alpha\text{-glu})]^+$	1	0.1 MPa	benzene		14:86		
13	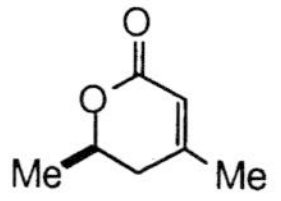 99.2%ee (R)	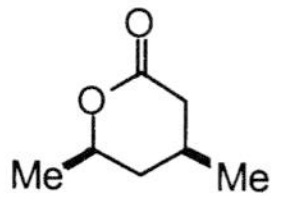 (4S,6R)	$Ru(OAc)_2((S)\text{-3,5-xylyl-biphep})$	0.2	60 bar	i-PrOH		80:20		11
14	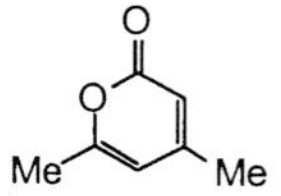	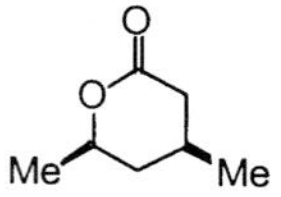	$Ru(OAc)_2((S)\text{-3,5-xylyl-biphep})$	0.2	60 bar	i-PrOH		92:8	25	11

Table 21.2 Diastereoselective hydrogenation of ketones and imines.

Substrate	Major diastereomer		Catalyst	mol%	P_{H_2}	Solvent	Temp. [°C]	Diastereo-selectivity	TOF	Reference
1 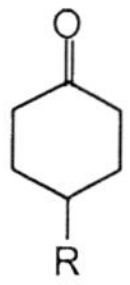	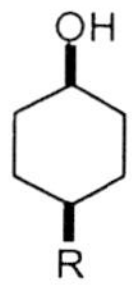	R = Me	Ru	0.2	4 atm	*i*-PrOH	28	92:8	485	12
		R = Ph	Ru	0.2	4 atm	*i*-PrOH	28	96:4	500	12
		R = *t*-Bu	Ru	0.2	4 atm	*i*-PrOH	28	98.4:1.6	500	12
2 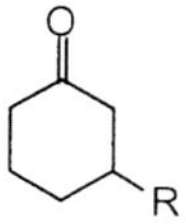	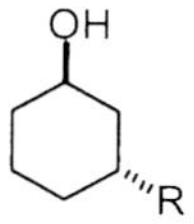	R = Me	Ru	0.2	4 atm	*i*-PrOH	28	96:4	500	12
3 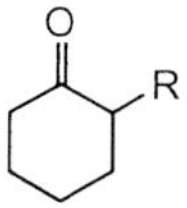	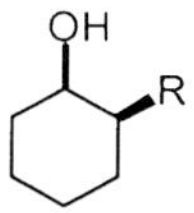	R = Me	Ru	0.2	4 atm	*i*-PrOH	28	98:2	475	12
		R = *t*-Bu	Ru	0.2	4 atm	*i*-PrOH	28	>99.8:0.2	500	12
4 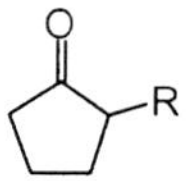	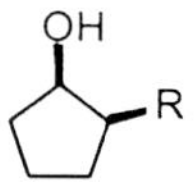	R = Me	Ru	0.2	4 atm	*i*-PrOH	28	99:1	500	12
5 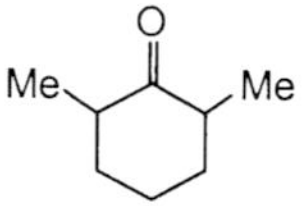	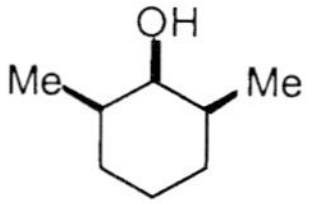		Ru	0.2	4 atm	*i*-PrOH	28	98.7:1.1:0.2		12

Table 21.2 (continued)

Substrate	Major diastereomer	Catalyst	mol%	P_{H_2}	Solvent	Temp. [°C]	Diastereo-selectivity	TOF	Reference
6	*syn*	R = Me Ru-a	0.2	4 atm	*i*-PrOH	28	96:4		12
		R = Me Ru-b	0.2	4 atm	*i*-PrOH	28	95:5		12
		R = Me Ru	0.2	4 atm	*i*-PrOH	28	86:14	480	12
		R = *n*-Bu Ru	0.2	4 atm	*i*-PrOH	28	93:7	480	12
		R = Ph Ru	0.2	4 atm	*i*-PrOH	28	98:2	480	12
7 (*R*)-carvone		Ru	0.2	4 atm	*i*-PrOH	28	81:19	66.4	13
		$RuCl_2$((*S*)-binap)(dmf)m-(*R*,*R*)-DPEN-KOH/*i*-PrOH	0.2	4 atm	*i*-PrOH	28	100:0	143	13
		$RuCl_2$((*R*)-binap)(dmf)m-(*S*,*S*)-DPEN-KOH/*i*-PrOH	0.2	8 atm	*i*-PrOH	28	34:66	29.4	13
8 (*R*)-pulegone		$RuCl_2$((*S*)-binap)(dmf)m-(*S*,*S*)-DPEN-KOH/*i*-PrOH	0.4	8 atm	*i*-PrOH	28	98:2	15.1	13
		$RuCl_2$((*R*)-binap)(dmf)m-(*R*,*R*)-DPEN-KOH/*i*-PrOH	0.4	8 atm	*i*-PrOH	28	95:5	14	13
9 (*R*)	(*R*,*R*)	[RhCl(diphos-3)]Cl	2	1000 psi	MeOH	r.t.	91:9		14
		[RhCl((*S*,*S*)-bdpp)]Cl	2	1000 psi	MeOH	r.t.	99.7:0.3		
		[RhCl((–)-diop]Cl	2	1000 psi	MeOH	r.t.	93:7		

Table 21.2 (continued)

	Substrate	Major diastereomer	Catalyst	mol%	P_{H_2}	Solvent	Temp. [°C]	Diastereo-selectivity	TOF	Reference
10	Ph, N, Me, Ph, Me (*S*)	Ph, HN, Me, Ph, Me (*S*,*S*)	[RhCl((*S*,*S*)-bdpp)]Cl	2	1000 psi	MeOH	r.t.	93.8:6.2		14
11	Ph, N, Me, o-MeOC$_6$H$_4$, Me (*R*)	Ph, HN, Me, o-MeOC$_6$H$_4$, Me (*R*,*R*)	[RhCl((*S*,*S*)-bdpp)]Cl	2	1000 psi	MeOH	r.t.	98.1:1.9		14
12	Ph, N, Me, PhCH$_2$CH$_2$, Me (*R*)	Ph, HN, Me, PhCH$_2$CH$_2$, Me (*R*,*R*)	[RhCl(diphos-3)]Cl	2	1000 psi	MeOH	r.t.	67:33		
			[RhCl((*S*,*S*)-bdpp)]Cl	2	1000 psi	MeOH	r.t.	84:16		14

Ru = $RuCl_2(PPh_3)_3$-EN-KOH; Ru-a = $RuCl_2(P(C_6H_4\text{-p-OMe})_3)_3$-EN-KOH; Ru-b = $RuCl_2(P(C_6H_4\text{-p-Me})_3)_3$-EN-KOH

sure. The reactivity of rhodium complexes in homogeneous hydrogenation is sensitive to the degree of substitution on the olefinic bond, and trisubstituted or tetrasubstituted alkenes are intact in the hydrogenation, whereas iridium catalysts can hydrogenate the multi-substituted olefins under atmospheric hydrogen pressure. The unreactivity of trisubstituted alkenes towards rhodium complexes is overcome by increasing the hydrogen pressure, and the selectivity is comparable to (or a little better than) that obtained with iridium catalysts under atmospheric pressure [5]. The results of the hydrogenation of alkenes with lanthanide, ruthenium and iridium catalysts are listed in Table 21.1 [6–8]. In these reductions, the catalyst approaches the face of the double bond from the less-hindered side, and selection of the diastereoface of the substrate is straightforward when the stereogenic center is disposed adjacent to the double bond [6]. In the hydrogenation of double bonds in steroidal compounds, diastereoselectivity induced by iridium catalysts is very high [8]. In the reduction of dehydroamino acid derivatives with a chiral unit, an achiral rhodium catalyst resulted in stereorandom products (Table 21.1, entries 9 and 10) [9]. In the reaction of pyrone, the hydrogenation does not stop at the dihydropyrone stage, and *cis*-lactone is obtained in high diastereoselectivity, whereas hydrogenation of (*R*)-dihydropyrone afforded *cis*-lactone in lower diastereoselectivity, suggesting the complex character of the second hydrogenation step (Table 21.1, entries 13 and 14) [11]. Ruthenium–phosphine complexes combined with a diamine ligand effectively reduced ketones with a stereogenic center, and high diastereoselectivities are obtained (Table 21.2, entries 1–6) [12, 13]. The apparent effect of an adjacent stereogenic center was also observed in the reduction of imines [14]. For high stereoinduction, the proximity of the aromatic ring to the C=N bond seems to be essential.

21.3 Substrate-Directive Diastereoselective Hydrogenation

21.3.1 Hydrogenation of Cyclic Alcohols with Endo- or Exo-Cyclic Olefinic Bond

In the diastereoselective hydrogenation of olefinic, keto or imino double bonds, enhanced diastereoface differentiation will be possible by utilizing the interaction of functional groups in the substrates with the metal or with the ligand of the complex. Heteroatom(s) in the functional group would ligate to the metal and serve to fix the coordination mode of the substrate onto the catalyst. The key factor in the enantioselective hydrogenation of dehydroamino acids and esters is the formation of the rhodium–enamide chelate complex in which olefinic and amidocarbonyl units ligate to the metal [15]. Also in the diastereoselective hydrogenation, chelation of the substrate would serve to control the course of the hydrogenation. This potentiality of functional group-directed hydrogenation was first disclosed by Thompson and McPherson in 1974. In the hydrogenation of a cyclic homoallyl alcohol derivative by $RhCl(PPh_3)_3$, the reaction did not even proceed under forced conditions (100 psi at 50 °C), but the substrate was hydrogenated by converting the alcohol to the potassium alkoxide **1** to afford the *cis*-product predominantly [Eq. (1)] [16].

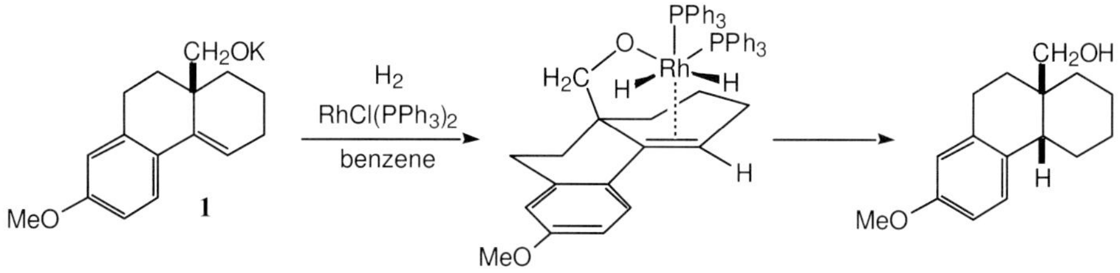

(1)

It was proposed that the chloride ion is displaced by the alkoxide ion from the coordination sphere of the dihydride complex, while delivery of the hydride to the unsaturated bond is controlled to afford the *cis*-product. In the case of the alcohol form, the chloride ion is not displaced, and the substrate is strongly resistant to hydrogenation by $RhCl(PPh_3)_3$. The heteroatom-directive hydrogenation would generally require vacant sites on the metal complex for the binding of H_2, the olefin unit and the directing heteroatom to the metal under hydrogenation conditions; thus, an active catalyst would be of 12-electron structure. Cationic $[Rh(diphosphine)(cod)]^+$ complex and cationic iridium complex (e.g., $[Ir(PCy_3)(py)(nbd)]^+$: Crabtree's catalyst [18]) and ruthenium complexes (e.g., $Ru(binap)(OAc)_2$) could hydrogenate the olefinic substrate with a directing heteroatom moiety to cause diastereoselective hydrogenation. Using $[Rh(diphosphine)(cod)]^+$ or $[Ir(PCy_3)(py)(nbd)]^+$ complexes, coordinating dienes are easily reduced by treatment with H_2, and 12-electron species are easily formed in a non-coordinating solvent such as dichloromethane (DCM). Brown reported the selective hydrogenation of acyclic allylic alcohols to produce chiral acyclic alcohols diastereoselectively using $[Rh(diphos\text{-}4)]^+$ catalyst in 1982 [19, 20]. Crabtree and Stork reported the highly diastereoselective hydrogenation of diverse types of cyclic alcohols in 1983 (allyl and homoallyl alcohols), using the cationic iridium complex ($[Ir(PCy_3)(py)(nbd)]PF_6$: Ir^+) [21, 22]. The data provided in Tables 21.3, 21.4 and 21.5 indicate the hydrogenation of diverse types of cyclic allylic and homoallylic alcohols.

In these reactions, the major diastereomer is formed by the addition of hydrogen *syn* to the hydroxyl group in the substrate. The cationic iridium catalyst $[Ir(PCy_3)(py)(nbd)]^+$ is very effective in hydroxy-directive hydrogenation of cyclic alcohols to afford high diastereoselectivity, even in the case of bishomoallyl alcohols (Table 21.4, entries 10–13) [5, 34, 35]. An intermediary dihydride species is not observed in the case of rhodium complexes, but iridium dihydride species are observed and the interaction of the hydroxyl unit of an unsaturated alcohol with iridium is detected spectrometrically through the presence of diastereotopic hydrides using NMR spectroscopy [21].

Table 21.3 Diastereoselective hydrogenation of cyclic allyl alcohols.

	Substrate	Major diastereomer	Catalyst	mol%	P_{H_2}	Solvent	Temp. [°C]	Diastereo-meric ratio	TOF	Reference(s)
1			Rh^+	3.5	375 psi			290:1	28	23
			Ir^+	20	15 psi	CH_2Cl_2	r.t.	98.5:1.5	>2.5	22
				20	15 psi	CH_2Cl_2	r.t.	98:2	>2.5	24
				2.5	15 psi	CH_2Cl_2	r.t.	140 ~ 150:1	20, 80	24, 25
2			Ir^+	2.5	1 atm	CH_2Cl_2	r.t.	99:1	80	25
3			Ir^+	2.5	1 atm	CH_2Cl_2	r.t.	940:1	80	25
4			Ir^+	2.5	1 atm	CH_2Cl_2	r.t.	98.5:1.5	80	25
5			Ir^+	2.5	1 atm	CH_2Cl_2	r.t.	96:4	80	25
6			Rh^+	5	800 psi	THF		98.6:1.4		5
			Ir^+	30	1 atm	CH_2Cl_2		highly selective		26

Table 21.3 (continued)

	Substrate	Major diastereomer	Catalyst	mol%	P_{H_2}	Solvent	Temp. [°C]	Diastereo-meric ratio	TOF	Reference(s)
7	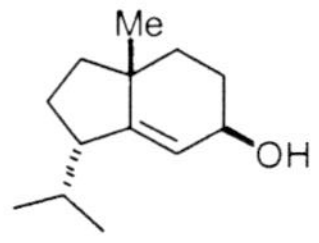	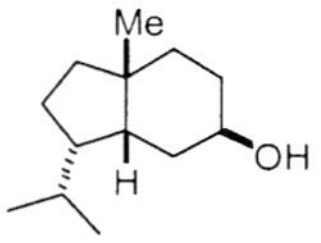	Ir^+	20	1 atm	CH_2Cl_2		100:0		26
8	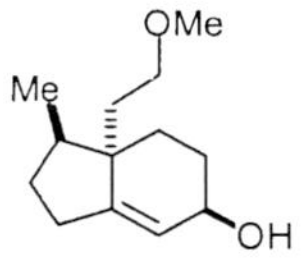	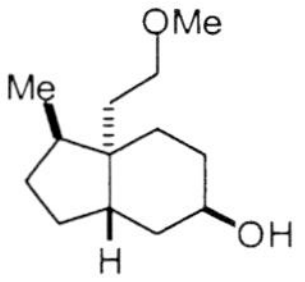	Rh^+	5	55 atm	THF		95:5	8	27
9	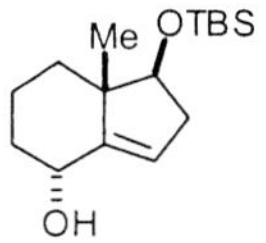	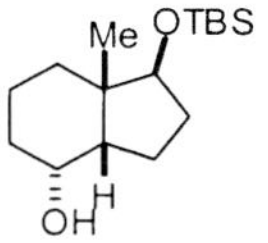	Ir^+	20	40 psi	CH_2Cl_2		highly selective	0.16	28
10	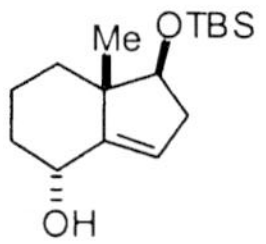	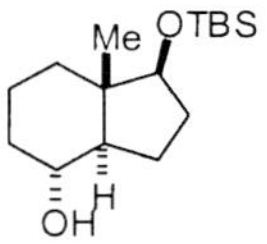	Rh	25	40 psi	benzene		75:25	0.15	28
11	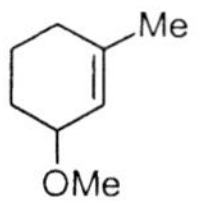	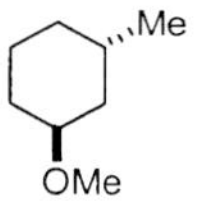	Ir^+	2.5	1 atm	CH_2Cl_2		>99.9:0.1	80	25

Table 21.3 (continued)

	Substrate	Major diastereomer	Catalyst	mol%	P_{H_2}	Solvent	Temp. [°C]	Diastereo-meric ratio	TOF	Reference(s)
12	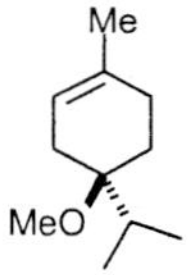 Me; MeO	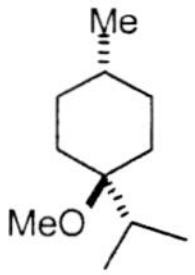 Me; MeO	Ir^+	2.5	1 atm	CH_2Cl_2		>99.9:0.1	80	25
13	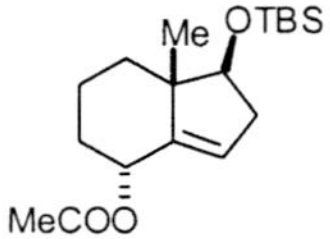 Me, OTBS; MeCOO	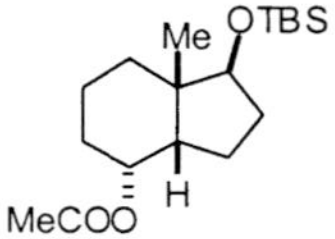 Me, OTBS; H; MeCOO	Ir^+	20	40 psi	CH_2Cl_2		100:0	0.13	28
14	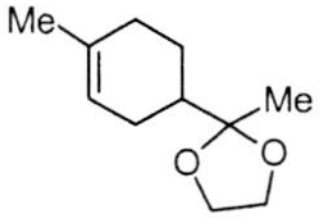 Me; Me; O, O	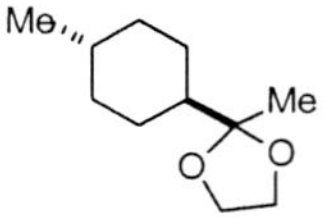 Me; Me; O, O	Ir^+	2.5	15 psi	CH_2Cl_2		62:38	36	25
15	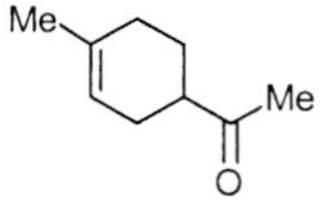 Me; Me; O	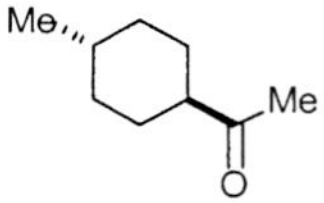 Me; Me; O	Ir^+	2.5	15 psi	CH_2Cl_2		99:1	80	25

Rh = $RhCl(PPh_3)_3$; Rh^+ = $[Rh(diphos\text{-}4)(cod)]^+$; Ir^+ = $[Ir(PCy_3)(py)(nbd)]^+$

Table 21.4 Diastereoselective hydrogenation of cyclic homoallyl and bishomoallyl alcohols.

	Substrate	Major diastereomer		Catalyst	mol%	P_{H_2}	Solvent	Temp. [°C]	Diastereomeric ratio	TOF	Reference
1	OH, Me	OH, Me		Ir^+	20	15 psi	CH_2Cl_2	r.t.	96.5:3.5	>2.5	22
2	HO, R, Me	HO, R, Me	R = Me	Ir^+	20	15 psi	CH_2Cl_2	r.t.	>100:1	>2.5	22
			R = ^{i}Pr	Ir^+	2.3	1 atm	CH_2Cl_2	0	1000:1	28.7	21
3	OH, Me	OH, Me		Rh^+	10	640 psi	CH_2Cl_2	r.t.	98.5:1.5	10	5
				Ir^+	20	15 psi	CH_2Cl_2	r.t.	97:3	>2.5	24
					2.5	15 psi	CH_2Cl_2	r.t.	98.1:1.9	20	24
4	OH, Me	OH, Me		Ir^+	20	15 psi	CH_2Cl_2	r.t.	86:14	>2.5	22
5	Me, OH, O	Me, OH, O, H		Ir^+	20	15 psi	CH_2Cl_2		96:4	<0.21	22

Table 21.4 (continued)

	Substrate	Major diastereomer	Catalyst	mol%	P_{H_2}	Solvent	Temp. [°C]	Diastereo-meric ratio	TOF	Reference
6	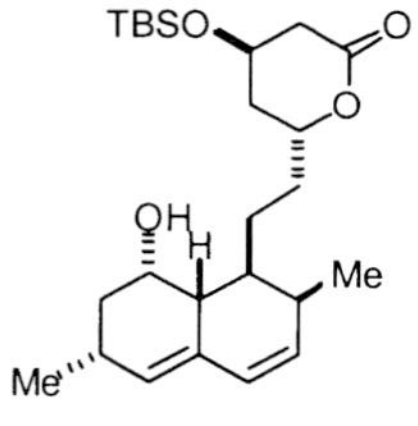	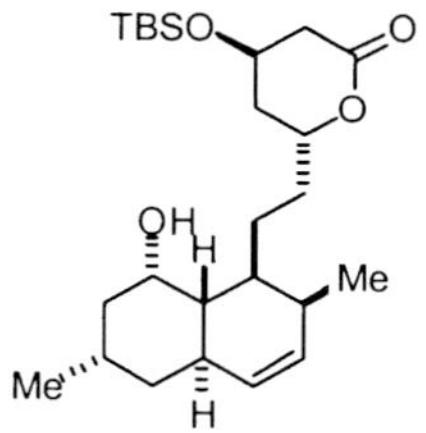	Ir^+	2.5	1 atm	CH_2Cl_2/ ^{i}PrOH		highly selective	1.9	30
7	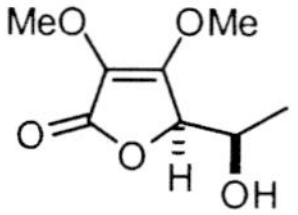	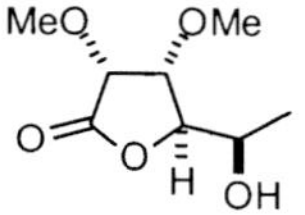	Rh^+	20	1850 psi	CH_2Cl_2		100:0		31
8	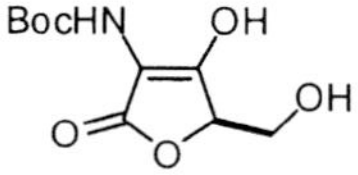	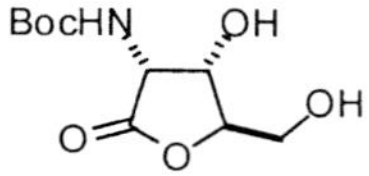	Rh^+	20	130 atm	CH_2Cl_2	r.t.	91:9	0.25	32
9	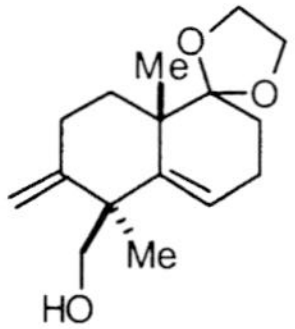	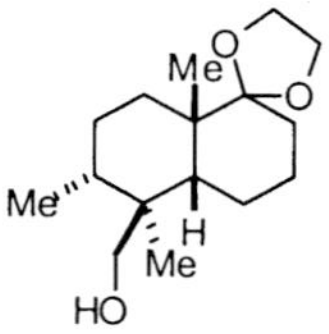	Ir^+	0.11	1000 psi	CH_2Cl_2		100:0	6.3	33

Table 21.4 (continued)

	Substrate	Major diastereomer	Catalyst	mol%	P_{H_2}	Solvent	Temp. [°C]	Diastereomeric ratio	TOF	Reference
10	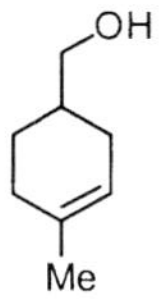	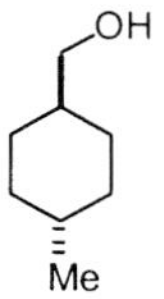	Rh^+	10	1000 psi	CH_2Cl_2	r.t.	95:5	10	5
			Ir^+	20	15 psi	CH_2Cl_2	r.t.	96.2:3.8	>2.5	5
11	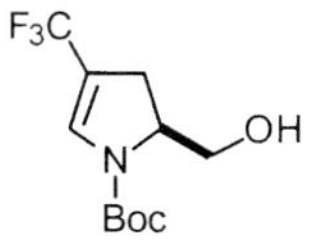	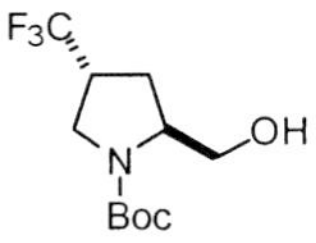	Ir^+	15	15 psi	CH_2Cl_2	r.t.	95:5		34
				2	15 psi	CH_2Cl_2	r.t.	99.4:0.6	12.5	34
12	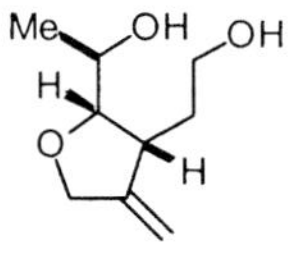	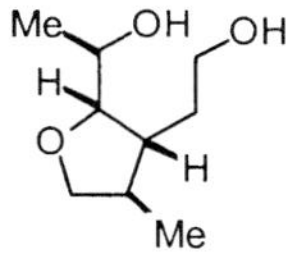	Ir^+	1	15 psi	CH_2Cl_2		>99:1		35
			Rh^+	6		CH_2Cl_2		90:10		
			Rh	100		toluene		6:94		
13	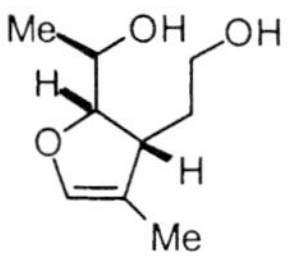	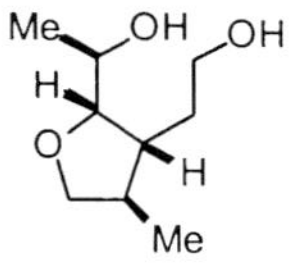	Ir^+	0.5	15 psi	CH_2Cl_2		>99:1		35

Rh = $RhCl(PPh_3)_3$; Rh^+ = $[Rh(diphos\text{-}4)(cod)]^+$; Ir^+ = $[Ir(PCy_3)(py)(nbd)]^+$

Table 21.5 Diastereoselective hydrogenation of alcohols with exocyclic double bond.

	Substrate	Major diastereomer	Catalyst	mol%	P_{H_2}	Solvent	Temp. [°C]	Diastereomeric ratio	TOF	Reference
1	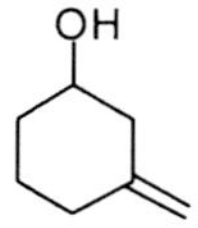 OH	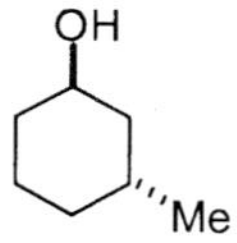 OH, Me	Rh^+	2	15 psi			>98:2		36
2	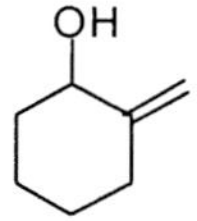 OH	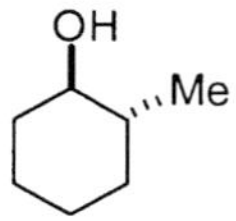 OH, Me	Rh^+	3.5	500 psi			75:25		23
			Rh^+	2	15 psi			45:55		36
3	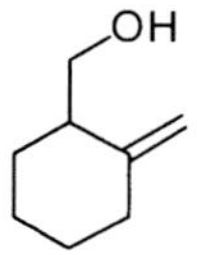 OH	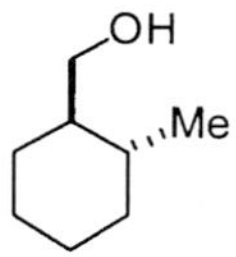 OH, Me	Rh^+	2	15 psi			47:53		36
4	OH, NHBz, OBz	OH, Me, NHBz, OBz	Rh^+	35	15 psi	CH_2Cl_2	r.t.	100:0	0.12	37
			Ir^+	17.5	15 psi	CH_2Cl_2	r.t.	72:28		37

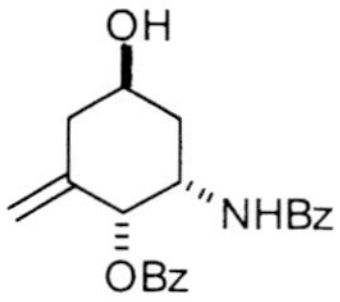

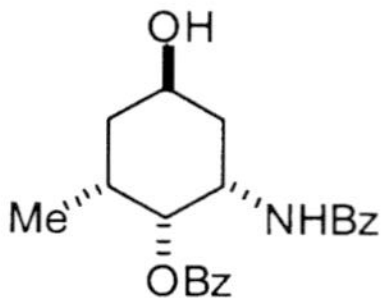

Table 21.5 (continued)

	Substrate	Major diastereomer	Catalyst	mol%	P_{H_2}	Solvent	Temp. [°C]	Diastereo-meric ratio	TOF	Reference
5	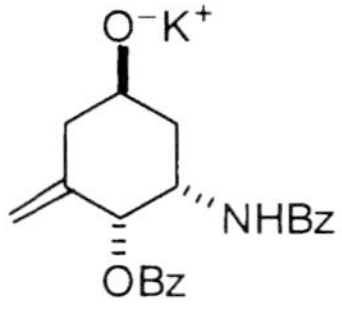 O⁻K⁺, NHBz, OBz	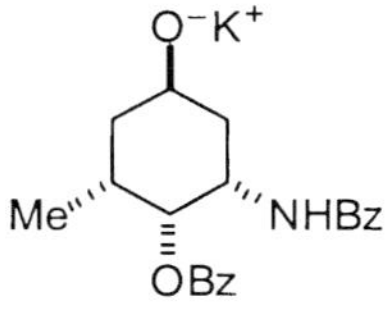 O⁻K⁺, Me, NHBz, OBz	Rh	100	15 psi	CH_2Cl_2	r.t.	95:5		37
				3.6	15 psi	CH_2Cl_2	r.t.	no reaction		37
6	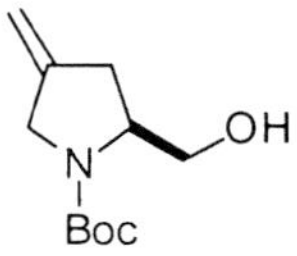 OH, N, Boc	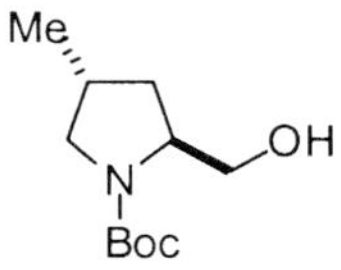 Me, OH, N, Boc	Ir^+	3	15 psi	CH_2Cl_2	r.t.	>97.5:2.5	1.67	38
7	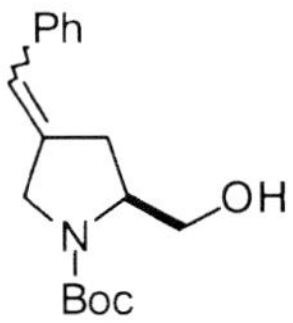 Ph, OH, N, Boc	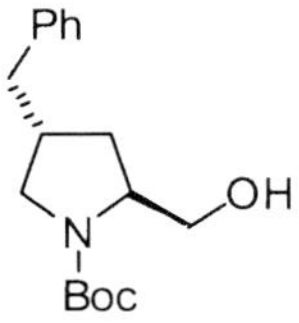 Ph, OH, N, Boc	Ir^+	3	15 psi	CH_2Cl_2	r.t.	>97.5:2.5	1.7	38
8	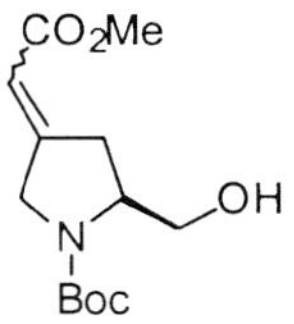 CO_2Me, OH, N, Boc	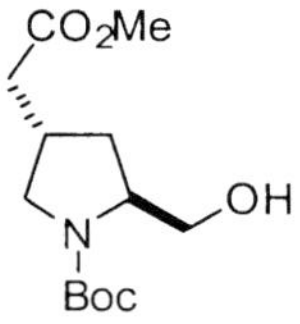 CO_2Me, OH, N, Boc	Ir^+	3	15 psi	CH_2Cl_2	r.t.	94:6	1.85	38

Table 21.5 (continued)

	Substrate	Major diastereomer	Catalyst	mol%	P_{H_2}	Solvent	Temp. [°C]	Diastereo-meric ratio	TOF	Reference
9			Rh^+	2	15 psi	CH_2Cl_2		95:5	10	29
			Ir^+	2	15 psi	CH_2Cl_2		99.7:0.3	1500	39
						CH_2Cl_2		99.7:0.3	6000	29
10	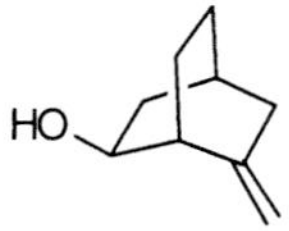	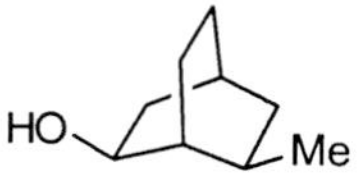	Ir^+	2	15 psi	CH_2Cl_2		55:45	16.7	29
11			Rh^+	2	15 psi	CH_2Cl_2		86.5:14.5	600	29
			Ir^+	2	15 psi	CH_2Cl_2		97.4:2.6	1000	29

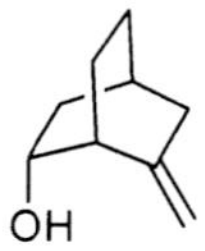

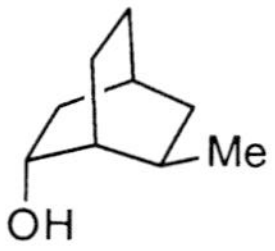

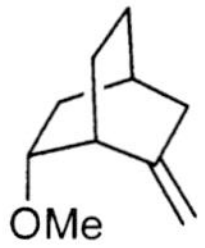

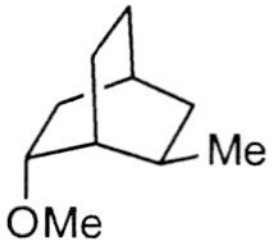

Rh = $RhCl(PPh_3)_3$; Rh^+ = $[Rh(diphos\text{-}4)(cod)]^+$; Ir^+ = $[Ir(PCy_3)(py)(nbd)]^+$

$[Ir(cod)(py)(PPh_3)]^+/H_2$ (2)

$\delta(H_A) = -29.4$, $\delta(H_B) = -9.6$ ppm

Thus, coordination of the OH-group controls the selection of the face to be reduced, and in some cases may also induce higher reactivity compared with the simple olefins not bearing OH-groups. Even in the presence of exceptional levels of steric congestion which are disposed to override the directivity of hydroxyl group, the addition of hydrogen occurs predominantly on the diastereoface bearing the hydroxyl group (Table 21.3, entry 6) [5, 26]. The concentration of the iridium catalyst strongly affects the stereoselectivity: a low concentration of the iridium complex is necessary to effect high diastereoselectivity; the decreases in stereoselectivity at a higher iridium concentration is ascribed to the formation of more than one complex, including trinuclear cluster complexes [24] in solution (Table 21.3, entry 1 and Table 21.4, entries 2 and 3). Cationic rhodium complexes such as $[Rh(nbd)(diphos\text{-}4)]^+$ are also effective for directive-hydrogenation of cyclic unsaturated alcohols, although in terms of selectivity they are somewhat inferior to the iridium catalysts. In entry 9 of Table 21.3, an attempt to obtain the *trans*-hydroindane product using a cationic iridium catalyst (Crabtree's catalyst) unexpectedly produced *cis*-hydroindanol highly selectively [28]. This suggests direction by the ether linkage of the TBSO unit. In contrast, hydrogenation by 25 mol.% of Wilkinson's catalyst in benzene (entry 10) afforded the *trans*-hydroindanol as a 3:1 mixture by the approach of rhodium catalyst from the less-hindered face of the double bond [28]. Even an exocyclic hydroxyl group can direct high diastereoselectivity with a rhodium or iridium catalyst in entries 7 and 9 of Table 21.4 [31, 33], though in entry 8 the directivity of the hydroxyl group is depressed by competition with the carbamate unit [32]. The hydroxyl group in bishomoallyl type alcohols also serves to direct the hydrogenation course to afford high diastereoselectivity. In entries 10–13 of Table 21.4, a cationic iridium complex can induce high diastereoselectivity by adding hydrogen from the face bearing the hydroxyl group [5, 34, 35]. With the Wilkinson catalyst, however, the reversed diastereoselectivity was again observed by the approach of the complex from the less-hindered side of the double bond without interaction with a hydroxyl unit (Table 21.4, entry 12).

Cationic iridium and rhodium catalysts are also effective for the hydrogenation of exocyclic olefinic alcohols (see Table 21.5), except for 2-exomethylenecyclohexanol and 2-methylenecyclohexanemethanol (entries 2 and 3). In entry 4, a cationic rhodium catalyst gave a single product whilst a cationic iridium catalyst induced only modest selectivity (72:28).

This lowering of the selectivity may be attributed to competitive binding between the hydroxyl and amide groups to iridium [37]. In entries 6, 7 and 8, the directivity of the hydroxyl group at the bishomoallylic position effectively overrides the effect of the carbamate unit [38]. In the hydrogenation of methylenebicyclo[2.2.2]octan-2-ol, the exo hydroxyl group does not serve as a directive group, and stereorandom hydrogenation proceeds contrary to the hydrogenation of methylenebicyclo[2.2.2]octan-2-ol with an endo hydroxyl group (entries 9 and 10) [29]. In entry 11 of Table 21.5, the methylenebicyclo[2.2.2]octane compound is hydrogenated in lower selectivity but at a higher rate than the parent alcohol with rhodium catalyst [29].

3 → (20 mol% $[Rh(nbd)(diphos\text{-}4)]^+BF_4^-$, H_2 (800 psi); KH, THF (68%), 3 days, TOF (h^{-1}) = 0.047) → 100 : 0 (3)

4 → (20 mol% $[Rh(nbd)(diphos\text{-}4)]^+BF_4^-$, H_2 (800 psi); NaH, THF (53%), 17 h, TOF (h^{-1}) = 0.16) → 100 : 0 (4)

As exemplified by Thompson for the case of tricyclic alcohol **1** [Eq. (1)], alkoxide has a strong coordinative ability to the metal, and high diastereoselectivity is realized in spite of the presence of a proximal bulky substituent in dihydrofuran derivatives [Eqs. (3) and (4)] [17].

Other functional groups which have a heteroatom rather than a hydroxyl group capable of directing the hydrogenation include alkoxyl, alkoxycarbonyl, carboxylate, amide, carbamate, and sulfoxide. The alkoxy unit efficiently coordinates to cationic iridium or rhodium complexes, and high diastereoselectivity is induced in the reactions of cyclic substrates (Table 21.3, entries 11–13) [25, 28]. An acetal affords much lower selectivity than the corresponding unsaturated ketone (Table 21.3, entries 14 and 15) [25].

Table 21.6 indicates the hydrogenation results of substrates with ester and carboxyl functionalities. An ester functionality also serves well as a directive unit, and high selectivity is reported for β,γ-unsaturated esters with both rhodium and iridium catalysts (Table 21.6, entries 1 and 2) [40, 41]. Directivity of the alkoxycarbonyl unit of an γ,δ-unsaturated ester is slightly diminished (entry 3) [25, 40, 41], and an acyloxy unit in the homoallylic position does not direct apparent stereoselectivity (entry 4) [41]. In the Wilkinson catalyst, hydroxyl, ether, esters or amide units cannot displace chloride ion from the metal, but carboxylate – being a better nucleophile – may be able to replace the chloride ion. In the hydrogenation of β,γ-carboxylic acids (Table 21.6, entries 8 and 9), the carboxylates are generated *in situ* by the addition of triethylamine, and the hydrogenation proceeds cleanly under 60 psi. As a consequence, one diastereoisomer is formed predominantly with a high selectivity of more than 99% diastereomeric

Table 21.6 Diastereoselective hydrogenation of double bond in cyclic esters and carboxylic acids.

	Substrate	Major diastereomer	Catalyst	mol%	P_{H_2}	Solvent	Temp. [°C]	Diastereomeric ratio	TOF	Reference
1	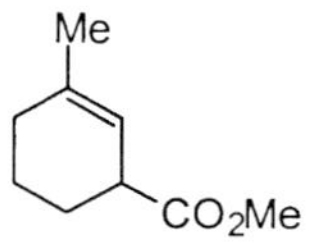 Me, CO_2Me	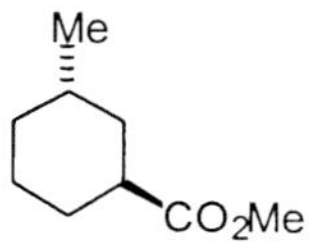 Me, CO_2Me	Rh^+	2	1 atm	CH_2Cl_2	r.t.	97:3	>1.8	40
			Ir^+	2	1 atm	CH_2Cl_2	r.t.	99.9:0.1	300	40
2	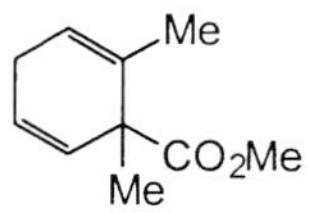 Me, CO_2Me, Me	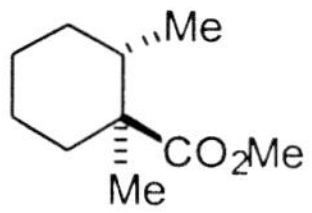 Me, CO_2Me, Me	Ir^+	5	1 atm	CH_2Cl_2	r.t.	99:1	4.5 ~ 9	41
3	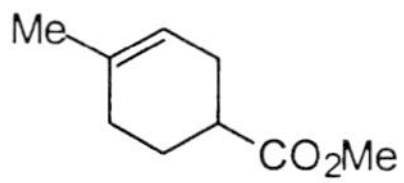 Me, CO_2Me	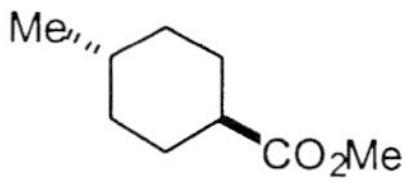 Me, CO_2Me	Rh^+	2	1 atm	CH_2Cl_2	r.t.	88:12	<0.23	40
			Ir^+	2	1 atm	CH_2Cl_2	r.t.	89:11	200	40
			Ir^+	2.5	1 atm	CH_2Cl_2	r.t.	95:5	77	25
			Ir^+	5	1 atm	CH_2Cl_2	r.t.	97.6:2.4	4.5–9	41
4	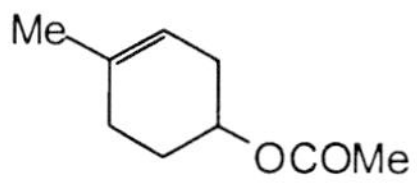 Me, OCOMe	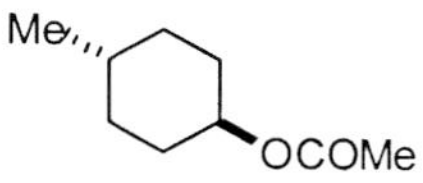 Me, OCOMe	Ir^+	5	1 atm	CH_2Cl_2	r.t.	50:50		41
5	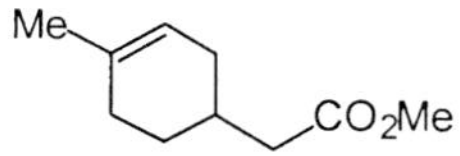 Me, CO_2Me	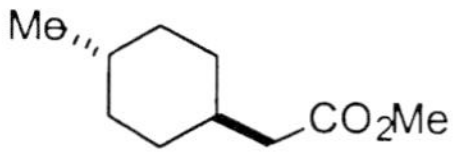 Me, CO_2Me	Ir^+	5	1 atm	CH_2Cl_2	r.t.	54.5:45.5		41
6	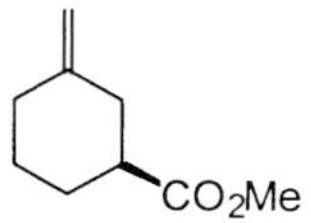 CO_2Me	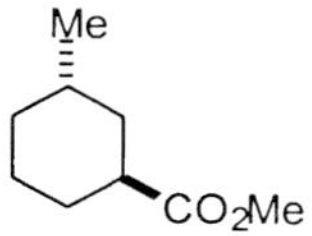 Me, CO_2Me	Rh^+	2	1 atm	CH_2Cl_2	r.t.	90:10	0.88	40
			Ir^+	2	1 atm	CH_2Cl_2	r.t.	81:19	300	40

Table 21.6 (continued)

	Substrate	Major diastereomer	Catalyst	mol%	P_{H_2}	Solvent	Temp. [°C]	Diastereo-meric ratio	TOF	Reference
7	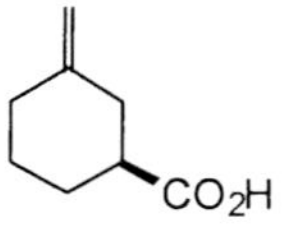 CO_2H	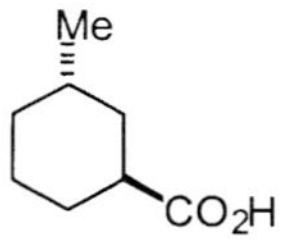 Me, CO_2H	Rh^+	2	1 atm	CH_2Cl_2	r.t.	50:50	<0.33	40
			Ir^+	2	1 atm	CH_2Cl_2	r.t.	88:12	8.3	40
8	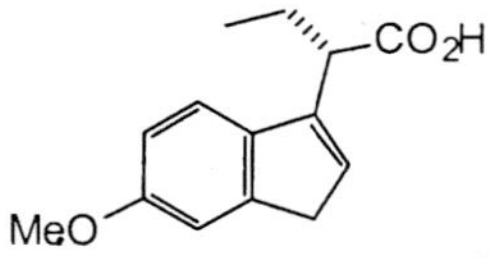 CO_2H, MeO	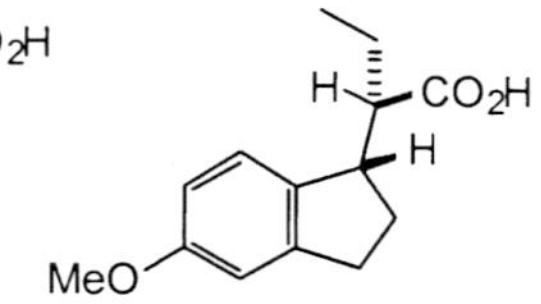 H, CO_2H, H, MeO	Rh	5	60 psi	THF/EtOH(1/9) (1.5 eq TEA)		>99.5:0.5	1.7	42
			Rh	5	60 psi	THF/EtOH(1/9) (0 eq TEA)		94:6	1.0	42
9	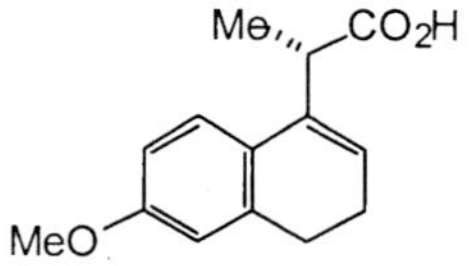 Me, CO_2H, MeO	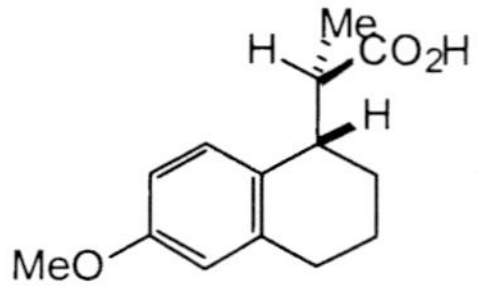 H, Me, CO_2H, H, MeO	Rh	5	60 psi	THF/EtOH(1/9) (1.5 eq TEA)		>99.5:0.5		42

Rh = $RhCl(PPh_3)_3$; Rh^+ = $[Rh(diphos\text{-}4)(cod)]^+$; Ir^+ = $[Ir(PCy_3)(py)(nbd)]^+$

excess (d.e.). The carboxylate anion binds to the rhodium complex, and discrimination of the diastereoface is caused by minimizing interaction of the stereogenic center with the peri-aromatic proton [42]. Even in the absence of amine, hydrogenation using the Wilkinson catalyst induces moderate reactivity and selectivity (88% d.e.) (entry 8), because part of the carboxylic acid undergoes dissociation to afford carboxylate in polar tetrahydrofuran (THF)/EtOH solution. In entry 7, the hydrogenation proceeds incompletely in DCM under the standard conditions. This implies the formation of inactive carboxylate complex in dichloromethane.

The amide group shows a prominent directivity in the hydrogenation of cyclic unsaturated amides by a cationic iridium catalyst, and much higher diastereoselectivity is realized than in the corresponding ester substrates (Table 21.7). In the case of β,γ-unsaturated bicyclic amide (entry 3), the stereoselectivity surpasses 1000:1 [41]. An increase of the distance between the amide carbonyl and olefinic bond causes little decrease in the selectivity (δ,ε-unsaturated amide, entry 6) compared with the case of the less-basic ester functionality (Table 21.6, entry 5).

In the case of cyclopentenyl carbamate in which a directive group is present at the homoallyl position, the cationic rhodium $[Rh(diphos\text{-}4)]^+$ or iridium $[Ir(PCy_3)(py)(nbd)]^+$ catalyst cannot interact with the carbamate carbonyl, and thus approaches the double bond from the less-hindered side. This affords a *cis*-product preferentially, whereas with the chiral rhodium–duphos catalyst, directivity of the carbamate unit is observed (Table 21.7, entry 7). The presence of a hydroxyl group at the allyl position induced hydroxy-directive hydrogenation, and higher diastereoselectivity was obtained (entry 8) [44].

21.3.2
Hydrogenation of Acyclic Allyl and Homoallyl Alcohols

The hydrogenation of acyclic allyl alcohols with a 1,1-disubstituted olefinic bond are listed as entries 1 to 5 of Table 21.8. The reduction of (α-hydroxyalkyl)acrylates proceeds stereoselectively with the cationic rhodium catalyst $[Rh(diphos\text{-}4)(nbd)]^+$, and 1,2-*anti*-compounds are obtained as the major product by the direction of the hydroxyl group. Under these reaction conditions, isomerization of the olefinic unit occurs to afford about 20% of the corresponding methyl ketone. If the isomerization occurs prior to the hydrogenation, diminished stereoselectivity would be observed, even if the individual reduction mode were to occur discriminately [5]. With a cationic iridium catalyst, the degree of isomerization is greater than with the rhodium catalyst, and the hydrogenation occurs with lower diastereoselectivity (Table 21.8, entries 5–7). This is in contrast to the high stereoinduction ability of iridium catalysts for cyclic unsaturated compounds, but the isomerization can be suppressed by increasing the hydrogen pressure. The concentration of iridium catalyst strongly affects the stereoselectivity, and at higher concentration stereorandom hydrogenation almost occurs

Table 21.7 Amido-directive diastereoselective hydrogenation.

	Substrate	Major diastereomer	Catalyst	mol%	P_{H_2}	Solvent	Diastereo-meric ratio	TOF	Reference
1	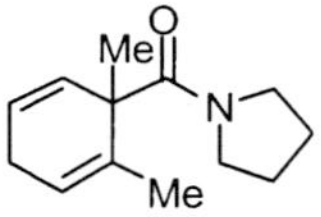	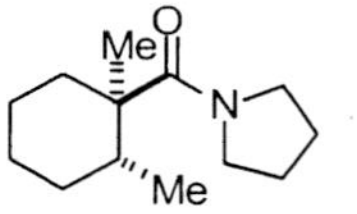	Ir^+	5	1 atm	CH_2Cl_2	170:1	4 ~ 8	41
2	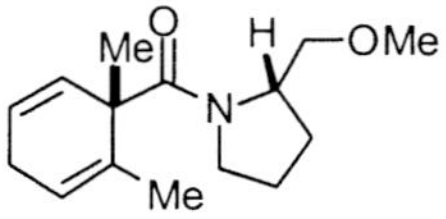	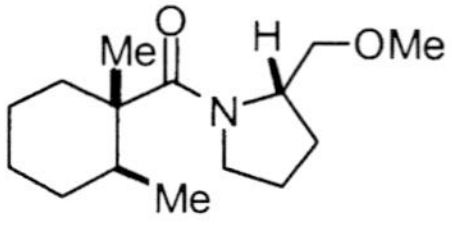	Ir^+	5	1 atm	CH_2Cl_2	530:1	4.5 ~ 9	41
3	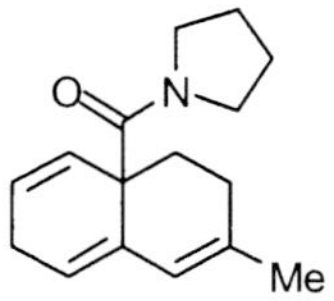	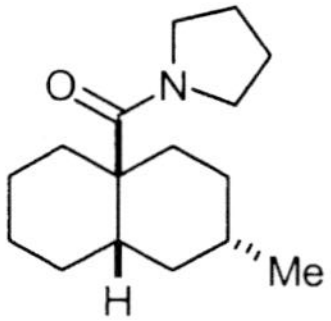	Ir^+	5	1 atm	CH_2Cl_2	>1000:1	4.5 ~ 9	41
4	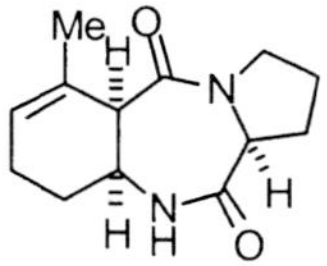	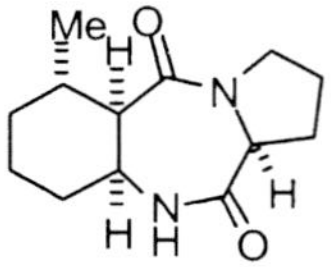	Ir^+ Ir^+	5 5	1 atm 1 atm	CH_2Cl_2 CH_2Cl_2	>99:1 >99:1	4.5 ~ 9 3.3	41 43
5	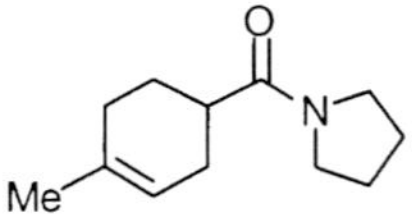	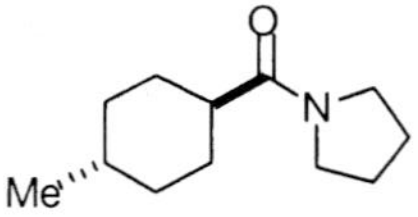	Ir^+	5	1 atm	CH_2Cl_2	130:1	4.5 ~ 9	41

Table 21.7 (continued)

	Substrate	Major diastereomer	Catalyst	mol%	P_{H_2}	Solvent	Diastereo-meric ratio	TOF	Reference
6	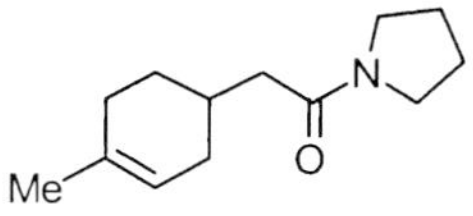 Me, O, N	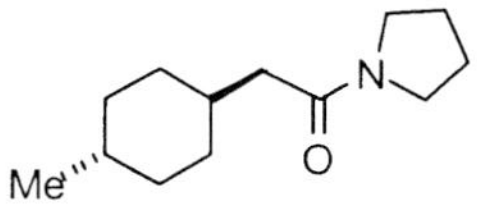 Me, O, N	Ir^+	5	1 atm	CH_2Cl_2	>100:1	4.5 ~ 9	41
7	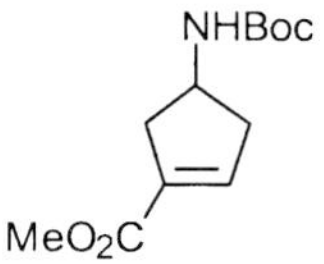 NHBoc, MeO_2C	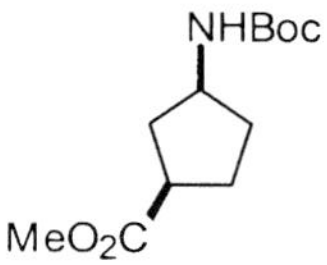 NHBoc, MeO_2C	Ir^+	1	5 atm	CH_2Cl_2	92.5:7.5	7.1	44
			Rh^+	1	5 atm	MeOH	78.5:21.5	7.1	44
			$[Rh((R,R)\text{-Me-Duphos})]^+$	1	5 atm	MeOH	20:80	7.1	44
			$[Rh((S,S)\text{-Me-Duphos})]^+$	1	5 atm	MeOH	9:91	7.1	44
8	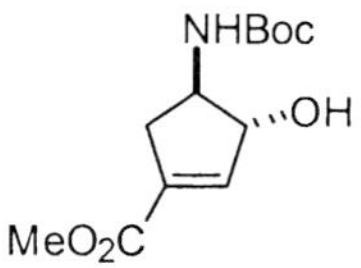 NHBoc, OH, MeO_2C	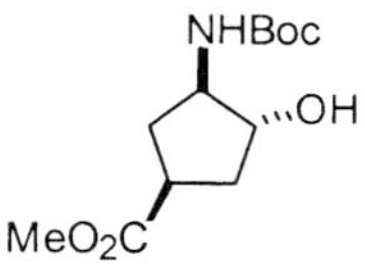 NHBoc, OH, MeO_2C	Ir^+	1	5 atm	CH_2Cl_2	97:3	7.1	44
			Rh^+	1	5 atm	MeOH	80:20	7.1	44
			$[Rh((R,R)\text{-Me-Duphos})]^+$	1	5 atm	MeOH	4:96	7.1	44
			$[Rh((S,S)\text{-Me-Duphos})]^+$	1	5 atm	MeOH	33.5:66.5	7.1	44

$Rh^+ = [Rh(diphos\text{-}4)(cod)]^+$; $Ir^+ = [Ir(PCy_3)(py)(nbd)]^+$

Table 21.8 Hydroxy-directed hydrogenation of acyclic allyl aclohols.

	Substrate	Major diastereomer	Catalyst	mol%	P_{H_2}	Solvent	Temp. [°C]	Diastereo-meric ratio	TOF	Reference
1	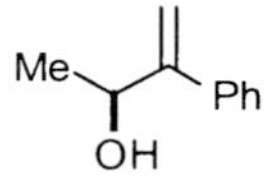	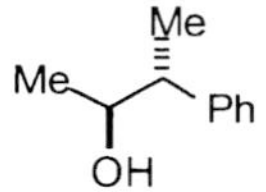	Rh^+	2	15 psi	CH_2Cl_2	0	97:3		19
2	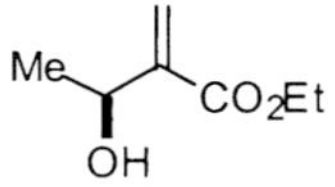	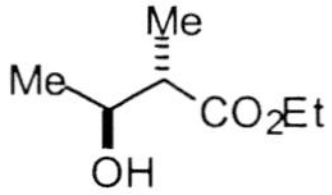	Rh^+	1	15 psi	CH_2Cl_2		100:1	50	45
						MeOH		100:1	50	45
3	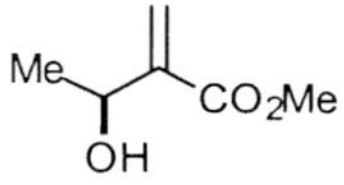	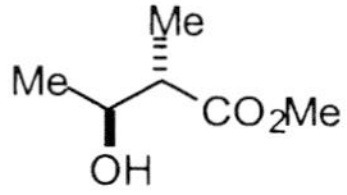	Rh(OAc)$_2$((*S*)-binap)		4 atm	MeOH	25	>23:1		46
			Rh(OAc)$_2$((*R*)-binap)		4 atm	MeOH	25	>23:1		46
4	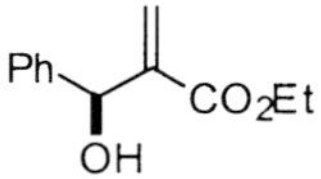	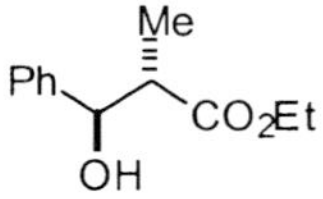	Rh^+	1	15 psi	CH_2Cl_2	20	100:1	50	45
						MeOH	20	97.5:2.5		
5	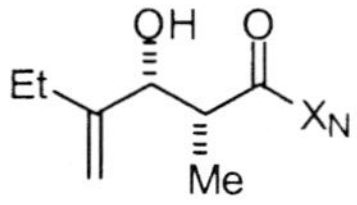	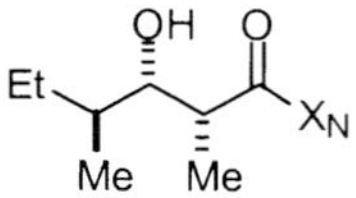	Rh^+	17.5	640 psi	CH_2Cl_2	25	93:7	>2.8	5
			Ir^+	20	15 psi	CH_2Cl_2	25	57:43	>2.5	24
			Ir^+	2.5	15 psi	CH_2Cl_2	25	85:15	>20	24
6	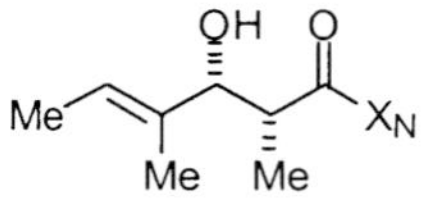	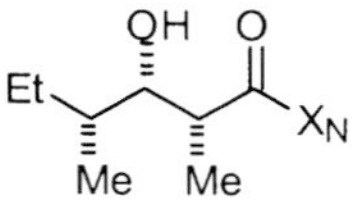	Rh^+	17.5	640 psi	CH_2Cl_2	25	91:9	>2.8	5
			Ir^+	20	15 psi	CH_2Cl_2	25	43:57	>2.5	24
			Ir^+	2.5	15 psi	CH_2Cl_2	25	73:27	>20	24

Table 21.8 (continued)

	Substrate	Major diastereomer	Catalyst	mol%	P_{H_2}	Solvent	Temp. [°C]	Diastereo-meric ratio	TOF	Reference
7			Rh^+	17.5	640 psi	CH_2Cl_2	25	94:6	>2.8	24
			Ir^+	2.5	15 psi	CH_2Cl_2	25	48:52	<0.4	24
8	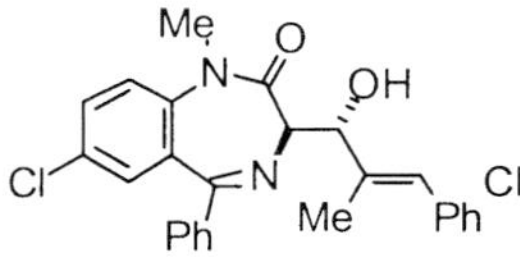	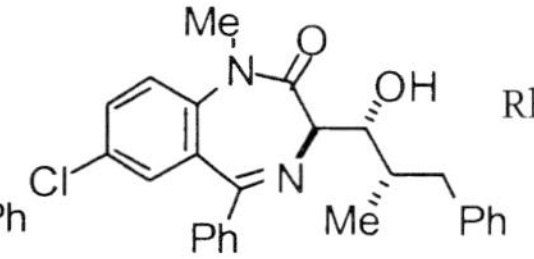	Rh^+	10	45 bar	CH_2Cl_2		75:25	1.1	47
9	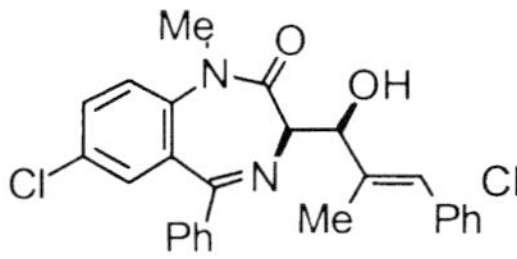	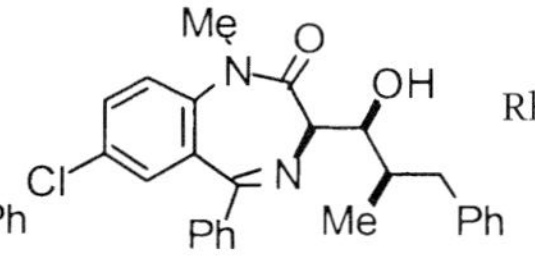	Rh^+	10	45 bar	CH_2Cl_2		80:20	1.3	47
10	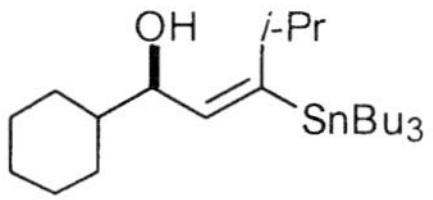	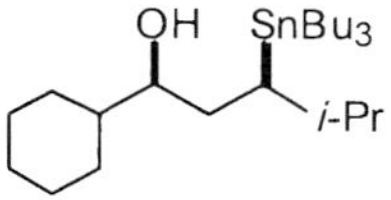	Rh^+	5	1500 psi	CH_2Cl_2	r.t.	300:1	0.47	48

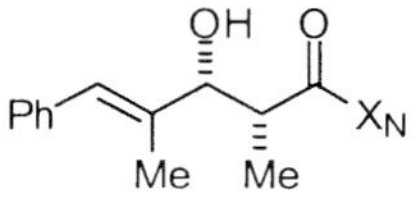

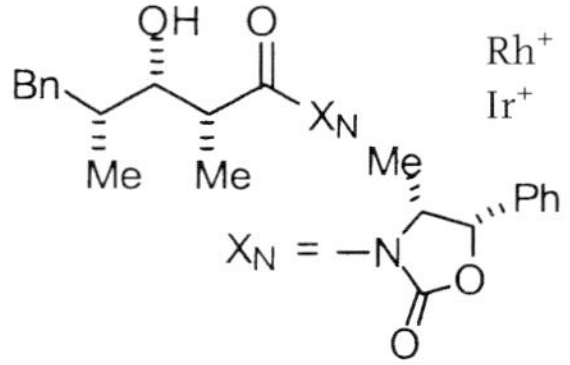

Table 21.8 (continued)

	Substrate	Major diastereomer	Catalyst	mol%	P_{H_2}	Solvent	Temp. [°C]	Diastereomeric ratio	TOF	Reference
11	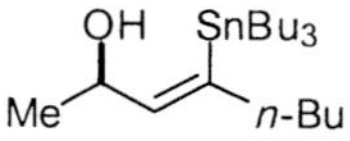 OH, $SnBu_3$, Me, *n*-Bu	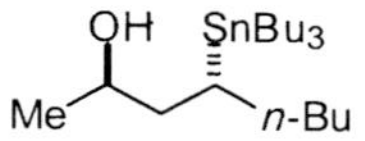 OH, $SnBu_3$, Me, *n*-Bu	Rh^+	5	1500 psi	CH_2Cl_2	r.t.	>100:1	0.2	48
12	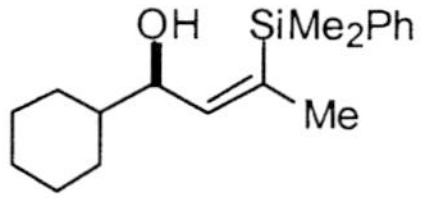 OH, $SiMe_2Ph$, Me	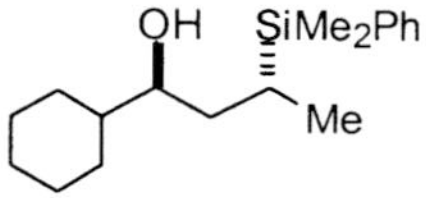 OH, $SiMe_2Ph$, Me	Rh^+	5	1500 psi	CH_2Cl_2	r.t.	>500:1	0.4	48
13	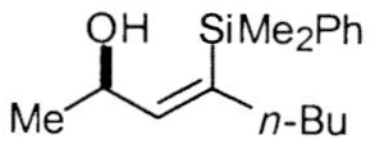 OH, $SiMe_2Ph$, Me, *n*-Bu	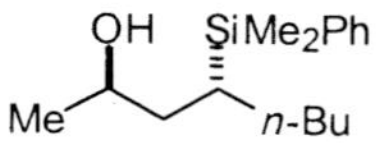 OH, $SiMe_2Ph$, Me, *n*-Bu	Rh^+	5	1500 psi	CH_2Cl_2	r.t.	>500:1	0.4	48
14	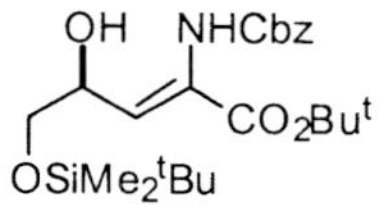 OH, NHCbz, CO_2Bu^t, $OSiMe_2{}^tBu$	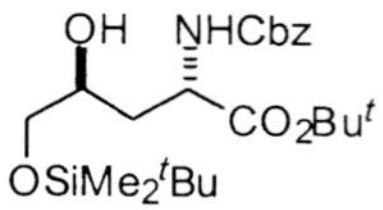 OH, NHCbz, CO_2Bu^t, $OSiMe_2{}^tBu$	$[Rh((R,R)\text{-dipamp})]^+$	0.7	3 bar	MeOH		highly selective	11.8	49

$Rh^+ = [Rh(diphos\text{-}4)(cod)]^+$; $Ir^+ = [Ir(PCy_3)(py)(nbd)]^+$

(Table 21.8, entries 5 and 6) [5, 24], similarly to the reduction of cyclic compounds.

The configuration of the product in diastereoselective hydrogenation – whether 1,2-*syn* or 1,2-*anti* – is related to the substitution pattern of the starting alkene. The allyl alcohol with a 1,1-disubstituted olefin unit affords the *anti*-product, while the *syn*-product is formed from the allyl alcohol with a trisubstituted olefinic bond (Table 21.8, entries 6–9). The complementarity in diastereoselective hydrogenation of di- and tri-substituted olefins may be rationalized based on the conformation analysis of the intermediary complex (Scheme 21.1) [23].

In entries 10–13 (Table 21.8) of trisubstituted alkenes, very high diastereoselectivity is realized by the use of a cationic rhodium catalyst under high hydrogen pressure, and the 1,3-*syn*- or 1,3-*anti*-configuration naturally corresponds to the (*E*)- or (*Z*)-geometry of the trisubstituted olefin unit [48, 49]. The facial selectivity is rationalized to be controlled by the A(1,3)-allylic strain at the intermediary complex stage (Scheme 21.2) [48].

Entries 8–13 in Table 21.9 illustrate the effect of S–O coordination on the hydrogenation of allyl alcohols. The hydrogenation of (*α*-hydroxyalkyl)vinyl sulfones follows the same stereochemical course as the corresponding acrylate via HO coordination (entries 8 and 9). However, the hydrogenation of (*α*-hydroxyalkyl)vinyl sulfoxides is directed by S–O coordination, which overrides the HO-participation in the stereochemical course (entries 10–13) [56]. The directing power of S–O may be limited to vinylic examples, as compounds having the S–O and double bond in an allylic relationship failed to reduce under the standard conditions.

The hydrogenation of acyclic homoallylic alcohols with a 1,1-disubstituted olefinic bond by cationic $[Rh(diphos\text{-}4)]^{+}$ catalyst proceeds in modest to moderate stereoselectivity, generally forming 1,3-*anti* compounds (Table 21.10, entries 1, 4 and 5), and the effect of the stereogenic center at the allylic position overrides the directivity of hydroxyl group. The 1,3-*syn* product is then observed though in poor selectivity (entry 3) [19, 57, 58]. Inspection of the hydrogenation prod-

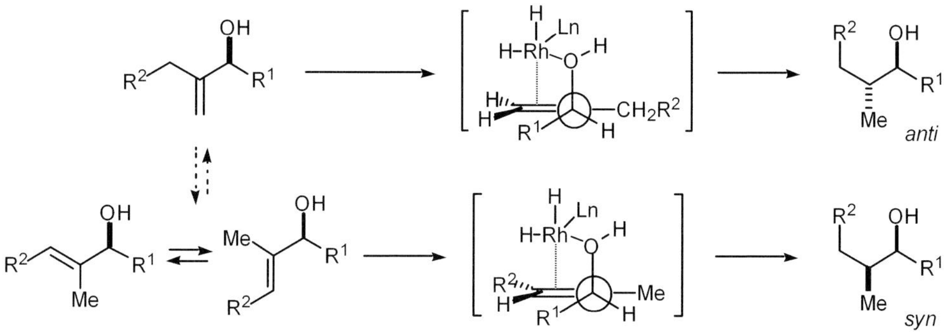

Scheme 21.1

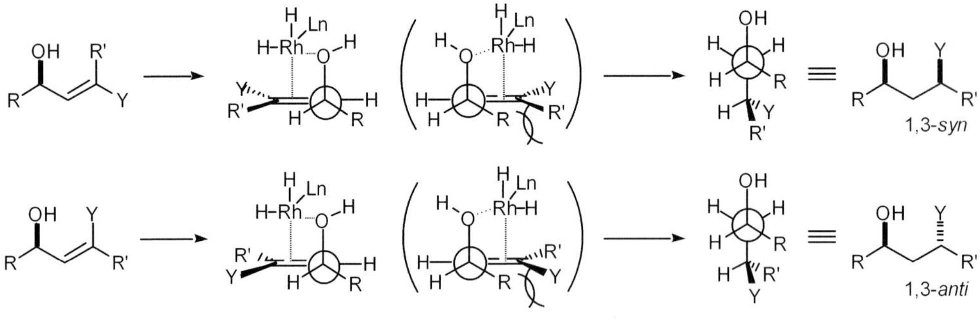

Scheme 21.2

ucts indicates that the substituent at the allylic position dictates the stereochemistry to afford the 1,2-*syn* product preferentially. The observed stereochemistry in 1,1-disubstituted homoallylic alcohols can be explained by considering the conformational analysis of the alkene complexes (Scheme 21.3). In conformation A, unfavorable A(1,2) interactions between R and R^1 are minimized and through A and C the 1,3-*anti* product will be formed. In the case of homoallylic alcohols without an allylic substituent, the group at the homoallylic carbon will adapt the pseudoequatorial orientation to afford the 1,3-*anti* product. In homoallylic alcohols with a stereogenic center at the allyl position, the group in the allylic position will also prefer the pseudoequatorial orientation so that the A(1,2) strain will be minimized [23, 57].

In the case of tri-substituted alkenes, the 1,3-*syn* products are formed in moderate to high diastereoselectivities (Table 21.10, entries 6~12). The stereochemistry of hydrogenation of homoallylic alcohols with a trisubstituted olefin unit is governed by the stereochemistry of the homoallylic hydroxy group, the stereogenic center at the allyl position, and the geometry of the double bond (Scheme 21.4). In entries 8 to 10 of Table 21.10, the product of 1,3-*syn* structure is formed in more than 90% d.e. with a cationic rhodium catalyst. The stereochemistry of the products in entries 10 to 12 shows that it is the stereogenic center at the allylic position which dictates the sense of asymmetric induction

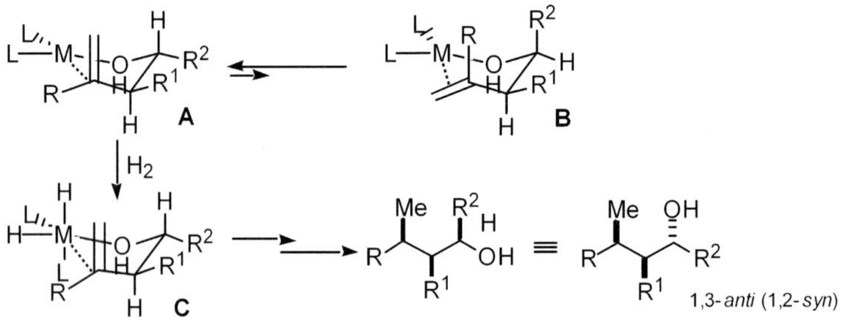

Scheme 21.3 Conformation of 1,1-disubstituted alkenes in the hydrogenation.

Table 21.9 Functinoal group-directed hydrogenation.

	Substrate	Major diastereomer	Catalyst	mol%	P_{H_2}	Solvent	Diastereo-meric ratio	TOF	Reference
1			$[RhCl(PPh_3)_3]$		50 bar	benzene	99:1		50
2			$RhCl(PPh_3)_3$		15 bar	benzene	>99:1		51
3			$Ru(OAc)_2$((*R*)-tol-binap)	0.2	1 atm	MeOH	99.9:0.1	10.4	52
			$Ru(OAc)_2$((*S*)-tol-binap)	0.2	1 atm	MeOH	22:78	10.4	52
			$RuBr_2$((*R*)-MeO-biphep)		15 bar		>99:1		53
4	100% *E*		$[RuCl((R)\text{-binap})]_2NEt_3$	2	70 atm	MeOH	99:1	1.07	54
5			$[Rh((S,S)\text{-dipamp})]^+$	0.23	3 bar	MeOH	80:20	28.6	49
			$[Rh((R,R)\text{-dipamp})]^+$	0.23	3 bar	MeOH	20:80		49

Table 21.9 (continued)

	Substrate	Major diastereomer	Catalyst	mol%	P_{H_2}	Solvent	Diastereo-meric ratio	TOF	Reference
6	Bn2N, NHCHO, Me, CO2Et	Bn2N, NHCHO, Me, CO2Et	[Rh(diphos-2)]$^+$	10	25 atm	MeOH	88:12	0.25	55
7	Bn2N, NHCHO, CO2Et, OH	Bn2N, NHCHO, CO2Et, OH	[Rh(diphos-2)]$^+$	10	85 atm	MeOH rt	90:10	0.24	55
8	Me, SO2Ph, OH	Me, Me, SO2Ph, OH	Rh$^+$	1	15 psi	CH_2Cl_2 MeOH	99.85:0.15 99.5:0.5		56
9	Ph, SO2Ph, OH	Me, Ph, SO2Ph, OH	Rh$^+$	1	15 psi	CH_2Cl_2 MeOH	99.85:0.15 400:1		56
10	Me, S, Ph, OH, O (S*,Rs*)	Me, Me, S, Ph, OH, O (S*,S*,Ss*)	Rh$^+$	1	15 psi	$Cl(CH_2)_2Cl$ MeOH	99.5:0.5 97.5:2.5		56
11	Me, S, Ph, OH, O (R*,Rs*)	Me, Me, S, Ph, OH, O (R*,S*,Ss*)	Rh$^+$	1	15 psi	$Cl(CH_2)_2Cl$ MeOH	99:1 80:20		56

Table 21.9 (continued)

	Substrate	Major diastereomer	Catalyst	mol%	P_{H_2}	Solvent	Diastereomeric ratio	TOF	Reference
12	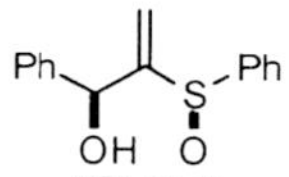 (*S**,*R*S*)	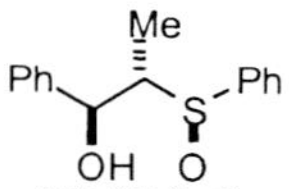 (*S**,*S**,*S*S*)	Rh^+	1	15 psi	CH_2Cl_2 MeOH	99.5:0.5 92.5:7.5		56
13	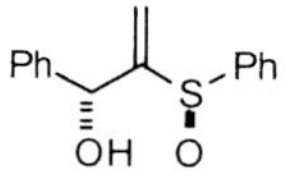 (*R**,*R*S*)	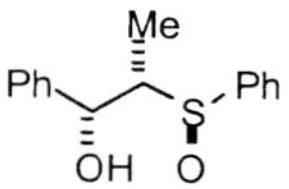 (*R**,*S**,*S*S*)	Rh^+	1	15 psi	CH_2Cl_2 MeOH	98.5:1.5 90.5:9.5		56

Rh^+ = [Rh(diphos.4)(cod)]$^+$; Ir^+ = [Ir(PCy$_3$)(py)(nbd)]$^+$

Table 21.10 Hydroxy-directed hydrogenation of acyclic homoallyl alcohols.

	Substrate	Major diastereomer	Catalyst	mol%	P_{H_2}	Solvent	Diastereo-selectivity	TOF	Reference
1	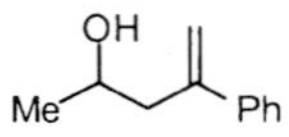	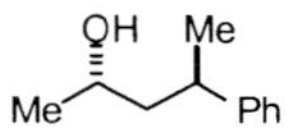	Rh^+	2	15 psi	THF	88:12		19
2	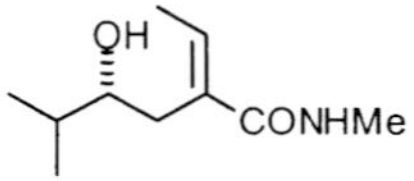	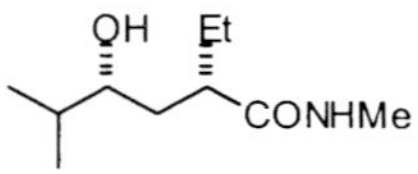	Rh^+	5	15 psi	MeOH	89:11	1	57
3	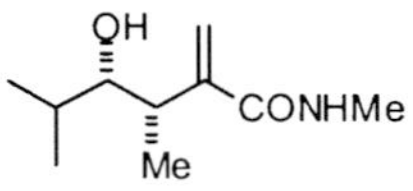	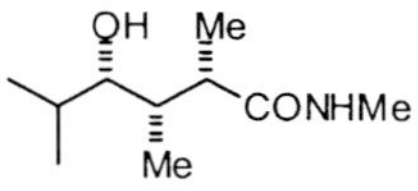	Rh^+	5	15 psi	MeOH	67:33	1	57
4	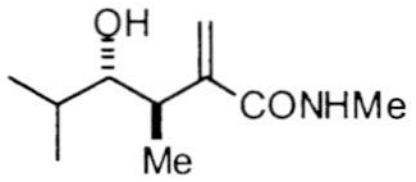	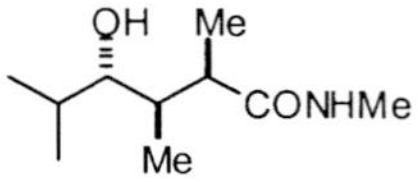	Rh^+	5	15 psi	MeOH	91:9	1	57
5	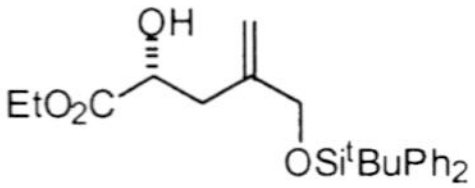	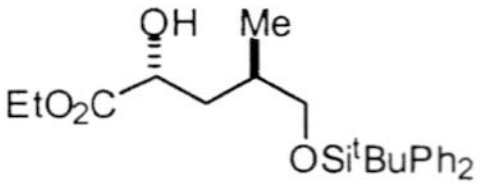	Rh^+		1000 psi	CH_2Cl_2	60:40		58
			$[Rh((S,S)\text{-Et-Duphos})]^+$		1000 psi	CH_2Cl_2	95:5		58
			$[Rh((R,R)\text{-Et-Duphos})]^+$		1000 psi	CH_2Cl_2	40:60		58
			$[Rh(R)\text{-phanephos})]^+$		1000 psi	CH_2Cl_2	25:75		58
			Ir^+		1000 psi	CH_2Cl_2	35:65		58
6	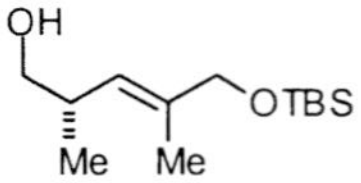	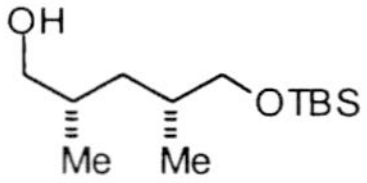	Rh^+	5	15 psi	CH_2Cl_2	95:5		59
			Ir^+	2.5	15 psi	CH_2Cl_2	73:27		59

Table 21.10 (continued)

	Substrate	Major diastereomer	Catalyst	mol%	P_{H_2}	Solvent	Diastereo-selectivity	TOF	Reference
7	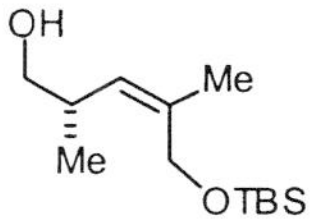	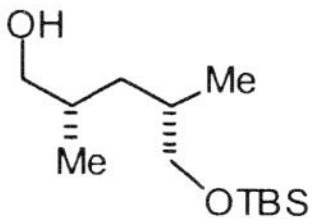	Rh^+	5	15 psi	CH_2Cl_2	91:9		59
8	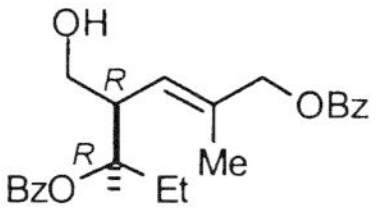	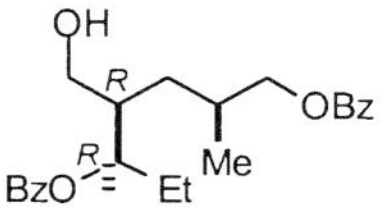	Rh^+	20	15 psi	CH_2Cl_2	97:3		59
			Ir^+	2.5	15 psi	CH_2Cl_2	97:3		59
9			Rh^+	20	15 psi	CH_2Cl_2	99:1		59
			Ir^+	2.5	15 psi	CH_2Cl_2	97:3		59
10			Rh^+	20	15 psi	CH_2Cl_2	97:3		59
			Ir^+	2.5		CH_2Cl_2	94:6		59
11			Rh^+	20	15 psi	CH_2Cl_2	89:11		59
			$[Rh((+)\text{-binap})]^+$		1000 psi	CH_2Cl_2	97:3		59
			$[Rh((-)\text{-binap})]^+$		1000 psi	CH_2Cl_2	92:8		59

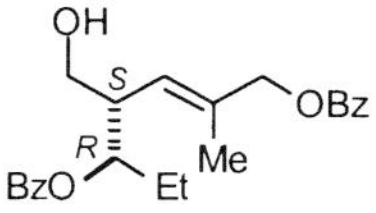

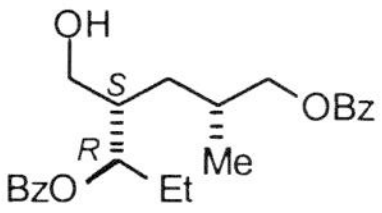

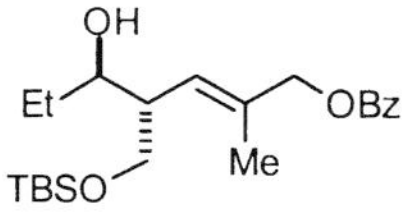

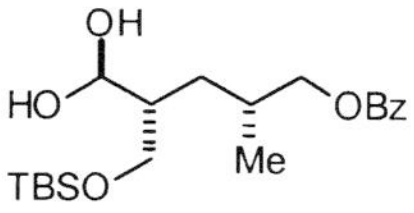

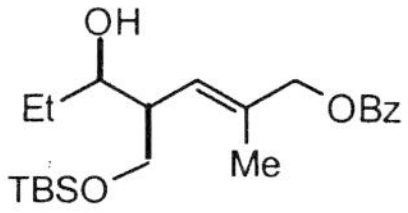

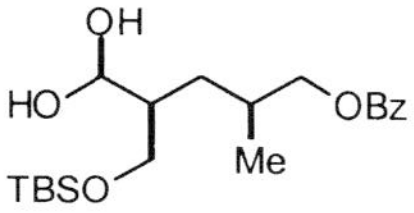

Table 21.10 (continued)

	Substrate	Major diastereomer	Catalyst	mol%	P_{H_2}	Solvent	Diastereo-selectivity	TOF	Reference
12	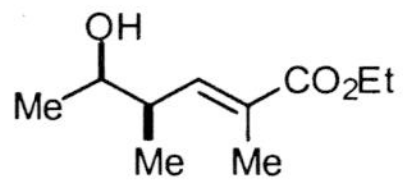	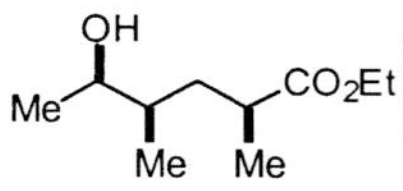	Rh^+		1000 psi	CH_2Cl_2	85:15		60
			$[Rh((+)\text{-binap})]^+$		1000 psi	CH_2Cl_2	98:2		59
13	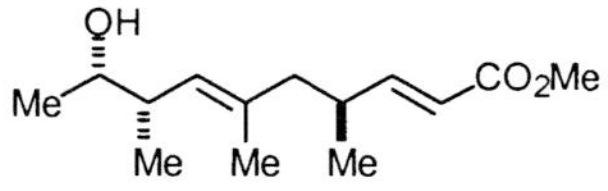	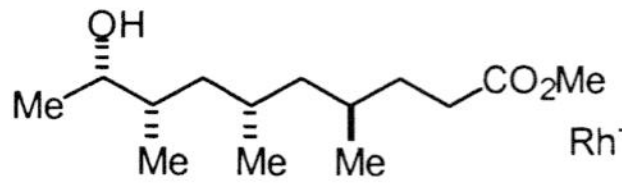 Rh^+							
			Rh^+	5	15 psi	CH_2Cl_2	94:6	1.6	61
14	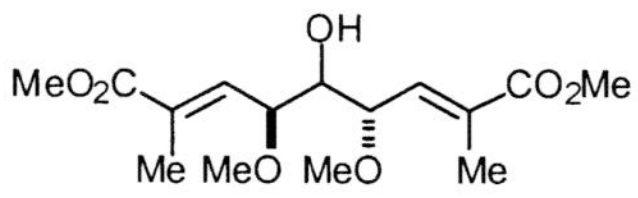	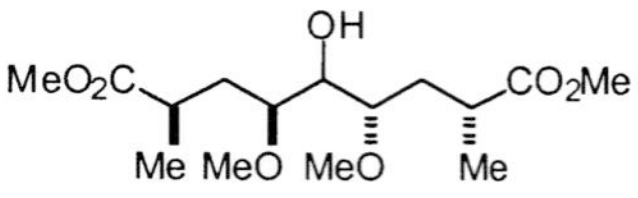							
			[Rh(diphos-4)(nbd)]$^+$	16	1000 psi	CH_2Cl_2	19:2.2:1	1.56	62
15	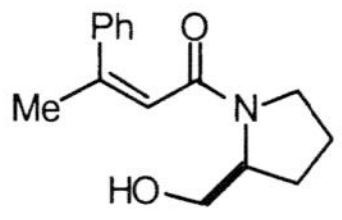	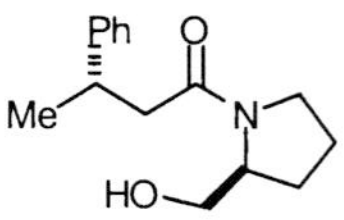	[Rh(diphos-4)(nbd)]$^+$	8	640 psi		80:20 1,5-asymmetric induction		23

Rh^+ = [Rh(diphos-4)]$^+$; Ir^+ = [Ir(PCy_3)(py)(nbd)]$^+$

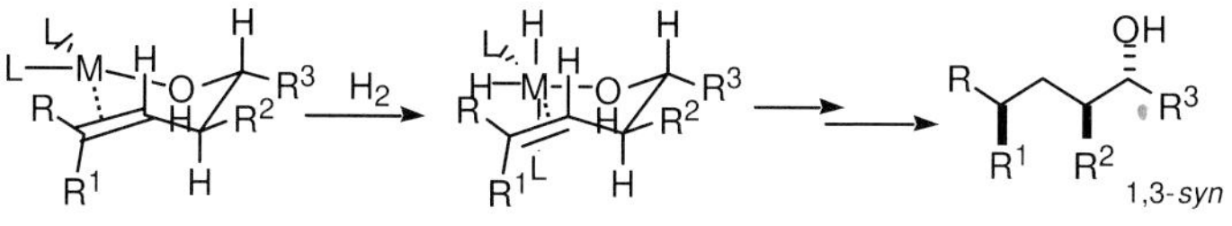

Scheme 21.4 Conformation of trisubstituted alkenes in the hydrogenation.

in combination with the direction by the hydroxyl group [59, 60]. In entries 6 and 7 of Table 21.10, the alkyl group at the allylic stereogenic center is small and the diastereoselectivity is ca. 80～90% d.e. In the directive hydrogenation of a 5-hydroxy-4,6-dimethoxy-2,7-nonadienedioic acid derivative, which was part of the synthesis of the C_{10}-C_{19} fragment of the immunosuppressive agent FK-506 (entry 14), the diastereoselectivity is controlled by the methoxy units at the allylic positions, and not by the hydroxy group, to afford a product with two 1,3-*syn* disposition in the structure [62].

In some cases of enantioselective hydrogenation of dehydroamino acids with a chiral cationic rhodium catalyst, the less-stable substrate–metal complex (minor species) reacts with hydrogen far more rapidly (～1000-fold faster) than the more stable complex (major species), and the stereochemistry of the predominant enantiomer is determined by the reaction of the minor species [63]. In these cases, the stereochemical outcome is not related to the initial equilibrium of the substrate–metal complex, and a higher hydrogen pressure and rise in reaction temperature suppress the enantioselectivity. In contrast to the enantioselective hydrogenation of dehydroamino acids, olefinic alcohols are hydrogenated with higher diastereoselectivity at higher hydrogen pressure and at lower reaction temperature [24, 48]. This implies that the major substrate–metal complex determines the stereochemical outcome of the hydrogenation.

In entry 15 of Table 21.10, it is noted that even a remote hydroxyl group directed hydrogenation by the cationic $[Rh(diphos\text{-}4)(nbd)]^+$ catalyst to afford a moderate diastereoselectivity (80:20) [23]. This is an interesting example of long-range 1,5-asymmetric induction.

21.3.3
Ester Unit- or Amide-Directive Hydrogenation

The diastereoselective hydrogenation of itaconate derivatives by cationic rhodium catalysts are listed in Table 21.11. The observed high diastereoselectivity indicates a strong directivity by the alkoxycarbonyl unit in the reduction of acyclic systems (entries 1 and 2). Decrease of stereoselectivity in entry 3 indicates the definite effect of the ether functionality on the sense of asymmetric induction, competing with the directivity exerted by the alkoxycarbonyl group [64]. An amide or a carbamate group also serves as an admirable directing group in acrylic acid derivatives (entries 4–6) [65–67], although the hydrogenation of an amine or its corresponding trifluoroacetate salt was quite unselective (entry 7) [65].

Table 21.11 Directive hydrogenation of acrylic acid derivatives.

	Substrate	Major diastereomer	Catalyst	mol%	P_{H_2}	Solvent	Diastereo-meric ratio	TOF	Reference
1	MeO_2C, CO_2Me, Me	MeO_2C, Me, CO_2Me, Me	Rh^+	2	1 atm	MeOH	99.6:0.4	50	64
2	MeO_2C, CO_2Me, Ph	MeO_2C, Me, CO_2Me, Ph	Rh^+	2	1 atm	MeOH	99.5:0.5		64
3	MeO_2C, CO_2Me, OMe	MeO_2C, Me, CO_2Me, OMe	Rh^+	2	1 atm	MeOH	98:2		64
4	MeO_2C, Me, NHCOMe	MeO_2C, Me, Me, NHCOMe	Rh^+	5		CH_2Cl_2	99:1		65
			Rh^+	5	1 atm	MeOH	100:0		65
5	MeO_2C, Ph, $NHCO_2Me$	MeO_2C, Me, Ph, $NHCO_2Me$	[Rh(diphos-2)]$^+$	1	30 atm	THF	95:5	2.1	66
			[Rh(diphos-2)]$^+$	1	30 atm	MeOH	98:2	2.1	66
			Ru(TFA)$_2$(PPh$_3$)$_2$	1	30 atm	THF	58:42	0.8	66
			Ru(TFA)$_2$(PPh$_3$)$_2$	1	75 atm	MeOH	99:1:0.9	2.5	66
			[Rh((*S*)-skewphos)]$^+$	1	30 atm	MeOH	93:7	28.3	67
			[Rh((*S*)-chiraphos)]$^+$	1	30 atm	MeOH	96:4	5.9	67
			[Rh((*R*)-binap)]$^+$	1	30 atm	MeOH	71:29	5.7	67

Table 21.11 Directive hydrogenation of acrylic acid derivatives.

	Substrate	Major diastereomer	Catalyst	mol%	P_{H_2}	Solvent	Diastereomeric ratio	TOF	Reference
6	MeO_2C–C(=CH_2)–CH(Me)–$NHCO_2Bu^t$	MeO_2C–CH(Me)–CH(Me)–$NHCO_2Bu^t$	Rh^+	5		CH_2Cl_2	99:1		65
			Rh^+	5	1 atm	MeOH	100:0		65
7	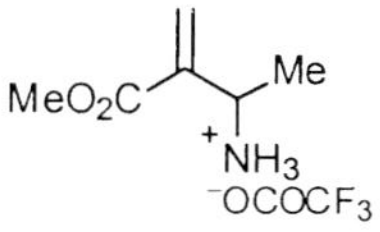 MeO_2C–C(=CH_2)–CH(Me)–$^+NH_3$ $^-OCOCF_3$	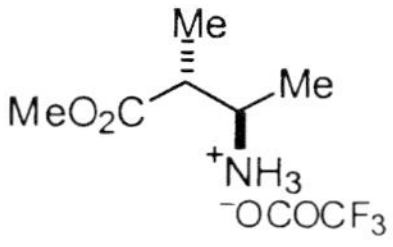 MeO_2C–CH(Me)–CH(Me)–$^+NH_3$ $^-OCOCF_3$	Rh^+	5		CH_2Cl_2	50:50		65
8	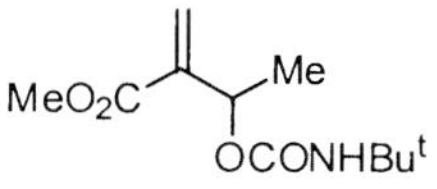 MeO_2C–C(=CH_2)–CH(Me)–$OCONHBu^t$	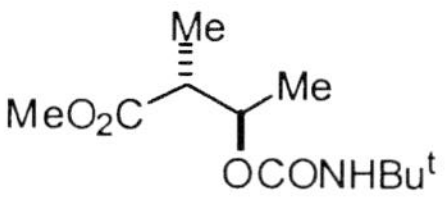 MeO_2C–CH(Me)–CH(Me)–$OCONHBu^t$	Rh^+	5		CH_2Cl_2	94:6		65

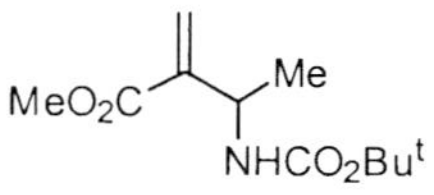

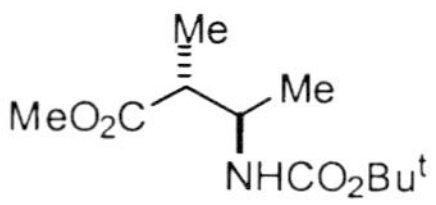

Rh^+ = $[Rh(diphos\text{-}4)(cod)]^+$

As described hitherto, diastereoselectivity is controlled by the stereogenic center present in the starting material (intramolecular chiral induction). If these chiral substrates are hydrogenated with a chiral catalyst, which exerts chiral induction intermolecularly, then the hydrogenation stereoselectivity will be controlled both by the substrate (substrate-controlled) and by the chiral catalyst (catalyst-controlled). On occasion, this will amplify the stereoselectivity, or suppress the selectivity, and is termed "double stereo-differentiation" or "double asymmetric induction" [68]. If the directions of substrate-control and catalyst-control are the same this is a matched pair, but if the directions of the two types of control are opposite then it is a mismatched pair.

H_2, Rh cat (5) [60]

cat	mol%	P_{H2}	solvent	diastereo. ratio
$RhCl(PPh_3)_3$	20	15 psi	CH_2Cl_2	83 : 15
$[Rh((+)\text{-binap})]^+$		1000 psi	CH_2Cl_2	98 : 2
$[Rh((-)\text{-binap})]^+$		1000 psi	CH_2Cl_2	67 : 33

H_2, Rh cat (6) [59]

cat	P_{H2}	solvent	diastereo. ratio
$[Rh((+)\text{-binap})]^+$	1000 psi	CH_2Cl_2	97 : 3
$[Rh((-)\text{-binap})]^+$	1000 psi	CH_2Cl_2	92 : 8

H_2, Ru cat (7) [52]

cat	mol%	P_{H2}	solvent	diastereo. ratio
$Ru(OAc)_2((R)\text{-tol-binap})$	0.2	1 atm	MeOH	99.9 : 0.1
$Ru(OAc)_2((S)\text{-tol-binap})$	0.2	1 atm	MeOH	22 : 78

Striking examples of this phenomenon are presented for allyl and homoallyl alcohols in Eqs. (5) to (7). The stereodirection in Eq. (5) is improved by a chiral (+)-binap catalyst and decreased by using the antipodal catalyst [60]. In contrast, in Eq. (6) both antipode catalysts induced almost the same stereodirection, indicating that the effect of catalyst-control is negligible when compared with the directivity exerted by the substrate [59]. In Eq. (7), the sense of asymmetric induction was inversed by using the antipode catalysts, where the directivity by chiral catalyst overrides the directivity of substrate [52]. In the case of chiral dehydroamino acids, where both double bond and amide coordinate to the metal, the effect of the stereogenic center of the substrate is negligibly small and diastereoface discrimination is unsuccessful with an achiral rhodium catalyst (see Table 21.1, entries 9 and 10) [9].

21.4
Hydrogenation of Dehydrooligopeptides

The hydrogenation of dehydrodipeptides and -tripeptides is a versatile method for the synthesis of oligopeptides of various compositions. In the homogeneous hydrogenation of dehydrodipeptide derivatives, coordination of the olefin and the amidocarbonyl oxygen to the metal is also anticipated (similar to the reduction of dehydroamino acid derivatives), and this presents the question of whether the reaction proceeds by substrate-control or by catalyst-control. In general, the chiral center in the dehydrodipeptide was found to have little influence on the stereoselectivity, and this small degree of substrate-control enables the synthesis of dipeptides or tripeptides having a desired configuration at will, with the newly forming chiral center being controlled by the external effect of the chiral catalysts [70, 71, 77, 78]. The structure of dehydrodipeptides also influences stereoselectivity, and dehydrodipeptides of the RCO-ΔAA-AA-OR′ type can be hydrogenated with high stereoselectivities, while those of the RCO-Gly-ΔAA-OR′ type could be converted with only moderate to good stereoselectivities. Ojima conducted the diastereoselective hydrogenation of various dehydrodipeptides and -tripeptides, and succeeded in preparing Leu-enkephalin analogues in high diastereoselectivity by coupling the dipeptide and tripeptide formed via hydrogenation using Rh-(*R*,*R*)-dipamp and Rh-Ph-CAPP catalysts [70, 73]. In these hydrogenations of dehydrodipeptides, the direction of asymmetric induction generally turned out to be the same as that observed in the asymmetric hydrogenation of (*Z*)-*α*-acylaminocinnamic acid. Kagan reported the diastereoselective synthesis of Leu-enkephalin analogues by the Rh-catalyzed hydrogenation of dehydroenkephalins Cbz-(O)Bn-(*S*)Tyr-(Gly)2-ΔPhe-(*S*)Leu-OMe **8** or Cbz-(O)Bn-ΔTyr-$(Gly)_2$-(*S*)Phe-(*S*)Leu-OMe **9**, with a diastereoselectivity up to 98:2 ((*S*,*S*,*S*):(*S*,*R*,*S*)) using $[Rh((R,R)\text{-dipamp})]^+$ [79] [Eqs. (8) and (9)]. The substitution of a glycyl residue by (*R*)alanyl in the above dehydropentapeptide had minimal effect on the stereochemical course of the hydrogenation (diastereoselectivity was 89:11).

BnO; Cbz-HN CONH CONH CONH CONH CO_2Me **8** → H_2, Rh cat. → Cbz-HN CONH CONH CONH CONH CO_2Me

$Rh((R,R)\text{-dipamp})^+$ 20 mol% 10 atm MeOH > 98 : 2 TOF = 0.1

(8) [79]

BnO; Cbz-HN CONH CONH CONH CONH CO_2Me **9** → H_2, Rh cat. → Cbz-HN CONH CONH CONH CONH CO_2Me

$Rh((R,R)\text{-dipamp})^+$ 20 mol% 10 atm MeOH > 93 : 7 TOF = 0.1

(9) [79]

In the enantioselective hydrogenation of dehydroamino acids, many diphosphine ligands are reported to give high enantioselectivity (often >95% ee). However, in the diastereoselective hydrogenation of dehydrodipeptides, many ligands – with very few exceptions – induce a somewhat lower stereoselectivity, especially in the case of mismatched pairs. Dipamp and Et-Duphos ligands retain their high chiral induction ability in the reduction of dehydrodipeptides [69, 70, 73, 76, 86, 87] (Table 21.12).

In the hydrogenation of dehydrodipeptides possessing a free carboxyl unit, the chiral ligands with an amine moiety would form an ion pair with the substrate which is expected to amplify the stereodifferentiation in the hydrogenation. Yamagishi realized high stereoinduction between 90 and 98% d.e. in the hydrogenation of dehydrodipeptides of the Ac-ΔPhe-AA-OH type in ethanol, using chiral diphosphinite ligands containing a dimethylamino moiety ((*S*,*S*)-POP-AEs) (Eq. (10)) [80]. In these systems, the chiral induction is governed by the chiral center of the substrate (substrate-control) and (*S*,*S*)- and (*R*,*R*)-products are formed highly selectively (Eq. (11)) [81].

Ac-HN CONH COOH **10** → H_2 (1 atm), $[Rh((S,S)\text{-MeO-POP-AE})]^+$, EtOH → Ac-HN CONH COOH
> 98 %de (*S*,*S*), TOF = 200

(10)

Ac-HN CONH COOH **11** → H_2 (1 atm), $[Rh((S,S)\text{-MeO-POP-AE})]^+$, EtOH → Ac-HN CONH COOH
86 %de (*R*,*R*), TOF = 2.8

Ar_2PO, Ar_2PO, N, NMe_2
(*S*,*S*)-MeO-POP-AE (Ar = $MeOC_6H_4$)

(11)

These diphosphinite ligands are not effective for the hydrogenation of Ac-ΔPhe-AA-OR-type substrates. This striking substrate-controlled behavior is also observed in the hydrogenation of RCO-ΔPhe-AA-OH-type substrates using an *achiral* diphos-3 ligand with a 2-dimethylaminoethyl unit at the 2-position (DPP-AE ligand). The $[Rh(I)(DPP\text{-}AE)]^+$ catalyst induced high diastereoselectivity (up to 96% d.e.) in the hydrogenation of Ac-ΔPhe-AA-OH in alcoholic solvents [82]. The kinetic parameters ($\Delta\Delta S^\ddagger$ and $\Delta\Delta H^\ddagger$) indicate that $\Delta\Delta G^\ddagger$ is governed by the $T\Delta\Delta S^\ddagger$ term, and not by the $\Delta\Delta H^\ddagger$ in the reactions where electrostatic interaction is possible. Moreover, the effect of solvent polarity and of the added amine on stereoselectivity also support the contribution of the attractive electrostatic interaction to the stereodifferentiation. NMR and circular dichroism spec-

Table 21.12 Diastereoselective hydrogenation of dehydrodipeptides.

	Substrate	Major diastereomer	Catalyst	mol%	P_{H_2}	Solvent	Diastereo-meric ratio	TOF	Reference
1	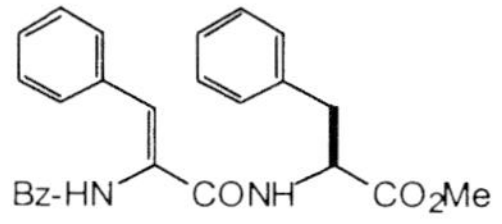 Bz-HN CONH CO_2Me	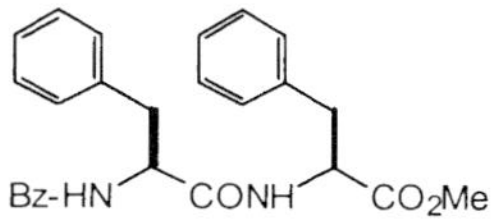 Bz-HN CONH CO_2Me	[Rh(diphos-4)]$^+$	1	1 atm	EtOH	62.2:37.8	17	69
			[Rh((*R*,*R*)-dipamp)]$^+$	1	10 atm	EtOH	97.8:2.2	6.7	70
			[Rh((+)-diop)]$^+$	1	5 atm	EtOH	83.6:16.4	6.7	70
			[Rh((–)-diop)]$^+$	1	5 atm	EtOH	84.1:15.9	6.7	70
			[Rh(Br-Ph-CAPP)]$^+$	1	1 atm	EtOH	0.8:99.2	33	69
2	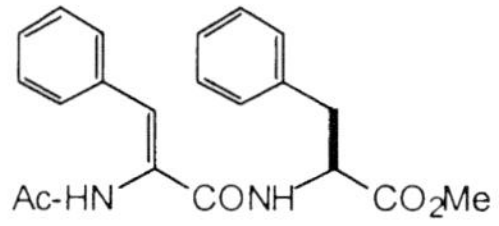 Ac-HN CONH CO_2Me	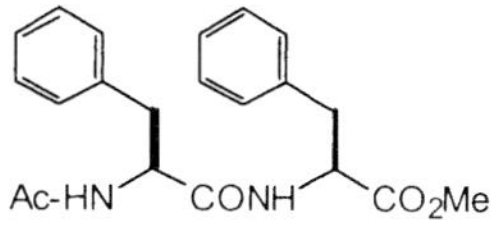 Ac-HN CONH CO_2Me	[Rh((+)-diop)]$^+$	1	5 atm	EtOH	92.7:7.3	4.2	70
			RhCl((+)-diop)	1	1 atm	MeOH	95:5		71
			RhCl((–)-diop)	1	1 atm	MeOH	10:90		71
			[Rh(Br-Ph-CAPP)]$^+$	1	5 atm	EtOH	2:98	4.2	70
3	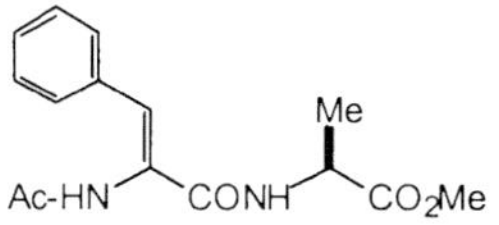 Ac-HN CONH Me CO_2Me	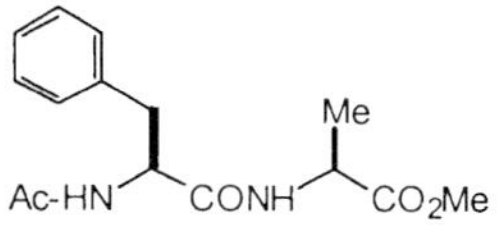 Ac-HN CONH Me CO_2Me	[Rh(Br-Ph-CAPP)]$^+$	1	5 atm	EtOH	1:99	5.5	70
			[Rh((+)-diop))$^+$	1	5 atm	EtOH	91.4:8.6	10.6	70
			[RhCl((+)-diop)]	3.5	1 atm	MeOH	85.6:14.4		72
			[RhCl((–)-diop)]	3.5	1 atm	MeOH	10.2:89.8		72
4	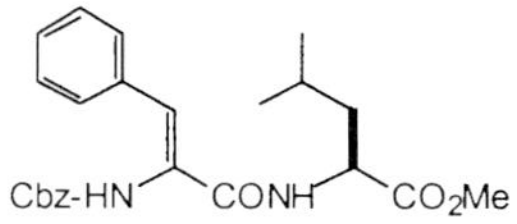 Cbz-HN CONH CO_2Me	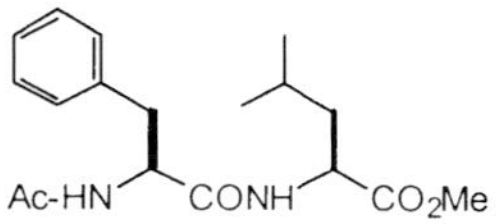 Ac-HN CONH CO_2Me	[Rh((*R*,*R*)-dipamp)]$^+$	2	20 atm	EtOH	97.8:2.2	2.5	73
5	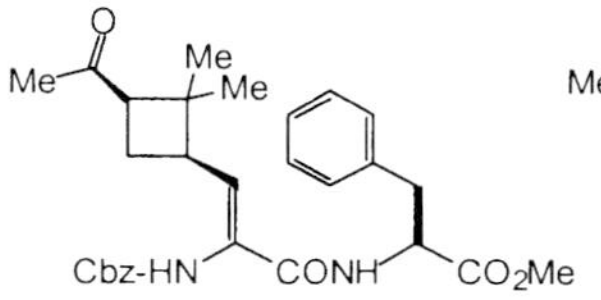 Me O Me Me Cbz-HN CONH CO_2Me	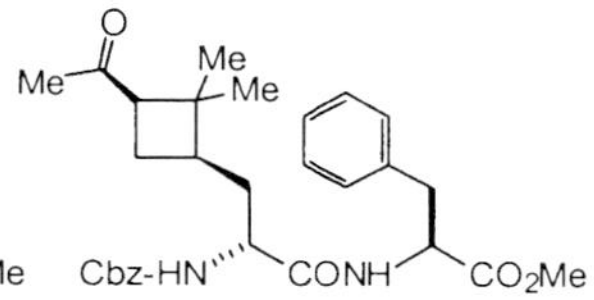 Me O Me Me Cbz-HN CONH CO_2Me	[Rh((*R*,*R*)-Et-Duphos))$^+$	4	2 atm	MeOH	>99.5:0.5	0.9	74
			[Rh((*S*,*S*)-Et-Duphos)]$^+$	4	2 atm	MeOH	5:95	0.4	74

Table 21.12 (continued)

	Substrate	Major diastereomer	Catalyst	mol%	P_{H_2}	Solvent	Diastereo-meric ratio	TOF	Reference
6	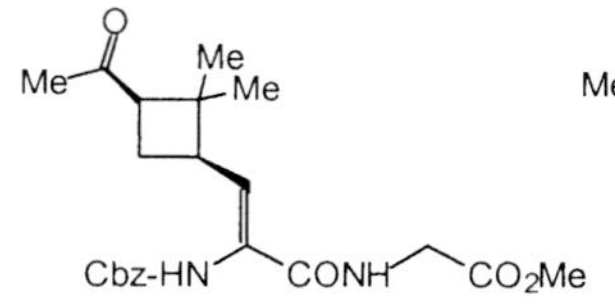 Me, O, Me, Me, Cbz-HN, CONH, CO_2Me	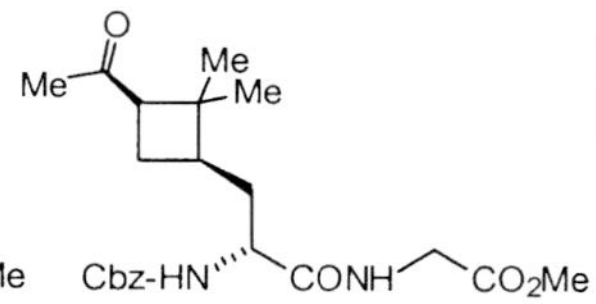 Me, O, Me, Me, Cbz-HN, CONH, CO_2Me	[Rh((*R,R*)-Et-Duphos)]$^+$	4	2 atm	MeOH	91:9	1	74
			[Rh((*S,S*)-Et-Duphos)]$^+$	4	2 atm	MeOH	11:89	0.9	74
7	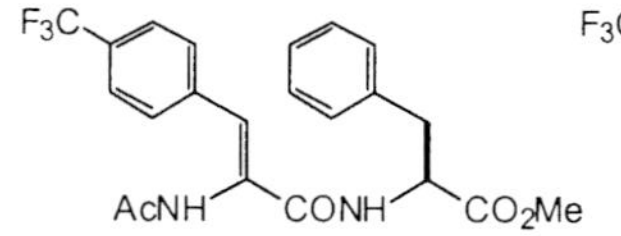 F_3C, AcNH, CONH, CO_2Me	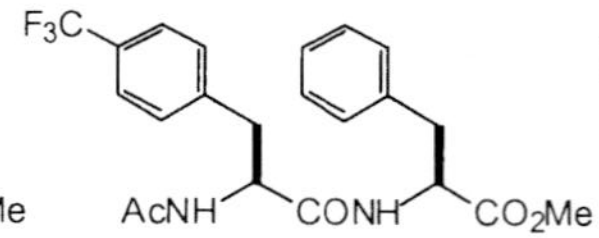 F_3C, AcNH, CONH, CO_2Me	[Rh((*S*)-Pindophos)]$^+$		0.1 MPa	MeOH	95.5:4.5		75
8	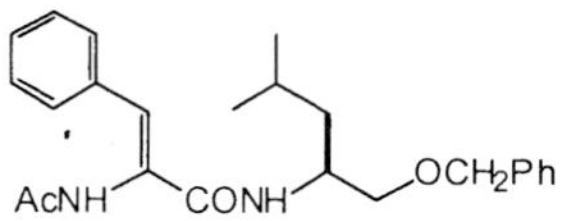 AcNH, CONH, OCH_2Ph	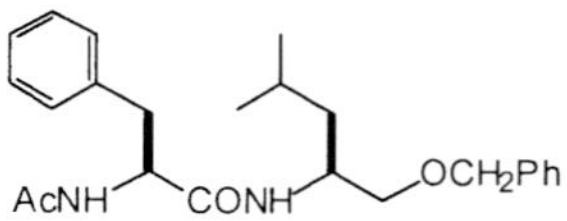 AcNH, CONH, OCH_2Ph	[Rh(diphos-4)]$^+$	0.33	5 atm	EtOH	58.7:41.3	7.6	76
			[Rh((+)-BPPM)]$^+$	1	5 atm		99.1:0.9	2.5	76
			[Rh(Ph-CAPP)]$^+$	1	5 atm		1.3:98.7	2.4	76
9	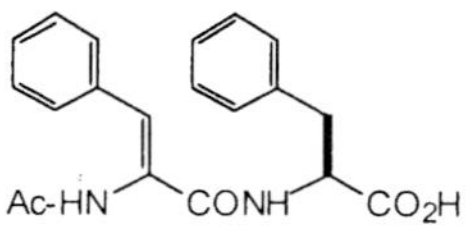 Ac-HN, CONH, CO_2H	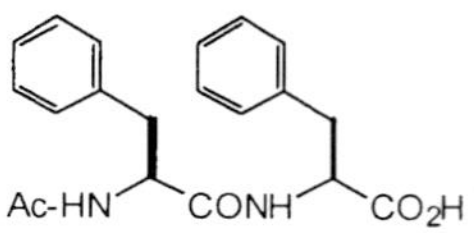 Ac-HN, CONH, CO_2H	[Rh(diphos-4)]$^+$	1	10 atm	EtOH	65.9:34.1	5.9	69
			[Rh((*R,R*)-dipamp)]$^+$	1	5 atm	EtOH	98.6:1.4	4.5	69
			[Rh((*R,R*)-bppm)]$^+$	1	10 atm	EtOH	99.4:0.6	4.9	69
			[Rh(DIOXOP)]$^+$	4	1 atm	EtOH	93:7	1	77
			[Rh((*S,S*)-Chiraphos)]$^+$	1	10 atm	EtOH	39.1:60.9	1	69
			[Rh((*S,S*)-MeO-POP-AE)]$^+$	2	1 atm	EtOH	>99:1	200	80
			[Rh(DPP-AE)]$^+$	2	1 atm	MeOH	97:3	100	82

Table 21.12 (continued)

	Substrate	Major diastereomer	Catalyst	mol%	P_{H_2}	Solvent	Diastereo-meric ratio	TOF	Reference
10	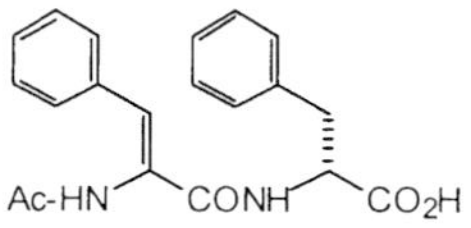 Ac-HN CONH CO_2H	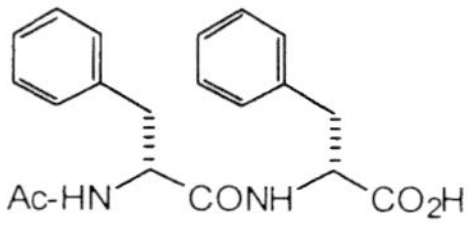 Ac-HN CONH CO_2H	$[Rh((S,S)\text{-MeO-POP-AE})]^+$	2	1 atm	EtOH	93:7	2.8	81
			$[Rh((S,S)\text{-POP-AE})]^+$	2	1 atm	EtOH	93:7	2.8	81
11	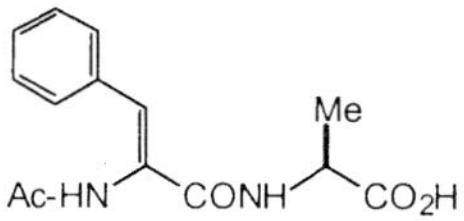 Ac-HN CONH Me CO_2H	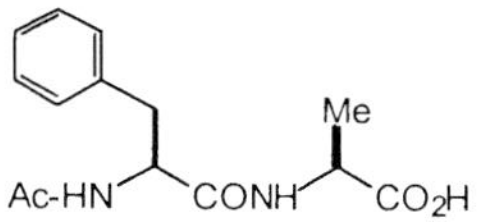 Ac-HN CONH Me CO_2H	$[Rh((S,S)\text{-MeO-POP-AE})]^+$	2	1 atm	EtOH	98:2	100	80
			$[Rh(DIOXOP)]^+$	4	1 atm	EtOH	86:14	1	77
			$[Rh(DPP\text{-AE})]^+$	2	1 atm	MeOH	94:6	63	82
			$[Rh((-)\text{-diop})]^+$	1	1 atm	EtOH/ C_6H_6	9:91		78
12	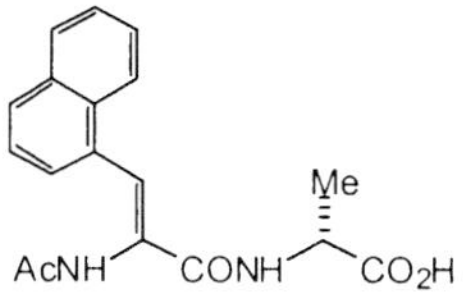 AcNH CONH Me CO_2H	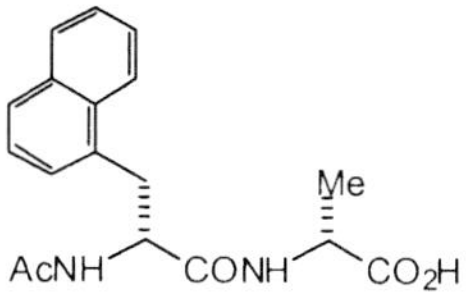 AcNH CONH Me CO_2H	$[Rh(DPP\text{-AE})]^+$	2	1 atm	MeOH	94.5:5.5	50	83
			$[Rh((-)\text{-bppm})]^+$	2	1 atm	MeOH	>99:1	50	83
13	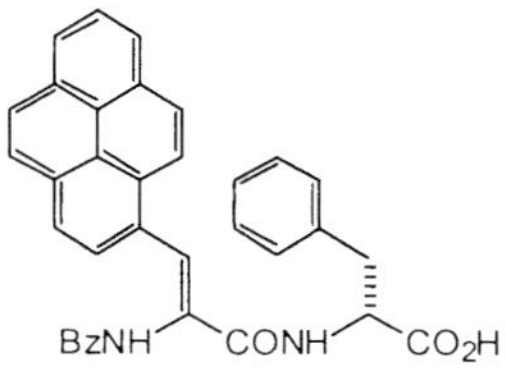 BzNH CONH CO_2H	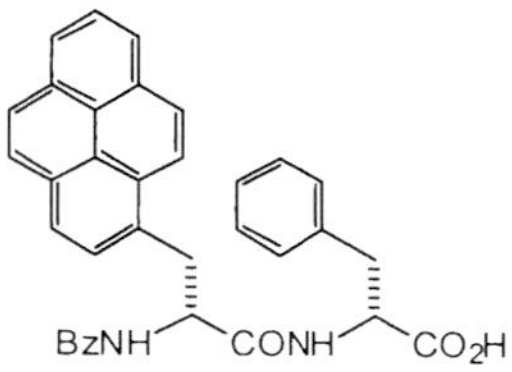 BzNH CONH CO_2H	$[Rh(diphos\text{-}3)]^+$	2	1 atm	MeOH	59:41	>1.7	83
			$[Rh(DPP\text{-AE})]^+$	2	1 atm	MeOH	94.5:5.5	150	83
			$[Rh((-)\text{-bppm})]^+$	2	1 atm	MeOH	94.5:5.5	5	83

troscopy suggest that the change of catalyst conformation occurs by electrostatic interaction between the (*S*)-substrate and the achiral ligand (induced fitting) to form a complex of the λ-conformation preferentially. Without electrostatic interaction, the rhodium complex exists as a 1:1 mixture of δ- and λ-conformations [82]. The effect of attractive electrostatic interactions on chiral induction was also reported by Hayashi, for the enantioselective hydrogenation of acrylic acid derivatives. A ferrocenyldiphosphine ligand having a dimethylaminoalkyl moiety induced a high enantioselectivity of more than 95% by utilizing electrostatic interactions with the substrate [84].

Kagan reported tandem asymmetric syntheses from achiral bisdehydrodipeptides by the sequential hydrogenation of two prochiral units [85]. With the $[Rh((R,R)\text{-dipamp})]^+$ catalyst, a high diastereoselectivity ratio of 98:2 ((*RR* and *SS*)/(*RS* and *SR*) ratio) was reported (Table 21.13, entries 1–3). In this case, the major diastereomer has a high ee-value (97.6% (*S*,*S*)), and the minor diastereomer has a negligible ee (15% (*S*,*R*)). This result indicates that, in each step of the reaction, the same stereoselectivity is realized. In the symmetrical bis(dehydroamino acid) derivatives, similar high diastereo-differentiation of 95–99% d.e. is realized using dipamp or duphos ligands, and in some cases the enantioselectivity of the major diastereomer reaches 100% (Table 21.13, entries 7 and 8) [86, 87]. Because of the high chiral induction ability of the catalyst, almost all of the minor monohydrogenated enantiomer is converted to the *meso*-product in the second hydrogenation step, and this results in an extremely high enantioselectivity of the major product.

Several attempts towards the asymmetric reduction of *N*-(α-ketoacyl)-α-amino acid derivatives by rhodium catalysts are reported to give chiral depsipeptide building blocks, *N*-(α-hydroxyacyl)-α-amino acid derivatives (Table 21.14). Using rhodium catalysts containing an electron-rich chiral diphosphine (Cydiop) or a chiral diphosphinite (Cy-POP-AE) ligand, the hydrogenation proceeds under atmospheric hydrogen pressure affording moderate to good diastereoselectivity (entries 2, 3 and 6) [89, 90], while the catalysts based on diop or bppm required high hydrogen pressure and only low selectivity is obtained [88].

21.5 Diastereoselective Hydrogenation of Keto-Compounds

For the hydrogenation of keto-compounds, ruthenium–phosphine catalysts are efficient in obtaining compounds with a stereogenic hydroxyl unit, though the reduction usually requires high hydrogen pressure. Ruthenium–diphosphine–diamine catalyst plus strong base, as was first reported by Noyori, functions well at lower pressures in the hydrogenation of keto-compounds [91, 92]. For enantio- and diastereoselective hydrogenation of keto-compounds, atropisomeric diphosphines such as binap [93], bichep [94], biphemp or biphep [95] are used effectively.

Table 21.13 Tandem asymmetric hydrogenation of didehydrodipeptides with two double bonds.

	Substrate	Major diastereomer	Catalyst	mol%	P_{H_2}	Solvent	Diastereo-meric ratio	%ee of major diastereo-mer	TOF	Refer-ence
1	AcNH CONH CO_2Me	AcNH CONH CO_2Me	[Rh((*R,R*)-di-pamp)]$^+$	3.5	1 atm	MeOH	98:2	99 (*S,S*)		85
			[Rh((*S,S*)-bppm)]$^+$	3.5	1 atm	MeOH	74.8:25.2	85 (*R,R*)		85
			[Rh(+)-(*R,R*)-diop)]$^+$	3.5	1 atm	MeOH	55:45	60 (*R,R*)		85
2	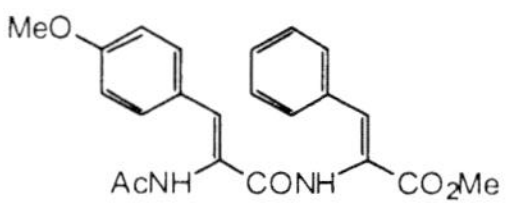 MeO AcNH CONH CO_2Me	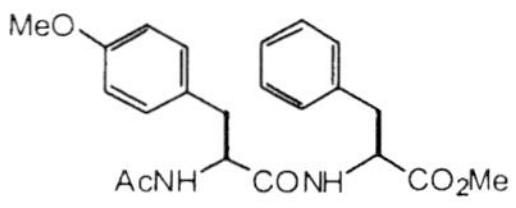 MeO AcNH CONH CO_2Me	[Rh((*R*)-di-pamp)]$^+$	3.5	1 atm	MeOH	98:2	97.6 (*S,S*)		85
3	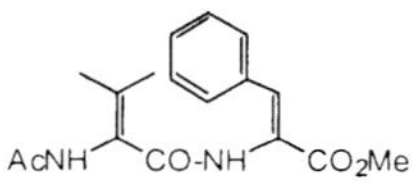 AcNH CO-NH CO_2Me	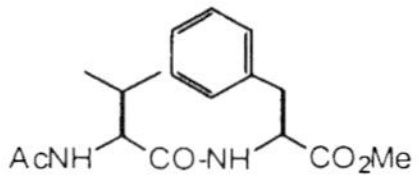 AcNH CO-NH CO_2Me	[Rh((*R,R*)-di-pamp)]$^+$	3.5	79 atm	MeOH	68:32	90.9 (*S,S*)	0.71	85
4	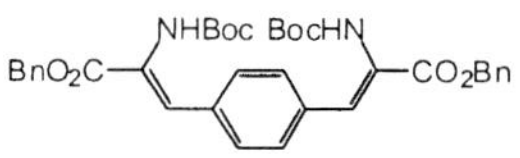 BnO_2C NHBoc BocHN CO_2Bn	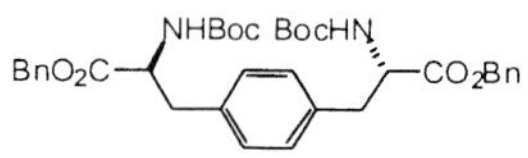 BnO_2C NHBoc BocHN CO_2Bn	[Rh(cod)((*R,R*)-dipamp)]$^+$	2	2.8 atm	MeOH	>99:1	>98 (*S,S*)	0.52	86
5	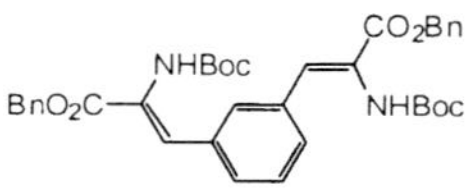 BnO_2C NHBoc CO_2Bn NHBoc	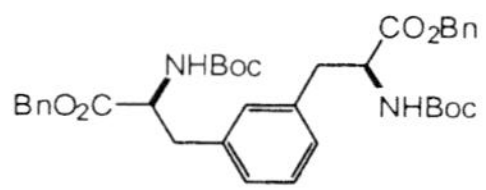 BnO_2C NHBoc CO_2Bn NHBoc	[Rh(cod)(*S,S*)-Me-Duphos)]$^+$	1.8	2.7 atm	MeOH	>99:1	>98 (*S,S*)	12.3	86

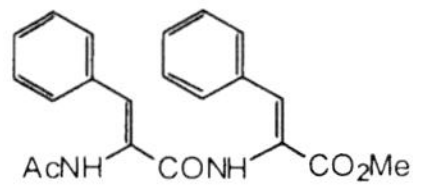

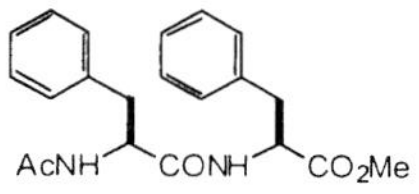

Table 21.13 (continued)

	Substrate	Major diastereomer	Catalyst	mol%	P_{H_2}	Solvent	Diastereomeric ratio	%ee of major diastereomer	TOF	Reference
6	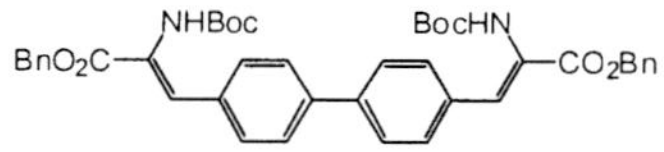	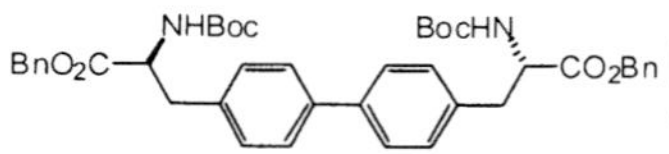	[Rh(cod)((*R,R*)-dipamp)]$^+$	3.6	3.5 atm	MeOH/THF	>99:1	>98 (*S,S*)	0.31	86
7	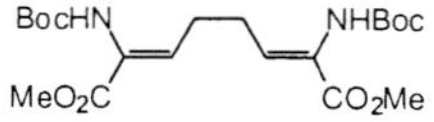	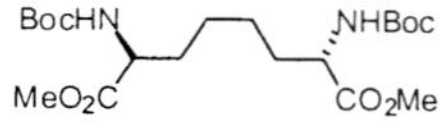	[Rh(cod)((*S,S*)-Et-Duphos)]$^+$	0.84	60 psi		99.5:0.5	100 (*S,S*)	6.6	87
			[Rh(cod)((*S,S*)-Me-Duphos)]$^+$		60 psi		98.5:1.5	100 (*S,S*)		87
			[Rh(cod)((*R,R*)-dipamp)]$^+$		60 psi		97.5:2.5	100 (*S,S*)		87
			$[RuCl_2(binap)]_2$ (TEA)		60 psi		85:15			87
			[Rh(cod)((*R,R*)-Chiraphos)]$^+$	0.84	60 psi		69.4:30.6	86 (*R,R*)	6.6	87
8	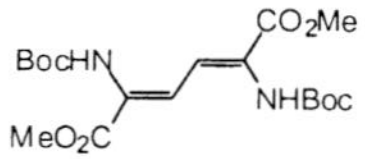	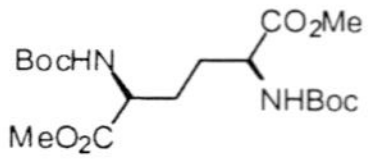	[Rh(cod)((*S,S*)-Et-Duphos)]$^+$	0.84	60 psi		98.5:1.5	100 (*S,S*)	6.6	87

Table 21.14 Diastereoselective hydrogenation of dehydrodepsipeptide.

	Substrate	Major diastereomer	Catalyst	mol%	P_{H_2}	Solvent	Diastereo-meric ratio	TOF	Reference
1	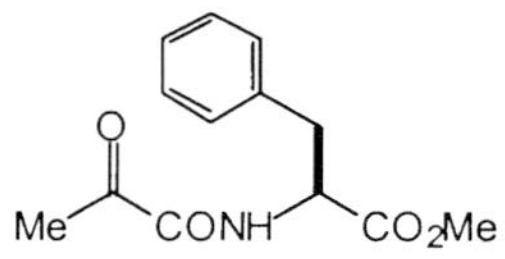 O, Me, CONH, CO_2Me	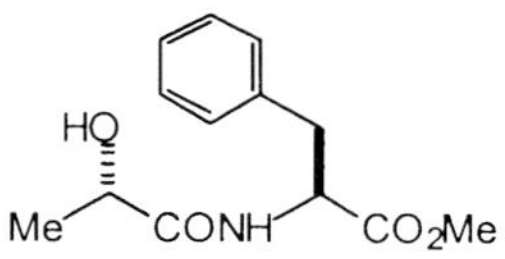 HO, Me, CONH, CO_2Me	$RhCl(PPh_3)_3$	1	50 atm	C_6H_6	60:40	5	88
			RhCl((+)-diop)	1	50 atm	C_6H_6	63:37	5	88
			RhCl((+)-bppm)	1	50 atm	C_6H_6	64:36	5	88
			RhCl((–)-cydiop)	5	1 atm	THF	73:27	0.8	89
			RhCl((+)-Cydiop)	5	1 atm	THF	26:74	0.7	89
2	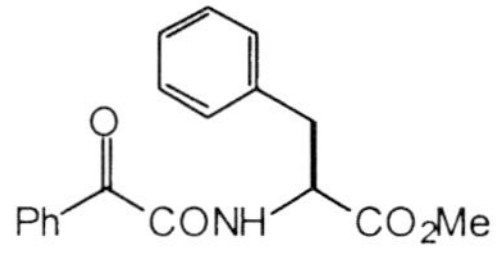 O, Ph, CONH, CO_2Me	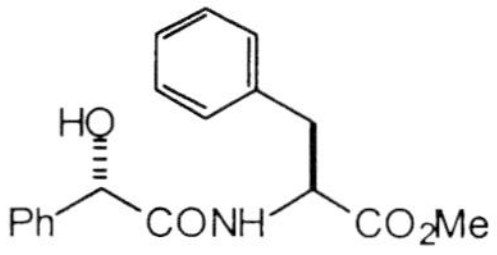 HO, Ph, CONH, CO_2Me	RhCl((–)-Cydiop)	5	1 atm	THF	86:14	1	89
			RhCl((+)-Cydiop)	5	1 atm	THF	16:84	1	89
3	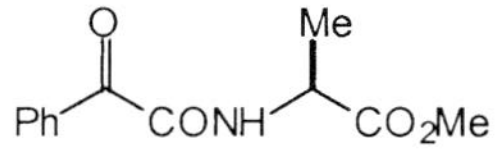 O, Me, Ph, CONH, CO_2Me	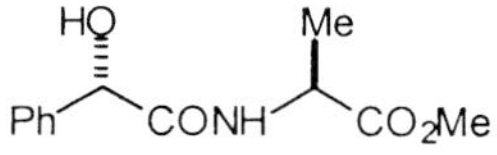 HO, Me, Ph, CONH, CO_2Me	RhCl((–)-Cydiop)	5	1 atm	THF	83:17	1	89
			RhCl((+)-Cydiop	5	1 atm	THF	18:82	1	89
			RhCl((*S*,*S*)-Cy-POP-AE)	2	1 atm	MeOH	33.5:66.5	2.1	90
4	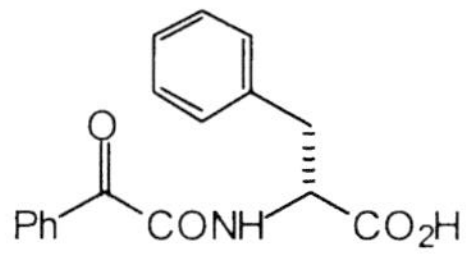 O, Ph, CONH, CO_2H	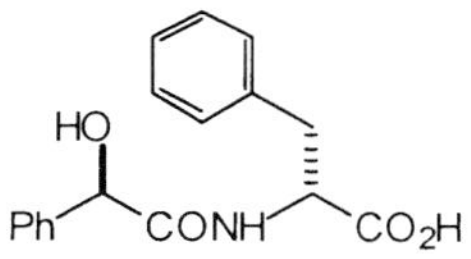 HO, Ph, CONH, CO_2H	RhCl((*S*,*S*)-Cy-POP-AE)	2	1 atm	MeOH	75:25	2.1	90
			$[Rh((S,S)\text{-Cy-POP-AE})]^+$	2	1 atm	MeOH	76.5:23.5	2.1	90

Table 21.14 (continued)

	Substrate	Major diastereomer	Catalyst	mol%	P_{H_2}	Solvent	Diastereomeric ratio	TOF	Reference
5	Ph–CO–CONH–CH(CH_2Ph)–CO_2H	Ph–CH(OH)–CONH–CH(CH_2Ph)–CO_2H	[Rh((*S*,*S*)-Cy-POP-AE)]$^+$	2	1 atm	MeOH	73.5:26.5	2.1	90
6	Ph–CO–CONH–CH(Me)–CO_2H	Ph–CH(OH)–CONH–CH(Me)–CO_2H	RhCl((*S*,*S*)-Cy-POP-AE)	2	1 atm	MeOH	83.5:16.5	2.1	90

21.5.1 Substrate-Directive Hydrogenation of Keto-Compounds

Several examples of substrate-directive reduction (hydroxyl, alkoxyl, carbamate or sulfoxide groups) have been reported in the hydrogenation of keto-compounds with ruthenium catalysts. In the reduction of β-keto ester derivatives, the γ- or δ-stereogenic center in the substrates significantly affects the degree of diastereoselection (Table 21.15). The hydrogenation of a β-keto ester with a carbamate unit at the γ-position ((*S*)-substrate) in the presence of Ru–(*R*)-binap (matched pair) exclusively afforded the *threo* product with the (3*S*,4*S*) configuration, whereas in the presence of Ru–(*S*)-binap (mismatched pair) the (3*R*,4*S*)-product was formed preferentially (Table 21.15, entry 1) [96]. The β-keto esters with a pyrrolidine unit showed similar behavior (entry 4) [97]. A silyloxy group at the δ-position could dictate the sense of asymmetric induction, and high diastereoselectivity is induced by using an achiral ruthenium catalyst, whereas the chiral ruthenium (*R*)-binap catalyst and the (*S*)-binap catalyst (matched and mismatched pairs, respectively) afford the same diastereomer, albeit in different selectivity (entry 7). On the other hand, the directivity of a hydroxy group was overwhelmed by the chirality of the catalyst, and a different diastereomer was formed preferentially by the antipode ruthenium catalysts (entry 8) [99].

In the hydrogenation of chiral sulfoxide **12**, the sulfoxide unit exerts strong stereodirectivity and, with axially stereogenic diphosphine ligands, high diastereoselectivity is realized whilst the antipode ligand affords lower diastereoselectivity (Scheme 21.5) [100]. The sense of asymmetric induction is rationalized by the favored conformation **E**, where the *p*-tolyl unit is disposed at a quasiequatorial position with a sulfoxide oxygen ligating to ruthenium [100].

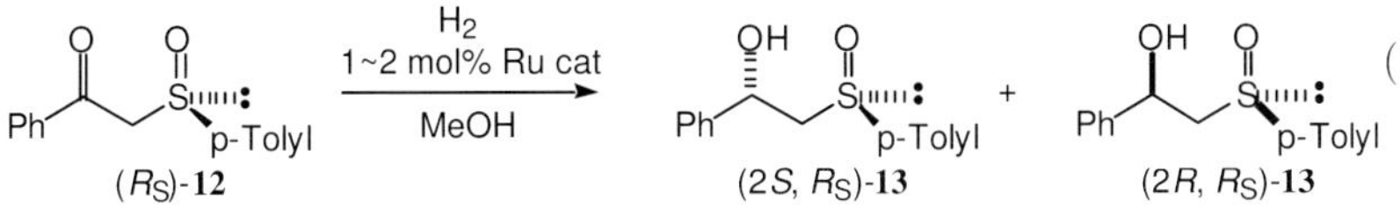

(12)

Catalyst	H_2(bar)	temp(ÞC)	time(h)	yield	(2*S*, R_S) : (2*R*, R_S)	TOF
$RuBr_2$(diphos-4)	50	50	24	12	78 : 22	0.25
$RuBr_2(PPh_3)_3$	50	20	64	99	80 : 20	1.6
$RuBr_2$((*S*)-MeO-biphep)	50	20	63	70	> 99 : 1	0.55
$RuBr_2$((*R*)-MeO-biphep)	50	20	63	95	10 : 90	0.75

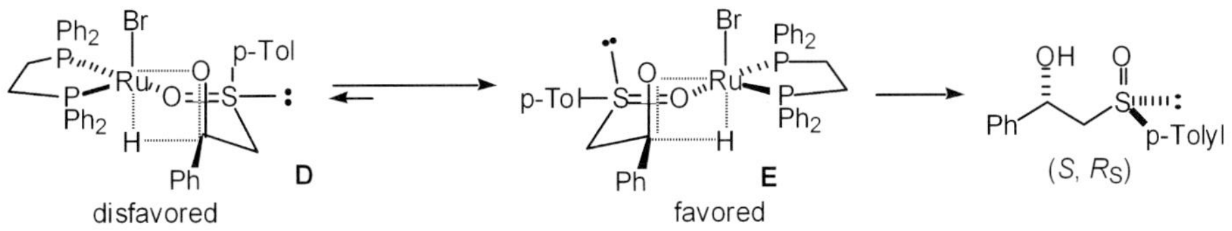

Scheme 21.5

Table 21.15 Hydroxy-directed hydrogenation of β-keto esters.

	Substrate	Major diastereomer	R	Catalyst	mol%	P_{H_2}	Solvent	Diastereomeric ratio	%ee (*syn*)	TOF	Reference
1	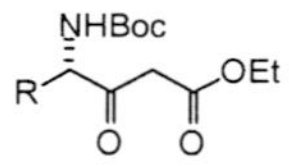	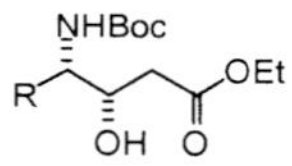	$PhCH_2$	$RuBr_2$((*R*)-binap)	0.2	100 atm	EtOH	>99:1	99	3.6	96
				$RuBr_2$((*S*)-binap)	0.2	100 atm	EtOH	9:91	>99		96
			CH_2CHMe_2	$RuBr_2$((*R*)-binap)	0.2	100 atm	EtOH	>99:1	97		96
			c-$C_6H_{11}CH_2$	$RuBr_2$((*R*)-binap)	0.2	100 atm	EtOH	>99:1	100		96
2	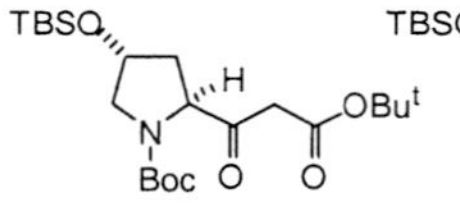	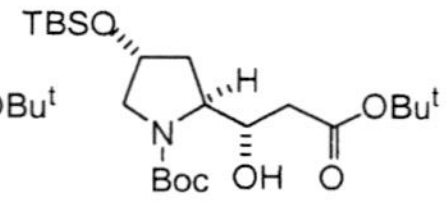		$RuCl_2$((*R*)-binap)1/2NEt_3	0.1	5 atm	THF/MeOH	96:4		19.2	97
				$RuCl_2$((*S*)-binap)1/2NEt_3	0.1	5 atm	THF/MeOH	ca 1:1		<2	97
3	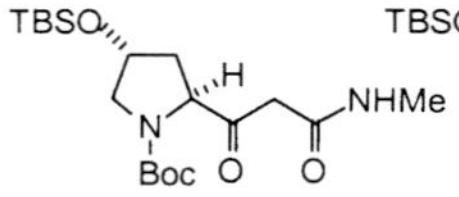	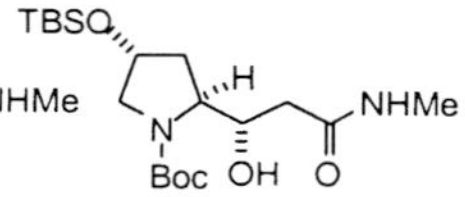		$RuCl_2$((*S*)-binap)1/2NEt_3	0.1	5 atm	THF/MeOH	34.5:65.5		32.9	97
4	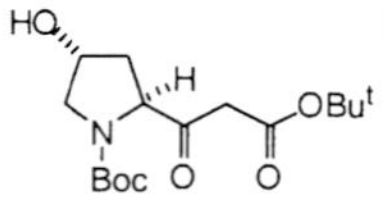	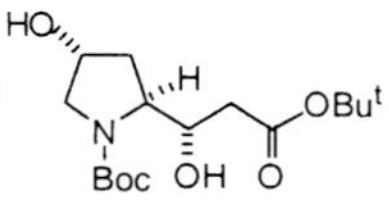		$RuCl_2$((*R*)-binap)Et_2NH	0.25	150 psi	MeOH 0.75 mol% HCl	>99:<1			97
				$RuCl_2$((*S*)-binap)Et_2NH	0.25	150 psi	MeOH 0.75 mol% HCl	12:88			97
5	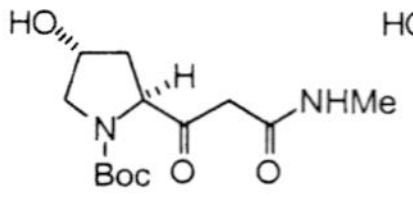	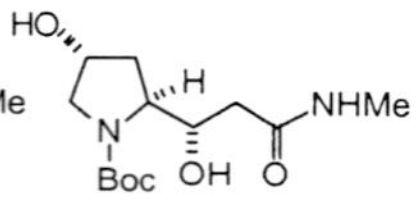		$RuCl_2$((*S*)-binap)1/2NEt_3	0.1	5 atm	THF/MeOH	23.5:76.5		5.5	98

Table 21.15 (continued)

	Substrate	Major diastereomer	R	Catalyst	mol%	P_{H_2}	Solvent	Diastereo-meric ratio	%ee (*syn*)	TOF	Refer-ence
6	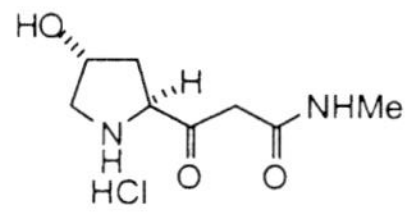	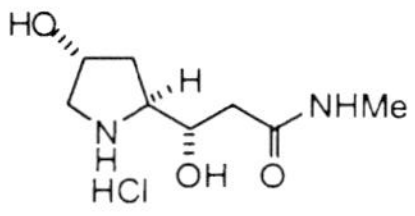		$RuCl_2$((*S*)-binap)1/2NEt_3	0.1	5 atm	THF/MeOH	1:99		35.4	98
7	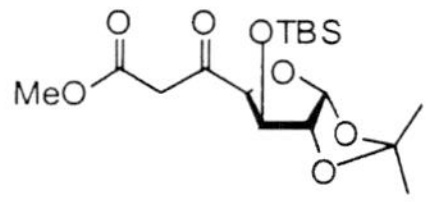	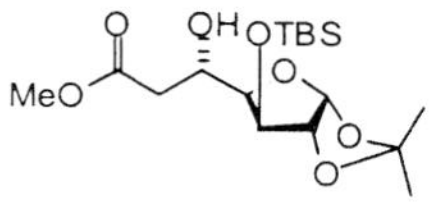		$RuBr_2$(diphos-2)	2	1 atm	MeOH	95:5		0.5	99
				$RuBr_2$((*R*)-binap)	2	1 atm	MeOH	>99:1		2.1	99
				$RuBr_2$((*S*)-binap)	2	1 atm	MeOH	82:18		1.5	99
8	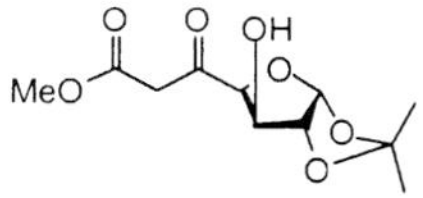	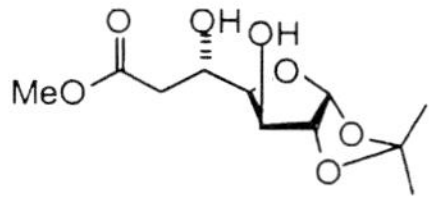		$RuBr_2$((*R*)-binap)	2	1 atm	MeOH	85:15		2.1	99
				$RuBr_2$((*S*)-binap)	2	1 atm	MeOH	10:90		1.0	99
9	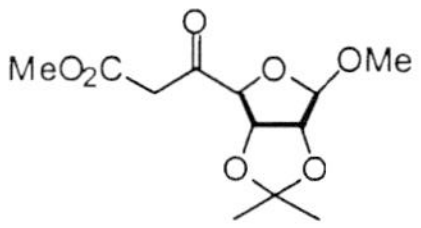	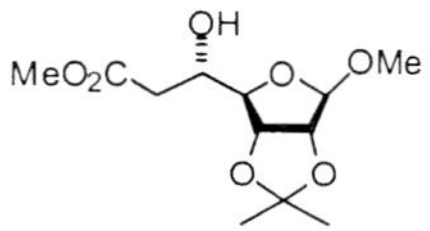									99
				$RuBr_2$((*R*)-binap)	2	1 atm	MeOH	90:10		2.1	99
				$RuBr_2$((*S*)-binap)	2	1 atm	MeOH	5:95		2.1	

21.5.2
Hydrogenation of Diketo Esters and Diketones

Reduction of *β,δ*-diketo esters using various atropisomeric diphosphine ligands afforded generally the *anti*-3,5-dihydroxy products in moderate to good diastereoselectivity, and in high enantioselectivity (Table 21.16) [101–104]. This suggests that stereocontrol in the hydrogenation of diketo esters is, in general, very similar to that found in the hydrogenation of diketones with Ru–binap (*vide infra*) [105, 106].

In these cases, using an (*S*)-axially chiral ligand, (3*R*)-3-hydroxy-5-oxoalkanoate is exclusively formed as monohydrogenation product, while the second hydrogenation step is a hydroxy-directed reduction by the Ru–catalyst bearing an (*S*)-axially chiral ligand to afford the (3*R*,5*S*)-dihydroxy product preferentially (Scheme 21.6) [103a]. With Ru–(*R*)-binap, this (3*R*)-3-hydroxy-5-oxoalkanoate is mainly converted to the (3*R*,5*R*)-*syn* dihydroxy compound. Formation of the 3-hydroxy-5-oxoalkanoate intermediate is also supported by the reduction experiment of (5*R*)-5-hydroxy-3-oxoalkanoate by Ru–(*R*)-binap to afford the (3*R*,5*R*)-*syn* diol as the major product [102]. There is a competitive ligation of functionalities to the Ru atom (Scheme 21.7). Because of the intervention of an enolic structure in **14** and the high final *anti*-selectivity, it is plausibly assumed that hydrogenation of the C-3 carbonyl unit of **14** arises mainly from a *β*-diketone chelated intermediate, which gives preferentially the (3*R*)-enantiomer using Ru–(*S*)-binap as catalyst. This is in contrast to the results with simple *β*-keto esters, which afford the (3*S*)-hydroxy product upon use of a Ru–(*S*)-axially chiral diphosphine catalyst [106, 107, 109].

In the consecutive hydrogenation of *β,δ*-diketo esters (Table 21.16), selection of the chiral ligand can determine the sense of diastereoselection, and the 3,5-*syn* dihydroxy product was formed predominantly upon use of a Ru–(*S*)-aminophosphinephosphinite-((*S*)-AMPP) catalyst, although the enantioselectivity of the *syn*-product is poor (Table 21.16, entry 7) [103a]. *Syn* 3,5-diol formation

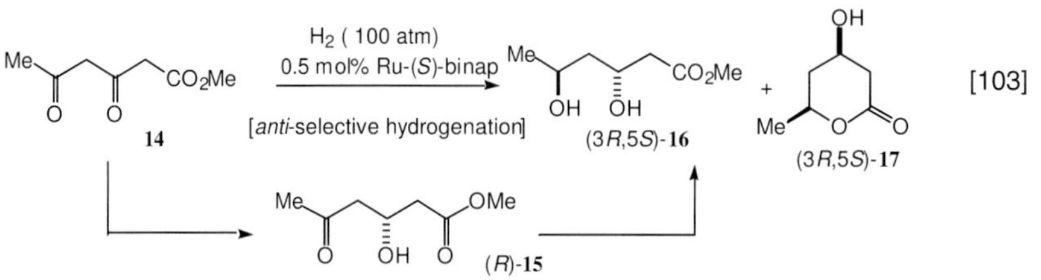

cat	mol%	P_{H2}	solvent	temp (ÞC)	**15** selec.	**16 + 17** selec.	*anti* : *syn*	%ee(*anti*)	TOF(h^{-1})
$RuBr_2$((*S*)-tol-binap)	0.5	100 atm	CH_2Cl_2	60	27	73	87 : 13	93 (3*R*,5*S*)	1.1
$RuBr_2$((*S*)-MeO-biphep)	0.5	100 atm	CH_2Cl_2	60	7	86	89 : 11	95 (3*R*,5*S*)	1.8
$RuBr_2$((*S*)-tol-binap)	0.5	100 atm	CH_2Cl_2	40	96	4	nd		5
$RuBr_2$((*S*)-MeO-biphep)	0.5	100 atm	CH_2Cl_2	40	96	4	nd		4.2

Scheme 21.6

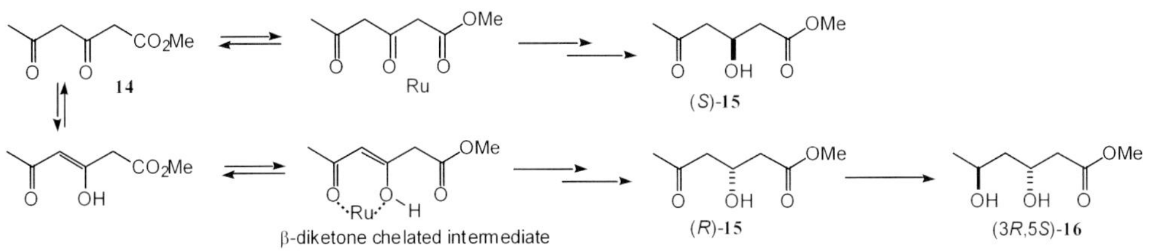

14 · Ru · (S)-15 · β-diketone chelated intermediate · (R)-15 · (3R,5S)-16

Scheme 21.7 Competitive chelation mode onto a Ru-(S)-binap type catalyst.

would lead to the synthesis of inhibitors of HMG-coenzyme A reductase [102]. In this case, an almost complete reversal of the diastereoselectivity was observed by changing the solvent from DCM to a polar solvent consisting of 1:1 DCM-methanol. In pure DCM, the Ru–(*S*)-AMPP catalyst induces the *syn*-rich diol (max. 92% d.e.), while in the DCM-methanol mixture the same catalyst leads to the formation of *anti*-rich product (72–84% d.e.). This reversal of diastereoselectivity was also observed in the reduction of *β*-diketones (Table 21.17, entry 13) [103 b].

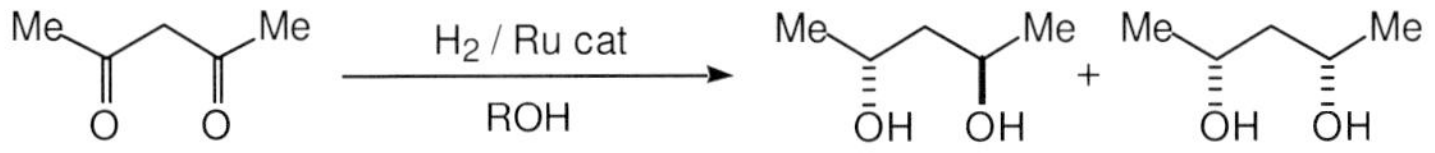

20 → (H_2 / Ru cat, ROH) (13)

cat	mol%	P_{H2}	*anti* : *syn*	ee%(*anti*-)	TOF	
Ru_2Cl_4((*R*)-binap)$_2$(NEt_3)	0.2	50 atm	99 : 1	> 99 (*R,R*)	25	[105]
$RuCl_2$((*R*)-binap)	0.05	72 atm	99 : 1	100 (*R,R*)	22.5	[106]
$RuBr_2$((*S*)-MeO-biphep)	2	20 atm	> 99.5 : 0.5	> 99 (*S,S*)	0.8	[110]

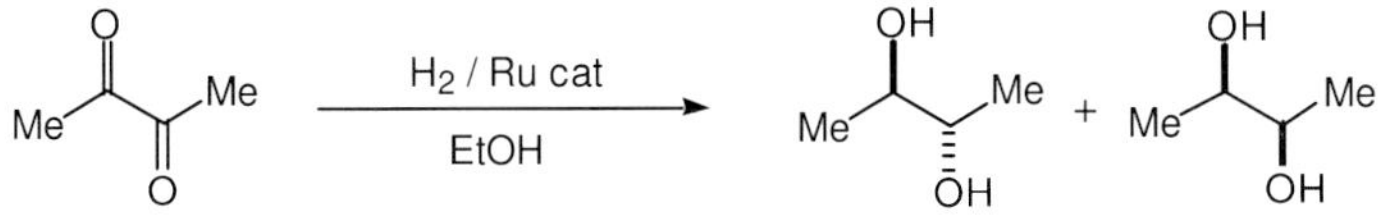

21 → (H_2 / Ru cat, EtOH) (14)

cat	mol%	P_{H2}	*anti* : *syn*	ee%(*anti*-)	TOF	
$RuBr_2$((*S*)-binap)	0.15	80 atm	26 : 74	92 (*S,S*)	11.1	[106]

In the hydrogenation of diketones by Ru–binap-type catalysts, the degree of *anti*-selectivity is different between *α*-diketones and *β*-diketones [Eqs (13) and (14)]. A variety of *β*-diketones are reduced by Ru-atropisomeric diphosphine catalysts to indicate admirable *anti*-selectivity, and the enantiopurity of the obtained *anti*-diol is almost 100% (Table 21.17) [105, 106, 110–112]. In this two-step consecutive hydrogenation of diketones, the overall stereochemical outcome is determined by both the efficiency of the chirality transfer by the catalyst (catalyst-control) and the structure of the initially formed hydroxyketones having a stereogenic center (substrate-control). The hydrogenation of monohydrogenated product ((*R*)-hydroxy ketone) with the antipode catalyst ((*S*)-binap catalyst) (mis-

Table 21.16 Diastereoselective hydrogenation of β,δ-diketo esters.

Substrate	Major diastereomer	Catalyst	mol%	P_{H_2}	Solvent	Diastereo-meric ratio anti : syn	%e.e. of major diastereomer	TOF	Reference
1		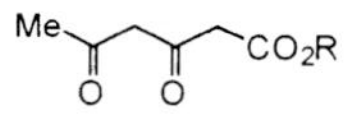 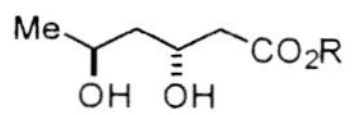 $RuCl_2(PPh_3)_3$	0.5	100 atm	CH_2Cl_2	73:27	0	2.6	101
		$[RuCl_2((S)$-binap$)]_2NEt_3$	0.5	100 atm	CH_2Cl_2	83:17	94 (3*R*,5*S*)	8.7	101
		$[RuCl_2((S)$-binap$)]_2NEt_3$	0.5	100 atm	MeOH	76:24	96 (3*R*,5*S*)	8.7	103
		$RuBr_2((S)$-Tolbinap)	0.5	100 atm	CH_2Cl_2	83:17	94 (3*R*,5*S*)	9.5	101
		$RuBr_2((S)$-MeO-biphep)	0.5	100 atm	CH_2Cl_2	84:16	98 (3*R*,5*S*)	5.1	101
		$[RuCl_2((S)$-binap$)]_2NEt_3$	0.1	100 atm	MeOH	81:19	78 (3*R*,5*S*)	10.4	102
		$[RuCl_2((R)$-binap$)]_2$-NEt_3	0.1	100 atm	MeOH	80:20	77 (3*S*,5*R*)	10.4	102
2		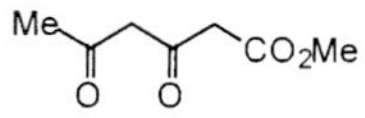 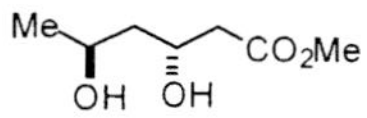 RhCl((*S*)-MeO-biphep)	0.5	50 atm	toluene	54:$6	79 (3*R*,5*S*)	0.26	101
		RhCl((*S,S*)-bppm)	0.5	50 atm	toluene	54:46	58 (3*R*,5*S*)	0.78	101
		RhCl((*S*)-Cy,Cy-oxoProNOP)	0.5	50 atm	toluene	65:35	80 (3*R*,5*S*)	0.43	101
		RhCl((*R,R*)-Me-Duphos)	0.5	50 atm	toluene	46:54		0.07	101
3		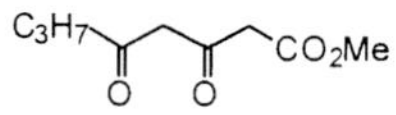 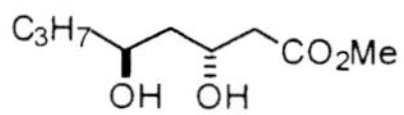 $[RuCl_2((S)$-binap$)]_2(NEt_3)$	0.1	100 atm	MeOH	76:24	78 (3*R*,5*S*)	10.4	102
		$[RuCl_2((R)$-binap$)]_2(NEt_3)$	0.1	100 atm	MeOH	78:22	77 (3*S*,5*R*)	10.4	102
4		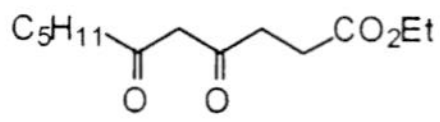 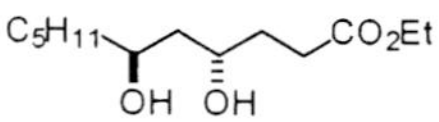 [RuCl((*S*)-binap)(p-cymene)]Cl		40 atm	MeOH	95:5 ~99:1	93 ~ 95 (4*S*,6*S*)		104

Table 21.16 (continued)

	Substrate	Major diastereomer		Catalyst	mol%	P_{H_2}	Solvent	Diastereo-meric ratio	%ee of major diastereomer *anti*:*syn*	TOF	Reference
			15 ee (%)								
5	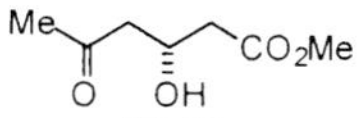 (*R*)-**15**	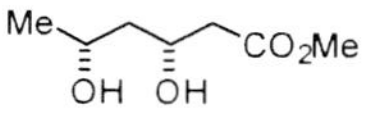	66 (*R*)	$Ru(TFA)_2((R)$-binap)	2	100 atm	CH_2Cl_2	10:90	75 (3*R*,5*R*)	2.4	103
			67 (*R*)	$Ru(TFA)_2((R)$-toalbinap)	2	100 atm	CH_2Cl_2	19:81	67 (3*R*,5*R*)	2.4	103
			67 (*R*)	$Ru(TFA)_2((R)$-MeO-biphep)	2	100 atm	CH_2Cl_2	14:86	72 (3*R*,5*R*)	2.2	103
6	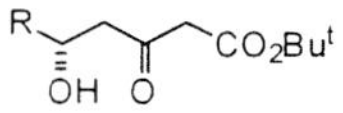 (*R*)-**18**	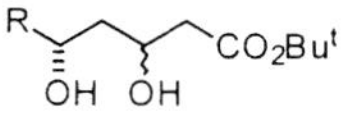 (*S*,*R*) or (*R*,*R*)	R = Me	$Ru_2Cl_4((S)$-binap$)_2(Et_3)$	0.2	100 atm	MeOH	95:5			102
			R = C_3H_7	$Ru_2Cl_4((R)$-binap$)_2(Et_3)$	0.2	100 atm	MeOH	40:60			102
				$Ru_2Cl_4((S)$-binap$)_2(Et_3)$	0.2	100 atm	MeOH	78:22			102
				$Ru_2Cl_4((R)$-binap$)_2(Et_3)$	0.2	100 atm	MeOH	43:57			102

Table 21.16 (continued)

	Substrate	Major diastereomer	Catalyst	mol%	P_{H_2}	Solvent	Diastereomeric ratio	%ee of major diastereomer *anti*:*syn*	TOF	Reference
7	Me, O, O, CO_2Me	Me, OH, OH, CO_2Me	Ru((*S*)-AMPP)(methallyl)$_2$	0.5	100 atm	CH_2Cl_2	13:87	14 (3*R*,5*R*)	0.85	103
			Ru((*S*)-AMPP)(TFA)$_2$	0.5		CH_2Cl_2	28:72	40 (3*R*,5*R*)	1.4	103
			Ru((*S*)-AMPP)((*R*)-MTPA)$_2$	0.5		CH_2Cl_2	4:96	<5 (3*R*,5*R*)	11.1	103
			Ru((*S*)-AMPP)((*S*)-MTPA)$_2$	0.5		CH_2Cl_2	8:92	5 (3*R*,5*R*)	2.9	103
			Ru((*S*)-AMPP)(TFA)$_2$	0.5		CH_2Cl_2/MeOH (1/1)	86:14	12 (3*S*,5*R*)	1.9	103
			Ru((*S*)-AMPP)((*R*)-MTPA)$_2$	0.5		CH_2Cl_2/MeOH (1/1)	92:8	5 (3*S*,5*R*)	0.85	103
			Ph, O-P-Ph, Ru(RCO_2)$_2$, N-P, Ph, Ph, O, cat 19							

Table 21.17 Consecutive hydrogenation of diketones.

	Substrate	Diastereome	Catalyst	mol%	P_{H_2}	Solvent	Temp. [°C]	Diastereo-meric ratio *anti:syn*	%ee *anti*	TOF	Refer-ence
1	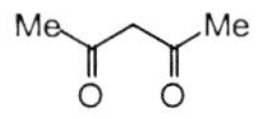	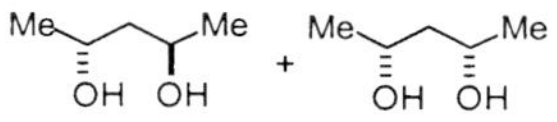	$RuCl_2((R)$-binap)	0.05	72 atm	EtOH		99:1	>99 (*R,R*)	22.5	106
2	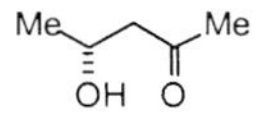 (*R*)-11	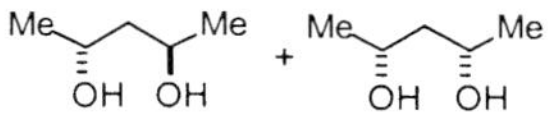	$RuCl_2((S)$-binap)	0.05	72 atm	EtOH		15:85			106
	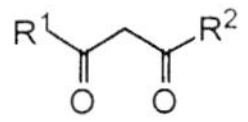	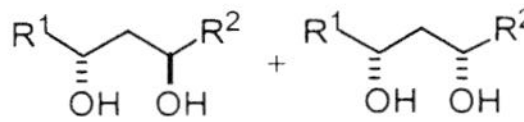 R^1 R^2									
3		Me Me	$Ru_2Cl_4((R)$-binap$)_2(NEt_3)$	0.2	50 atm	MeOH		99:1	>99 (*R,R*)	25	105
4		Me Et				MeOH		94:6	94 (*R,R*)	22	105
5		Me *i*Pr				MeOH		97:3	98 (*S,R*)	23	105
6		Me *i*Bu				MeOH		91:9	98 (*S,R*)	21	105
7		Et Et				MeOH		98:2	96 (*R,R*)	23	105
8		Me Me	$RuBr_2((R,R)$-Me-duphos)	2	70 atm	MeOH	80	97:3	93 (*R,R*)	0.8	110
9		C_6H_{11} C_6H_{11}	$RuBr_2((R)$-MeO-biphep)	2	100 atm	MeOH	r.t.	>99.5:0.5	>99 (*S,S*)	2.1	110
10		Bn Bn		2	30 atm	MeOH	r.t.	>97.5:2.5	>95 (*R,R*)	1.25	110
11	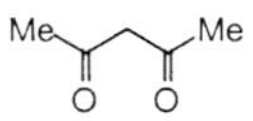		$RuBr_2((S)$-MeO-biphep)	2	20	MeOH	r.t.	>99.5:0.5	>99 (*S,S*)	0.8	110
			$RuCl_2(PPh_3)((S)$-biphep)	0.05	100 atm	MeOH	50	94:6	>99 (*S,S*)	133	111
					100 atm	EtOH	50	99.4:0.6	>99 (*S,S*)	83.3	112

Table 21.17 (continued)

	Substrate	Diastereome	Catalyst	mol%	P_{H_2}	Solvent	Temp. [°C]	Diastereo-meric ratio *anti*:*syn*	%ee *anti*	TOF	Reference
12	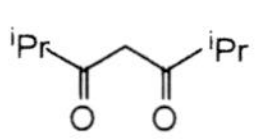	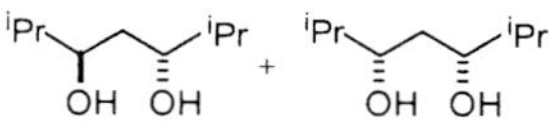	$RuBr_2$((*S*)-MeO-bi-phep)	2	20 atm	MeOH	r.t.	>99.5:0.5	>99 (*R*,*R*)	0.8	110
13	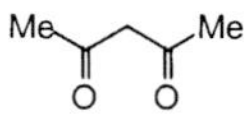	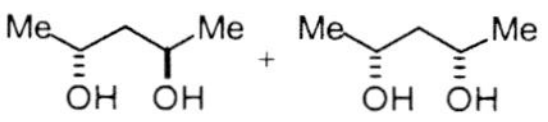 (*R*,*R*)-*anti* + *meso*	**19a** (R = CF_3)	0.5	100 atm	CH_2Cl_2		40:60	93 (*R*,*R*)	8.3	103
				0.5	100 atm	CH_2Cl_2/MeOH (1/1)		88:12	20 (*R*,*R*)	8.3	103
			19b (R = (*R*)-PhC(CF_3)(OMe))	0.5	100 atm	CH_2Cl_2		8:92	86 (*R*,*R*)	2	103
			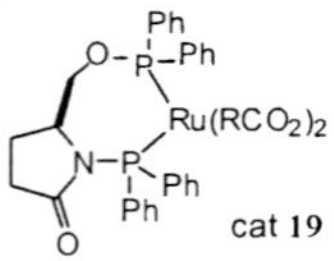	0.5	100 atm	CH_2Cl_2/MeOH (1/1)		92:8	14 (*R*,*R*)	3.5	103
			cat 19								
14	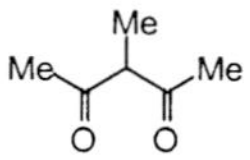	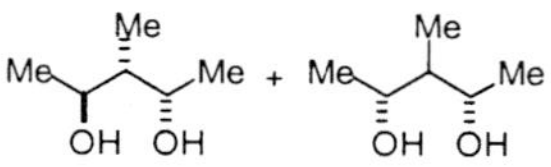	$RuCl_2$((*S*)-(binap)	0.05	94 atm	EtOH		99:1	>99 (*R*,*R*)	35.5	106

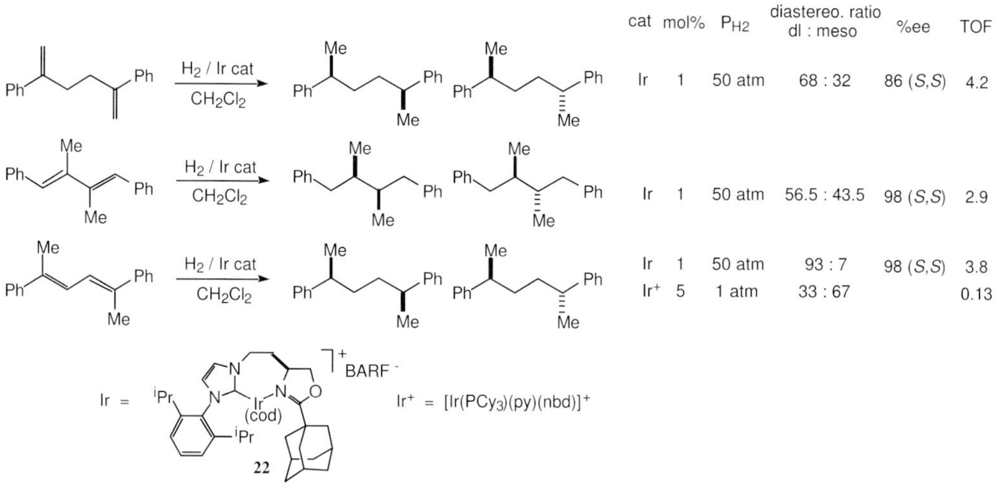

cat	mol%	P_{H2}	diastereo. ratio dl : meso	%ee	TOF
Ir	1	50 atm	68 : 32	86 (*S*,*S*)	4.2
Ir	1	50 atm	56.5 : 43.5	98 (*S*,*S*)	2.9
Ir	1	50 atm	93 : 7	98 (*S*,*S*)	3.8
Ir^+	5	1 atm	33 : 67		0.13

$Ir^+ = [Ir(PCy_3)(py)(nbd)]^+$

Scheme 21.8 Consecutive hydrogenation of symmetrical dienes [111].

matched pair) affords a *meso*-diol exclusively (Table 21.17, entry 1). This indicates that the catalyst-control in the second step is much more dominant over the substrate-control favoring *anti*-diol formation; thus, the high enantiomeric purity of the *anti*-diol is the result of a double stereodifferentiation [68]. In the reaction of α-diketones, substrate-control in the second hydrogenation step favors *meso*-diol formation, while minor *anti*-diol is obtained with high enantiomeric purities [Eq. (14)] [106]. This *dl*- and *meso*-products formation is also observed in the hydrogenation of symmetrical dienes [113] by an iridium carbene catalyst **22** (Scheme 21.8), or in the symmetrical bis(dehydroamino acids) by rhodium diphosphine catalysts (see Table 21.13, entries 4–8) [86, 87].

21.6
Kinetic Resolution to Selectively Afford Diastereomers and Enantiomers

In the hydrogenation of a compound with a stereogenic center by a chiral catalyst, the two possible stereocombinations – the matched pair and the mismatched pair – often afford different degrees of stereoinduction. The reaction rate of the hydrogenation is different for these two combinations (k_R and k_S, $k_R \neq k_S$), and kinetic chemical resolution is possible starting from racemic substrate using chiral catalyst by controlling the chemical conversion [114]. Kinetic resolution is now recognized as a viable tool for obtaining certain optically active compounds. In the homogeneous hydrogenation of unsaturated alcohols with a stereogenic center, various chiral catalysts were applied for their kinetic resolution. Using Rh–(*R*,*R*)-dipamp, racemic methyl (α-hydroxyethyl)acrylate **23** was resolved in THF at 0 °C to afford the (*S*)-substrate in 93% ee and the *anti*-product as the major product with a $k_R:k_S$ ratio of 6.5:1 at 75% conversion [Eq.

(15)] [45]. With an increase of conversion, the enantiopurity of unreacted (*S*)-substrate increases and the diastereoselectivity of the product decreases. Using Ru–((*S*)-binap)$(OAc)_2$, unreacted (*S*)-substrate was obtained in more than 99% ee and a 49:1 mixture of *anti*-product (37% ee (2*R*,3*R*)) at 76% conversion with a higher $k_R : k_S$ ratio of 16:1 [46]. In the case of a racemic cyclic allyl alcohol **24**, high enantiopurity of the unreacted alcohol was obtained using Ru–binap catalyst with a high $k_R : k_S$ ratio of more than 70:1 [Eq. (16)] [46]. In these two cases, the transition state structure is considered to be different since the sense of diastereoface selection with the (*S*)- or the (*R*)-catalysts is opposite if a similar OH/C=C bond spatial relationship is assumed.

rac. **23** → (H_2 / chiral cat, kinetic resolution) → (2*R*,3*R*) major diastereomer + (*S*) (15)

cat.	mol%	P_{H2}	solvent	temp (°C)	conv.	diastereo. ratio	recov. **23**	$k_R : k_S$	
$[Rh(dipamp)]^+$		1 atm	THF	0	75%		93 (*S*)	6.5 : 1	[45]
$[Ru(OAc)_2$((*S*))-binap)]	0.06	4 atm	MeOH	25	76%	98 : 2	> 99 (*S*)	16 : 1	[46]
$[Ru(OAc)_2$((*S*))-binap)]	0.1	50 atm	MeOH	20	63%	96.9 : 3.1	97 (*S*)	14 : 1	

rac-**24** → (H_2 / Ru cat, kinetic resolution) → (1*R*,3*R*) major diastereomer + (*S*) (16)

cat.	mol%	P_{H2}	solvent	temp (°C)	conv.	diastereo. ratio	recov. **24**	$k_R : k_S$	
$[Ru(OAc)_2$((*R*))-binap)]	0.17	4 atm	MeOH	26	54%	300 : 1	> 99 (*S*)	74 : 1	[46]
$[Ru(OAc)_2$((*R*))-binap)]	0.055	100 atm	MeOH	26	51%	98.5 : 1.5	95 (*S*)	76 : 1	

In the case of (*α*-acylaminoethyl)acrylate or (*α*-carbamoylethyl)acrylate, amido or carbamate functional groups work well in the direction of the kinetic resolution (Table 21.18). Reduction proceeded rapidly but then slowed markedly after consumption of 55–60% of the theoretical amount of H_2, indicating the large difference between the values of k_R and k_S (entries 1–3) [115]. The degree of enantiomer differentiation is considerably influenced by the hydrogen pressure, and higher k_f/k_s values were obtained at lower pressure with many substrates. The high selectivity should be noted, since binding of the substrate to rhodium through the olefin and amido units gives a fairly flexible chelate complex. In contrast, the reduction of (*α*-methoxyethyl)acrylate by the rhodium catalyst indicated a poor kinetic resolution (entry 4), in contrast to the effective OMe-directed hydrogenation with the iridium catalyst (see Table 21.3, entries 11 and 12) [25].

Table 21.18 Kinetic resolution of acrylic acid derivatives.

	Substrate	Major product	Recovered substrate	Catalyst	mol%	P_{H_2}	Solvent	Conversion [%]	Diastereomeric ratio	%ee recov.	k_f/k_s	TOF	Reference
1	CH_2, MeO_2C, Me, $NHCO_2Bu^t$	Me, MeO_2C, Me, $NHCO_2Bu^t$	CH_2, MeO_2C, Me, $NHCO_2Bu^t$	$[Rh(nbd)(R,R)$-dipamp$)]^+$	4	1 atm	MeOH	56	highly sel.	87 (*S*)	15	14	115
2	CH_2, MeO_2C, Me, $NHCO_2Bu^t$	Me, MeO_2C, Me, $NHCO_2Bu^t$	CH_2, MeO_2C, Me, $NHCO_2Bu^t$	$[Rh(nbd)(R,R)$-dipamp$)]^+$	4	1 atm	MeOH	60	highly sel.	98 (*S*)	22	10	115
3	CH_2, MeO_2C, Me, NHCOMe	Me, MeO_2C, Me, NHCOMe	CH_2, MeO_2C, Me, NHCOMe	$[Rh(nbd)(R,R)$-dipamp$)]^+$	4	1 atm	MeOH	56	highly sel.	96 (*S*)	21	17.4	115
4	CH_2, MeO_2C, Me, OMe	Me, MeO_2C, Me, OMe	CH_2, MeO_2C, Me, OMe	$[Rh(nbd)(R,R)$-dipamp$)]^+$	4	1 atm	MeOH	50	80:20	16 (*S*)	1.5	0.3	115
5	MeO_2C, CO_2Me, R	Me, MeO_2C, CO_2Me, R	MeO_2C, CO_2Me, R										
			R = Et	$[Rh((nbd)(R,R)$-dipamp$)]^+$	2	1 atm	MeOH	52.7		81 (*S*)	16		64
			R = Ph		2								
			R = OMe	$[Rh((nbd)(R,R)$-dipamp$)]^+$	2	1 atm	MeOH	62.3		82 (*S*)	7.2		64
				$[Rh((nbd)(R,R)$-dipamp$)]^+$		1 atm	MeOH	62.2		93 (*S*)	11.5		64

In the hydrogenation of 3-substituted itaconate ester derivatives by rhodium–dipamp, the alkoxycarbonyl group at the stereogenic center also exerts a powerful directing effect, comparable to that induced by OH in the kinetic resolution of (α-hydroxyethyl)acrylate, leading to a high enantiomer-discriminating ability up to $k_R:k_S=16:1$ (Table 21.18, entry 5) [64].

21.7
Kinetic Resolution of Keto- and Imino-Compounds

Kinetic resolution results of ketone and imine derivatives are indicated in Table 21.19. In the kinetic resolution of cyclic ketones or keto esters, ruthenium atropisomeric diphosphine catalysts **25** induced high enantiomer-discriminating ability, and high enantiopurity is realized at near 50% conversion [116, 117]. In the case of a bicyclic keto ester, the presence of hydrogen chloride in methanol served to raise the enantiomer-discriminating ability of the Ru–binap catalyst (entry 1) [116].

Racemic 2,5-disubstituted 1-pyrrolines were kinetically resolved effectively by hydrogenation with a chiral titanocene catalyst **26** at 50% conversion, which indicates a large difference in the reaction rate of the enantiomers (Table 21.19, entries 4 and 5), while 2,3- or 2,4-disubstituted 1-pyrrolines showed moderate selectivity in the kinetic resolution (entries 6 and 7) [118]. The enantioselectivity of the major product with *cis*-configuration was very high for all disubstituted pyrrolidines. The high selectivity obtained with 2,5-disubstituted pyrrolines can be explained by the interaction of the substituent at C5 with the tetrahydroindenyl moieties of the catalyst [Eq. (17)].

H_2, 5 mol% Ti cat, 50% conv.

99 %ee (*S,S*) + 99 %ee (*R*) (17)

Ti cat **26** (X_2 = 1,1'-binaphthyl-2,2'-diolate)

In the kinetic resolution of acyclic chiral imines derived from α-methylbenzylamine and acetophenone derivatives, Rh(I)–(2*S*,4*S*)-bdpp catalyst forming a six-membered chelate ring exerted good kinetic resolution results. Catalysts with 2-carbon bridged diphosphines resulted in low reactivity and low selectivity (Table 21.19, entry 8). The hydrogenation of (*R*)-**27** by Rh(I)–(2*S*,4*S*)-bdpp gives an extremely high diastereomeric ratio of *RR*:*SR*=333:1 with *threo* stereochemistry, while in the reduction of (*S*)-**27** by Rh(I)–(2*S*,4*S*)-bdpp, the *threo* product is also formed in *SS*:*RS*=15.2:1 ratio, indicating strong substrate-controlled selectivity [14]. Under kinetic resolution conditions, however, the Rh–bdpp catalyst re-

Table 21.19 Kinetic resolution of ketones and imines.

Substrate	Major diastereomer	Recovered substrate	Catalyst	mol%	HCl (mol%)	P_{H_2}	Solvent	Conversion [%]	Diastereomeric ratio	%ee recov.	k_R/k_S	TOF	Reference
1 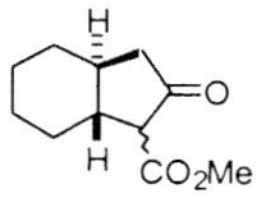	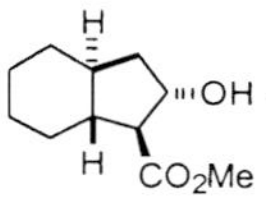 +	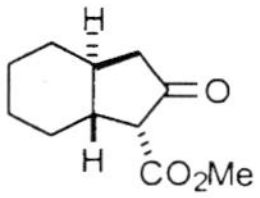	$RuCl_2$((*S*)-binap)	1.2	13.0 9.5 8.1	52 psi	MeOH	98.4 43.5 33.0	72:28 98:2 100:0				116
2 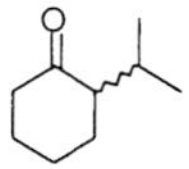	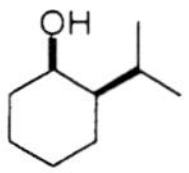 + (1*R*,2*R*)	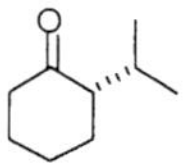 (2*S*)	(*S*,*SS*)-**25**	0.05		8 atm	*i*-PrOH	53	highly sel.	91 (*S*)	28		117
3 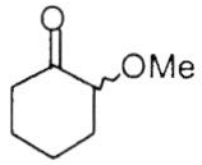	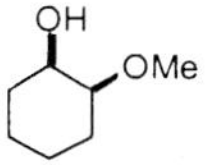 + (1*R*,2*S*)	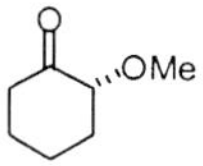 (2*R*)	(*S*,*SS*)-**25** 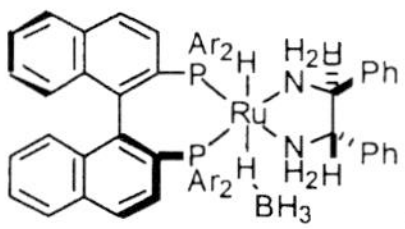 (*S*,*SS*)-**25** Ar = 3,5-$(CH_3)_2C_6H_3$	0.05		8 atm	*i*-PrOH	53	highly sel.	94 (*R*)	38		117
4 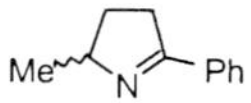	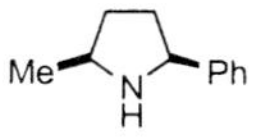 +	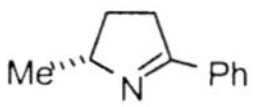	Titanocene cat **26**	5		80 psi	THF	50	highly sel. 99% ee	99 (*R*)			118

Table 21.19 (continued)

	Substrate	Major diastereomer	Recovered substrate	Catalyst	mol%	HCl (mol%)	P_{H_2}	Solvent	Conversion [%]	Diastereomeric ratio	%ee recov.	k_R/k_S	TOF	Reference
5	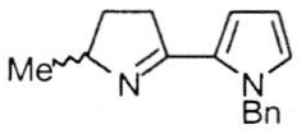 Me N N Bn	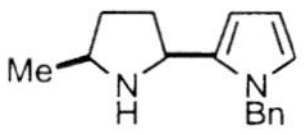 Me N H N Bn	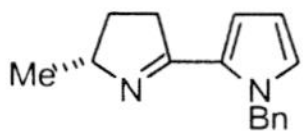 + Me N N Bn	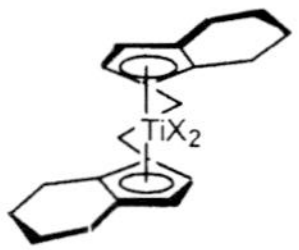 TiX_2 Titanocene cat **26**	5		80 psi	THF	50	highly sel. 98% ee	96 (*R*)			118
6	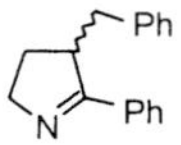 Ph N Ph	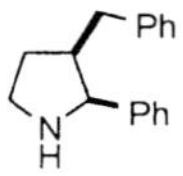 Ph N H Ph	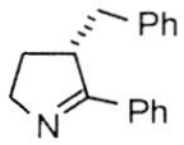 + Ph N Ph	Titanocene cat	5		80 psi	THF	50	85:15 >95% ee	75 (*R*)			118
7	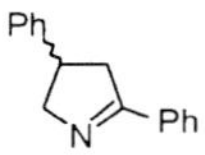 Ph N Ph	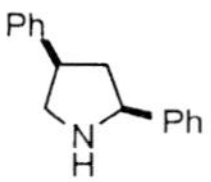 Ph N H Ph	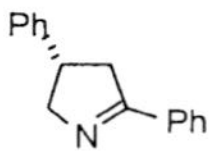 + Ph N Ph	Titanocene cat	5		80 psi	THF	50	75:25 99% ee	49 (*R*)			118
8	Ph N Me Ph Me rac-**27**	Ph HN Me Ph Me (*R*,*R*)-**28**	+ Ph N Me Ph Me (*S*)-**27**	[RhCl((*S*,*S*)-bdpp)]Cl	2		1000 psi	MeOH	67		83 (*S*)	5.7		14
				[RhCl((*S*,*S*)-chiraphos)]Cl	2		1000 psi	MeOH	82		7 (*S*)			14
9	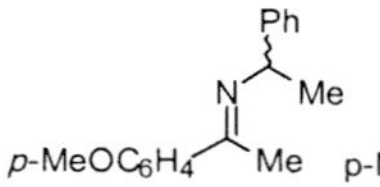 Ph N Me *p*-MeOC$_6$H$_4$ Me rac	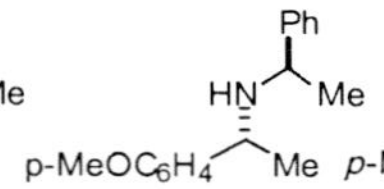 Ph HN Me p-MeOC$_6$H$_4$ Me (*R*,*R*)	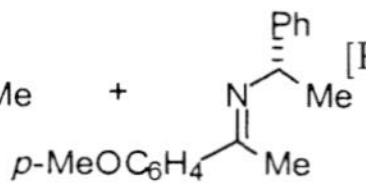 + Ph N Me *p*-MeOC$_6$H$_4$ Me (*S*)	[RhCl((*S*,*S*)-bdpp)]Cl	2		1000 psi	MeOH	72		98 (*S*)			14

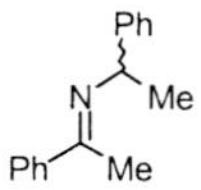

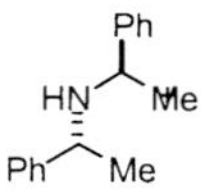

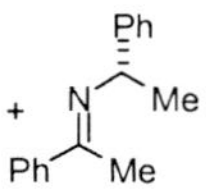

quired relatively high conversion in order to obtain (*S*)-chiral imine of high enantiomeric excess (entries 8 and 9) [14].

21.8 Dynamic Kinetic Resolution

In the kinetic resolution, the yield of desired optically active product cannot exceed 50% based on the racemic substrate, even if the chiral-discriminating ability of the chiral catalyst is extremely high. In order to obtain one diastereomer selectively, the conversion must be suppressed to less than 50%, while in order to obtain one enantiomer of the starting material selectively, a higher than 50% conversion is required. If the stereogenic center is labile in the racemic substrate, one can convert the substrate completely to gain almost 100% yield of the diastereomer formation by utilizing dynamic stereomutation.

In 1989, Noyori reported the first example of dynamic kinetic resolution in the enantioselective hydrogenation of *α*-substituted *β*-keto esters [119]. If the racemization of enantiomeric keto esters is rapid enough with respect to hydrogenation of the *β*-keto unit and the chiral discriminating ratio ($k_R : k_S$) is high, the hydrogenation will afford only one diastereomer out of four possible stereoisomeric hydroxy esters. The efficiency of the dynamic kinetic resolution and the sense of diastereoselection and enantioselection are strongly dependent on the structure of the substrate and the reaction conditions, including the solvent. The results of the dynamic kinetic resolution of *β*-keto esters are presented in Tables 21.20 to 21.22.

In the hydrogenation of cyclic *β*-keto esters (ketones substituted with an alkoxycarbonyl moiety), Ru(II)–binap reduced a racemic substrate in DCM with high *anti*-diastereoselectivity to give a 99:1 mixture of the *trans*-hydroxy ester (92% ee) and the *cis*-hydroxy ester (92% ee), quantitatively [Eq. (18)] [119, 120].

R = Me, Et — H_2 / Ru-(*R*)-binap, CH_2Cl_2, *anti*-selective → (1*R*,2*R*) + (1*S*,2*R*) 99 : 1 (18)

On the other hand, racemic *β*-keto esters with an amide or carbamate group in the *α*-position were reduced with high *syn*-diastereoselectivity (99:1) and with high enantioselectivity, leading to threonine-type products [Eq. (19)]. In polar methanol solution, the diastereoselectivity diminished to 71:29 (Table 21.20, entry 1) [123]. Results obtained using isotope-labeling experiments suggest that the hydrogenation proceeds via the ketone, and not via the enol [64]. The origin of the *syn* selectivity directed by an amide or a carbamate group was explained by a transition state stabilized by hydrogen bonding between the CONH and ester OR units [119]. *Anti*-selectivity in the case of cyclic *β*-keto esters is rationalized by the steric constraint of the cyclic ketone moiety (Scheme 21.10).

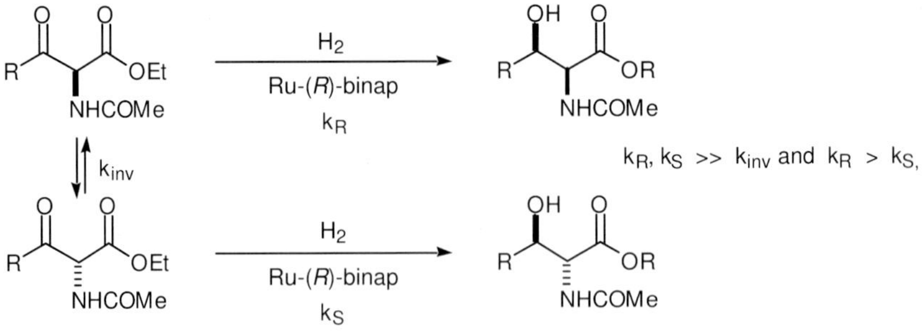

$k_R, k_S >> k_{inv}$ and $k_R > k_S$,

Scheme 21.9 Dynamic kinetic resolution of β-keto ester.

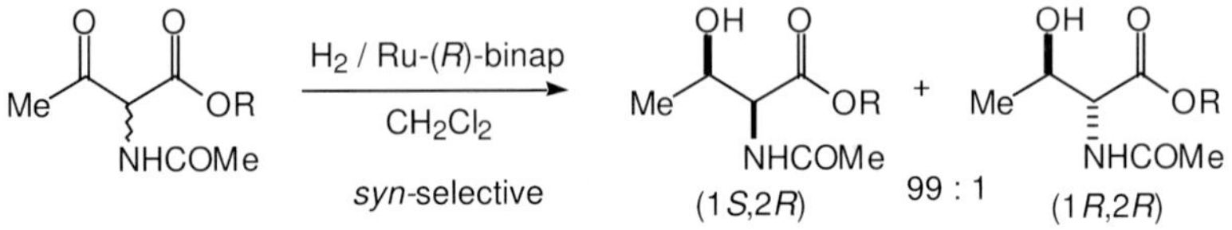

(19)

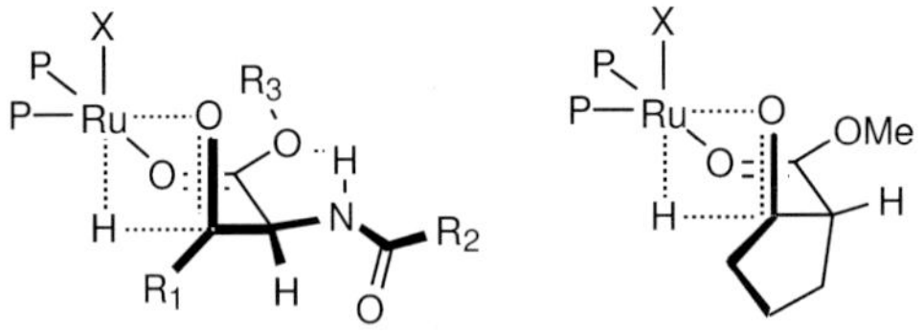

Scheme 21.10

Genêt also reported the dynamic kinetic resolution of α-substituted β-keto esters using several atropisomeric diphosphine ligands [121, 122, 124, 125] with high diastereo- and enantioselectivity. The *syn*:*anti* preference is lower using 10 mol.% of catalyst than in the presence of 1 mol.% of catalyst in MeOH [120]. This supported the fact that the hydrogenation needs to be slower than the racemization of the chiral center in order to achieve high stereoselectivities. For effective dynamic kinetic resolution, high hydrogen pressure should generally be avoided. With a rhodium catalyst, β-keto esters were reduced in moderate to good diastereoselectivities, but with low enantioselectivities [121].

The sense of diastereoselectivity in the dynamic kinetic resolution of 2-substituted β-keto esters depends on the structure of the keto ester. The ruthenium catalyst with atropisomeric diphosphine ligands (binap, MeO-biphep, synphos, etc.) induced *syn*-products in high diastereomeric and enantiomeric selectivity in the dynamic kinetic resolution of β-keto esters with an α-amido or carbamate moiety (Table 21.21) [119–121, 123, 125–127]. In contrast to the above examples of α-amido-β-keto esters, the TsOH or HCl salt of β-keto esters with an α-amino unit were hydrogenated with excellent *anti*-selectivity using ruthenium-atropiso-

Table 21.20 Dynamic kinetic resolution of β-keto esters.

	Substrate	Major diastereomer	Minor diastereomer		Catalyst	mol%	P_{H_2}	Solvent	Temp. [°C]	Diastereo-meric ratio [*anti*:*syn*]	%ee (*anti*)	TOF	Refer-ence
1	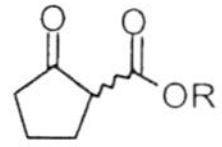	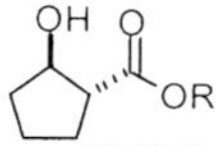 + (2*R*,3*R*)	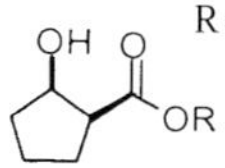	R = Et	[RuCl(C_6H_6)((*R*)-binap)]Cl	0.085	100 atm	CH_2Cl_2	50	99:1	92 (2*R*,3*R*)	16.7	119
					$RuBr_2$((*R*)-binap)	1	20 atm	MeOH	80	97:3	94 (2*R*,3*R*)	25	122
					$RuBr_2$((*R*)-binap)	1	100 atm	EtOH	80	96:4	85 (2*R*,3*R*)	2.1	133
				R = Me	[RuC((C_6H_6)((*R*)-binap)]Cl	0.085	100 atm	CH_2Cl_2	50	99:1	92 (2*R*,3*R*)	16.7	120
2	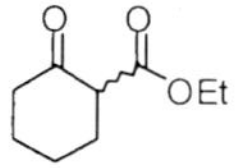	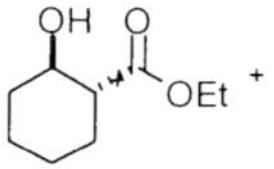 + (2*R*,3*R*)	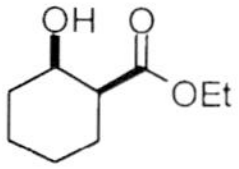		[Rucl(C_6H_6)((*R*)-binap)]Cl	0.085	100 atm	CH_2Cl_2	50	95:5	90 (2*R*,3*R*)	16.7	120
					$RuBr_2$((*R*)-binap)	1	20 atm	CH_2Cl_2	80	76.5:23.5	91 (2*R*,3*R*)	33	122
					$RuBr_2$((*R*)-binap)	1		MeOH		46:54	88 (2*R*,3*R*)	50	122
					$RuBr_2$((*S*)-binap)	1	20 atm	EtOH	80	73.5:26.5	91 (2*S*,3*S*)	33	133
3	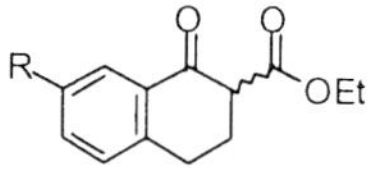	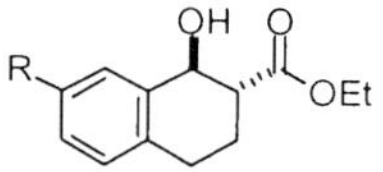 + (2*R*,3*R*)	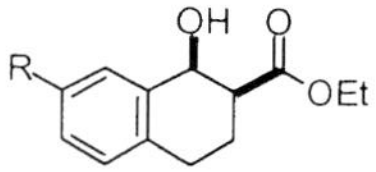										
				R = H	$RuBr_2$((*R*)-MeO-biphep)		10 atm	EtOH	80	98.5:1.5	95 (2*R*,3*R*)	1.2	133
				R = OMe	$RuBr_2$((*R*)-binap)	3	10 atm	CH_2Cl_2	80	98.5:1.5	96 (2*R*,3*R*)	0.7	122
					$RuBr_2$((*R*)-MeO-biphep)	3	10 atm	MeOH	80	98.5:1.5	95 (2*R*,3*R*)	0.6	122
4	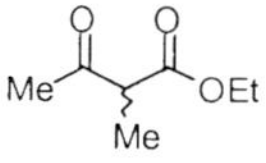	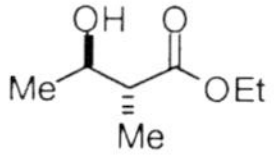 +	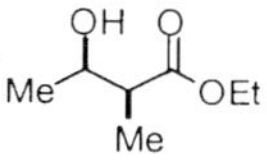		$RuBr_2$((*R*)-binap)	0.083	100 atm	EtOH	25	49:51	97 (2*R*,3*R*)	30	120
					[RuCl(C_6H_6)((*R*)-binap)]Cl	0.085	100 atm	CH_2Cl_2	50	68:32	94 (2*R*,3*R*)	20	107

Table 21.20 (continued)

	Substrate	Major diastereomer	Minor diastereomer	Catalyst	mol%	P_{H_2}	Solvent	Temp. [°C]	Diastereomeric ratio [*anti*:*syn*]	%ee (*anti*)	TOF	Reference
5	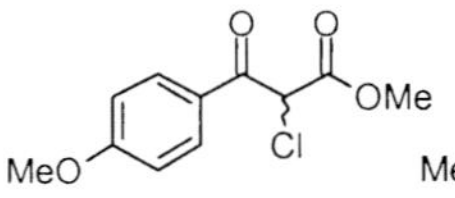	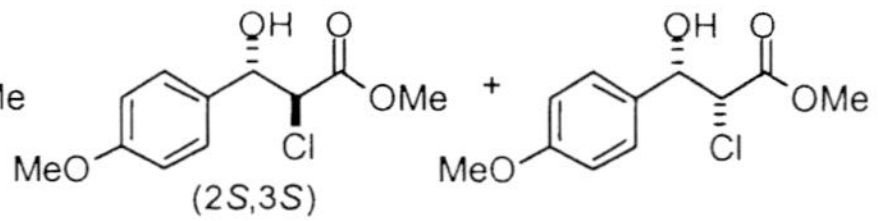 (2*S*,3*S*)										
				Ru(allyl)$_2$((*S*)-binap)	1	80 atm	CH_2Cl_2	80	96:4	94 (2*S*,3*S*)	5.3	124
				Ru(allyl)$_2$((*R*-binap)	0.5	5atm	MeOH	50	8.5:91.5	77 (2*R*,3*R*)	12.5	124
				[Ru((*S*)-MeO-biphep)]	1	60 bar	CH_2Cl_2	80	97.5:2.5	94 (2*S*,3*S*)	2.0	134
6	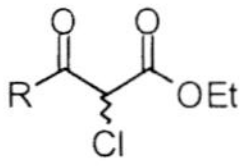	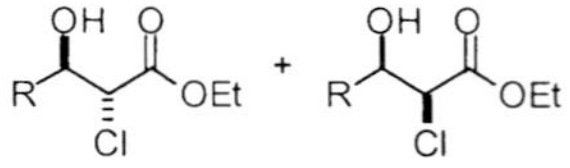 (2*R*,3*R*)										
			R = Me	Ru(allyl)$_2$((*R*)-binap)	0.5	90 atm	CH_2Cl_2	80	99:1	99 (2*R*,3*R*)	20	124
					1	30 atm	EtOH	27	52:48	93 (2*R*,3*R*)	5	124
				[Ru((*R*)-MeO-biphep)]	1	80 bar	CH_2Cl_2	80	99.5:0.5	98 (2*R*,3*R*)		133
			P = Rh	[Ru((*S*)-binap)]	0.5	86 atm	CH_2Cl_2	80	96:4	83 (2*S*,3*S*)	42.7	124
				RuBr$_2$((*R*)-binap)	1	30 atm	EtOH	27	9.5:90.5	31 (2*R*,3*R*)	5	124

meric diphosphine catalysts to afford *anti-β*-hydroxy-*α*-amino acids (Table 21.22, entries 1–4) [128–131]. In this case, a five-membered transition state is envisioned. High *anti*-selectivity was also observed in the reaction of an *α*-phthalimido-*β*-keto ester in methanol (Table 21.22, entry 5) [132], and in the reduction of keto esters having an *α*-chloro group (Table 21.20, entries 5 and 6) [124, 133, 134]. *β*-Keto esters with a cyclic keto unit also afforded *anti*-products selectively, as described [53, 119, 120, 122, 123, 133].

Dynamic kinetic resolution of *β*-keto phosphonic esters containing an *α*-amido group was examined using Ru–(*R*)-binap, and resulted in a phosphonate analogue of *α*-amino acids with (*R*,*R*)-*syn* configuration and very high selectivity ($k_R/k_S = 39$; $k_{inv}/k_S = 31$) [135]. The overall stereochemical outcome was explained on the basis of the Felkin-Anh model (Scheme 21.11), wherein *β*-keto phosphonic esters with *α*-bromo substituent were hydrogenated with high *syn*-selectivity [136].

In the hydrogenation of simple 2-alkyl-3-oxobutanoates, the interconversion between the enantiomers is relatively slow, and very poor resolution results are observed (see Table 21.20, entry 4) [107, 119, 120, 123]. Efficient dynamic kinetic resolution of 2-alkyl *β*-keto esters was first observed in the asymmetric hydrogenation of *α*-alkyl-*β*-keto esters which are derived from (*S*)-proline bearing a stereogenic center at the *γ*-position. The type of *N*-protecting group played a dramatic role, and the *N*-Boc substrate **29b** afforded naturally occurring (2*R*,3*R*)-dolaproine with *syn*-configuration, while *β*-keto esters *N*-protected as an amine hydrogen chloride salt **29a** afforded an *anti*-adduct in moderate to high diastereoselectivity (Scheme 21.12) [137, 138].

Dynamic kinetic resolution is possible for *α*-alkyl or *α*-alkoxy cyclic ketones in the presence of KOH, which causes mutation of the stereogenic center; *syn*-alcohols were obtained selectively with high enantioselectivity using ruthenium–3,5-xyl-binap. Dynamic kinetic resolution of 2-arylcycloalkanones also proceeded with extremely high *syn*-selectivity and with high enantioselectivity using ruthenium–binap-diamine as catalyst (Table 21.23) [12, 139, 140].

21.9 Conclusions

By linking the steric factor of the stereogenic center of substrates with the chiral-inducing ability of properly designed ligands, homogeneous diastereoselective hydrogenation can attain levels of stereoselectivity that will enable the industrial

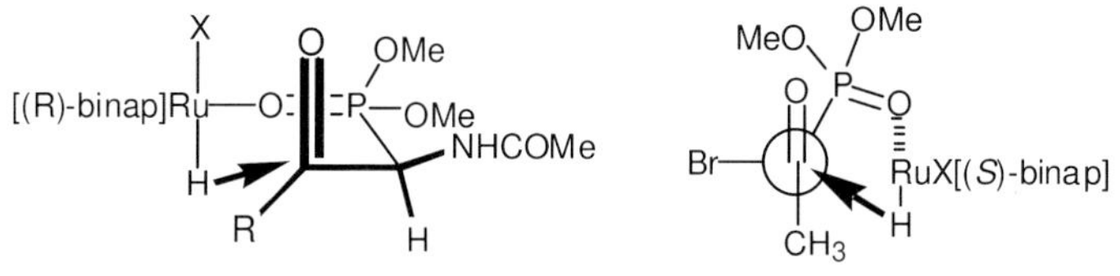

Scheme 21.11 Felkin-Anh model for the hydrogenation of keto phosphate.

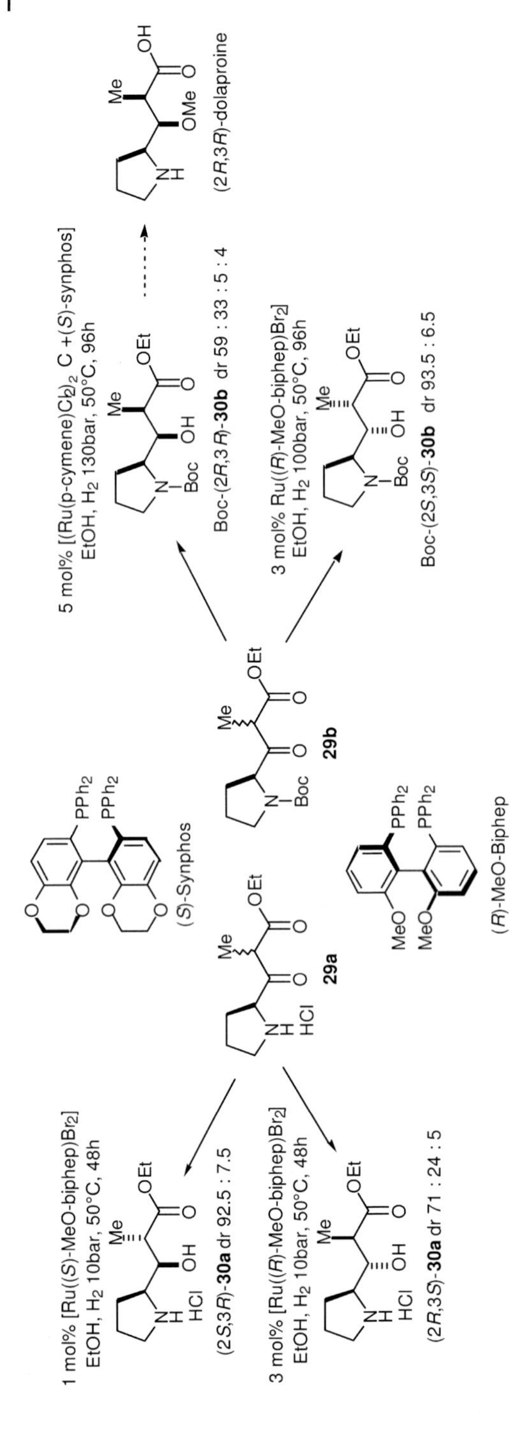

Scheme 21.12 Effect of *N*-protecting groups on the dynamic kinetic resolution.

Table 21.21 Dynamic kinetic resolution of α-subsituted β-keto esters.

	Substrate	Major diastereomer	Minor diastereomer	Catalyst	mol%	P_{H_2}	Solvent	Diastereo-meric ratio [*anti*:*syn*]	%ee (*syn*)	TOF	Refer-ence
1	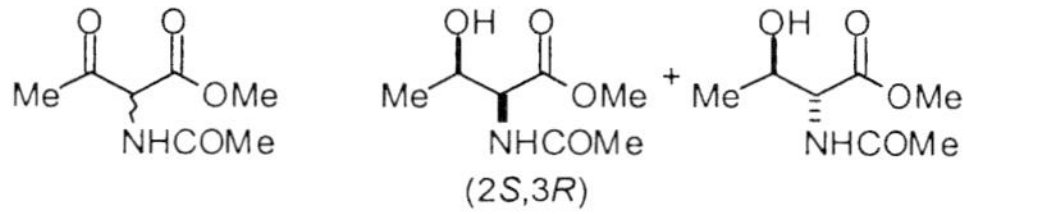	(2*S*,3*R*)		$Ru(O_2CCF_3)_2$[(*R*)-binap]	1	90 atm	CH_2Cl_2	95:5	51	0.43	121
				$RuBr_2$[(–)-chiraphos]	1	90 atm	CH_2Cl_2	97:3	85	0.83	121
				[Rh(nbd)((–)-dipamp)]$^+$	1	70 atm	THF	94:6	39	1.0	121
				$RuBr_2$[(*R*)-binap]	0.4	100 atm	CH_2Cl_2	99:1	98	5	119
				$RuBr_2$[(*R*)-binap]	0.4	100 atm	MeOH	71:29	90		123
2	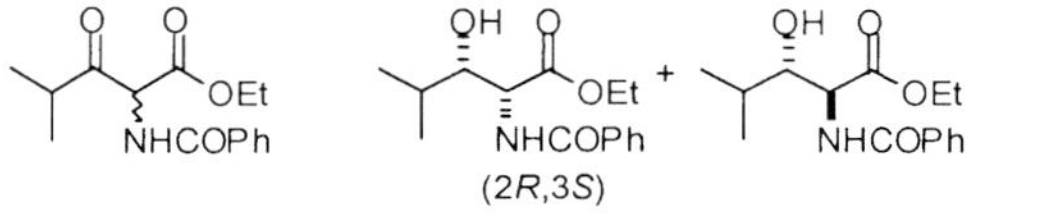	(2*R*,3*S*)		$RuCl_2$((*S*)-binap)$(dmf)_m$	2	100 atm	CH_2Cl_2	100:0	99	1.0	131
				$RuBr_2$((*S*)-synphos)	2	130 bar	CH_2Cl_2	99.5:0.5	97	0.5	127
3	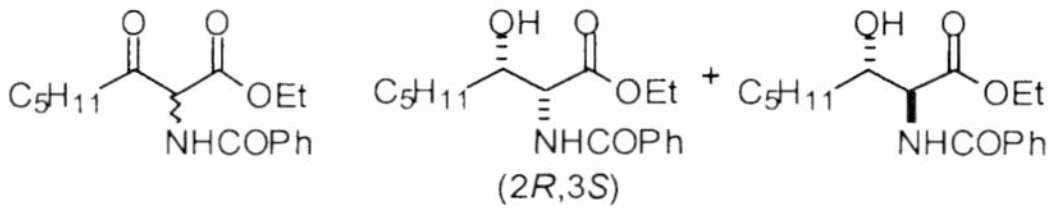	(2*R*,3*S*)		$RuBr_2$((*S*)-synphos)	2	130 bar	CH_2Cl_2	99:1	99	0.4	131
4	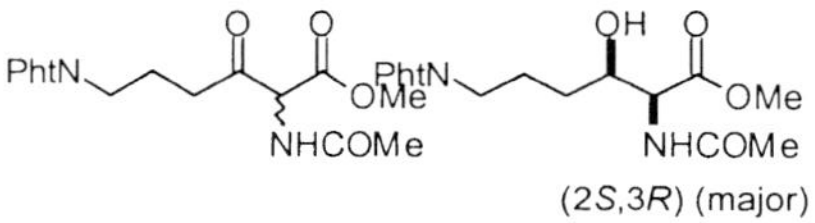	(2*S*,3*R*) (major)		$RuBr_2$((*R*)-MeO-biphep)	0.5	100 bar	CH_2Cl_2	>99:1		1.85	125
				$RuBr_2$((*R*)-MeO-biphep)	1	100 bar	MeOH	65:35		1.1	125
5	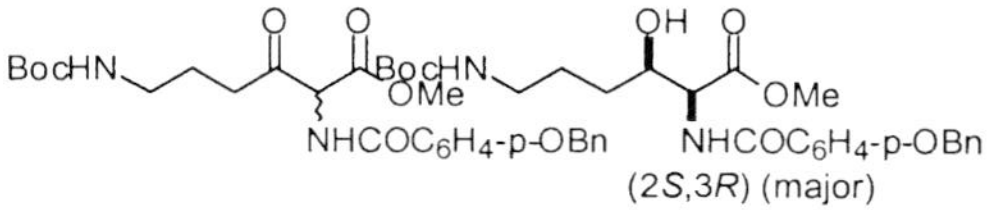	(2*S*,3*R*) (major)		$RuBr_2$((*R*)-MeO-biphep)	1	130 bar	CH_2Cl_2	96:5:3.5	94	1.1	126

Table 21.21 (continued)

	Substrate	Major diastereomer	Minor diastereomer	Catalyst	mol%	P_{H_2}	Solvent	Diastereomeric ratio [*anti*:*syn*]	%ee (*syn*)	TOF	Reference
6	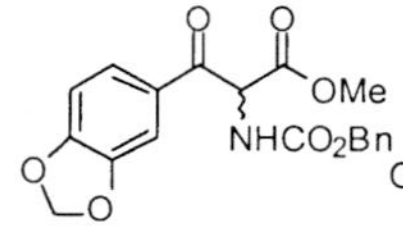 (structure: OMe, NHCO$_2$Bn)	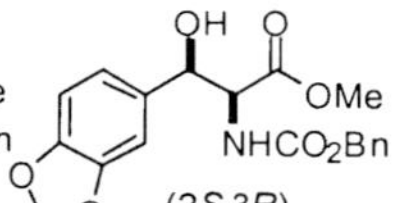 (2*S*,3*R*) (structure: OH, OMe, NHCO$_2$Bn) +	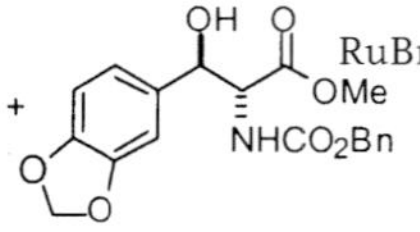 (structure: OH, OMe, NHCO$_2$Bn)	RuBr$_2$[(*R*)-binap]	0.4	100 atm	CH_2Cl_2	99:1	92	2.4	119
7	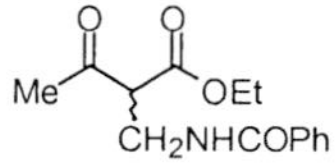 (structure: Me, OEt, CH$_2$NHCOPh)	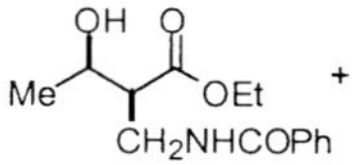 (2*S*,3*R*) (structure: OH, Me, OEt, CH$_2$NHCOPh) +	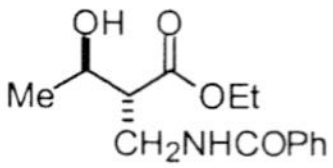 (structure: OH, Me, OEt, CH$_2$NHCOPh)	[RuCl$_2$((*R*)-binap)]$_2$NEt$_3$	1	100 atm	CH_2Cl_2	94:6	98	5	119
				[RuCl((*R*)-3,5-Bu$^t_2$-binap)(*p*-cymene)]	0.1	50 atm	CH_2Cl_2/MeOH	99:1	99	13.8	123
				[RuBr((*R*)-binap)(benzene)]Br	1	50 atm	CH_2Cl_2/H_2O	89.5:10.5	98	2.3	123
				[RuBr((*R*)-binap)(benzene]Br	1	50 atm	MeOH	54.5:45.5	80	2.4	123
8	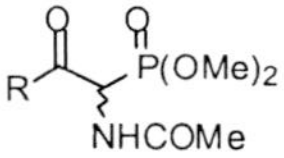 (structure: R, P(OMe)$_2$, NHCOMe)	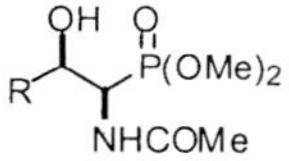 (1*S*,2*R*) (structure: OH, R, P(OMe)$_2$, NHCOMe) +	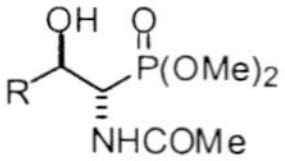 (structure: OH, R, P(OMe)$_2$, NHCOMe)								
			R = Ph	[RuCl$_2$(*R*)-binap](dmf)$_n$	1	4 atm	MeOH	98:2	95	0.8	135
			R = Me	[RuCl$_2$(*R*)-binap](dmf)$_n$	0.17	4 atm	MeOH	97:3	>98	9.0	135
							$k_R/k_S = 39$ and $k_{inv}/k_S = 31$				
9	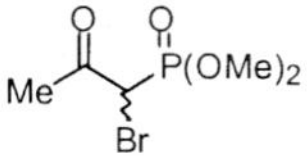 (structure: Me, P(OMe)$_2$, Br)	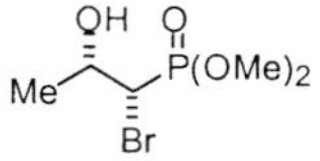 (1*R*,2*S*) (structure: OH, Me, P(OMe)$_2$, Br) +	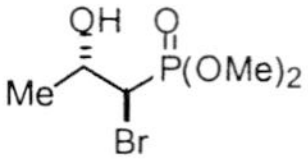 (structure: OH, Me, P(OMe)$_2$, Br)	[RuCl$_2$(*S*)-binap](dmf)$_n$	0.05	4 atm	MeOH	90:10	98	0.8	136
							$k_R/k_S = 13$ and $k_{inv}/k_S = 11.5$				

Table 21.22 Dynamic kinetic resolution of α-substituted β-keto esters.

	Substrate	Major diastereomer	Minor diastereomer	Catalyst	mol%	P_{H_2}	Solvent	Diastereomeric ratio [*anti*:*syn*]	%ee (*anti*)	TOF	Reference
1	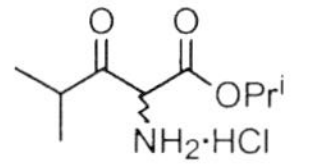 O O OPr^i NH_2·HCl	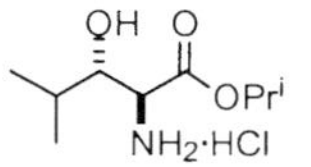 OH O OPr^i NH_2·HCl (2*S*,3*S*) +	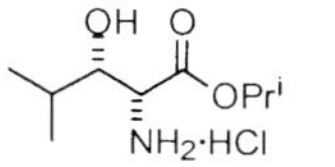 OH O OPr^i NH_2·HCl	[RuCl$_2$(*S*)-binap](dmf)$_n$	4	100 atm	CH_2Cl_2	98:2	92	0.5	128
				RuBr$_2$((*S*)-MeO-biphep)	1	12 bar	EtOH	99:1	87	4.2	131
				RuBr$_2$((*S*)-synphos)	2	12 bar	CH_2Cl_2/2-PrOH	99.5:0.5	97	1.9	130
2	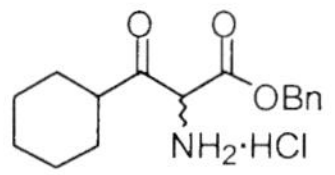 O O OBn NH_2·HCl	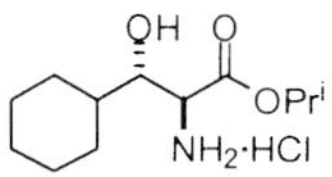 OH O OPr^i NH_2·HCl (2*S*,3*S*) +	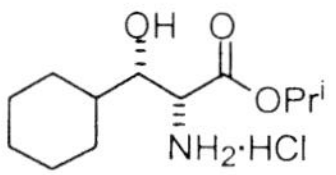 OH O OPr^i NH_2·HCl	[RuCl(*S*)-binap](dmf)$_n$	4	100 atm	CH_2Cl_2	>99:1	97	0.44	128
3	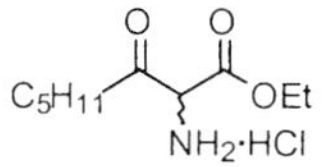 O O C_5H_{11} OEt NH_2·HCl	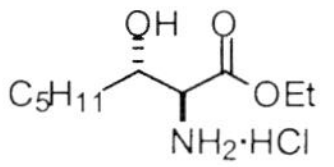 OH O C_5H_{11} OEt NH_2·HCl +	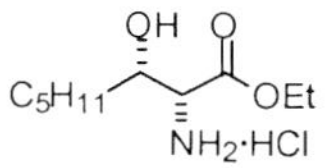 OH O C_5H_{11} OEt NH_2·HCl	RuBr$_2$((*S*)-synphos)	2	12 bar	CH_2Cl_2 (EtOH)	96.5:3.5	91	1.8	130
4	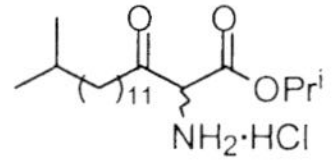 O O ( )$_{11}$ OPr^i NH_2·HCl	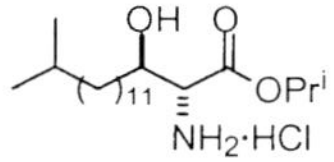 OH O ( )$_{11}$ OPr^i NH_2·HCl +	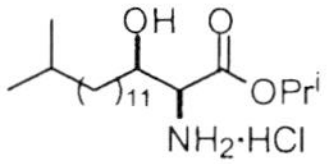 OH O ( )$_{11}$ OPr^i NH_2·HCl	RuBr$_2$((*R*)-MeO-biphep)	2	20 bar	MeOH	98:2	41	0.7	129
				RuBr$_2$((*R*)-MeO-biphep)	2	20 bar	CH_2Cl_2	90:10	88	0.7	129
				RuBr$_2$((*R*)-synphos)	2	12 bar	CH_2Cl_2/MeOH	96:4	92	1.7	129
5	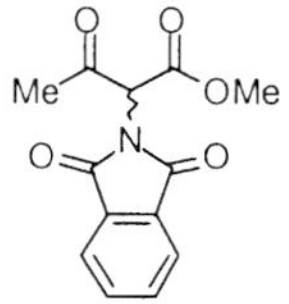 O O Me OMe O N O	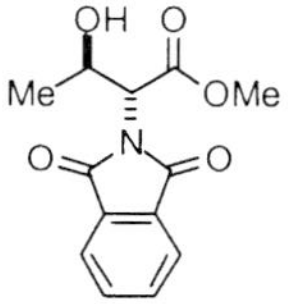 OH O Me OMe O N O +	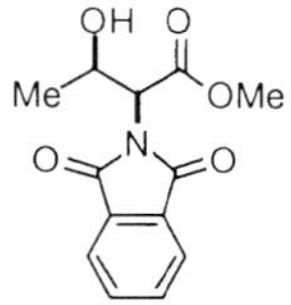 OH O Me OMe O N O	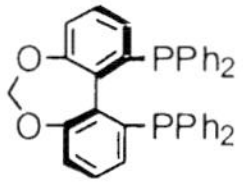 O O PPh$_2$ PPh$_2$ (*R*)-C$_3$-Tunephos [Ru(*S*)-C$_3$-Tunephos]	2	100 bar	MeOH	>97:3	>99	0.7	132

Table 21.23 Dynamic kinetic resolution of ketones.

	Substrate	Diastereomers	Catalyst	mol%	P_{H_2}	Solvent	Diastereomeric ratio [*syn*:*anti*]	%ee (*syn*)	TOF	Reference
1	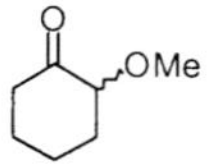 OMe	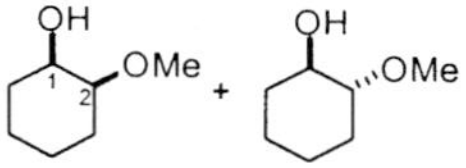 OMe (1*R*,2*S*) + OMe	[RuCl$_2$((*S*)-binap)]$_2$(NEt$_3$)-(*S*,*S*)-DPEN-KOH	0.1	50 atm	*i*-PrOH	98.5:1.5	92 (*R*,*S*)	1000	139
			[RuCl$_2$((*S*)-3,5-xylyl-binap)]$_2$(NEt$_3$)-(*S*,*S*)-DPEN-KOH	0.1	50 atm	*i*-PrOH	99.5:0.5	99 (*R*,*R*)	50	139
2	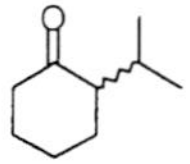	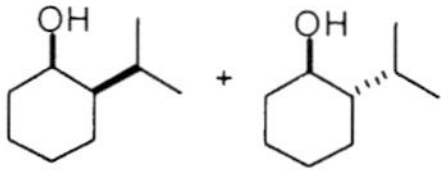 +	RuCl$_2$((*S*)-binap)(dmf)m-(*R*,*R*)-DPEN-KOH	0.2	4 atm	*i*-PrOH	99.8:0.2	93 (*R*,*R*)	45.5	12
3	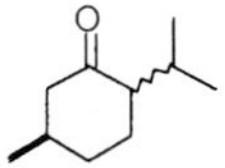	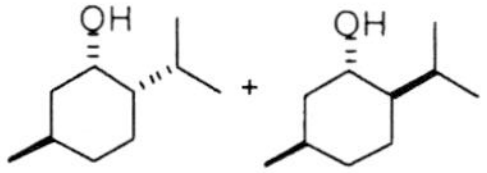 +	RuCl$_2$((*R*)-binap)(dmf)m-(*S*,*S*)-DPEN-KOH	0.2	4 atm	*i*-PrOH	highly selective			12
4	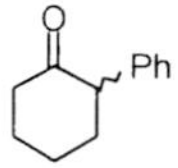 Ph	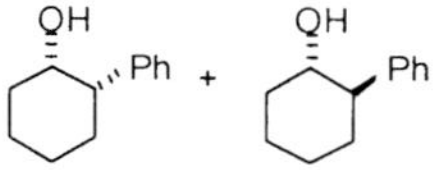 Ph + Ph	RuCl$_2$((*R*)-tol-binap)(dmf)m-(*S*,*S*)-DPEN-^{t}BuOK	0.05	8 atm	*i*-PrOH	100:0	97 (*S*,*S*)	500	140
5	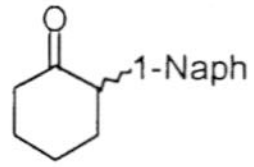 1-Naph	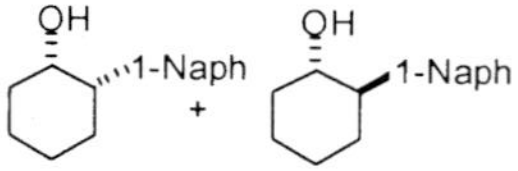 1-Naph + 1-Naph	RuCl$_2$((*R*)-tol-binap)(dmf)m-(*R*,*R*)-DPEN-^{t}BuOK	0.05	8 atm	*i*-PrOH	98:2	93 (*S*,*S*)	125	140

Table 21.23 (continued)

	Substrate	Diastereomers	Catalyst	mol%	P_{H_2}	Solvent	Diastereo-meric ratio [*syn*:*anti*]	%ee (*syn*)	TOF	Reference
6	O Ph	OH Ph + OH Ph	$RuCl_2$((*R*)-tol-binap)(dmf)m -(*S*,*S*)-DPEN-^{t}BuOK	0.05	8 atm	*i*-PrOH	100:0	95 (*S*,*S*)	83	140
7	O OMe N C_3H_7	OH OMe N C_3H_7 + OH OMe N C_3H_7	$RuCl_2$((*S*)-tol-binap)(dmf)m -(*S*,*S*)-DPEN-^{t}BuOK	0.2	8 atm	*i*-PrOH	>99:1	97 (*S*,*R*)	20.8	140

preparation of optically active compounds with several stereogenic centers. Functional group-directed hydrogenation led to good results *via* the interaction of a heteroatom in the substrate with the metal or with the ligand, whilst by selecting the catalyst, diastereoselective hydrogenation can induce excellent stereoselectivity *via* double stereo-differentiation (matched pair). However, in general the activity of hydrogenation catalysts remains poor, especially for those substrates with bulky groups proximal to the reaction site. Consequently, it will be necessary to develop more efficient catalysts to produce higher turnover frequencies.

Abbreviations

d.e.	diastereomeric excess
DCM	dichloromethane
ee	enantiomeric excess
TBSO	tert-butyldimethylsilyloxy
THF	tetrahydrofuran

References

1 (a) Knowles, W.S., Sabacky, M.J., *J. Chem. Soc., Chem. Commun.* **1968**, 1445; (b) Horner, L., Siegel, H., Buthe, H., *Angew. Chem., Int. Ed. Engl.* **1968**, *7*, 942; (c) Kagan, H.B., Chiral Ligands for Asymmetric Catalysis, in: Morrison, J.D. (Ed.), *Asymmetric Synthesis*, Academic Press, Inc.: Orlando, FL, **1985**, Vol. 5, p. 1; (d) Koenig, K.E., The Applicability of Asymmetric Homogeneous Catalytic Hydrogenation, in: *Asymmetric Synthesis*; Morrison, J.D. (Ed.), Academic Press, Inc., Orlando, FL, **1985**, Vol. 5, p. 71; (e) Brunner, H., *Top. Stereochem.* **1988**, *18*, 129; (f) Arntz, D., Schaefer, A., Asymmetric Hydrogenation, in: Noels, A.F., Graziani, M., Hubert, A.J. (Eds.), *Metal Promoted Selectivity in Organic Synthesis*, Kluwer Academic, Dordrecht, **1991**, p. 161; (g) Takaya, H., Ohta, T., Noyori, R., Asymmetric Hydrogenation, in: Ojima, I. (Ed.), *Catalytic Asymmetric Synthesis*, VCH Publishers, Inc., New York, NY, **1993**, p. 1.

2 Jardine, F.H., *Prog. Inorg. Chem.* **1984**, *431*, 265.

3 Halpern, J., Harrod, J.F., James, B.R., *J. Am. Chem. Soc.* **1961**, *83*, 753.

4 Crabtree, R.H., Felkin, H., Morris, G.E., *J. Organomet. Chem.* **1977**, *141*, 205.

5 Evans, D.A., Morrissey, M.M., *J. Am. Chem. Soc.* **1984**, *106*, 3866.

6 (a) Molander, G.A., Winterfeld, J., *J. Organomet. Chem.* **1996**, 524, 275; (b) Molander, G.A., Dowdy, E.D., *Top. Organomet. Chem.* **1999**, *2*, 119.

7 Jenke, T., Süss-Fink, G., *J. Organomet. Chem.* **1991**, *405*, 383.

8 Suggs, J.W., Cox, S.D., Crabtree, R.H., Quirk, J.M., *Tetrahedron Lett.* **1981**, *22*, 303.

9 Debenham, S.D., Debenham, J.S., Burk, M.J., Toone, E.J., *J. Am. Chem. Soc.* **1997**, *119*, 9897.

10 Berens, U., Fischer, C., Selke, R., *Tetrahedron: Asymm.* **1995**, *6*(5), 1105.

11 Fehr, M.J., Consiglio, G., Scalone, M., Schmid, R., *New J. Chem.* **1998**, 1499.

12 Ohkuma, T., Ooka, H., Yamakawa, M., Ikariya, T., Noyori, R., *J. Org. Chem.* **1996**, *61*, 4872.

13 Ohkuma, T., Ikehira, H., Ikariya, T., Noyori, R., *Synlett* **1997**, 467.

14 Lensink, C., de Vries, J.G., *Tetrahedron: Asymm.* **1993**, *4*, 215.

15 Morrison, J. D. (Ed.), *Asymmetric Synthesis*, Vol. 5, Academic Press, New York, 1985.
16 Thompson, H.W., McPherson, E., *J. Am. Chem. Soc.* **1974**, *96*, 6232.
17 (a) Paquette, L.A., Peng, X., Bondar, D., *Org. Lett.* **2002**, *4*(6), 937; (b) Peng, X., Bondar, D., Paquette, L.A., *Tetrahedron* **2004**, *60*, 9589.
18 Crabtree, R.H., Demou, P.C., Eden, D., Mihelcic, J.M., Parnell, C.A., Quirk, J.M., Morris, G.E., *J. Am. Chem. Soc.* **1982**, *104*, 6994.
19 Brown, J.M., Naik, R.G., *J. Chem. Soc., Chem. Commun.* **1982**, 348.
20 Brown, J.M., *Angew. Chem. Int. Ed. Engl.* **1987**, *26*, 190.
21 Crabtree, R.H., Davis, M.W., *Organometallics* **1983**, *2*, 681.
22 Stork, G., Kahne, D.E., *J. Am. Chem. Soc.* **1983**, *105*, 1072.
23 Hoveyda, A.H., Evans, D.A., Fu, G.C., *Chem. Rev.* **1993**, *93*, 1307.
24 Evans, D.A., Morrissey, M.M., *Tetrahedron Lett.* **1984**, *25*(41), 4637.
25 Crabtree, R.H., Davis, M.W., *J. Org. Chem.* **1986**, *51*, 2655.
26 Corey, E.J., Engler, T.A., *Tetrahedron Lett.* **1984**, *25*, 149.
27 Sakurai, K., Kitahara, T., Morñi, K., *Tetrahedron* **1990**, *46*(3), 761.
28 Fernández, B., Martínez Pérez, J.A., Granja, J.R., Castedo, L., Mouriño, A., *J. Org. Chem.* **1992**, *57*, 3173.
29 Brown, J.M., Hall, S.A., *Tetrahedron* **1985**, *41*, 4639.
30 DeCamp, A.E., Verhoeven, T.R., Shinkai, I., *J. Org. Chem.* **1989**, *54*, 3207.
31 Poss, A.J., Smyth, M.S., *Tetrahedron Lett.* **1988**, *29*(45), 5723.
32 Kawai, A., Hara, O., Hamada, Y., Shioiri, T., *Tetrahedron Lett.* **1988**, *29*(48), 6331.
33 Watson, A.T., Park, K., Wiemer, D.F., *J. Org. Chem.* **1995**, *60*, 5102.
34 Del Valle, J.R., Goodman, M., *Angew. Chem. Int. Ed. Engl.* **2002**, *41*, 1600.
35 Bueno, J.M., Coterón, J.M., Chiara, J.L., Fernández-Moyaralas, A., Fiandor, J.M., Valle, N., *Tetrahedron Lett.* **2000**, *41*, 4379.
36 Brown, J.M., Hall, S.A., *Tetrahedron Lett.* **1984**, *25*, 1393.
37 Machado, A.S., Olesker, A., Castillon, S., Lukacs, G., *J. Chem. Soc., Chem. Commun.* **1985**, 330.
38 Del Valle, J.R., Goodman, M., *J. Org. Chem.* **2003**, *68*, 3923.
39 Brown, J.M., Derome, A.E., Hall, S.A., *Tetrahedron* **1985**, *41*, 4647.
40 Brown, J.M., Hall, S.A., *J. Organomet. Chem.* **1985**, *285*, 333.
41 Schultz, A.G., McCloskey, P.J., *J. Org. Chem.* **1985**, *50*, 5905.
42 (a) Zhang, M., Zhu, L., Ma, X., *Tetrahedron: Asymm.* **2003**, *14*, 3447; (b) Zhang, M., Zhu, L., Ma, X., Dai, M., Lowe, D., *Org. Lett.* **2003**, *5*(9), 1587.
43 Schultz, A.G., McCloskey, P.J., Court, J.J., *J. Am. Chem. Soc.* **1987**, *109*, 6493.
44 Smith, M.E.B., Derrien, N., Lloyd, M.C., Taylor, S.J.C., Chaplin, D.A., McCague, R., *Tetrahedron Lett.* **2001**, *42*, 1347.
45 Brown, J.M., Cutting, I., *J. Chem. Soc., Chem. Commun.* **1985**, 578.
46 Kitamura, M., Kasahara, I., Manabe, K., Noyori, R., Takaya, H., *J. Org. Chem.* **1988**, *53*, 708.
47 Hameršak, Z., Gašo, D., Kovač, S., Hergold-Brundić, A., Vicković, I., Šunjić, V., *Helv. Chim. Acta* **2003**, *86*(6), 2247.
48 (a) Lautens, M., Zhang, C.H., Crudden, C.M., *Angew. Chem. Int. Ed. Engl.* **1992**, *31*(2), 232; (b) Lautens, M., Zhang, C.H., Goh, B.J., Crudden, C.M., Johnson, M.J.A., *J. Org. Chem.* **1994**, *59*, 6208.
49 Schmidt, U., Stäbler, F., Lieberknecht, A., *Synthesis* **1992**, 482.
50 Paterson, I., Bower, S., Tillyer, R.D., *Tetrahedron Lett.* **1993**, *34*(27), 4393.
51 Paterson, I., Bower, S., McLeod, M.D., *Tetrahedron Lett.* **1995**, *36*, 175.
52 Kitamura, M., Nagai, K., Hsiao, Y., Noyori, R., *Tetrahedron Lett.* **1990**, *31*, 549.
53 Genêt, J.-P., *ACS Symposium Series* **1996**, *641* (Reductions in Organic Synthesis), 31.
54 Terada, M., Sayo, N., Mikami, K., *Synlett* **1995**, 411.
55 Reetz, M.T., Kayser, F., *Tetrahedron: Asymm.* **1992**, *3*, 1377.
56 Ando, D., Bevan, C., Brown, J.M., Price, D.W., *J. Chem. Soc., Chem. Commun.* **1992**, 592.
57 Birtwistle, D.H., Brown, J.M., Herbert, R.H., James, A.P., Lee, K.-F., Taylor,

R. J., *J. Chem. Soc., Chem. Commun.* **1989**, 194.

58 Wehn, P. M., Du Bois, J., *J. Am. Chem. Soc.* **2002**, *124*, 12950.

59 Evans, D. A., Morrissey, M. M., Dow, R. L., *Tetrahedron Lett.* **1985**, *26*(49), 6005.

60 Evans, D. A., DiMare, M., *J. Am. Chem. Soc.* **1986**, *108*, 2476.

61 Evans, D. A., Dow, R. L., *Tetrahedron Lett.* **1986**, *27*(9), 1007.

62 Villalobos, A., Danishefsky, S. J., *J. Org. Chem.* **1990**, *55*, 2776.

63 Landis, C. R., Halpern, J., *J. Am. Chem. Soc.* **1987**, *109*, 1746.

64 Brown, J. M., James, A.P., *J. Chem. Soc., Chem. Commun.* **1987**, 181.

65 Brown, J. M., Cutting, I., James, A. P., *Bull. Soc. Chim. Fr.* **1988**, (2), 211.

66 (a) Yamamoto, K., Takagi, M., Tsuji, J., *Bull. Chem. Soc. Jpn.* **1988**, *61*, 319; (b) Yamamoto, K., *Yuki Gousei Kagaku Kyokaishi* **1989**, *47*(2), 123.

67 Takagi, M., Yamamoto, K., *Tetrahedron* **1991**, *47*(2), 8869.

68 Masamune, S., Choy, W., Petersen, J. S., Sita, L. R., *Angew. Chem. Int. Ed. Engl.* **1985**, *24*, 1.

69 Ojima, I., Kogure, T., Yoda, N., Suzuki, T., Yatabe, M., Tanaka, T., *J. Org. Chem.* **1982**, *47*, 1329.

70 Ojima, I., Suzuki, T., *Tetrahedron Lett.* **1980**, *21*, 1239.

71 Meyer, D., Poulin, J.-C., Kagan, H. B., Levin-Pinto, H., Morgat, J.-L., Fromageot, P., *J. Org. Chem.* **1980**, *45*, 4680.

72 El-Baba, S., Poulin, J.-C., Kagan, H. B., *Tetrahedron* **1984**, *40*, 4275.

73 (a) Ojima, I., Yoda, N., Yatabe, M., *Tetrahedron Lett.* **1982**, *23*, 3917; (b) Ojima, I., Yoda, N., Yatabe, M., Tanaka, T., Kogure, T., *Tetrahedron* **1984**, *40*, 1255.

74 Aguado, G. P., Moglioni, A. G., Brousse, B. N., Ortuño, R.M., *Tetrahedron: Asymm.* **2003**, *14*, 2445; Aguado, G. P., Moglioni, A. G., García-Expósito, E., Branchadell, V., Ortuño, R. M., *J. Org. Chem.* **2004**, *69*, 7971.

75 Kreuzfeld, H.-J., Döbler, C., Schmidt, U., Krause, H.W., *Chirality* **1998**, *10*, 535.

76 Ojima, I., Yatabe, M., *Chem. Lett.* **1982**, 1335.

77 Sinou, D., Lafont, D., Descotes, G., *J. Organomet. Chem.* **1981**, *217*, 119.

78 Onuma, K., Ito, T., Nakamura, A., *Chem. Lett.* **1980**, 481.

79 Hammadi, A., Nuzillard, J.M., Poulin, J. C., Kagan, H. B., *Tetrahedron: Asymm.* **1992**, *3*(10), 1247.

80 (a) Yatagai, M., Zama, M., Yamagishi, T., Hida, M., *Chem. Lett.* **1983**, 12036; (b) Yatagai, M., Zama, M., Yamagishi, T., Hida, M., *Bull. Chem. Soc. Jpn.* **1984**, *57*, 739.

81 Yatagai, M., Yamagishi, T., Hida, M., *Bull. Chem. Soc. Jpn.* **1984**, *57*, 823.

82 Yamagishi, T., Ikeda, S., Yatagai, M., Yamaguchi, M., Hida, M., *J. Chem. Soc., Perkin Trans. 1* **1988**, 1787.

83 Ikeda, S., Yamagishi, T., Yamaguchi, M., Hida, M., *Chemistry Express* **1990**, *5*(1), 29.

84 Hayashi, T., Kawamura, N., Ito, Y., *J. Am. Chem. Soc.* **1987**, *109*, 7876.

85 (a) Poulin, J.-C., Kagan, H. B., *J. Chem. Soc., Chem. Commun.* **1982**, 1261; (b) El-Baba, S., Poulin, J.-C., Kagan, H. B., *Bull. Soc. Chim. Fr.* **1994**, *131*, 525.

86 Ritzén, A., Basu, B., Chattopadhyay, S. K., Dossa, F., Frejd, T., *Tetrahedron: Asymm.* **1998**, *9*, 503.

87 Hiebl, J., Kollmann, H., Rovenszky, F., Winkler, K., *J. Org. Chem.* **1999**, *64*, 1947.

88 Ojima, I., Tanaka, T., Kogure, T., *Chem. Lett.* **1981**, 823.

89 Tani, K., Tanigawa, E., Tatsuno, Y., Otsuka, S., *Chem. Lett.* **1986**, 737.

90 Yamagishi, T., Ikeda, S., Egawa, T., Yamaguchi, M., Hida, M., *Bull. Chem. Soc. Jpn.* **1990**, *63*, 281.

91 (a) Ohkuma, T., Ooka, H., Hashiguchi, S., Ikariya, T., Noyori, R., *J. Am. Chem. Soc.* **1995**, *117*, 2675; (b) Ohkuma, T., Ooka, H., Ikariya, T., Noyori, R., *J. Am. Chem. Soc.* **1995**, *117*, 10417; (c) Doucet, H., Ohkuma, T., Murata, K., Yokozawa, T., Kozawa, M., Katayama, E., England, A. F., Ikariya, T., Noyori, R., *Angew. Chem., Int. Ed.* **1998**, *37*(12), 1703.

92 Abdur-Rashid, K., Lough, A. J., Morris, R. H., *Organometallics* **2001**, *20*, 1047.

93 Miyashita, A., Takaya, H., Souchi, T., Noyori, R., *Tetrahedron* **1984**, *40*(8), 1245.

94 Miyashita, A., Karino, H., Shimamura, J., Chiba, T., Nagano, K., Nohira, H., Takaya, H., *Chem. Lett.* **1989**, 1849.

95 Schmid, R., Cereghetti, M., Heiser, B., Schönholzer, P., Hansen, H.-J., *Helv. Chim. Acta* **1988**, *71*(4), 897; Schmid, R., Foricher, J., Cereghetti, M., Schönholzer, P., *Helv. Chim. Acta* **1991**, *74*(2), 370.

96 Nishi, T., Kitamura, M., Ohkuma, T., Noyori, R., *Tetrahedron Lett.* **1988**, *29*(48), 6327.

97 Armstrong, III, J. D., Keller, J. L., Lynch, J., Liu, T., Hartner, Jr., F. W., Ohtake, N., Okada, S., Imai, Y., Okamoto, O., Ushijima, R., Nakagawa, S., Volante, R.P., *Tetrahedron Lett.* **1997**, *38*(18), 3203.

98 Ohtake, N., Jona, H., Okada, S., Okamoto, O., Imai, Y., Ushijima, R., Nakagawa, S., *Tetrahedron: Asymm.* **1997**, *8*(17), 2939.

99 (a) Drouillat, B., Poupardin, O., Bourdreux, Y., Greck, C., *Tetrahedron Lett.* **2003**, *44*, 2781; (b) Thomassigny, C., Greck, C., *Tetrahedron: Asymm.* **2004**, *15*, 199.

100 Duprat De Paule, S., Piombo, L., Ratovelomanana-Vidal, V., Greck, C., Genêt, J.-P., *Eur. J. Org. Chem.* **2000**, 1535.

101 Blandin, V., Carpentier, J.-F., Mortreux, A., *Eur. J. Org. Chem.* **1999**, 1787.

102 Shao, L., Kawano, H., Saburi, M., Uchida, Y., *Tetrahedron* **1993**, *49*, 1997.

103 (a) Blandin, V., Carpentier, J.-F., Mortreux, A., *Eur. J. Org. Chem.* **1999**, 3421; (b) Blandin, V., Carpentier, J.-F., Mortreux, A., *New. J. Chem.* **2000**, *24*, 309.

104 Schulz, S., *Chem. Commun.* **1999**, 1239.

105 Kawano, H., Ishii, Y., Saburi, M., Uchida, Y., *J. Chem. Soc., Chem. Commun.* **1988**, 87.

106 Kitamura, M., Ohkuma, T., Inoue, S., Sayo, N., Kumobayashi, H., Akutagawa, S., Ohta, T., Takaya, H., Noyori, R., *J. Am. Chem. Soc.* **1988**, *110*, 629.

107 Noyori, R., Ohkuma, T., Kitamura, M., Takaya, H., Sayo, N., Kumobayashi, H., Akutagawa, S., *J. Am. Chem. Soc.* **1987**, *109*, 5856.

108 Burk, M. J., Harper, T. G. P., Kalberg, C.S., *J. Am. Chem. Soc.* **1995**, *117*, 4423.

109 Genêt, J.-P., Pinel, C., Ratovelomanana-Vidal, V., Mallart, S., Pfister, X., Bischoff, L., Caño de Andrade, M. C., Darses, S., Galopin, C., Laffitte, J. A., *Tetrahedron: Asymm.* **1994**, *5*, 675.

110 Blanc, D., Ratovelomanana-Vidal, V., Marinetti, A., Genêt, J.-P., *Synlett* **1999**, (4), 480.

111 Mezzetti, A., Consiglio, G., *J. Chem. Soc., Chem. Commun.* **1991**, 1675.

112 Mezzetti, A., Tschumper, A., Consiglio, G., *J. Chem. Soc., Dalton Trans.* **1995**, 49.

113 Cui, X., Ogle, J.W., Burgess, K., *Chem. Commun.* **2005**, 672.

114 Kagan, H. B., Fiaud, J. C., *Top. Stereochem.* **1988**, *18*, 249.

115 Brown, J. M., James, A. P., Prior, L. M., *Tetrahedron Lett.* **1987**, *28*, 2179.

116 Taber, D. F., Wang, Y., *J. Am. Chem. Soc.* **1997**, *119*, 22.

117 Ohkuma, T., Koizumi, M., Muñiz, K., Hilt, G., Kabuto, C., Noyori, R., *J. Am. Chem. Soc.* **2002**, *124*, 6508.

118 Viso, A., Lee, N. E., Buchwald, S. L., *J. Am. Chem. Soc.* **1994**, *116*, 9373.

119 Noyori, R., Ikeda, T., Ohkuma, T., Widhalm, M., Kitamura, M., Takaya, H., Akutagawa, S., Sayo, N., Saito, T., Taketomi, T., Kumobayashi, H., *J. Am. Chem. Soc.* **1989**, *111*, 9134.

120 Kitamura, M., Ohkuma, T., Tokunaga, M., Noyori, R., *Tetrahedron: Asymm.* **1990**, *1*, 1.

121 Genêt, J.-P., Pinel, C., Mallart, S., Juge, S., Thorimbert, S., Laffitte, J. A., *Tetrahedron: Asymm.* **1991**, *2*(7), 555.

122 Genêt, J.-P., Pfister, X., Ratovelomanana-Vidal, V., Pinel, C., Laffitte, J. A., *Tetrahedron Lett.* **1994**, *35*, 4559.

123 (a) Mashima, K., Matsumura, Y., Kusano, K., Kumobayashi, H., Sayo, N., Hori, Y., Ishizaki, T., Akutagawa, S., Takaya, H., *J. Chem. Soc., Chem. Commun.* **1991**, 609; (b) Takaya, H., Ohta, T., Mashima, K., *Advances in Chemistry Series* **1992**, *230*, 123.

124 Genêt, J.-P., Caño de Andrade, M. C., Ratovelomanana-Vidal, V., *Tetrahedron Lett.* **1995**, *36*, 2063.

125 Coulon, E., Caño de Andrade, M.C., Ratovelomanana-Vidal, V., Genêt, J.-P., *Tetrahedron Lett.* **1998**, *39*, 6467.

126 Phansavath, P., Duprat de Paule, S., Ratovelomanana-Vidal, V., Genêt, J.-P., *Eur. J. Org. Chem.* **2000**, 3903.

127 Makino, K., Okamoto, N., Hara, O., Hamada, Y., *Tetrahedron: Asymm.* **2001**, *12*, 1757.

128 Makino, K., Goto, T., Hiroki, Y., Hamada, Y., *Angew. Chem. Int. Ed. Engl.* **2004**, *43*, 882.

129 Labeeuw, O., Phansavath, P., Genêt, J.-P., *Tetrahedron: Asymm.* **2004**, *15*, 1899.

130 Mordant, C., Dünkelmann, P., Ratovelomanana-Vidal, V., Genêt, J.-P., *Chem. Commun.* **2004**, 1296.

131 Mordant, C., Dünkelmann, P., Ratovelomanana-Vidal, V., Genêt, J.-P., *Eur. J. Org. Chem.* **2004**, 3017.

132 Lei, A., Wu, S., He, M., Zhang, X., *J. Am. Chem. Soc.* **2004**, *126*, 1626.

133 Ratovelomanana-Vidal, V., Genêt, J.-P., *J. Organomet. Chem.* **1998**, *567*, 163.

134 Mordant, C., Caño de Andrade, C., Touati, R., Ratovelomanana-Vidal, V., Hassine, B. B., Genêt, J.-P., *Synthesis* **2003**, (15), 2405.

135 Kitamura, M., Tokunaga, M., Pham, T., Lubell, W. D., Noyori, R., *Tetrahedron Lett.* **1995**, *36*(32), 5769.

136 Kitamura, M., Tokunaga, M., Noyori, R., *J. Am. Chem. Soc.* **1995**, *117*, 2931.

137 Genêt, J.-P. *Pure Appl. Chem.* **2002**, *74*, 77.

138 Mordant, C., Reymond, S., Ratovelomanana-Vidal, V., Genêt, J.-P., *Tetrahedron* **2004**, *60*, 9715.

139 Matsumonto, T., Murayama, T., Mitsuhashi, S., Miura, T., *Tetrahedron Lett.* **1999**, *40*, 5043.

140 Ohkuma, T., Li, J., Noyori, R., *Synlett* **2004**, (8), 1383.

22
Hydrogen-Mediated Carbon–Carbon Bond Formation Catalyzed by Rhodium

Chang-Woo Cho and Michael J. Krische

22.1
Introduction and Mechanistic Considerations

The development of direct catalytic methods for reductive carbon–carbon bond formation has emerged as the subject of intensive investigation [1–10]. The catalytic hydrometallative reductive coupling of alkenes [1], alkynes [2, 3], allenes [4], conjugated enones [5–7], conjugated dienes [8–10] and conjugated enynes [11] to carbonyl partners and imines has been achieved using silanes, stannanes, boranes and alanes as terminal reductant. The use of such terminal reductants mandates stoichiometric byproduct generation. Related hydrogen-mediated transformations would proceed with complete levels of atom economy [12]. However, while metal catalysts capable of reversible transfer hydrogenation have been applied to the development of C–C bond formations predicated on dehydrogenation-trapping-rehydrogenation [13], true hydrogen-mediated reductive C–C bond formations only have been achieved for processes involving migratory insertion of carbon monoxide, for example, alkene hydroformylation and the Fischer-Tropsch reaction [14, 15].

The question persists as to whether the organometallic intermediates that appear transiently during the course of catalytic hydrogenation can be intercepted and re-routed to products of C–C bond formation. In the case of rhodium-catalyzed alkene hydroformylation, a key feature appears to be the involvement of mono-hydride-based catalytic cycles, wherein the formation of (alkyl)(hydrido)metal intermediates occurs *subsequent* to C–C bond formation. In contrast, conventional dihydride-based hydrogenation cycles generally afford (alkyl)(hydrido)metal intermediates in *advance* of potential C–C bond formation. For such dihydride-based hydrogenation cycles, the capture of hydrogenation intermediates is likely untenable due to rapid C–H reductive elimination. This may account, in part, for the exceptional rarity of hydrogen-mediated C–C bond formation in the absence of carbon monoxide [15].

Recent studies from our laboratory demonstrate the feasibility of hydrogen-mediated C–C bond formation under "CO-free conditions." Here, at least two

The Handbook of Homogeneous Hydrogenation.
Edited by J. G. de Vries and C. J. Elsevier
Copyright © 2007 WILEY-VCH Verlag GmbH & Co. KGaA, Weinheim
ISBN: 978-3-527-31161-3

distinct mechanistic pathways potentially operate. Initial studies on hydrogen-mediated reductive aldol coupling demonstrate that conventional hydrogenation pathways are suppressed through the use of cationic rhodium precatalysts in the presence of a mild base. Such conditions are believed to promote heterolytic hydrogen activation (H_2+M–X → M–H+HX) [16, 20]. Monohydride-mediated hydrometallation should furnish organometallic species that do not possess hydride ligands, disabling direct C–H reductive elimination manifolds and extending the lifetimes of the organometallic intermediates obtained upon hydrometallation to facilitate their capture. Hence, one strategy for hydrogen-mediated C–C bond formation involves the hydrogenation of reactants using catalysts that operate *via* monohydride-based catalytic cycles. A second strategy for hydrogen-mediated C–C bond formations takes advantage of the fact that hydrogen activation can be quite slow for certain conventional hydrogenation catalysts. Here, oxidative coupling of the reacting partners prior to hydrogenation activation becomes feasible [17] (Scheme 22.1).

Among homogeneous hydrogenation catalysts, those based on rhodium are especially well studied [18–20, 28]. Whereas neutral rhodium(I)-complexes such as Wilkinson's catalyst induce *homolytic* hydrogen activation [18, 19], the use of cationic rhodium(I) complexes in conjunction with basic additives is believed to promote *heterolytic* activation pathways [20]. Heterolytic hydrogen activation by cationic rhodium complexes presumably is owed to the enhanced acidity of the cationic dihydrides that result upon oxidative addition in comparison to their neutral counterparts [21]. Thus, heterolytic hydrogen activation is believed to occur through a two-stage process involving hydrogen oxidative addition followed by base-induced H–X reductive elimination [22] (Scheme 22.2).

Hydrogen activation is rate-determining for enantioselective hydrogenations employing cationic rhodium catalysts [28]. This observation is significant given that closely related cationic rhodium(I) complexes are known to catalyze a variety of C–C bond formations believed to proceed through the initial oxidative coupling of π-unsaturated partners to furnish metallocyclic intermediates [17]. Accordingly, tandem oxidative coupling-metallocycle hydrogenolysis strategies toward hydrogen-mediated C–C bond formation have proven fruitful (*vide supra*).

Here, a comprehensive overview of hydrogen-mediated C–C bond formation under CO-free conditions is presented [23]. This emergent family of reductive couplings now encompasses:

- the intra- and intermolecular reductive coupling of enone and enal pronucleophiles with aldehyde and ketone partners [24];
- the intermolecular reductive coupling of 1,3-cyclohexadiene with α-ketoaldehydes [25];
- the intermolecular reductive coupling of 1,3-enynes and 1,3-diynes with α-ketoaldehydes and iminoacetates [26]; and
- the reductive cyclization of 1,6-diynes and 1,6-enynes [27].

These results establish catalytic hydrogenation as a powerful and mechanistically novel means of catalytic C–C bond formation, and support the feasibility

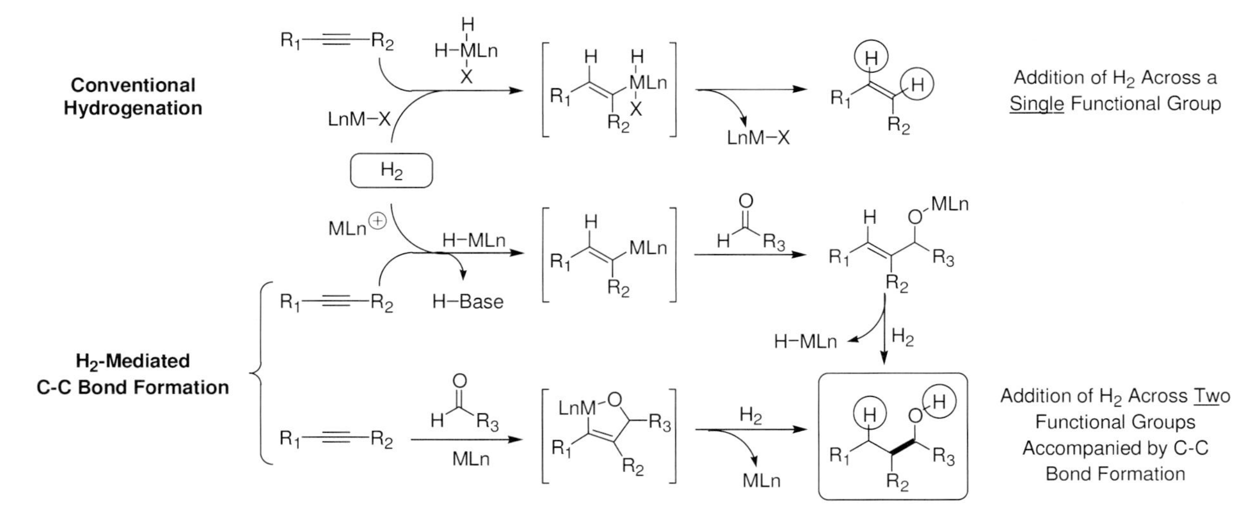

Scheme 22.1 Potential mechanistic pathways for hydrogen-mediated C–C bond formation.

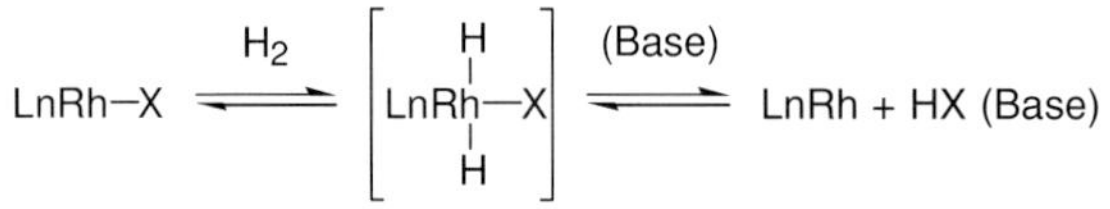

Scheme 22.2 Formal heterolytic hydrogen activation *via* deprotonation of a dihydride intermediate.

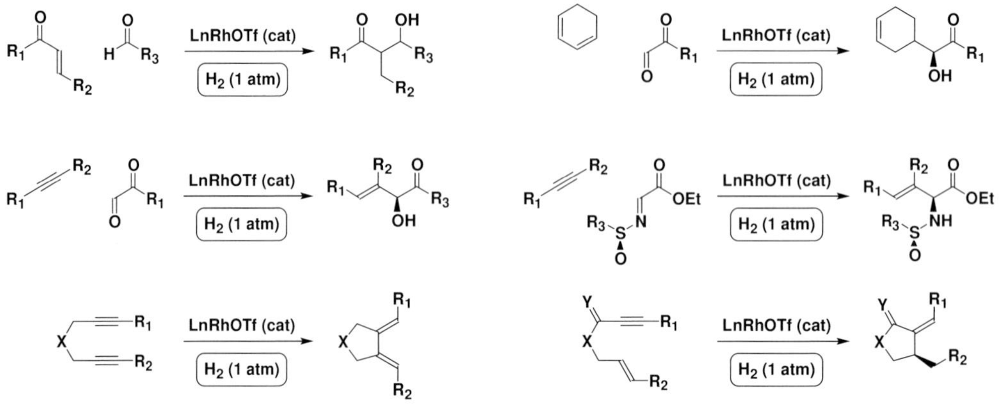

Scheme 22.3 Hydrogen-mediated C–C bond formations catalyzed by rhodium.

of developing a broad new class of catalytic reductive C–C bond formations (Scheme 22.3).

22.2
Reductive Coupling of Conjugated Enones and Aldehydes

22.2.1
Intramolecular Reductive Aldolization

Initial studies pertaining to the Rh-catalyzed aldol cycloreduction under hydrogenation conditions are consistent with the bifurcated catalytic mechanism depicted in Schemes 22.1 and 22.4 [24 a]. Catalytic hydrogenation of the indicated enone-aldehyde using the neutral complex $Rh(PPh_3)_3Cl$ provides only trace quantities of the aldol product due to competitive 1,4-reduction via conventional hydrogenation. In contrast, rhodium salts that embody increased cationic character, such as $Rh^I(COD)_2OTf$, provide almost equal proportions of aldol and 1,4-reduction products. Finally, when $Rh^I(COD)_2OTf$ is used in conjunction with substoichiometric quantities of the mildly basic additive potassium acetate, the proportion of aldol product is increased such that simple 1,4-reduction manifolds are nearly fully suppressed. The observed *syn*-diastereoselectivity suggests intermediacy of a *Z*-enolate and a Zimmerman-Traxler-type transition state. These optimized conditions

Table 22.1 Partitioning of aldolization and 1,4-reduction pathways depends critically on the use of cationic Rh-complexes and mildly basic additives.[a)]

Catalyst (10 mol%), Ligand (24 mol%)
H_2 (1 atm), Additive, DCE
25 °C, 3 h

Substrate	Catalyst	Ligand	Additive (mol%)	Yield Aldol (Syn-Anti)	Yield 1,4-Reduction
n=2, R=Ph	$Rh(PPh_3)Cl$	–	–	1% (99:1)	95%
n=2, R=Ph	$Rh(COD)_2OTf$	PPh_3	–	21% (99:1)	25%
n=2, R=Ph	$Rh(COD)_2OTf$	PPh_3	KOAc (30%)	59% (58:1)	21%
n=2, R=Ph	$Rh(COD)_2OTf$	$(p\text{-}CF_3Ph)_3P$	–	57% (14:1)	22%
n=2, R=Ph	**$Rh(COD)_2OTf$**	**$(p\text{-}CF_3Ph)_3P$**	**KOAc (30%)**	**89% (10:1)**	**0.1%**
n=2, R=*p*-MeOPh	$Rh(COD)_2OTf$	$(p\text{-}CF_3Ph)_3P$	KOAc (30%)	74% (5:1)	3%
n=2, R=2-Naphthyl	$Rh(COD)_2OTf$	$(p\text{-}CF_3Ph)_3P$	KOAc (30%)	90% (10:1)	1%
n=2, R=2-thiophenyl	$Rh(COD)_2OTf$	$(p\text{-}CF_3Ph)_3P$	KOAc (30%)	76% (19:1)	2%
n=2, R=2-Furyl	$Rh(COD)_2OTf$	$(p\text{-}CF_3Ph)_3P$	KOAc (30%)	70% (6:1)	10%
n=1, R=Ph	$Rh(COD)_2OTf$	$(p\text{-}CF_3Ph)_3P$	KOAc (30%)	71% (24:1)	1%
n=2, R=CH_3	$Rh(COD)_2OTf$	$(p\text{-}CF_3Ph)_3P$	KOAc (30%)	65% (1:5)	–

a) As product ratios were found to vary with surface to volume ratio of the reaction mixture, all transformations were conducted on 1.48 mmol scale in 50 mL round bottomed flasks.

proved general for the cycloreduction of aromatic, heteroaromatic and aliphatic enone substrates to form five- and six-membered ring products (Table 22.1).

The pronounced effect of basic additives on partitioning of the aldolization and 1,4-reduction manifolds suggests that enolate-hydrogen reductive elimination pathways are disabled through deprotonation of the (hydrido)metal intermediates $LnRh^{III}X(H)_2$ or (enolato)$Rh^{III}X(H)Ln$. Thus, as proposed by Osborn and Schrock [20], deprotonation shifts the catalytic mechanism from a dihydride-based cycle to a monohydride-based cycle. In the former case, 1,4-reduction products would predominate. In the latter case, owing to the absence of (alkyl)(hydrido)rhodium intermediates, capture of the rhodium enolate through its addition to the appendant aldehyde is facilitated. The following control experiments were performed. Exposure of the simple 1,4-reduction product to the reaction conditions does not result in aldolization. Conversely, re-exposure of the aldol product to the reaction conditions does not result in retroaldolization. Finally, exposure of the substrate to standard reaction conditions in the *absence* of hydrogen does not afford products of Morita-Baylis-Hillman cyclization. For the sake of clarity, the catalytic mechanism indicated in Scheme 22.4 has been simplified. For example, equilibria involving association of the substrate to the catalyst prior to hydrogen activation are omitted, although such equilibria are known to be an important feature of the enantioselective hydrogenation of dehydroamino acids employing cationic rhodium catalysts [28] (Scheme 22.4).

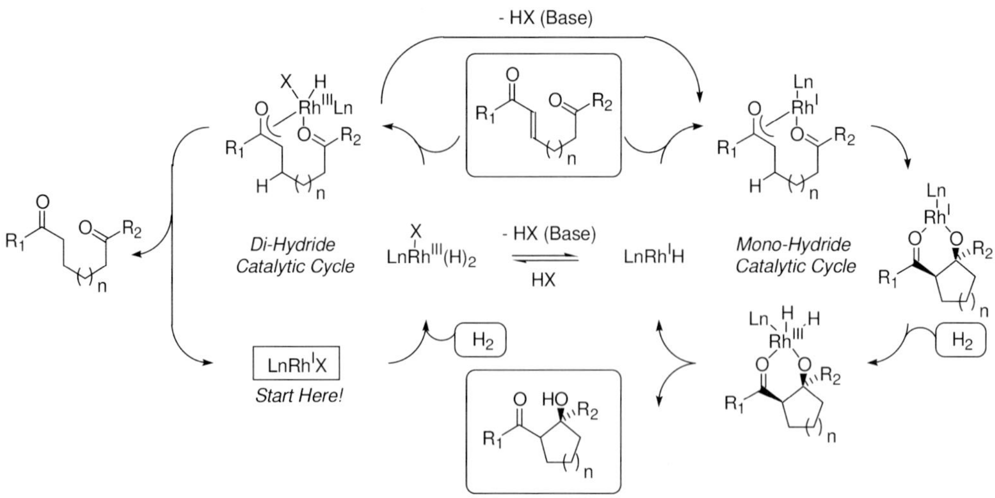

Scheme 22.4 A bifurcated mechanism accounting for the effect of basic additives.

In order to explore the scope of this hydrogen-mediated aldol addition methodology, additions to ketone acceptors were explored. Because ketones are less electrophilic than aldehydes, competitive conventional hydrogenation was anticipated to be problematic. Indeed, upon exposure of keto-enones to basic hydrogenation conditions, the formation of five- and six-membered ring aldol products is accompanied by substantial quantities of conventional hydrogenation products. As retro-aldolization does not occur upon resubmission of the aldol products to the reaction conditions, enone hydrogenation must occur prior to carbonyl addition. Nevertheless, serviceable yields of the ketone aldol products are obtained. Moreover, very high levels of *syn*-diastereoselectivity are observed, which again are attributed to the intermediacy of a *Z*-enolate and a Zimmerman-Traxler-type transition state. While aldolization proceeds readily at ambient temperature, more reproducible ratios of aldol and 1,4-reduction product are observed at 80 °C [24 b] (Table 22.2).

In order to gain further insight into the reaction mechanism, the indicated oxygen-tethered keto-enone was subjected to basic hydrogenation conditions under 1 atmos. elemental deuterium. Deuterium incorporation is observed at the former enone β-position exclusively. In addition to mono-deuterated material (81% composition), doubly-deuterated (8% composition) and non-deuterated materials (11% composition) are observed. These data suggest reversible hydrometallation in the case of keto-enone substrates. Consistent with the mechanism depicted in Scheme 22.4, deuterium is not incorporated at the α-position of the aldol product [24 b] (Scheme 22.5).

For the cycloreduction of keto-enones, competitive 1,4-reduction in response to the reduced electrophilicity of the carbonyl partner is observed. Diones are more susceptible to addition by viture of inductive effects and the relief of di-

Table 22.2 Catalytic hydrogen-mediated reductive aldol cyclization of keto-enones.

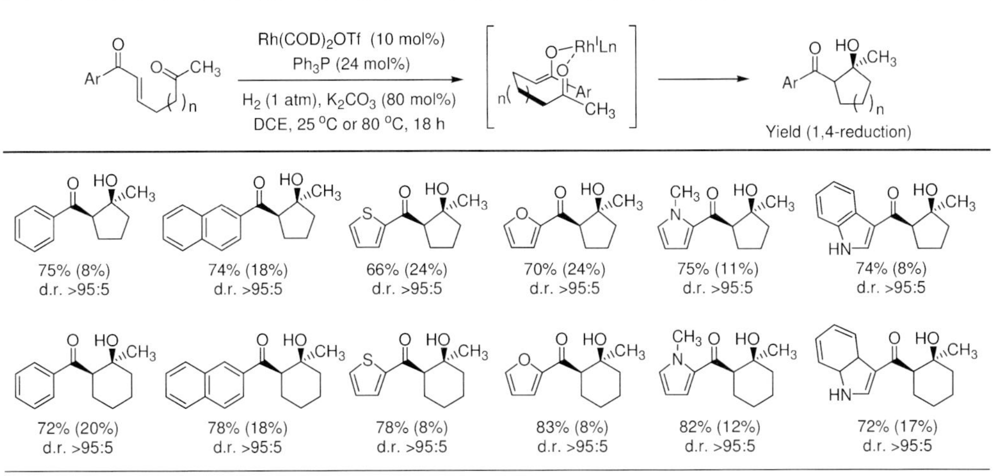

Rh(COD)$_2$OTf (10 mol%)
Ph$_3$P (24 mol%)
H$_2$ (1 atm), K$_2$CO$_3$ (80 mol%)
DCE, 25 °C or 80 °C, 18 h

Yield (1,4-reduction)

75% (8%) d.r. >95:5	74% (18%) d.r. >95:5	66% (24%) d.r. >95:5	70% (24%) d.r. >95:5	75% (11%) d.r. >95:5	74% (8%) d.r. >95:5
72% (20%) d.r. >95:5	78% (18%) d.r. >95:5	78% (8%) d.r. >95:5	83% (8%) d.r. >95:5	82% (12%) d.r. >95:5	72% (17%) d.r. >95:5

a) As product ratios were found to vary with surface-to-volume ratio of the reaction mixture, all transformations were conducted on 1.48 mmol scale in 13×100-mm sealed test tubes.

pole-dipole interactions. Accordingly, catalytic hydrogenation of dione-containing substrates affords the corresponding aldol products in good yield and with excellent *syn*-diastereoselectivity for both benzoyl- and acetyl-containing enones. Simple 1,4-reduction only accompanies formation of the strained *cis*-decalone ring system [24 b] (Scheme 22.6).

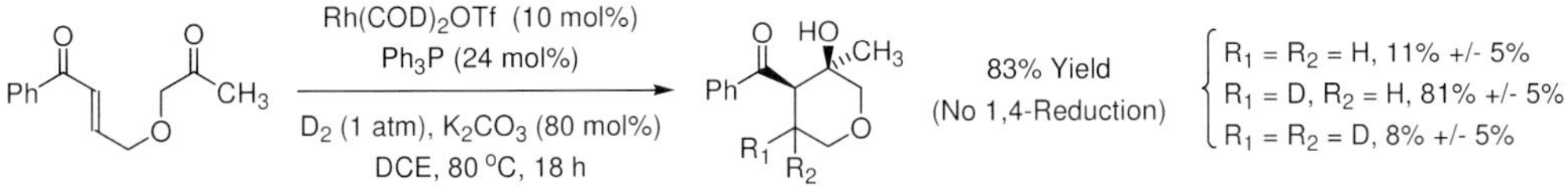

Rh(COD)$_2$OTf (10 mol%)
Ph$_3$P (24 mol%)
D$_2$ (1 atm), K$_2$CO$_3$ (80 mol%)
DCE, 80 °C, 18 h

83% Yield
(No 1,4-Reduction)

R$_1$ = R$_2$ = H, 11% +/- 5%
R$_1$ = D, R$_2$ = H, 81% +/- 5%
R$_1$ = R$_2$ = D, 8% +/- 5%

Scheme 22.5 Deuterium-labeling studies suggest reversible hydrometallation for keto-enone substrates.

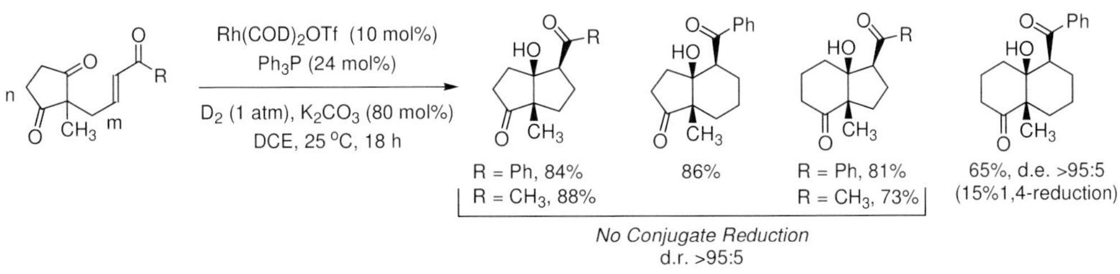

Rh(COD)$_2$OTf (10 mol%)
Ph$_3$P (24 mol%)
D$_2$ (1 atm), K$_2$CO$_3$ (80 mol%)
DCE, 25 °C, 18 h

R = Ph, 84%
R = CH$_3$, 88%

86%

R = Ph, 81%
R = CH$_3$, 73%

65%, d.e. >95:5
(15%1,4-reduction)

No Conjugate Reduction
d.r. >95:5

Scheme 22.6 Catalytic hydrogen-mediated reductive aldol cyclization of enone-diones.

Rh(COD)$_2$OTf (10 mol%)
(2-furyl)$_3$P (24 mol%)
H_2 (1 atm), K_2CO_3 (100 mol%)
THF, 40 °C, 1-3 h

72% (syn:anti, 2:1) (16% 1,4-reduction); 73% (10:1) (21%); 63% (5:1) (30%); 59% (4:1) (29%)

n = 1, 63% (1:3) (21%); n = 2, 61% (2:1) (20%); 63% (2:1) (21%); 67% (2:1) (20%); 61% (5:1) (20%)

Scheme 22.7 Catalytic addition of metallo-aldehyde enolates to ketones.

Perhaps the most elusive variant of the aldol reaction involves the addition of metallo-aldehyde enolates to ketones. A single stoichiometric variant of this transformation is known [29]. As aldolization is driven by chelation, intramolecular addition to afford a robust transition metal aldolate should bias the enolate-aldolate equilibria toward the latter [30, 31]. Indeed, upon exposure to basic hydrogenation conditions, keto-enal substrates provide the corresponding cycloaldol products, though competitive 1,4-reduction is observed (Scheme 22.7) [24 d].

22.2.2 Intermolecular Reductive Aldolization

In principle, the presumed rhodium(I) enolates that occur transiently during the course of enone hydrogenation may: (i) engage in aldolization; or (ii) hydrogenolytically cleave *via* oxidative addition of hydrogen followed by reductive elimination. In principle, intermolecular capture of such hydrogenation intermediates should suffer due to increasingly competitive conventional hydrogenation. In practice, only a modest excess of vinyl ketone is required to offset conventional hydrogenation manifolds. For example, hydrogenation of phenyl vinyl ketone (PVK) (150 mol%) in the presence of aromatic and heteroaromatic aldehydes (100 mol%) provides good yields of the corresponding aldol products. As PVK is prone toward anionic polymerization, these results are especially noteworthy. Consistent with the bifurcated mechanism depicted in Scheme 22.4, the addition of potassium acetate significantly increases the yield of aldol product (Table 22.3) [24 a].

In the case of methyl vinyl ketone (MVK), similar reactivity is observed. Exposure of MVK (150 mol%) and *p*-nitrobenzaldehyde to basic hydrogenation conditions provides the corresponding aldol product in good yield, though poor diastereoselectivity is observed [24 a]. Remarkably, upon use of *tris*(2-furyl)phosphine as ligand and Li_2CO_3 as basic additive, the same aldol product is formed with high levels of *syn*-selectivity [24 e]. Addition of MVK to activated ketones such as 1-(3-bromophenyl)propane-1,2-dione is accomplished under similar con-

Table 22.3 Use of phenyl vinyl ketone (PVK) in intermolecular hydrogen-mediated reductive aldol coupling.

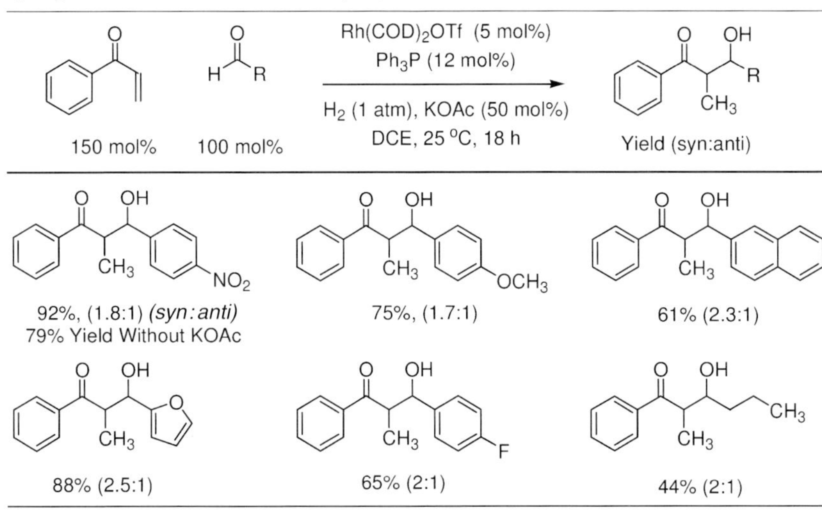

a) As product ratios were found to vary with surface-to-volume ratio of the reaction mixture, all transformations were conducted on 1.0 mmol scale in 50-mL round-bottomed flasks.

ditions [24e]. Notably, neither reduction of the nitro group or aryl bromide is observed (Scheme 22.8).]

Intermolecular cross aldolization of metallo-aldehyde enolates typically suffers from polyaldolization, product dehydration and competitive Tishchenko-type processes [32]. While such cross-aldolizations have been achieved through amine catalysis and the use of aldehyde-derived enol silanes [33], the use of aldehyde enolates in this capacity is otherwise undeveloped. Under hydrogenation conditions, acrolein and crotonaldehyde serve as metallo-aldehyde enolate precursors, participating in selective cross-aldolization with α-ketoaldehydes [24c]. The resulting β-hydroxy-γ-ketoaldehydes are highly unstable, but may be trapped *in situ* through the addition of methanolic hydrazine to afford 3,5-disubstituted pyridazines (Table 22.4).

To corroborate the proposed mechanism, the catalytic reductive aldol coupling of acrolein with phenyl glyoxal monohydrate was performed under 1 atmos. elemental deuterium. Exposure of the aldol product to excess hydrazine *in situ* results in formation of the pyridazine, which incorporates precisely one deuterium atom in a manner consistent with the general mechanism proposed in Scheme 22.4 (Scheme 22.9).

Thus far, the use of acrylates and related acyl derivatives as nucleophilic partners in hydrogen-mediated reductive aldol coupling has been unsuccessful due to competitive conventional hydrogenation. Although the mechanistic basis of these results remains unclear, it may be speculated that for acrylates and struc-

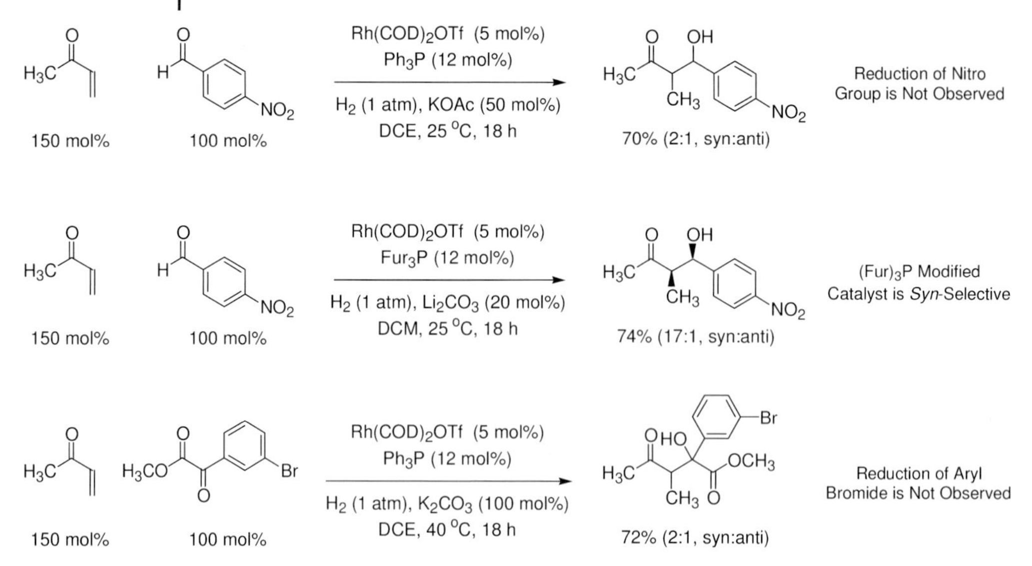

a) As product ratios were found to vary with surface-to-volume ratio of the reaction mixture, all transformations were conducted on 1.0 mmol scale in 50-mL round-bottomed flasks.

Scheme 22.8 Use of methyl vinyl ketone (MVK) in intermolecular hydrogen-mediated reductive aldol coupling.[a]

Table 22.4 Use of acrolein and crotonaldehyde in intermolecular hydrogen-mediated reductive aldol coupling.

Acrolein/crotonaldehyde (R_1)	Aldehyde (R_2)	Conditions	Intermediate	Cyclization	Product
(500 mol%)	(100 mol%)	$Rh(COD)_2OTf$ (1-5 mol%), Ph_3P (2.4-12 mol%), H_2 (1 atm), KOAc (100 mol%), DCE, 25 °C, 3 h	[aldol adduct]	H_2NNH_2, - 3 H_2O, 1 h	pyridazine

Product	Yield
4-methyl-6-phenylpyridazine	62%
4-methyl-6-(N-methylpyrrol-2-yl)pyridazine	30%
4-methyl-6-(2-naphthyl)pyridazine	59%
4-methyl-6-(2-thienyl)pyridazine	31%
4-ethyl-6-(2-naphthyl)pyridazine	50%
4-ethyl-6-phenylpyridazine	47%

a) As product ratios were found to vary with surface-to-volume ratio of the reaction mixture, all transformations were conducted on 1.0 mmol scale in 50-mL round-bottomed flasks.

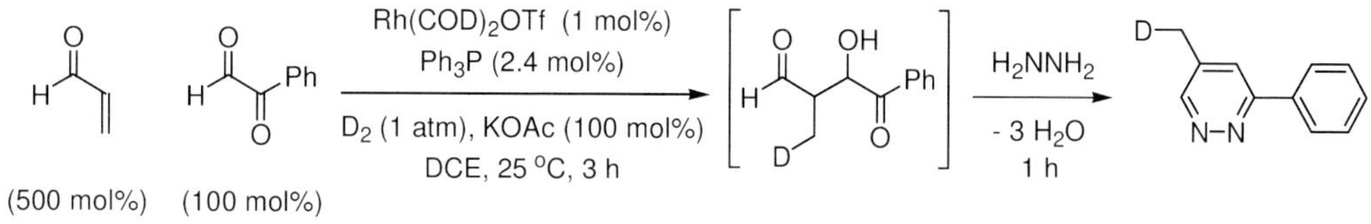

a) As product ratios were found to vary with surface-to-volume ratio of the reaction mixture, all transformations were conducted on 1.0 mmol scale in 50-mL round-bottomed flasks.

Scheme 22.9 Intermolecular reductive aldol coupling of acrolein and phenyl glyoxal under a D_2 atmosphere.[a)]

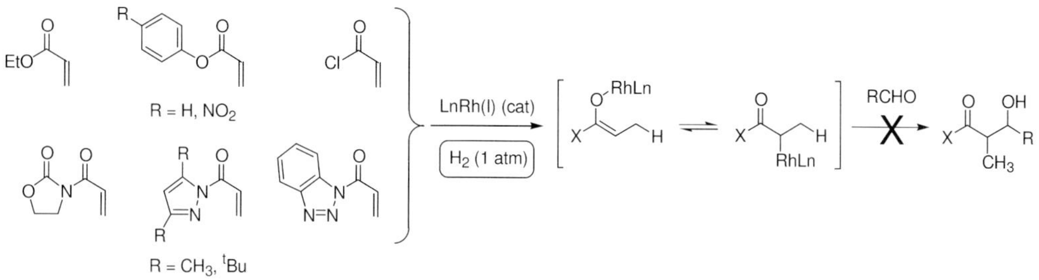

Scheme 22.10 Attempted reductive aldol coupling of ethyl acrylate and related acyl derivatives.

tural relatives possessing heteroatom substitution at the acyl position, the haptomeric equilibrium pertaining to the *O*-bound and *C*-bound forms of the rhodium(I) enolate is biased toward the latter. As aldol addition should occur by way of the *O*-bound enolate in accordance with the Zimmerman-Traxler model, the intervention of a *C*-bound enolate may diminish the rate of aldolization to the point that competitive conventional hydrogenation predominates.

22.3 Reductive Coupling of 1,3-Cyclohexadiene and α-Ketoaldehydes

Given the structural homology of conjugated enones and 1,3-dienes, the reductive coupling of 1,3-cyclohexadiene and phenyl glyoxal was examined under hydrogenation conditions [22]. Optimization studies again reveal the requirement of cationic rhodium catalysts. Whereas hydrogenation of 1,3-cyclohexadiene and phenyl glyoxal using Wilkinson's catalyst provides products of simple reduction, a 61% yield of reductive coupling product is obtained using $Rh(COD)_2OTf$ with PPh_3 as ligand. When $(p\text{-}CH_3OPh)_3P$ is employed as ligand, the yield of coupling product increases to 77%. Related cationic complexes, such as $Rh(COD)_2BF_4$, exhibit similar efficiencies when used in conjunction with $(p\text{-}CH_3OPh)_3P$. Under optimized conditions, the catalytic reductive coupling of 1,3-cyclohexadiene with di-

Table 22.5 Catalytic reductive coupling of 1,3-cyclohexadiene with alkyl, aryl and heteroaryl α-ketoaldehydes.

Catalyst, Phosphine
H_2 (1 atm), DCE
25 °C, 18 h
(100 mol%) (500 mol%)

Entry	Catalyst (mol%)	Ligand (mol%)	Yiedl (%)
1	$Rh(PPh_3)_3Cl$ (10%)	–	–
2	$Rh(COD)_2OTf$ (10%)	Ph_3P (20%)	61
3	$Rh(COD)_2OTf$ (10%)	$(p\text{-}CF_3Ph)_3P$ (20%)	24
4	$Rh(COD)_2OTf$ (10%)	$(p\text{-}CH_3OPh)_3P$ (20%)	77
5	$Rh(COD)_2BF_4$ (10%)	$(p\text{-}CH_3OPh)_3P$ (20%)	79
6	**$Rh(COD)_2OTf_4$ (5%)**	**$(p\text{-}CH_3OPh)_3P$ (10%)**	**76**

76% (1:1.6) 77% (1:1.5) 74% (1:1.4)

78% (1:1.3) 73% (1:1.4) 71% (1:1.6)

verse α-ketoaldehydes was examined. Aryl, heteroaryl and aliphatic α-ketoaldehydes provide reductive coupling products in good yield. Notably, basic additives are not required, suggesting that heterolytic hydrogen activation may not be operative (Table 22.5).

Deuterium-labeling studies reveal that the reductive coupling of 1,3-cyclohexadiene with α-ketoaldehydes occurs through a mechanism very different than that postulated for related enone-aldehyde couplings. Reductive coupling of 1,3-cyclohexadiene with 2-naphthyl glyoxal under an atmosphere of D_2(g) results in the incorporation of precisely two deuterium atoms as an equimolar distribution of 1,2- and 1,4-regioisomers. The relative stereochemistry of the deuterated materials could not be assigned (Scheme 22.11).

$Rh(COD)_2OTf$ (5 mol%)
$(p\text{-}MeOPh)_3P$ (10 mol%)
D_2 (1 atm), DCE
25 °C, 18 h
(100 mol%) (500 mol%) (1:1)

Scheme 22.11 Reductive coupling of 1,3-cyclohexadiene and 2-naphthyl glyoxal under an atmosphere of D_2(g).

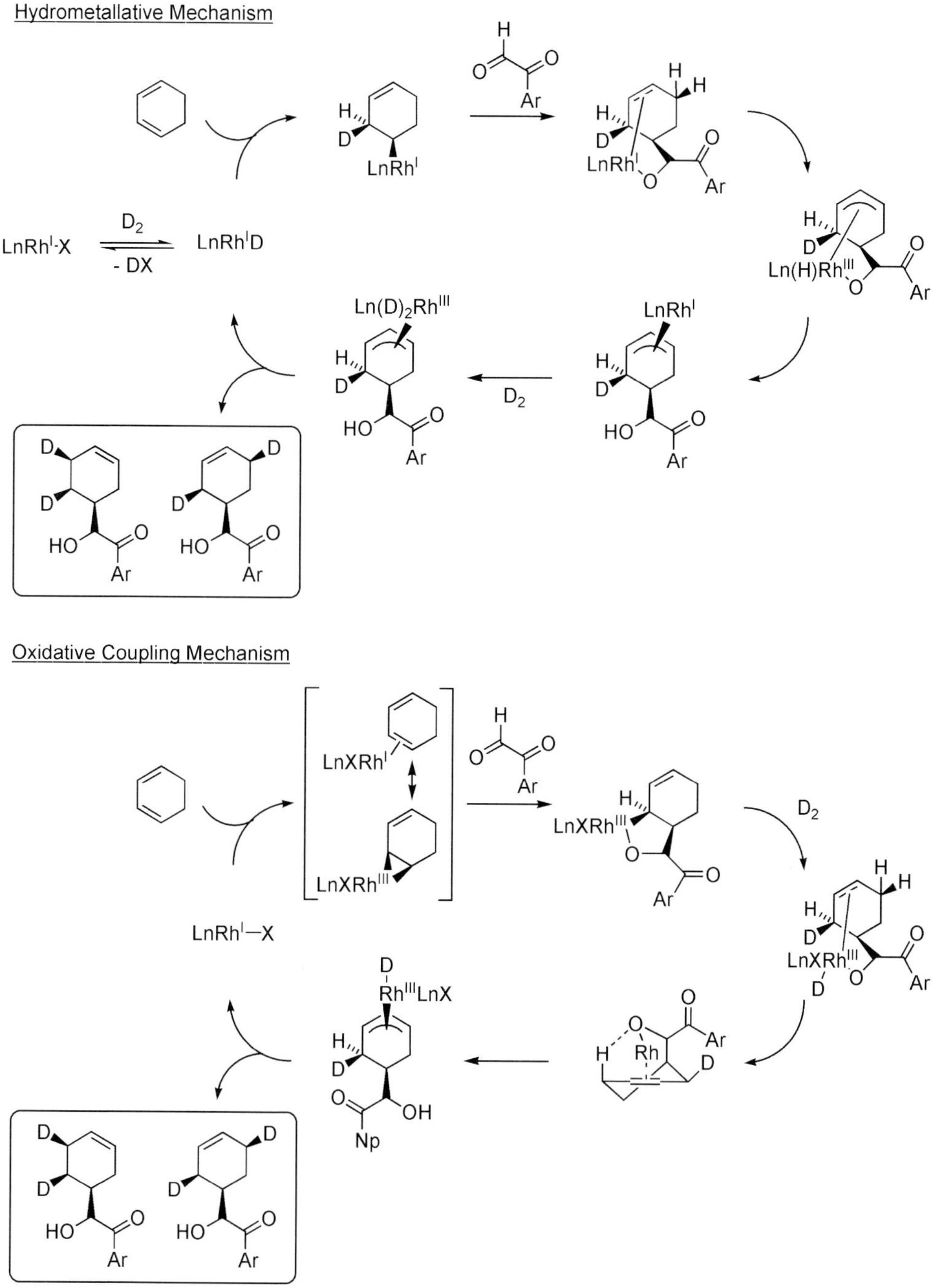

Hydrometallative Mechanism

Oxidative Coupling Mechanism

Scheme 22.12 Possible mechanisms for the reductive coupling of 1,3-cyclohexadiene and 2-naphthyl glyoxal under an atmosphere of D_2(g).

The mechanism initially proposed involves diene hydrometallation from a monohydride derived via heterolytic hydrogen activation (Scheme 22.12, upper). Here, diene deuterometallation gives rise to a homo-allyl rhodium intermediate, which engages in carbonyl addition to afford a rhodium alkoxide. The indicated regiochemistry of C–C bond formation is consistent with that observed by Loh in the nickel-catalyzed reductive coupling of 1,3-cyclohexadiene with aldehydes [9e]. Additionally, as observed by Mori, the presence of 1,3-cyclohexadiene induces 1,4-regiochemistry in nickel-promoted diene-aldehyde cyclizations [8b]. Allylic C–H insertion provides a rhodium(III) π-allyl, which upon O–H reductive elimination gives rise to a rhodium(I) π-allyl. Similar allylic C–H insertions are observed in metal-catalyzed alkene isomerization [34]. Finally, oxidative addition of elemental deuterium followed by C–D reductive elimination completes the catalytic cycle. The intermediacy of rhodium a π-allyl is required to account for the incorporation of precisely two deuterium atoms as an equimolar distribution of 1,2- and 1,4-regioisomers (Scheme 22.12, upper).

A related mechanistic proposal involves diene-glyoxal oxidative coupling (Scheme 22.12, lower). Here, complexation by low-valent rhodium(I) confers nucleophilic character to the bound diene *via* backbonding, as suggested by the Dewar-Chatt-Duncanson model for alkene coordination [35]. For low-valent *early* transition metals, such "back-bonding" is driven by the stability associated with a d^0-electronic configuration. For example, as demonstrated by the Kulinkovich reaction, complexation of olefins by Ti(II) causes them to behave as vicinal dianions: Ti(II)(olefin)$\leftrightarrow$Ti(IV)(metallocyclopropane) [36]. For *late* transition metals, the driving force associated with attaining a noble gas electronic configuration is absent, perhaps accounting for the requirement of highly activated electrophilic partners such as α-ketoaldehydes. In any case, addition of the diene to the glyoxal provides the formal product of oxidative coupling. This oxarhodacycle may react with deuterium via sigma bond metathesis to afford a rhodium alkoxide, which abstracts an allylic hydrogen to provide a rhodium π-allyl complex. Subsequent C–D reductive elimination delivers the dideuterated products as an equimolar distribution of regioisomers (Scheme 22.12, lower). Recently, this diene-glyoxal coupling was performed under an atmosphere of HD(g) as the terminal reductant. The coupling product was found to incorporate a single molecule of deuterium, distributed over the same three carbons found when D_2(g) was used as reductant. These data disqualify the initially disclosed hydrometallative mechanism, and strongly support the latter mechanism involving direct oxidative coupling.

22.4 Reductive Coupling of Conjugated Enynes and Diynes with Activated Aldehydes and Imines

The reductive coupling of 1,3-cyclohexadiene and α-ketoaldehydes, which occurs without over-reduction of the olefinic product, suggests the feasibility of utilizing more highly unsaturated pronucleophiles in the form of 1,3-enynes. In the

Catalyst (5 mol%)
BIPHEP (5 mol%)
H_2 (1 atm), DCE
25 °C, 18 h

(200 mol%) (100 mol%)

Catalyst	Yield
$[Rh(COD)Cl]_2$	---%
$Rh(COD)_2OTf$	86%
$Rh(COD)_2BF_4$	87%

Scheme 22.13 Reductive coupling of 1-phenyl but-3-en-1-yne with phenyl glyoxal.

event, the reaction conditions optimized for the cyclohexadiene-α-ketoaldehyde couplings, which employ $(p\text{-MeOPh})_3P$ as ligand, proved ineffective at promoting the coupling of 1-phenylbut-3-en-1-yne and phenyl glyoxal. However, under otherwise identical conditions, the use of bidentate ligands such as BIPHEP provides the products of reductive coupling in excellent yield. Under optimized conditions, the coupling proceeds smoothly to afford diene-containing products as single regio- and stereoisomers. Over-reduction of the diene-containing products is not observed. Presumably, upon complete consumption of glyoxal, excess enyne nonproductively coordinates rhodium, dramatically retarding the rate of any further reduction. As for the diene couplings, basic additives are not required. Additionally, reductive coupling fails upon use of neutral Rh(I) precatalysts, such as $[Rh(COD)Cl]_2$ (Scheme 22.13, Table 22.6).

Catalytic reductive coupling of 1-phenyl but-3-en-1-yne with phenyl glyoxal conducted under 1 atmos. elemental deuterium provides the *mono*-deuterated product in 85% yield. It is instructive to compare hydrometallative and oxidative coupling mechanisms involving both heterolytic and homolytic deuterium activation. For the hydrometallative mechanism involving heterolytic deuterium activation, direct alkyne deuterometallation to afford the vinyl rhodium intermediate is followed by carbonyl addition and hydrogenolytic cleavage of the resulting

Table 22.6 Reductive coupling of assorted 1,3-enynes with alkyl, aryl and heteroaryl α-ketoaldehydes.

$Rh(COD)_2OTf$ (5 mol%)
BIPHEP (5 mol%)
H_2 (1 atm), DCE
25 °C, 18 h

(200 mol%) (100 mol%)

Ar	R	Yield
Ph	Ph	86%
Ph	2-Nap	89%
Ph	*t*-Bu	61%
Ph	N-methylpyrrol-2-yl	75%
Ph	2-furyl	70%
Ph	2-thienyl	78%
p-CH_3OPh	Ph	89%
p-NO_2Ph	Ph	61%
2-furyl	Ph	75%
2-thienyl	Ph	70%
Ph (CH_3-substituted enyne)	Ph	80%
Ph (H_3C-substituted enyne)	Ph	80%

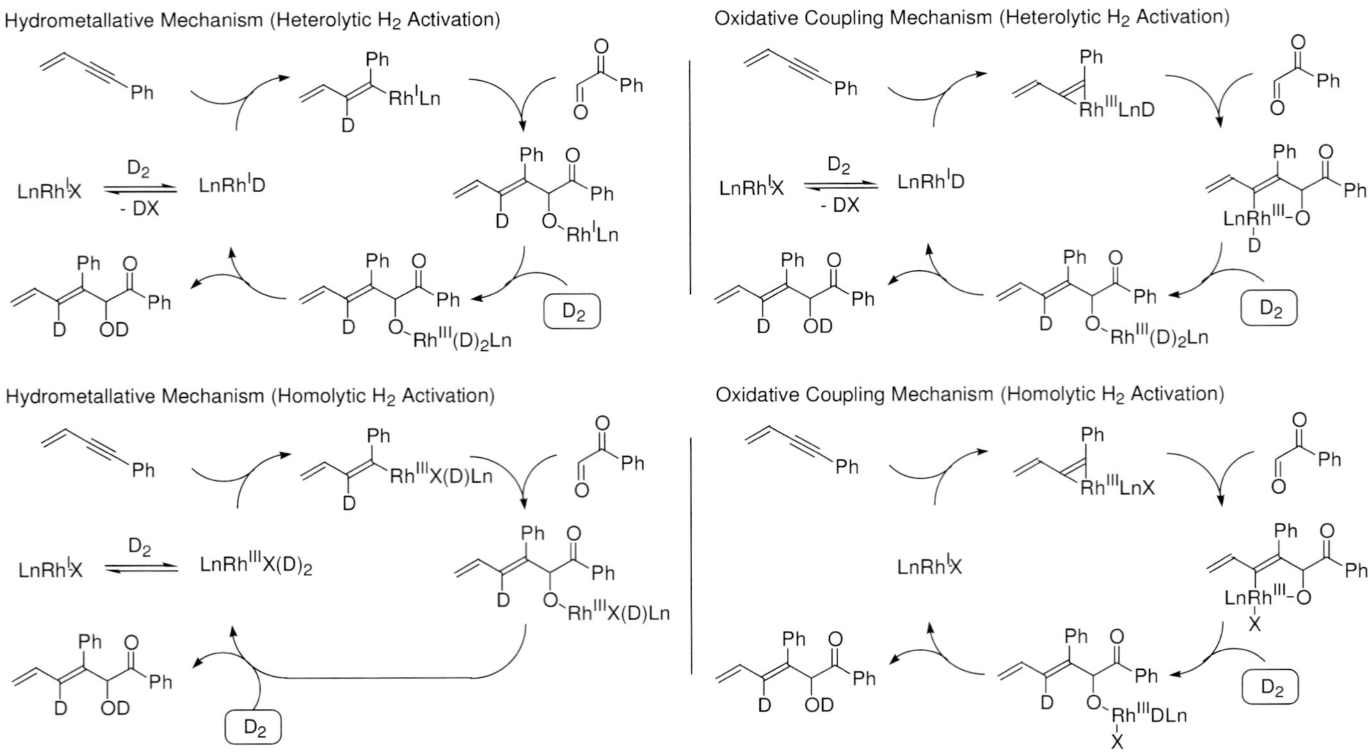
Hydrometallative Mechanism (Heterolytic H2 Activation)
Oxidative Coupling Mechanism (Heterolytic H2 Activation)
Hydrometallative Mechanism (Homolytic H2 Activation)
Oxidative Coupling Mechanism (Homolytic H2 Activation)
LnRhIX
LnRhID
LnRhIIIX(D)2
D2
- DX
Rh^I Ln
Rh^III LnD
Rh^III X(D)Ln
Rh^III LnX
Rh^III(D)2Ln
Rh^III DLn
Ph
OD
D

Scheme 22.14 Plausible mechanisms for the reductive coupling of 1-phenyl but-3-en-1-yne with phenyl glyoxal under an atmosphere of D_2(g).

Rh(I)-alkoxide. Notably, this mechanism requires deuterometallation to occur with complete regioselection (Scheme 22.14, top-left). The corresponding hydrometallative mechanism involving homolytic deuterium activation also requires completely regioselective deuterometallation. Moreover, carbonyl addition must compete favorably with C–D reductive elimination (Scheme 22.14, bottom-left). A mechanism involving heterolytic hydrogen activation followed by oxidative coupling of the reactants is plausible. Here, C–D reductive elimination of the resulting (hydrido)Rh(III)-oxametallocyclopentene followed by hydrogenolytic cleavage of rhodium alkoxide completes the catalytic cycle (Scheme 22.14, top-right). Finally, direct oxidative coupling of the reactants with subsequent hydrogenolytic cleavage of the resulting metallocycle may be envisaged. This latter mechanism requires deuterium activation by a rhodium(III) intermediate. An increasing body of evidence supports participation of organorhodium(III) complexes in σ-bond metathesis pathways [37], including reactions with hydrogen [37c]. In the case of the hydrometallative mechanisms, C–H bond formation precedes C–C bond formation. In the case of the oxidative coupling mechanisms, the converse is true. The oxidative coupling mechanism better accounts for the regiochemistry of reductive coupling and is preferred on the basis of related mechanistic studies (*vide supra*). In principle, it should be possible to discriminate between heterolytic and homolytic hydrogen activation modes *via* H_2–D_2 and HD isotope crossover experiments. However, rapid exchange of the hydroxylic protons and deuterons under the reaction conditions renders this prospect untenable (Scheme 22.14).

1,3-Diynes also participate in highly regio- and stereoselective reductive couplings to aryl, heteroaryl and aliphatic glyoxals under catalytic hydrogenation conditions [26b]. Unlike the corresponding reaction of 1,3-enynes, both *mono*- and *bis*(phosphines) may serve as ligands. Consistent with the requirement of cationic rhodium(I) catalysts, $Rh(COD)_2OTf$ and $Rh(COD)_2BF_4$ are viable catalysts, while $Rh(PPh_3)_3Cl$ is not. Remarkably, formation of the highly unsaturated 1,3-enyne products is not accompanied by over-reduction. As previously stated, it would appear that upon complete consumption of glyoxal, excess enyne nonproductively coordinates rhodium, retarding the rate of further reduction. Reductive coupling performed under an atmosphere of D_2 provides the indicated mono-deuterated product. This result may be interpreted on the basis of the mechanisms outlined in Scheme 22.14 (Scheme 22.15).

Ph, Ph (200 mol%) — O, O, Ph (100 mol%)

Catalyst (5 mol%)
Ph_3P (10 mol%)
H_2 (1 atm) or D_2 (1 atm)
DCE, 25 °C, 18 h

Ph, Ph, O, Ph, (D), OH

Catalyst	Yield
$Rh(PPh_3)_3Cl$	---%
$Rh(COD)_2OTf$	79%
$Rh(COD)_2BF_4$	80%

Scheme 22.15 Reductive coupling of diphenylbutadiyne with phenyl glyoxal.

A highly enantioselective variant of this transformation has been developed using the commercially available chiral *bis*(phosphine) (*R*)-Cl-MeO-BIPHEP. Optimization studies pertaining to the enantioselective transformation reveal that high levels of asymmetric induction are critically dependent upon the dihedral angle of the diphenylphosphino moieties of the ligand. Under optimized conditions, coupling products are produced in 71–77% yield and 86–95% enantiomeric excess. Notably, highly enantioselective C–C bond formation is achieved at ambient temperature and pressure (Table 22.7).

Under optimum conditions identified for enantioselective coupling, non-symmetric 1,3-diynes react with marked levels of regioselectivity. Specifically, for 1-phenyl-4-alkyl 1,3-diynes, coupling occurs preferentially at the aromatic terminus (Table 22.8). Competition experiments provide some insight into the mechanistic basis for such regioselectivity. Catalytic hydrogenation of phenyl glyoxal in the presence of equimolar quantities of 1,4-diphenylbutadiene and 1,4-diphenylbut-3-en-1-yne results in coupling to the more highly unsaturated enyne partner. Similarly, catalytic hydrogenation of phenyl glyoxal in the presence of equimolar quantities of 1,4-diphenylbut-3-en-1-yne and 1,4-diphenylbutadiyne

Table 22.7 Enantioselective catalytic reductive coupling of 1,3-diynes with alkyl, aryl and heteroaryl α-ketoaldehydes.

Rh(COD)$_2$OTf (5 mol%)
Ligand (5 mol%)
H_2 (1 atm), PhCOCHO
Solvent (0.1 M), 25 °C, 18 h

Entry	Ligand	Solvent	Yield (%)	e.e. (%)
1	(*R*)-BINAP	DCE	72	47
2	(*R*)-Phanephos	DCE	79	67
3	(*R*)-Cl-OMe-BIPHEP	DCE	74	76
4	(*R*)-Cl-OMe-BIPHEP	EtOH	69	80
5	(*R*)-Cl-OMe-BIPHEP	THF	67	82
6	**(*R*)-Cl-OMe-BIPHEP**	**PhH**	**74**	**91**

Ph, 74%, 91% ee
2-Nap, 77%, 90% ee
t-Bu, 72%, 95% ee
N-methylpyrrol-2-yl (CH_3), 71%, 86% ee
2-furyl, 76%, 91% ee
2-thienyl, 74%, 93% ee

Table 22.8 Regioselective reductive coupling of non-symmetric 1,3-diynes to various α-ketoaldehydes.

Product	ee	Yield (ratio)	Regioisomer	ee
H_3C, Ph, O, Ph, OH	83% ee	80% Yield (5.2:1)	Ph, CH_3, O, Ph, OH	88% ee
Bu^t, Ph, O, Ph, OH	98% ee	57% Yield (>99:1)	Ph, ^{t}Bu, O, Ph, OH	(not observed)
n-Pr, Ph, O, Ph, OH	85% ee	62% Yield (4:1)	Ph, n-Pr, O, Ph, OH	89% ee
H_3C, Ph, O, OH	90% ee	65% Yield (12:1)	Ph, CH_3, O, OH	93% ee

results in coupling to the more highly unsaturated diyne partner. Chemoselective coupling to the more highly unsaturated pronucleophile suggests preferential coordination of low-valent rhodium to the most π-acidic reactant. Moreover, for non-symmetric 1,3-diynes, regioselective C–C bond formation occurs at the terminus of the diyne that embodies the lowest LUMO energy. Hence, the chemo- and regioselectivity of reductive coupling may be explained by the Dewar-Chatt-Duncanson model for alkyne coordination [35] – that is, coordination of low-valent rhodium is driven by backbonding, with nucleophilic character developing at positions where backbonding occurs most effectively (Scheme 22.16).

Rhodium-catalyzed hydrogenation of 1,3-enynes and 1,3-diynes in the presence of ethyl (*N-tert*-butanesulfinyl)iminoacetate and ethyl (*N*-2,4,6-triisopropylbenzenesulfinyl)iminoacetate, respectively, results in reductive coupling to afford unsaturated α-amino acid esters in good to excellent yields with exceptional levels of stereocontrol [26c] (Tables 22.9 and 22.10). Remarkably, the reductive coupling of non-symmetric 1,3-diynes to iminoacetates occurs with complete levels of regioselectivity. Further hydrogenation of the diene side chain using Wilkinson's catalyst provides the corresponding β,γ-unsaturated amino acid esters. Exhaustive hydrogenation of both the diene- and enyne-containing side chains

Ph, Ph (100 mol%) + Ph, Ph (100 mol%) + O, Ph, O (100 mol%) → [$Rh(COD)_2OTf$ (5 mol%), BIPHEP (5 mol%), H_2 (1 atm), DCE, 25 °C, 18 h] → Ph, Ph, O, Ph, OH (Not Formed) + Ph, Ph, O, Ph, OH (46% Yield)

Ph, Ph (100 mol%) + Ph, Ph (100 mol%) + O, Ph, O (100 mol%) → [$Rh(COD)_2OTf$ (5 mol%), BIPHEP (5 mol%), H_2 (1 atm), DCE, 25 °C, 18 h] → Ph, Ph, O, Ph, OH (Not Formed) + Ph, Ph, O, Ph, OH (59% Yield)

Scheme 22.16 Competition experiments reveal reductive coupling occurs chemoselectively with the strongest π-acid.

Table 22.9 Reductive coupling of 1,3-enynes with ethyl (*N-tert*-butanesulfinyl)iminoacetate.

Enyne (200 mol%) + ethyl (*N-tert*-butanesulfinyl)iminoacetate (100 mol%); Rh(COD)$_2$OTf (5 mol%), BIPHEP (5 mol%), H_2 (1 atm), DCM, 25-35 °C, 3 h

R	Coupling product	After a)
Ph	92% (>95:5 d.r.)	74%
n-butyl (H_3C chain)	66% (>95:5 d.r.)	93%
CH_2OTBS	68% (>95:5 d.r.)	80%
CH_2NHBoc	94% (>95:5 d.r.)	76%

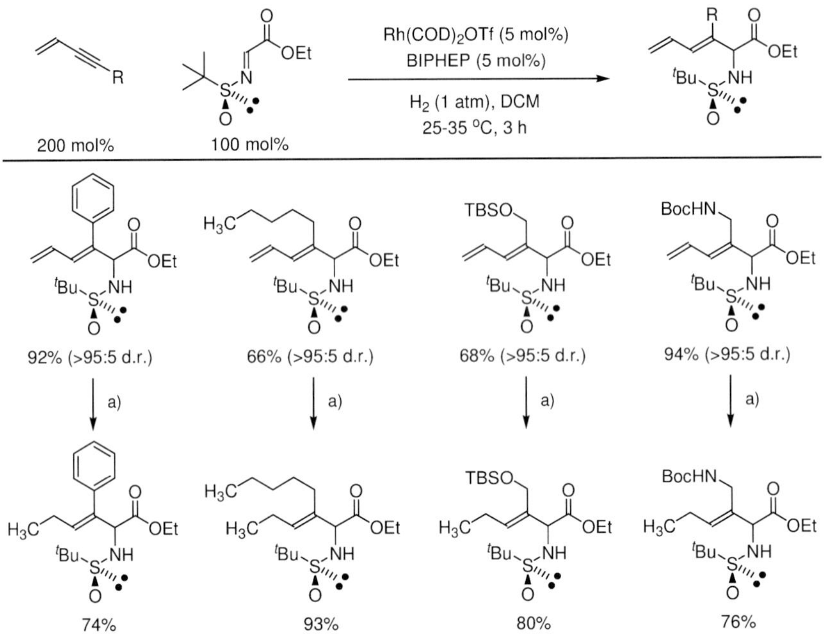

a) Rh(PPh$_3$)$_3$Cl (10 mol%), H_2 (1 atm), toluene, 25 °C, 18 h.

Table 22.10 Reductive coupling of 1,3-diynes with ethyl (*N*-2,4,6-triisopropylbenzenesulfinyl)iminoacetate.

Diyne R_1–C≡C–C≡C–R_2 (200 mol%) + ethyl (*N*-2,4,6-triisopropylbenzenesulfinyl)iminoacetate (100 mol%); Rh(COD)$_2$OTf (5 mol%), BIPHEP (5 mol%), H_2 (1 atm), DCM, 25-35 °C, 3 h

R_1	R_2	Yield
H_3C	Ph	86% (>95:5 d.r.)
H_3C	2-furyl	92% (>95:5 d.r.)
CH_2NHBoc	Ph	96% (>95:5 d.r.)
CH_2OH	CH_2OH	78% (>95:5 d.r.)

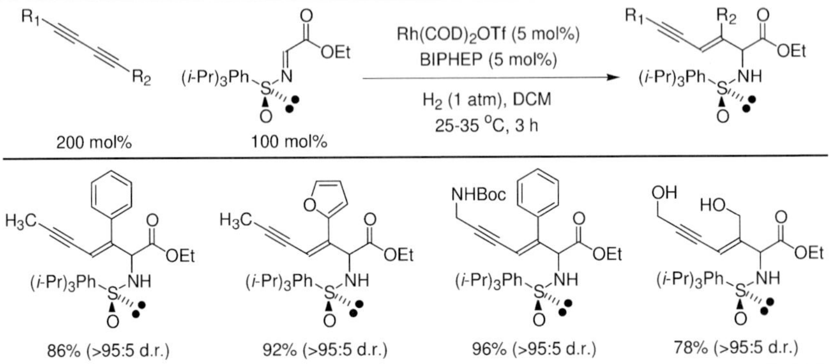

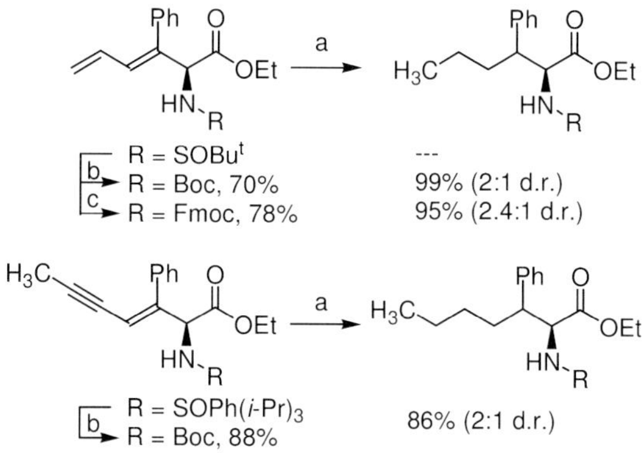

R = SOBu^t → ---
b → R = Boc, 70% → 99% (2:1 d.r.)
c → R = Fmoc, 78% → 95% (2.4:1 d.r.)

R = SOPh(*i*-Pr)$_3$
b → R = Boc, 88% → 86% (2:1 d.r.)

Scheme 22.17 Exhaustive hydrogenation of diene- and enyne-containing reductive coupling products using Crabtree's catalyst.

using Crabtree's catalyst also proceeds readily. However, the *N-tert*-butanesulfinyl residue must be exchanged for a carbamate protecting group (Scheme 22.17).

22.5 Reductive Cyclization of 1,6-Diynes and 1,6-Enynes

Hydrogenation of 1,6-diynes using cationic rhodium precatalysts promotes reductive cyclization to afford 1,2-dialkylidenecycloalkane products. As for all hydrogen-mediated C–C bond formations described in this account, cationic rhodium precatalysts are required. Whereas reductive cyclization proceeds readily using $Rh(COD)_2OTf$ and $Rh(COD)_2BF_4$, the neutral precatalyst $[Rh(COD)Cl]_2$ is ineffective (Scheme 22.18) [27, 38]. Under optimized conditions, reductive cyclization proceeds smoothly across a range of 1,6-diynes. Near-identical hydrogenation conditions are effective for the reductive cyclization of 1,6-enynes [39, 40]. Notably, conformationally predisposed substrates possessing geminal substitution in the tether are not necessary (Table 22.11).

Enantioselective hydrogenation of 1,6-enynes using chirally modified cationic rhodium precatalysts enables enantioselective reductive cyclization to afford alkylidene-substituted carbocycles and heterocycles [27b, 41, 42]. Good to excellent yields and exceptional levels of asymmetric induction are observed across a structurally diverse set of substrates. For systems that embody 1,2-disubstituted alkenes, competitive β-hydride elimination *en route* to products of cycloisomerization is observed. However, related enone-containing substrates cannot engage in β-hydride elimination, and undergo reductive cyclization in good yield (Table 22.12).

The products of reductive cyclization incorporate two non-exchangeable hydrogen atoms. Homolytic and heterolytic hydrogen activation pathways may now be discriminated on the basis of hydrogen-deuterium crossover experiments. Reductive cyclization of the indicated nitrogen-tethered enyne under a mixed atmosphere

Table 22.11 Reductive cyclization of assorted 1,6-diynes and 1,6-enynes.

1,6-Diynes: $Rh(COD)_2OTf$ (3 mol%), Phosphine (3 mol%), H_2 (1 atm), DCE, 25 °C, 3 h			1,6-Enynes: $Rh(COD)_2OTf$ (5 mol%), Phosphine (5 mol%), H_2 (1 atm), DCE, 25 °C, 2 h		
85% (*rac*-BINAP)	90% (BIPHEP)	79% (BIPHEP)	89% (BIPHEP)	75% (*rac*-BINAP)	49% (BIPHEP)
78% (*rac*-BINAP)	89% (*rac*-BINAP)	68% (BIPHEP)	65% (BIPHEP)	80% (*rac*-BINAP)	82% (BIPHEP)
51% (BIPHEP)	73% (*rac*-BINAP)	62% (*rac*-BINAP)	91% (BIPHEP)	79% (*rac*-BINAP)	79% (*rac*-BINAP)

Catalyst (3 mol%)
BIPHEP (3 mol%)
H_2 (1 atm), DCE
25 °C, 3 h

Catalyst	Yield
$[Rh(COD)Cl]_2$	---%
$Rh(COD)_2OTf$	77%
$Rh(COD)_2BF_4$	81%

Scheme 22.18 Reductive cyclization of 1,6-diynes.

of H_2 and D_2 or under an atmosphere of DH does not provide crossover products, in accordance with homolytic hydrogen activation. Interestingly, exposure of related systems incorporating a *cis*-1,2-disubstituted alkene to identical conditions under a D_2 atmosphere does not induce reductive cyclization. Rather, products of cycloisomerization are formed. A hydrometallative mechanism for cycloisomerization would be initiated by D_2 oxidative addition and propagated by rhodium hydrides derived upon β-hydride elimination from intermediate **A**. Deuterium incorporation should occur in the first turnover of the catalytic cycle, yet deuterium incorporation is not observed, even at stoichiometric catalyst loadings. The extent of deuterium incorporation for the isotopically labeled reaction products is determined by electrospray ionization-mass spectrometry (ESI-MS) analysis with isotopic correction and is corroborated by ^{1}H-NMR analysis (Scheme 22.19).

The acquisition of cycloisomerization products without deuterium incorporation is inconsistent with a hydrometallative mechanism. Furthermore, the ab-

Table 22.12 Enantioselective reductive cyclization of 1,6-enynes.

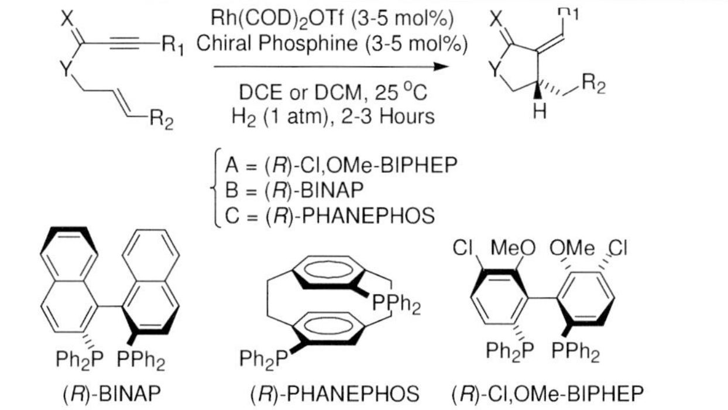

$Rh(COD)_2OTf$ (3-5 mol%), Chiral Phosphine (3-5 mol%)

DCE or DCM, 25 °C

H_2 (1 atm), 2-3 Hours

A = (*R*)-Cl,OMe-BIPHEP
B = (*R*)-BINAP
C = (*R*)-PHANEPHOS

(*R*)-BINAP (*R*)-PHANEPHOS (*R*)-Cl,OMe-BIPHEP

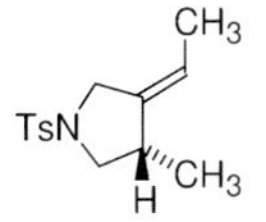 A: 69%, 94% e.e. B: 79%, 93% e.e. C: multiple products	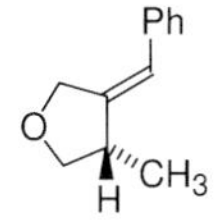 A: 84%, 95% e.e. B: 80%, 97% e.e. C: multiple products	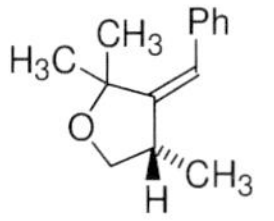 A: 85%, 84% e.e. B: 76%, 82% e.e. C: multiple products
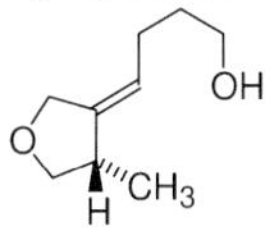 A: 77%, 95% e.e. B: 75%, 95% e.e. C: multiple products	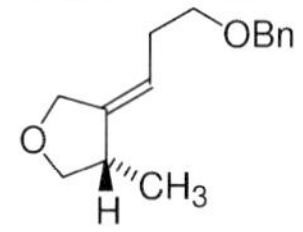 A: 75%, 98% e.e. B: 77%, 98% e.e. C: multiple products	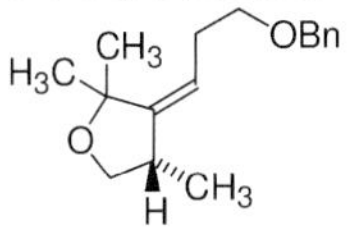 A: 82%, 98% e.e. B: 64%, 98% e.e. C: multiple products
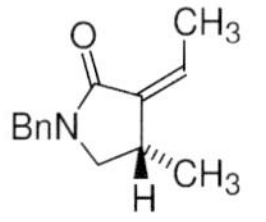 A: 40%, 26% e.e. B: 24%, 10% e.e. C: 73%, 91% e.e.	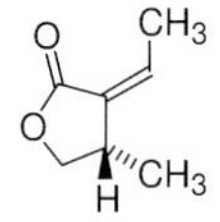 A: 89%, 6% e.e. B: 79%, 20% e.e. C: 73%, 94% e.e.	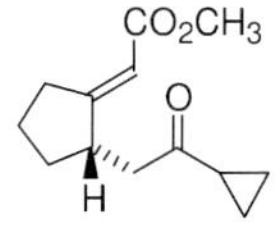 A: 68%, 98% e.e. B: 63%, 90% e.e. C: complete reduction
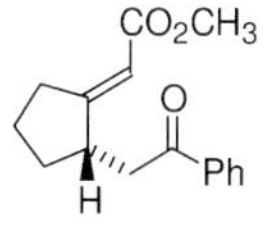 A: 78%, 88% e.e. B: 66%, 96% e.e. C: complete reduction	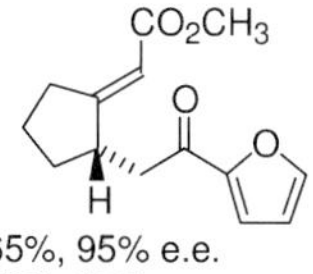 A: 65%, 95% e.e. B: 72%, 96% e.e. C: complete reduction	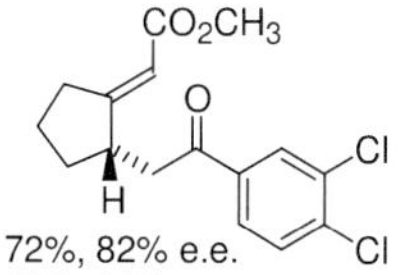 A: 72%, 82% e.e. B: 63%, 94% e.e. C: complete reduction

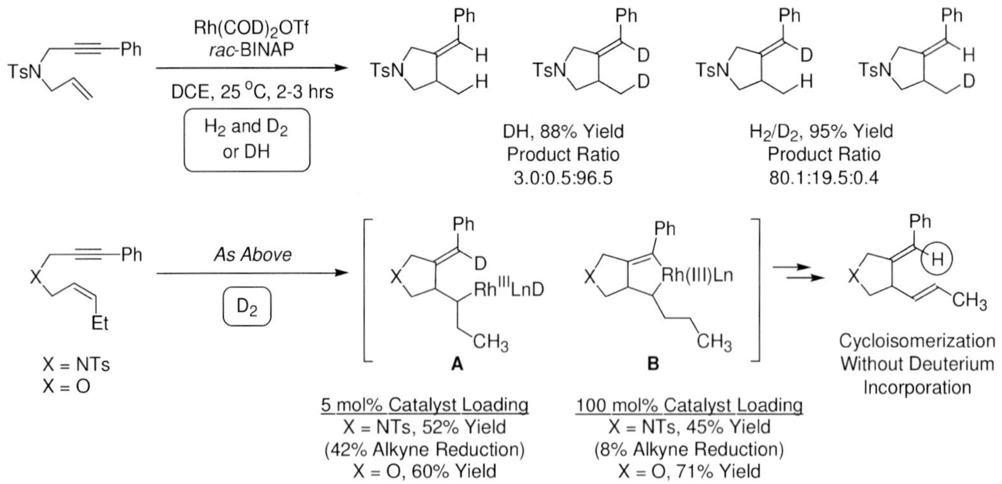

Rh(COD)$_2$OTf
rac-BINAP
DCE, 25 °C, 2-3 hrs
H_2 and D_2 or DH

DH, 88% Yield
Product Ratio
3.0:0.5:96.5

H_2/D_2, 95% Yield
Product Ratio
80.1:19.5:0.4

As Above
D_2

X = NTs
X = O

5 mol% Catalyst Loading
X = NTs, 52% Yield
(42% Alkyne Reduction)
X = O, 60% Yield

100 mol% Catalyst Loading
X = NTs, 45% Yield
(8% Alkyne Reduction)
X = O, 71% Yield

Cycloisomerization
Without Deuterium
Incorporation

Scheme 22.19 Mechanistic studies involving hydrogen-deuterium crossover experiments, along with the observance of non-conjugated cycloisomerization products, suggest that rhodium(III) metallocyclopentene formation occurs in advance of hydrogen activation.

sence of conjugated cycloisomerization products suggests β-hydride elimination from metallocycle **B**. If indeed oxidative cyclization occurs initially to form metallocycle **B** [17], subsequent hydrogenolytic cleavage must occur via: (i) hydrogen oxidative addition or (ii) hydrogen activation via σ-bond metathesis [37c]. Whereas hydrogen oxidative addition to a Rh(III) metallocycle would afford a Rh(V) intermediate, hydrogen activation via σ-bond metathesis would not. In either case, it would appear that C–C bond formation occurs in advance of hydrogen activation. Hydrogen oxidative addition followed by rhodium(V) metallocycle formation is unlikely and, to our knowledge, without precedent. Finally, it is worth noting that hydrogen oxidative addition is rate-determining for asymmetric hydrogenations catalyzed by cationic rhodium complexes, which suggests that the oxidative cyclization manifold may compete favorably with hydrogen oxidative addition [28].

22.6 Conclusion

From the seminal studies of Sabatier [43] and Adams [44] to the more recent studies of Knowles [45] and Noyori [46], catalytic hydrogenation has been regarded as a method of reduction. The results herein demonstrate the feasibility of transforming catalytic hydrogenation into a powerful and atom-economical method for reductive C–C bond formation. Given the profound socioeconomic impact of alkene hydroformylation, the development of catalysts for the hydrogen-mediated coupling of basic feedstocks, such as α-olefins and styrenes with aldehyde partners, represents a paramount scientific challenge. The mechanistic underpinnings of these transformations have begun to unfold. Hydrogen-mediated enone couplings require mild basic additives and, hence, likely involve heterolytic hydrogen activation by way of an intermediate dihydride. In contrast, the mechanistic data linked to the reductive coupling of conjugated dienes, enynes, and diynes to α-ketoaldehydes and related iminoacetates, along with data pertaining to the reductive cyclizations of 1,6-diynes and 1,6-enynes, suggest oxidative coupling to form metallacyclic intermediates which are then hydrogenolytically cleaved. This oxidative coupling-hydrogenolysis motif should play a key role in the design of related hydrogen-mediated couplings. It is the authors hope that the emergent mechanistic principles gleaned from these initial studies will ultimately spawn a broad new class of hydrogen-mediated C–C couplings.

Acknowledgments

The authors acknowledge the Robert A. Welch Foundation (F-1466), the Research Corporation Cottrell Scholar Award, the National Science Foundation CAREER Award, the Alfred P. Sloan Foundation, the Camille and Henry Dreyfus Foundation, Eli Lilly, and Merck.

Abbreviations

MVK methyl vinyl ketone
PVK phenyl vinyl ketone

References

1 For use of alkenes as nucleophilic partners in catalytic reductive couplings, see: (a) N. M. Kablaoui, S. L. Buchwald, *J. Am. Chem. Soc.* **1995**, *117*, 6785; (b) W. E. Crowe, M. J. Rachita, *J. Am. Chem. Soc.* **1995**, *117*, 6787; (c) N. M. Kablaoui, S. L. Buchwald, *J. Am. Chem. Soc.* **1996**, *118*, 3182.

2 For use of alkynes as nucleophilic partners in catalytic hydrometallative reductive couplings to aldehydes, see: (a) E. Oblinger, J. Montgomery, *J. Am. Chem. Soc.* **1997**, *119*, 9065; (b) X.Q. Tang, J. Montgomery, *J. Am. Chem. Soc.* **1999**, *121*, 6098; (c) W.-S. Huang, J. Chen, T. F. Jamison, *Org. Lett.* **2000**, *2*, 4221; (d) E.A. Colby, T. F. Jamison, *J. Org. Chem.* **2003**, *68*, 156; (e) K. M. Miller, W.-S. Huang, T. F. Jamison, *J. Am. Chem. Soc.* **2003**, *125*, 3442; (f) K. Takai, S. Sakamoto, T. Isshiki, *Org. Lett.* **2003**, *5*, 653; (g) J. Chan, T. F. Jamison, *J. Am. Chem. Soc.* **2003**, *125*, 11514; (h) E.A. Colby, K.C. O'Brien, T. F. Jamison, *J. Am. Chem. Soc.* **2004**, *126*, 998; (i) J. Chan, T. F. Jamison, *J. Am. Chem. Soc.* **2004**, *126*, 10682; (j) K. M. Miller, T. F. Jamison, *J. Am. Chem. Soc.* **2004**, *126*, 15342.

3 For use of alkynes as nucleophilic partners in catalytic hydrometallative reductive couplings to imines, see: S.J. Patel,

T.F. Jamison, *Angew. Chem. Int. Ed.* **2003**, *42*, 1364.
4 For use of alkynes as nucleophilic partners in catalytic hydrometallative reductive couplings to imines, see: (a) M.V. Chevliakov, J. Montgomery, *J. Am. Chem. Soc.* **1999**, *121*, 11139; (b) J. Montgomery, M. Song, *Org. Lett.* **2002**, *4*, 4009; (c) K.K.D. Amarasinghe, J. Montgomery, *J. Am. Chem. Soc.* **2002**, *124*, 9366.
5 For use of conjugated enones as nucleophilic partners in catalytic intermolecular reductive couplings to aldehydes, see: (a) A. Revis, T.K. Hilty, *Tetrahedron Lett.* **1987**, *28*, 4809; (b) S. Isayama, T. Mukaiyama, *Chem. Lett.* **1989**, 2005; (c) I. Matsuda, K. Takahashi, S. Sato, *Tetrahedron Lett.* **1990**, *31*, 5331; (d) S. Kiyooka, A. Shimizu, S. Torii, *Tetrahedron Lett.* **1998**, *39*, 5237; (e) T. Ooi, K. Doda, D. Sakai, K. Maruoka, *Tetrahedron Lett.* **1999**, *40*, 2133; (f) S.J. Taylor, J.P. Morken, *J. Am. Chem. Soc.* **1999**, *121*, 12202; (g) S.J. Taylor, M.O. Duffey, J.P. Morken, *J. Am. Chem. Soc.* **2000**, *122*, 4528; (h) C.-X. Zhao, M.O. Duffey, S.J. Taylor, J.P. Morken, *Org. Lett.* **2001**, *3*, 1829; (i) C.-X. Zhao, J. Bass, J.P. Morken, *Org. Lett.* **2001**, *3*, 2839.
6 For use of conjugated enones as nucleophilic partners in catalytic intramolecular reductive couplings to aldehydes, see: (a) T.-G. Baik, A.L. Luis, L.-C. Wang, M.J. Krische, *J. Am. Chem. Soc.* **2001**, *123*, 5112; (b) D. Emiabata-Smith, A. McKillop, C. Mills, W.B. Motherwell, A.J. Whitehead, *Synlett* **2001**, 1302; (c) L.-C. Wang, H.-Y. Jang, Y. Roh, V. Lynch, A.J. Schultz, X. Wang, M.J. Krische, *J. Am. Chem. Soc.* **2002**, *124*, 9448; (d) R.R. Huddleston, D.F. Cauble, M.J. Krische, *J. Org. Chem.* **2003**, *68*, 11; (e) M. Freiria, A.J. Whitehead, D.A. Tocher, W.B. Motherwell, *Tetrahedron* **2004**, *60*, 2673.
7 For use of conjugated enones as nucleophilic partners in catalytic intermolecular reductive couplings to aldimines, see: (a) J.A. Townes, M.A. Evans, J. Queffelec, S.J. Taylor, J.P. Morken, *Org. Lett.* **2002**, *4*, 2537; (b) T. Muraoka, S.-I. Kamiya, I. Matsuda, K. Itoh, *Chem. Commun.* **2002**, 1284.
8 For use of conjugated dienes as nucleophilic partners in catalytic intramolecular reductive couplings to aldehydes, see: (a) Y. Sato, M. Takimoto, K. Hayashi, T. Katsuhara, K. Tagaki, M. Mori, *J. Am. Chem. Soc.* **1994**, *116*, 9771; (b) Y. Sato, Y. Takimoto, M. Mori, *Tetrahedron Lett.* **1996**, *37*, 887; (c) Y. Sato, T. Takanashi, M. Hoshiba, M. Mori, *Tetrahedron Lett.* **1998**, *39*, 5579; (d) Y. Sato, N. Saito, M. Mori, *J. Am. Chem. Soc.* **2000**, *122*, 2371; (e) K. Shibata, M. Kimura, M. Shimizu, Y. Tamaru, *Org. Lett.* **2001**, *3*, 2181; (f) Y. Sato, N. Saito, M. Mori, *J. Org. Chem.* **2002**, *67*, 9310; (g) Y. Sato, T. Takanishi, M. Hoshiba, M. Mori, *J. Organomet. Chem.* **2003**, *688*, 36.
9 For use of conjugated dienes as nucleophilic partners in catalytic intermolecular reductive couplings to aldehydes, see: (a) M. Kimura, A. Ezoe, K. Shibata, Y. Tamaru, *J. Am. Chem. Soc.* **1998**, *120*, 4033; (b) M. Kimura, H. Fujimatsu, A. Ezoe, K. Shibata, M. Shimizu, S. Matsumoto, Y. Tamaru, *Angew. Chem. Int. Ed.* **1999**, *38*, 397; (c) M. Kimura, K. Shibata, Y. Koudahashi, Y. Tamaru, *Tetrahedron Lett.* **2000**, *41*, 6789; (d) M. Kimura, A. Ezoe, S. Tanaka, Y. Tamaru, *Angew. Chem. Int. Ed.* **2001**, *40*, 3600; (e) T.-P. Loh, H.-Y. Song, Y. Zhou, *Org. Lett.* **2002**, *4*, 2715; (f) Y. Sato, R. Sawaki, N. Saito, M. Mori, *J. Org. Chem.* **2002**, *67*, 656; (g) L. Bareille, P. Le Gendre, C. Moise, *Chem. Commun.* **2005**, 775.
10 For use of conjugated dienes as nucleophilic partners in catalytic intermolecular reductive couplings to aldimines, see: M. Kimura, A. Miyachi, K. Kojima, S. Tanaka, Y. Tamaru, *J. Am. Chem. Soc.* **2004**, *126*, 14360.
11 For use of conjugated enynes as nucleophilic partners in catalytic intermolecular reductive couplings to aldehydes, see: K.M. Miller, T. Luanphaisarnnont, C. Molinaro, T.F. Jamison, *J. Am. Chem. Soc.* **2004**, *126*, 4130.
12 For reviews on atom economy, see: (a) B.M. Trost, *Science* **1991**, *254*, 1471; (b) B.M. Trost, *Angew. Chem. Int. Ed. Engl.* **1995**, *34*, 259.
13 (a) P.J. Black, M.G. Edwards, J.M.J. Williams, *Tetrahedron* **2005**, *61*, 1363;

(b) M. G. Edwards, R. F. R. Jazzar, B. M. Paine, D. J. Shermer, M. K. Whittlesey, J. M. J. Williams, D. D. Edney, *Chem. Commun.* **2004**, 90; (c) G. Cami-Kobeci, J. M. J. Williams, *Chem. Commun.* **2004**, 1072; (d) M. G. Edwards, J. M. J. Williams, *Angew. Chem. Int. Ed.* **2002**, *41*, 4740; (e) P. J. Black, W. Harris, J. M. J. Williams, *Angew. Chem. Int. Ed.* **2001**, *40*, 4475.

14 For recent reviews on alkene hydroformylation and the Fischer-Tropsch reaction, see: (a) B. Breit, *Acc. Chem. Res.* **2003**, *36*, 264; (b) B. Breit, W. Seiche, *Synthesis* **2001**, 1; (c) W. A. Herrmann, *Angew. Chem. Int. Ed. Engl.* **1982**, *21*, 117; (d) C.-K. Rofer-Depoorter, *Chem. Rev.* **1981**, *81*, 447.

15 Prior to our work, two examples of hydrogen-mediated C–C bond formation under CO-free conditions are reported: (a) G. A. Molander, J. O. Hoberg *J. Am. Chem. Soc.* **1992**, *114*, 3123; (b) K. Kokubo, M. Miura, M. Nomura, *Organometallics* **1995**, *14*, 4521.

16 For reviews on the heterolytic activation of elemental hydrogen, see: (a) P. J. Brothers, *Prog. Inorg. Chem.* **1981**, *28*, 1; (b) G. Jeske, H. Lauke, H. Mauermann, H. Schumann, T. J. Marks, *J. Am. Chem. Soc.* **1985**, *107*, 8111.

17 Rh(III)-metallocycles derived from 1,6-enynes are postulated as reactive intermediates in catalytic [4+2] and [5+2] cycloadditions, Pauson-Khand reactions and cycloisomerizations: P. Cao, B. Wang, X. Zhang, *J. Am. Chem. Soc.* **2000**, *122*, 64901 and references cited therein.

18 (a) C. A. Tolman, P. Z. Meakin, D. L. Lindner, J. P. Jesson, *J. Am. Chem. Soc.* **1974**, *96*, 2762; (b) J. Halpern, T. Okamoto, A. Zakhariev, *J. Mol. Catal.* **1976**, *2*, 65.

19 For a review, see: L. Marko, *Pure Appl. Chem.* **1979**, *51*, 2211.

20 Monohydride formation by deprotonation of a dihydride intermediate is known for cationic Rh-complexes: (a) R.R. Schrock, J.A. Osborn, *J. Am. Chem. Soc.* **1976**, *98*, 2134; (b) R.R. Schrock, J.A. Osborn, *J. Am. Chem. Soc.* **1976**, *98*, 2143; (c) R. R. Schrock, J. A. Osborn, *J. Am. Chem. Soc.* **1976**, *98*, 4450.

21 For a review of the acidity of metal hydrides, see: J. R. Norton, in: A. Dedieu (Ed.), *Transition Metal Hydrides*. New York, **1992**, pp. 309.

22 Direct heterolytic activation of hydrogen by $RhCl(CO)(PPh_3)_2$ has been suggested, but likely involves an intermediate dihydride: D. Evans, J.A. Osborn, G. Wilkinson, *J. Chem. Soc. A* **1968**, 3133.

23 For earlier reviews encompassing aspects of this work, see: (a) H.-Y. Jang, M. J. Krische, *Eur. J. Org. Chem.* **2004**, 3953; (b) H.-Y. Jang, M. J. Krische, *Acc. Chem. Res.* **2004**, *37*, 653; (c) H.-Y. Jang, R. R. Huddleston, M. J. Krische, *Chemtracts* **2003**, *16*, 554.

24 (a) H.-Y. Jang, R. R. Huddleston, M. J. Krische, *J. Am. Chem. Soc.* **2002**, *124*, 15156; (b) R. R. Huddleston, M. J. Krische, *Org. Lett.* **2003**, *5*, 1143; (c) G. A. Marriner, S. A. Garner, H.-Y. Jang, M. J. Krische, *J. Org. Chem.* **2004**, *69*, 1380; (d) P. K. Koech, M. J. Krische, *Org. Lett.* **2004**, *6*, 691; (e) S. A. Garner, C.-K. Jung, M. J. Krische, *Org. Lett.* **2006**, *8*, 519.

25 H.-Y. Jang, R. R. Huddleston, M. J. Krische, *Angew. Chem. Int. Ed.* **2003**, *42*, 4074.

26 (a) H.-Y. Jang, R. R. Huddleston, M. J. Krische, *J. Am. Chem. Soc.* **2004**, *126*, 4664; (b) R. R. Huddleston, H.-Y. Jang, M. J. Krische, *J. Am. Chem. Soc.* **2003**, *125*, 11488; (c) J.-R. Kong, C.-W. Cho, M. J. Krische, *J. Am. Chem. Soc.* **2005**, *127*, 11269.

27 (a) H.-Y. Jang, M. J. Krische, *J. Am. Chem. Soc.* **2004**, *126*, 7875; (b) H.-Y. Jang, F. W. Hughes, H. Gong, J. Zhang, J. S. Brodbelt, M. J. Krische, *J. Am. Chem. Soc.* **2004**, *127, 6174*.

28 For excellent reviews, see: (a) J. Halpern, *Asymm. Synth.* **1985**, *5*, 41; (b) C. R. Landis, T. W. Brauch, *Inorg. Chim. Acta* **1998**, *270*, 285; (c) I. Gridnev, T. Imamoto, *Acc. Chem. Res.* **2004**, *37*, 633.

29 A method for the stoichiometric addition of metallo-aldehyde enolates to ketones has recently been reported: K. Yachi, H. Shinokubo, K. Oshima, *J. Am. Chem. Soc.* **1999**, *121*, 9465.

30 E. M. Arnett, F. J. Fisher, M. A. Nichols, A. A. Ribeiro, *J. Am. Chem. Soc.* **1989**, *111*, 748.

31 The failure of *tris*(dialkylamino)sulfonium enolates to react with aldehydes is attributed to unfavorable enolate-aldolate equilibria: (a) R. Noyori, J. Sakata, M. Nishizawa, *J. Am. Chem. Soc.* **1980**, *102*, 1223; (b) R. Noyori, I. Nishida, J. Sakata, *J. Am. Chem. Soc.* **1981**, *103*, 2106; (c) R. Noyori, I. Nishida, J. Sakata, *J. Am. Chem. Soc.* **1983**, *105*, 1598.

32 (a) C.H. Heathcock, in: B.M. Trost, I. Fleming, C.H. Heathcock (Eds.), *Comprehensive Organic Synthesis: Additions to C-X Bonds Part 2.* Pergamon Press, New York, pp. 181; (b) B. Alcaide, P. Almendros, *Angew. Chem. Int. Ed.* **2003**, *42*, 858.

33 (a) S. Denmark, S.K. Ghosh, *Angew. Chem. Int. Ed.* **2001**, *40*, 4759; (b) A.B. Northrup, D.W.C. MacMillan, *J. Am. Chem. Soc.* **2002**, *124*, 6798; (c) C. Pidathala, L. Hoang, N. Vignola, B. List, *Angew. Chem. Int. Ed.* **2003**, *42*, 2785.

34 For a recent discussion, see: T.C. Morrill, C.A. D'Souza, *Organometallics* **2003**, *22*, 1626 and references therein.

35 (a) M.J.S. Dewar, *Bull. Soc. Chim. Fr.* **1951**, *18*, C71; (b) J. Chatt, L.A. Duncanson, *J. Chem. Soc.* **1953**, 2939.

36 For a review, see: F. Sato, H. Urable, S. Okamoto, *Chem. Rev.* **2000**, *100*, 2835.

37 For σ-bond metathesis involving Rh(III) intermediates, see: (a) J.F. Hartwig, K.S. Cook, M. Hapke, C.D. Incarvito, Y. Fan, C.E. Webster, M.B. Hall, *J. Am. Chem. Soc.* **2005**, *127*, 2538; (b) C. Liu, R.A. Widenhoefer, *Organometallics* **2002**, *21*, 5666; (c) F. Hutschka, A. Dedieu, W. Leitner, *Angew. Chem. Int. Ed. Engl.* **1995**, *34*, 1742.

38 For metal-catalyzed cyclization of 1,6- and 1,7-diynes, see: (a) B.M. Trost, D.C. Lee, *J. Am. Chem. Soc.* **1988**, *110*, 7255; (b) K. Tamao, K. Kobayashi, Y. Ito, *J. Am. Chem. Soc.* **1989**, *111*, 6478; (c) B.M. Trost, F.J. Fleitz, W.J. Watkins, *J. Am. Chem. Soc.* **1996**, *110*, 5146; (d) M. Lautens, N.D. Smith, D. Ostrovsky, *J. Org. Chem.* **1997**, *62*, 8970; (e) S.-Y. Onozawa, Y. Hatanaka, M. Tanaka, *Chem. Commun.* **1997**, 1229; (f) S.-Y. Onozawa, Y. Hatanaka, N. Choi, M. Tanaka, *Organometallics* **1997**, *16*, 5389; (g) I. Ojima, J. Zhu, E.S. Vidal, D.F. Kass, *J. Am. Chem. Soc.* **1998**, *120*, 6690; (h) T. Muraoka, I. Matsuda, K. Itoh, *Tetrahedron Lett.* **1998**, *39*, 7325; (i) S. Gréau, B. Radetich, T.V. RajanBabu, *J. Am. Chem. Soc.* **2000**, *122*, 8579; (j) J.W. Madine, X. Wang, R.A. Widenhoefer, *Org. Lett.* **2001**, *3*, 385; (k) X. Wang, H. Chakrapani, J.W. Madine, M.A. Keyerleber, R.A. Widenhoefer, *J. Org. Chem.* **2002**, *67*, 2778; (l) T. Muraoka, I. Matsuda, K. Itoh, *Organometallics* **2002**, *21*, 3650; (m) C. Liu, R.A. Widenhoefer, *Organometallics* **2002**, *21*, 5666; (n) T. Uno, S. Wakayanagi, Y. Sonoda, K. Yamamoto, *Synlett* **2003**, 1997; (o) B.M. Trost, M.T. Rudd, *J. Am. Chem. Soc.* **2003**, *125*, 11516.

39 For reviews encompassing the Pd-catalyzed cycloisomerization and reductive cyclization of 1,6-enynes, see: (a) B.M. Trost, *Acc. Chem. Res.* **1990**, *23*, 34; (b) I. Ojima, M. Tzamarioudaki, Z. Li, R.J. Donovan, *Chem. Rev.* **1996**, *96*, 635; (c) B.M. Trost, M.J. Krische, *Synlett* **1998**, 1; (d) C. Aubert, O. Buisine, M. Malacria, *Chem. Rev.* **2002**, *102*, 813.

40 For selected examples of the cycloisomerization of 1,6-enynes catalyzed by metals other than palladium, see: (a) Titanium: S.J. Sturla, N.M. Kablaoui, S.L. Buchwald, *J. Am. Chem. Soc.* **1999**, *121*, 1976; (b) Rhodium: P. Cao, B. Wang, X. Zhang, *J. Am. Chem. Soc.* **2000**, *122*, 6490; (c) Nickel-Chromium: B.M. Trost, J.M. Tour, *J. Am. Chem. Soc.* **1987**, *109*, 5268; (d) Ruthenium: M. Nishida, N. Adachi, K. Onozuka, H. Matsumura, M. Mori, *J. Org. Chem.* **1998**, *63*, 9158; (e) B.M. Trost, F.D. Toste, *J. Am. Chem. Soc.* **2000**, *122*, 714; (f) J. LaPaih, D.C. Rodriguez, S. Derien, P.H. Dixneuf, *Synlett* **2000**, 95; (g) Cobalt: A. Ajamian, J.L. Gleason, *Org. Lett.* **2003**, *5*, 2409; (h) Iridium: N. Chatani, H. Inoue, T. Morimoto, T. Muto, S. Muria, *J. Org. Chem.* **2001**, *66*, 4433.

41 For an excellent review covering enantioselective metal-catalyzed cycloisomerization of 1,6- and 1,7-enynes, see: I.J.S. Fairlamb, *Angew. Chem. Int. Ed.* **2004**, *43*, 1048.

42 For rhodium-catalyzed enantioselective enyne cycloisomerization and hydrosilylation-cyclization, see: (a) C. Ping, X. Zhang, *Angew. Chem. Int. Ed.* **2000**, *39*, 4104; (b) A. Lei, M. He, S. Wu, X.

Zhang, *Angew. Chem. Int. Ed.* **2002**, *41*, 3457–3460; (c) A. Lei, J. P. Waldkirch, M. He, X. Zhang, *Angew. Chem. Int. Ed.* **2002**, *41*, 4526; (d) A. Lei, M. He, X. Zhang, *J. Am. Chem. Soc.* **2003**, *125*, 11472; (e) H. Chakrapani, C. Liu, R. A. Widenhoefer, *Org. Lett.* **2003**, *5*, 157.

43 For a biographical sketch of Paul Sabatier, see: A. Lattes, *C. R. Acad. Sci. Ser. IIC: Chemie* **2000**, *3*, 705.

44 For a biographical sketch of Roger Adams, see: D.S. Tarbell, A.T. Tarbell, *J. Chem. Ed.* **1979**, *56*, 163.

45 (a) W. S. Knowles, *Prix Nobel* **2001**, *2002*, 160; (b) W. S. Knowles, *Angew. Chem. Int. Ed.* **2002**, *41*, 1998; (c) W. S. Knowles, *Adv. Synth. Catal.* **2003**, *345*, 3.

46 (a) R. Noyori, *Prix Nobel* **2001**, *2002*, 186; (b) R. Noyori, *Angew. Chem. Int. Ed.* **2002**, *41*, 2008; (c) R. Noyori, *Adv. Synth. Catal.* **2003**, *345*, 15.

Part IV
Asymmetric Homogeneous Hydrogenation

The Handbook of Homogeneous Hydrogenation.
Edited by J.G. de Vries and C.J. Elsevier
Copyright © 2007 WILEY-VCH Verlag GmbH & Co. KGaA, Weinheim
ISBN: 978-3-527-31161-3

23
Enantioselective Alkene Hydrogenation: Introduction and Historic Overview

David J. Ager

23.1
Introduction

This chapter describes, from an historic perspective, the development of ligands and catalysts for enantioselective hydrogenations of alkenes. There is no in-depth discussion of the many ligands available as the following chapters describe many of these, as well as their specific applications. The purpose here is to provide an overall summary and perspective of the area. By necessity, a large number of catalyst systems have not been mentioned. The discussion is also limited to the reductions of carbon–carbon unsaturation. In almost all cases, rhodium is the transition metal to catalyze this type of reduction. In order to help the reader, the year of the first publication in a journal has been included in parentheses under each structure.

Before 1968, attempts to perform enantioselective hydrogenations had either used a chiral auxiliary attached to the substrate [1] or a heterogeneous catalyst that was on a chiral support, usually derived from Nature [2]. Since the disclosure of chiral phosphine ligands to bring about enantioselective induction in a hydrogenation, many systems have been developed, as evidenced in this book. The evolution of these transition-metal catalysts has been discussed in a number of reviews [3–12].

In addition to academic curiosity, enantioselective hydrogenation catalysts have enjoyed an extra impetus for their development. The early commercial successes of Knowles with the Dopa process (*vide infra*), followed by the related applications of BINAP-based catalysts, have led many companies to develop their own ligand systems if not based on a completely new scaffold, then at least sufficiently different to allow patent protection and freedom to operate. This has resulted in a wide range of catalysts and ligand systems to perform the same, or very similar, reactions. In this chapter, ligands have been included if they have a familiar name. Acronyms are given in upper case, while the names of ligands that are not based on acronyms are given lower case with the parts of the name denoted by an upper-case letter.

The Handbook of Homogeneous Hydrogenation.
Edited by J. G. de Vries and C. J. Elsevier
Copyright © 2007 WILEY-VCH Verlag GmbH & Co. KGaA, Weinheim
ISBN: 978-3-527-31161-3

23.2
Development of CAMP and DIPAMP

During the late 1960s, Horner et al. [13] and Knowles and Sabacky [14] independently found that a chiral monodentate tertiary phosphine, in the presence of a rhodium complex, could provide enantioselective induction for a hydrogenation, although the amount of induction was small [15–20]. The chiral phosphine ligand replaced the triphenylphosphine in a Wilkinson-type catalyst [10, 21, 22]. At about this time, it was also found that $[Rh(COD)_2]^+$ or $[Rh(NBD)_2]^+$ could be used as catalyst precursors, without the need to perform ligand exchange reactions [23].

Knowles found that the monophosphine CAMP (**1a**) could provide an ee-value of up to 88% for the reduction of dehydroamino acids. CAMP was an extension of PAMP (**1b**) that provides ee-values of 50–60% in analogous reactions [24]. At this time, Kagan showed that DIOP (**2**) (*vide infra*), where the stereogenic centers are not at phosphorus, could also provide enantioselective induction in an hydrogenation. DIOP also showed that a bisphosphine need not have the chirality at phosphorus, and that good stereoselectivity might result from a C_2-symmetric ligand. Knowles then developed the C_2-symmetric ligand DIPAMP (**3**) [22, 25]. The use of Rh-DIPAMP for the synthesis of L-Dopa is well known and is still practiced today [12, 22, 27]. A number of variations of the DIPAMP structure were investigated for the synthesis of this important pharmaceutical, but the parent remains the best ligand in this class [22].

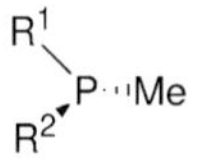

1 (1972)

a R^1 = Ph, R^2 = *o*-An (PAMP)
b R^1 = *c*-C_6H_{11}, R^2 = *o*-An (CAMP)

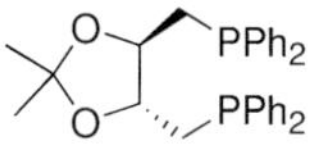

2 (DIOP) (1971)

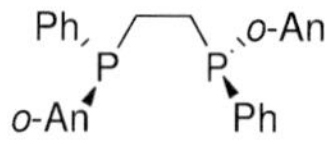

3 (DIPAMP) (1977)

It is interesting to note that a few "rules of thumb" and myths came out of these early studies. Many of these have been perpetuated for decades, and the myths are only just being put to rest. Knowles showed that only two phosphorus ligands were needed on the metal to achieve reduction, and not three as in Wilkinson's catalyst [10]. The success of DIPAMP and DIOP led to the belief that bisphosphine ligands were required for high enantioselective induction, and that C_2-symmetry was also desirable. These hypotheses molded the design of new ligands for many years. We now know that monodentate ligands, as well as asymmetric bidentate ligands, can provide high ee-values.

Another trend that arose from Knowles' results was that a wide range of enamides (dehydroamino acids) could be reduced to amino acids [22, 25, 27, 29]. This was in contrast to the enzymatic reactions known then, where enzymes were believed to be very substrate specific. As we now know, there is no general catalytic system to perform asymmetric hydrogenations and even within a small class of substrates, some ligand variation is required to achieve optimal results.

The results obtained with the Knowles' catalyst system have led to a number of useful tools that have helped with the development of other ligand families. The low ee-values obtained with simple unsaturated acids as compared to the enamides of dehydroamino acid derivatives show that the oxygen atoms of the amide is a key to complex formation with the metal center. Knowles also proposed a quadrant model that has been adapted for many reactions [5, 22]. The mechanism of the reaction has been investigated, and it is known that the addition of the substrate to the metal is regioselective and that competing catalytic cycles can occur [5, 10, 22, 25, 27, 30–46].

Perhaps the one major drawback with DIPAMP is the long synthetic sequence required for its preparation, though shorter and cheaper methods are now available [12]. The ligand continues to be a player for the synthesis of amino acid derivatives at scale, including L-Dopa, as mentioned above [12, 25, 27–29]. Its continued use is a testament to the power of the initial discoveries, as well as showing that a chemical catalyst can achieve selectivities only previously seen with enzymes.

The difficulty in preparing *P*-chiral ligands is a large barrier to entry for this class of ligand. It has taken over twenty years to see new, useful ligands of this class appear. BisP* (**4**) provides good selectivity for the reductions of dehydroamino acids [47, 48], enamides [49], *E*-β-acylaminoacrylates [50], and α,β-unsaturated-α-acyloxyphosphonates [51], but rates can be slow. Other ligands of this type are MiniPhos (**5**) [47, 48, 52] and the unsymmetrical **6** [53, 54], TangPhos (**7**) is also a member of the class [55–58], as are BIPNOR (**8**) [59, 60] and ^{i}Pr-BeePhos (**9**) [61]. Mention should also be made of the DuPhos-type hybrid **10** that works well for the reductions of itaconic acids [62]. A recent addition to the general class is trichickenfootphos (**11**); this has been developed for the reduction of enamides, dehydroamino acids and α,β-unsaturated nitriles [63, 64]. (Throughout this chapter, R in generic structures denotes an alkyl group unless otherwise stated.)

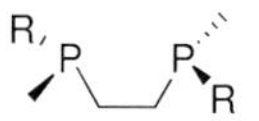

4 R = *t*-Bu (BisP*)
(1998)

R P P R

5 R = *t*-Bu (MiniPhos)
(1998)

R P P R^1

6 R =Ad, R^1 = *t*-Bu (unsymmetrical BisP*)
(2001)

H P H P *t*-Bu *t*-Bu

7 (TangPhos)
(2002)

Ph P Ph P Ph Ph

8 (BIPNOR) (1997)

P P *i*-Pr *i*-Pr

9 (^{i}Pr-BeePHOS) (1997)

HO HO P Ph P OMe

10 (2003)

P P

11 (Trichickenfootphos)
(2003)

23.3 DIOP

Kagan and colleagues found that DIOP (**2**) could provide significant enantioselective induction in a hydrogenation, and this finding led to Knowles' development of the DIPAMP system [65–67]. Certainly, at the time when the results were reported the ee-values were considered high, though this would not be the case today. DIOP is prepared from tartaric acid, and has the stereogenic centers in the carbon backbone rather than at the phosphorus atoms. The use of two phosphorus groups within the same molecule provided the move to the current plethora of bisphosphine ligands. The second key finding was the use of stereogenic centers in the backbone, as these are much easier to introduce than obtaining a chiral phosphorus with high enantiopurity.

DIOP has not found widespread usage after the initial investigations, presumably due to the lower selectivity.

During the 1980s, Achiwa and colleagues examined a number of derivatives of DIOP, and found that MOD-DIOP (**12c**) allowed for the enantioselective hydrogenation of itaconic acid derivatives with >96% ee [68–75].

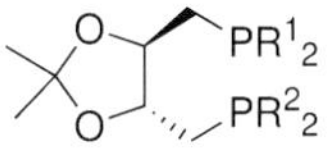

12 (1987)

a $R^1 = R^2$ = Cy (Cy-DIOP)
b R^1 = Cy. R^2 = Ph (DIOCP)
c $R^1 = R^2$ = 3,5-Me_2-4-$MeOC_6H_2$ (MOD-DIOP)

At the turn of the millennium, Zhang and RajanBabu have independently returned to derivatives of DIOP and found that the introduction of α-methyl groups as in DIOP* (**13**) greatly increases enantioselectivity [76–79]. Zhang attributes this improvement to a reduction in the conformational flexibility of the backbone within the metal complex [3, 76]. It is interesting to note that Kagan himself prepared the *S,S,S,S*-isomer of DIOP*, but enantioselectivities were lower than with DIOP itself [80].

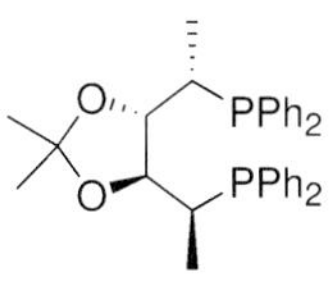

13 (DIOP*)
(2000)

The system can also be made rigid by modification of the ketal portion, as illustrated by **14** and SK-Phos (**15**). These ligands provided high stereoselection for reductions of enamides and MOM-protected β-hydroxy enamides [81].

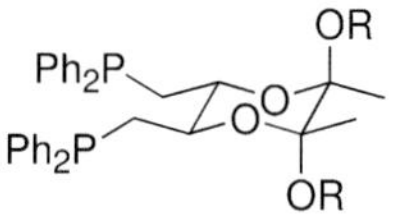

OR
Ph_2P
Ph_2P
O O
OR

14 (2002)

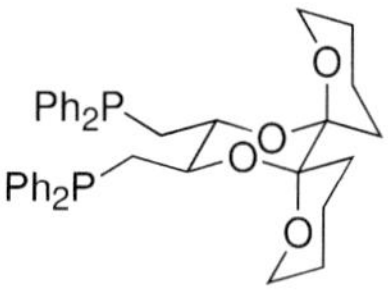

Ph_2P
Ph_2P
O O O O

15 (SK-Phos)
(2002)

As already stated, DIOP led the way for a number of ligand systems that were built on a carbon framework containing stereogenic centers. Some of these ligands followed closely on the heels of DIOP, such as ChiraPhos (**16**) where the chelate ring is five-membered [82, 83]. Even one stereogenic center in the backbone, as in ProPhos (**17**), provides reasonable selectivity [83, 84]. The main problem with these systems is that of slow reactions.

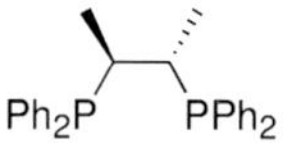

Ph_2P PPh_2

16 (ChiraPhos)
(1977)

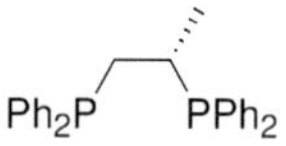

Ph_2P PPh_2

17 (ProPhos)
(1977)

Other early variations on the theme led to BPPM (**18a**) [85] and CBD (**19**) [86].

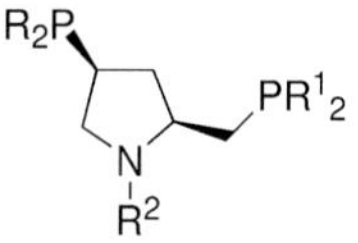

R_2P
PR^1_2
N
R^2

18 (1976)

a $R = R^1$ = Ph, R^2 = Boc (BPPM)
b $R = R^1$ = Ph, R^2 = H (PPM)
c $R = R^1$ = Ph, R^2 = CO-*t*-Bu (PPPM)
d $R = R^1$ = 3,5-Me_2-4-$MeOC_6H_2$, R^2 = Boc (MOD-BPPM)
e R = Cy, R^1 = Ph, R^2 = CO_2-*t*-Bu (BCPM)
f R = Cy, R^1 = 3,5-Me_2-4-$MeOC_6H_2$, R^2 = CO_2-*t*-Bu (MOD-BCPM)
g R = Cy, R^1 = Ph, R^2 = CONHMe (MCCPM)

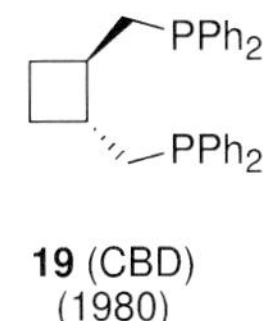

19 (CBD)
(1980)

BPPM is derived from 4-hydroxyproline [85]. As "unnatural" amino acids are often the target product for enantioselective reductions, Knowles' comment in his 1983 review is interesting: "BPPM, like DIOP, is a seven-membered chelator derived from natural (2*R*,4*R*)-hydroxyproline. It can give high efficiency at very fast rates. Unfortunately, it gives unwanted D-amino acids..." [22]. Although this may not be seen as a shortfall today, derivatives (**18**) of BPPM have been developed, mainly through different nitrogen substituents, and derivatives such as PPPM (**18c**) still give D-amino acids [85, 87–98]. The ligands are also useful for the reduction of itaconic acid derivatives [3].

One of the branches in ligand design was provided by Kumada and his introduction of the ferrocene backbone for BPPFA [99–101] (**20a**) and BPPOH [102] (**20b**). This development leads us to the next class of ligands – ferrocene-based. Other variations for development include changes in the backbone and incorporation of the phosphorus into a phospholane (see Section 23.6).

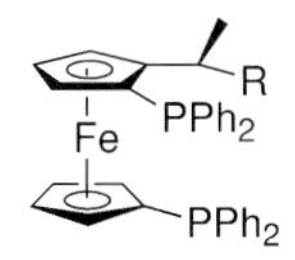

20 (1976)
a R = NMe_2 (BPPFA)
b R = OH (BPPOH)

The electronic properties around the phosphorus atom can be varied by manipulation of the groups on that atom. MOD-DIOP (**12c**) was developed by Achiwa and used to reduce itaconic acids [68–72, 75, 103]. Some variations built on BCPM (**18e**), itself a variant of BPPM, such as the MOD-BCPM (**18f**) and MCCPM (**18g**) [88–93, 95–98, 104]. Other variants are PYRPHOS (**21a**; also called DeguPHOS) [105, 106], DPCP (**22**) [107], NorPhos (**23**) [108], BDPP (**24a**) (also called SkewPHOS) [109–111], and PPCP (**25**) [112].

Ar_2P PAr_2
N
R

21 (1984)
a Ar = Ph, R = CH_2Ph (PYRPHOS, DeguPHOS)
b Ar = 3,5-Me_2-4-$MeOC_6H_2$, R = CH_2Ph (MOD-DeguPHOS)

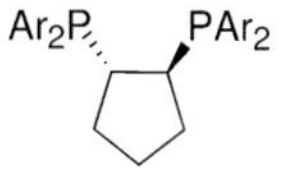

Ar_2P PAr_2

22 (DPCP)
(1983)

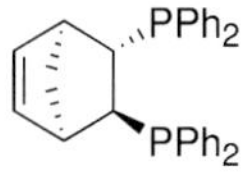

PPh_2
PPh_2

23 (NorPhos)
(1981)

PAr_2 PAr_2

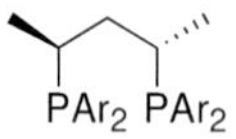

24 (1981)

a Ar = Ph (BDPP, SkewPhos)
b Ar = 3,5-Me_2-4-$MeOC_6H_2$ (MOD-BDPP)

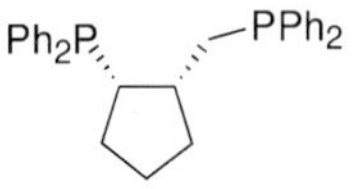

Ph_2P PPh_2

25 (PPCP)
(1991)

One ligand system which was developed during the late 1990s, and has proven quite versatile for the reduction of a wide variety of unsaturations, including α-and β-dehydroamino acids, arylenamides and MOM-protected β-hydroxy enamides, is the rigid BICP (**26**) [113–118].

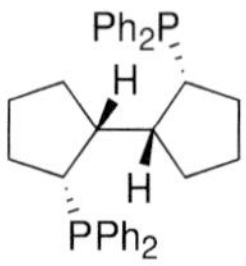

Ph_2P H H PPh_2

26 (BICP)
(1997)

The BDPMI (**27**) system can also be considered to be a DIOP variant, as an imidazole ring forms the rigid backbone [119–121]. Excellent stereoselectivity is seen with this system for the reductions of arylenamides.

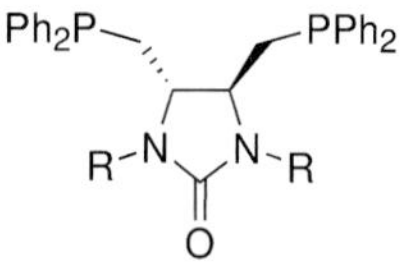

Ph_2P PPh_2 R N N R O

27 (BDPMI)
(2002)

An aromatic system can also provide a rigid backbone, as seen with PhanePhos (**28a**) [122–124].

PAr_2

PAr_2

28 (1997)

a, Ar = Ph (PhanePhos)
b, Ar = 3,5-$Me_2C_6H_3$ (Xyl-PhanePhos)

23.4 Ferrocene Ligands

Kumada's use of a ferrocene moved away from the C_2-symmetrical motive, as planar chirality can result from the two ferrocene rings having different substituents. The development of this class of ligand is well documented [5, 125–127]. The best-known uses of these ligands are for reductions of carbon–heteroatom multiple bonds, as in the synthesis of the herbicide, Metolachlor^{TM} [128, 129].

The key access compound to the early members of the class, and indeed some later ones, is the Ugi amine (**29**) and its relationship to PPFA (**30**) and BPPFA (**20a**) can be clearly seen [100, 130, 131].

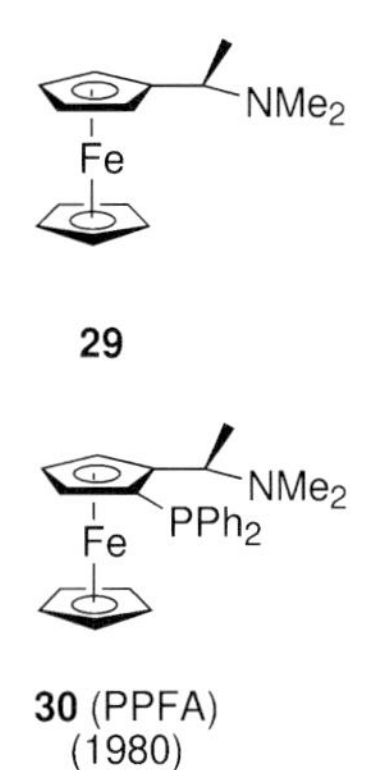

29

30 (PPFA)
(1980)

Analogues of BPPFA and BPPFOH have been prepared, but for many applications these two ligands still prove to be the best for enantioselective hydrogenations [125]. The introduction of another functional group into the side chain, as in **31**, provided the first catalysts capable of hydrogenating the tetra-substituted α,β-unsaturated acids with high enantioselectivity, even though the activity was very low (turnover frequency, TOF, $\sim 2\ h^{-1}$) [132, 133].

31 (1987)

a $R_2 = (CH_2)_4$
b R = Bu

Developments after these Ugi derivatives have taken a number of pathways. The MandyPhos family of ligands (**32**) have been used to reduce enamides to α-amino acids as well as an enol acetate to produce an α-hydroxy ester [134–140]. The substituents R and R^1 can be used for the fine-tuning of a specific substrate. Many of the family have R^1 as a secondary amine, relating the family back to PPFA. For confusion, MandyPhos has also been called FerriPhos, while the derivative **32** ($R = R^1 = Et$) is known as FerroPhos.

32 (MandyPhos) (1989)
$R = R^1 = Et$ (FerroPhos)

Perhaps the first successful variation of the PPFA framework was the development of the JosiPhos family of ligands (**33**) [125, 131, 141, 142]. Here, the two phosphorus groups are attached to the same cyclopentenyl ring rather than one to each of the rings. The C_2-symmetry model is now a distant memory for these ligand families.

The *R,S*-family **33**, and of course its enantiomer, provide high enantioselectivities and activities for the reductions of itaconic and dehydroamino acid derivatives as well as imines [141]. The JosiPhos ligands have found industrial applications for reductions of the carbon–carbon unsaturation within α,β-unsaturated carbonyl substrates [125, 127, 131, 143–149]. In contrast, the *R,R*-diastereoisomer of **30** does not provide high stereoselection in enantioselective hydrogenations [125, 141].

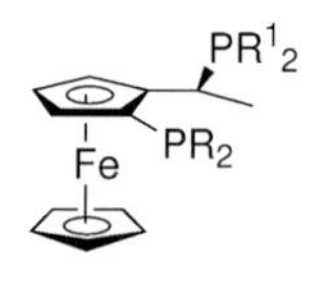

33 (JosiPhos)
(1994)

The recent introduction of the TaniaPhos ligands (**34**) provides another excellent catalyst system for the reduction of dehydroamino acid derivatives and enol acetates [150–153]. One surprise is that the sense of induction in the product is opposite that observed with a JosiPhos-derived catalyst [125].

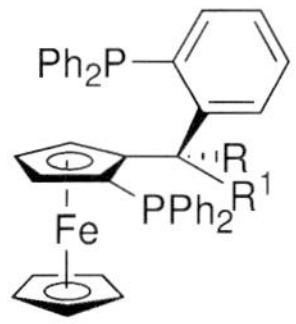

34 (TaniaPhos)
(1999)

Another ligand, the potential of which has only recently been exploited, is BoPhoz (**35**). This ligand is an aminophosphine as well as a phosphine (see also Section 23.7). It has shown high selectivities and activities with enamides and itaconates [154, 155].

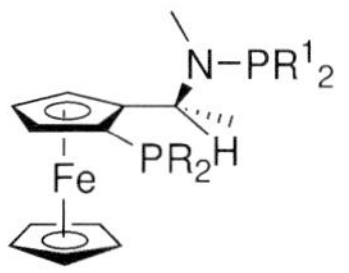

With ferrocenes, an alternative approach has been to attach the phosphorus moieties only to side chains. The WalPhos family (**36**) forms an eight-membered chelate with the metal. Members of this family provide good selectivity and reactivity for the reductions of dehydroamino and itaconic acid derivatives as well as α,β-unsaturated carboxylic acids [145, 156].

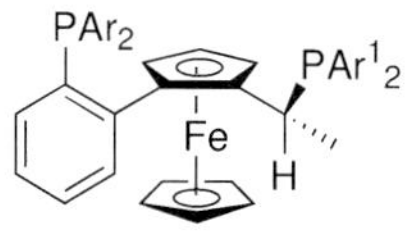

36 (WalPhos)
(2001)

A different variation on this theme has been developed by Ito, where the TRAP ligands (**37**) form a nine-membered metallocycle [157–162]. The ruthenium catalysts seem to function best at low pressures, but highly functionalized dehydroamino esters can be reduced with high degrees of asymmetric induction [157, 159–164], as well as indoles [165].

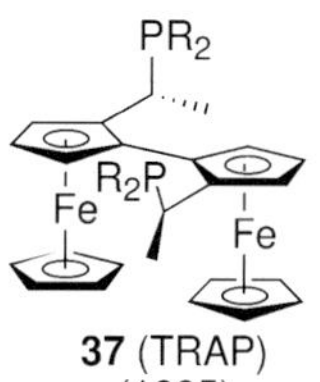

37 (TRAP)
(1995)

where R = alkyl or aryl

It is possible to use other metallocenes as the backbone, as illustrated by the rhenium complex (**38**) [166].

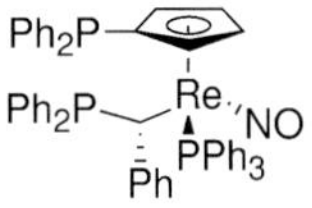
Ph2P
Ph2P
Re
NO
PPh3
Ph

38 (2002)

23.4.1
Ferrocene Hybrids

The rhodium complexes of the ferrocene derivatives **39** have shown useful characteristics for the reduction of itaconates as well as dehydroamino acid derivatives [15, 167–170]. These compounds are hybrids between ferrocene-based ligands and the various other types. The *P*-chiral compounds, which in some ways are DIPAMP hybrids, showed tolerance for the reduction of *N*-methyl enamides to produce *N*-methyl-*α*-amino acid derivatives [169–171].

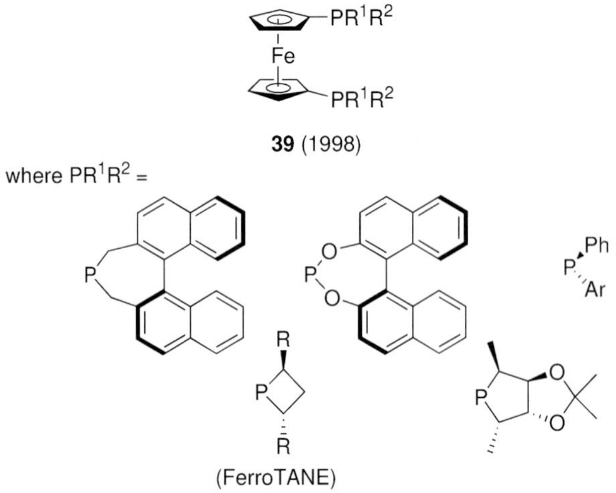
PR1R2
Fe
PR1R2
39 (1998)
where PR1R2 =
P
P
O
O
Ph
P
Ar
R
P
R
(FerroTANE)
P
O
O

23.5
Atropisomeric Systems

BINAP (**40a**) was first reported as a ligand in an enantioselective hydrogenation in 1980 [172], and provides good selectivity for the reductions of dehydroamino acid derivatives [173], enamides, allylic alcohols and amines, and *α*,*β*-unsaturated acids [4, 9, 11, 12, 174, 175]. The fame of the ligand system really came with the reduction of carbonyl groups with ruthenium as the metal [11, 176]. The Rh-BINAP systems is best known for the enantioselective isomerizations used for the industrial-scale synthesis of menthol and other terpenes [12, 177–181], rather than enantioselective hydrogenations.

The use of atropisomeric ligands for carbon–carbon bond reductions was, however, the jumping-off point for variations such as BICHEP (**41a**) [182–185], BIPHEP (**41b**) [127, 186, 187], MeO-BIPHEP (**41c**) [179, 187, 188], and Cl-MeOBIPHEP (**41d**) [189]. A slightly different approach was taken by Achiwa with the BIMOP (**41e**) [190], FUPMOP (**41f**) [191], and MOC-BIMOP (**41g**) [192], with more substituents on the aryl rings. Again, most of the applications and high reactivities are seen for carbon–heteroatom unsaturation hydrogenations [12].

40 (1980)

a Ar = Ph (BINAP)
b Ar = *p*-MeC_6H_4 (TolBINAP)
c Ar = 3,5-$(Me)_2C_6H_3$ (XylBINAP)

41 (1988)

where	R^1	R^2	R^3	R^4	R^5	R^6	R^7	R^8	R^9	R^{10}	
a	Cy	Cy	H	H	H	Me	Me	H	H	H	(BICHEP)
b	Ph	Ph	H	H	H	Me	Me	H	H	H	(BIPHEP)
c	Ph	Ph	H	H	H	OMe	OMe	H	H	H	(MeO-BIPHEP)
d	Ph	Ph	H	H	Cl	OMe	OMe	Cl	H	H	(Cl-MeOBIPHEP)
e	Ph	Ph	H	Me	OMe	Me	Me	OMe	Me	H	(BIMOP)
f	Ph	Ph	H	CF_3	H	CF_3	Me	OMe	Me	H	(FUPMOP)
g	Cy	Cy	H	Me	OMe	Me	Me	OMe	Me	H	(MOC-BIMOP)
h	Ph	Ph	Ph	OMe	OMe	OMe	OMe	OMe	OMe	Ph	(*o*-Ph-HexaMeO-BIPHEP)

The success of BINAP and the associated ligands families has led to many variations, and most have shown improved properties for specific applications (Fig. 23.1). Examples include derivatives of the naphthyl system of BINAP, as well as those derived from BIPHEP.

BINAP itself has been shown to be effective for the reduction of α,β-unsaturated carboxylic acids [8, 36, 177, 215–220], but H_8-BINAP often provides higher ee-values [193, 194]. The ruthenium complex with P-Phos provides high selectiv-

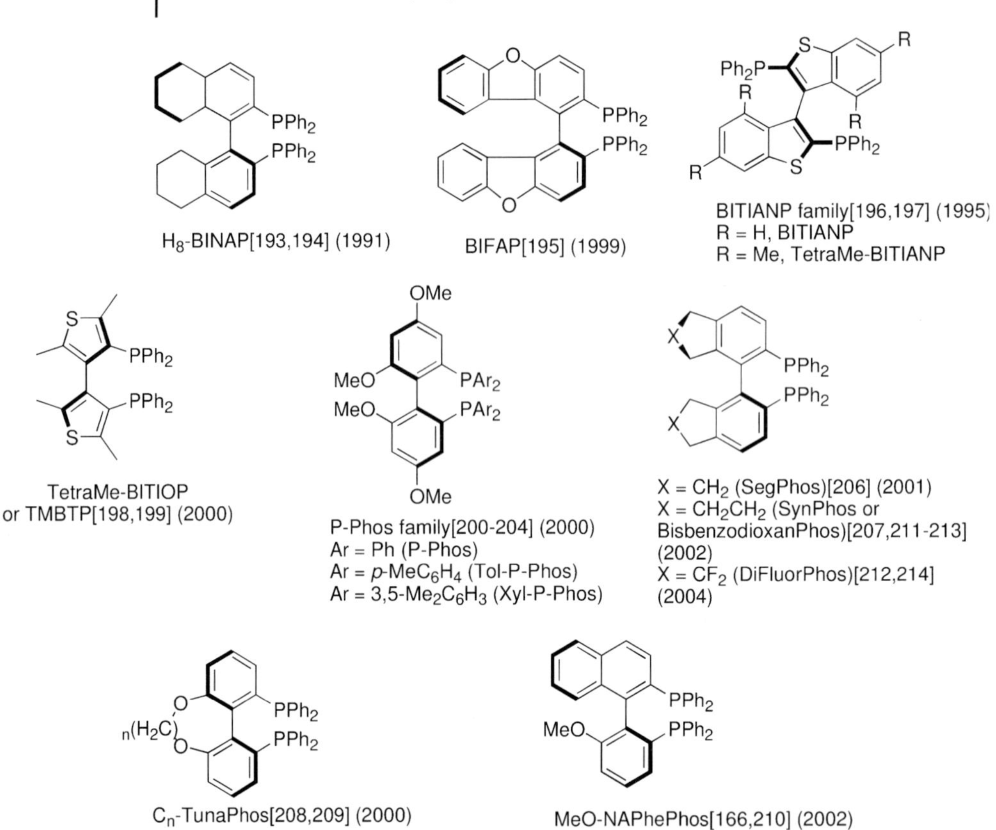

Fig. 23.1 Ligands based on atropisomeric systems.

ities and reactivities with *α*-arylacrylic acids [200–202]. The rhodium complex with *o*-Ph-HexaMeO-BIPHEP works well with cyclic enamides [205].

Many of these ligands have been modified to make them water-soluble, usually by the addition of a sulfate group, or attached to a support [3, 221].

23.6 DuPhos

The development of the next major class of ligands occurred during the 1990s, with Burk's DuPhos (**42**) family of phospholane ligands [222, 223]. (An individual member of the family is named after the substituent R; in Me-DuPhos, R=Me.) This structure could be considered an improvement on the DIOP-derived ligands, where the stereogenic centers are now closer to phosphorus. In addition to the aromatic spacer of DuPhos, there is also the related BPE (**43**) family, where the spacer between the two phosphorus atoms is less rigid. In both series the phosphorus is now part of a five-membered ring that has adjacent stereogenic centers [224, 225]. DuPhos has been shown to be useful for the preparation of α-amino acid derivatives [222, 226–241], including β-branched examples that are not accessible with DIPAMP [222, 242, 243]. The catalyst system is also successful with enamides, enol acetates, unsaturated carboxylic, and itaconic acids [7, 115, 222, 244–250].

42 (DuPhos) (1990)

43 (BPE) (1990)

As the chirality with the DuPhos ligands is within the phospholane rings, a wide variety of backbones can be used ranging from ferrocene to heterocycles [61, 62, 77, 222, 251–262].

Variations have also been made on the DuPhos theme by changing the nature of the phospholane ring (Fig. 23.2). These ligands retain the high selectivity of

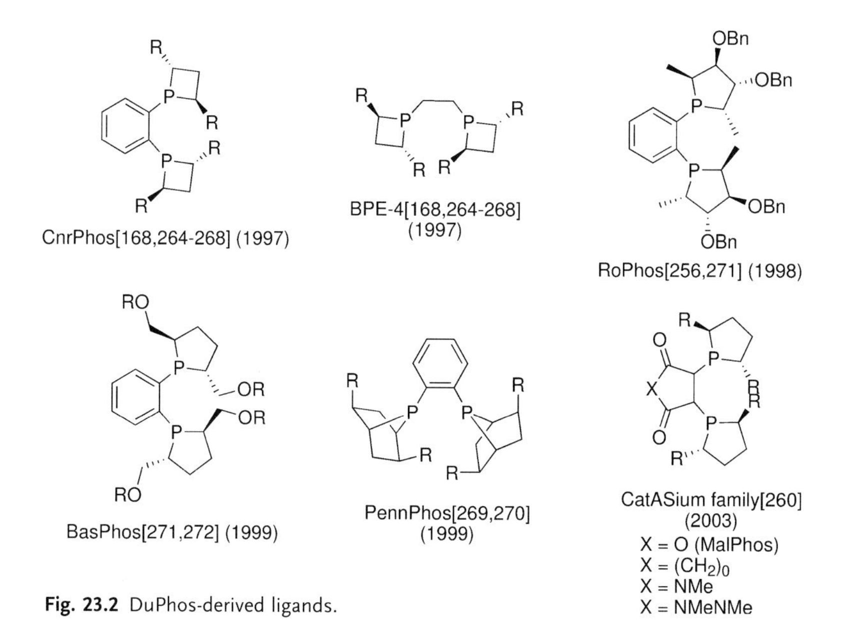

Fig. 23.2 DuPhos-derived ligands.

DuPhos. The exception for rhodium-catalyzed reductions are CnrPhos and BPE-4 [168, 264–268]. MalPhos has proven useful for the reductions of *β*-acylamino-acrylates [260]. The ferrocene hybrid (FerroTANE) was referred to earlier (see Section 23.4.1) [167, 222]. The PennPhos ligand is useful for the reductions of cyclic enamides and enol acetates; both classes of compounds are difficult for DuPhos itself to reduce with high selectivity [269, 270].

Neither of the phosphorus atoms needs not be in an asymmetric phospholane ring, as illustrated by both Saito and Pringle with UCAPs (**44**) [273, 274].

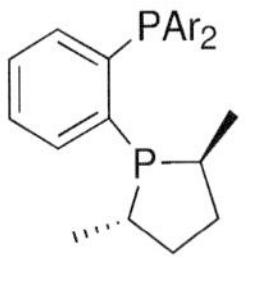
PAr2
P

44 (UCAPs)

23.7
Variations at Phosphorus

In addition to the use of a phosphine ligands, other types of phosphorus moieties have also been used. In some cases, carbohydrates have formed the basis of the backbone, as illustrated by CarboPhos [275–280]. Some of these ligands have been available for almost twenty years. The electronic effects of the ligands can be very important with this class of compound. Other ligands of this class are variations on backbones established for bisphosphane ligands (Fig. 23.3). Most of these ligands have been employed for the reductions of dehydroamino acids; for example, the Phenyl-*β*-Glup ligand has been employed in a process involving L-Dopa [275–277].

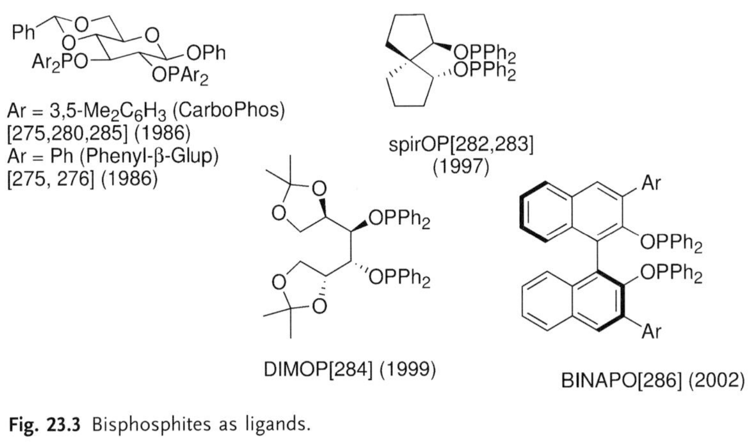
Ph
O
O
O
Ar2PO
OPh
OPAr2
Ar = 3,5-Me2C6H3 (CarboPhos)
[275,280,285] (1986)
Ar = Ph (Phenyl-β-Glup)
[275, 276] (1986)
OPPh2
OPPh2
spirOP[282,283]
(1997)
O
O
OPPh2
OPPh2
O
O
DIMOP[284] (1999)
Ar
OPPh2
OPPh2
Ar
BINAPO[286] (2002)

Fig. 23.3 Bisphosphites as ligands.

In addition to oxygen, the phosphorus can be tied to the backbone through nitrogen. Indeed, this was one of the earliest variations of the DIOP family (see Section 23.3) with PNNP (**45**) [22, 280, 287]. Care must be taken to avoid hydrolysis of the labile P–N bonds [22, 288]. Other examples of nitrogen-linked compounds are BDPAB (**46** and **47**) and its derivatives (Fig. 23.4). These ligands are clearly variations of the BINAP series [289–291].

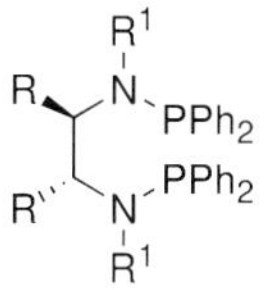

45
R = Ph, R^1 = H (PNNP)
(1979)

46 (1998)
Ar = Ph (BDPAB)
Ar = 3,5-$Me_2C_6H_3$ (Xyl-BDPAB)

47 (H_8-BDPAB) (1998)

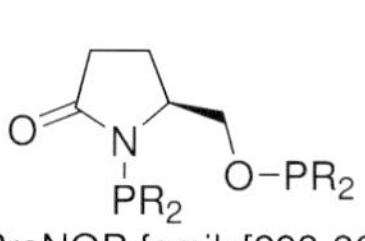

oxoProNOP family[293-300] (1993)
R = Cy (Cy,Cy-oxoProNOP)
R = Cp (Cp,Cp-oxoProNOP)

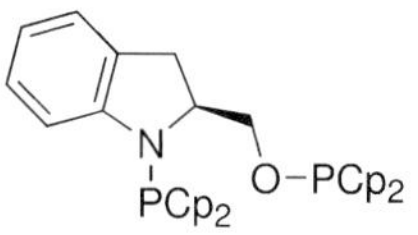

Cp,Cp-IndoNOP[301]
(1996)

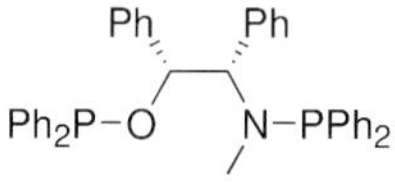

DPAMPP[302,303]
(2000)

where Cp = cyclopentadienyl

Fig. 23.4 Ligands with phosphorus attached to the backbone through oxygen and nitrogen.

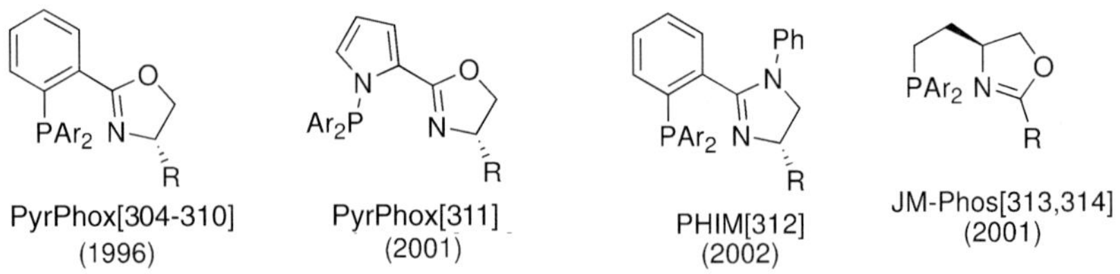
PAr2 N
R
PyrPhox[304-310]
(1996)
Ar2P N
R
PyrPhox[311]
(2001)
Ph
N
PAr2 N
R
PHIM[312]
(2002)
PAr2 N
O
R
JM-Phos[313,314]
(2001)

Fig. 23.5 Monophosphorus ligands.

The linkers may be nitrogen and carbon, as in BoPhoz – that is, a ferrocene-type ligand, as has already been mentioned. An example of a phosphine–phosphoramidite is provided by QuinaPhos (**48**) [292].

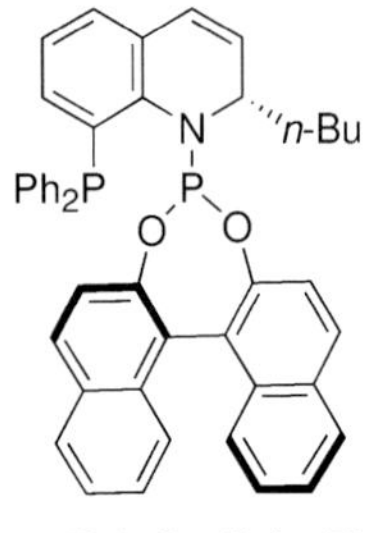
N
n-Bu
Ph2P
P
O
O

48 (*n*-Bu-QuinaPhos)
(2000)

In some cases, oxygen and nitrogen have been used as linkers to the backbones that are variations of those described previously (Fig. 23.4).

23.8
Monophosphorus Ligands

Just as Wilkinson's catalyst gave rise to the bisphosphine ligands, Crabtree's catalyst [304] spawned the family of phosphorus–nitrogen ligands for simple alkenes. Subsequently, Pfaltz developed the Phox family, which provides high ee-values with nonfunctionalized alkenes [305–310]. Other analogues are also illustrated in Figure 23.5.

23.9
A Return to Monodentate Ligands

The observation by Knowles that bisphosphines gave better selectivity for asymmetric hydrogenations resulted in a gap of over twenty-five years before monodentate ligands were investigated in detail and became useful ligands. There are several classes of these monodentate ligands (Fig. 23.6), all of which were introduced within a surprisingly short time period [314]. BINOL has proven to be a very successful backbone with this class of ligand, and covers phosphites (**49**),

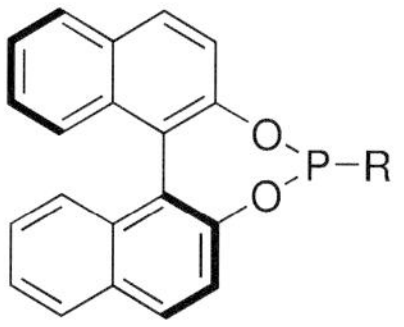

49[275,315]
(2000)

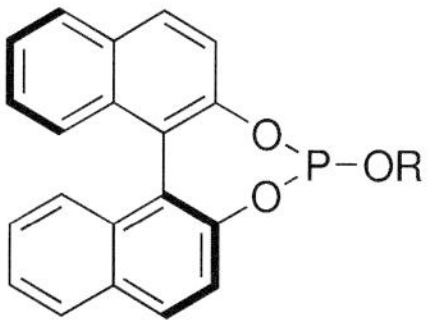

50[316,317]
(2000)

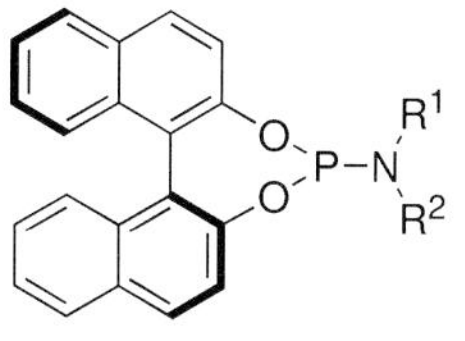

51[318-323]
(2000)
$R^1 = R^2$ = Me (MonoPhos)

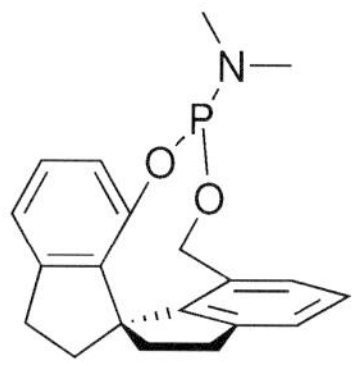

SiPhos[324-326]
(2002)

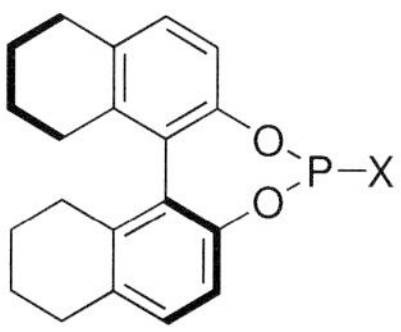

52[316,320,327-330]
(2000)
where X = OR or NR^1R^2

where R = alkyl or aryl

Fig. 23.6 Examples of monodentate ligands.

phosphinites (**50**), and phosphoramidites (**51**). Other variations include SiPhos, as well as reduced aromatic systems (**52**).

These ligands can be used to reduce a wide variety of carbon–carbon unsaturations, and have the advantage that their simple preparation can be used for rapid ligand library synthesis and screening [314, 318, 319, 330, 331].

23.10 Summary

The development of ligands and catalytic systems for the enantioselective hydrogenation of carbon–carbon unsaturation has been rapid, but was given an astounding start by the studies of Knowles, followed by the key findings of Kagan and colleagues. Many derivatives of these ligands have since found a place in the arsenal of the synthetic chemist. Both, Togni's ferrocene-based ligands and Burk's DuPhos family, have expanded the substrate potential of the approach as well as providing for higher selectivities and reactivities. However, recent studies have seen a return to *P*-chiral ligands, while another fairly recent contribution has come from the monodentate ligands of Pringle, Reetz, Feringa, Minnaard, and de Vries. With studies in other areas also beginning to bear fruit, it is possible that we will shortly see ligands based on alkenes making a significant impact. Although there is no "general" catalyst, there is still room for improvement with many potential substrate classes; clearly, the number of ligands appearing in the literature will continue to increase.

References

1 Harada, K. Asymmetric heterogeneous catalytic hydrogenation. In: Morrison, J.D. (Ed.), *Asymmetric Synthesis.* Academic Press, Inc., Orlando, FL, **1985**, *Vol. 5*, p. 345.

2 Blaser, H. *Tetrahedron: Asymmetry* **1991**, *2*, 843.

3 Tang, W., Zhang, Z. *Chem. Rev.* **2003**, *103*, 3029.

4 Takaya, H., Ohta, T., Noyori, R. Asymmetric hydrogenation. In: Ojima, I. (Ed.), *Catalytic Asymmetric Synthesis.* VCH Publishers, Inc., New York, NY, **1993**, p. 1.

5 Koenig, K.E. The applicability of asymmetric homogeneous catalytic hydrogenation. In: Morrison, J.D. (Ed.), *Asymmetric Synthesis.* Academic Press, Inc., Orlando, FL, **1985**; *Vol. 5*, p. 71.

6 Ojima, I., Clos, N., Bastos, C. *Tetrahedron* **1989**, *45*, 6901.

7 Nugent, W.A., RajanBabu, T.V., Burk, M.J. *Science* **1993**, *259*, 479.

8 Noyori, R. *Science* **1990**, *248*, 1194.

9 Noyori, R. *Tetrahedron* **1994**, *50*, 4259.

10 Knowles, W.S. *Angew. Chem., Int. Ed.* **2002**, *41*, **1998**.

11 Noyori, R. *Angew. Chem., Int. Ed.* **2002**, *41*, 2008.

12 Laneman, S.A. In: Ager, D.J. (Ed.), *Handbook of Chiral Fine Chemicals*, 2nd edition. CRC, Taylor Francis, Boca Raton, **2005**, p. 186.

13 Horner, L., Siegel, H., Buthe, H. *Angew. Chem., Int. Ed.* **1968**, *7*, 942.

14 Knowles, W.S., Sabacky, M.J. *J. Chem. Soc., Chem. Commun.* **1968**, 1445.

15 Marinetti, A., Carmichael, D. *Chem. Rev.* **2002**, *102*, 201.

16 Kagan, H.B. Chiral ligands for asymmetric catalysis. In: Morrison, J.D. (Ed.), *Asymmetric Synthesis.* Academic Press, Inc., Orlando, FL, **1985**, *Vol. 5*, p. 1.

17 Pietrusiewicz, K.M., Zablocka, M. *Chem. Rev.* **1994**, *94*, 1375.

18 Holz, J., Quirmbach, M., Börner, A. *Synthesis* **1997**, 983.

19 Ohff, M., Holz, J., Quirmbach, M., Börner, A. *Synthesis* **1998**, 1391.

20 Laurenti, D., Santelli, M. *Org. Prep. Proc. Int.* **1999**, *31*, 245.

21 Osborn, J.A., Jardine, F.H., Young, J.F., Wilkinson, G. *J. Chem. Soc. A* **1966**, 1711.

22 Knowles, W.S. *Acc. Chem. Res.* **1983**, *16*, 106.

23 Green, M., Kuc, T.A., Taylor, S.H. *J. Chem. Soc., Chem. Commun.* **1970**, 1553.

24 Knowles, W.S., Sabacky, M.J., Vineyard, B.D. *J. Chem. Soc., Chem. Commun.* **1972**, 10.

25 Vineyard, B.D., Knowles, W.S., Sabacky, M.J., Bachman, G.L., Weinkauff, D.J. *J. Am. Chem. Soc.* **1977**, *99*, 5946.

26 Knowles, W.S. *J. Chem. Ed.* **1986**, *63*, 222.

27 Knowles, W.S. Asymmetric hydrogenations – The Monsanto L-Dopa process. In: Blaser, H.-U., Schmidt, E. (Eds.), *Asymmetric Catalysis on Industrial Scale.* Wiley-VCH, Weinheim, **2004**, p. 23.

28 Ager, D.J., Laneman, S.A. The synthesis of unnatural amino acids. In: Blaser, H.-U., Schmidt, E. (Eds.), *Asymmetric Catalysis on Industrial Scale.* Wiley-VCH, Weinheim, **2004**, p. 259.

29 Laneman, S.A., Froen, D.E., Ager, D.J. The preparation of amino acids via Rh(DIPAMP)-catalyzed asymmetric hydrogenations. In: Herkes, F.E. (Ed.), *Catalysis of Organic Reactions.* Marcel Dekker, New York, **1998**, p. 525.

30 Chan, A.S.C., Pluth, J.J., Halpern, J. *J. Am. Chem. Soc.*, **1980**, *102*, 5953.

31 Halpern, J., Riley, D.P., Chan, A.S.C., Pluth, J.J. *J. Am. Chem. Soc.* **1977**, *99*, 8055.

32 Chan, A.S.C., Halpern, J. *J. Am. Chem. Soc.* **1980**, *102*, 838.

33 Chan, A.S.C., Pluth, J.J., Halpern, J. *J. Am. Chem. Soc.* **1980**, *102*, 5952.

34 Halpern, J. *Science* **1982**, *217*, 401.

35 Landis, C.R., Halpern, J. *J. Am. Chem. Soc.* **1987**, *109*, 1746.

36 Ashby, M.T., Halpern, J. *J. Am. Chem. Soc.* **1991**, *113*, 589.

37 Brown, J.M., Chaloner, P.A. *J. Chem. Soc., Chem. Commun.* **1978**, 321.

38 Brown, J.M., Chaloner, P.A. *Tetrahedron Lett.* **1978**, *19*, 1877.

39 Brown, J.M., Chaloner, P.A. *J. Chem. Soc., Chem. Commun.* **1979**, 613.

40 Brown, J. M., Chaloner, P. A. *J. Chem. Soc., Chem. Commun.* **1980**, 344.
41 Brown, J. M., Chaloner, P. A., Glaser, R., Geresh, S. *Tetrahedron* **1980**, *36*, 815.
42 Alcock, N. W., Brown, J. M., Derome, A., Lucy, A. R. *J. Chem. Soc., Chem. Commun.* **1985**, 575.
43 Brown, J. M., Chaloner, P. A., Morris, G. A. *J. Chem. Soc., Perkin Trans. II* **1987**, 1583.
44 Brown, J. M. *Chem. Soc. Rev.* **1993**, *22*, 25.
45 Halpern, J. Asymmetric catalytic hydrogenation: Mechanism and origin of enantioselection. In: Morrison, J. D. (Ed.), *Asymmetric Synthesis.* Academic Press, Orlando, **1985**, *Vol. 5*, p. 41.
46 Koenig, K. E., Sabacky, M. J., Bachman, G. L., Christopfel, W. C., Barnstorff, H. D., Friedman, R. B., Knowles, W. S., Stults, B. R., Vineyard, B. D., Weinkauff, D. J. *Ann. N. Y. Acad. Sci.* **1980**, *333*, 16.
47 Imamoto, T., Watanabe, J., Wada, Y., Masuda, H., Yamada, H., Tsuruta, H., Matsukawa, S., Yamaguchi, K. *J. Am. Chem. Soc.* **1998**, *120*, 1635.
48 Gridnev, I. D., Yamanoi, Y., Higashi, N., Tsuruta, H., Yasutake, M., Imamoto, T. *Adv. Synth. Catal.* **2001**, *343*, 118.
49 Gridnev, I. D., Yasutake, M., Higashi, N., Imamoto, T. *J. Am. Chem. Soc.* **2001**, *123*, 5268.
50 Yasutake, M., Gridnev, I. D., Higashi, N., Imamoto, T. *Org. Lett.* **2001**, *3*, 1701.
51 Gridnev, I. D., Higashi, N., Imamoto, T. *J. Am. Chem. Soc.* **2001**, *123*, 4631.
52 Yamanoi, Y., Imamoto, T. *J. Org. Chem.* **1999**, *64*, 2988.
53 Ohashi, A., Imamoto, T. *Org. Lett.* **2001**, *3*, 373.
54 Ohashi, A., Kikuchi, S.-I., Yasutake, M., Imamoto, T. *Eur. J. Org. Chem.* **2002**, 2535.
55 Tang, W., Zhang, X. *Angew. Chem., Int. Ed.* **2002**, *41*, 1612.
56 Tang, W., Zhang, X. *Org. Lett.* **2002**, *4*, 4159.
57 Tang, W., Liu, D., Zhang, X. *Org. Lett.* **2003**, *5*, 205.
58 Cong-Dung Le, J., Pagxenkopf, B. L. *J. Org. Chem.* **2004**, *69*, 4177.
59 Robin, F., Mercier, F., Ricard, L., Mathey, F., Spagnol, M. *Chem. Eur. J.* **1997**, *3*, 1365.
60 Mathey, F., Mercier, F., Robin, F., Ricard, L. *J. Organometal. Chem.* **1998**, *577*, 117.
61 Shimizu, H., Saito, T., Kumobayashi, H. *Adv. Synth. Catal.* **2003**, *345*, 185.
62 Carmichael, D., Doucet, H., Brown, J. M. *J. Chem. Soc., Chem. Commun.* **1999**, 261.
63 Hoge, G., Wu, H.-P., Kissel, W. S., Pflum, D. A., Greene, D. J., Bao, J. *J. Am. Chem. Soc.* **2004**, *126*, 5966.
64 Burk, M. J., De Konig, P. D., Grote, T. M., Hoekstra, M. S., Hoge, G., Jennings, R. A., Kissel, W. S., Le, T. V., Lennon, I. C., Mulhern, T. A., Ramsden, J. A., Wade, R. A. *J. Org. Chem.* **2003**, *68*, 5731.
65 Kagan, H. B., Langlois, N., Dang, T. P. *J. Organometal. Chem.* **1975**, *90*, 353.
66 Kagan, H. B., Dang, T. P. *J. Chem. Soc., Chem. Commun.* **1971**, 481.
67 Kagan, H. B., Dang, T. P. *J. Am. Chem. Soc.* **1972**, *94*, 6429.
68 Chiba, M., Takahashi, H., Takahashi, H., Morimoto, T., Achiwa, K. *Tetrahedron Lett.* **1987**, *28*, 3675.
69 Morimoto, T., Chiba, M., Achiwa, K. *Tetrahedron Lett.* **1988**, *29*, 4755.
70 Morimoto, T., Chiba, M., Achiwa, K. *Tetrahedron Lett.* **1989**, *30*, 735.
71 Morimoto, T., Chiba, M., Achiwa, K. *Chem. Pharm. Bull.* **1989**, *37*, 3161.
72 Morimoto, T., Chiba, M., Achiwa, K. *Heterocycles* **1990**, *30*, 363.
73 Yoshikawa, K., Inoguchi, K., Morimoto, T., Achiwa, K. *Heterocycles* **1990**, *31*, 261.
74 Morimoto, T., Chiba, M., Achiwa, K. *Tetrahedron Lett.* **1990**, *31*, 261.
75 Morimoto, T., Chiba, M., Achiwa, K. *Chem. Pharm. Bull.* **1993**, *41*, 1149.
76 Li, W., Zhang, X. *J. Org. Chem.* **2000**, *65*, 5871.
77 Yan, Y.-Y., RajanBabu, T. V. *J. Org. Chem.* **2000**, *65*, 900.
78 Yan, Y.-Y., RajanBabu, T. V. *Org. Lett.* **2000**, *2*, 199.
79 Yan, Y.-Y., RajanBabu, T. V. *Org. Lett.* **2000**, *2*, 4137.
80 Kagan, H. B., Fiaud, J. C., Hoornaert, C., Meyer, D., Poulin, J. C. *Bull. Soc. Chim. Belg.* **1979**, *88*, 923.
81 Li, W., Waldkirch, J. P., Zhang, X. *J. Org. Chem.* **2002**, 7618.
82 Fryzuk, M. B., Bosnich, B. *J. Am. Chem. Soc.* **1979**, *101*, 3043.

83 Fryzuk, M. B., Bosnich, B. *J. Am. Chem. Soc.* **1977**, *99*, 6262.

84 Fryzuk, M. D., Bosnich, B. *J. Am. Chem. Soc.* **1978**, *100*, 5491.

85 Achima, K. *J. Am. Chem. Soc.* **1976**, *98*, 8265.

86 Glaser, R., Geresh, S., Twaik, M. *Isr. J. Chem.* **1980**, *20*, 102.

87 Ojima, I., Kogure, T., Yoda, N. *J. Org. Chem.* **1980**, *45*, 4728.

88 Takahashi, H., Hattori, M., Chiba, M., Morimoto, T., Achiwa, K. *Tetrahedron Lett.* **1986**, *27*, 4477.

89 Takahashi, H., Morimoto, T., Achiwa, K. *Chem. Lett.* **1987**, 855.

90 Takahashi, H., Achiwa, K. *Chem. Lett.* **1987**, 1921.

91 Inoguchi, K., Morimoto, T., Achiwa, K. *J. Organomet. Chem.* **1989**, *370*, C9.

92 Takahashi, H., Yamamoto, N., Takeda, H., Achiwa, K. *Chem. Lett.* **1989**, 559.

93 Takahashi, H., Sakuraba, S., Takeda, H., Achiwa, K. *J. Am. Chem. Soc.* **1990**, *112*, 5876.

94 Sakuraba, S., Achiwa, K. *Synlett* **1991**, 689.

95 Sakuraba, S., Nakajima, N., Achiwa, K. *Synlett* **1992**, 829.

96 Sakuraba, S., Nakajima, N., Achiwa, K. *Tetrahedron: Asymmetry* **1993**, *7*, 1457.

97 Sakuraba, S., Takahashi, H., Takeda, H., Achiwa, K. *Chem. Pharm. Bull.* **1995**, *43*, 738.

98 Takeda, H., Hosokawa, S., Aburatani, M., Achiwa, K. *Synlett*, **1991**, 193.

99 Hayashi, T., Mise, T., Mitachi, S., Yamamoto, K., Kumada, M. *Tetrahedron Lett.* **1976**, 1133.

100 Hayashi, T., Mise, T., Fukushima, M., Kagotani, M., Nagashima, N., Hamada, Y., Matsumoto, A., Kawakami, S., Monishi, M., Yamomoto, K., Kumada, M. *Bull. Chem. Soc. Jpn.* **1980**, *53*, 1138.

101 Hayashi, T., Kumada, M. *Acc. Chem. Res.* **1982**, *15*, 395.

102 Hayashi, T., Mise, T., Kumada, M. *Tetrahedron Lett.* **1976**, 4351.

103 Yoshikawa, K., Inoguchi, K., Morimoto, T., Achiwa, K. *Heterocycles* **1990**, *31*, 261.

104 Sakuraba, S., Achiwa, K. *Synlett* **1991**, 689.

105 Andrade, J. G., Prescher, G., Schaefer, A., Nagel, U. In: *Chem. Ind.*, Kosik, J. (Ed.), Marcel Dekker, **1990**, p. 33.

106 Nagel, U. *Angew. Chem., Int. Ed. Engl.* **1984**, *23*, 435.

107 Allen, D. L., Gibson, V. C., Green, M. L. H., Skinner, J. F., Bashkin, J., Grebenik, P. D. *J. Chem. Soc., Chem. Commun.* **1983**, 895.

108 Brunner, H., Pieronczyk, W., Schönhammer, B., Streng, K., Bernal, I., Korp, J. *Chem. Ber.* **1981**, *114*, 1137.

109 Bakos, J., Toth, I., Markó, L. *J. Org. Chem.* **1981**, *46*, 5427.

110 NacNeil, P. A., Roberts, N. K., Bosnich, B. *J. Am. Chem. Soc.* **1981**, *103*, 2273.

111 Bakos, J., Tóth, I., Heil, B., Markó, L. *J. Organometal. Chem.* **1985**, *279*, 23.

112 Inoguchi, K., Achiwa, K. *Synlett* **1991**, 49.

113 Zhu, G., Cao, P., Jiang, Q., Zhang, X. *J. Am. Chem. Soc.* **1997**, *119*, 1799.

114 Zhu, G., Zhang, X. *J. Org. Chem.* **1998**, *63*, 9590.

115 Zhu, G., Casalnuovo, A. L., Zhang, X. *J. Org. Chem.* **1998**, *63*, 8100.

116 Zhu, G., Zhang, X. *Tetrahedron: Asymmetry* **1998**, *9*, 2415.

117 Zhu, G., Chen, Z., Zhang, X. *J. Org. Chem.* **1999**, *64*, 6907.

118 Cao, P., Zhang, X. *J. Org. Chem.* **1999**, *64*, 2127.

119 Lee, S.-G., Zhang, Y. J., Song, C. E., Lee, J. K., Choi, J. H. *Angew. Chem., Int. Ed.* **2002**, *41*, 847.

120 Lee, S.-G., Zhang, Y. J. *Org. Lett.* **2002**, *4*, 2429.

121 Lee, S.-G., Zhang, Y. J. *Tetrahedron: Asymmetry* **2002**, *13*, 1039.

122 Pye, P. J., Rossen, K., Reamer, R. A., Tsou, N. N., Volante, R. P., Reider, P. J. *J. Am. Chem. Soc.* **1997**, *119*, 6207.

123 Pye, P. J., Rossen, K., Reamer, R. A., Volante, R. P., Reider, P. J. *Tetrahedron Lett.* **1998**, *39*, 4441.

124 Burk, M. J., Hems, W., Herzberg, D., Malan, C., Zanotti-Gerosa, A. *Org. Lett.* **2000**, *2*, 4173.

125 Blaser, H.-U., Lotz, M., Spindler, F. In: *Handbook of Chiral Fine Chemicals*, 2nd edition. Ager, D. J. (Ed.), CRC: Taylor Francis, Boca Raton, **2005**, p. 287.

126 Colacot, T. J. *Chem. Rev.* **2003**, *103*, 3101.

127 Blaser, H.-U., Malan, C., Pugin, B., Spindler, F., Steiner, H., Studer, M. *Adv. Synth. Catal.* **2003**, *345*, 103.

128 Spindler, F., Pugin, B., Jalett, H.-P., Buser, H.-P., Pittelkow, U., Blaser, H.-U. Catalysis of Organic Reactions. In *Chem. Ind.*, Malz, J. (Ed.), Dekker: New York, **1996**; *Vol. 68*, p. 153.

129 Blaser, H.-U., Spindler, F. *Chimia* **1997**, *51*, 297.

130 Hayashi T. In: *Ferrocenes*. Togni, A., Hayashi T. (Eds.), VCH, Weinheim, **1995**, p. 105.

131 Togni, A., Breutel, C., Schnyder, A., Spindler, F., Landert, H., Tijani, A. *J. Am. Chem. Soc.* **1994**, *116*, 4062.

132 Hayashi, T., Kawamura, N., Ito, Y. *Tetrahedron Lett.* **1988**, *29*, 5969.

133 Hayashi, T., Kawamura, N., Ito, Y. *J. Am. Chem. Soc.* **1987**, *109*, 7876.

134 Almena Perea, J., Lotz, M., Knochel, P. *Tetrahedron: Asymmetry* **1999**, *10*, 375.

135 Hayashi, T., Yamamoto, A., Hojo, M., Ito, Y. *J. Chem. Soc., Chem. Commun.* **1989**, 495.

136 Schwink, L. *Tetrahedron Lett.* **1996**, *37*, 25.

137 Kang, J., Lee, J. H., Ahn, S. H., Choi, J. S. *Tetrahedron Lett.* **1998**, *39*, 5523.

138 Kang, J., Lee, J. H., Kim, J. B., Kim, G. J. *Chirality* **2000**, *12*, 378.

139 Almena Perea, J., Börner, A., Knochel, P. *Tetrahedron Lett.* **1998**, *39*, 8073.

140 Lotz, M., Ireland, T., Almena Perea, J., Knochel, P. *Tetrahedron: Asymmetry* **1999**, *10*, 1839.

141 Blaser, H. U., Brieden, W., Pugin, B., Spindler, F., Studer, M., Togni, A. *Topics in Catalysis* **2002**, *19*, 3.

142 Blaser, H. U., Buser, H. P., Coers, K., Hanreich, R., Jalett, H. P., Jelsch, E., Pugin, B., Schneider, H. D., Spindler, F., Wegmann, A. *Chimia* **1999**, *53*, 275.

143 Blaser, H. U., Spindler, F., Studer, M. *Appl. Catal. A: General* **2001**, *221*, 119.

144 Blaser, H. U. *Adv. Synth. Catal.* **2002**, *344*, 17.

145 Sturm, T., Xiao, L., Weissensteiner, W. *Chimia* **2001**, *55*, 688.

146 Sturm, T., Weissensteiner, W., Spindler, F., Mereiter, K., López-Agenjo, A. M., Manzano, B. R., Jalón, F. A. *Organometallics* **2002**, *21*, 1766.

147 Bader, R. R., Baumeister, P., Blaser, H. U. *Chimia* **1996**, *30*, 9.

148 Imwinkelried, R. *Chimia* **1997**, *51*, 300.

149 McGarrity, J. F., Brieden, W., Fuchs, R., Mettler, H.-P., Schmidt, B., Werbitzky, O. Liberties and constraints in the development of asymmetric hydrogenations on a technical scale. In: *Asymmetric Catalysis on Industrial Scale*. Blaser, H. U., Schmidt, E. (Eds.), Wiley-VCH, Weinheim, **2004**, Chapter III.3, p. 283.

150 Ireland, T., Tappe, K., Grossheimann, G., Knochel, P. *Chem. Eur. J.* **2002**, *8*, 843.

151 Lotz, M., Polborn, K., Knochel, P. *Angew. Chem., Int. Ed.* **2002**, *41*, 4708.

152 Ireland, T., Grossheimann, G., Wieser-Jeunesse, C., Knochel, P. *Angew. Chem., Int. Ed.* **1999**, *38*, 3212.

153 Tappe, K., Knochel, P. *Tetrahedron: Asymmetry* **2004**, *15*, 91.

154 Boaz, N. W., Debenham, S. D., Mackenzie, E. B., Large, S. E. *Org. Lett.* **2002**, *14*, 2421.

155 Boaz, N. W., Debenham, S. D., Large, S. E., Moore, M. K. *Tetrahedron: Asymmetry* **2003**, *14*, 3575.

156 Sturm, T., Weissensteiner, W., Spindler, F. *Adv. Synth. Catal.* **2003**, *345*, 160.

157 Kuwano, R., Sato, K., Kurokawa, T., Karube, D., Ito, Y. *J. Am. Chem. Soc.* **2000**, *122*, 7614.

158 Kuwano, R., Sawamura, M., Ito, Y. *Bull. Chem. Soc. Jpn.* **2000**, *73*, 2571.

159 Kuwano, R., Sawamura, M., Ito, Y. *Tetrahedron: Asymmetry* **1995**, *6*, 2521.

160 Kuwano, R., Ito, Y. *J. Org. Chem.* **1999**, *64*, 1232.

161 Kuwano, R., Okuda, S., Ito, Y. *J. Org. Chem.* **1998**, *63*, 3499.

162 Kuwano, R., Okuda, S., Ito, Y. *Tetrahedron: Asymmetry* **1998**, *9*, 2773.

163 Sawamura, M., Hamashima, H., Sugawara, M., Kuwano, N., Ito, Y. *Organometallics* **1995**, *14*, 4549.

164 Sawamura, M., Kuwano, R., Ito, Y. *J. Am. Chem. Soc.* **1995**, *117*, 9602.

165 Kuwano, R., Sato, K., Kurokawa, T., Karube, D. Ito, Y. *J. Am. Chem. Soc.*, **2000**, *122*, 7614.

166 Kromm, K., Osburn, P. L., Gladysz, J. A. *Organometallics* **2002**, *21*, 4275.

167 Berens, U., Burk, M. J., Gerlach, A., Hems, W. *Angew. Chem., Int. Ed.* **2000**, *39*, 1981.

168 Marinetti, A., Genet, J.-P., Jus, S., Blanc, D., Ratovelamanana-Vidal, V. *Chem. Eur. J.* **1999**, *5*, 1160.
169 Stoop, R. M., Mezzetti, A., Spindler, F. *Organometallics* **1998**, *17*, 668.
170 Maienza, F., Wörle, M., Steffanut, P., Mezzetti, A., Spindler, F. *Organometallics* **1999**, *18*, 1041.
171 Nettekoven, U., Kamer, P. C. J., van Leeuwen, P. W. N. M., Widhalm, M., Spek, A. L., Lutz, M. *J. Org. Chem.* **1999**, *64*, 3996.
172 Miyashita, A., Yasuda, A., Takaya, H., Toriumi, K., Ito. T., Souchi, T., Noyori, R. *J. Am. Chem. Soc.* **1980**, *102*, 7932.
173 Miyashita, A., Takaya, H., Souchi, T., Noyori, R. *Tetrahedron* **1984**, *40*, 1245.
174 Chan, A. S. C. *CHEMTECH* **1993**, *March*, 46.
175 Takaya, H., Ohta, T., Sayo, N., Kumobayashi, H., Akutagawa, S., Inoue, S., Kasahara, I., Noyori, R. *J. Am. Chem. Soc.* **1987**, *109*, 1596.
176 Noyori, R., Ohkuma, T. *Angew. Chem., Int. Ed.* **2001**, *40*, 40.
177 Noyori, R., Takaya, H. *Acc. Chem. Res.* **1990**, *23*, 345.
178 Akutagawa, S., Tani, K. Asymmetric isomerization of allylamines. In: *Catalytic Asymmetric Synthesis.* Ojima, I. (Ed.), VCH Publishers, Inc.: New York, **1993**, p. 41.
179 Noyori, R., Hasiguchi, S., Yamano, T. Asymmetric synthesis. In: *Applied Homogeneous Catalysis with Organic Compounds.* Herrmann, B. C. W. A. (Ed.), Wiley-VCH: Weinheim, **2002**; Vol. 1, p. 557.
180 Otsuka, S., Tani, K. In: *Asymmetric Catalytic Isomerization of Functionalized Olefins.* Morrison, J. D. (Ed.), Academic Press, Inc.: Orlando, FL, **1985**; Vol. 5, p. 171.
181 Kagan, H. B. *Bull. Soc. Chim. Fr.* **1988**, 846.
182 Miyashita, A., Karino, H., Shimamura, J., Chiba, T., Nagano, K., Nohira, H., Takaya, H. *Chem. Lett.* **1989**, 1007.
183 Miyashita, A., Karino, H., Shimamura, J., Chiba, T., Nagano, K., Nohira, H., Takaya, H. *Chem. Lett.* **1989**, 1849.
184 Chiba, T., Miyashita, A., Nohira, H. *Tetrahedron Lett.* **1991**, *32*, 4745.
185 Chiba, T., Miyashita, A., Nohira, H., Takaya H. *Tetrahedron Lett.* **1993**, *34*, 2351.
186 Schmid, R., Cereghetti, M., Heiser, B., Schönholzer, P., Hansen, H.-J. *Helv. Chim. Acta* **1988**, *71*, 897.
187 Schmid, R., Foricher, J., Cereghetti, M., Schonholzer, P. *Helv. Chim. Acta* **1991**, *74*, 370.
188 Gautier, I., Ratovelomanana-Vidal, V., Savignac, P., Genet, J.-P. *Tetrahedron Lett.* **1996**, *37*, 7721.
189 Gerlach, A., Scholz, U. *Spec. Chem. Magazine* **2004**, 37.
190 Yamamoto, N., Murata, M., Morimoto, T., Achiwa, K. *Chem. Pharm. Bull.* **1991**, *39*, 1085.
191 Murata, M., Morimoto, T., Achiwa, K. *Synlett* **1991**, 827.
192 Yoshikawa, K., Yamamoto, N., Murata, M., Awano, K., Morimoto, T., Achiwa, K. *Tetrahedron: Asymmetry* **1992**, *3*, 13.
193 Zhang, X., Mashima, K., Koyano, K., Sayo, N., Kumobayashi, H., Akutagawa, S., Takaya, H. *Tetrahedron Lett.* **1991**, *32*, 7283.
194 Zhang, X., Mashima, K., Koyano, K., Sayo, N., Kumobayashi, H., Akutagawa, S., Takaya, H. *J. Chem. Soc., Perkin Trans. I* **1994**, 2309.
195 Sollewijn Gelpke, A. E., Kooijman, H., Spek, A. L., Hiemstra, H. *Chem. Eur. J.* **1999**, *5*, 2472.
196 Benincori, T., Brenna, E., Sannicolò, F., Trimarco, L., Antognazza, P., Cesarotti, E. *J. Chem. Soc., Chem. Commun.* **1995**, 685.
197 Benincori, T., Brenna, E., Sannicolò, F., Trimarco, L., Antognazza, P., Cesarotti, E. Demartin, F., Pilati, T. *J. Org. Chem.* **1996**, *61*, 6244.
198 Beninicori, T., Cesarotti, E., Piccolo, O., Sannicolò, F. *J. Org. Chem.* **2000**, *65*, 2043.
199 Banzinger, M., Cercus, J., Hirt, H., Laumen, K., Malan, C., Spindler, F., Struber, F., Troxler, T. *Tetrahedron: Asymmetry* **2003**, *14*, 3469.
200 Pai, C.-C., Lin, C.-W., Lin, C.-C., Chen, C.-C., Chan, A. S. C. *J. Am. Chem. Soc.* **2000**, *122*, 11513.

201 Wu, J., Wai, H. K., Kim, H. L., Zhong, Y. Z., Yeung, C. H., Chan, A. S. C. *Tetrahedron Lett.* **2002**, 1539.

202 Wu, J., Chen, X., Guo, R., Yeung, C. H., Chan, A. S. C. *J. Org. Chem.* **2003**, *68*, 2490.

203 Wu, J., Chen, H., Zhou, Z.-Y., Yeung, C. H., Chan, A. S. C. *Synlett* **2001**, 1050.

204 Wu, J., Pai, C.-C., Kwok, W., Guo, R., Au-Yeung, T. T.-L., Yeung, C.H., Chan, A.S.C. *Tetrahedron: Asymmetry* **2003**, *14*, 987.

205 Tang, W., Chi, Y., Zhang, X. *Org. Lett.* **2002**, *4*, 1695.

206 Saito, T., Yokozawa, T., Ishizaki, T., Moroi, T., Sayo, N., Miura, T., Kumobayashi, H. *Adv. Synth. Catal.* **2001**, *343*, 264.

207 Pai, C.-C., Li, Y.-M., Zhong, Y. Z., Chan, A.S.C. *Tetrahedron Lett.* **2002**, *43*, 2789.

208 Zhang, Z., Qian, H., Longmire, J., Zhang, X. *J. Org. Chem.* **2000**, *65*, 6223.

209 Wu, S., Wang, W., Tang, W., Lin, M., Zhang, X. *Org. Lett.* **2002**, *4*, 4495.

210 Michaud, G., Bulliard, M., Ricard, L., Genêt, J.-P., Marinetti, A. *Chem. Eur. J.* **2002**, *8*, 3327.

211 Duprat de Paule, S., Jeulin, S., Ratovelomanana-Vidal, V., Genêt, J.-P., Champion, N., Dellis, P. *Tetrahedron Lett.* **2003**, *44*, 823.

212 Jeulin, S., Duprat de Paul, S., Ratovelomanana-Vidal, V., Genêt, J.-P., Champion, N., Dellis, P. *Proc. Nat. Acad. Sci. USA* **2004**, *101*, 5799.

213 Duprat de Paul, S., Jeulin, S., Ratovelomanana-Vidal, V., Genêt, J.-P., Champion, N., Dellis, P. *Eur. J. Org. Chem.* **2003**, 1931.

214 Jeulin, S., Duprat de Paul, S., Ratovelomanana-Vidal, V., Genêt, J.-P., Champion, N., Dellis, P. *Angew. Chem., Int. Ed.* **2004**, *43*, 320.

215 Ager, D. J., Babler, S., Froen, D. E., Laneman, S. A., Pantaleone, D. P., Prakash, I., Zhi, B. *Org. Proc. Research. Develop.* **2003**, *7*, 369.

216 Mashima, K., Kusano, K., Ohta, T., Noyori, R., Takaya, H. *J. Chem. Soc., Chem. Commun.* **1989**, 1208.

217 Noyori, R. *Chem. Soc. Rev.* **1989**, 187.

218 Takaya, H., Ohta, T., Mashima, K., Noyori, R. *Pure Appl. Chem.* **1990**, *62*, 1135.

219 Uemura, T., Zhang, X., Matsumura, K., Sayo, N., Kumobayashi, H., Ohta, T., Nozaki, K., Takaya, H. *J. Org. Chem.* **1996**, *61*, 5510.

220 Zhang, X., Uemura, T., Matsumura, K., Sayo, N., Kumobayashi, H., Tayaya, H. *Synlett* **1994**, 501.

221 Wan, K. T., Davis, M. E. *Nature* **1994**, *370*, 449.

222 Burk, M. J., Ramsden, J. A. In: *Handbook of Chiral Fine Chemicals*, 2nd edition. Ager, D. J. (Ed.), CRC: Taylor Francis, Boca Raton, **2005**, p. 249.

223 Cobley, C. J., Johnson, N. B., Lennon, I. C., McCague, R., Ramsden, J. A., Zanotti-Gerosa, A. In: *Asymmetric Catalysis on Industrial Scale.* Blaser, H. U., Schmidt, E. (Eds.), Wiley-VCH, Weinheim, **2004**, Chapter III.2, p. 269.

224 Burk, M. J., Feaster, J. E., Harlow, R. L. *Organometallics* **1990**, *9*, 2653.

225 Burk, M. J., Harlow, R. L., *Angew. Chem., Int. Ed.* **1990**, *29*, 1462.

226 Burk, M. J., Feaster, J. E., Nugent, W. A., Harlow, R. L. *J. Am. Chem. Soc.* **1993**, *115*, 10125.

227 Burk, M. J., Bienewald, F. In: *Transition Metals for Organic Synthesis and Fine Chemicals.* Bolm, C., Beller, M. (Eds.), VCH Publishers: Weinheim, Germany, **1998**, Vol. 2, p. 13.

228 Stammers, T. A., Burk, M. J. *Tetrahedron Lett.* **1999**, *40*, 3325.

229 Masquelin, T., Broger, E., Mueller, K., Schmid, R., Obrecht, D. *Helv. Chim. Acta* **1994**, *77*, 1395.

230 Jones, S. W., Palmer, C. F., Paul, J. M., Tiffin, P. D. *Tetrahedron Lett.* **1999**, *40*, 1211.

231 Debenham, S. D., Debenham, J. S., Burk, M. J., Toone, E. J. *J. Am. Chem. Soc.* **1997**, *119*, 9897.

232 Debenham S. D., Cossrow, J., Toone, E. J. *J. Org. Chem.* **1999**, *64*, 9153.

233 Xu, X., Fakha, G., Sinou, D. *Tetrahedron* **2002**, *58*, 7539.

234 Rizen, A., Basu, B., Chattopadhyay, S. K., Dossa, F., Frejd, T. *Tetrahedron: Asymmetry* **1998**, *9*, 503.

235 Hiebl, J., Kollmann, H., Rovenszky, F., Winkler, K. *J. Org. Chem.* **1999**, *64*, 1947.
236 Maricic, S., Ritzén, A., Berg, U., Frejd, T. *Tetrahedron*, **2001**, *57*, 6523.
237 Shieh, W.-C., Xue, S., Reel, N., Wu, R., Fitt, J., Repic, O. *Tetrahedron: Asymmetry* **2001**, *12*, 2421.
238 Wang, W., Yang, J., Ying, J., Xiong, C., Zhang, J., Cai, C., Hruby, V. J. *J. Org. Chem.* **2002**, *67*, 6353.
239 Wang, W., Cai, M., Xiong, C., Zhang, J., Trivedi, D., Hruby, V. J. *Tetrahedron* **2002**, *58*, 7365.
240 Burk, M. J., Allen, J. G., Kiesman, W. F. *J. Am. Chem. Soc.* **1998**, *120*, 657.
241 Teoh, E., Campi, E. M., Jackson, W. R., Robinson, A. J. *J. Chem. Soc., Chem. Commun.* **2002**, 978.
242 Hoerrner, R. S., Askin, D., Volante, R. P., Reider, P. J. *Tetrahedron Lett.* **1998**, *39*, 3455.
243 Burk, M. J., Gross, M. F., Martinez, J. P. *J. Am. Chem. Soc.* **1995**, *117*, 9375.
244 Burk, M. J. *Acc. Chem. Res.* **2000**, *33*, 363.
245 Burk, M. J., Wang, Y. M., Lee, J. R. *J. Am. Chem. Soc.* **1996**, *118*, 5142.
246 Burk, M. J. *J. Am. Chem. Soc.* **1991**, *113*, 8518.
247 Boaz, N. W. *Tetrahedron Lett.* **1998**, *39*, 5505.
248 Burk, M. J., Stammers, T. A., Straub, J. A. *Org. Lett.* **1999**, *1*, 387.
249 Burk, M. J., Kalberg, C. S., Pizzano, A. *J. Am. Chem. Soc.* **1998**, *120*, 4345.
250 Burk, M. J., Bienewald, F., Harris, M., Zanotti-Gerosa, A. *Angew. Chem., Int. Ed.* **1998**, *37*, 1931.
251 Morimoto, T., Ando, N., Achiwa, K. *Synlett* **1996**, 1211.
252 Dierkes, P., Ramdeehul, S., Barloy, L., De Cian, A., Fischer, J., Kamer, P. C. J., van Leeuwen, P. W. N. M., Osborn, J. A. *Angew. Chem., Int. Ed.* **1998**, *37*, 3116.
253 Schmid, R., Broger, E. A., Cereghetti, M., Crameri, Y., Foricher, J., Lalonde, M., Muller, R. K., Scalone, M., Schoettel, G., Zutter, U. *Pure Appl. Chem.* **1996**, *68*, 131.
254 Burk, M. J., Pizzano, A., Martin, J. A., Liable-Sands, L., Rheingold, A. L. *Organometallics* **2000**, *19*, 250.
255 Burk, M. J., Gross, M. F. *Tetrahedron Lett.* **1994**, *35*, 9363.
256 Holz, J., Quirmbach, M., Schmidt, U., Heller, D., Stürmer, R., Börner, A. *J. Org. Chem.* **1998**, *63*, 8031.
257 RajanBabu, T. V., Yan, Y.-Y., Shin, S. *J. Am. Chem. Soc.* **2001**, *123*, 10207.
258 Li, W., Zhang, Z., Zhang, X. *J. Org. Chem.* **2000**, *65*, 3489.
259 Holz, J., Stürmer, R., Schmidt, U., Drexler, H.-J., Heller, D., Krimmer, H.-P., Börner, A. *Eur. J. Org. Chem.* **2001**, 4615.
260 Holz, J., Monsees, A., Jiao, H., You, J., Komarov, I. V., Fischer, C., Drauz, K., Borner, A. *J. Org. Chem.* **2003**, *68*, 1701.
261 Fernandez, E., Gillon, A., Heslop, K., Horwood, E., Hyett, D. J., Orpen, A. G., Pringle, P. G. *J. Chem. Soc., Chem. Commun.* **2000**, 1663.
262 Landis, C. R., Wiechang, J., Owen, J. S., Clark, T. P. *Angew. Chem., Int. Ed.* **2001**, *40*, 3432.
263 Matsumura, K., Shimizu, H., Saito, T., Kumobayashi, H. *Adv. Synth. Catal.* **2003**, *345*, 180.
264 Marinetti, A., Kruger, V., Buzin, F.-X. *Tetrahedron Lett.* **1997**, *38*, 2947.
265 Marinetti, A., Labrue, F., Genêt, J.-P. *Synlett* **1999**, 1975.
266 Marinetti, A., Jus, S., Genêt, J.-P. *Tetrahedron Lett.* **1999**, *40*, 8365.
267 Marinetti, A., Jus, S., Genêt, J.-P., Ricard, L. *Tetrahedron* **2000**, *56*, 95.
268 Marinnetti, A., Jus, S., Genêt, J.-P., Ricard, L. *J. Organometal. Chem.* **2001**, *624*, 162.
269 Zhang, Z., Zhu, G., Jiang, Q., Xiao, D., Zhang, X. *J. Org. Chem.* **1999**, *64*, 1774.
270 Jiang, Q., Xiao, D., Zhang, Z., Cao, P., Zhang, X. *Angew. Chem., Int. Ed.* **1999**, *38*, 516.
271 Holz, J., Stürmer, R., Schmidt, U., Drexler, H.-J., Heller, D., Krimmer, H.-P., Börner, A. *Eur. J. Org. Chem.* **2000**, 4615.
272 Holz, J., Heller, D., Stürmer, R., Börner, A. *Tetrahedron Lett.* **1999**, *40*, 7059.
273 Matsumura, K., Shimizu, H., Saito, T., Kumobayashi, H. *Adv. Synth. Catal.* **2003**, *345*, 180.
274 Claver, C., Fernandez, E., Gillon, A., Heslop, K., Hyett, D. J., Martorell, A., Orpen, A. G., Pringle, P. G. *J. Chem. Soc., Chem. Commun.* **2000**, 961.

275 Selke, R., Pracejus, H. *J. Mol. Catal.* **1986**, *37*, 213.

276 Vocke, W., Hanel, R., Flother, F.-U. *Chem. Technol.* **1987**, *39*, 123.

277 de Vies, J. G. In: *Encyclopedia of Catalysis.* Horvath, I. (Ed.), Wiley, New York, **2003**, Vol. 3, p. 295.

278 Selke, R. *J. Organometal. Chem.* **1989**, *370*, 249.

279 Selke, R., Facklam, C., Foken, H., Heller, D. *Tetrahedron: Asymmetry* **1993**, *4*, 369.

280 RajanBabu, T. V., Ayers, T. A., Casalnuovo, A. L. *J. Am. Chem. Soc.* **1994**, *116*, 4101.

281 RajanBabu, T. V., Ayers, T. A., Halliday, G. A., You, K. K., Calabrese, J. C. *J. Org. Chem.* **1997**, *62*, 6012.

282 Chan, A. S. C., Hu, W., Pai, C.-C., Lau, C.-P., Jiang, Y., Mi, A., Yan, M., Sun, J., Lou, R., Deng, J. *J. Am. Chem. Soc.* **1997**, *119*, 9570.

283 Hu, W., Yan, M., Lau, C.-P., Yang, S. M., Chan, A. S. C., Jiang, Y., Mi, A. *Tetrahedron Lett.* **1999**, *40*, 973.

284 Chen, Y., Li, X., Tong, S.-K., Choi, M. C. K., Chan, A. S. C. *Tetrahedron Lett.* **1999**, *40*, 957.

285 Selke, R. *J. Organometal. Chem.* **1989**, *370*, 249.

286 Zhou, Y.-G., Zhang, X. *Chem. Commun.* **2002**, 1124.

287 Fiorini, M., Giongo, G. M. *J. Mol. Catal.* **1979**, *5*, 303.

288 Pracejus, G., Pracejus, H. *Tetrahedron Lett.* **1977**, *39*, 3497.

289 Zhang, F.-Y., Pai, C.-C., Chan, A. S. C. *J. Am. Chem. Soc.* **1998**, *120*, 5808.

290 Zhang, F.-Y., Kwok, W. H., Chan, A. S. C. *Tetrahedron: Asymmetry* **2001**, *12*, 2337.

291 Guo, R., Li, X., Wu, J., Kwok, W. H., Chen, J., Choi, M. C. K., Chan, A. S. C. *Tetrahedron Lett.* **2002**, *43*, 6803.

292 Franciò, G., Faraone, F., Leitner, W. *Angew. Chem., Int. Ed.* **2000**, *39*, 1428.

293 Roucoux, A., Agbossou, F., Mortreux, A., Petit, F. *Tetrahedron: Asymmetry* **1993**, *4*, 2279.

294 Agbossou, F., Carpentier, J.-F., Hatat, C., Kokel, N., Mortreux, A. *Organometallics* **1995**, *14*, 2480.

295 Roucoux, A., Devocelle, M., Carpentier, J.-F., Agbossou, F., Mortreux, A. *Synlett.* **1995**, 358.

296 Roucoux, A., Thieffry, L., Carpentier, J.-F., Devocelle, M., Méliet, C., Agbossou, F., Mortreux, A. *Organometallics* **1996**, *15*, 2440.

297 Devocelle, M., Agbossou, F., Mortreux, A. *Synlett* **1997**, 1306.

298 Carpentier, J.-F., Mortreux, A. *Tetrahedron Asymmetry* **1997**, *8*, 1083.

299 Pasquier, C., Naili, S., Pelinski, L., Brocard, J., Mortreux, A., Agbossou, F. *Tetrahedron: Asymmetry* **1998**, *9*, 193.

300 Agbossou, F., Carpentier, J.-F., Hapiot, F., Suisse, I., Mortreux, A. *Coord. Chem. Rev.* **1998**, *178–180*, 1615.

301 Kreuzfeld, H.-J., Schmidt, U., Döbler, C., Krause, H. W. *Tetrahedron: Asymmetry* **1996**, *7*, 1011.

302 Xie, Y., Lou, R., Li, Z., Mi, A., Jiang, Y. *Tetrahedron: Asymmetry* **2000**, *11*, 1487.

303 Lou, R., Mi, A., Jiang, Y., Qin, Y., Li, Z., Fu, F., Chan, A. S. C. *Tetrahedron* **2000**, *56*, 5857.

304 Crabtree, R. H. *Acc. Chem. Res.* 1979, *12*, 331.

305 Helmchen, G., Kudis, S., Sennhenn, P., Steinhagen, H. *Pure Appl. Chem.* **1997**, *69*, 513.

306 Pfaltz, A. *Acta Chem. Scand. B* **1996**, *50*, 189.

307 Helmchen, G., Pfaltz, A. *Acc. Chem. Res.* **2000**, *33*, 336.

308 Lightfoot, A., Schnider, P., Pfaltz, A. *Angew. Chem., Int. Ed.* **1998**, *37*, 2897.

309 Blackmond, D.G., Lightfoot, A., Pfaltz, A., Rosner, T., Schnider, P., Zimmermann, N. *Chirality* **2000**, *12*, 442.

310 Pfaltz, A., Blankenstein, J., Hilgraf, R., Hörmann, E., McIntyre, S., Menges, F., Schönleber, M., Smidt, S. P., Wüstenberg, B., Zimmermann, N. *Adv. Synth. Catal.* **2003**, *345*, 33.

311 Cozzi, P. G., Zimmermann, N., Hilgraf, R., Schaffner, S., Pfaltz, A. *Adv. Synth. Catal.* **2001**, *343*, 450.

312 Menges, F., Neuburger, M., Pfaltz, A. *Org. Lett.* **2002**, *4*, 4713.

313 Hou, D.-R., Reibenspies, J., Colacot, T. J., Burgess, K. *Chem. Eur. J.* **2001**, *7*, 5391.

314 de Vries, J. G. In: *Handbook of Chiral Fine Chemicals*, 2nd edition. Ager, D. J. (Ed.), CRC: Taylor Francis, Boca Raton, **2005**, p. 269.

315 Reetz, M. T., Sell, T. *Tetrahedron Lett.* **2000**, *41*, 6333.

316 Reetz, M. T., Mehler, G. *Angew. Chem., Int. Ed.* **2000**, *39*, 3889.

317 Reetz, M. T., Mehler, G., Meiswinkel, A., Sell, T. *Tetrahedron Lett.* **2002**, *43*, 7941.

318 de Vries, J. G., de Vries, A. H. M. *Eur. J. Org. Chem.* **2003**, 799.

319 van den Berg, M., Minnaard, A. J., Schudde, E. P., van Esch, J., de Vries, A. H. M., de Vries, J. G., Feringa, B. L. *J. Am. Chem. Soc.* **2000**, *122*, 11539.

320 van den Berg, M., Minnaard, A. J., Haak, R. M., Leeman, M., Schudde, E. P., Meetsma, A., Feringa, B. L., de Vries, A. H. M., Maljaars, C. E. P., Willans, C. E., Hyett, D., Boogers, J. A. F., Henderickx, H. J. W., de Vries, J. G. *Adv. Synth. Catal.* **2003**, *345*, 308.

321 Jia, X., Guo, R., Li, X., Yao, X., Chan, A. S. C. *Tetrahedron Lett.* **2002**, *43*, 5541.

322 van den Berg, M., Haak, R. M., Minnaard, A. J., de Vries, A. H. M., de Vries, J. G., Feringa, B. L. *Adv. Synth. Catal.* **2002**, *344*, 1003.

323 Ager, D., van den Berg, M., Minnaard, A. J., Feringa, B. L., de Vries, A. H. M., Willans, C. E., Boogers, J. A. F., de Vries, J. G. An affordable catalyst for the production of amino acids. In: *Methodologies in Asymmetric Catalysis*. Malhotra, S. V. (Ed.), ACS Symposium Series 880, American Chemical Society, Washington DC, **2004**, pp. 115.

324 Fu, Y., Xie, J.-H., Hu, A.-G., Zhou, H., Wang, L.-X., Zhou, Q.-L. *J. Chem. Soc., Chem. Commun.* **2002**, 480.

325 Hu, A.-G., Fu, Y., Xie, J.-H., Zhou, H., Wang, L.-X., Zhou, Q.-L. *Angew. Chem., Int. Ed.* **2002**, *41*, 2348.

326 Zhu, S.-F., Fu, Y., Xie, J.-H., Liu, B., Xing, L., Zhou, Q.-L. *Tetrahedron: Asymmetry* **2003**, *14*, 3219.

327 Gergely, I., Hegedüs, C; Gulyás, H., Szöllősy, Á., Monsees, A., Riermeier, T., Bakos, J., *Tetrahedron: Asymmetry* **2003**, *14*, 1087.

328 Hannen, P., Millitzer, H.-C., Vogl, E.M., Rampf, F.A. *J. Chem. Soc., Chem. Commun.* **2003**, 2210.

329 Zeng, Q., Liu, H., Cui, X., Mi, A., Jiang, Y., Li, X., Choi, M. C. K., Chan, A.S.C. *Tetrahedron: Asymmetry* **2002**, *13*, 115.

330 Lefort, L., Boogers, J. A. F., de Vries, A. H. M., de Vries, J. G. *Org. Lett.* **2004**, *6*, 1733.

331 Peña, D., Minnaard, A. J., Boogers, J. A. F., de Vries, A. H. M., de Vries J. G., Feringa, B. L. *Org. Biomol. Chem.* **2003**, *1*, 1087.

24
Enantioselective Hydrogenation: Phospholane Ligands

Christopher J. Cobley and Paul H. Moran

24.1
Introduction and Extent of Review

The ability to efficiently synthesize enantiomerically enriched materials is of key importance to the pharmaceutical, flavor and fragrance, animal health, agrochemicals, and functional materials industries [1]. An enantiomeric catalytic approach potentially offers a cost-effective and environmentally responsible solution, and the assessment of chiral technologies applied to date shows enantioselective hydrogenation to be one of the most industrially applicable [2]. This is not least due to the ability to systematically modify chiral ligands, within an appropriate catalyst system, to obtain the desired reactivity and selectivity. With respect to this, phosphorus(III)-based ligands have proven to be the most effective.

Amongst the hundreds of chiral phosphorus-based ligands developed since the seminal studies of Knowles and Horner [3], only a select few ligand families have had a revolutionary impact on the field. The highly modular chiral C_2-symmetric phospholane ligands (DuPhos™ and BPE), developed by Burk and coworkers at DuPont, are one such example. As a result, much effort has been directed towards building on this breakthrough discovery and extending both the design and application of this ligand class.

In this chapter, we review the growing family of phospholane-based chiral ligands, and specifically examine their applications in the field of enantioselective hydrogenation. In general, this ligand class has found its broadest applicability in the reduction of prochiral olefins and, to a significantly lesser extent, ketones and imines; this is reflected in the composition of the chapter. Several analogous phosphacycle systems have also been included, where appropriate.

Whilst trying to be comprehensive, we have also intended to introduce a strong applied flavor to this summary. In the industrial case, catalyst performance is critically judged on overall efficiency, namely catalyst productivity and activity as well as enantioselectivity. As a result, turnover numbers (TONs) and turnover frequencies (TOFs) have been included or calculated whenever possible and meaningful.

The Handbook of Homogeneous Hydrogenation.
Edited by J. G. de Vries and C. J. Elsevier
Copyright © 2007 WILEY-VCH Verlag GmbH & Co. KGaA, Weinheim
ISBN: 978-3-527-31161-3

However, the reader should be aware of the danger of comparing systems tested under nonequivalent conditions (e.g., *in situ* versus preformed catalysts or alternative solvents). It is also worth noting that as this chapter is dedicated to applications in enantioselective hydrogenation, there may be many examples of phospholane-containing ligands that do not feature. Since this is by no means the first review of this type [2, 4], hopefully those reviews dealing with more general enantioselective applications will capture these aspects [5].

24.2 Phospholane Ligands: Synthesis and Scope

24.2.1 Early Discoveries and the Breakthrough with DuPhos and BPE

The first reported application of phospholane-based ligands for enantiomeric hydrogenation was described by Brunner and Sievi in 1987 [6]. Unfortunately, these *trans*-3,4-disubstituted phospholanes (**1–3**) were derived from tartaric acid, and proved to be relatively unselective for the rhodium-catalyzed hydrogenation of (*Z*)-*α*-(*N*-acetamido)cinnamic acid (6.6–16.8% ee). This was, presumably, due to the remoteness of the chiral centers from the metal coordination sphere failing to impart a significant influence. This was also found to be the case with several other bi- and tridentate analogues [7].

The fundamental discovery by Burk et al. that the analogous *trans*-2,5-disubstituted phospholanes formed a more rigid steric environment led to the introduction of the DuPhos and BPE ligand classes (Fig. 24.1) [8–13]. Subsequently, these ligands have been successfully employed in numerous enantiomeric catalytic systems [4a, 5], the most fruitful and prolific being Rh-catalyzed hydrogenations. The reduction of *N*-substituted *α*- and *β*-dehydroamino acid derivatives,

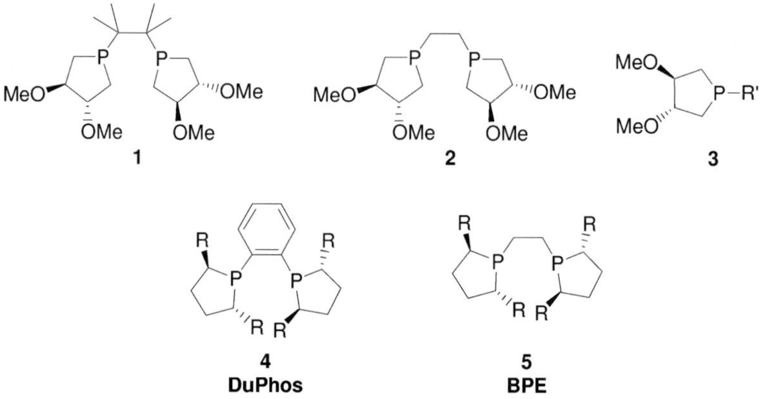

Fig. 24.1 The first phospholanes to be used for enantiomeric hydrogenation.

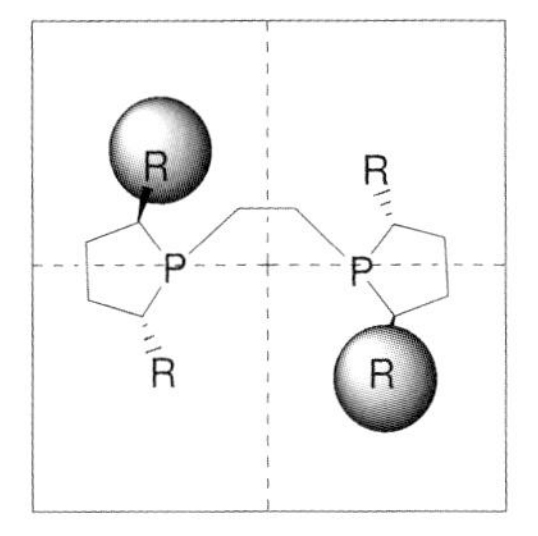

Fig. 24.2 The steric quadrant model for 2,5-disubstituted phospholanes.

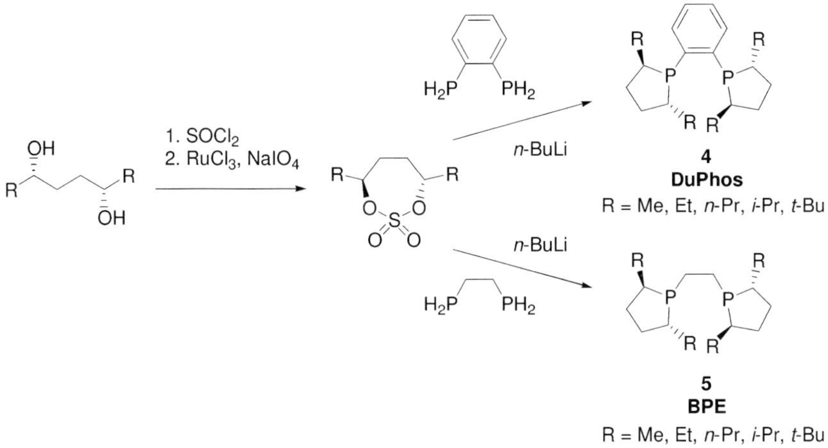

1. $SOCl_2$
2. $RuCl_3$, $NaIO_4$

H_2P PH_2

n-BuLi

4
DuPhos
R = Me, Et, *n*-Pr, *i*-Pr, *t*-Bu

n-BuLi

H_2P PH_2

5
BPE
R = Me, Et, *n*-Pr, *i*-Pr, *t*-Bu

Scheme 24.1 The synthesis of *trans*-2,5-dialkyl-phospholanes, DuPhos.

β-dehydroamino alcohols, *N*-acylhydrazones, *N*-substituted enamides, enol esters, *α,β*-unsaturated acid and *β*-keto ester derivatives have all been achieved in exceptionally high enantioselectivity [4a, 14–21]. Furthermore, the combination of robustness, high activity, and excellent selectivity has rendered these ligands suitable for commercial-scale industrial applications [2d, 4b, 22, 23]. A simplistic, qualitative guide to explaining the high degree of selectivity observed has been provided in part by the quadrant model (Fig. 24.2) [4a]. By having two of the four phospholane substituents project into the open coordination plane of the metal, steric interactions influence the reaction pathway, though some dispute as to the validity of this model has recently been raised [24].

The conventional synthesis of *trans*-2,5-dialkyl phospholanes starting from a chiral 1,4-diol is shown in Scheme 24.1. Originally, these 1,4-diols were obtained via electrochemical Kolbe coupling of single enantiomer *α*-hydroxy acids [25], but this method proved to be commercially impracticable and has since been replaced by more viable biocatalytic routes [26]. Reaction of the chiral 1,4-diol with thionyl chloride followed by ruthenium-catalyzed oxidation with so-

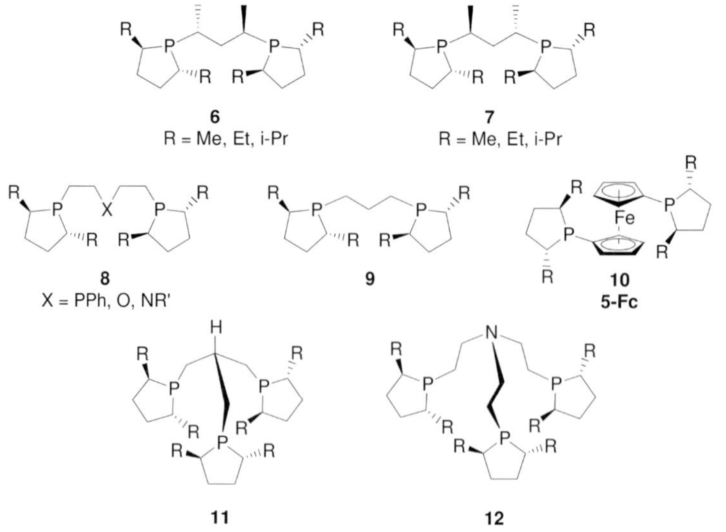

6
R = Me, Et, i-Pr

7
R = Me, Et, i-Pr

8
X = PPh, O, NR'

9

10
5-Fc

11

12

Fig. 24.3 Multidentate 2,5-disubstituted phospholanes displaying wide backbone diversity.

dium periodate yields the cyclic sulfate [27]. Treatment with 2 equiv. of a strong base, such as BuLi, and addition of a primary phosphine affords the tertiary phospholane with net inversion of stereochemistry. Practical methods have been developed for the large-scale manufacture of these ligands [28]. An alternative method via lithium phosphides was originally applied, but this was handicapped by excessive P–P bond formation [8], in addition to partial racemization of the phospholane chiral centers [11, 29]. Clearly, multidentate ligands may be obtained if a moiety containing more than one primary phosphine is used, and indeed numerous examples with a wide diversity of backbones were reported by Burk et al. (Fig. 24.3) [8, 10, 11, 29, 30].

Unfortunately, *trans*-2,5-diaryl phospholanes cannot be prepared using the traditional method described above for the alkyl derivatives; the basic conditions employed tend to induce elimination reactions with the corresponding cyclic aryl sulfate or dimesylate [31]. In 1991, Fiaud and co-workers reported a route to single enantiomer *trans*-1,2,5-triphenylphospholane oxide via epimerization of the previously reported *cis*-isomer and liquid chromatographic separation of the racemate [32]. Later, an alternative approach was developed using a chelotropic reaction between 1,4-diphenylbuta-1,3-diene and a dichloroaminophosphine (Scheme 24.2) [31]. After reduction, epimerization and hydrolysis, a diastereomeric salt resolution of the resulting racemic *trans*-2,5-diphenyl phospholanic acid could now be achieved, yielding the enantiomerically pure phospholane synthon **13**. This was ultimately converted to a series of monodentate 2,5-triphenylphospholane ligands (**14**), and shown to give reasonable to high enantioselectivities for the hydrogenation of (*Z*)-methyl-2-acetamidocinnamate (MAC), itaconic acid and esters, and *N*-acetyl enamides [31, 33]. This procedure has since

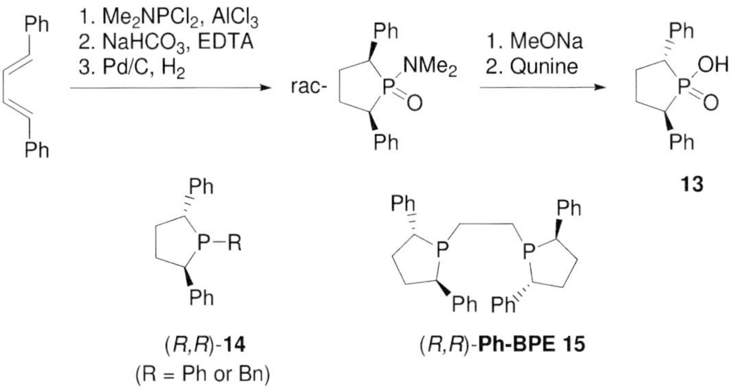

Scheme 24.2 Preparation, resolution and resulting ligands from 2,5-diaryl phospholanic acid.

been used to prepare the bidentate bisphospholane, Ph-BPE **15** (Scheme 24.2) [34]. This has been shown to have excellent levels of selectivity and activity for the hydrogenation of a range of olefinic substrates when compared to the dialkyl analogues [34, 35].

Although the DuPhos and BPE family of ligands have been shown to form active asymmetric hydrogenation catalysts with a range of transition metals (namely Ru, Ir, Pt, Pd and Au), none has shown the high degrees of selectivity and activity typically reported for the Rh-based catalysts. On the whole, the most successful results have been obtained with preformed mononuclear, cationic complexes employing the diolefin co-ligands 1,5-cyclooctadiene (COD) or norbornadiene (NBD). There has been some debate regarding COD precatalysts being uneconomic for use in industrial processes when compared with the NBD analogues [36]. A study performed at high catalyst loadings (molar substrate to catalyst ratio (SCR) = 100) showed there to be a rate difference for some substrates due to the NBD precatalyst forming the active species faster, but with no difference in enantioselectivity. However, when more industrially practical conditions were applied (SCR 2000 to 10 000), this effect became insignificant to the catalyst's overall productivity; furthermore, it was shown to be substrate-dependent [37]. In fact, the experimental conditions (e.g., stirring rate) were found to have a far more dramatic effect than the choice of precatalyst. For this class of reaction, hydrogen mass transfer into solution is the most important individual process parameter to affect the overall reaction rate [38].

In general, the choice of counteranion has a minor effect on catalyst performance, with typical examples being selected from BF_4^-, OTf^-, PF_6^-, or $BARF^-$. In one example, however, it was noted that [(*R,R*)-Et-DuPhos Rh COD]OTf gave superior selectivity for the reduction of *β*-*β*disubstituted *α*-dehydroamino acid derivatives than the corresponding BARF complex when performed in a range of solvents, including supercritical carbon dioxide [39].

In recent years, considerable effort has been made to immobilize homogeneous hydrogenation catalysts because of the obvious potential advantages, such as improved separation and catalytic performance [4b, 40]. Although beyond the remit of this chapter, it is worth mentioning that significant success has been achieved with several examples involving catalysts based on phospholane ligands [41].

Unsurprisingly, the immense success of DuPhos and BPE has created considerable interest in this ligand class, resulting in a vast number of variants appearing over the past few years. This has partly been driven by a desire to circumvent the original patents, but also by others in an attempt to explore certain mechanistic or design theories. On the whole, these ligands display similar properties to DuPhos and BPE with variable degrees of selectivity and activity when applied to enantioselective hydrogenation. This expansion has been partly facilitated by the modular nature of these ligands [4a], with modifications to the backbones, phospholane substituents, and second chelating site. A summary of these ligands concludes this section.

24.2.2 Modifications to the Backbone

The structures depicted in Figure 24.4 all display alterations to the original DuPhos and BPE backbones, and a concomitant variation in the ligand bite angle. In general, these have been prepared using the traditional cyclic sulfate method with the corresponding primary diphosphine. Pringle et al. have successfully applied a chiral *trans*-1,2-diphosphinocyclopentane to the synthesis of matching and mis-matching bidentate phospholanes **16** [24]. Hydrogenation of MAC was achieved with 77% to 98% ee, depending on the relative chirality of the backbone and 2,5-positions of the phospholane rings, with the overall stereochemis-

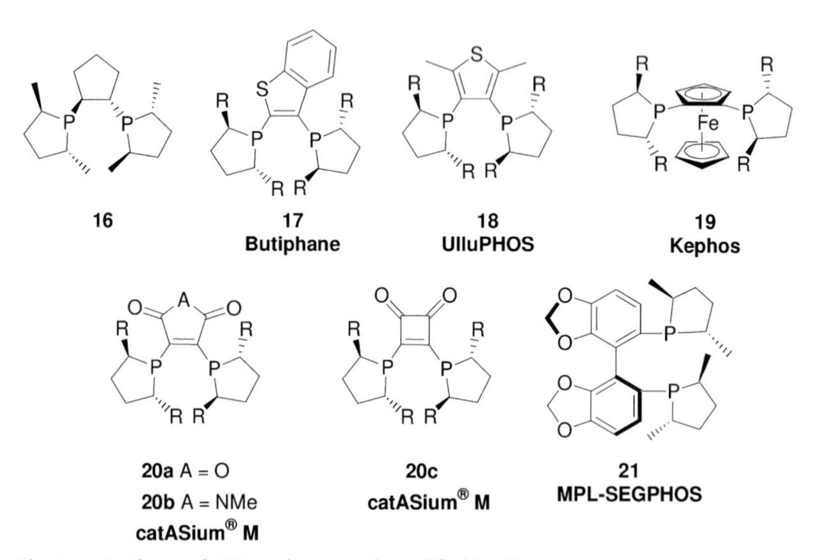

Fig. 24.4 DuPhos and BPE analogues with modified backbones.

try of the product being determined by the phospholane moieties. The sulfur heterocycle-based ligands, Butiphane (**17**) [42, 43] and UlluPHOS (**18**) [43, 44], have both been reported to be applicable for the Rh-catalyzed enantiomeric reduction of simple α-dehydroamino acid and itaconate derivatives, giving comparable results to Me-DuPhos in each case. Interestingly, the synthesis of Butiphane, together with that of several other benzo[b]thiophene-based ligands, was facilitated by the use of *N,N*-dialkyl-aminophosphine-containing intermediates acting as directing groups in the *ortho*-lithiation of the backbone. Several ferrocenyl-1,2-diphosphines, including Kephos (**19**), have also recently been reported to be effective for the reduction of several standard model substrates [45].

One exception to the use of primary phosphines is in the reported syntheses of the catASium® M class of ligands **20** [46–49]. In one report, reaction of the cyclic sulfate with $P(TMS)_3$ yields the TMS-protected secondary phospholane, which could then be reacted with the appropriate 1,2-dichloro species [46]. An alternative procedure to the same intermediate involves preparation of 1-phenylphospholane via the bismesylate, subsequent lithium-induced P–Ph cleavage, and quenching with TMSCl [49]. The ligand based on 2,3-dichloromaleic anhydride (**20a**; originally referred to as MalPHOS [46]) has been shown to be effective for the chiral reduction of α- and β-dehydroamino acid derivatives and itaconate derivatives.

An interesting approach to investigating the relationship between the position of enantiodescriminating sites in a number of chiral ligands and enantioselectivity in enantioselective hydrogenation has been proposed by Saito et al. [50]. In this report, (*aS,S,S*)-MPL-SEGPHOS (**21**) was used for the reduction of MAC, albeit in 75% ee.

24.2.3
Modifications to the Phospholane Substituents

In recent years, numerous DuPhos and BPE analogues have been introduced that contain structural variations at the 2,5-positions of the phospholane segments and/or additional stereogenic centers (Fig. 24.5).

Several groups have independently reported the synthesis of D-mannitol-derived phospholanes with either ketal, ether or hydroxy substituents in the 3,4-positions. The earliest ligand class, Rophos containing either a 1,2-benzene (**22**) or 1,2-ethane backbone (**23**), was introduced by Börner, Holz and co-workers in 1998 [51]. By taking advantage of the difference in reactivity between the primary and secondary alcohols, the mannitol framework could be manipulated to prepare the cyclic sulfate and, ultimately, the desired diphosphine. These were applied to the Rh-catalyzed hydrogenation of a range of olefinic substrates, all with excellent enantioselectivity. The research groups of Zhang [52] and RajanBabu [53] have both reported the synthesis of the *iso*-propylidene ketal bisphospholane **24** (R = Me or Et) and the tetrahydroxy bisphospholane **25** (R = Me or Et). Surprisingly, whilst ligand **24** (KetalPhos) was described as being inactive for the hydrogenation of dehydroamino acid derivatives when the catalyst was

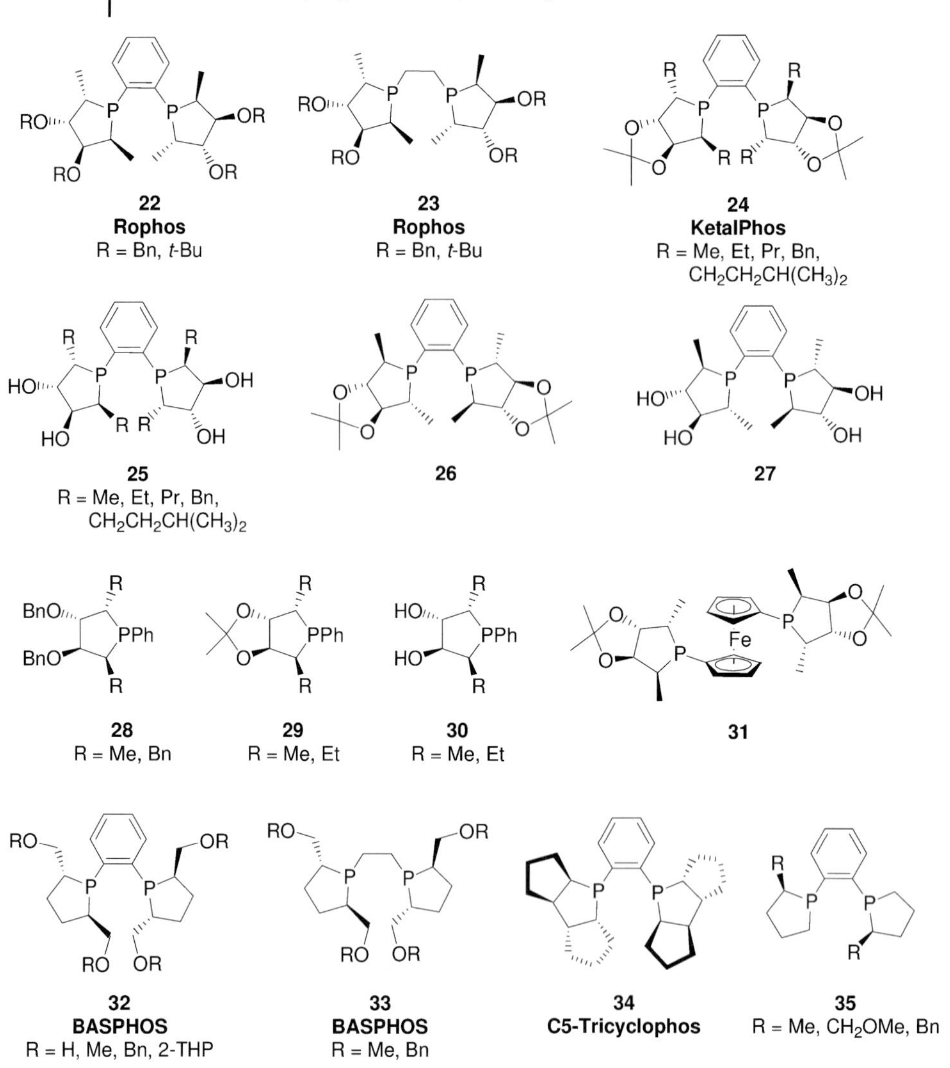

Fig. 24.5 Ligands with modifications to the phospholane substituents.

prepared *in situ* with $[Rh(COD)_2]X$ ($X = BF_4$, SbF_6, PF_6 and OTf) [52a, b], the isolated precatalyst, $[(\mathbf{24})\ Rh(COD)]BF_4$, was shown to be active and very selective (>90% ee) [53b]. A mannitol-derived cyclic sulfate has also been employed in the synthesis of monodentate phospholanes **28–30** [52a, b] and the ferrocenyl-based diphosphine **31** [54]. Although enantioselective hydrogenation with **28–30** has not been reported, **31** has been shown to be extremely active (TON 10000; TOF >800 h^{-1}) and selective (89.8–99.9% ee) for the Rh-catalyzed reduction of a range of functionalized olefins. By preparing a diasteromeric bisepoxide pair from D-mannitol (Scheme 24.3), RajanBabu and Yan have also accessed the dia-

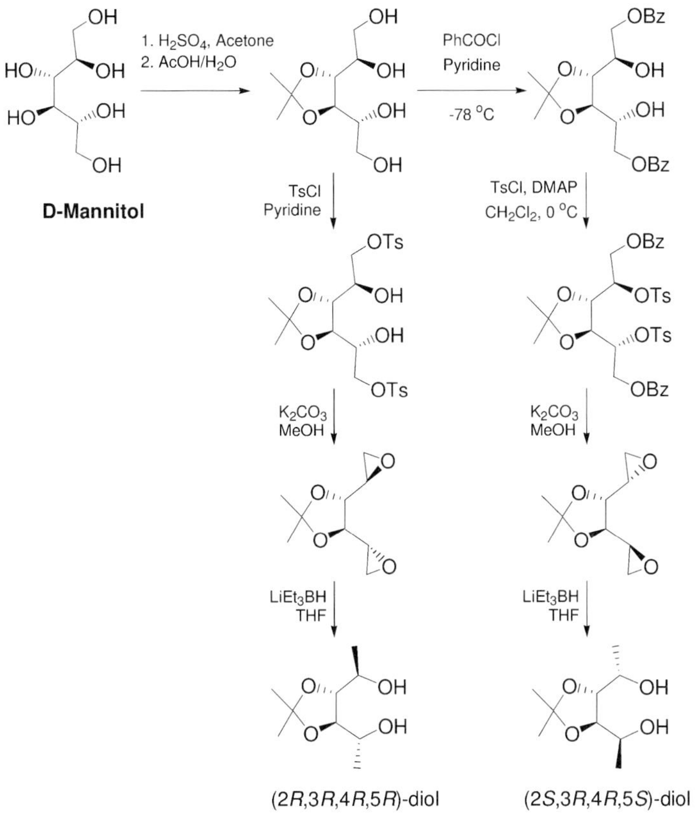

Scheme 24.3 The preparation of diastereomeric diols from D-mannitol.

stereomeric 3,4-disubstituted phospholanes **26** and **27** [53a]. When comparing **24** (R=Me) with **26**, as expected the opposite enantiomer was obtained for the hydrogenation of methyl 2-acetamidoacrylate (MAA), but interestingly **26** gave a slight improvement on the level of selectivity [97.4% ee (*R*) versus 90.5% ee (*S*)] [53b]. Rieger et al. extended the range of substituents at the 2,5-positions of ligands **24** and **25** (R=Et, *n*-Pr, isoamyl and Bn) by means of copper-catalyzed coupling of the appropriate Grignard reagent to the mannitol-derived bisepoxide [55]. Testing this series of ligands against the hydrogenation of α-methylcinnamic acid and itaconic acid showed high selectivity in every case (96–99% ee).

Several methods have been described to liberate the hydroxyl groups from **24** to produce the water-soluble, tetrahydroxyl bidentate ligand **25** [52, 53b]. Water-soluble ligands are of interest due to the prospect of recycling the catalyst into an aqueous phase, ideally without loss of performance. The enantiomeric hydrogenation of itaconic acid was performed in aqueous methanol over a range of solvent compositions (MeOH: H_2O, 9:1 to 3:97), with consistently excellent levels of performance (100% conversion, 99% ee, SCR 100, 12 h) [52b]. Interestingly, when applied to the reduction of MAA under comparable conditions, an increase in the percentage of water was found to have a deleterious effect on selectivity [53b]. However, at equal volumes of methanol and water, Rh complexes of **25** and analogous tetrahydroxy phospholanes could be recycled (up to five runs at SCR 100) by extracting the product into ether, with no significant losses in enantioselectivity.

Another family of mannitol-derived bisphospholanes was introduced by Holz and Börner. Removal of the hydroxy groups at the 3 and 4 positions leads to a key intermediate that ultimately produces 2,5-disubstituted phospholanes BASPHOS **32** and **33**. The water-soluble, tetrahydroxyl-substituted variant **32** (R=H) displayed excellent selectivities for the Rh-catalyzed hydrogenation of 2-acetamido acrylic acid and the corresponding methyl ester in water (99.6% and 93.6% ee, respectively) [56a]. RajanBabu and co-workers confirmed this and showed that the catalyst could be recycled up to four times, with no loss in selectivity (SCR 100) [53b]. An interesting feature of the synthesis of this ligand is the protection of the air-sensitive phosphine groups as the rhodium complex prior to liberation of the hydroxyl groups (tetrahydropyranyl group removal), saving two borane protection–deprotection steps. The corresponding 2,5-bis(alkoxymethylene)-substituted ligands **32** and **33** (R=Me, Bn) have also been tested for the Rh-catalyzed hydrogenation of α- and β-dehydroamino acid derivatives, itaconates and an unsaturated phosphonate, together with Ru-catalyzed reduction of prochiral β-keto esters [56b–e]. The wide range of enantioselectivities obtained (8 to 99% ee) was found to depend strongly on both the phospholane substituent and the backbone used.

A unique tricyclic bisphospholane ligand, C5-Tricyclophos (**34**), has been described in a patent by Zhang [57]. Derived from resolved bicyclopentyl-2,2′-diol (originally used in the preparation of the chiral diphosphine, BICP [58]), this li-

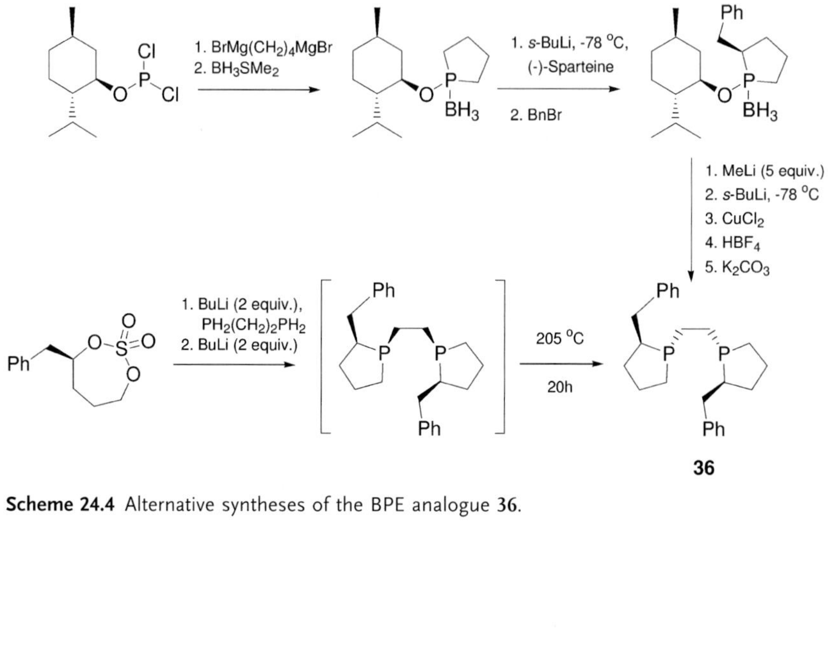

Scheme 24.4 Alternative syntheses of the BPE analogue **36**.

gand has shown moderate enantioselectivity for the reduction of α-acetamidocinnamate (53% ee) and MAC (78% ee). An interesting class of P-chirogenic monosubstituted phospholanes, **35** and **36**, has recently been introduced by Hoge [59]. Originally, the 1,2-ethane variant **36** was prepared using menthol as a chiral auxiliary for the directed selective benzylation of the phospholane ring and subsequent phosphorus methylation with stereochemical retention (Scheme 24.4) [59a]. Oxidative homo-coupling and deboronation completed the synthesis. A more versatile method was subsequently published, via the traditional cyclic sulfate route, for the preparation of BPE, DuPhos and monodentate analogues with alternative phospholane substituents (R=Me and CH_2OMe) [59b]. These ligands have been successfully applied to the hydrogenation of α- and β-dehydroamino acid derivatives and a pharmaceutically important precursor to Pregabalin [59a, b, d], giving results comparable to BPE and DuPhos [60, 61].

24.2.4
Other Phospholane-Containing Ligands

In addition to direct DuPhos and BPE analogues, several other ligands containing five-membered phosphacycles have been reported (Fig. 24.6). As early as 1991, non-C_2-symmetric phospholane-containing phosphines **37–39** were reported by Brunner and Limmer [7]. These were prepared by base-induced addition of the secondary phospholane to the appropriate diphenylphosphino-substituted olefin. As for the symmetrical 3,4-disubstituted bisphospholanes, enantioselectivities for the Rh-catalyzed reduction of α-acetamidocinnamate were poor.

Brown et al. [62] prepared a family of unsymmetrical diphosphine ligands **40** by the conjugate addition of racemic borane-protected *o*-anisylphenyl phosphide to diethylvinylphosphonate followed by deprotection, reduction and phospholane formation with the appropriate cyclic sulfate (2,5-hexanediol- or L-mannitol-derived). The diastereomers of the mannitol-derived phosphines could be separated chromatographically and converted to their dihydroxyl analogue, whereas the disubstituted-phospholane required medium-pressure liquid (flash) chromatography (MPLC). Rh-catalyzed hydrogenations with these ligands gave moderate enantioselectivities for several standard substrates and, whilst some significant matching and mis-matching effects were observed, the chirality of the product was determined primarily by the phospholane moiety.

Since this report, several research groups have replaced one phospholane ring of Me-DuPhos with a diaryl phosphine group. Stelzer et al. [63] described the synthesis of ligand **41** (R=Me, Ar=Ph) by treating the standard cyclic sulfate with a mixed primary-tertiary diphosphine. The ligand was purified via its dihydrochloride salt, liberating the free diphosphine quantitatively by treatment with $NaHCO_3$. Independently, Saito [64] and Pringle [65] reported the use of **41** (R=Me, Ar=Ph) in Rh-catalyzed asymmetric hydrogenation of a range of olefins, with particularly good results being obtained for prochiral enamides. Saito and co-workers made a small family of this class of ligand, UCAPs, and demonstrated that adjusting the diaryl-substituted phosphine could lead to higher selectivities than Me-DuPhos for a trisubstituted enamide [64]. The structurally related P,N ligand, DuPHAMIN **42** has also been prepared by Brauer and coworkers [66]. Remarkably, large matching and mis-matching effects were observed for the Rh-catalyzed hydrogenation of MAC, with the (*R,R,R*) ligand giving complete conversion at 20 °C (96% ee), but the (*S,R,R*) ligand being inactive. Pringle also synthesized the ferrocenyl-based **43** [65], but showed this to be less efficient than the phenylene-linked analogue.

The sterically bulky and conformationally rigid bicyclic ligand PennPhos (**44**) was developed by Zhang [57, 67]. The synthesis uses chiral 1,4-cyclohexanediols, converting them to the dimesylate to enable cyclization with 1,2-diphosphinobenzene under basic conditions. This has given high selectivity in Rh-catalyzed hydrogenation of both aryl and alkyl methyl ketones [57, 67b], cyclic enol acetates [67c, d], enol ethers [67d], cyclic enamides [67d, e] and *α*-dehydroamino acid derivatives [57]. Under certain conditions, the selectivities obtained for cyclic enamides are superior to those achieved with Me-DuPhos, but inferior for acyclic enamides. The bulky monodentate ligand **45** has been described in a patent by Börner, but gave poor enantioselectivities for *α*- and *β*-amino acid derivates, dimethylitaconate (DMI) and itaconic acid [68].

The research group of Zhang has also introduced two rigid P-chiral bisphospholane ligands, TangPhos **46** and DuanPhos **47** (Scheme 24.5), both of which contain two chiral phosphorus centers and two chiral carbon centers. Since the synthesis of TangPhos employs an enantioselective deprotonation of 1-*t*-butylphospholane sulfide with a butyllithium–sparteine complex, only one enantiomer is readily accessible [69]. On the other hand, either enantiomer of DuanPhos can be obtained enantiomerically pure by resolution of the corresponding bisoxide with either L- or D-dibenzoyl tartaric acid [70] (Scheme 24.5). Both ligands have been found to be very efficient in the Rh-catalyzed hydrogenation of a range of olefinic substrates such as *α*-acetamidoacrylate derivatives, *α*-arylenamides [69, 70], *β*-acetamidoacrylates [69b, 70, 71], itaconic acids, and enol acetates [69b, 70, 72]. DuanPhos has also been reported to give high rates (TON 4500; TOF 375 h^{-1}) and excellent enantioselectivities (93–99% ee) for a range of *β*-secondary-amino ketone salts [73].

The 1-*t*-butylphospholane sulfide intermediate to TangPhos was also used to prepare the P,N ligands **48** by reacting the lithium complex with CO_2 and then oxazoline formation with a range of chiral amino alcohols [69b, 74]. The Ir complexes of these ligands have been successfully used in the reduction of *β*-methylcinnamic esters (80–99% ee) and methylstilbene derivatives (75–95% ee), a particularly challenging class of unfunctionalized olefins [4c].

The BeePHOS (**49**) and mBeePHOS (**50**) classes of ligands introduced by Saito [75] are prepared by reacting the appropriate primary phosphine with a mesylated alkylalcohol-substituted aryl halide. Although a single diastereomer is obtained, the absolute configuration is unknown. Whilst trials of Ru-catalyzed hydrogenation of MAC and methyl *α*-hydroxymethylacrylate were disappointing, the Rh-catalyzed reactions were more active. On the whole, selectivities were lower than those obtained with Me-DuPhos under the same conditions. Ligand

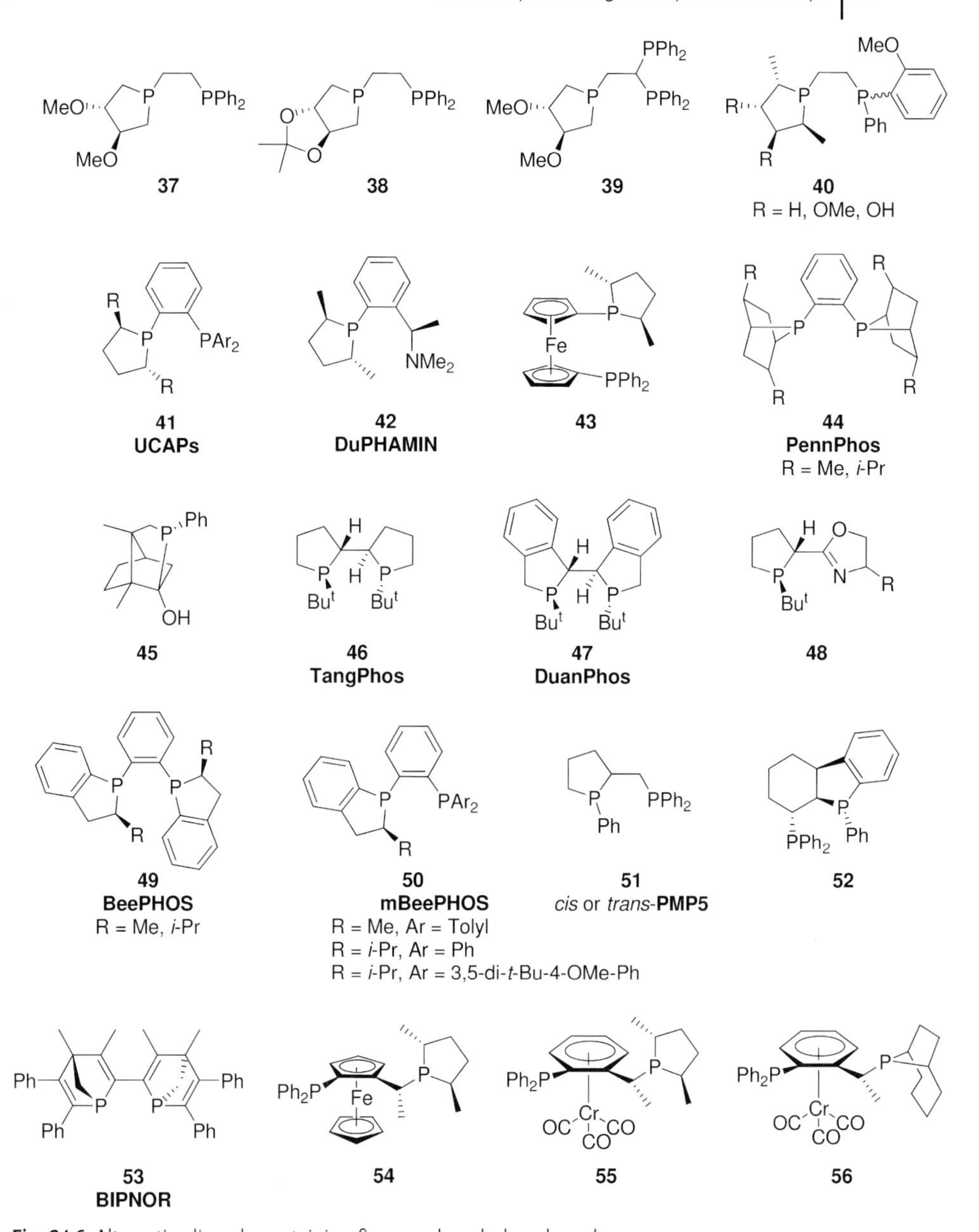

Fig. 24.6 Alternative ligands containing five-membered phosphacycles.

51 has recently appeared in separate patents from both Kobayashi and Schmid [76]. Given the name *cis* and *trans*-PMP5 by Schmid et al., the Rh complexes have been reported to be active catalysts for the reduction of several standard substrates, α-enol acetates and β-ketoacid derivatives, with variable enantioselectivities [76 b]. In general, the *cis* isomer is more selective than the *trans*.

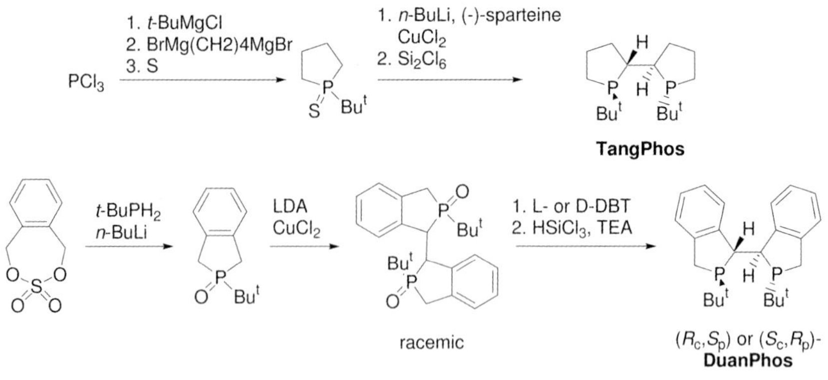

Scheme 24.5 The syntheses of TangPhos and DuanPhos.

The rigid bicyclic diphosphine **52** was prepared by Knochel and co-workers by the radical cyclization of a bromophosphine oxide, itself obtained from a double [2,3]-sigmatropic shift of an intermediate phosphinite [77]. Unfortunately, this ligand only gave moderate enantioselectivity (21–58% ee) against standard model substrates under normal screening conditions (MeOH, room temperature, 10 atm, SCR 100). The P-chiral diphosphine BIPNOR (**53**) was synthesized by Mathey et al. via a [4+2] cycloaddition of tetramethyl-1,1′-bisphospholyl and tolan, the crucial intermediate arising from a double [1,5] shift of each phosphole around the ring [78]. With the phosphorus atoms being located at the bridgehead of a bicyclic system, none of the usual racemization pathways potentially observed for P-homochiral phosphines can occur (Berry pseudorotation and edge inversion). Both *meso* and *rac* diastereomers and the resulting racemic mixture are separated by chromatography of their Pd complexes. Enantioselectivities for the Rh-catalyzed hydrogenation of α-acetamidocinnamic acid and itaconic acid are comparable to those achieved with DuPhos-based catalysts.

The group of Salzer has recently reported phospholanes **54**–**56** based on chiral half-sandwich complexes [79]. These were obtained by treatment of the appropriately substituted complex with a secondary phospholane, itself accessed via the cyclic sulfate or dimesylate and PH_3. These were tested against a range of substrates with C=C, C=O and C=N bonds, with variable results [80].

24.2.5 Related Phosphacycle-Based Ligands

Although strictly not phospholanes, several other noteworthy P-heterocycle-containing ligands have been applied to asymmetric hydrogenation (Fig. 24.7). The first optically active phosphetanes to be used in catalysis were described by Marinetti and Ricard [81]. Although active in Pd-catalyzed hydrosilylation, these monodentate ligands (**57**) proved to be very poor for the hydrogenation of MAC [81 b, c]. More recently, Burk and co-workers [82] and the groups of Marinetti and Genêt

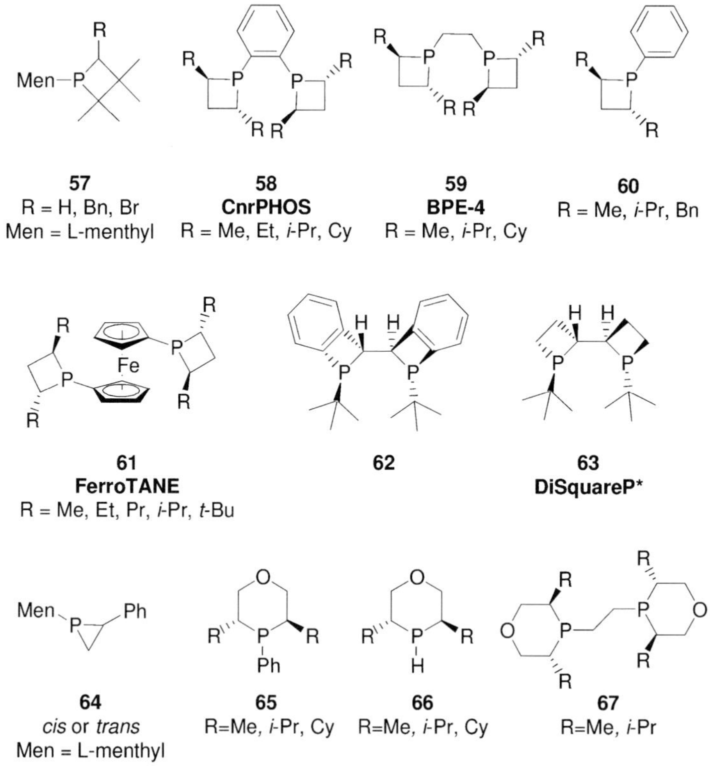

57
R = H, Bn, Br
Men = L-menthyl

58
CnrPHOS
R = Me, Et, *i*-Pr, Cy

59
BPE-4
R = Me, *i*-Pr, Cy

60
R = Me, *i*-Pr, Bn

61
FerroTANE
R = Me, Et, Pr, *i*-Pr, *t*-Bu

62

63
DiSquareP*

64
cis or *trans*
Men = L-menthyl

65
R=Me, *i*-Pr, Cy

66
R=Me, *i*-Pr, Cy

67
R=Me, *i*-Pr

Fig. 24.7 Related P-heterocyclic ligands.

[83] have independently prepared and examined several enantiomerically pure ferrocenyl-based 2,4-disubstituted phosphetanes as ligands for asymmetric hydrogenation. These ligands were prepared from the appropriate primary phosphines and a range of chiral 2,4-diols using the traditional cyclic sulfate methodology (*vide supra*). Although not as enantioselective as their bisphospholane analogues in reducing dehydroamino acid derivatives [83 a, d], Genêt's CnrPHOS (**58**) and BPE-4 (**59**) are reported to be extremely selective in the Ru-catalyzed hydrogenation of several β-ketoesters (73–98% ee) [83 d, e]. Interestingly, reasonable levels of enantioselectivity were achievable with the monodentate ligands **60** against α-acetamidocinnamic acid (10 to 86% ee) [83 f]. The ferrocenyl-based bisphosphetanes, FerroTANE (**61**), have been shown to be exceptional for the Rh-catalyzed reduction of (E)-β-dehydroamino acids [56 c, 84] as well as a number of itaconic and succinamide derivatives [82 a, 85, 86], outperforming DuPhos in both cases. Albeit less selective, FerroTANE has also been examined for Ru-catalyzed hydrogenation of β-ketoesters [83 d] and Rh-catalyzed hydrogenation of α-dehydroamino acids [83 c], and is a precursor to the potent anticonvulsant (S)-Pregabalin [61].

In recent years, the research group of Imamoto has been very active in the area of C_2-symmetric P-stereogenic phosphine ligands [87]. Two such ligands,

62 and DiSquareP* **63**, were prepared using the same strategy, the key being an oxidative homocoupling of the corresponding benzophosphetene or phosphetane, respectively [87]. Both ligands have been applied to the Rh-catalyzed hydrogenation of MAC, but in particular DiSquareP* has displayed excellent activity (TON 50000; TOF ~1100 h^{-1}) and selectivity (99% ee) for this and other *α*-dehydroamino acid derivatives. Interestingly, **63** is also an extremely selective ligand for the reduction of *α*-substituted enamides, but does not perform well on either substrate class when *β*,*β*-disubstitution is present.

The three-membered phosphirane **64** was studied by Marinetti et al. for the Rh-catalyzed hydrogenation of MAA, MAC, and itaconic acid with, in general, poor enantioselectivities [88]. Since ring-opened oxidized phosphorus species were observed at the end of the reactions, some doubt was voiced as to the exact nature of the catalytic species. The oxaphosphinanes **65**, were synthesized by Helmchen via reaction of the diol ether mesylates with dilithiophenylphosphine [89]. Since these showed poor performance for the Rh-catalyzed reduction of *α*-dehydroamino acids and itaconate derivatives, the corresponding secondary phosphinanes **66** were prepared by cleavage of the P–Ph bond with lithium. These were then converted through to the bidentate analogue **67**. Interestingly, both **66** and **67** performed well against these standard substrates (80–98% ee), but gave the opposite sense of stereochemical induction for a number of products [89].

24.3
Enantioselective Hydrogenation of Alkenes

24.3.1
Enantioselective Hydrogenation of *α*-Dehydroamino Acid Derivatives

The pivotal role of natural *α*-amino acids among a myriad of biologically active molecules is widely appreciated, and is of particular importance in the pharmaceutical industry. Unnatural *α*-amino acids also have a prominent position in the development of new pharmaceutical products. It has been shown that substitution of natural *α*-amino acids for unnatural amino acids can often impart significant improvements in physical, chemical and biological properties such as resistance to proteolytic breakdown, stability, bioavailability, and efficacy. One of the many synthetic methods available for the production of enantiomerically enriched *α*-amino acids is the metal-catalyzed enantioselective reduction of *α*-dehydroamino acid derivatives [90].

The parent DuPhos and BPE ligands exhibit excellent enantioselectivities routinely in excess of 95% with the majority of model *α*-dehydroamino acid substrates (Table 24.1) [4a, 8, 12, 13, 20, 90]. High molar SCRs (in the order of >1000:1), as well as TOFs in excess of 1000 h^{-1}, are indicative of the high catalyst activity and productivity typically found with DuPhos and BPE systems with these simple substrates. Burk reported that in the enantiomeric hydrogenation of MAA and MAC, with alkyl-DuPhos–Rh catalysts, optimal enantioselectivity could be achieved with the *n*-Pr-DuPhos ligand over other alkyl DuPhos or BPE ligands [13]. Cationic rhodium catalysts derived from Ph-BPE, the first aryl member of the diphospholane ligand class, are reported to be significantly more reactive and selective than the analogous alkyl-BPE ligands in the hydrogenation of various model substrates [34]. Experimental and computational mechanistic studies using Me-DuPhos-Ir and Me-DuPhos–Rh respectively revealed that an "anti-lock and key" reaction pathway also operates with DuPhos; consequently, the facial selectivity of the more reactive minor diastereoisomer is the source of enantioselectivity in the final product [91–93]. A small number of experimental and theoretical investigations of the use of DuPhos–Rh catalysts in supercritical CO_2 have been reported, with some notable differences with standard substrates being observed [39, 94].

The mannitol-derived phospholane systems from Zhang, Rajanbabu and Börner (ligands **22–27**, **31–33**) have been extensively tested with model substrates. In general, these ligands have been shown to hydrogenate a similar range of simple α-dehydroamino acid substrates to DuPhos and BPE, and are able to replicate their high enantioselectivities and reactivities. Furthermore, despite the further elaboration of the phospholane ring systems of several of the mannitol ligands, the stereochemical outcome of the reported reactions is identical to that of DuPhos and BPE ligands with the same spatial arrangement. As expected, the diastereomeric hydroxylated ligands (*S*,*S*,*S*,*S*)-**25** and (*R*,*S*,*S*,*R*)-**27** gave the opposite sense of stereoinduction in the hydrogenation of MAA with ee-values of >99% (*S*) and 97% (*R*) respectively. This indicates that the spatial orientation of the 2,5-positions of the phospholane is the principal factor in defining the stereochemical outcome of MAA hydrogenation. Interestingly, the ketal variants (*S*,*S*,*S*,*S*)-**24** and (*R*,*S*,*S*,*R*)-**26** showed a more marked difference in the hydrogenation of MAA, with ee-values of 90.5% (*S*) and 97.4% (*R*), respectively [53b]. Zhang's mannitol-derived ferrocenyl phospholane Me-f-KetalPhos system facilitates the hydrogenation of MAA in 99.4% ee [54], whereas the parent Me- and Et-5-Fc ligands achieve only 64 and 83% ee, respectively [30]. Zhang has gone on to show that the hydrogenation of an extensive range of simple aromatic, substituted aromatic and heteroaromatic β-substituted α-dehydroamino acid systems can be achieved with his mannitol-derived systems with excellent enantioselectivity, albeit it under standard screening conditions and typically with high catalyst loadings [52b]. Börner and co-workers noted that the significant degree of structural variation possible within the BASPHOS ligand family imparts a greater degree of substrate sensitivity than their DuPhos and BPE counterparts, and thus results in more variable enantioselectivities over a wide range of simple substrates [56b].

UlluPHOS [43, 44], catASium M [48, 95], Kephos [45, 96] and Butiphane [97] – four ligand systems which possess larger P–Rh–P bite angles than DuPhos [44, 46, 97] – all achieved enantioselectivities >95% when used in the hydrogenation of some model substrates. Much importance has been attached to P–Rh–P bite angles larger than the parent DuPhos system. It is believed that the pos-

Table 24.1 Phospholanes reported to hydrogenate model α-dehydroamino acid derivatives in >95% ee.

$R^1CH=C(NHAc)CO_2R^2 \xrightarrow[H_2]{\text{Rh-catalyst}} R^1CH_2CH(NHAc)CO_2R^2$

68 R^1 = H, R^2 = Me **69** R^1 = H, R^2 = H
70 R^1 = Ph, R^2 = Me **71** R^1 = Ph, R^2 = H

Substrate	Ligand	SCR	Reaction conditions [a]	TON	TOF [h^{-1}]	% ee (config.)	Reference(s)
68	(*S*,*S*)-Me-DuPhos [b]	1000	MeOH, 20 °C, 2 atm, 1 h	1000	>1000	99	13, 27
68	(*S*,*S*)-Et-BPE [b]	1000	MeOH, 20 °C, 2 atm, 1 h	1000	>1000	98	13 ,27
68	(*R*,*R*)-Ph-BPE	5000	MeOH, 25 °C, 9.9 atm, 1 h	5000	–	>99	34
68	(*R*,*R*)-Me–**16**	1000	MeOH, rt, 2 atm, 1–16 h	1000	–	95	24
68	(*R*,S_p,S_p,*R*)-Bn-**35** [b]	100	MeOH, rt, 2 atm, 15 min	100	>400	98 (*S*)	59b
68	(*R*,*R*)-UlluPHOS	1000	MeOH, 27 °C, 2.8 atm, 1 h	1000	>1000	98 (*S*)	43, 44
68	(*S*,*S*,*S*,*S*)-Me-**25** [b]	100	MeOH, rt, 3 atm, 9 h	100	–	98 (*S*)	52a, b
68	(*R*,*S*,*S*,*R*)-**26**	100	MeOH, rt, 2.8 atm, 7 h [c]	–	–	97 (*R*)	53b
68	(*S*,*S*,*S*,*S*)-Me-f-Ketal-Phos	10000	THF, rt, 3 atm, 12 h	10000	833	99 (*S*)	54
68	(*R*,*R*,*S*,*S*)-DuanPhos	10000	MeOH, rt, 1.4 atm, 2 h	10000	5000	99 (*R*)	70
68	(*S*,*S*)-DiSquareP*	100	MeOH, rt, 1 atm, 1 h	100	100	99	87b
69	(*R*,*R*)-H-Ph-BASPHOS	–	H_2O	–	–	>99 (*S*)	56a
69	(*R*,S_p,S_p,*R*)-Bn-**35** [b]	100	MeOH, rt, 2 atm, 15 min	100	>400	97 (*S*)	59b
69	(*S*,*S*,*S*,*S*)-Me-**25**	100	MeOH, rt, 3 atm, 9 h	100	–	>99 (*S*)	52a, b
69	(*R*,*R*)-Me-**67**	1000	MeOH, 20 °C, 1.1 atm, 24 h	1000	–	97 (*R*)	89
70	(*R*,*R*)-Ph-BPE	3000	MeOH, 28 °C, 10 atm, 1.25 h	3000	>2400	99	34
70	(*R*,*R*)-*n*-Pr-DuPhos [b]	1000	MeOH, 20 °C, 2 atm, 1 h	1000	>1000	99	13, 27
70	(*R*,*R*)-Me–**16**	1000	MeOH, rt, 2 atm, 1–16 h	1000	–	98	24
70	(+)-*i*-Pr-BeePHOS	200	MeOH, 30 °C, 4 atm, 14–16 h	200	–	98	75
70	(*R*,*R*,*R*)-DuPHAMIN	100	Toluene, 20 °C, 5 atm, 12 h [f]	95	7.9	96	66
70	(*R*,*R*)-Me-Ph-BASPHOS [b]	100	MeOH, 25 °C, 1 atm, 15 min	100	400	99 (*S*)	56b, e

Table 24.1 (continued)

Substrate	Ligand	SCR	Reaction conditions [a]	TON	TOF [h^{-1}]	% ee (config.)	Reference(s)
70	(*R*,*R*)-Bn-Et-BASPHOS	100	MeOH, 25 °C, 1 atm, 5 h	100	20	96 (*S*)	56b
70	(*R*,S_p,S_p,*R*)-Bn-**35** [b]	100	MeOH, rt, 2 atm, 15 min	100	>400	95 (*S*)	59b
70	(*R*,*R*)-*cis*-PMP5	1000	MeOH, rt, 1.5 atm, 2 h [e]	750	375	98 (*S*)	76b
70	(*S*,*S*,*S*,*S*)-Me-**25**	100	MeOH, rt, 3 atm, 12 h	100	–	>99 (*S*)	52a
70	(*S*,*S*,*S*,*S*)-Et-**25**	100	MeOH, rt, 3 atm, 12 h	100	–	>99 (*S*)	52b
70	(*S*,*S*,*S*,*S*)-*t*-Bu-Rophos **23** [b]	100	MeOH, rt, 1 atm, 2 h [d]	50	25	98 (*S*)	53b
70	(*S*,*S*,*S*,*S*)-Bn-Rophos **22**	100	MeOH, rt, 1 atm, 48 min [d]	50	63	96 (*S*)	53b
70	(*S*,*S*,*S*,*S*)-Me-f-Ketal-Phos	100	THF, rt, 1 atm, 1 h	100	–	99 (*S*)	54
70	[(*R*,*R*)-catASium M	200	THF, 25 °C, 1.5 atm, 2 h	200	100	96 (*R*)	48, 95
70	(*S*,*S*)-Me-Kephos	1000	MeOH, rt, 1 atm	1000	300	97	45, 96
70	(*R*,*R*)-Et-Butiphane	1000	MeOH, rt, 1 atm	1000	550	99	97
70	(*R*,*R*)-**14**	100	MeOH, 20 °C, 1 atm, 24 min	100	250	93 (*S*)	33
70	(*S*,*S*,*R*,*R*)-TangPhos	10000	MeOH, rt, 1.4 atm	10000	–	>99 (*R*)	69
70	(*R*,*R*,*S*,*S*)-DuanPhos	100	MeOH, rt, 1.4 atm, 12 h	100	–	99 (*R*)	70
70	(*S*,*S*)-**62**	1000	MeOH, 20 °C, 2 atm, 1 h [c]	–	–	96	87a
70	(*R*,*R*)-Me-FerroTANE	100	MeOH, 50 °C, 1 atm, 24 h [c]	–	–	96 (*R*)	83d
70	(*S*,*S*)-DiSquareP*	50000	MeOH, rt, 6 atm, 43 h	50000	1163	99	87b
71	(*R*,*R*)-Me-DuPhos	1000	EtOH, 27 °C, 2 atm, 1 h	1000	>1000	95 (*S*)	44
71	(*R*,S_p,S_p,*R*)-Bn-**35**	100	MeOH, rt, 2 atm, 15 min	100	>400	96 (*S*)	59b
71	(*R*,*R*)-UlluPHOS	1000	EtOH, 27 °C, 2 atm, 1 h	1000	>1000	99 (*S*)	43, 44
71	(*S*,*S*,*S*,*S*)-Me-**25** [b]	244	MeOH, rt, 1.3 atm, 20 h	244	–	99 (*S*)	52, 55
71	(*S*,*S*,*S*,*S*)-Et-**25**	100	MeOH, rt, 3 atm, 12 h	100	–	>99 (*S*)	52b, 55

Table 24.1 (continued)

Substrate	Ligand	SCR	Reaction conditions [a)]	TON	TOF $[h^{-1}]$	% ee (config.)	Reference(s)
71	(*S,S,S,S*)-*t*-Bu-Rophos **23** [b)]	100	MeOH, rt, 1 atm, 2 h [d)]	50	26	97 (*S*)	51
71	(*S,S,R,R*)-TangPhos	100	MeOH, rt, 1.4 atm, 12 h [c)]	–	–	99 (*R*)	69

a) Complete conversion unless otherwise stated.
b) Similar high enantioselectivities have also been obtained with several other ligands of this class.
c) No conversion given.
d) 50% conversion.
e) 75% conversion.
f) 95% conversion.

session of a wider P–Rh–P angle places the substrate in closer proximity to the metal, resulting in a more intimate contact between the substrate and catalyst, which could impart a greater degree of selectivity. However, this potentially oversimplifies the differences in performance of certain ligands in enantioselective hydrogenations. It is likely that the outcome of enantiomeric hydrogenations is governed by a variety of stereoelectronic factors, as well as reaction parameters such as pressure, temperature, and solvent. Sannicolo et al. have studied the reaction rates of Me-DuPhos and UlluPHOS in the hydrogenation of 2-acetamidocinnamic acid under identical conditions. The resulting kinetic rate data revealed that the UlluPHOS–Rh catalyst hydrogenated the substrate more quickly than the Me-DuPhos catalyst, with $k_{\mathrm{UlluPHOS}}/k_{\mathrm{Me\text{-}DuPhos}}=7.73$; this was in part attributed to the greater electron density of the thiophene-based ligand [44]. Et-Butiphane and Me-Kephos both hydrogenated MAC in high enantioselectivity, with SCRs of 1000:1; however, the respective TOFs of 550 and 300 h^{-1} were somewhat lower than the value of >2400 h^{-1} reported for the Ph-BPE ligand at a SCR of 3000:1 and with the same substrate under near-identical conditions [35, 96, 97]. Pringle's 1,2-diphospholano-cyclopentane ligand **16** can be synthesized as both λ and δ conformers, though only the δ conformer is reported to achieve high enantioselectivities [24]. The corresponding λ conformer achieves enantioselectivities approximately 20% lower than the δ conformer, indicating a strong matching/mismatching effect between the chirality of the phospholane rings and chiral backbone.

The non-*trans*-2,5-disubstituted phospholanes from Hoge (**35**), Takasago (BeePHOS family) and Zhang (TangPhos and DuanPhos) are all capable of achieving high enantioselectivities with standard α-dehydroamino acid substrates (see Table 24.1) Hoge's 1,2-phenylene system (**35**) generally gave higher enantioselectivities with model substrates than its related 1,2-ethylene system (**36**) [59c]. Takasago's BeePHOS family showed variable performance when hydrogenating MAC in that the ee-values ranged from 47 to 98% [75]. Zhang demonstrated that both the TangPhos and DuanPhos systems are able of performing hydrogenations with high enantioselectivities at economical catalyst loadings [69, 70]. In comparison to other non-*trans*-2,5-disubstituted phospholanes, only TangPhos so far has shown broad applicability akin to DuPhos and BPE. Moreover, Hoge's mono-substituted **35** and Takasago's BeePHOS show better utility with other substrate classes.

Monophospholanes or bidentate ligands containing a single phospholane moiety have also been successfully applied in the hydrogenation of standard *α*-dehydroamino acid substrates, though they have not yet been shown to be useful beyond the standard substrates. Remarkably, Fiaud's monophospholane **14** achieves 93% ee with MAC, whereas the bidentate monophospholano ligands (*R,R,R*)-DuPHAMIN (**42**) [66] and (*R,R*)-*cis*-PMP5 (**51**) [76b] hydrogenate MAC in 96% and 98% ee, respectively. Unsurprisingly, the TOFs for DuPHAMIN are too low to be of industrial use. Other related bidentate-monophospholane systems (**37–41**, **43**, **52**, **54–56**) have generally been found to give moderate to low ee-values with model substrates under normal screening conditions [7, 62, 65, 75, 77, 80].

Standard *α*-dehydroamino acid substrates have been hydrogenated in high enantioselectivities by phosphetanes and phosphinanes. Imamoto's phosphetanes **62** and DiSquareP* (**63**) achieve excellent enantioselectivities; furthermore, DiSquareP* achieves a SCR of 50000:1 with MAC, indicating exceptional catalyst productivity and stability [87]. The ferrocenyl system (*R,R*)-Me-FerroTANE hydrogenates MAC in 96% ee [83d]. Other phosphetanes, such as Genêt's Cy-BPE-4, *i*-Pr-CnrPHOS [83a], Berens' monophosphetanes [82b] or Takasago's IPT-SEGPHOS [50] have to date been found to give only moderate enantioselectivities with model substrates. Helmchen's bidentate oxa-phosphinane **67**, remarkably hydrogenates 2-acetamidoacrylic acid in 97.4% ee, while several analogous mono-oxa-phosphinanes have demonstrated high enantioselectivity (>90% ee) in the reduction of MAC and are currently the only reported examples of chiral phosphinanes which are highly selective [89].

β-Substituted *α*-dehydroamino acids are frequently synthesized as mixtures of (*E*/*Z*)-isomers [90], and several studies have shown that the geometry of *β*-substituted *α*-dehydroamino acid substrates can have a profound effect on both enantioselectivity and hydrogenation rates [98]. In some exceptional cases the opposite enantiomer can be produced when hydrogenating the (*E*) or (*Z*)-olefin with a single catalyst enantiomer [99]. Burk demonstrated, in a series of experiments using *n*-Pr-DuPhos–Rh with the isomerically pure (*E*) and (*Z*)-methyl-2-acetamido-2-butenoate, that both geometrical isomers of the alkene could be hydrogenated with almost identical high enantioselectivity and the same sense of facial selectivity regardless of the alkene geometry (see Scheme 24.6). Deuterium-labeling studies of the reduction of methyl 2-acetamido-2-pentenoate showed that the origin of the DuPhos–Rh catalysts' high enantioselectivity with (*E*/*Z*)-*α*-dehydroamino acid mixtures was not the result of alkene isomerization [13]. Reduction of (*E*) and (*Z*)-isomers of methyl 2-acetamido-2-pentenoate with D_2 and *n*-Pr-DuPhos–Rh gave rise to diastereomerically pure isotopomers

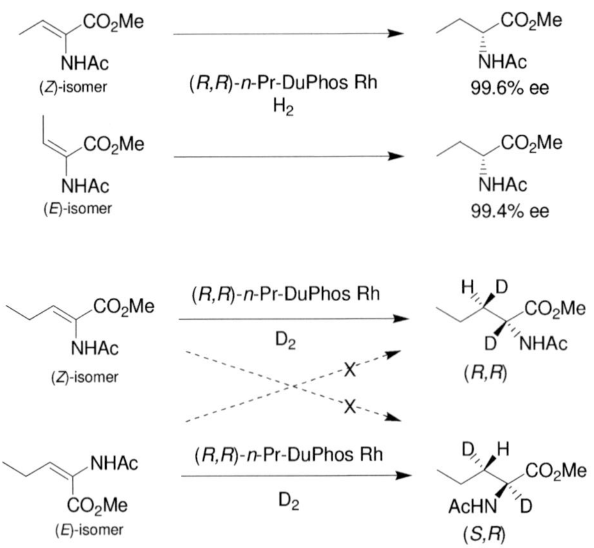

Scheme 24.6 Hydrogenation of (*E*) and (*Z*)-*α*-dehydroamino acid derivatives.

which, together with a 1:1 ratio of deuterium incorporation in the *α* and *β*-positions, excludes an (*E*/*Z*)-isomerization mechanism.

The synthetic utility of phospholane-derived catalysts has been directly extended to a broad range of simple, non-standard *α*-dehydroamino acid substrates, with enantioselectivities in excess of >95% ee being readily achieved [20, 100–107] (Scheme 24.7). The commercially available anti-fungicide (*R*)-metalaxyl has a MAA-related structure, and in a study by Blaser et al. conducted to assess the viability of an enantioselective hydrogenation approach to the active ingredient, the Me-DuPhos–Rh system produced the desired *α*-amino acid in high enantioselectivity (95.6% ee) and with extremely high productivity and activity (TON 50000; TOF 5200 h^{-1}) [108, 109]. Some simple phenoxycarbonyl-protected cyclic *α*-dehydroamino acid substrates have been hydrogenated. In the case of five- and six-membered systems only low or modest enantioselectivities could be obtained, whereas seven-, eight-, nine-, thirteen-, and sixteen-atom ring systems gave 86 to 97% ee [110]. It has also been shown that even a polymer-supported dehydrophenylalanine substrate is readily hydrogenated by Me-DuPhos–Rh with high ee and de values [111].

Tandem processes consisting of enantioselective hydrogenation and cross-coupling have been shown to provide a useful approach for generating a diverse range of substituted aromatic *α*-amino acids, the corresponding *α*-dehydroamino acid precursors of which are not easily prepared (Scheme 24.8). Burk and Hruby have exploited the ability of DuPhos–Rh catalysts to hydrogenate various halogen- and boronic acid-substituted *β*-aromatic and *β*-heteroaromatic *α*-dehydroamino acids with high enantiomeric excesses. The resulting aromatic halides or boronic acids can be coupled with a variety of vinyl, aryl, and heteroaryl-groups

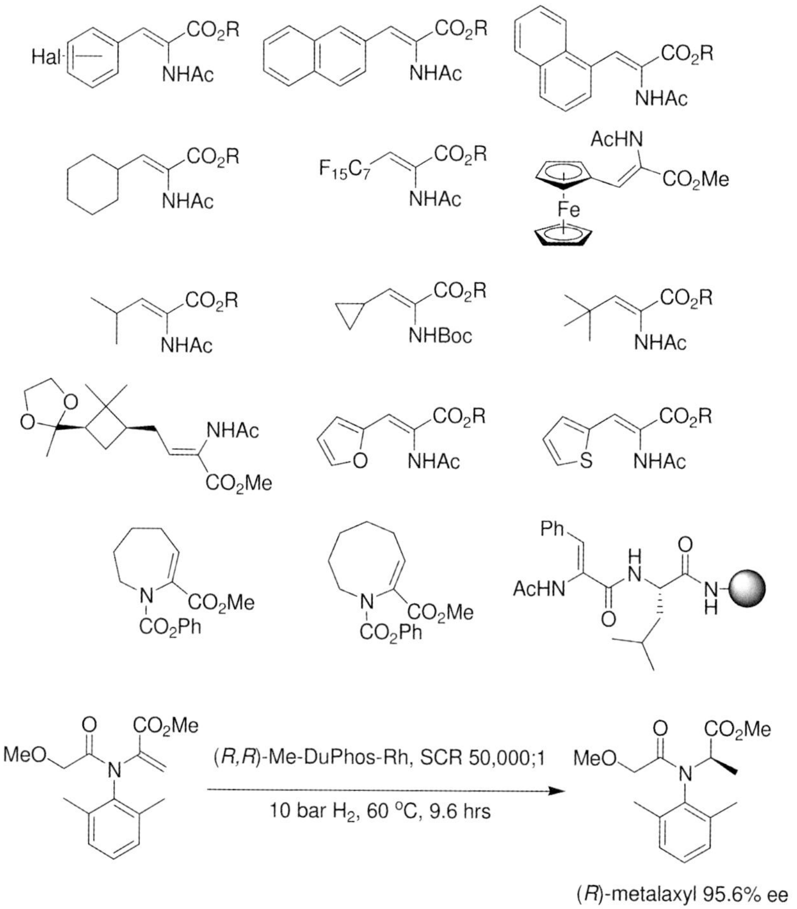

Hal, CO_2R, NHAc, CO_2R, NHAc, CO_2R, NHAc, CO_2R, NHAc, $F_{15}C_7$, CO_2R, NHAc, AcHN, CO_2Me, Fe, CO_2R, NHAc, CO_2R, NHBoc, CO_2R, NHAc, O, O, NHAc, CO_2Me, CO_2R, O, NHAc, CO_2R, S, NHAc, N, CO_2Me, CO_2Ph, N, CO_2Me, CO_2Ph, Ph, AcHN, H, N, O, O, N, H

MeO, O, CO_2Me, N — (R,R)-Me-DuPhos-Rh, SCR 50,000;1 / 10 bar H_2, 60 °C, 9.6 hrs → MeO, O, CO_2Me, N

(R)-metalaxyl 95.6% ee

Scheme 24.7 Non-standard α-dehydroamino acid derivatives reduced by Rh-phospholane-based catalysts.

to produce a diverse range of new unnatural α-amino acids [20, 90, 100, 112–114]. Hruby has used this approach to great effect to generate a number of novel χ^2-constrained α-amino acids [112, 114].

A further key factor in the success of phospholane-derived catalysts is their ability to hydrogenate a variety of α-dehydroamino acids possessing functional groups, which in theory could either inhibit or adversely interfere with the selectivity of a hydrogenation process or, indeed, themselves be hydrogenated. These include strongly donating groups (e.g., heteroatoms, heterocycles, sulfides) or unsaturated groups (e.g., olefinic, ketonic and nitro groups). A large number of heteroaryl-α-amino acids have been prepared via asymmetric hydrogenation with chiral phospholane-modified catalysts. Zhang et al. reported that TangPhos and Et-**25** have been used in the preparation of a 2-thiophenylalanine

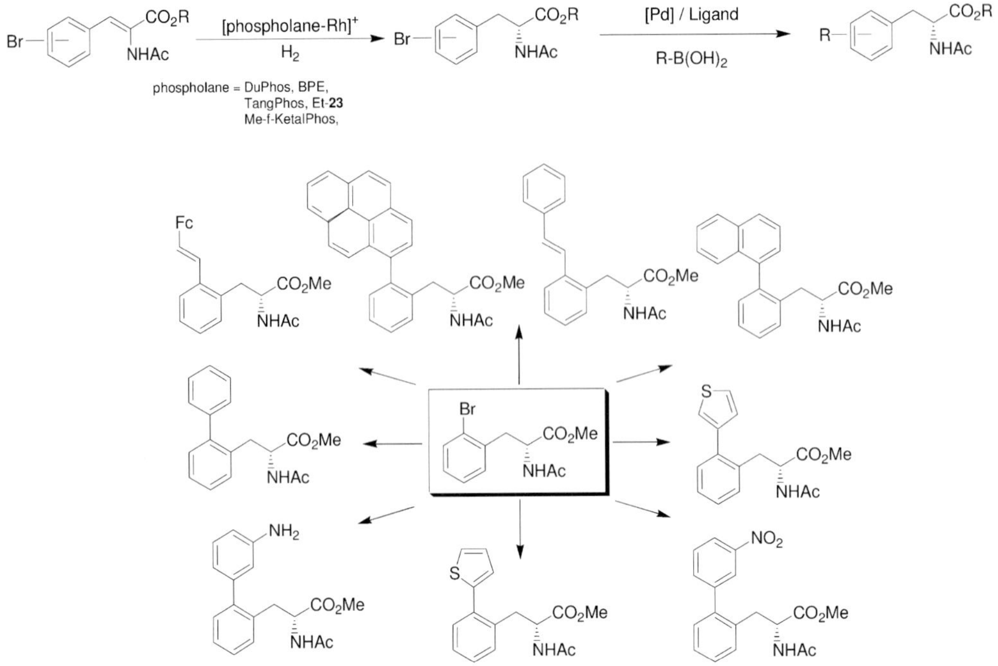
[phospholane-Rh]+
H2
[Pd] / Ligand
R-B(OH)2
phospholane = DuPhos, BPE, TangPhos, Et-23 Me-f-KetalPhos,
Fc
CO2Me
NHAc
Br
S
NH2
NO2

Scheme 24.8 Unnatural amino-acid derivatives accessed via tandem catalysis.

derivative in >99% ee at low pressure, without interference from the thiophene moiety [52b, 69a]. However, DuPhos, in particular, has been applied most extensively in this field. Simple thiophenyl [4a, 13, 20, 100, 115], furanyl [4a, 20, 100, 115], pyrroyl [115], pyrrolidyl [116], coumaryl [117] and a diverse range of tryptophanyl-α-amino acids [113, 114, 118–120] have all been synthesized with high enantioselectivities by means of enantiomeric hydrogenation of the requisite α-dehydroamino acids (see Fig. 24.8). In a number of cases, prolonged reaction times and molar catalysts loadings in the range of 1 to 3% were required to effect complete conversion. Moody generated di- and tri-peptide fragments of stephanotic acid using Et-DuPhos–Rh in the key asymmetric step [119]. In a remarkable piece of work, Carlier demonstrated that all five unnatural regioisomers of tryptophan derivatives could be accessed via enantioselective hydrogenation of the requisite α-dehydroamino acids with Et-DuPhos–Rh, and in no less than 96.7% ee in each case [118]. Substantially more complex and highly functionalized tryptophanyl-substrates have been prepared by Feldman et al. (Fig. 24.8), albeit with low enantioselectivities [121].

Pyridyl- and quinolyl-substrates are significantly more challenging to hydrogenate, due to the greater donating power of the nitrogen in these systems and, in general, modified hydrogenation protocols are necessary. A 2-quinolyl-alanine derivative was prepared by enantioselective hydrogenation with [Et-DuPhos–Rh]$^+$, in the presence of HBF_4, as the N-protonated species in 94% ee (see

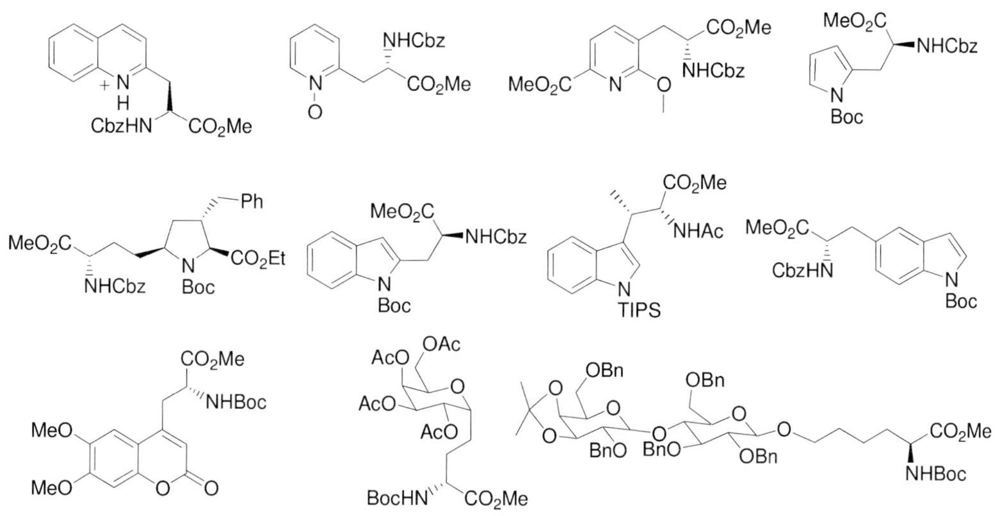

Fig. 24.8 Heteroalanines, tryptophan derivatives and glycosylated α-amino acid derivatives.

Fig. 24.8) [122]. Whilst this protocol can be used to prepare 3-pyridyl-alanine derivatives [22], the corresponding 2-pyridyl-alanine cannot be made [122]. However, Adamczyk has prepared several 2-pyridyl-alanine analogues through hydrogenation of the pyridine-*N*-oxide substrates in 80–83% ee (see Fig. 24.8) [123]. In general, only when the 2- and 6-positions of the pyridine ring are occupied can 2-, 3- or 4-pyridyl-alanine derivatives be prepared, without nitrogen modification, via hydrogenation with $[\text{phospholane–Rh}]^+$ catalysts [122–124].

Numerous examples exist of simple heteroatom-substituted substrates which have been hydrogenated by $[\text{phospholane–Rh}]^+$ including, amongst others, sulfide substrates [13,14], (*E*/*Z*)-isomers of *N*,*N*′-protected 2,3-diaminopropanoic and 2,3-diaminobutanoic acid derivatives [125], ε-NO_2-substituted α-dehydroamino acid [126], 4-piperidinylglycine precursor [127], and a ketonic substrate [14]. Heavily functionalized glycosylated α-amino acid derivatives have also been prepared using DuPhos–Rh catalysts [128]. Diastereoisomers of structurally complex and functionalized dipeptides have been prepared by Ortuño; a matching/mismatching effect is clear between the chiral substrate and the respective catalyst enantiomers, where (*R*,*R*)-Et-DuPhos–Rh gave >99% de and (*S*,*S*)-Et-DuPhos–Rh resulted in only 90% de, though the distal ketone moiety was not reduced [129].

A number of di- and tri-α-dehydroamino acid substrates have been shown to be hydrogenated in high ee and de with DuPhos–Rh catalysts, despite the potential for the initial chiral centers formed to interfere in subsequent stereodiscriminating steps [122,130] (Fig. 24.9). Interestingly, a number of these substrates are orthogonally protected at the acid or the amide functional groups, which is apparently not a barrier to high ee- and de-values [130a–d]. Hruby used this approach to synthesize a series of novel rigid dipeptide β-turn mimetics via the reduction of symmetrical di-α-dehydroamino acids [131].

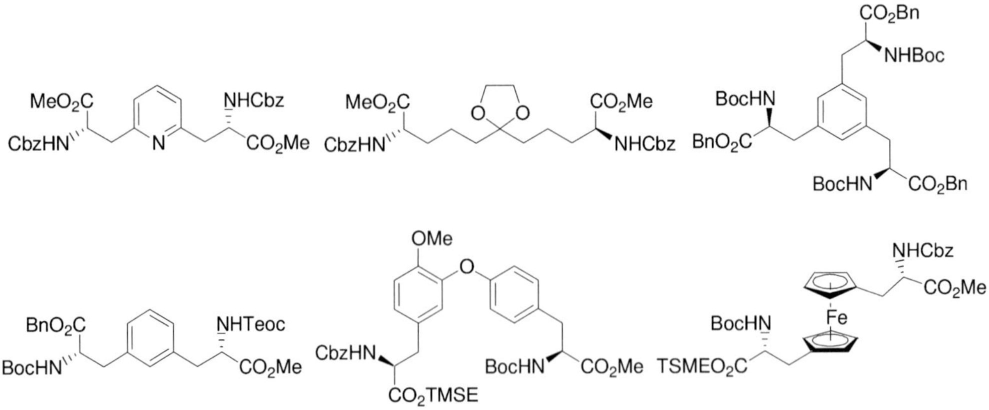

Fig. 24.9 Di and tri α-amino acid derivatives.

Phospholane-modified catalysts have shown the ability to discriminate between olefinic bonds in conjugated/nonconjugated β- and β,β-disubstituted α-didehydroamino acids, in both a highly regio- and enantioselective fashion [13, 90, 132, 133] (Scheme 24.9). The latter group involves the hydrogenation of tetrasubstituted alkenes with concomitant formation of two stereogenic centers. In general, the more reactive functionalized enamide bond is selectively reduced over simple unfunctionalized bonds. However, over-reduction can be observed, particularly when the reaction times are prolonged, as in the case of highly substituted olefins or in reactions near completion and the reduction of the distal bond becomes more favorable [132]. Over-reduction can be tempered by careful monitoring of the hydrogen uptake, solvent screening, lowering the hydrogen pressure, and reducing catalyst loadings [132, 133].

The choice of catalyst can have a significant effect on enantioselectivity and, in certain cases, the regioselectivity and activity. With nonconjugated α-didehydroamino acids, such as 2-acetamido-trideca-2,7-dienoic acid methyl ester or 2-acetamido-tetradeca-2,13-dienoic acid methyl ester, the proximal olefin is easily reduced at low catalyst loading with *n*-Pr-DuPhos–Rh in >99% ee and essentially complete regioselectivity [13]. However, in the case of conjugated α-didehydroamino acids, such as the 2-acetamido-6-(*tert*-butyl-dimethyl-silanyloxy)-hexa-2,4-dienoic acid methyl ester, the wrong choice of catalyst can lead to significant undesired over-reduction. 2-Acetamido-6-(*tert*-butyl-dimethyl-silanyloxy)-hexa-2,4-dienoic acid methyl ester can be hydrogenated in high ee and regioselectivity with Et-DuPhos–Rh (>99% ee and <0.5% over-reduction), whereas *i*-Pr-DuPhos–Rh gives only 87.8% ee and demonstrated little or no regioselectivity [133]. It has generally been observed that the DuPhos– and BPE–Rh systems can reduce tri-substituted $\alpha,\beta,\gamma,\delta$-didehydroamino acids with a remarkable degree of chemoselectivity in all but a few cases studied. Where the distal C=C bond is also activated to a certain extent, as in the case of styryl-dienamides, the degree of over-reduction is routinely <2%.

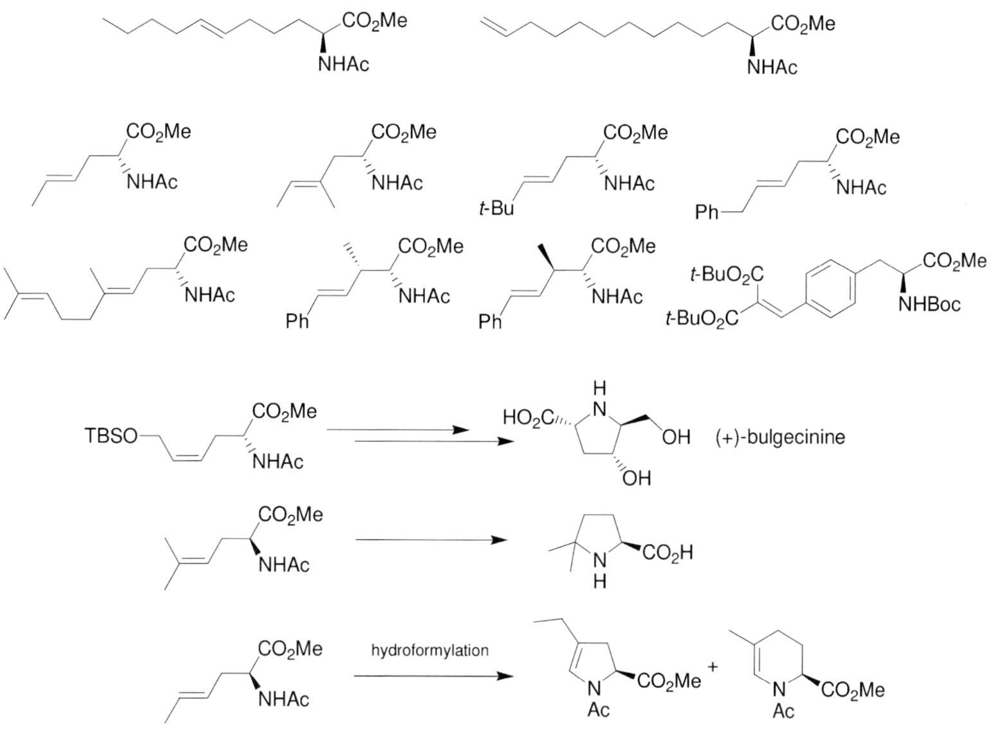

Scheme 24.9 Unsaturated α-amino acid derivatives prepared via chemoselective asymmetric hydrogenation.

In cases where the proximal double bond is highly substituted, such as tetrasubstituted α-didehydroamino acids, selective reduction of the proximal double bond becomes increasingly difficult. Burk found, with a series of β,β-disubstituted $\alpha,\beta,\gamma,\delta$-didehydroamino acids, that only the smaller, sterically less-congested catalysts Me-BPE–Rh and Me-DuPhos–Rh were able to achieve high reactivity and selectivities [132]. The overall lower reactivity of these highly substituted systems generally requires higher catalyst loadings and more forcing conditions to achieve high or full conversions. In many cases complete conversion could not be achieved, and over-reduction approached 10%. Reduction of the (2*Z*,4*E*)-isomers in comparison to the (2*E*,4*E*)-dienamides has also been studied; unsurprisingly, the (2*Z*,4*E*)-isomer is more readily reduced in contrast to the (2*E*,4*E*)-dienamides.

The reduction of dienamides with [phospholane–Rh]$^+$ catalysts has been applied in the synthesis of a number of biologically interesting targets (Scheme 24.9). Burk and co-workers synthesized (+)-bulgecinine from 2-acetamido-6-(*tert*-butyl-dimethyl-silanyloxy)-hexa-2,4-dienoic acid methyl ester utilizing [(*R*,*R*)-Et-DuPhos–Rh]$^+$ [133], while Boehringer Ingelheim used (*R*,*R*,*S*,*S*)-TangPhos–Rh and (*S*,*S*)-Et-DuPhos–Rh to generate key intermediates in protease inhibitors [134], and 5,5-dimethylproline has been generated using (*S*,*S*)-Et-DuPhos–Rh

[135]. Garbay reported the chemoselective reduction of a α-dehydrophenylalanine substrate bearing a *p*-acrylate moiety [105]. Robinson et al. have also used a tandem, one-pot asymmetric hydrogenation-hydroformylation-cyclization approach to generate six- to eight-membered cyclic α-amino acids [136].

The enantioselective hydrogenation of β,β-disubstituted α-dehydroamino acids by means of [diphosphine-Metal]$^+$ catalysts is challenging in terms of enantioselectivity, chemoselectivity (*vide supra*), and reactivity. Few ligand systems have been tested with β,β-disubstituted substrates, and the reported results indicate that variable ee-values and high catalyst loadings are commonplace for many of the [diphosphine-Metal]$^+$ systems with tetrasubstituted olefins [137]. However, DuPhos and BPE catalysts demonstrate the capacity to consistently hydrogenate a wide range of β,β-disubstituted α-dehydroamino acid substrates with excellent ee-values and industrially applicable loadings [14, 20, 39, 90, 125, 138] (Fig. 24.10). Furthermore, the ability to hydrogenate (*E*) or (*Z*) α-dehydroamino acids with high enantioselectivity means that, with dissimilar substituents, two stereogenic centers can be created with high enantio- and diastereoselectivity when the (*E*) and (*Z*) α-dehydroamino acid is hydrogenated with both catalyst enantiomers.

In the majority of cases reported, optimal stereoselectivity and reactivity in sterically congested tetrasubstituted alkenes can be achieved with sterically less cumbersome Me-DuPhos and Me-BPE ligands. This is most graphically highlighted with the model substrate 2-acetamido-3-methyl-but-2-enoic acid methyl ester, where both Me-BPE and Me-DuPhos–Rh catalysts hydrogenate the substrate in >95% ee, whereas Et-DuPhos achieves 74% ee, *n*-Pr-DuPhos 45% ee, and *i*-Pr-DuPhos merely 14% ee [14]. This trend has been found over a broad range of substrates [14, 132], although there are some exceptional cases where Et-DuPhos–Rh achieves high enantio- and diastereoselectivity [39, 138]. Burk also made the observation that benzene was the optimal solvent for the majority of cases, however supercritical CO_2 (scCO_2) has also been shown to be a suitable medium for the enantioselective hydrogenation of β,β-disubstituted α-dehydroamino acids [39]. Zhang's phospholane catalyst Me-f-KetalPhos–Rh hydrogenated the model substrate 2-acetamido-3-methyl-but-2-enoic acid methyl

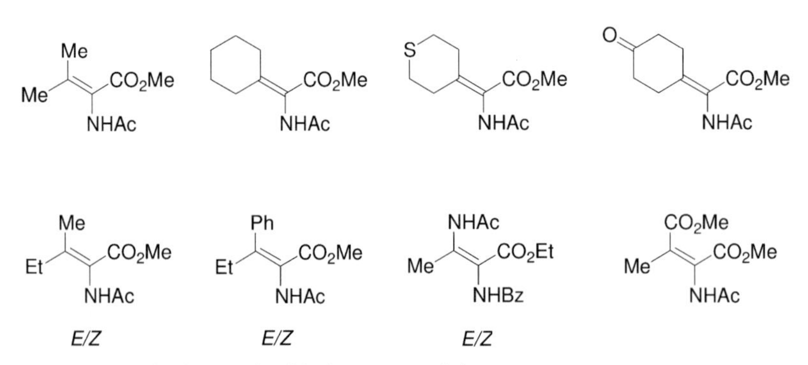

Fig. 24.10 β,β-disubstituted α-dehydroamino acid derivatives.

ester in 87.3% ee, whereas the phosphetanes DiSquareP* [87b], *i*-Pr-CnrPHOS, and Cy-BPE-4 [83a] have been shown to give only low to moderate ee-values with a few model substrates. (*S,S*)-*i*-Pr-CnrPHOS–Rh demonstrated a remarkable inversion of facial selectivity with 2-acetamido-3-phenyl-but-2-enoic acid methyl ester, giving the (2*S*,3*R*)-product in 38% ee at 10 bar H_2 and the (2*R*,3*S*)-isomer in 80% ee at 100 bar H_2.

Several biologically interesting targets have been synthesized via phospholane–Rh catalyzed hydrogenation of β,β-disubstituted α-dehydroamino acid substrates, including all four diastereoisomers of *N,N'*-protected 2,3-diaminobutanoic acid derivatives [125] (*vide supra*), a 4-piperidinylglycine derivative used as metalloproteinase and thrombin inhibitors (*vide supra*) [127a], as well as a sterically congested β-methyltryptophan derivative [120].

24.3.2 Enantioselective Hydrogenation of β-Dehydroamino Acid Derivatives

Enantiomerically pure β-amino acids and their derivatives are an important class of compounds due to of their use as chiral building blocks in the synthesis of both biologically active molecules [139] and novel peptidomimetics [140]. Their incorporation is partly a result of the unusual secondary structures they can create, and also because they are frequently resistant to proteolysis. Currently, the principal methods used for their preparation involve chiral auxiliaries in stoichiometric reactions and to a lesser extent enantioselective catalysis [141]. One of the most promising and industrially viable methodologies involves the enantiomeric hydrogenation of an appropriate β-dehydroamino acid derivative with a homogeneous metal catalyst [142]. Owing to its simplicity, this approach has seen rapid development in recent years, with the most successful catalysts typically being Rh and Ru complexes containing phosphorus-based ligands, including several diphospholanes. The most selective examples (>95% ee), achieved with the most commonly used β-dehydroamino acid-derived substrates (R^1 = Me or Ph), are collected in Table 24.2, together with several results useful for comparison purposes.

As can be seen from Table 24.2, these rhodium catalysts are in general extremely active under very mild reaction conditions (H_2 pressure 1–20 atm, room temperature), albeit at catalyst loadings typical of screening studies (SCR 100). Although rare exceptions are known [143], the hydrogenation of (*E*)-β-dehydroamino acid esters generally proceeds with considerably higher enantioselectivity than the corresponding (*Z*)-isomers. It is worth mentioning, however, that Rh-TangPhos is reported to perform remarkably well against either stereoisomer [69b, 71]. This is important since in the synthesis of dehydroamino acids, the (*E/Z*)-isomeric mixtures obtained can be difficult to separate, especially in the case of β-aryl substitution [71, 143, 148]. Furthermore, the (*Z*)-isomer is predominantly formed due to stabilizing hydrogen bonds [149]. Whether or not this additional bonding retards coordination to the metal center and concomitantly lowers selectivity is arguable, especially in protic media. However, use of the

Table 24.2 Phospholanes reported to hydrogenate model β-dehydroamino acid derivatives in >95% ee.

AcHN, CO_2R^2, R^1 —(Rh-catalyst, H_2)→ AcHN, CO_2R^2, R^1

72 R^1 = Me, R^2 = Et
73 R^1 = Me, R^2 = Me
74 R^1 = Ph, R^2 = Et
75 R^1 = Ph, R^2 = Me

Substrate	Ligand	SCR	Reaction conditions [a)]	TON	TOF [h^{-1}]	% ee (config.)	Reference(s)
(*E*)-**72**	(*R*,*R*)-Me-DuPhos	100	Toluene, rt, 2.7 atm, 24 h[c]	100	–	99	143
(*Z*)-**72**	(*R*,*R*)-Me-DuPhos	100	Toluene, rt, 20 atm, 24 h[c]	100	–	88	143
(*E*)-**72**	(*S*,*S*)-Et-DuPhos	100	THF, 25 °C, 2 atm, 1 h[c]	100	–	99	144
(*Z*)-**72**	(*S*,*S*)-Et-DuPhos	100	THF, 25 °C, 2 atm, 1 h[c]	100	–	89	144
(*E*)-**72**	(*R*,*R*)-*i*-Pr-DuPhos	100	TFE, rt, 9.7 atm, <2 min	100	>3000	99	145
(Z)-**72**	(*R*,*R*)-*i*-Pr-DuPhos	100	TFE, rt, 9.7 atm, <2 min	100	>3000	92	145
(*E*/*Z*)-**72**	(*R*,*R*)-*i*-Pr-DuPhos	1000	TFE, rt, 9.7 atm, 40 min	1000	1500	95	145
(*E*)-**72**	(*R*,*R*)-Et-BPE	100	THF, 40 °C, 2 atm, 1 h[c)]	100	–	99	144
(*Z*)-**72**	(*R*,*R*)-Et-BPE	100	THF, 40 °C, 2 atm, 1 h[c)]	100	–	90	144
(*E*)-**72**	(*R*,*R*,*S*,*S*)-DuanPhos	100	MeOH, rt, 1.4 atm[b)]	–	–	99 (*R*)	70
(*Z*)-**72**	(*R*,*R*,*S*,*S*)-DuanPhos	100	MeOH, rt, 1.4 atm[b)]	–	–	97 (*R*)	70
(*E*)-**73**	(*R*,*R*)-Me-DuPhos	100	Toluene, rt, 2.7 atm, 24 h[c)]	100	–	99	143
(*Z*)-**73**	(*R*,*R*)-Me-DuPhos	100	Toluene, rt, 2.7 atm, 24 h[c)]	100	–	64	143
(*E*)-**73**	(*S*,*S*)-Me-DuPhos	100	MeOH, 25 °C, 1 atm, 1 h[c)]	100	–	98	146a
(*Z*)-**73**	(*S*,*S*)-Me-DuPhos	100	MeOH, 25 °C, 1 atm, 1 h[c)]	100	–	88	146a
(*E*)-**73**	(*S*,*S*)-Et-DuPhos	100	MeOH, 25 °C, 1 atm[b)]	100	–	98	146
(*Z*)-**73**	(*S*,*S*)-Et-DuPhos	100	MeOH, 25 °C, 1 atm[b)]	100	–	88	146
(*E*/*Z*)-**73**	(*S*,*S*)-Et-DuPhos	100	MeOH, 25 °C, 1 atm[b)]	100	–	92	146
(*E*)-**73**	(*R*,*R*)-Et-DuPhos	100	THF, 40 °C, 2 atm, 1 h[c)]	100	–	95	144
(*Z*)-**73**	(*R*,*R*)-Et-DuPhos	100	THF, 40 °C, 2 atm, 1 h[c)]	100	–	86	144

Table 24.2 (continued)

Substrate	Ligand	SCR	Reaction conditions [a]	TON	TOF [h^{-1}]	% ee (config.)	Reference(s)
(*E*)-**73**	(*R*,*R*)-Et-BPE	100	THF, 40 °C, 2 atm, 1 h [c]	100	–	98	144
(*Z*)-**73**	(*R*,*R*)-Et-BPE	100	THF, 40 °C, 2 atm, 1 h [c]	100	–	82	144
(*E*)-**73**	(*R*,*R*)-Me-Ph-BASPHOS	100	MeOH, 25 °C, 1 atm [b]	100	–	99 (*S*)	56b
(*Z*)-**73**	(*R*,*R*)-Me-Ph-BASPHOS	100	MeOH, 25 °C, 1 atm [b]	100	–	70 (*S*)	56b
(*E*)-**73**	(S,R_p,S,R_p)-**36**	100	THF, rt, 1.4 atm, 5 min	100	1200	96 (*R*)	59c
(*Z*)-**73**	(S,R_p,S,R_p)-**36**	100	THF, rt, 1.4 atm, 45 min	100	133	89 (*R*)	59c
(*E*)-**73**	(R,S_p,S_p,R)-**35**	100	THF, rt, 1.4 atm, 15 min	100	400	96 (*S*)	59c
(*Z*)-**73**	(R,S_p,S_p,R)-**35**	100	THF, rt, 1.4 atm, 1 h 15 min	100	80	83 (*S*)	59c
(*E*)-**73**	(*R*,*R*)-catASium M	100	MeOH, 25 °C, 1 atm, 3 h [c]	100	–	98 (*R*)	95
(*Z*)-**73**	(*R*,*R*)-catASium M	100	MeOH, 25 °C, 1 atm, 3 h [c]	100	–	83 (*R*)	95
(*E*)-**73**	(*R*,*R*)-Et-FerroTANE	100	MeOH, 25 °C, 1 atm [b]	100	–	99 (*S*)	56d, 147
(*Z*)-**73**	(*R*,*R*)-Et-FerroTANE	100	MeOH, 25 °C, 1 atm [b]	100	–	28 (*S*)	56d, 147
(*E*)-**73**	(*S*,*S*,*R*,*R*)-TangPhos	200	THF, rt, 1.4 atm, 24 h [c]	200	–	>99	69b, 71
(*Z*)-**73**	(*S*,*S*,*R*,*R*)-TangPhos	200	THF, rt, 1.4 atm, 24 h [c]	200	–	99	69b, 71
(*E*/*Z*)-**73**	(*S*,*S*,*R*,*R*)-TangPhos	200	THF, rt, 1.4 atm, 24 h [c]	200	–	>99	69b, 71
(*Z*)-**73**	(*S*)-Me-Butiphane	200	MeOH, 25 °C, 5 atm [b]	200	–	98	42
(*E*)-**74**	(*R*,*R*)-Et-FerroTANE	100	MeOH, 25 °C, 1 atm [b]	100	–	99 (*S*)	56d, 147
(*E*)-**75**	(*R*,*R*)-Et-FerroTANE	100	MeOH, 25 °C, 1 atm [b]	100	–	99 (*S*)	56d, 147
(*E*/*Z*)-**75**	(*S*,*S*,*R*,*R*)-TangPhos	200	THF, rt, 1.4 atm, 24 h [c]	–	–	94 (*S*)	69b, 71

a) Complete conversion unless otherwise stated.
b) No reaction time given.
c) No conversion given

strongly polar solvent 2,2,2-trifluoroethanol (TFE) had a beneficial effect on enantioselectivities when a Rh-*i*-Pr-DuPhos catalyst was applied [145]. Limited solvent studies have been performed with a number of phospholane-based ligand systems [65, 71, 143, 146a], but in general alcoholic solvents are the most suitable, together with tetrahydrofuran (THF) and dichloromethane (CH_2Cl_2).

Several accounts have described (*Z*)-dehydroamino acid esters as being less active than the corresponding (*E*)-isomer [59c, 143–145]. In fact, Bruneau and Demonchaux reported that when reduction of an (*E*/*Z*)-mixture of **73** with Rh-Et-DuPhos in THF was not complete, only unreacted (*Z*)-**73** was detected. These findings conflict, however, with results obtained in MeOH [56d], where the ligand structure was also found to be significant to the relative reactivity of each stereoisomer. As for *α*-dehydroamino acid derivatives, preformed metal–diphosphine complexes generally perform in superior fashion to those prepared *in situ* [56d].

Zhang et al. reported that for Rh-catalyzed enantiomeric hydrogenation with either BICP or Me-DuPhos, the (*Z*)-isomers were generally less reactive, and required higher pressures for complete conversion [143]. Enantioselectivities for the (*E*)-isomer were shown to be unaffected by increased pressure. On the other hand, Heller and co-workers showed that operating at low hydrogen pressures in a polar solvent had a significantly beneficial effect on enantioselectivities when hydrogenating (*Z*)-**73** with [Et-DuPhos Rh(COD)]BF_4 (e.g., 35% ee at 45 atm and 87% ee at 1 atm), albeit at the expense of reaction rates [146a]. This was also found to be the case with several other ligand systems [56d]. In light of these findings, a tentative mechanistic concept has been proposed which provides evidence that the reaction proceeds via an "unsaturated route" with the prochiral olefin coordinating to the metal center prior to oxidative addition of hydrogen [150].

In general, the same sense of chiral induction is obtained with either geometrical stereoisomer, which facilitates the use of (*E*/*Z*)-isomeric mixtures. An exception to this was recently reported by Heller and Börner [56d]. Remarkably, hydrogenation of methyl (*Z*)-*β*-acetylamino pentenoate with [(*S*,*S*)-Et-DuPhosRh (COD)]BF_4 at 1 bar gave the (*R*)-enantiomer of product in 31% ee, whereas the same reaction at 30 bar resulted in an inversion of configuration and the (*S*)-product in 77% ee.

The effects of temperature on enantioselectivities have been examined using a Rh-Et-DuPhos catalyst in both MeOH [56d] and THF [144]. With *β*-dehydroamino acid derivative **73** in MeOH, an increase in temperature was found to have a slight beneficial effect for both (*E*) and (*Z*)-isomers over a 70 °C range, with maximum values being observed between 0 °C and 25 °C. In THF, however, the effect is much more pronounced, especially for the (*Z*)-isomer which varies in selectivity from 65% ee at 10 °C to 86% ee at 25 °C. Interestingly, when substrate **72** was reduced with a Rh-Et-BPE catalyst in THF, this temperature dependence on enantioselectivity for the (*Z*)-isomer was most apparent, the selectivities varying from 43% ee (10 °C) to 90% ee (40 °C). Examination of these results also seemed to indicate that the hydrogenation of *β*-dehydroamino acid derivatives follows an unsaturated pathway (*vide supra*) [144].

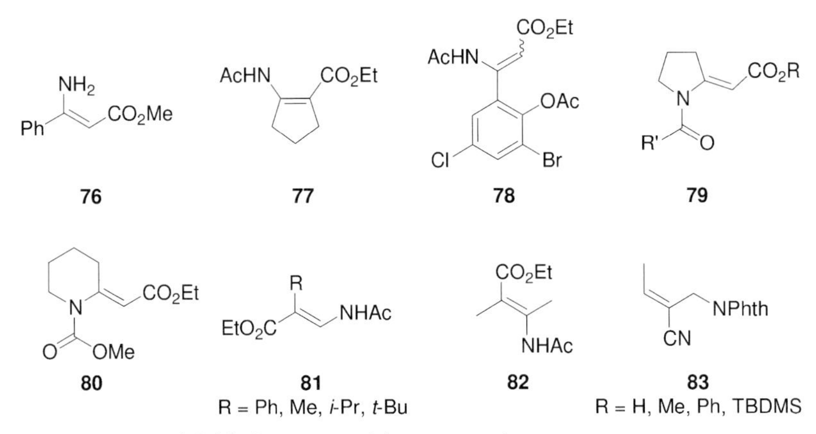

Fig. 24.11 Unusual β-dehydroamino acid derivatives to have been reduced with phospholane-based catalysts.

Several phospholane-based ligands have shown a wide substrate scope beyond the standard examples represented in Table 24.2. Both Et-FerroTANE **61** [147] and TangPhos **46** [69b, 71] have been successfully applied to a diverse range of methyl and ethyl β-aryl-dehydroamino acids containing various aromatic substituents, whilst catASium M **20a** [95] has been used for the reduction of numerous β-alkyl-dehydroamino acid esters.

In addition to the standard substrates described above, the enantioselective hydrogenations of several other β-dehydroamino acid derivatives using phospholane-based or related ligands are worthy of note (Fig. 24.11). Using TFE as solvent, a screen of commercially available catalysts showed that the unprotected β-dehydroamino acid ester (*Z*)-**76** could be partially reduced (77% conversion) in 88% ee with [(*R,R*)-Et-FerroTANE Rh(COD)]BF_4, but ultimately, an *in-situ*-prepared Rh-Josiphos-type catalyst gave superior results [151]. Interestingly, preliminary deuterium-labeling studies suggested that the hydrogenation of (*Z*)-**76** proceeds through the imine tautomer in an analogous fashion to β-ketoester hydrogenations. Enantioselective reduction of the tetrasubstituted cyclic β-dehydroamino acid ester (*E*)-**77** was recently reported by Zhang et al., and although both *in-situ*-prepared Ru catalysts of Me-DuPhos and TangPhos were found to give complete conversion (SCR 20) in moderate enantioselectivities (71% ee and 57% ee, respectively), atropisomeric biaryl-based ligands were more selective (e.g., C2- to C5-TunePhos all gave 99% ee) [152].

As one of several routes investigated for the preparation of a key intermediate for a $\alpha_v\beta_3$ integrin antagonist, the enantiomeric hydrogenation of olefin **78** was examined [153]. Despite investigating several different catalysts under multiple reaction conditions on various derivatives of **78**, a viable method was not forthcoming. The best result obtained was 70% ee with [(*S,S*)-Et-DuPhos Rh(COD)]OTf in CH_2Cl_2 at 5 atm H_2. Lee and co-workers examined the enantioselective synthesis of homoproline derivatives via Rh-catalyzed reduction of the cyclic substrate (*E*)-**79** with a range of different ligands [154]. Although, (*R,R*)-Me-DuPhos was the most selective (>99% ee (*R*)), the conversion was only 37%. The chiral bidentate phosphine Me-BDMI proved to be the overall ligand of choice. Reduction of the structurally related *β*-dehydroamino acid ester (*E*)-**80** was recently described in a patent by Solvay, with [(*R,R*)-Me-DuPhos Rh(COD)]OTf giving complete conversion to the (*R*)-product in 95.5% ee (5 atm H_2, 25 °C, SCR 100) [155].

Finally, an interesting variation on the standard substitution pattern shown above is the regioisomeric β^2-amino acids, the substituent being *α* to the carboxylic functionality. Two approaches for their synthesis, which adopts enantiomeric hydrogenation with a phospholane-based ligand, have been described. Robinson, Jackson and colleagues reported the preparation of a range of substrates of type (*E*)-**81** and subsequent Rh-catalyzed enantioselective hydrogenation with either Me or Et-DuPhos and Me-BPE. Through modification of the amide group, the ligand and the solvent, moderate enantioselectivities up to 67% ee where attainable [156]. The *α,β*-disubstituted *β*-dehydroamino acid ester (*E*)-**82** was also examined with BPE and DuPhos catalyst systems; Rh-Me-BPE was the most selective, giving complete conversion in 65% ee over 72 h in benzene (4 atm, room temperature). An alternative approach to β^2-amino acids was also explored through the reduction of *α,β*-unsaturated nitriles **83**, again using Rh-Me-BPE or Rh-Et-DuPhos catalysts (up to 48% ee), and the amino acids being obtained after hydrolysis and phthalimide deprotection [157]. Conversion of the nitrile to the corresponding methyl ester and switching to a Ru-BINAP-based system increased the enantioselectivities to 84% ee.

24.3.3
Enantioselective Hydrogenation of Enamides

Chiral amines constitute an important class of compounds that have been extensively employed as resolving agents, chiral auxiliaries, and pharmaceutical intermediates. Traditionally, classical resolution or biocatalysis have been typically chosen as the preferred methods for industrial manufacturing, though the enantioselective hydrogenation of enamides or imines (*vide infra*) has recently received much attention, with several phospholane-based ligands proving to be applicable. Table 24.3 details the most selective phospholanes (≥95% ee) to have been used in the Rh-catalyzed hydrogenation of frequently used simple enamides, namely *N*-(1-phenyl-vinyl)-acetamide **84** and *N*-(1-phenyl-propenyl)-acetamide **85**. In general, the conditions employed are mild, with high reactivities being observed at low temperatures and pressures. For example, TONs as high as 5000 to 10 000 have been achieved with Ph-BPE **15** [34] or the P-chiral phospholane, TangPhos **46** [69]. In the case of the *β*-branched enamide **85**, (*S,S,R,R*)-DiSquareP* **63** has been reported to be extremely selective for reduction of the (*E*)-isomer (>99% ee), but with the enantioselectivity being much lower for the corresponding (*Z*)-isomer (37% ee) [87 b]. However, this is not the case for the diphospholanes BPE and TangPhos, with excellent levels of selectivity being obtained even when an (*E/Z*)-isomeric mixture is applied.

Table 24.3 Phospholanes reported to hydrogenate model enamide substrates in >95% ee.

Rh-catalyst, H_2

84 R = H
85 R = Me

Substrate	Ligand	SCR	Reaction conditions[a)]	TON	TOF $[h^{-1}]$	% ee (config.)	Reference(s)
84	(*S,S,R,R*)-DiSquareP*	100	MeOH, rt, 1 atm, 1 h	100	>100	>99 (*R*)	87b
84	(*S,S,S,S*)-Et-**16**	100	MeOH, rt, 10 atm, 24 h	100	–	96 (*S*)	52b, c
84	(*R,R*)-Me-Ph-UCAP	100	MeOH, rt, 5 atm, 3 h	100	>33	94 (*R*)	65, 67
84	(*S,S,R,R*)-TangPhos	10 000	MeOH, rt, 1.4 atm[d)]	10 000	–	99 (*R*)	69
84	(*R,R,S,S*)-DuanPhos	100	MeOH, rt, 1.4 atm[d)]	10 000	–	>99 (*R*)	70
84	(*R,R*)-Ph-BPE	5 000	MeOH, 25 °C, 10 atm[d]	5 000	–	99	34
84	(*R,R*)-Me-BPE	500	MeOH, 22 °C, 4 atm, 15 h	500	>33	95	17
(*E*)-85	(*S,S,R,R*)-DiSquareP*	100	MeOH, rt, 1 atm, 1 h	100	>100	>99 (*R*)	87b
(*Z*)-85	(*S,S,R,R*)-DiSquareP*	100	MeOH, rt, 2 atm, 1 h	100	>100	37 (*R*)	87b
(*Z*)-85[c)]	(*S,S*)-Me-DTBM-UCAP	500	MeOH, 30 °C, 4 atm, 15 h	500	>33	99 (*S*)	64
(*E/Z*)-85	(*S,S,R,R*)-TangPhos	100	MeOH, rt, 1.4 atm, 12 h	100	–	98 (*R*)	69
(*E/Z*)-85	(*R,R*)-Me-BPE	100	MeOH, rt, 1.4 atm, 12 h	100	–	95 (*R*)	17

a) Complete conversion unless otherwise stated.
b) Similar high enantioselectivities have also been obtained with several other ligands of this class.
c) *N*-benzoyl instead of *N*-acetyl.
d) No reaction time given.

In addition to these simple model substrates, several phospholane-containing ligands have shown broad functional group tolerance when applied to the Rh-catalyzed hydrogenation of aromatically substituted α-arylenamides. Burk et al. reported the use of DuPhos and BPE ligands for the reduction of α-arylenamides containing alkyl, halogen, thio, alkoxy, aromatic, and heteroaromatic substituents, commenting that the enantioselectivities tended to increase with decreasing steric demand of the phospholane moiety at the 2,5-positions [17]. This was later extended to include esters, ketones and cyano groups, showing that these systems are also chemoselective [35]. Similar levels of tolerance have been demonstrated, with several phospholanic ligands originating from the group of Zhang, including the tetrahydroxy diphosphine **25** (R = Me) [52 b, 53 b], TangPhos [69] and, to a lesser

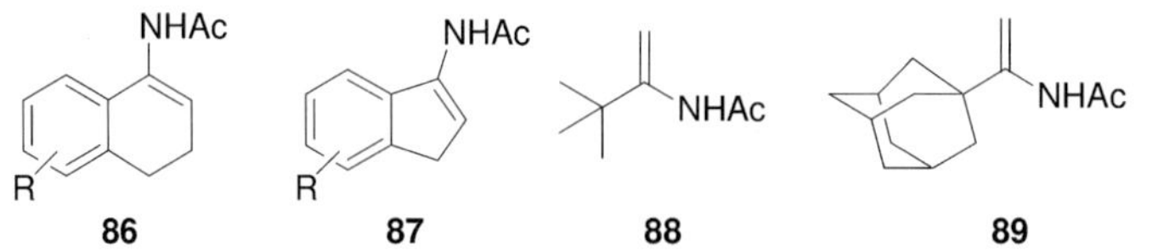

Fig. 24.12 General enamide classes to have been reduced with phospholane-based catalysts.

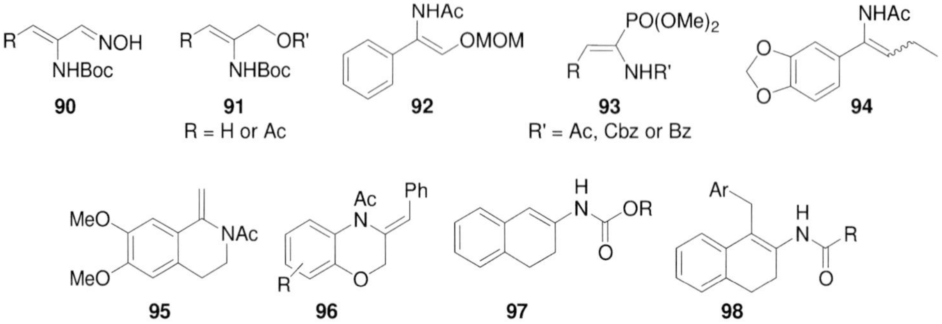

Fig. 24.13 Unusual enamides to have been reduced with phospholane-based catalysts.

extent, DuanPhos [70]. The effective hydrogenation of (*Z*)-**85** and the *N*-benzoyl analogue were independently reported by Pringle [65] and Saito [64] respectively, using a range of unsymmetrical diphosphines, UCAPs **41**.

Both (*S*,*S*)-Me-BPE [19] and (*R*,*S*,*R*,*S*)-Me-PennPhos **44** [67 d, e] have been successfully applied to the Rh-catalyzed hydrogenation of a range of cyclic enamides derived from α-tetralones and α-indolones (**86** and **87**). Under mild conditions, both ligands achieved good to excellent enantioselectivities (71–99% ee), with PennPhos giving reasonable levels of catalyst reactivity (TON up to 2000; TOF $\sim 100\ h^{-1}$), even for tetra-substituted enamides. Interestingly, PennPhos gives lower selectivities for acyclic enamides when compared to BPE catalysts, whereas BPE requires lower temperatures to attain high enantioselectivities with cyclic enamides (e.g., 71% ee at 20 °C and 92% ee at 0 °C for unsubstituted **86**).

Alkyl enamides, such as *N*-(1-*tert*-butyl-vinyl)-acetamide **88** and *N*-(1-adamantyl-vinyl)-acetamide **89**, can also be hydrogenated in high enantioselectivity (>99% ee) and activity (TON 5000; TOF $>625\ h^{-1}$) with Rh-Me-DuPhos [19]. Remarkably, these bulky alkyl enamides are reduced with the opposite sense of induction, a phenomenon also observed when the bisphospholane DiSquareP* **63** was applied [87 b]. A computational modeling study by Landis and Feldgus suggested that the reduction of α-alkyl and α-arylenamides involves different coordination pathways [93, 158].

In addition to standard cyclic and acyclic enamides, the effective hydrogenation of several more unusual enamides has been reported (Fig. 24.13). A concise method for the synthesis of chiral β-amino alcohols, amino oximes and chiral 1,2-diamines has been described by Burk et al. via the enantioselective hydrogenation of **90** or **91** using Rh catalysts of Me or Et-DuPhos [18]. In general, the enantioselectivities were high (91–99% ee), with reactions proceeding smoothly to completion within 12 h at SCR 1000. Zhang and co-workers have reported the hydrogenation of a series of MOM-protected β-hydroxy-α-arylenamides **92** as a mixture of (*E/Z*) isomers [159]. Although BICP–Rh and Me-DuPhos–Rh complexes were both found to be excellent catalysts for this transformation (90–99% ee), Rh-Me-DuPhos displayed higher enantioselectivity over a broader substrate range. Ultimately, the products could be converted to chiral α-arylglycinols by *O*-MOM and *N*-acetyl deprotection under acidic conditions. By screening a range of diphosphines, including five phospholane-based ligands, Pagenkopf et al. extended the scope of this reaction to include *o*-alkoxy-substituted enamides [160]. Me-BPE and Me-DuPhos were found to be the most selective ligands (92–98% ee), with better results being achieved with the isolated [(P-P) Rh COD]OTf precatalysts over *in-situ* preparation. The size of the *o*-substituent was not found to have any significant effect on selectivity.

By using a Rh-catalyst containing a ligand from either the DuPhos or BPE family, Burk and co-workers successfully hydrogenated a range of phosphonated enamides **93** (R′=Ac or Cbz) in moderate to high enantioselectivities (57–95% ee), with aryl-substituted examples giving lower selectivity than alkyl analogues [161]. In contrast to the reduction of several other substrate classes with Rh-DuPhos or Rh-BPE complexes [12, 13, 15, 18, 21], a strong dependence on olefin geometry was observed, with (*E*)-isomers being significantly more selective. Börner and Holz also reported the Rh-catalyzed reduction of a phosphonated enamide **93** (R=Ph, R′=Bz) with two DuPhos/BPE type ligand systems, Rophos (**22** and **23**) [51a] and BASPHOS (**32** and **33**) [56e]. Although enantioselectivities with Rophos were generally higher than with BASPHOS (up to 99% ee versus 79% ee), activities were slightly lower (TOF 6 h^{-1} versus 25 h^{-1}). The enantioselective hydrogenation of enamide **94** using [(*R,R*)-Me-DuPhos Rh(COD)]BF_4 was reported by Storace et al. [162]. This intermediate to the leukocyte elastase inhibitor, DMP 777, was prepared quantitatively in 96.5% ee at SCR 1800 under mild conditions (2 atm, room temperature) in MeOH. Although a single crystallization afforded the optically pure amide in 86% yield and >99% ee, ultimately enantiomeric hydrogenation was not chosen as the preferred method for manufacturing.

As well as endo-cyclic enamides (*vide supra*), phospholane-based Rh catalysts have also been applied to the enantiomeric reduction of exo-cyclic enamides. Zhang reported TangPhos to hydrogenate **95** in 97% ee [69], while Zhou showed Me-DuPhos to be extremely efficient for a broad range of substituted dihydrobenzoxazines **96** (92–99% ee) [163]. Finally, the hydrogenation of a series of trisubstituted ene carbamates **97** and tetrasubstituted enamides **98** was found to be catalyzed by Ru complexes of either Me-DuPhos and Me-BPE [164]. The catalysts were formed *in situ* by reacting the diphosphine with [Ru(COD)(methallyl)$_2$] in the presence of HBF_4 or triflic acid. Notably, the use of atropisomeric ligands, BINAP and BIPHEMP, led to no activity and similarly, hydrogenation did not occur with the more commonly used [(diphosphine)Rh(COD)]BF_4 precatalysts.

24.3.4
Enantioselective Hydrogenation of Unsaturated Acid and Ester Derivatives

The enantioselective hydrogenation of α,β- or β,γ-unsaturated acid derivatives and ester substrates including itaconic acids, acrylic acid derivatives, butenolides, and dehydrojasmonates, is a practical and efficient methodology for accessing, amongst others, chiral acids, chiral α-hydroxy acids, chiral lactones and chiral amides. These are of particular importance across the pharmaceutical and the "flavors and fragrances" industries.

The enantioselective hydrogenation of itaconic acid derivatives in particular has received much interest; one area of significance is the generation of succinate compounds for use as peptidomimetics [82 a]. As indicated in Table 24.4, a variety of phospholanes are suitable ligands for hydrogenating the commonly used model substrates, itaconic acid and its dimethyl ester (DMI), with high enantioselectivities. Moreover, phosphetanes and phosphinanes have demonstrated high enantioselectivities with both substrates. In several cases exceptionally high catalyst activity has been demonstrated; Ph-BPE and catASium M are reported to hydrogenate DMI with TOFs of 60000 and 40000 h^{-1}, respectively [34, 47]. Although high enantioselectivities are also reported for the parent itaconic acid substrate, catalyst activities are substantially lower with this substrate in comparison to DMI. Surprisingly, the parent Me-DuPhos ligand is neither particularly selective nor active with the diacid [44], and a number of ligands are reported with superior selectivity and activity. BASPHOS ligands have demonstrated a significant degree of substrate sensitivity with DMI and itaconic acid; enantioselectivities ranging from 8.1% to 97.9% are reported for (*R,R*)-Me-Et-BASPHOS (**33**) and (*R,R*)-Me-Ph-BASPHOS (**32**), respectively [56 b]. Pressure and solvent effects have been observed in the hydrogenation of DMI using a catASium M ligand, where higher pressures (7.9 atm) and CH_2Cl_2 as solvent gave superior results [46–48]. Itaconic acid has been hydrogenated by **25** in various MeOH/H_2O mixtures, ranging from 9:1 to 3:97, without variance or loss of enantioselectivity. The three-carbon bridged tridentate ligand (*S,S*)-Me-**11** hydrogenated DMI in 94% ee, albeit with protracted reaction times, whereas the related bidentate ligand (*R,R*)-Me-**7** surprisingly gave only 78% ee [11]. Remarkably, Corma et al. have reported the first use Me-DuPhos-based Pt and Au catalysts for the reduction of simple itaconic acid derivatives in high enantioselectivity (3 to 95% ee) and extremely high rates (TOF up to 10200 h^{-1}) [166]. Unfortunately, these high activities were achieved at the expense of selectivity.

β-Substituted and β,β-disubstituted itaconic acid substrates, generated via the Stobbe condensation and resulting in mono 1-esters, provide a more structurally diverse and challenging set of substrates. Currently, only DuPhos, BPE and – to a lesser extent TangPhos and catASium M – have been shown to achieve high enantioselectivities across this broad range of itaconate substrates (Fig. 24.14). The (*E*/*Z*)-mixtures typically formed in Stobbe condensations can be tolerated by phospholane-based catalysts without loss of performance [21, 72]. Performing itaconate hydrogenations at temperatures of around 0 °C has been found to be

Table 24.4 Phospholanes reported to hydrogenate model itaconic acid substrates in >95% ee.

RO_2C–C(=CH_2)–CH_2–CO_2R —(Rh-catalyst, H_2)→ RO_2C–C*H(CH_3)–CH_2–CO_2R

99 R = Me
100 R = H

Substrate	Ligand	SCR	Reaction conditions [a]	TON	TOF $[h^{-1}]$	% ee (config.)	Reference(s)
99	(*S,S*)-Et-DuPhos	1000	MeOH, 20 °C, 5.4 atm, 2.8 h	1000	– [b]	>97 (*R*)	21
99	(*R,R*)-Et-BPE	1000	MeOH, 20 °C, 5.4 atm, 1 h	1000	1000	97 (*R*)	34
99	(*R,R*)-Ph-BPE	10000	MeOH, 28 °C, 9.9 atm, 10 min	10000	60000	99 (*R*)	34
99	(*R,R*)-Me-Ph-BASPHOS	100	MeOH, 28 °C, 1 atm, 3 h	100	33.3	97 (*R*)	56b
99	(*R,R*)-UlluPHOS	1000	MeOH, 27 °C, 2 atm, 2.8 h	1000	–	>99 (*S*)	44
99	(*R,R*)-Me-DuPhos	1000	MeOH, 27 °C, 2 atm, 2.8 h	1000	–	>99 (*S*)	44
99	(*S,S,S,S*)-*t*-Bu-**23**	100	MeOH, rt, 1 atm, 8 min	50 [c]	375	99 (*R*)	51
99	(*S,S,S,S*)-Bn-**22**	100	MeOH, rt, 1 atm, 28 min	50 [c]	107	98 (*R*)	51
99	catASium M	– [a]	CH_2Cl_2, 25 °C, 7.9 atm, 15 min	10000	40000	99	47
99	(*S,S,R,R*)-TangPhos	5000	THF, rt, 1.4 atm	5000	– [b]	99 (*S*)	72
99	(*R,R,S,S*)-DuanPhos	100	THF, rt, 1.4 atm	–	– [b]	99 (*S*)	70
99	(*Rp,Rc,R,R*)-**54**	100	MeOH, 25 °C, 1 atm, 1 h	100	100	>99 (*S*)	80
99	(*R,R*)-Et-FerroTANE	200	MeOH, 25 °C, 5.4 atm, 1 h	200	>200	98 (*S*)	82, 83
99	(*S,S*)-*n*-Pr-FerroTANE	200	MeOH, 20 °C, 5.4 atm, 1 h	200	>200	97 (*R*)	82, 83
100	(*S,S*)-Et-DuPhos	100	MeOH, rt, 17.8 atm, 2 h	95 [d]	47.5	96 (*R*)	165
100	(*R,R*)-Me-Ph-BASPHOS	100	MeOH, 25 °C, 1 atm, 20 min	100	300	97 (*R*)	56b
100	(*R,R*)-*cis*-PMP5	1000	MeOH, rt, 1.5 atm, 2 h	998 [c]	499	97 (*R*)	76b
100	(*S,S,S,S*)-Me-**25**	192	MeOH, rt, 1.3 atm, 20 h	192	–	>99 (*R*)	55
100	(*S,S,S,S*)-Bn-**23**	100	MeOH, rt, 1 atm, 24 min	50	125	98 (*R*)	51
100	(*S,S,S,S*)- Bn-**22**	100	MeOH, rt, 1 atm, 10 min	50	300	98 (*R*)	51
100	(*S,S,S,S*)-**31**	100	MeOH, rt, 5.4 atm, 12 h	100	8.3	>99 (*R*)	54

Table 24.4 (continued)

Substrate	Ligand	SCR	Reaction conditions [a]	TON	TOF $[h^{-1}]$	% ee (config.)	Reference(s)
100	catASium M	– [a]	MeOH, 25 °C, 3.9 atm, 3 h	500	– [b]	97	47
100	(*S,S,R,R*)-TangPhos	200	THF, rt, 1.4 atm	–	– [b]	99 (*S*)	72
100	(*R,R*)-*ent*-Cy-**66**	500	*i*-PrOH, 20 °C, 1 atm, 24 h	–	– [b]	96 (*S*)	89

a) No catalyst loading given.
b) Insufficient data on loading, time or yield to calculate TOF.
c) Reaction not gone to completion.

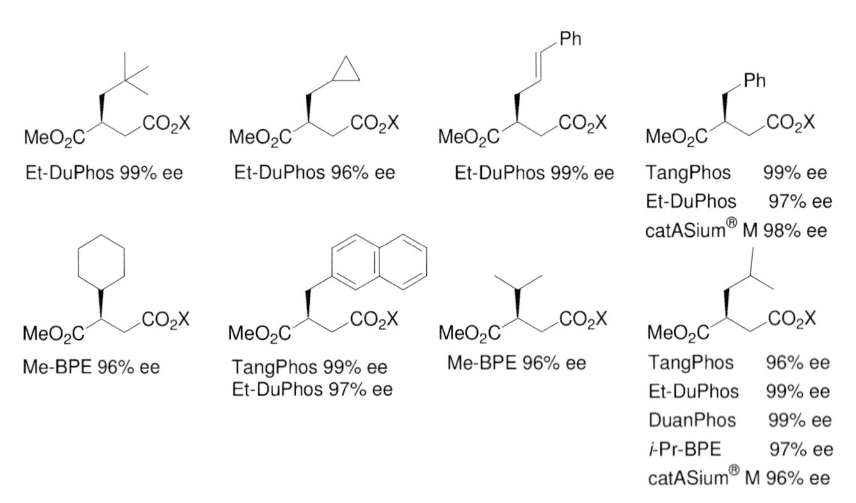

Fig. 24.14 Succinic acid derivatives produced by phospholane–Rh-catalyzed hydrogenation (X=H, Na or R_3NH).

beneficial in terms of enantioselectivity. The nature of the secondary binding group can play a critical role in terms of reaction rates and enantioselectivity with this substrate class; the hydrogenation of 2-isopropylidenesuccinic acid 1-methyl ester with (*R,R*)-Me-BPE at SCR 300:1 gave only 33% conversion and 88% ee, whereas the *tert*-butylamine salt of the acid gave complete conversion at SCR 500:1 with 95% ee under comparable conditions [167]. The use of a salt form is reported to enhance both reactivity and selectivity, enabling industrially viable catalyst loadings of SCR >4000:1 to be readily achieved [21]. Amine and alkali metal salts of itaconic acid are generally employed, though optimal results seem to be best achieved with preformed and purified amine salts [168]. The enhanced performance of the itaconate acid salts is possibly a result of both the enhanced binding ability of the carboxylate group and higher substrate purity.

β,β-Disubstituted itaconic acid substrates require higher catalyst loadings and hydrogen pressures to achieve reasonable reaction rates, which is unsurprising given the level of steric congestion around the olefinic bond. The most effective hydrogenation was attained when using the sterically less cumbersome Me-BPE ligand; indeed, when used in conjunction with substrates as the amine salt, enantioselectivities of 96% could be realized [21].

Inverse itaconate derivatives (4-itaconic acid derivatives) have been studied to a lesser extent than the 1-itaconic acid derivatives. In the limited number of reported hydrogenations of itaconic acid 4-esters, the parent DuPhos ligands have performed poorly with model substrates. For example, in the hydrogenation of itaconic acid 4-methyl ester in MeOH, both Me- and Et-DuPhos gave less than 3% yield and poor to moderate ee-values (41 to 74%) [82c]. However, catASium M is reported to achieve 99% ee and a TOF of 8000 h^{-1} in CH_2Cl_2 with itaconic acid 4-methyl ester [47]. The related Me- and Et-5-Fc ferrocenyl phospholanes hydrogenate itaconic acid 4-methyl ester with complete conversion at SCR 2000:1, though the enantioselectivities remain low at <45% [82c]. FerroTANE based catalysts reduce itaconic acid 4-methyl ester and 2-pentylidene-succinic acid 4-methyl ester in >94% ee at SCRs of between 1000:1 and 2000:1 [82c]. Moreover, a series of 2-alkylsuccinic acids 4-*tert*-butyl ester targets have been generated in high ee using Et-FerroTANE and Et-DuPhos. This approach has been used in synthesizing the MMP-3 inhibitor UK-370,106 101 [169] (Scheme 24.10).

Inverse amido-itaconates also proved to be challenging substrates for phospholane-based catalysts, and only limited success has been achieved to date. 2-Methylenesuccinamic acid (**102**) has been reported to be reduced by Et-DuPhos–Rh in 96% ee at SCR 100000:1, with an average TOF of 13000 h^{-1}. The removal of a trace chloride-containing contaminant was found to be crucial in obtaining high enantioselectivities and reaction rates [85]. An isoquinuclidine containing inverse amido-itaconate **103**, which is currently being evaluated in preparations for the treatment of diabetes, has been prepared using Rh-Et-DuPhos in 98.7% ee [170]. Whilst phospholane systems have achieved only moderate success in this substrate class, the FerroTANE family of ligands has been reported broadly to outperform all ligand classes, with superior enantioselectivities

t-BuO2C CO2H Ph → t-BuO2C CO2H Ph

(*S,S*)-Et-FerroTANE 96% ee
(*S,S*)-Et-DuPhos 96% ee

→ HO2C N O O N H Ph O Ph

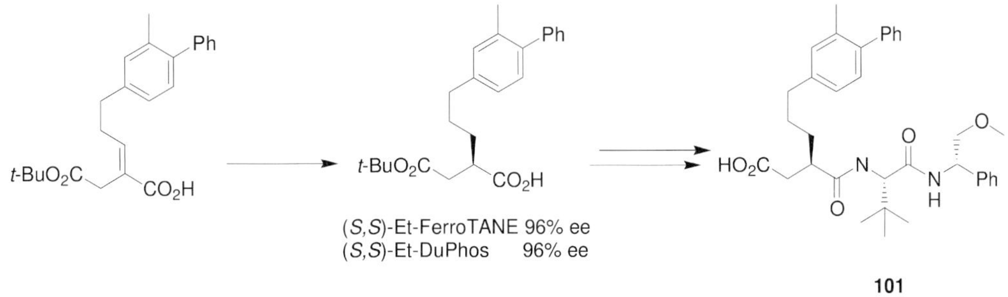

101
UK-370,106

Scheme 24.10 Inverse itaconate approach to a protease inhibitor.

and industrially viable catalyst activities. Burk tested an array of inverse amido-itaconates (e.g., **104**) where both the amide fragment and β-substitution on the olefin were varied, and found most enantioselectivities to be >95% with TOFs in the range of 1000 to 6000 h^{-1} when Et-FerroTANE was used as the chiral ligand [82 a, c].

Although the efficient enantioselective reduction of α,β-unsaturated acids and lactones is typically achieved using ruthenium-biaryldiphosphine-based catalysts [2], several reports have been made of phospholane-ruthenium, -rhodium and -iridium systems that perform this hydrogenation with comparably high enantioselectivities. Burk has reported that the ubiquitous model substrate for Ru-biarylphosphine catalysts, tiglic acid, can been reduced with up to 94% ee with a Ru-*i*-Pr-DuPhos catalyst [27, 171 a, 172 a]. However, the array of substrate structures reported for this class is diverse and bears little similarity to the model substrates (Fig. 24.15). Simple acrylic acid derivatives such as methyl α-hydroxymethacrylate and 2-[[(phenylmethoxy)amino]methyl]-2-hexenoic acid methyl ester **105** have been hydrogenated by both Me-DuPhos and TangPhos–Rh catalysts in 90% and 96–98% ee [75, 173]. (*E*)-β-methylcinnamates **106** have been reduced using Zhang's mixed phospholane-oxazoline catalyst Ir-**48**, giving results comparable to Pfaltz's established iridium phosphonite-oxazoline systems [74]. Trisubstituted aryl/heteroaryl-sulfonylated acrylic acid derivatives (e.g., **107**) have been hydrogenated in remarkably high enantioselectivities with Me-DuPhos–Rh [174]. A challenging α-(γ-amino)-β-imidazolyl-acrylic acid substrate **108** was reduced using *i*-Pr-5-Fc–Rh in the presence of quinidine to give the target product in high ee using a combined enantioselective hydrogen/classical resolution approach [175]. A series of diastereomeric scaffolds (e.g., **109**) were synthesized from functionalized chiral

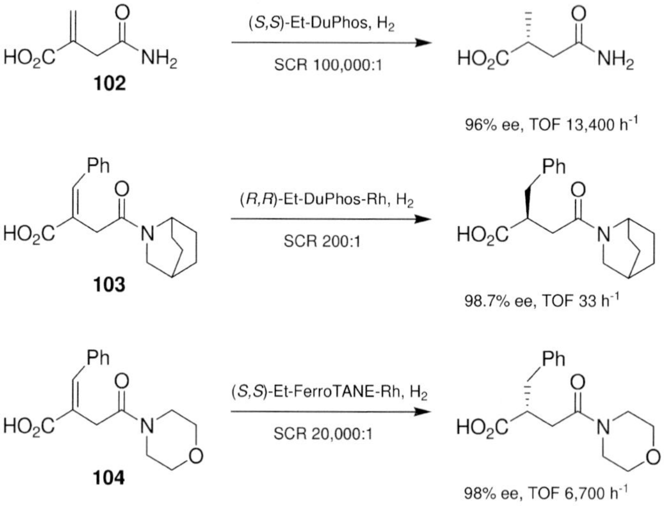

Scheme 24.11 Enantioselective reduction of inverse amido-itaconates.

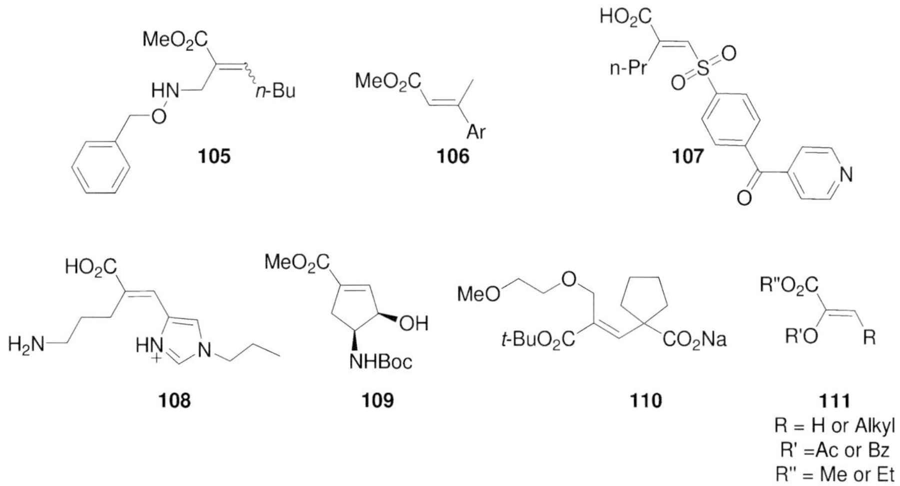

105 106 107 108 109 110 111
R = H or Alkyl
R' = Ac or Bz
R'' = Me or Et

Fig. 24.15 Diverse *α,β*-unsaturated acid derivatives reduced with phospholane catalysts.

cyclopentene rings bearing an acrylate substructure, while facial selectivity could be controlled via the judicious use of DuPhos or BPE–Rh catalysts or Crabtree's catalyst [176]. Sterically congested *α,β*-unsaturated lactone substrates (butenolides) bearing potentially inhibiting heterocycles have been hydrogenated with Me-DuPhos–Rh in 80% ee [177]. A key glutarate component **110** in the atrial natriuretic factor (ANF) potentiator Candoxatril has been synthesized in high ee using both a MeDuPhos–Rh and Ph-BPE–Rh catalyst [34, 178]. The use of a Rh-phospholane catalyst circumvented the problem of the generation of unreactive enol ethers as side products, which was a major issue when Ru-BINAP was used [178]. A PennPhos–Ru species reduced 3-(*p*-fluorobenzylidene) valerolactam in 70% ee, a target molecule for producing 3-alkylpiperidines as pharmacophores; however, BDPP-Ir was found to be a better system for reducing the *exo*-cyclic olefin [179].

A wide range of *α*-(acetyloxy)- and *α*-(benzoyloxy)acrylates **111**, with both alkyl and aryl *β*-substituents, have been successfully hydrogenated with cationic Rh-DuPhos complexes [180]. In particular, the reduction could be performed on an (*E*/*Z*)-isomeric mixture, with selectivities generally being greater than 97% ee. A brief solvent study showed MeOH, *i*-PrOH, or CH_2Cl_2 to be the best solvents in terms of both catalyst activity and selectivity, and benzene to inhibit the reaction by formation of a stable adduct. Increased hydrogen pressure had a negligible influence on selectivity. Ultimately, the hydrogenation products were converted to enantiomerically enriched *α*-hydroxy esters and 1,2-diols, without any loss in optical purity. An example of a (*Z*)-*α*-(phenoxy)-*β*-alkyl-acrylate has been reported to have been reduced in 86 and 89% ee using Et-FerroTANE–Ru and *i*-Pr-DuPhos–Ru catalysts; Me-f-KetalPhos–Ru, whilst catalytically active, produced an essentially racemic product, and a SynPhos–Ru catalyst proved to be the optimal catalyst screened [181].

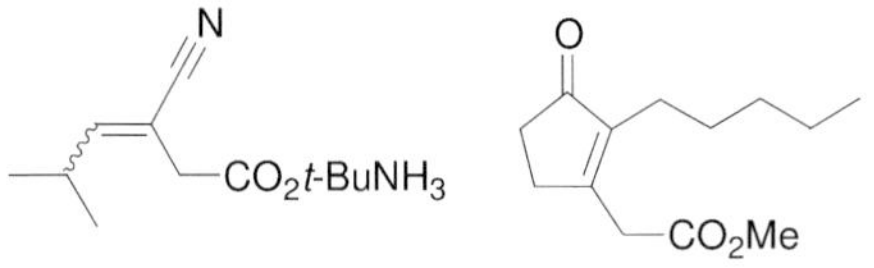

Fig. 24.16 β,γ-unsaturated acids reduced with MeDuPhos–Rh and Ru catalysts.

The hydrogenation of a β,γ-unsaturated acid **112** has been key to the enantioselective synthesis of a potent anticonvulsant (*S*)-(+)-3-aminomethyl-5 methylhexanoic acid, Pregabalin [60, 61] (see Fig. 24.16). (*E*/*Z*)-mixtures of three possible precursors were examined for reactivity and selectivity. The ester substrate was found to be not particularly useful in terms of both reactivity and selectivity towards a number of DuPhos, BPE and FerroTANE catalysts; remarkably, using (*R,R*)-Me-DuPhos at room temperature or at 55 °C resulted in a reversal of facial selectivity, albeit with modest enantioselectivities in each case. The *t*-$BuNH_3^+$ and K^+ salts of the acid were found to be superior in both reactivity and enantioselectivity, and whilst both salts performed comparably well, the hydrogenation was optimized using the *t*-$BuNH_3^+$ salt as a result of substrate quality concerns [60]. An optimized procedure using (*R,R*)-Me-DuPhos–Rh has been used at the kilogram scale to give the desired product in 97.7% ee at SCR 2700 : 1 and 4.4 atm. Hoge's C_1-symmetric phospholanes (**35** and **36**) are also selective towards this substrate: at low pressure (2 atm) and SCR 100 : 1, **35** was moderately more selective than **36**, giving 96% and 92% ee, respectively. Hoge demonstrated with **36** that high enantioselectivities at lower catalyst loadings could only be achieved with a concomitant increase in H_2 pressure; subsequently, 97% ee could be achieved at 13.5 atm [59 a, b, d]. Jasmonoid compounds have indicated numerous phtyobiological activities and olfactory properties, the *cis*-jasmonate compounds being of particular interest to the perfume industry (Fig. 24.16). Direct reduction of the olefinic double bond of dehydrojasmonate **113** via *syn*-addition of H_2 is the most direct route to the *cis*-isomers. However, traditional cationic Rh-phospholanes are not electrophilic enough to hydrogenate the tetrasubstituted double bond. Bergens developed a more electrophilic, coordinatively unsaturated 16-electron cationic ruthenium-hydride-phosphine system that was not only capable of hydrogenating the double bond but, when modified with Me-DuPhos and provided with enantiomeric excesses up to 60% and a >99 : 1 *cis*/*trans* ratio, furnished the desired enantiomerically enriched *cis*-isomer. However, a Josiphos variant operated with better enantioselectivity [182]. It is not clear whether the pendant ester group plays a secondary binding role during the hydrogenation.

24.3.5
Enantioselective Hydrogenation of Unsaturated Alcohol Derivatives

One method of accessing chiral alcohols is through the enantioselective hydrogenation of the corresponding enol acetates. Despite this substrate class having similar structures to enamides, far fewer successful examples of this reaction have been reported. It has been argued that this may be in part due to the enol acetate having a weaker binding acyl group than the analogous enamide [4c]. Notwithstanding this, Burk was successful in hydrogenating a range of simple α-substituted enol acetates **114** in good to excellent enantioselectivities (89–99% ee) with either Rh-DuPhos or Rh-BPE catalysts [12]. High enantioselectivities have also been obtained when applying Rh-Me-DuPhos for the reduction of 1-alkenyl or 1-alkynyl enol acetates, **115** and **116** [183]. In the case of **116**, the triple bond is reduced to a double bond with (*Z*)-configuration after reduction of the enol acetate. Interestingly, the judicious choice of **115** or **116** as substrate allows access to either (*E*) or (*Z*)-α,β-unsaturated acetates. The analogous saturated straight-chain derivatives were found to hydrogenate smoothly, but with only moderate selectivity (64–77% ee). [Me-DuPhos Rh(COD)]OTf has also been reported to catalyze the hydrogenation of 2-acetyloxy-1,1,1-trifluorododec-2-ene in 92% ee, but higher selectivities were obtained with Ru-based catalysts of atropisomeric biaryl ligands, BINAP and BIPHEMP.

Burk et al. also showed the Rh-complexes of Me and Et-DuPhos to be effective catalysts for the enantioselective reduction of several phosphonated enol acetates of type **117** (86–96% ee with TOF $\sim 10\ h^{-1}$ at 25 °C, 4 atm), providing an efficient route to enantiomerically enriched alkyl-substituted α-hydroxy phosphonites [161]. Remarkably, these catalysts were inactive against an aryl-substituted analogue, and although partial conversion could be obtained with Me-BPE, both the selectivity and reactivity were much lower (70% ee, TOF $1.5\ h^{-1}$). Recently, the research group of Zhang has reported several phospholane-based ligands to be effective for the Rh-catalyzed reduction of acyclic, α-aryl substituted enol esters of type **118** (PennPhos [67c,d], TangPhos [69b, 72], DuanPhos [70] and **48** (R = Me) [52b]). In general, high to excellent enantioselectivities are achieved (81–99% ee) under mild reaction conditions, albeit at high catalyst loadings (SCR 100). The *in-situ*-prepared Rh-PennPhos catalyst [67c,d] has also been applied to the reduction of five- and six-membered cyclic enol acetates, **119** and **120**, derived from substituted 1-indanone and 1-tetralone, respectively.

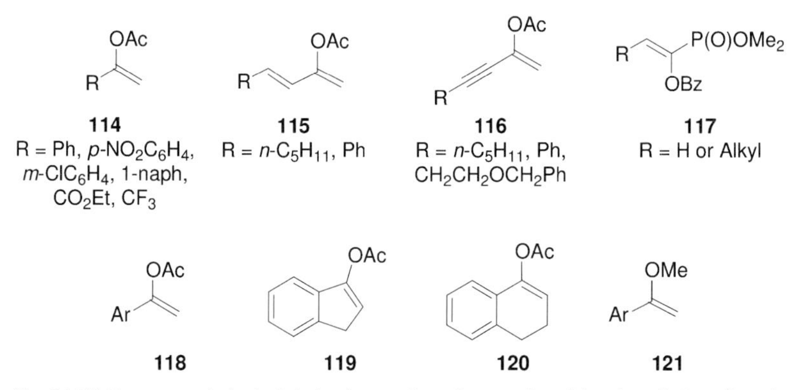

Fig. 24.17 Unsaturated alcohol derivatives to have been reduced by phospholane ligands.

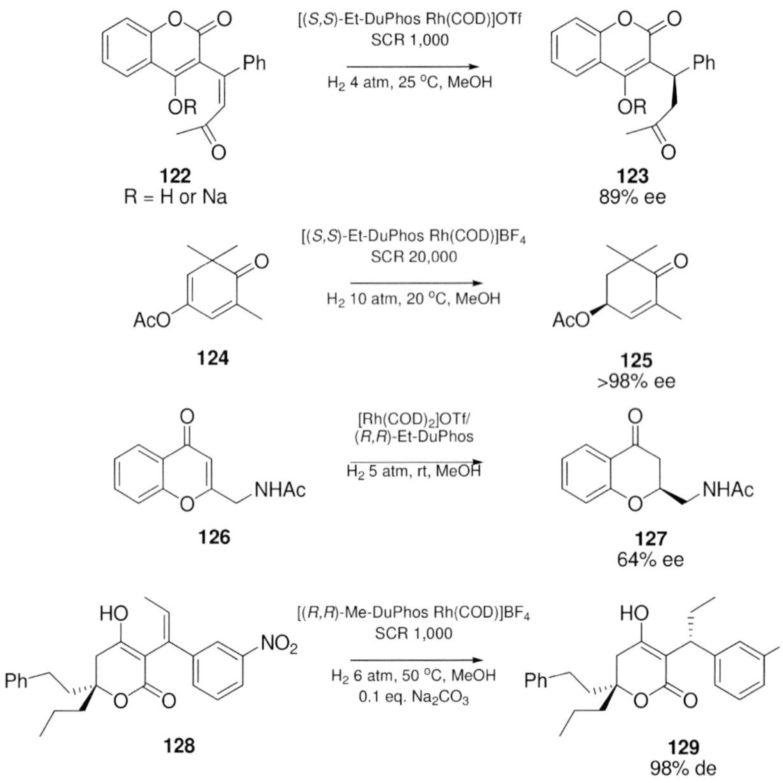

Scheme 24.12 Several pharmaceutically active compounds to have been synthesized via Rh-phospholane-mediated asymmetric hydrogenation.

Catalyst performance was far superior to the corresponding BINAP or Me-DuPhos systems, with both conversions and selectivities being higher. The hydrogenation of enol ethers using Rh-PennPhos catalysts has been reported in a patent by Zhang [67d]. Under mild conditions, high enantioselectivities were obtained (73–94% ee) for 1-aryl-1-methoxy-ethene derivatives **121**, compared to Me-DuPhos (40–73% ee) and BINAP (46–48% ee).

The enantioselective reduction of unsaturated alcohol derivatives has been applied to the synthesis of several biologically active compounds (Scheme 24.12). Warfarin (**123**, R=H) is an important anticoagulant that is normally prescribed as the racemate, despite the enantiomers having dissimilar pharmacological profiles. One of the earliest reported uses of DuPhos was in the development of a chiral switch for this bioactive molecule, facilitating the preparation of (*R*)- and (*S*)-warfarin [184]. Although attempted reduction of the parent hydroxycoumarin **122** (R=H) led to formation of an unreactive cyclic hemiketal, hydrogenation of the sodium salt proceeded smoothly with Rh-Et-DuPhos in 86–89% ee.

Hoffman la Roche have reported the synthesis of an intermediate to zeaxanthin via the enantioselective reduction of cyclic enolacetate **124** [185]. Using a Rh-Et-DuPhos catalyst, excellent levels of selectivity (98% ee) could be obtained at extremely low catalyst loadings (TON 20000; TOF 5000 h^{-1}). Chroman derivatives, such as **127**, have been reported by Merck to affect the central nervous system [186]. As part of a wide ligand screen, an *in-situ*-prepared Rh-Et-DuPhos complex has been reported to be amongst the most selective for the preparation of **127**, albeit in 64% ee. Finally, an enantioselective approach to **129**, a key intermediate to the HIV protease inhibitor tipranavir (PNU-140710), was developed by Chirotech for Pharmacia & Upjohn [187]. The use of [(*R*,*R*)-Me-DuPhos–Rh(COD)]BF_4 and Na_2CO_3 as a co-catalyst gave quantitative conversion in 93% de (SCR 1000, 6 atm, 50–60 °C). Once again, (*E*/*Z*)-mixtures could be tolerated together with high chemoselectivity with regards to over-reduction of the nitro functional group. Both UlluPHOS **18** [43, 44] and catASium M **20a** (A=O, R=Me) [188] have also been applied to this reaction.

24.3.6 Enantioselective Hydrogenation of Miscellaneous C=C Bonds

Through continued exploration of the applicability of enantiomeric hydrogenation, phospholane-based catalysts have been reported to be efficient for the reduction of several atypical olefinic substrates (Fig. 24.18).

Due to the lack of an ordered chelate complex provided by the substrate containing a secondary binding site, the enantioselective reduction of unfunctionalized olefins remains a challenging area where only limited success has been achieved. Noyori et al. reported Me-DuPhos–Ru catalysts to be effective for the hydrogenation of α-ethylstyrenes **130** [189]. By activating the precatalyst with an alkoxide base in 2-propanol, respectable enantioselectivities (71–89% ee) and activity (TON up to 2600; TOF 160 h^{-1}) could be obtained under mild reaction conditions. Using the cationic iridium complexes of a class of phospholane-oxazoline ligands, **48**, Zhang and co-workers successfully reduced methylstilbene derivatives **131** in 75–91% ee [74]. These selectivities are comparable to those obtained with the best ligand systems for this substrate class [4b,c, 190]. Although outperformed by biaryl-based ligands such as BINAP, Me-DuPhos has

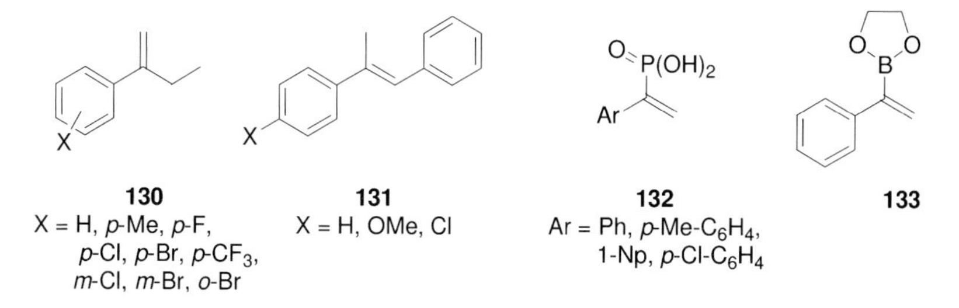

Fig. 24.18 Unusual olefins to have been reduced with phospholane-based catalysts.

also shown to be active in the Ru-catalyzed reduction of *α*-aryl-substituted ethylphosphonates **132** (16–37% ee) [191] and ethanediol 1-phenylethenylboronic ester **133** (42% ee) [192].

24.4
Enantioselective Hydrogenation of C=O and C=N Bonds

24.4.1
Enantioselective Hydrogenation of Ketones

The enantioselective hydrogenation of ketones using Rh or Ru diphosphine catalysts is the most efficient method for the synthesis of chiral alcohols. Although in general, atropisomeric biaryl-based chiral ligands have proven to be the most versatile for this substrate class [2, 4b, c], significant success has been achieved with phospholane-containing systems (Fig. 24.19). As early as 1991, Burk et al. reported the use of Rh-phospholane **7** catalysts for the reduction of methyl acetoacetate [11], albeit with poor enantioselectivity (20–27% ee). Switching to a Ru-based precatalyst greatly enhanced both the activities and selectivities obtained [172], with [*i*-Pr-BPE $RuBr_2$] being found to be broadly effective for a range of *β*-keto esters **134** (76–99% ee, TON 500, TOF $>15\ h^{-1}$) at low hydrogen pressures [172a]. Interestingly, amongst other applications, this was used for the preparation of enantiomerically pure 1,4-dicyclohexyl-1,4-butanediol and, ultimately, the synthesis of a new phospholane ligand, Cy-BPE; a rare case of "ligand self-generation". In general, the DuPhos class of phospholanes (and analogues [43, 44]) shows much lower activities for the reduction of *β*-keto esters than the more basic BPE ligands, with higher pressures, temperatures and reaction times being required [193]. Despite these harsher conditions, excellent enantioselectivities (96–99% ee) were obtained for the reduction of **134** (R = Me or $MeO_2C(CH_2)_2$, R′ = Me or Et) with the tetramethoxy ligand, BASPHOS **32** (R = Me) [56b].

The groups of Marinetti and Genêt have shown that several bisphosphetane-derived ligands (**58**, **59**, **61**) form effective Ru-based catalysts for the hydrogena-

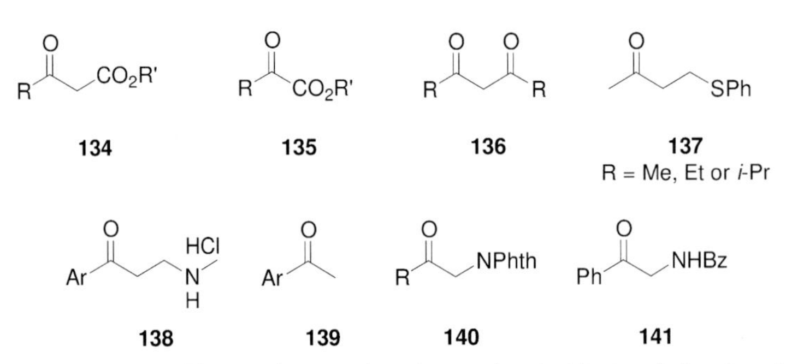

Fig. 24.19 General ketone classes to have been reduced with phospholane-based catalysts.

tion of *β*-keto esters, but once again, higher temperatures and pressures were generally required for reasonable activity [83 b–e]. Overall, moderate to high enantioselectivities were observed, with the bulkier 2,4-disubstituted phosphetanes (*i*-Pr > Et > Me) being more selective [83 d, e]. The bisphosphetanes CnrPHOS **58** (R = *i*-Pr or Cy) have also been applied to the Ru-catalyzed enantiomeric reduction of *α*-keto ester **135** (R = Ph, R′ = Et), *β*-diketones **136**, and *β*-thioketone **137**. In the case of substrates **135** and **136**, excellent selectivities were observed (98% and 97% ee, respectively), despite the high temperatures employed, with **136** also being obtained with high diastereoselectivity (94%). Cy-BPE-4 **59** has also been shown to be extremely selective for the Ru-catalyzed reduction of *β*-diketone **136** (R = Me), giving 98% ee and 95% de [83 b].

One application of phospholane-containing ligands for the enantiomeric reduction of a ketone recently appeared in a patent application from Lonza [194]. In the presence of NaOMe, both Me-DuPhos and Me-KetalPhos (**24**) gave reasonable selectivities (71.3% and 79.8% ee, respectively) for the Rh-catalyzed reduction of a *β*-amino ketone **138** (Ar = 2-thienyl), an intermediate for the pharmaceutically active drug duloxetine. Zhang and co-workers have also demonstrated the use of Rh-DuanPhos (**47**) for the reduction of a range of *β*-secondary-amino ketone hydrochlorides, including precursors to the drugs (*S*)-fluoxetine and (*S*)-duloxetine [73]. In general, remarkably high enantioselectivities were obtained (93–99% ee) with high TONs (> 4500) and TOF (375 h^{-1}) when using the secondary amino group, whilst the corresponding tertiary amine was unreactive.

The Rh-catalyzed hydrogenation of methyl pyruvate, **135** (R,R′ = Me), was studied by Burk et al. with bisphospholanes containing a chiral backbone, **6** and **7** (Fig. 24.20). Significant matching and mismatching effects were observed (43% versus 75% ee), with the matched system being ligand **7** [29]. Genêt also studied the reduction of *α*-keto esters, showing Me-DuPhos to form an effective Ru-catalyst when tested against **135** (R = Ph, R′ = Me) (80% ee) [172 b].

As for unfunctionalized olefinic substrates, the enantioselective hydrogenation of unfunctionalized ketones is considerably more challenging due to the absence of a secondary chelating moiety. Undoubtedly, until now the catalyst of choice for this substrate class is the *trans*-[$RuCl_2$(diphosphine)(diamine)] catalyst developed by Noyori [195], with numerous examples being reported with high enantioselectivities (> 95% ee) and activities (TON up to 2 400 000; TOF 259 000 h^{-1}) [196]. Although far less active (TOF < 50 h^{-1}), Zhang has reported significant selectivities with a Rh-PennPhos (**44**) system [57, 67 b]. When com-

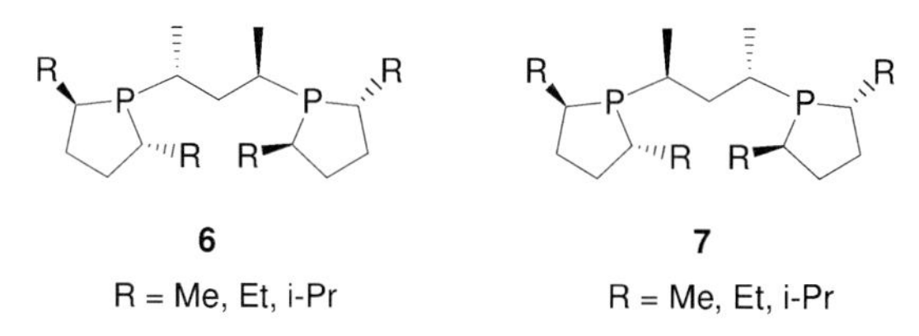

Fig. 24.20 Matching and mismatching bisphospholanes.

bined with either 2,6-lutidine or KBr additives, moderate to high enantioselectivities (55–99% ee) are achieved for a range of simple ketones, including dialkyl-substituted ketones (up to 92% ee), a class against which the Noyori system is much less effective. A ruthenium complex of the P-chiral diphosphine BIPNOR (**53**) has also been reported to reduce unfunctionalized ketones of class **139** with moderate enantioselectivity (57–81% ee) [78b, d].

Finally, the group of Zhou has recently published the first Pd-catalyzed enantiomeric reduction of ketones using Me-DuPhos [197]. By performing the reaction in TFE, a series of α-phthalimido ketones **140** were reduced in high yield and 75–92% ee, albeit at high catalyst loadings (SCR 50), reaction times (12 h) and pressures (13.7 atm). This procedure was extended to include ketones **134** (R = Ph, R′ = Et), **139** (Ar = Ph), and **141**.

24.4.2
Enantioselective Hydrogenation of Imines and C=N–X Bonds

Enantiomerically pure amines are extremely important building blocks for biologically active molecules, and whilst numerous methods are available for their preparation, the catalytic enantioselective hydrogenation of a C=N bond potentially offers a cheap and industrially viable process. The multi-ton synthesis of (*S*)-metolachlor fully demonstrates this [108]. Although phospholane-based ligands have not proven to be the ligands of choice for this substrate class, several examples of their effective use have been reported.

Indeed, the imine intermediate **142** in the synthesis of metolachlor has been reduced in 97% ee using an iridium complex of the phospholane-containing ligand **55** [80].

A *trans*-[$RuCl_2$(diphosphine)(1,2-diamine)] complex with (*R,R*)-Et-DuPhos and (*R,R*)-1,2-diaminocyclohexane as the ligand combination has been found to be effective for the hydrogenation of imine **143**, with up to 94% ee being obtained under the standard basic conditions employed for this catalytic system [198]. Unfortunately, the optimum combination of chiral diphosphine and diamine was found to be substrate-dependent, with only 40% ee being obtained for 2-methylquinoxaline **144** with Et-DuPhos.

Corma et al. have recently demonstrated the hydrogenation of **145** using a binuclear gold complex of (*R,R*)-Me-DuPhos [166]. Reasonable rates were observed (TOF 1005 h^{-1}), with the enantioselectivity being higher (75% ee) than that obtained with Pt- and Ir-based catalysts (15% ee in each case).

Burk et al. showed the enantioselective hydrogenation of a broad range of *N*-acylhydrazones **146** to occur readily with [Et-DuPhos Rh(COD)]OTf [14]. The reaction was found to be extremely chemoselective, with little or no reduction of alkenes, alkynes, ketones, aldehydes, esters, nitriles, imines, carbon-halogen, or nitro groups occurring. Excellent enantioselectivities were achieved (88–97% ee) at reasonable rates (TOF up to 500 h^{-1}) under very mild conditions (4 bar H_2, 20 °C). The products from these reactions could be easily converted into chiral amines or α-amino acids by cleavage of the N–N bond with samarium diiodide.

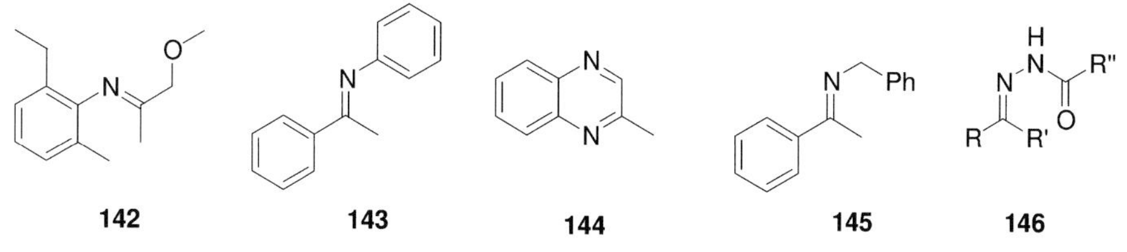

Fig. 24.21 C=N bonds to have been reduced with phospholane-based catalysts.

24.5 Concluding Remarks

Since the breakthrough introduction of the DuPhos and BPE family of ligands by Burk, the use of phospholanes in asymmetric hydrogenation has witnessed an explosion of interest, with many new and imaginative analogues emerging. This intense activity has extended the applicability of this important class of ligands beyond the standard substrates and towards the synthesis of a diverse range of chiral intermediates. This in turn has led to the realization of their commercial potential with multi-kilogram catalyst sales and applications on the industrial scale. We are only now beginning to witness the true synthetic utility of this technology and, with its increased adoption, it is anticipated that many more large-scale applications will be reported in the future.

Abbreviations

ANF	atrial natriuretic factor
BPE	1,2-bis(trans-2,5-dialkylphospholano)ethane
COD	1,5-cyclooctadiene
de	diastereomeric excess
DMI	dimethylitaconate
DuPhos	1,2-bis(trans-2,5-dialkylphospholano)benzene
ee	enantiomeric excess
LC	liquid chromatography
MAA	methyl 2-acetamidoacrylate
MAC	(*Z*)-methyl-2-acetamidocinnamate
MPLC	medium-pressure liquid (flash) chromatography
NBD	norbornadiene
SCR	substrate:catalyst ratio
TFE	2,2,2-trifluoroethanol
THF	tetrahydrofuran
TOF	turnover frequency
TON	turnover number

References

1 (a) R. A. Sheldon, *Chiral Technologies*, Marcel Dekker Inc., New York, **1993**; (b) A. N. Collins, G. N. Sheldrake, J. Crosby, *Chirality in Industry*, John Wiley & Sons, New York, **1992**.

2 (a) I. Ojima, *Catalytic Asymmetric Synthesis*, VCH, New York, **2000**; (b) R. Noyori, *Asymmetric Catalysis in Organic Synthesis*, Wiley, New York, **1994**; (c) E. N. Jacobsen, A. Pfaltz, H. Yamamoto, *Comprehensive Asymmetric Catalysis*, Springer, Heidelberg, Volumes 1–3, **1999**; (d) H.-U. Blaser, F. Spindler, M. Studer, *App. Catal. A: General*, **2001**, *221*, 119.

3 For a historical account, see H. B. Kagan, in: *Comprehensive Asymmetric Catalysis*, Springer, Heidelberg, vol. 1, **1999**, chapter 2, p. 14.

4 (a) M. J. Burk, *Acc. Chem. Res.*, **2000**, *33*, 363–372; (b) H.-U. Blaser, C. Malan, B. Pugin, F. Spindler, H. Steiner, M. Studer, *Adv. Synth. Catal.*, **2003**, *345*, 103; (c) W. Tang, X. Zhang, *Chem. Rev.*, **2003**, *103*, 3029.

5 T. P. Clark, C. R. Landis, *Tetrahedron: Asymm.*, **2004**, *15*, 2123.

6 H. Brunner, R. Sievi, *J. Organomet. Chem.*, **1987**, *328*, 71.

7 H. Brunner, S. Limmer, *J. Organomet. Chem.*, **1991**, *413*, 55.

8 M. J. Burk, J. E. Feaster, R. L. Harlow, *Organometallics*, **1990**, *9*, 2653.

9 M. J. Burk (E. I. du Pont de Nemours and Company), US 5008457, **1991**.

10 M. J. Burk, R. L. Harlow, *Angew, Chem. Int. Ed. Engl.*, **1990**, *29*, 1462.

11 M. J. Burk, J. E. Feaster, R. L. Harlow, *Tetrahedron: Asymm.*, **1991**, *2*, 569.

12 M. J. Burk, *J. Am. Chem. Soc.*, **1991**, *113*, 8518.

13 M. J. Burk, J. E. Feaster, W. A. Nugent, R. L. Harlow, *J. Am. Chem. Soc.*, **1993**, *115*, 10125.

14 M. J. Burk, M. F. Gross, J. P. Martinez, *J. Am. Chem. Soc.*, **1995**, *117*, 9375.

15 (a) M. J. Burk, J. E. Feaster, *J. Am. Chem. Soc.*, **1992**, *114*, 6266; (b) M. J. Burk (E. I. du Pont de Nemours and Company), US 5250731, **1993**; (c) M. J. Burk, J. P. Martinez, J. E. Feaster, N. Cosford, *Tetrahedron*, **1994**, *50*, 4399.

16 M. J. Burk, *Chemtracts-Org. Chem.*, **1998**, *11*, 787.

17 M. J. Burk, Y. M. Wang, J. R. Lee, *J. Am. Chem. Soc.*, **1996**, *118*, 5142.

18 M. J. Burk, N. B. Johnson, J. R. Lee, *Tetrahedron Lett.*, **1999**, *40*, 6685.

19 (a) M. J. Burk, G. Casy, N. B. Johnson, *J. Org. Chem.*, **1998**, *63*, 6804; (b) N. B. Johnson, M. J. Burk, G. Casy (Chirotech Technology Ltd.), WO 99/18065, **1999**.

20 M. J. Burk, M. F. Gross, T. G. P. Harper, C. S. Kalberg, J. R. Lee, J. P. Martinez, *Pure & Appl. Chem.*, **1996**, *68*, 37.

21 M. J. Burk, F. Bienewald, M. Harris, A. Zanotti-Gerosa, *Angew. Chem. Int. Ed.*, **1998**, *37*, 1931.

22 C. J. Cobley, N. B. Johnson, I. C. Lennon, R. McCague, J. A. Ramsden, A. Zanotti-Gerosa, in: H.-U. Blaser, E. Schmidt (Eds.), *Asymmetric Catalysis on an Industrial Scale: Challenges, Approaches and Solutions*. Wiley-VCH, Weinheim, **2004**.

23 G. Casy, N. B. Johnson, I. C. Lennon, *Chim. Oggi, Chem. Today*, **2003**, *21*, 63.

24 E. Fernandez, A. Gillon, K. Heslop, E. Horwood, D.J. Hyett, A.G. Orpen, P.G. Pringle, *Chem. Commun.*, **2000**, 1663.

25 M. J. Burk (E. I. du Pont de Nemours and Company), US 5021131, **1991**.

26 For a historical account, see M.J. Burk, *ChemTracs-Org. Chem. I*, **1998**, *11*, 787.

27 M. J. Burk (E. I. du Pont de Nemours and Company), US 5171892, **1992**.

28 U. Berens (Chirotech Technology Ltd), US 6545183 B1, **2003**.

29 M. J. Burk, A. Pizzano, J. A. Martin, L. M. Liable-Sands, A.L. Rheingold, *Organometallics*, **2000**, *19*, 250.

30 M. J. Burk, M. F. Gross, *Tetrahedron Lett.*, **1994**, *35*, 9363.

31 F. Guillen, J.-C. Fiaud, *Tetrahedron Lett.*, **1999**, *40*, 2939.

32 J.-C. Fiaud, J.-Y. Legros, *Tetrahedron Lett.*, **1991**, *32*, 5089.

33 F. Guillen, M. Rivard, M. Toffano, J.-Y. Legros, J.-C. Darran, J.-C. Fiaud, *Tetrahedron*, **2002**, *58*, 5895.

34 C. J. Pilkington, A. Zanotti-Gerosa, *Org. Lett.*, **2003**, *5*, 1273.

35 P. Harrison, G. Meek, *Tetrahedron Lett.*, **2004**, *45*, 9277.

36 A. Börner, D. Heller, *Tetrahedron Lett.*, **2001**, *42*, 223.

37 (a) C.J. Cobley, I.C. Lennon, R. McCague, J.A. Ramsden, A. Zanotti-Gerosa, *Tetrahedon Lett.*, **2001**, *42*, 7481; (b) C.J. Cobley, I.C. Lennon, R. McCague, J.A. Ramsden, A. Zanotti-Gerosa, in: *Catalysis of Organic Reactions*. Marcel Dekker, New York, chapter 24, **2002**, p. 329.

38 Y. Sun, R.N. Landau, J. Wang, C. LeBlond, D.G. Blackmond. *J. Am. Chem. Soc.*, **1996**, *118*, 1348.

39 M.J. Burk, S. Feng, M.F. Gross, W. Tumas, *J. Am. Chem. Soc.*, **1995**, *117*, 8277.

40 H.-U. Blaser, B. Pugin, M. Studer, in: D.E. De Vos, I.F.J. Vankelecom, P.A. Jacobs (Eds.), *Chiral Catalyst Immobilization and Recycling*. Wiley-VCH, Weinheim, **2000**, p. 1.

41 For examples involving DuPhos-based catalysts, see (a) A. Crosman, W.F. Hoelderich, *J. Catal.*, **2005**, *232*, 43; (b) W.P. Hems, P. McMorn, S. Riddel, S. Watson, F.E. Hancock, G.J. Hutchings, *Org. Bio. Chem.* **2005**, *3*, 1547; (c) C. Simons, U. Hanefeld, I. Arends, R.A. Sheldon, *Chem. Eur. J.*, **2004**, *10*, 5829; (d) R.L. Augustine, P. Goel, N. Mahata, C. Reyes, S.K. Tanielyan, *J. Mol. Catal. A: Chem.* **2004**, *216*, 189; (e) R.L. Augustine, S.K. Tanielyan, N. Mahata, Y. Gao, A. Zsigmond, H. Yang, *App. Catal. A: General*, **2003**, *256*, 69; (f) J.A.M. Brandts, P.H. Berben, *Org. Proc. Res. Dev.*, **2003**, *7*, 393; (g) H.H. Wagner, H. Hausmann, W.F. Hoelderich, *J. Cat.*, **2001**, *203*, 150; (h) F.M. de Rege, D.K. Morita, K.C. Ott, W. Tumas, R.D. Broene, *Chem. Commun.*, **2000**, 1797; (i) I.F.J. Vankelecom, A. Wolfson, S. Geresh, M. Landau, M. Gottlieb, H.M. Moshe, *Chem. Commun.* **1999**, 2407; (j) R. Augustine, S. Tanielyan, S. Anderson, *Chem. Commun.*, **1999**, 1257; (k) K. De Smet, S. Aerts, E. Ceulemans, I.F.J. Vankelecom, P.A. Jacobs, *Chem. Commun.*, **2001**, 597.

42 U. Berens (Solvias AG), WO 03/031456, **2003**.

43 F. Sannicol, O. Piccolo, T. Benincori, M. Sada, A. Verrazzani, S. Tollis, E. Ullucci, L. de Ferra, S. Rizzo (Chemi S.P.A.), WO 03/074169 A2, **2003**.

44 T. Benincori, T. Pilati, S. Rizzo, F. Sannicol, M.J. Burk, L. de Ferra, E. Ullucci, O. Piccolo, *J. Org. Chem.*, **2005**, *70*, 5436.

45 M. Lotz, M. Kesselgruber, M. Thommen, B. Pugin (Solvias AG), WO 2005/056568, **2005**.

46 J. Holz, A. Monsees, H. Jiao, J. You, I.V. Komarov, C. Fischer, K. Drauz, A. Börner, *J. Org. Chem.*, **2003**, *68*, 1701.

47 J. Almena, A. Monsees, R. Kadyrov, T.H. Riermeier, B. Gotov, J. Holz, A. Börner, *Adv. Synth. Catal.*, **2004**, *346*, 1263.

48 A. Börner, J. Holz, A. Monsees, T. Riermeier, R. Kadyrov, C.A. Schneider, U. Dingerdissen, K. Drauz (Degussa AG), WO 03/084971 A1, **2003**.

49 T. Riermeier, A. Monsees, J.J. Almena Perea, R. Kadyrov, B. Gotov, W. Zeiss, N. Iris, A. Börner, J. Holz, K. Drauz (Degussa AG), WO 2005/049629 A1, **2005**.

50 H. Shimizu, T. Ishizaki, T. Fujiwara, T. Saito, *Tetrahedron Asymm.*, **2004**, *15*, 2169.

51 (a) J. Holz, M. Quirmbach, U. Schmidt, D. Heller, R. Stürmer, A. Börner, *J. Org. Chem.*, **1998**, *63*, 8031; (b) R. Stürmer, A. Börner, J. Holz (BASF AG), DE 19725796, **1998**.

52 (a) W. Li, Z. Zhang, D. Xiao, X. Zhang, *Tetrahedron Lett.*, **1999**, *40*, 6701; (b) W. Li, Z. Zhang, D. Xiao, X. Zhang, *J. Org. Chem.*, **2000**, *65*, 3489; (c) X. Zhang (The Penn State Research Foundation), WO 00/11008, **2000**.

53 (a) Y.-Y. Yan, T.V. RajanBabu, *J. Org. Chem.*, **2000**, *65*, 900; (b) T.V. RajanBabu, Y.-Y. Yan, S. Shin, *J. Am. Chem. Soc.*, **2001**, *123*, 10207.

54 (a) D. Liu, W. Li, X. Zhang, *Org. Lett.*, **2002**, *4*, 4471–4474; (b) X. Zhang (The Penn State Research Foundation), US 2003/0040629, **2003**.

55 A. Bayer, P. Murszat, U. Thewalt, B. Rieger, *Eur. J. Inorg. Chem.*, **2002**, 2614.

56 (a) J. Holz, D. Heller, R. Stürmer, A. Börner , *Tetrahedron Lett.*, **1999**, *40*, 7059; (b) J. Holz, R. Stürmer, U. Schmidt, H.-J. Drexler, D. Heller, H.-P. Krimmer, A. Börner, *Eur. J. Org. Chem.*, **2001**, 4615; (c) D. Heller, H.-J. Drexler, J. You, W. Baumann, K. Drauz, H.-P. Krimmer, A. Börner, *Chem. Eur. J.*, **2002**, *8*, 5196; (d) D. Heller, J. Holz, I. Komarov, H.-J. Drexler, J. You, K. Drauz, A. Börner,

Tetrahedron Asymm., **2002**, *13*, 2735; (e) R. Stürmer, A. Börner, J. Holz (BASF AG), DE19824121, **1999**.
57 X. Zhang (The Penn State Research Foundation), US 6037500, **2000**.
58 G. Zhu, P. Cao, Q. Jiang, X. Zhang, *J. Am. Chem. Soc.*, **1997**, *119*, 1799.
59 (a) G. Hoge, *J. Am. Chem. Soc.*, **2003**, *125*, 10219; (b) H. Hoge, *J. Am. Chem. Soc.*, **2004**, *126*, 9920; (c) G. Hoge, B. Samas, *Tetrahedron: Asymm.*, **2004**, *15*, 2155; (d) G.S. Hoge, O.P. Goel (Warner-Lambert Company), WO 02/48161, **2002**.
60 M.J. Burk, O.M. Prakash, M.S. Hoekstra, T.F. Mich, T.A. Mulhern, J.A. Ramsden (Warner-Lambert Company), US 2003/0212290 A1, **2003**.
61 M.J. Burk, P.D. de Koning, T.M. Grote, M.S. Hoekstra, G. Hoge, R.A. Jennings, W.S. Kissel, T.V. Le, I.C. Lennon, T.A. Mulhern, J.A. Ramsden, R.A. Wade, *J. Org. Chem.*, **2003**, *68*, 5731.
62 (a) D. Carmichael, H. Doucet, J.M. Brown, *Chem. Commun.*, **1999**, 261; (b) J.M. Brown, D. Carmichael, H. Doucet (ISIS Innovation Ltd.), WO 00/26220, **2000**.
63 K.W. Kottsiper, U. Kuehner, O. Stelzer, *Tetrahedron Asymm.*, **2001**, *12*, 1159.
64 (a) K. Matsumura, H. Shimizu, T. Saito, H. Kumobayashi, *Adv. Synth. Catal.*, **2003**, *345*, 180; (b) K. Matsumura, T. Saito (Takasago International Corporation), EP 1318156 A1, **2003**; (c) K. Matsumura, T. Saito (Takasago International Corporation), EP 1318155 A1, **2003**.
65 S. Basra, J.G. de Vries, D.J. Hyett, G. Harrison, K.M. Heslop, A.G. Orpen, P.G. Pringle, K. von der Luehe, *Dalton Trans.*, **2004**, 1901.
66 D.J. Brauer, K.W. Kottsieper, S. Roßenbach, O. Stelzer, *Eur. J. Inorg. Chem.*, **2003**, 1748.
67 (a) Z. Chen, Q. Jiang, G. Zhu, D. Xiao, P. Cao, X. Zhang, *J. Org. Chem.*, **1997**, *62*, 4521; (b) Q. Jiang, Y. Jiang, D. Xiao, P. Cao, X. Zhang, *Angew. Chem. Int. Ed.*, **1998**, *37*, 1100; (c) Q. Jiang, D. Xiao, Z. Zhang, P. Cao, X. Zhang, *Angew. Chem. Int. Ed.*, **1999**, *38*, 516; (d) X. Zhang (The Penn State Research Foundation), WO 99/59721, **1999**; (e) Z. Zhang, G. Zhu, Q. Jiang, D. Xiao, X. Zhang, *J. Org. Chem.*, **1999**, *64*, 1774.
68 I. Komarov, A. Börner (Institut für Organische Katalyseforschung an der Universität Rostock), DE 10223442 A1, **2002**.
69 (a) W. Tang, X. Zhang, *Angew. Chem. Int. Ed.*, **2002**, *41*, 1612; (b) X. Zhang, W. Tang (The Penn State Research Institute), WO 03/042135, **2003**.
70 (a) D. Liu, X. Zhang, *Eur. J. Org. Chem.*, **2005**, 646–649; (b) X. Zhang, W. Tang (The Penn State Research Foundation), US 2004/0229846 A1, **2004**.
71 W. Tang, X. Zhang, *Org. Lett.*, **2002**, *4*, 4159.
72 W. Tang, D. Liu, X. Zhang, *Org. Lett.*, **2003**, *5*, 205.
73 D. Liu, W. Gao, C. Wang, X. Zhang, *Angew. Chem. Int. Ed.*, **2005**, *44*, 1687.
74 W. Tang, W. Wang, X. Zhang, *Angew. Chem. Int. Ed.*, **2003**, *42*, 943.
75 (a) H. Shimizu, T. Saito, H. Kumobayashi, *Adv. Synth. Catal.*, **2003**, *345*, 185; (b) H. Shimizu, T. Saito (Takasago International Corp.), US2003/0144139 A1, **2003**.
76 (a) O. Kobayashi, JP 2002069086 A2, **2002**; (b) P. Osinski, K.M. Piertrusiewicz, R. Schmid (Hoffmann-La Roche Inc.), US 2004/0110975 A1, **2004**.
77 S. Demay, M. Lotz, K. Polborn, P. Knochel, *Tetrahedron Asymm.*, **2001**, *12*, 909.
78 (a) S. Lelièvre, F. Mercier, F. Matthey, *J. Org. Chem.*, **1996**, *61*, 3531; (b) F. Matthey, F. Mercier, F. Robin, L. Ricard, *J. Organomet. Chem.*, **1998**, *577*, 117; (c) M. Spagnol, F. Dallemer, F. Matthey, F. Mercier, V. Mouries (Rhodia Chimie), WO 99/47530, **1999**; (d) F. Matthey, F. Robin, F. Mercier, M. Spagnol (Rhone-Poulenc Chimie), WO 98/00375, **1998**.
79 W. Braun, B. Calmuschi, J. Haberland, W. Hummel, A. Liese, T. Nickel, O. Stelzer, A. Salzer, *Eur. J. Inorg. Chem.*, **2004**, 2235.
80 W. Braun, A. Salzer, F. Spindler, E. Alberico, *App. Catal. A: General*, **2004**, *274*, 191.
81 (a) A. Marinetti, L. Ricard, *Tetrahedron*, **1993**, *49*, 10291; (b) A. Marinetti, *Tetrahedron Lett.*, **1994**, *35*, 5861; (c) A. Marinetti, L. Ricard, *Organometallics*, **1994**, *13*, 3956.
82 (a) U. Berens, M. J. Burk, A. Gerlach, W. Hems, *Angew. Chem. Int. Ed.*, **2000**, *39*, 1981; (b) U. Berens (Chirotech Technology

Ltd.), US 6545183, **2003**; (c) U. Berens, M. J. Burk, A. Gerlach (Chirotech Technology Ltd.), US 6172249, **2001**.

83 (a) A. Marinetti, S. Jus, J.-P. Genêt, *Tetrahedron Lett.*, **1999**, *40*, 8365; (b) A. Marinetti, S. Jus, J.-P. Genêt, L. Ricard, *J. Organomet. Chem.*, **2001**, *624*, 162; (c) A. Marinetti, F. Labrue, J.-P. Genêt, *Synlett*, **1999**, *12*, 1975; (d) A. Marinetti, F. Labrue, B. Pons, S. Jus, L. Ricard, J.-P. Genêt, *Eur. J. Inorg. Chem.*, **2003**, 2583; (e) A. Marinetti, J.-P. Genêt, S. Jus, D. Blanc, V. Ratovelomanana-Vidal, *Chem. Eur. J.*, **1999**, *5*, 1160; (f) A. Marinetti, S. Jus, F. Labrue, A. Lemarchand, J.-P. Genêt, L. Ricard, *Synlett*, **2001**, *14*, 2095.

84 J. You, H.-J. Drexler, S. Zhang, C. Fischer, D. Heller, *Angew. Chem. Int. Ed.*, **2003**, *42*, 913.

85 C. J. Cobley, I. C. Lennon, C. Praquin, A. Zanotti-Gerosa, R. A. Appell, C. T. Goralski, A. C. Sutterer, *Org. Proc. Res. Dev.*, **2003**, *7*, 407.

86 A. M. Derrick, N. M. Thomson (Pfizer Ltd.), EP 1199301 A1, **2002**.

87 (a) T. Imamoto, K. V. L. Crépy, K. Katagiri, *Tetrahedron Asymm.*, **2004**, *15*, 2213; (b) T. Imamoto, N. Oohara, H. Takahashi, *Synthesis*, **2004**, *9*, 1353.

88 A. Marinetti, F. Mathey, L. Ricard, *Organometallics*, **1993**, *12*, 1207.

89 M. Ostermeier, J. Prieß, G. Helmchen, *Angew. Chem. Int. Ed.*, **2002**, *41*, 612.

90 M. J. Burk, F. Bienewald, in: *Transition Metals for Organic Synthesis*. Wiley-VCH, Weinheim, **1998**, chapter 2, p. 13.

91 S. K. Armstrong, J. M. Brown, M. J. Burk, *Tetrahedron Lett.*, **1993**, *34*, 879.

92 S. Feldgus, C. R. Landis, *J. Am. Chem. Soc.*, **2000**, *122*, 12714.

93 S. Feldgus, C. R. Landis, *Organometallics*, **2001**, *20*, 2374.

94 B. Guzel, M. A. Omary, J. P. Fackler, A. Akgerman, *Inorg. Chim. Acta*, **2002**, *35*, 45.

95 J. Holz, A. Monsees, H. Jiao, J. You, I. V. Kamarov, C. Fischer, K. Drauz, A. Börner, *J. Org. Chem.*, **2003**, *68*, 1701.

96 M. Kesselgruber, F. Spindler (Solvias AG), personal communication.

97 H-U. Blaser (Solvias AG), personal communication.

98 (a) J. W. Scott, D. D. Keith, G. Nix, D. R. Parrish, S. Remington, G. R. Roth, J. M. Townsend, D. Valentine, R. Yang, *J. Org. Chem.*, **1981**, *46*, 5086; (b) B. D. Vineyard, W. S. Knowles, M. J. Sabacky, G. L. Bachman, D. J. Weinkauff, *J. Am. Chem. Soc.*, **1977**, *99*, 5946.

99 A. Miyashita, H. Takaya, T. Souchi, R. Noyori, *Tetrahedron*, **1984**, *40*, 1245.

100 M. J. Burk, J. R. Lee, J. P. Martinez, *J. Am. Chem. Soc.*, **1994**, *116*, 10847.

101 T. A. Stammers, M. J. Burk, *Tetrahedron Lett.*, **1999**, *40*, 3325.

102 (a) G. P. Aguado, A. Alvarez-Larena, O. Illa, A. G. Moglioni, R. M. Ortuño, *Tetrahedron Asymm.*, **2001**, *12*, 25; (b) G. P. Aguado, A. G. Moglioni, E. García-Expósito, V. Branchadell, R. M. Ortuño, *J. Org. Chem.*, **2004**, *69*, 7971.

103 (a) N. W. Boaz, S. D. Debenham, S. E. Large, M. K. Moore, *Tetrahedron Asymm.*, **2003**, *14*, 3575; (b) S. D. Debenham, N. W. Boaz (Eastman Chemical Company), WO 02/26695 A2, **2002**.

104 X. Li, C. Yeung, A. S. C. Chan, D.-S. Lee, T.-K. Yang, *Tetrahedron Asymm.*, **1999**, *10*, 3863.

105 H. Chen, J.-P. Luzy, C. Garbay, *Tetrahedron Lett.*, **2005**, *46*, 3319.

106 L. Chen, J. W. Tilley, R. W. Guthrie, F. Mennona, T.-N. Huang, G. Kaplan, R. Trilles, D. Miklowski, N. Huby, V. Schwinge, B. Wolitzky, K. Rowan, *Bioorg. Med. Chem. Lett.*, **2000**, *10*, 729.

107 Y. Lee, R. B. Silvermann, *Tetrahedron*, **2001**, *57*, 5339.

108 H.-U. Blaser, F. Spindler, *Topics in Catalysis*, **1997**, *4*, 275.

109 F. Spindler, B. Pugin, H. Buser, H.-P. Jalett, U. Pittelkow, H.-U. Blaser, *Pesticide Science*, **1998**, *54*, 302.

110 (a) K. C. Nicolaou, G.-Q. Shi, K. Namoto, F. Bernal, *Chem. Commun.*, **1998**, 1757; (b) S. H. Lim, S. Ma, P. Beak, *J. Org. Chem.*, **2001**, *66*, 9056.

111 T. Doi, N. Fujimoto, J. Watanabe, T. Takakashi, *Tetrahedron Lett.*, **2003**, *44*, 2161.

112 W. Wang, J. Zhang, C. Xiong, V. J. Hruby, *Tetrahedron Lett.*, **2002**, *43*, 2137.

113 W. Wang, C. Xiong, J. Yang, V. J. Hruby, *Tetrahedron Lett.*, **2001**, *42*, 7717.

114 W. Wang, C. Xiong, J. Zhang, V.J. Hruby, *Tetrahedron*, **2002**, *58*, 3101.

115 T. Masquelin, E. Broger, K. Mueller, R. Schmid, D. Obrecht, *Helv. Chim. Acta*, **1994**, *77*, 1395.

116 W. Wang, J. Yang, J. Ying, C. Xiong, J. Zhang, C. Cai, V.J. Hruby, *J. Org. Chem.*, **2002**, *67*, 6353.

117 W. Wang, H. Li, *Tetrahedron Lett.*, **2004**, *45*, 8479.

118 P.R. Carlier, P.C.-H. Lam, D.M. Wong, *J. Org. Chem.*, **2002**, *65*, 6256.

119 D.J. Bentley, C.J. Moody, *Org. Biomol. Chem.*, **2004**, *2*, 3545.

120 R.S. Hoerner, D. Askin, R.P. Volante, P.J. Reider, *Tetrahedron Lett.*, **1998**, *39*, 3455.

121 K.S. Feldman, K.J. Eastman, G. Lessene, *Org. Lett.*, **2002**, *4*, 3525.

122 S.W. Jones, C.F. Palmer, J.M. Paul, P.D. Tiffin, *Tetrahedron Lett.*, **1999**, *40*, 1211.

123 M. Adamczyk, S.R. Akireddy, R.E. Reddy, *Org. Lett.*, **2002**, *3*, 3157.

124 (a) M. Adamczyk, S.R. Akireddy, R.E. Reddy, *Tetrahedron Asymm.*, **2001**, *12*, 2385–2387; (b) M. Adamczyk, S.R. Akireddy, R.E. Reddy, *Tetrahedron Lett.*, **2002**, *58*, 6951.

125 (a) A.J. Robinson, C.Y. Lim, L. He, P. Ma, H.-Y. Li, *J. Org. Chem.*, **2001**, *66*, 4141; (b) A.J. Robinson, P. Stanislawski, D. Mulholland, L. He, H.-Y. Li, *J. Org. Chem.*, **2001**, *66*, 4148.

126 J. Singh, D.R. Kronenthal, M. Schwinden, J.D. Godfrey, R. Fox, E.J. Vawter, B. Zhang, T.P. Kissick, B. Patel, O. Mneimne, M. Humora, C.G. Papaioannou, W. Szymanski, M.K.Y. Wong, C.K. Chen, J.E. Heikes, J.D. DiMarco, J. Qiu, R.P. Deshpande, J.Z. Gougoutas, R.H. Mueller, *Org. Lett.*, **2003**, *5*, 3155.

127 (a) W.-C. Shieh, S. Xue, N.M. Reel, J.J. Fitt (Novartis Corporation), US**2002**/0133014, **2002**.; (b) W.-C. Shieh, S. Xue, N. Reel, R. Wu, J. Fitt, O. Repiè, *Tetrahedron Asymm.*, **2001**, *12*, 2421.

128 (a) S.D. Debenham, J.S. Debenham, M.J. Burk, E.J. Toone, *J. Am. Chem. Soc.*, **1997**, *119*, 9897; (b) S.D. Debenham, J. Cossrow, E.J. Toone, *J. Org. Chem.*, **1999**, *64*, 9153; (c) J.R. Allen, C.R. Harris, S.J. Danishefsky, *J. Am. Chem. Soc.*, **2001**, *123*, 1890; (d) J.R. Allen, S.J. Danishefsky, *J. Prakt. Chem.*, **2000**, *342*, 736; (e) X. Xu, G. Fakha, D. Sinou, *Tetrahedron*, **2002**, *58*, 7539; (f) S. Liu, R.N. Ben, *Org. Lett.*, **2005**, *7*, 2385.

129 G.P. Aguado, A.G. Moglioni, B.N. Brousse, R.M. Ortuño, *Tetrahedron Asymm.*, **2003**, *12*, 2445.

130 (a) A. Ritzén, B. Basu, S.K. Chattopadhyay, F. Dossa, T. Frejd, *Tetrahedron Asymm.*, **1998**, *8*, 503; (b) K.B. Jørgensen, O.R. Gautun, *Tetrahedron*, **1999**, *55*, 10527; (c) S. Maricic, A. Ritzén, U. Berg, T. Frejd, *Tetrahedron*, **2001**, *57*, 6523; (d) J.M. Travins, F.A. Etzkorn, *J. Org. Chem.*, **1997**, *62*, 8387; (e) A. Ritzén, B. Basu, A. Wllberg, T. Frejd, *Tetrahedron Asymm.*, **1998**, *9*, 3491; (f) S. Maricic, U. Berg, T. Frejd, *Tetrahedron*, **2002**, *58*, 3085; (g) J. Hiebl, H. Kollmann, F. Rovensky, K. Winkler, *J. Org. Chem.*, **1999**, *64*, 1947.

131 W. Wang, C. Xiong, V.J. Hruby, *Tetrahedron Lett.*, **2001**, *42*, 3159.

132 M.J. Burk, K.M. Bedingfield, W.F. Kiesman, J.G. Allen, *Tetrahedron Lett.*, **1999**, *40*, 3093.

133 M.J. Burk, J.G. Allen, W.F. Kiesman, *J. Am. Chem. Soc.*, **1998**, *120*, 657.

134 (a) C.A. Busacca, N. Haddad, S.R. Kapadia, L. Smith Keenan, J.C. Lorenz, C.H. Senanayake (Boehringer Ingelheim Pharmaceuticals Inc.), WO 2005/044799 A1, **2005**; (b) A.-M. Faucher, M.D. Bailey, P.L. Beaulieu, C. Brochu, J.-S. Duceppe, J.-M. Ferland, E. Ghiro, V. Gorys, T. Halmos, S.H. Kawai, M. Poirier, B. Simoneau, Y.S. Tsantrizos, M. Llinás-Brunet, *Org. Lett.*, **2004**, *6*, 2901.

135 J. Elaridi, W.R. Jackson, A.J. Robinson, *Tetrahedron Asymm.*, **2005**, *16*, 2025.

136 (a) E. Teoh, E.M. Campi, W.R. Jackson, A.J. Robinson, *Chem. Commun.*, **2002**, 978; (b) E. Teoh, E.M. Campi, W.R. Jackson, A.J. Robinson, *New J. Chem.*, **2003**, *27*, 387; (c) E. Teoh, W.R. Jackson, A.J. Robinson, *Aus. J. Chem.*, **2005**, *58*, 63.

137 (a) M. Sawamura, R. Kuwano, Y. Ito, *J. Am. Chem. Soc.*, **1995**, *117*, 9602; (b) T. Imamoto, J. Watanabe, Y. Wada, H. Masuda, H. Yamada, H. Tsuruta,

S. Matsukawa, K. Yamaguchi, *J. Am. Chem. Soc.*, **1998**, *120*, 1635; (c) A. Ohashi, T. Imamoto, *Org. Lett.*, **2001**, *3*, 373; (d) D. A. Evans, F. E. Michael, J. S. Tedrow, K. R. Campos, *J. Am. Chem. Soc.*, **2003**, *125*, 3534; (e) X. Jiang, M. van den Berg, A. J. Minnaard, B. L. Feringa, J. G. de Vries, *Tetrahedron Asymm.*, **2004**, *15*, 2223.

138 G. J. Roff, R.C. Lloyd, N. J. Turner, *J. Am. Chem. Soc.*, **2004**, *126*, 4098.

139 For examples, see (a) V. Wehner, H. Blum, M. Kurz, H. U. Stilz, *Synthesis*, **2002**, *14*, 2032; (b) S. D. Bull, S. G. Davies, D. J. Fox, M. Gianotti, P. M. Kelly, C. Pierres, E. D. Savory, A. D. Smith, *J. Chem. Soc., Perkin Trans I*, **2002**, 1858; (c) E. Juaristi, H. Lopez-Ruiz, *Curr. Med. Chem.*, **1999**, *6*, 983; (d) I. Ojima, S. Lin, T. Wang, *Curr. Med. Chem.*, **1999**, *6*, 927; (e) G. Cardillo, C. Tomasini, *Chem. Soc. Rev.*, **1996**, 117.

140 For examples, see (a) K. Gadermann, T. Hintermann, J.V. Schreiber, *Curr. Med. Chem.*, **1999**, *6*, 905; (b) D. Seebach, S. Abele, K. Gadermann, G. Guichard, T. Hintermann, B. Jaun, J. L. Matthews, J. V. Schreiber, *Helv. Chim. Acta*, **1998**, *81*, 932; (c) S. H. Gellman, *Acc. Chem. Res.*, **1998**, *31*, 173; (d) D. Seebach, J. L. Matthews, *Chem. Commun.*, **1997**, 2015.

141 E. Juaristi, *Enantioselective Synthesis of Amino Acids*. Wiley-VCH, New York, **1997**.

142 For an excellent review of this area, see H.-J. Drexler, J. You, S. Zhang, C. Fischer, W. Baumann, A. Spannanberg, D. Heller, *Org. Proc. Res. Dev.*, **2003**, *7*, 355.

143 G. Zhu, Z. Chen, X. Zhang, *J. Org. Chem.*, **1999**, *64*, 6907.

144 T. Jerphagnon, J.-L. Renaud, P. Demonchaux, A. Ferreira, C. Bruneau, *Tetrahedron Asymm.*, **2003**, *14*, 1973.

145 C. J. Cobley, C. G. Malan (The Dow Chemical Company), WO 0316264A1, **2003**.

146 (a) D. Heller, J. Holz, H.-J. Drexler, J. Lang, K. Drauz, H.-P. Krimmer, A. Börner, *J. Org. Chem.*, **2001**, *66*, 6816; (b) H.-P. Krimmer, K. Drauz, J. Lang, A. Börner, D. Heller, J. Holz (Degussa AG), DE 10100971 A1, **2001**.

147 J. You, H.-J. Drexler, S. Zhang, C. Fischer, D. Heller, *Angew. Chem. Int. Ed.*, **2003**, *42*, 913.

148 (a) Y. G. Zhou, W. Tang, W.-B. Wang, W. Li, X. Zhang, *J. Am. Chem. Soc.*, **2002**, *124*, 4952; (b) S. Lee, Y.J. Zhang, *Org. Lett.*, **2002**, *4*, 2429.

149 W. D. Lubell, M. Kitamura, R. Noyori, *Tetrahedron Asymm.*, **1991**, *2*, 543.

150 H.-J. Drexler, W. Baumann, T. Schmidt, S. Zhang, A. Sun, A. Spannenberg, C. Fischer, H. Buschmann, D. Heller, *Angew. Chem. Int. Ed.*, **2005**, *44*, 1184.

151 Y. Hisao, N. R. Rivera, T. Rosner, S. W. Krska, E. Njolito, F. Wang, Y. Sun, J. D. Armstrong, E. J. J. Grabowski, R. D. Tillyer, F. Spindler, C. Malan, *J. Am. Chem. Soc.*, **2004**, *126*, 9918.

152 W. Tang, S. Wu, X. Zhang, *J. Am. Chem. Soc.*, **2003**, *125*, 9570.

153 J. D. Clark, G. A. Weisenburger, D. K. Anderson, P.-J. Colson, A. D. Edney, D. J. Gallagher, H. P. Kleine, C. M. Knable, M. K. Lantz, C. M. V. Moore, J. B. Murphy, T. E. Rogers, P. G. Ruminski, A. S. Shah, N. Storer, B. E. Wise, *Org. Proc. Res. Dev.*, **2004**, *8*, 51.

154 Y. J. Zhang, J. H. Park, S. Lee, *Tetrahedron Asymm.*, **2004**, *15*, 2209.

155 R. Callens, M. Larcheveque, C. Pousset (Solvay SA), FR 2853316 A1, **2003**.

156 J. Elaridi, A. Thaqi, A. Prosser, W.R. Jackson, A. J. Robinson, *Tetrahedron Asymm.*, **2005**, *16*, 1309.

157 D. Saylik, E. M. Campi, A. C. Donohue, W. R. Jackson, A.J. Robinson, *Tetrahedron Asymm.*, **2001**, *12*, 657.

158 The same effect has been observed and studied with the diphosphine *t*Bu-BisP*: I.D. Gridnev, M. Yasutake, N. Higashi, T. Imamoto, *J. Am. Chem. Soc.*, **2001**, *123*, 5268.

159 G. Zhu, A. L. Casalnuovo, X. Zhang, *J. Org. Chem.*, **1998**, *63*, 8100.

160 J. Cong-Dung Le, B. L. Pagenkopf, *J. Org. Chem.*, **2004**, *69*, 4177.

161 M. J. Burk, T. A. Stammers, J. A. Straub, *Org. Lett.*, **1999**, *1*, 387.

162 L. Storace, L. Anzalone, P. N. Confalone, W. P. Davis, J. M. Fortunak,

M. Giangiorano, J. J. Haley, K. Kamholz, H.-Y. Li, P. Ma, W. A. Nugent, R. L. Parsons, P. J. Sheeran, C. E. Silverman, R. E. Waltermire, C. C. Wood, *Org. Proc. Res. Dev.*, **2002**, 6, 54.
163 Y.-G. Zhou, P.-Y. Yang, X.-W. Han, *J. Org. Chem.*, **2005**, *70*, 1679.
164 (a) P. Dupau, A.-E. Hay, C. Bruneau, P. H. Dixneuf, *Tetrahedron Asymm.*, **2001**, *12*, 863; (b) P. Dupau, C. Bruneau, P. H. Dixneuf, *Adv. Synth. Catal.*, **2001**, *343*, 331.
165 P. M. Donate, D. Frederico, R. da Silva, M. G. Constantino, G. Del Ponte, P. S. Bonatto, *Tetrahedron Asymm.*, **2003**, 3253.
166 C. González-Arellano, A. Corma, M. Iglesias, F. Sánchez, *Chem. Commun.*, **2005**, 3451.
167 M. J. Burk, F. Bienewald, M. E. Fox, A. Zanotti-Gerosa (Chirotech Technology Limited), WO 99/31041.
168 M. J. Burk, F. Bienewald, A. Zanotti-Gerosa (Chirotech Technology Limited), WO 99/52852.
169 (a) A. M. Derrick, N. M. Thomson (Pfizer Ltd. and Pfizer Inc.), EP 1199301 A1; (b) C. P. Ashcroft, S. Challenger, A. M Derrick, R. Storey, N. M. Thomson, *Org. Proc. Res. Dev.*, **2003**, *7*, 362.
170 (a) Y. Toyama, A. Noda, F. Yamamoto, T. Toyama (Kotobuki Pharmaceutical Company Ltd.), JP 2003212874 A2; (b) H. Tomiyama, Y. Kobayashi, A. Noda (Kotobuki Pharmaceutical Company Ltd.), WO2002076981 A1.
171 (a) W. A. Nugent, T. V. Rajanbabu, M. J. Burk, *Science*, **1993**, *259*, 479; (b) O. M. Akotsi, K. Metera, R. D. Reid, R. McDonald, S. H. Bergens, *Chirality*, **2000**, *12*, 514.
172 (a) M. J. Burk, G. P. Harper, C. S. Kalberg, *J. Am. Chem. Soc.*, **1995**, *117*, 4423; (b) J.-P. Gent, C. Pinel, V. Ratovelomanana-Vidal, S. Mallart, X. Pfister, L. Bischoff, M.C. Caño de Andrade, S. Darses, C. Galopin, J. A. Laffitte, *Tetrahedron Asymm.*, **1994**, *5*, 675; (c) J.-P. Gent, V. Ratovelomanana-Vidal, M. C. Caño de Andrade, X. Pfister, P. Guerreiro, J. Y. Lenoir, *Tetrahedron Lett.*, **1995**, *36*, 4801.
173 M. Prashad, H.-Y. Kim, B. Hu. J. Slade, P. K. Kapa (Novartis AG), WO 2004/076053 A2.
174 J. P. Paul, C. Palmer (Darwin Discovery Ltd.), WO 99/15481.
175 I. Appleby, L. T. Boulton, C. J. Cobley, C. Hill, M. L. Hughes, P. D. de Koning, I. C. Lennon, C. Praquin, J. A. Ramsden, H. J. Samuel, N. Willis, *Org. Lett*, **2005**, *7*, 1931.
176 M. E. B. Smith, N. Derrien, M. C. Lloyd, S. J. C. Taylor, D. A. Chaplin, R. McCague, *Tetrahedron Lett.*, **2001**, *42*, 1347.
177 M. Scalone, U. Zutter (F. Hoffman-La Roche AG), EP 0974590 A1, **2000**.
178 M. J. Burk, F. Bienewald, S. Challenger, A. Derrick, J. A. Ramsden, *J. Org. Chem.*, **1999**, *64*, 3290.
179 T.-Y. Yue, W. A. Nugent, *J. Am. Chem. Soc.*, **2002**, *124*, 13692.
180 M. J. Burk, C. S. Kalberg, A. Pizzano, *J. Am. Chem. Soc.*, **1998**, *120*, 4345.
181 P. E. Maligres, S. H. Krska, G. R. Humphrey, *Org. Lett*, **2004**, *6*, 3147.
182 (a) D. A. Dobbs, K. P. M Vanhessche, E. Brazi, V. Rautenstrauch, J.-Y. Lenoir, J.-P. Genet, J. Wiles, S. H. Bergens, *Angew. Chem. Int. Ed.*, **2000**, *39*, 1992; (b) J. A. Wiles, S. H. Bergens, K. P. M. Vanhessche, D. A. Dobbs, V. Rautenstrauch, *Angew. Chem. Int. Ed.*, **2001**, *40*, 914; (c) D. A. Dobbs, K. P. M. Vanhessche, V. Rautenstrauch (Firmenich SA), US 6,455,460 B1.
183 (a) N. W. Boaz, *Tetrahedron Lett.*, **1998**, *39*, 5505; (b) N. W. Boaz (Eastman Chemical Company), EP 0872467 B1, **2003**.
184 A. Robinson, H.-Y. Li, J. Feaster, *Tetrahedron Lett.*, **1996**, *37*, 8321; (b) H.-Y. Li, A. J. Robinson (The DuPont Merck Pharmaceutical Company), US 5686631, **1997**.
185 (a) E. A. Broger, Y. Crameri, R. Schmid, T. Siegfried (F. Hoffman-La Roche AG), EP 691325, **1996**; (b) M. Scalone, R. Schmid, E. A. Broger, W. Burkart, M. Cereghetti, Y. Crameri, J. Foricher, M. Henning, F. Kienzie, F. Montavon, G. Schoettel, D. Tesauro, S. Wang, R. Zell, U. Zutter, Proceedings of the ChiraTech'97 Symposium, The Catalyst Group, Spring House, USA, **1997**.

186 H.-H Bokel, P. Mackert, C. Mürmann, N. Schweickert (Merck), WO 00/35901, **2000**.

187 (a) B. D. Hewitt, M. J. Burk, N. B. Johnson (Pharmacia & Upjohn Company), WO 00/55150, **2000**. (b) K. S. Fors, J. R. Gage, R. F. Heier, R. C. Kelly, W. R. Perrault, N. Wicnienski, *J. Org. Chem. Soc.*, **1998**, *63*, 7248; (c) J. R. Gage, R. C. Kelly, B. D. Hewitt (Pharmacia & Upjohn), WO 9912919, **1999**.

188 F. D. Klingler, M. Steigerwald, R. Ehlenz (Boehringer Ingelheim International GmbH), WO 2004/085427 A1, **2004**.

189 G. S. Forman, T. Ohkuma, W. P. Hems, R. Noyori, *Tetrahedron Lett.*, **2000**, *41*, 9471.

190 For examples, see (a) A. Pfaltz, J. Blankenstein, R. Hilgraf, E. Hormann, S. McIntyre, F. Menges, M. Schönleber, S. P. Smidt, B. Wüstenberg, N. Zimmermann, *Adv. Synth. Catal.*, **2003**, *345*, 33; (b) M. C. Perry, X. Cui, M. T. Powell, D.-. Hou, J. H. Reibenspies, K. Burgess, *J. Am. Chem. Soc.*, **2003**, *125*, 113.

191 (a) J.-C. Henry, D. Lavergne, V. Ratovelomanana-Vidal, J.-P. Gent, I. P. Beletskaya, T. M. Dolgina, *Tetrahedron Lett.*, **1998**, *39*, 3473; (b) N. S. Goulioukina, T. M. Dolgina, J.-C. Henry, D. Lavergne, V. Ratovelomanana-Vidal, J.-P. Gent, *Tetrahedron Asymm.* **2001**, *12*, 319.

192 M. Ueda, A. Saitoh, N. Miyaura, *J. Organomet. Chem.*, **2002**, *642*, 145.

193 J. Madec, X. Pfister, V. Ratovelomanana-Vidal, J.-P. Gent, *Tetrahedron*, **2001**, *57*, 2563.

194 Inventors unknown (Lonza AG), EP 1510517 A1, **2005**.

195 T. Ohkuma, H. Ooka, S. Hashiguchi, T. Ikariya, R. Noyori, *J. Am. Chem. Soc.*, **1995**, *117*, 2675.

196 R. Noyori, T. Ohkuma, *Angew. Chem. Int. Ed.*, **2001**, *40*, 40 and references therein.

197 Y.-Q. Wang, S.-M. Lu, Y.-G. Zhou, *Org. Lett.*, **2005**, *7*, 3235.

198 (a) C. J. Cobley, J. P. Henschke, *Adv. Synth. Catal.*, **2003**, *345*, 195; b) C. J. Cobley, J. P. Henschke, J. A. Ramsden, WO 0208169 A1, **2002**.

25
Enantioselective Hydrogenation of Alkenes with Ferrocene-Based Ligands

Hans-Ulrich Blaser, Matthias Lotz, and Felix Spindler

25.1
Introduction

Ferrocene as a (at the time rather exotic) backbone for chiral ligands was introduced by Kumada and Hayashi [1] based on Ugi's pioneering studies related to the synthesis of enantiopure ferrocenes (Fig. 25.1). Ppfa, as well as bppfa and bppfoh, proved to be effective ligands for a variety of asymmetric transformations. From this starting point, several ligand families with a range of structural variations have been developed during the past few years. In this chapter we will describe effective ligand structures developed over time, the main focus being on diphosphine derivatives (Fig. 25.2) and their application to the hydrogenation of alkenes. Three recently published reviews cover some of the same area, but from slightly different points of view. Colacot and Barbaro et al. [2] presented general overviews on ferrocene-based chiral ligands and their application to various asymmetric transformations, while Blaser et al. [3] and Tang and Zhang [4] reviewed the recent progress in the application of diphosphines for the enantioselective hydrogenation. These reviews can serve to put the present account into a broader perspective.

This chapter is organized according to the position of the phosphine groups **P**, as depicted in Fig. 25.2. It is important to realize that many of the ligands described here have both planar (C_p ring with two different substituents) as well

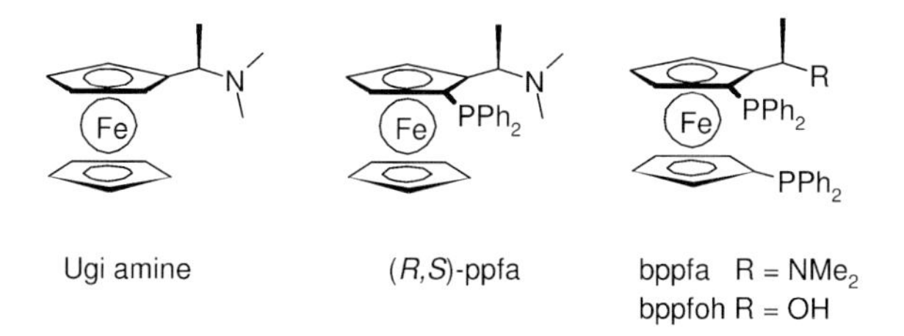

Fig. 25.1 Structures of Ugi's amine and the first ferrocene-based chiral phosphine ligands.

The Handbook of Homogeneous Hydrogenation.
Edited by J. G. de Vries and C. J. Elsevier
Copyright © 2007 WILEY-VCH Verlag GmbH & Co. KGaA, Weinheim
ISBN: 978-3-527-31161-3

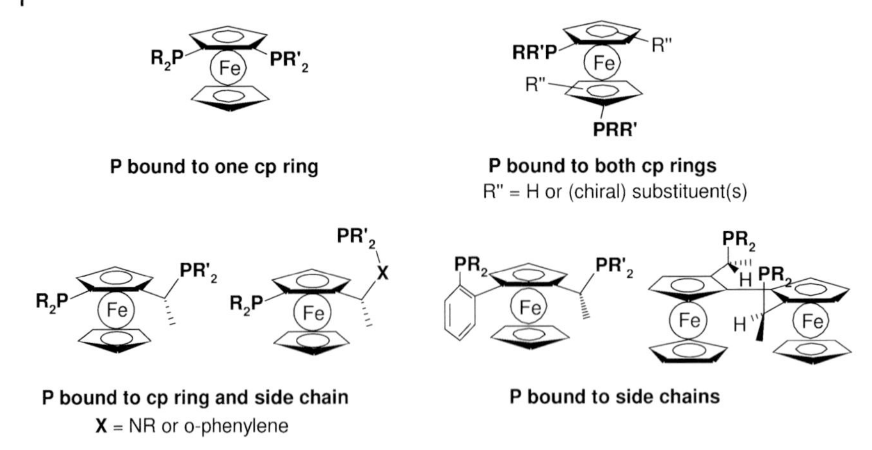

Fig. 25.2 Subclasses of ferrocenyl diphosphines.

as central (in the side chain) chirality. In most cases, the central chirality dominates the sense of induction, but strong matched-mismatched effects are very common. Except for selected cases we will not include the absolute configuration in the ligand names, but as a rule the central chirality is given first, followed by the planar and (if applicable) the axial chirality (e.g., see ppfa in Fig. 25.1).

When assessing catalytic results reported for new ligands, one must bear in mind that their quality and relevance differ widely. For most new ligands only experiments with selected model test substrates carried out under standard conditions are available, and very few have already been applied to industrially relevant problems. The test substrates for alkenes used most frequently are Acetamido Cinnamic Acid (ACA) or its methyl ester (MAC), Methyl Acetamido Acrylate (MAA), ITaconic Acid or DiMethyl ITaconate (ITA, DMIT) and selected aryl enamides (Fig. 25.3).

Especially for new ligands, reaction conditions are usually optimized for enantioselectivity, whereas catalyst productivity (given as turnover number, TON, or substrate/catalyst ratio, SCR) and catalyst activity (given as turnover frequency, TOF (h^{-1}), at high conversion) are often only a first indication of the potential of the ligand. The decisive test – namely the application of a new ligand to "real world problems" which will tell about the scope and limitations of a ligand (family) concerning tolerance to changes in the substrate structure and/or the presence of functional groups – will often come much later.

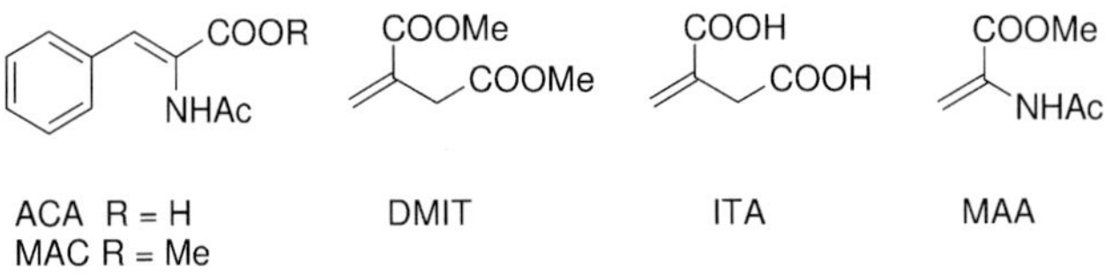

COOR NHAc — ACA R = H, MAC R = Me

COOMe COOMe — DMIT

COOH COOH — ITA

COOMe NHAc — MAA

Fig. 25.3 Structures and abbreviations of frequently used model test substrates.

25.2
Ligands with Phosphine Substituents Bound to One Cyclopentadiene Ring

Until now, only a few effective ligands of this type have been identified (Fig. 25.4). Kagan and co-workers [5] prepared one of the few chiral diphosphines with only planar chirality and obtained 95% ee for the hydrogenation of DMIT with **L1** (Table 25.1, entry 1.1.), but enantioselectivities for several enamide derivatives were below 82% ee (the best results were with the cyclohexyl analogue of **L1**). For the reactions with DMIT or MAC, the cationic Rh-kephos complex showed comparable or better performance than corresponding duphos catalysts.

25.3
Ligands with Phosphine Substituents Bound to both Cyclopentadiene Rings

As noted earlier, the first effective ligands were prepared by Kumada and Hayashi during the 1970s, starting from Ugi's amine. Depending on the reaction conditions, phosphine substituents were introduced either on one or on both cyclopentadiene rings. It transpired that only the diphosphines bppfa and bppfoh were useful for hydrogenation reactions. Only recently, new, usually C_2 symmetrical, ligand families were prepared with excellent catalytic properties for a variety of hydrogenation reactions (Fig. 25.5).

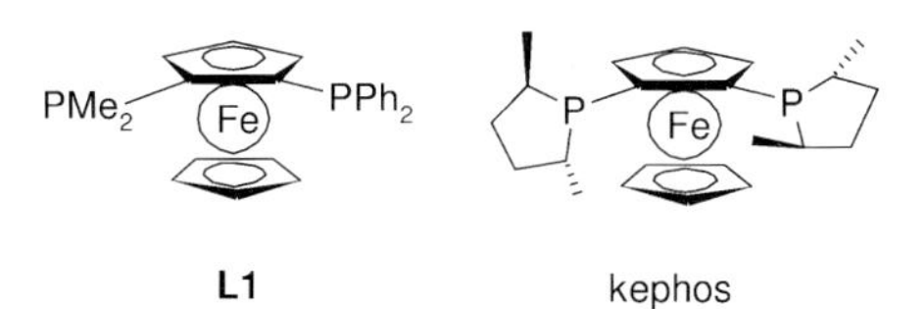

Fig. 25.4 Structures and abbreviations of diphosphines with P bound to one C_p ring.

Table 25.1 Selected results for Rh-catalyzed hydrogenations using diphosphines having both P bound to one C_p ring (for structures, see Fig. 25.4).

Entry	Ligand	Substrate	$p(H_2)$ [bar]	SCR	TOF [h^{-1}]	ee [%]	Comments	Reference
1.1	**L1**	DMIT	1	100	n.a.	95		5
1.2	kephos	DMIT	10	1000	1000	99.5	ee > 99.9% at SCR 200	6
1.3	kephos	MAC	10	1000	300	96	ee 98% at SCR 200	6

SCR = substrate:catalyst ratio; n.a. = not available.

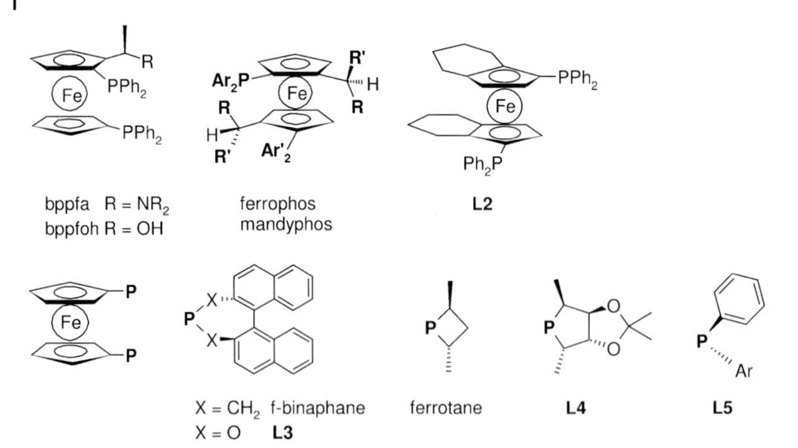

Fig. 25.5 Generic structures and names or numbers of ferrocene-based diphosphines with P bound to both C_p rings.

25.3.1 Bppfa, Ferrophos, and Mandyphos Ligands

Bppfoh and bppfa derivatives have been applied most successfully for the Rh-catalyzed hydrogenation of dehydro amino acid derivatives such as MAC (ee 97%) and of functionalized ketones [7]. The nature of the amino group has a significant effect on enantioselectivity and often also on activity, and is used to tailor the ligand for a particular substrate. Rh-bppfa complexes were among the first catalysts able to hydrogenate tetrasubstituted C=C bonds, albeit with relatively low activity (Table 25.2, entries 2.1–2.3). Ferrophos, one of the very few li-

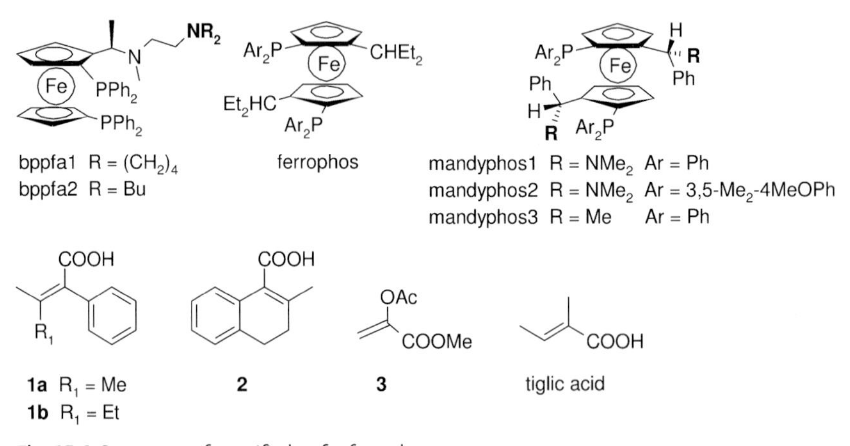

Fig. 25.6 Structures of specific bppfa, ferrophos, and mandyphos ligands and of test substrates.

Table 25.2 Selected results for Rh-catalyzed hydrogenations using bppf, ferrophos, and mandyphos derivatives (for structures, see Figs. 25.5 and 25.6)

Entry	Ligand	Substrate	$p(H_2)$ [bar]	SCR	TOF [h^{-1}]	ee [%]	Comments	Reference
2.1	bppfa1	**1a**	50	200 [a]	7 [a]	98.4		7
2.2	bppfa1	**1b**	50	200 [a]	2 [a]	97	*cis/trans* 97/3	7
2.3	bppfa2	**2**	50	100 [a]	2 [a]	87	*cis/trans* >99/1	7
2.4	ferrophos	ACA	2	100 [a]	~10–30	98.9	97.6% ee for MAC	8
2.5	Mandyphos1	pCl-MAC	1	100 [a]	n.a.	>99	98% ee for MAC	9c
2.6	Mandyphos2	MAC	1	20000	~3000	98.7	>99% ee at SCR 200	9d
2.7	Mandyphos2	tiglic acid	5	200 [a]	11 [a]	97	Ru complex	9d
2.8	Mandyphos3	MAC	1	100 [a]	≥600 [a]	98.6	97.9% ee for MAA	9b
2.9	Mandyphos3	**3**	1	100 [a]	8 [a]	95		9a

a) Standard test results, not optimized.

gands with only planar chirality, shows good ee-values but low activity for dehydroamino acid derivatives (Table 25.2, entry 2.4).

Mandyphos ligands [9] are highly modular bidentate analogues of ppfa where not only the phosphine moieties but also the R substituent have been used for fine-tuning purposes. Both C_2 (Ar=Ar′) as well as C_1 (Ar ≠ Ar′) symmetrical ligands have been prepared and tested in an extended screening [9d]. Even though the scope of this family is not yet fully explored, these test results indicate high enantioselectivities as well as high activity for the Rh-catalyzed hydrogenation of dehydroamino acid derivatives (Table 25.2, entries 2.5, 2.6, 2.8), tiglic acid (Ru complex, entry 2.7) and enol acetate **3** (entry 2.9). Mandyphos2 was shown to be the most versatile ligand, leading to very high TON and TOF for the MAC hydrogenation (entry 2.6). Both enantiomers of the mandyphos family are equally well accessible, and selected derivatives are now commercialized by Solvias in collaboration with Umicore (formerly OMG) [10].

25.3.2 Miscellaneous Diphosphines

A variety of C_2 symmetrical diphosphine ligands with a ferrocenyl backbone (see Fig. 25.5) have recently been described and tested, with sometimes quite impressive results. Interesting examples are f-binaphane [11], ferrotane [12], **L2**

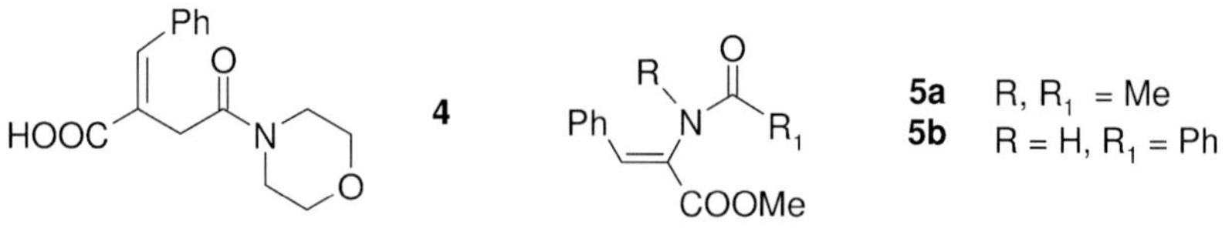

Ph O HOOC N O 4 R O Ph N R$_1$ COOMe 5a R, R$_1$ = Me 5b R = H, R$_1$ = Ph

Fig. 25.7 Structures of substrates listed in Table 25.3.

Table 25.3 Selected results for Rh-catalyzed hydrogenations using miscellaneous diphosphines (for structures of ligands, see Fig. 25.5; for structures of substrates, see Fig. 25.7).

Entry	Ligand	Substrate	$p(H_2)$ [bar]	SCR	TOF [h^{-1}]	ee [%]	Comments	Reference
3.1	Ferrotane [a]	4	5	20000	~7000	98–99	98% ee for DMIT	12
3.2	**L2**	MAC	1	1000 [b]	40 [b]	97	96% for MAA	13
3.3	**L3**	DMIT	1.3	5400	270	>99.5		14
3.4	**L4**	ITA	5	100 [b]	<10 [b]	99.5	ee 90% for DMIT	15
3.5	**L4**	MAA	3	10000	850	99.9	best solvent THF	15
3.6	**L4**	MAC [c]	1	100 [b]	100 [b]	>99.9	best solvent THF	15
3.7	**L5**	5a	1	200 [b]	15 [b]	97	Ar 2-anisyl	16b
3.8	**L5**	5b	2	100 [b]	15 [b]	98.5	Ar 9-phenanthryl	16c

a) Et-ferrotane.
b) Standard test results, not optimized.
c) Various substituted analogues were tested.

with only planar chirality [13], bisphosphonite **L3** [14], the sugar-based phospholane **L4** [15] and the P-chiral phosphines **L5** [16].

Rh complexes of ferrotanes showed very good performance for various amido itaconates, and achieved very high TONs and TOFs for substrate **4** (Table 25.3, entry 3.1). The planar chiral Rh-**L2** complex achieved up to 97% ee for MAC (entry 3.2) and bisphosphonite **L3** based on a binol or related moiety achieved very high ee-values and respectable TONs for the hydrogenation of itaconates (entry 3.3). The sugar-based ligand **L4** is excellent for dehydroamino and itaconic acid derivatives, with good TONs and very high ee-values (entries 3.4–3.6). Rh-**L5** complexes (with Ar=2-anisyl, 1-naphthyl or 9-phenanthryl) reduce MAC and ACA with ee-values of 95–99% (results not shown). In contrast to many other ligands, Rh-**L5** catalysts are quite tolerant towards changes in the structure of the amide moiety, showing high ee-values for N-methyl (**5a**, entry 3.7) or benzoyl derivatives (**5b**, entry 3.8).

25.4 Ligands with Phosphine Substituents Bound to a Cyclopentadiene Ring and to a Side Chain

The first successful variation of the ppfa structure was carried out by Togni and Spindler, who replaced the amino group at the stereogenic center of the side chain by a second phosphino moiety. Later, very effective ligands were also obtained when a further bridging group was introduced between the stereogenic center and the second phosphino group, as in bophoz or **L7** (Fig. 25.8).

25.4.1 Josiphos

The josiphos ligands arguably constitute the most versatile and successful ferrocenyl ligand family. Because the two phosphine groups are introduced in consecutive steps with very high yields (as shown in Scheme 25.1), a variety of ligands is readily available with widely differing steric and electronic properties. A comprehensive review on the catalytic performance of josiphos ligands has recently been published [17]. Until now, only the (*R*, *S*)-family (and its enantiomers) but not the (*R*, *R*) diastereomers have led to high enantioselectivities (the first descriptor stands for the stereogenic center, the second for the planar chirality). The ligands are technically developed, and available in commercial quanti-

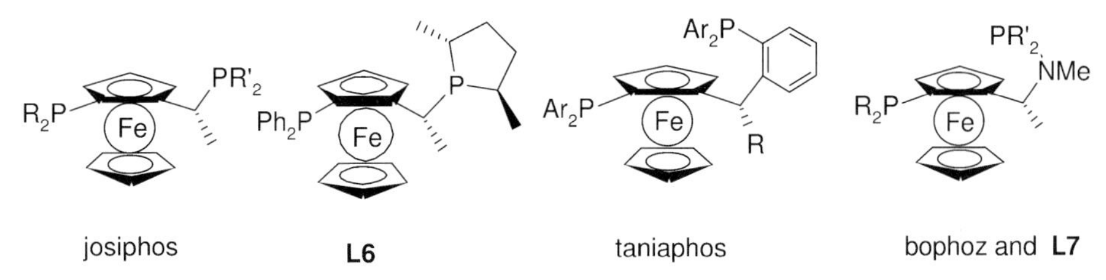

Fig. 25.8 Structures and names of the most important diphosphines with P bound to a C_p ring and a chiral side chain.

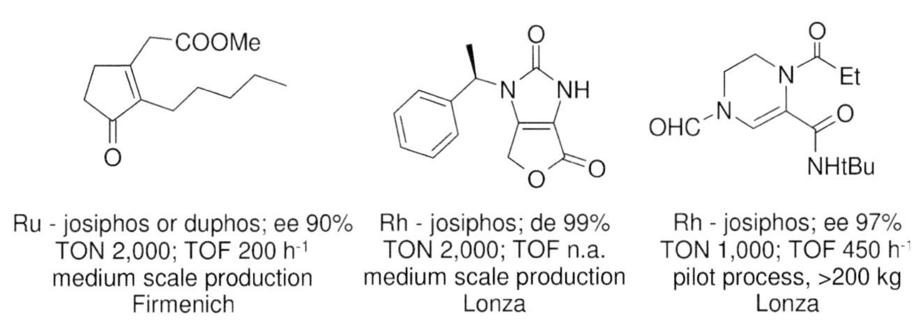

Fig. 25.9 Industrial applications of josiphos ligands for (for further information, see [19]).

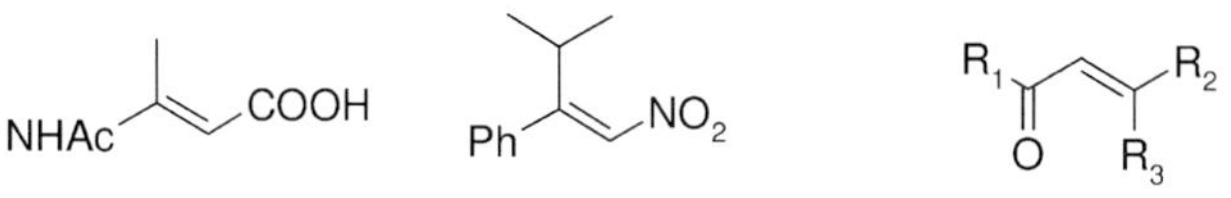
NHAc
COOH
Ph
NO2
R1
R2
O
R3

6 **7**

8a R_1 = Me, R_2 = $(CH_2)_4OR$, R_3 = Me
8b R_1 = Me, R_2 = Hex, R_3 = Me
8c R_1 = $Cy(CH_2)_2$, R_2 = $(CH_2)_4Cl$, R_3 = Me

Fig. 25.10 Structures of substrates listed in Table 25.4.

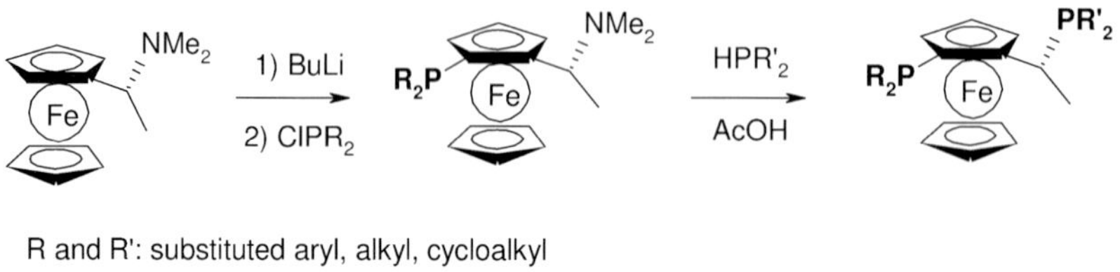
NMe2
Fe
1) BuLi
2) ClPR2
R2P
HPR'2
AcOH
PR'2
R and R': substituted aryl, alkyl, cycloalkyl

Scheme 25.1 Preparation of josiphos ligands starting from the Ugi amine.

ties from Solvias [10]. The most important application is undoubtedly the hydrogenation of C=N functions where the largest enantioselective process has been realized for the enantioselective production of the herbicide (*S*)-metolachlor [18] and of highly substituted C=C bonds. Several smaller productions and some pilot processes use josiphos ligands and important examples are shown in Fig. 25.9.

Table 25.4 Selected results for the Rh- and Cu-catalyzed hydrogenation using josiphos ligands (for structures, see Figs. 25.8 and 25.10).

Entry	Substrate	Metal-ligand (R, R′) [a)]	$p(H_2)$ [bar]	SCR	TOF [h^{-1}]	ee [%]	Reference
4.1	ACA	Rh-(3,5-$(CF_3)_2$Ph, Cy)	1	600	>600	99	6
4.2	MAA	Rh-(3,5-$(CF_3)_2$Ph, Cy)	1	600	>600	98	6
4.3	MAA	Rh-(Ph, Cy)	1	100 [b)]	330 [b)]	97	17
4.4	DMIT	Rh-(Ph, Cy)	1	100 [b)]	200 [b)]	99.9	17
4.5	DMIT	Rh-**L6**	1	200 [b)]	200 [b)]	99.5	20
4.6	**6**	Rh-(3,5-$(CF_3)_2$Ph, Cy)	1	100 [b)]	90 [b)]	92	4
4.7	**7**	Cu-(Ph, Cy)	c)	100 [b)]	~16 [b)]	94	21
4.8	**8a**	Cu-(Ph, Cy)	c)	1640	~75	98	22
4.9	**8b**	Cu-(Ph, Cy)	c)	100 [b)]	>10 [b)]	99	22
4.10	**8c**	Cu-(Ph, Cy)	c)	100 [b)]	>10 [b)]	99	22

a) See Fig. 25.8, Cy=cyclohexyl.
b) Standard test results, not optimized.
c) Reducing agent: polymethylhydrosiloxane/NaO*t*Bu.

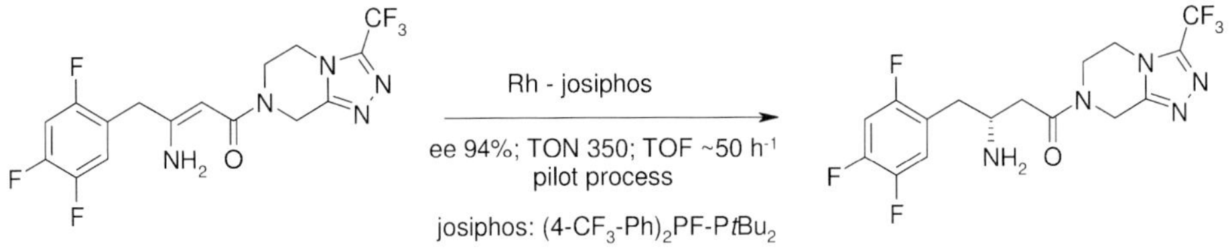

Rh - josiphos

ee 94%; TON 350; TOF ~50 h^{-1}
pilot process

josiphos: (4-CF_3-Ph)$_2$PF-P*t*Bu_2

Scheme 25.2 Hydrogenation of the β-dehydro amino acid amide intermediate for MK-0431.

As can be seen from Table 25.4, several Rh-josiphos complexes are excellent catalysts for the hydrogenation of α-dehydro amino acid derivatives and DMIT with ee-values of 97–99.9% (entries 4.1–4.5). While the reactions have not been optimized, satisfactory TONs and TOFs have been observed. Good ee-values are also obtained for a β-dehydro amino acid derivative (entry 4.6) and for the Cu-catalyzed reduction with PMHS of activated C=C bonds (entries 4.7–4.10), albeit with relatively low TOFs. Interestingly, in all cases the best ligands have unsubstituted or electron-deficient aryl groups on the ring phosphorus and a PCy_2 group at the side chain.

Recently, Merck chemists reported the Rh-josiphos-catalyzed hydrogenation of unprotected dehydro β-amino acids with ee-values up to 97%, but relatively low activity [23]. It was also shown that not only simple derivatives but also the complex intermediate for MK-0431 depicted in Scheme 25.2 can be hydrogenated successfully, and this has been produced on a >50 kg scale with ee-values up to 98%, albeit with low to medium TONs and TOFs [24].

25.4.2
Immobilized Josiphos and Josiphos Analogues

Several josiphos ligands were functionalized at the lower C_p ring and grafted to silica gel or a water-soluble group [25a] to give very active catalysts for the Ir-catalyzed MEA imine reduction; a Rh-josiphos complex grafted to several dendrimers (e.g., see Fig. 25.11) hydrogenated DMIT with ee-values up to 98.6% with similar activities as the mononuclear catalyst [25b]. Salzer and co-workers [20] prepared a number of josiphos analogues based on an arene chromium tricarbonyl scaffold (**L8**) and tested their Rh complexes on several alkenic substrates. With few exceptions, relatively low ee-values and catalyst activities were observed: ≤79% ee for DMIT, ≤87% for MAC, ≤91% for MAA, and ≤85% for the β-dehydro amino acid derivative **6**. Weissensteiner and co-workers [26] described two josiphos analogues **L9** and **L10** with restricted rotation of the side chain, and observed a strong decrease in enantioselectivity (best ee ≤91% for DMIT with **L10**, R′=Cy).

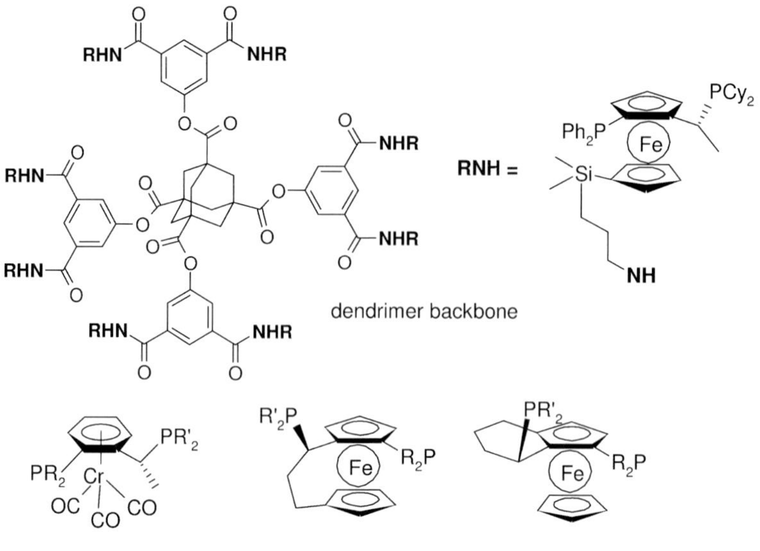

Fig. 25.11 Structure of immobilized josiphos and josiphos analogues.

25.4.3 Taniaphos

Compared to the josiphos ligands, taniaphos ligands have an additional phenyl ring inserted at the side chain of the Ugi amine. Whereas the effect of changing the two phosphine moieties has only been investigated with a few derivatives (Table 25.5, entries 5.3, 5.4, 5.5, 5.8–5.10), the nature of the substituent at the stereogenic center has a strong effect on the induction of stereochemistry for the Rh-catalyzed hydrogenation of MAC and DMIT. Rather surprisingly, a change of the substituent can even lead to a different sense of induction. For MAC, methyl or methoxy substituents lead to the opposite absolute configuration of the product compared to R=NMe_2, *i*-Pr or H (entries 5.1–5.4). Similar effects are also observed for DMIT (entries 5.7– 5.11) and for the hydrogenation of enol acetate **9** where ee-values up to 98% but low activities are achieved (entry 5.13). Interestingly, changing the absolute configuration of the stereogenic center only has an effect on the level of the ee but not on the sense of induction (compare entries 3/4 and 7/8). Enamides **10** are hydrogenated with high ee-values but low TOFs (entries 5.14 and 5.15). Currently, several taniaphos ligands are being marketed by Solvias in collaboration with Umicore (formerly OMG) [10].

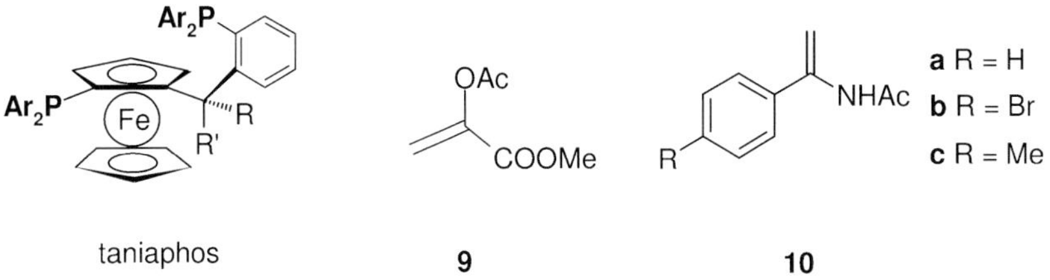

Fig. 25.12 Structure of taniaphos derivatives and substrates listed in Table 25.5.

Table 25.5 Selected results for Rh-catalyzed hydrogenations using taniaphos (Ar=Ph, pH_2 1 bar) (for structures, see Fig. 25.12).

Entry	Ligand (R, R′)	Substrate	$p(H_2)$ [bar]	TON [a]	TOF [a] $[h^{-1}]$	ee [%]	Comments	Reference(s)
5.1	(H, *i*Pr)	MAC	1	100	25	97 (*R*)	52% (*S*) for R′=Me!	27a
5.2	(H, NMe_2)	MAC	1	100	200	95 (*R*)	77% (*R*) for R′=H	27a
5.3	(H, OMe)	MAC	1	100	50	94 (*S*)	92% (*S*) for Ar=Xyl	27b
5.4	(OMe, H)	MAC	1	100	67	99 (*S*)	99% (*S*) for Ar=Xyl	27b
5.5	(H, NMe_2)	MAC	1	200	>200	99.5 (*S*)	Ar=3,5-Me_2-4-MeOPh	6, 27c
5.6	(H, NMe_2)	MAA	1	200	106	97	Ar=3,5-Me_2-4-MeOPh	6, 27c
5.7	(H, *i*Pr)	DMIT	1	100	25	98 (*S*)	19% (*R*) for R′=Me!	27a
5.8	(H, NMe_2)	DMIT	1	100	7	91 (*S*)	75% (*S*) for R′ = H [b]	27a
5.9	(H, NMe_2)	DMIT	1	200	>200	99.5 (*S*)	Ar=3,5-Me_2-4-MeOPh	6, 27c
5.10	(OMe, H)	DMIT	1	100	200	98 (*R*)	90% (*R*) for Ar=Xyl	27b
5.11	(H, OMe)	DMIT	1	100	40	95 (*R*)		27b
5.12	(H, NMe_2)	**6**	1	100	27	99.5 (*S*)	Ar = 3,5-Me_2-4-OMePh	6, 27c
5.13	(OMe, H)	**9**	1	100	5	98 (*S*)	80% (*S*) for (O, OMe)	27b
5.14	(OMe, H)	**10a**	1	100	7	96	97% ee for **10b**	27b
5.15	(OMe, H) [c]	**10a**	1	100	67	92	95% ee for **10c**	27b

a) Standard test results, not optimized.
b) At 10 bar, low conversion at 1 bar.
c) Ar=Xyl.

25.4.3 Various Ligands

Bophoz [28] and **L11** [29] are modular ligands with a PR_2 group on the C_p ring and an aminophosphine or a phosphoramidite, respectively, at the side chain. Bophoz ligands are air-stable and effective for the Rh-catalyzed hydrogenation of a variety of enamides and itaconates with high ee-values, TONs and TOFs (Table 25.6, entries 6.1–6.3); depending on the solvent the stability of the N–PR_2 bond might be a critical issue. As observed for several ligands forming seven-membered chelates, high activities can be reached (maximum TOFs up 68 000 h^{-1}) and TONs up to 10 000 have been achieved [28c]. A feasibility study for the

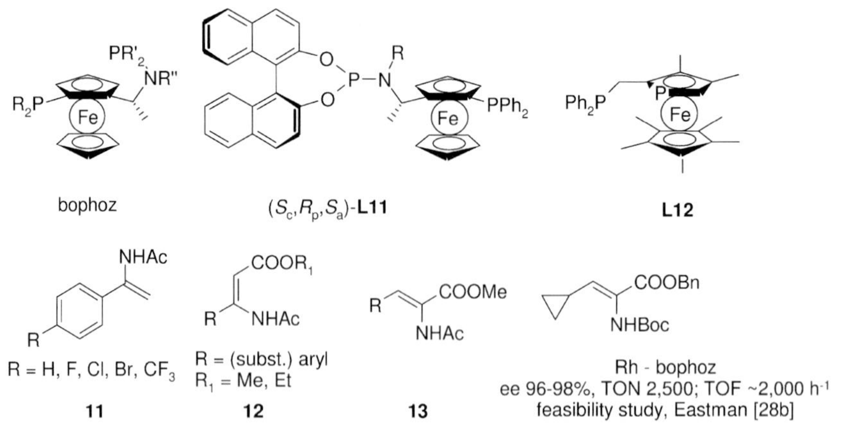

PR'2
NR"
R2P
Fe
bophoz
O
O
P–N
R
PPh2
(Sc,Rp,Sa)-L11
Ph2P
P
L12
NHAc
R
R = H, F, Cl, Br, CF3
11
COOR1
R
NHAc
R = (subst.) aryl
R1 = Me, Et
12
R
COOMe
NHAc
13
COOBn
NHBoc
Rh - bophoz
ee 96-98%, TON 2,500; TOF ~2,000 h-1
feasibility study, Eastman [28b]

Fig. 25.13 Structures of bophoz, **L11**, **L12** and of substrates listed in Table 25.6.

preparation of enantiopure cyclopropylalanine has been reported (Fig. 25.13). Ligands of the type **L11** have three elements of chirality (central, planar, and axial), and all combinations were actually prepared and tested for enamides **11**. As can be seen comparing entries 6.4–6.7 in Table 25.6, the absolute configuration of the binaphthol moiety determines the absolute configuration of the product; the relative configurations of the other chiral elements have a variable but usually very strong effect on the magnitude of the ee. Best results were achieved for the (S_c,R_p,S_a)-**L11** diastereomer shown in Fig. 25.13 (entry 6.8), which also shows very high enantioselectivities and very good TONs and TOFs for DMIT (entry 6.9). MAC was also hydrogenated very effectively, but only when 2 equiv. of ligand were added (entry 6.10); β-dehydroamino esters **12** are also good substrates for Rh/**L11** catalysts (entries 6.11, 6.12). Ligand **L12** with only planar chirality does not quite fit in this category since one of the P atoms is not attached to but is part of the C_p ring. The corresponding Rh complexes achieve respectable ee-values for several dehydroamino esters, but have very low activity (entry 6.13).

25.5 Ligands with Phosphine Substituents Bound only to Side Chains

Until now, only two families of ligands have been realized where both P groups are attached to side chains, probably because the resulting metal complexes have relatively large chelate rings which usually are not suitable for enantioselective catalysis. A cursory inspection of the ligands depicted in Fig. 25.14 shows that, due to steric bulk of the ferrocene backbone, both diphosphines probably have sufficiently restricted flexibility so that good stereocontrol is still possible.

The starting point for walphos was also the Ugi amine. Like josiphos, walphos ligands are modular but form eight-membered metallocycles due to the

Table 25.6 Selected results for Rh-catalyzed hydrogenation using bophoz, **L11**, and **L12** (for structures, see Fig. 25.13).

Entry	Ligand	Substrate	$p(H_2)$ [bar]	SCR	TOF [h^{-1}]	ee [%]	Comments	Reference
6.1	bophoz[a]	MAC	~1	10000	10000	97	99.4% ee for MAA	28c
6.2	bophoz[a]	subst MAC	~1	100[b)]	~100[b)]	97–99	various derivatives	28c
6.3	bophoz[c]	subst ITA	~3	2500	n.a.	94–99	various derivatives	28c
6.4	(S_c,R_p,S_b)-**L11**	**11** (R=H)	10	100[b]	100[b)]	99.6	(*R*)-product	29a
6.5	(S_c,R_p,R_b)-**L11**	**11** (R=H)	10	100[b)]	100[b)]	11	(*S*)-product	29a
6.6	(S_c,S_p,R_b)-**L11**	**11** (R=H)	10	100[b)]	100[b)]	99.6	(*S*)-product	29a
6.7	(S_c,S_p,S_b)-**L11**	**11** (R=H)	10	100[b)]	100[b)]	83	(*R*)-product	29a
6.8	(S_c,R_p,S_b)-**L11**	**11** (R=H)	10	5000	5000	99.3	for other R ee ~99%	29a
6.9	(S_c,R_p,S_b)-**L11**	DMIT	10	10000	20000	>99		29a
6.10	(S_c,R_p,S_b)-**L11**	MAC	10	10000	10000	>99	2 equiv. ligand	29a
6.11	(S_c,R_p,S_b)-**L11**	**12**	10	100[b)]	~80[b)]	97–>99	ee >99% for R=H	29b
6.12	(S_c,R_p,S_b)-**L11**	**12** (R=H)	10	5000	~4000	97	ee 98% at S/C 1000	29b
6.13	**L12**	**13** (R=Et)	1	20[b)]	<2[b)]	96	MAC 87% ee	30

a) R, R'=Ph, R''=H or Me at SCR 100, ee 99.1%.
b) Standard test results, not optimized.
c) R, R'=Ph, R''=H or Me.

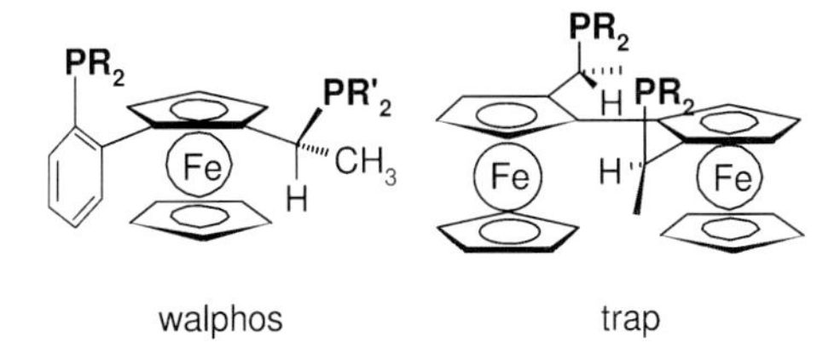

Fig. 25.14 Structures and names of walphos and trap ligands.

Table 25.7 Selected results for Rh- and Cu-catalyzed hydrogenation using (R,R′)-walphos and R-trap ligands (for ligand and substrates structures, see Figs. 25.14 and 25.16, respectively).

Entry	M – (R, R′)	Substrate	$p(H_2)$ [bar]	SCR[a]	TOF $[h^{-1}]^a$	ee [%]	Comments	Reference(s)
7.1	Rh-(Ph, Ar_1)[b)]	MAC	1	200[b)]	≥10b[b)]	95	ee 94% for Ar′=Ph	31b
7.2	Rh-(Ph, Ar_1)[b)]	DMIT	1	200[b)]	≥10b[b)]	92		31b
7.3	Cu-(Ph, Ar_2)[c)]	**8**	[c)]	100	>10	97	R_1 Me, R_2 $(CH_4)_2OR$, R_3 Et	22
7.4	Cu-(Ph, Ar_2)[c)]	**8**	[c)]	100	>10	97	R_1 *t*Bu, R_2 Bu, R_3 Me	22
7.5	Rh-Ph-trap[d)]	**14a**	50	100	50[c)]	94 (*R*)	ee 7% (*S*) without Cs_2CO_3	33a
7.6	Rh-Ph-trap[d)]	**14b**	50	100	200	95	ee 78% for Boc derivative	33a
7.7	Rh-Et-trap[f)]	MAA	0.5	100	50[e)]	96 (*R*)	ee 70/2%(!) at 1/100 bar	33b
7.8	Rh-*i*Bu-trap	MAC	1	100	4	92 (*S*)	R=Et[f)] ee 77% (*R*)!	33b
7.9	Rh-Et-trap[g)]	ITA	1	200	30	96	ee 68% for DMIT	33c
7.10	Rh-Pr-trap	**15**, **16a**	1	100	4	97	ee 82% for **16b**	33e,f
7.11	Rh-*i*Bu-trap	**17**	1	100	5	97	intermediate for indinavir	33d

a) Standard test results, not optimized.
b) Ar_1=3,5-Me_2-4-MeO-Ph.
c) Ar_2=3,5-$(CF_3)_2$-Ph, reducing agent PMHS/NaO*t*Bu.
d) In presence of Cs_2CO_3.
e) At 60 °C.
f) For R=Pr/Ph/*i*Pr, ee=85% (*R*)/21% (*S*)/5% (*S*), respectively.
g) For R=*i*Bu, ee=17%.

additional phenyl ring attached to the cyclopentadiene ring [31]. They also show promise for the enantioselective hydrogenation of dehydroamino and itaconic acid derivatives (Table 25.7, entries 7.1 and 7.2), and the Cu-catalyzed enantioselective reduction of *α,β*-unsaturated ketones **8** (entries 7.3 and 7.4). There are noticeable electronic effects, but the scope of this ligand family is still under investigation; several derivatives are available from Solvias on a technical scale [10]. The first industrial application has just been realized in collaboration with Speedel/Novartis for the hydrogenation of SPP100-SyA, a sterically demanding *α,β*-unsaturated acid intermediate of the renin inhibitor SPP100 (Fig. 25.15). The process has already been operated on a multi-100 kg scale.

The trap (*trans*-chelating phosphines) ligands developed by Ito and co-workers [33] form nine-membered metallocycles where trans-chelation is possible. However, it is not clear whether the *cis* isomer which has been shown to be present in small amounts or the major *trans* isomer is responsible for the catalytic activ-

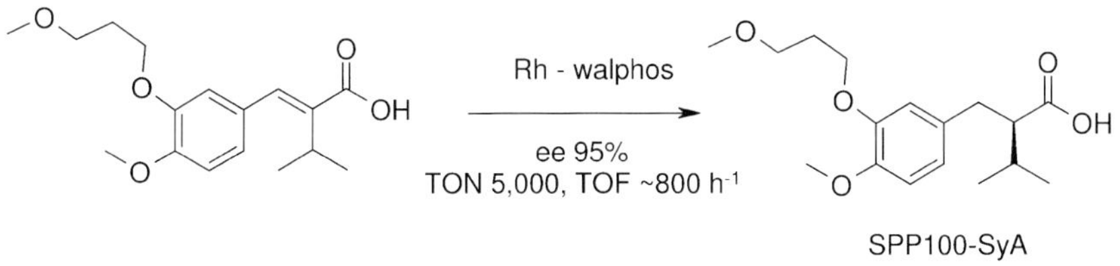

Rh - walphos

ee 95%

TON 5,000, TOF ~800 h^{-1}

SPP100-SyA

Fig. 25.15 Pilot-scale application of the walphos ligand (R=Ph, R′=3,5-$(CF_3)_2$-Ph) [32].

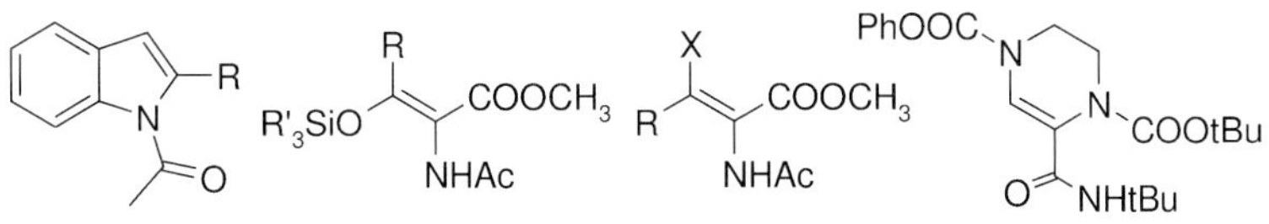

14a R = Me
14b R = COOEt

15

16a X = OOC*t*Bu
16b X = NHCbz

17

Fig. 25.16 Substrate structures listed in Table 25.7.

ity. Until now, only a few different PR_2 fragments have been tested, but it is clear that the choice of R strongly affects the level of enantioselectivity and sometimes even the sense of induction (e.g., see Table 25.7, entries 7.7 and 7.8). The Rh complexes function best at very low pressures of 0.5–1 bar, but often need elevated temperatures (e.g., entry 7.7). Effectively reduced are indole-derivatives **14** (entries 7.5, 7.6, the first examples of heteroaromatic substrates with high ee-values), dehydroamino (entries 7.7, 7.8; best ligand PEt_2-trap, unusual p and T effects), and itaconic acid derivatives (entry 7.9). β-Hydroxy–α-amino acids and α,β-diamino acids can be prepared via asymmetric hydrogenation of tetra-substituted alkenes **14–16** with respectable de-values of 99–100% and ee-values of 97% and 82%, respectively, but low catalyst activities (entry 7.10). Also described was the hydrogenation of an indinavir intermediate **17** (entry 7.11).

25.6 Major Applications of Ferrocene Diphosphine-Based Catalysts

As can be seen in the preceding section, ferrocene-based complexes are very versatile ligands for the enantioselective hydrogenation of a variety of alkenes. One reason for this is undoubtedly the modularity of most of the described ligand families which allows them to influence the activity and enantioselectivity in an extraordinarily broad range. In the following section a short overview is provided of substrates where ferrocene-based ligands define the state of the art not only for alkene hydrogenation but also for the enantioselective reduction of C=O and C=N groups. A comparison with other classes of ligands can be found in an above-mentioned review [3].

25.6.1 Hydrogenation of Substituted Alkenes

Rh complexes of ferrocene-based ligands are very effective for the hydrogenation of several types of α- and β-dehydroamino (Fig. 25.17, structures **18–22**), enamides (**23**) and enol acetates (**24**), as well as for itaconic acid derivatives (**25**, **26**) and α,β-unsaturated acids (**27**). Of particular interest are substrates which have unusual substituents (**19**, **21**) at the C=C moiety or are more sterically hindered than the usual model compounds (**20**, **27**). Cu complexes of ferrocenyl diphosphines are very effective for the reduction of nitroalkenes **28** and α,β-unsaturated ketones **29** with high chemoselectivity. Effective metal/ligand combinations with very high ee-values and often respectable TONs and TOFs are listed in Table 25.8. Several industrial applications have already been reported using Rh-josiphos and Ru-josiphos (see Fig. 25.10), as well as for Rh-bophoz (see Fig. 25.13) and Rh-walphos (see Fig. 25.14).

25.6.2 Hydrogenation of C=O and C=N Functions

Ferrocene-based complexes have some potential for the enantioselective reduction of ketones, but compared to other ligand classes this is relatively limited [3]. Rh complexes of bppfa, bophoz and josiphos are among the most selective catalysts for the hydrogenation of α-functionalized ketones (Table 25.9; Fig. 25.18, **30–32**). Ru complexes of walphos and ferrotane are quite effective for

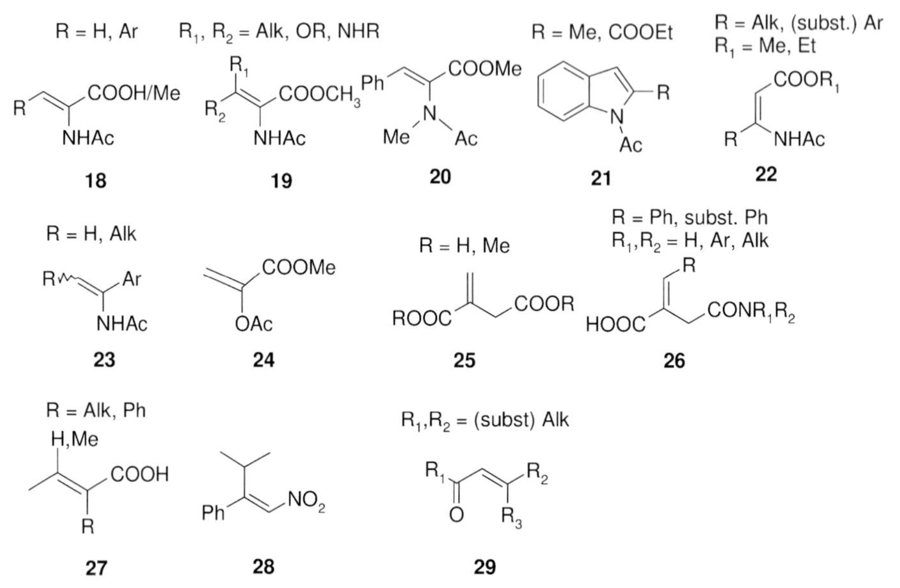

Fig. 25.17 Structures of substrates listed in Table 25.8.

Table 25.8 Best SCRs, TOF and ee-values for the reduction of selected functionalized alkenes (for substrates, see Fig. 25.17).

Substrate	Metal-ligand	TON	TOF [h^{-1}]	ee [%]
18	Rh-bophoz, Rh-josiphos, Rh-mandyphos, Rh-taniaphos, Rh-**L2**, Rh-**L4**	Up to 20000	Up to 10000	98–>99
19	Rh-trap	100 a)	4 a)	97
20	Rh-**L5**	200 a)	15 a)	96–98
21	Ru-taniaphos, Rh-trap	100–200 a)	25–200 a)	94–96
22	Rh-josiphos, Rh-taniaphos, Rh-**L11**	Up to 5000	Up to ~4000	92–>99
23	Rh-**L3**, Rh-taniaphos, Rh-**L11**	Up to 5000	Up to 6000	92–98
24	Rh-mandyphos, Rh-taniaphos, Rh-**L11**	Up to 10000	Up to 20000	95–98
25	Rh-josiphos, Rh-taniaphos, Rh-**L3**, Rh-**L4**	200 a)	>200 a)	97–99.9
26	Rh-ferrotane	Up to 20000	Up to 7000	98–99
27	Rh-bppfa, Rh-mandyphos, Rh-walphos	Up to 5000	Up to ~800	95
28	Cu-josiphos	100 a)	~16 b)	94
29	Cu-josiphos, Cu-walphos	1640	10–80 a)	97–98

a) Standard test results, not optimized.

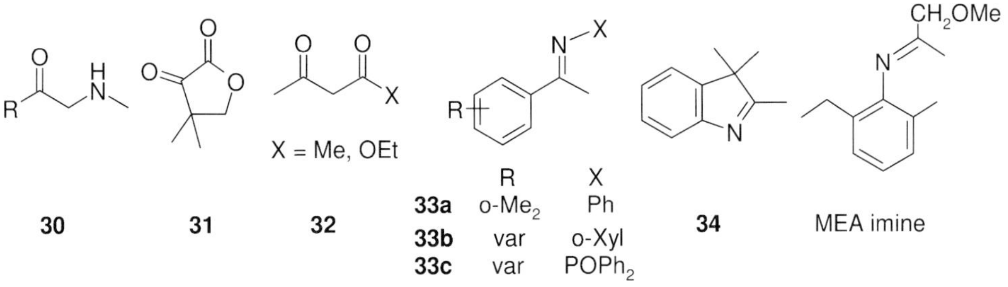

Fig. 25.18 Structures of ketone and imine substrates listed in Table 25.9.

β-keto esters and diketones **32**, though this is usually the domain of Ru-binap-type catalysts. Josiphos and f-binaphane however are the ligands of choice for the Ir-catalyzed hydrogenation of N-aryl imines such as **33** and **34**. Special mention should be made of the Ir-josiphos catalyst system which is able to hydrogenate MEA imine with TONs up to 2×10^6 [18].

Table 25.9 Best catalysts for the hydrogenation of C=O and C=N functions (for substrates, see Fig. 25.18).

Substrate	Metal-ligand (additive)	TON	TOF $[h^{-1}]$	ee [%]
30	Rh-bppfoh, bppfsh	200–2000	2–125	95–>99
31	Rh-bophoz, josiphos	100–200 [a)]	1 [a)]–>1000	97–99
32	Ru-walphos, ferrotane	5–200 [a)]	<1–25 [a]	95–99
33 a	Ir-josiphos/I^-/H^+	200 [a)]	n.a.	96
33 b	Ir-f-binaphane/I_2	100 [a)]	2 [a)]	>99
33 c	Rh-josiphos	500	500	99
34	Ir-josiphos/I^-/H^+	250 [a)]	56 [a)]	93
MEA imine	Ir-josiphos/I^-/H^+	2 000 000	>400 000	80

a) Standard test results, not optimized.
n.a. = data not available.

Abbreviations

ACA acetamido cinnamic acid
DMIT dimethyl itaconate
ee enantiomeric excess
ITA itaconic acid
MAA methyl acetamido acrylate
MAC methyl esters of acetamido cinnamic acid
SCR substrate:catalyst ratio
TOF turnover frequency
TON turnover number

References

1 For an account, see T. Hayashi, in: A. Togni, T. Hayashi (Eds.), *Ferrocenes.* VCH, Weinheim, **1995**, p. 105.
2 (a) T.J. Colacot, *Chem. Rev.* **2003**, *103*, 3101; (b) P. Barbaro, C. Bianchini, G. Giambastiani, S.L. Parisel, *Coord. Chem. Rev.* **2004**, *248*, 2131.
3 H.U. Blaser, Ch. Malan, B. Pugin, F. Spindler, H. Steiner, M. Studer, *Adv. Synth. Catal.* **2003**, *345*, 103.
4 W. Tang, X. Zhang, *Chem. Rev.* **2003**, *103*, 3029.
5 G. Argouarch, O. Samuel, O. Riant, J.-C. Daran, H.B. Kagan, *Eur. J. Org. Chem.* **2000**, 2893.
6 M. Kesselgruber, C. Malan, B. Pugin, F. Spindler (Solvias AG), unpublished results.
7 T. Hayashi, N. Kawamura, Y. Ito, *J. Am. Chem. Soc.* **1987**, 1, 7876; T. Hayashi, N. Kawamura, Y. Ito, *Tetrahedron Lett.* **1988**, *29*, 5969.
8 J. Kang, J.H. Lee, S.H. Ahn, J.S. Choi, *Tetrahedron Lett.* **1998**, *39*, 5523.
9 (a) M. Lotz, T. Ireland, J. Almena Perea, P. Knochel, *Tetrahedron: Asymmetry* **1999**, *10*, 1839; (b) J. Almena Perea, A. Börner, P. Knochel, *Tetrahedron Lett.* **1998**, *39*, 8073; (c) J. Almena Perea, M. Lotz, P. Knochel, *Tetrahedron: Asymmetry* **1999**, *10*, 375; d) F. Spindler, C. Malan,

M. Lotz, M. Kesselgruber, U. Pittelkow, A. Rivas-Nass, O. Briel, H.U. Blaser, *Tetrahedron: Asymmetry* **2004**, *15*, 2299.

10 For more information see www.solvias.-com/ligands. M. Thommen, H.U. Blaser, *PharmaChem*, July/August, **2002**, 33.

11 D. Xiao, X. Zhang, *Angew. Chem.* **2001**, *113*, 3533.

12 (a) U. Berens, M.J. Burk, A. Gerlach, W. Hems, *Angew. Chem.* **2000**, *112*, 2057; (b) A. Marinetti, J.-P. Genet, S. Jus, D. Blanc, V. Ratovelamanana-Vidal, *Chem. Eur. J.* **1999**, *5*, 1160; (c). A. Marinetti, D. Carmichael, *Chem. Rev.* **2002**, *102*, 201.

13 M.F. Reetz, E.W. Beuttenmüller, R. Goddard, M. Pasto, *Tetrahedron Lett.* **1999**, *40*, 4977.

14 M.T. Reetz, A. Gosberg, R. Goddard, S.-H. Kyung, *Chem. Commun.* **1998**, 2077; M. T. Reetz, A. Gosberg, WO 0014096, **1998** (assigned to Studiengesellschaft Kohle MBH).

15 D. Liu, W. Li, X. Zhang, *Org. Lett.* **2002**, *4*, 4471.

16 (a) For overviews, see M. Ohff, J. Holz, M. Quirmbach, A. Börner, *Synthesis*, **1998**, 1391; F. Maienza, F. Spindler, M. Thommen, B. Pugin, A. Mezzetti, *Chimia* **2001**, *55*, 694; (b) F. Maienza, M. Wörle, P. Steffanut, A. Mezzetti, F. Spindler, *Organometallics* **1999**, *18*, 1041; (c) U. Nettekoven, P.C.J. Kamer, P.W.N.M. van Leeuwen, M. Widhalm, A.L. Spek, M. Lutz, *J. Org. Chem.* **1999**, *64*, 3996.

17 For an overview, see H.U. Blaser, W. Brieden, B. Pugin, F. Spindler, M. Studer, A. Togni, *Topics in Catalysis* **2002**, *19*, 3.

18 H.U. Blaser, H.P. Buser, K. Coers, R. Hanreich, H.P. Jalett, E. Jelsch, B. Pugin, H.D. Schneider, F. Spindler, A. Wegmann, *Chimia* **1999**, *53*, 275.

19 H.U. Blaser, F. Spindler, M. Studer, *Applied Catal. A: General* **2001**, *221*, 119.

20 W. Braun, A. Salzer, F. Spindler, E. Alberico, *Appl. Catal. A: General* **2004**, *274*, 191.

21 C. Czekelius, E.M. Carreira, *Angew. Chem. Int. Ed.* **2003**, *42*, 4793; C. Czekelius, E.M. Carreira, *Org. Lett.* **2004**, *6*, 4575.

22 B.H. Lipshutz, J.M. Servesko, *Angew. Chem. Int. Ed.* **2003**, *42*, 4789.

23 Y. Hsiao, N.R. Rivera, T. Rosner, S.W. Krska, E. Njolito, F. Wang, Y. Sun, J.D. Armstrong, E.J.J. Grabowski, R.D. Tillyer, F. Spindler, C. Malan, *J. Am. Chem. Soc.* **2004**, *126*, 9918.

24 M. Rouhi, *Chem. Eng. News* **2004**, *82*(37), 28.

25 (a) B. Pugin, H. Landert, F. *Spindler*, H.-U. Blaser, *Adv. Synth. Catal.* **2002**, *344*, 974; (b) C. Köllner, B. Pugin, A. Togni, *J. Am. Chem. Soc.* **1998**, *120*, 10274.

26 T. Sturm, W. Weissensteiner, F. Spindler, K. Mereiter, A.M. Lopez-Agenjo, F.A. Jalon, *Organometallics* **2002**, *21*, 1766.

27 (a) T. Ireland, K. Tappe, G. Grossheimann, P. Knochel, *Chem. Eur. J.* **2002**, *8*, 843; (b) M. Lotz, K. Polborn, P. Knochel, *Angew. Chem. Int. Ed.* **2002**, *41*, 4708; (c) F. Spindler, C. Malan, M. Lotz, M. Kesselgruber, U. Pittelkow, A. Rivas-Nass, O. Briel, H.U. Blaser, *Tetrahedron: Asymmetry* **2004**, *15*, 2299.

28 (a) N.W. Boaz, S.D. Debenham, E.B. Mackenzie, S.E. Large, *Org. Lett.* **2002**, *4*, 2421; (b) N.W. Boaz, S.D. Debenham, S.E. Large, M.K. Moore, *Tetrahedron: Asymmetry* **2003**, *14*, 3575; (c) N.W. Boaz, E.B. Mackenzie, S.D. Debenham, S.E. Large, J.A. Ponasik, *J. Org. Chem.* **2005**, *70*, 1872.

29 (a) X-P. Hu, Z. Zheng, *Org. Lett.* **2004**, *6*, 3585; (b) X-P. Hu, Z. Zheng, *Org. Lett.* **2005**, *7*, 419.

30 S. Qiao, G.C. Fu, *J. Org. Chem.* **1998**, *63*, 4168.

31 (a) T. Sturm, L. Xiao, W. Weissensteiner, *Chimia* **2001**, *55*, 688; (b) W. Weissensteiner, T. Sturm, F. Spindler, *Adv. Synth. Catal.* **2003**, *345*, 160.

32 P. Herold, S. Stutz, T. Sturm, W. Weissensteiner, F. Spindler, WO 02/02500, **2002** (assigned to Speedel Pharma AG).

33 (a) R. Kuwano, K. Sato, T. Kurokawa, D. Karube, Y. Ito, *J. Am. Chem. Soc.* **2000**, *122*, 7614; (b) R. Kuwano, M. Sawamura, Y. Ito, *Bull. Chem. Soc. Jpn.* **2000**, *73*, 2571; (c) R. Kuwano, M. Sawamura, Y. Ito, *Tetrahedron: Asymmetry* **1995**, *6*, 2521; (d) R. Kuwano, Y. Ito, *J. Org. Chem.* **1999**, *64*, 1232; (e) R. Kuwano, S. Okuda, Y. Ito, *J. Org. Chem.* **1998**, *63*, 3499; (f) R. Kuwano, S. Okuda, Y. Ito, *Tetrahedron: Asymmetry* **1998**, *9*, 2773.

26
The other Bisphosphine Ligands for Enantioselective Alkene Hydrogenation

Yongxiang Chi, Wenjun Tang, and Xumu Zhang

26.1
Introduction

This chapter describes atropisomeric biaryl bisphosphine ligands; modified DIOP-type ligands; P-chiral bisphosphane ligands; other bisphosphane ligands; and their applications in the enantioselective hydrogenation of olefins.

26.2
Chiral Bisphosphine Ligands

26.2.1
Atropisomeric Biaryl Bisphosphine Ligands

In 1980, Noyori and Takaya reported an atropisomeric C_2-symmetric bisphosphine ligand, BINAP [1]. This ligand was first used in Rh-catalyzed enantioselective hydrogenation of α-(acylamino)acrylic acids, and high selectivities were reported for some substrates [2]. However, the significant impact of BINAP in asymmetric hydrogenation did not gain very much attention until it was applied in ruthenium chemistry. In 1986, Noyori and Takaya prepared a BINAP–Ru dicarboxylate complex for the asymmetric hydrogenation of various functionalized alkenes [3]. Subsequently, these authors discovered that the halogen-containing BINAP–Ru complexes were also efficient catalysts for enantioselective hydrogenation of a range of functionalized ketones [4]. During the mid-1990s, another major breakthrough was made on BINAP–Ru chemistry when Noyori discovered that the Ru–BINAP/diamine complexes were efficient catalysts for the enantioselective hydrogenation of some unfunctionalized ketones [5]. This advance addressed a long-standing challenging problem in enantioselective hydrogenation. Importantly, the catalytic system can selectively reduce ketones in the presence of carbon–carbon double or triple bonds [6]. Inspired by Noyori's studies on the BINAP chemistry, other research groups developed many excellent atropisomeric biaryl bisphosphine ligands. For example, Miyashima reported a BI-

The Handbook of Homogeneous Hydrogenation.
Edited by J. G. de Vries and C. J. Elsevier
Copyright © 2007 WILEY-VCH Verlag GmbH & Co. KGaA, Weinheim
ISBN: 978-3-527-31161-3

CHEP ligand, which was successfully applied in both Rh- and Ru-catalyzed enantioselective hydrogenation [7]. Schmid et al. reported BIPHEMP [8] and MeO-BIPHEP [9] ligands, both of which were successfully applied in many Ru-catalyzed hydrogenations. Achiwa also developed several atropisomeric ligands such as BIMOP [10], FUPMOP [11], and MOC-BIMOP (Fig. 26.1) [12].

Modification of the electronic and steric properties of BINAP, BIPHEMP, and MeO-BIPHEP can lead to the development of new efficient atropisomeric ligands (Fig. 26.1). In fact, Takaya has found that a modified BINAP ligand, H_8-BINAP, provides better enantioselectivity than BINAP in the Ru-catalyzed hydrogenation of unsaturated carboxylic acids [13]. Mohr has developed a bis-steroidal bisphosphine **1**, which has shown similar catalytic results to BINAP in the Ru-catalyzed enantioselective hydrogenation [14]. Hiemstra has developed a dibenzofuran-based bisphosphine BIFAP, which has shown excellent enantioselectivity in the Ru-catalyzed hydrogenation of methyl acetoacetate [15]. The dihedral angle of the biaryl backbone is expected to have a strong influence on the enantioselectivity. Another chiral biaryl bisphosphine ligand, SEGPHOS, was developed in Takasago. The ligand, which possesses a narrower dihedral angle than BINAP, has provided greater enantioselectivity than BINAP in the Ru-catalyzed hydrogenation of a wide variety of carbonyl compounds [16]. Chan [17a] and Genêt [17b,c] have reported a closely related ligand bisbenzodioxanPhos (SYNPHOS) independently. In order systematically to investigate the influence of the dihedral angle of biaryl ligands on enantioselectivity, Zhang has developed a series of TunePhos ligands with tunable dihedral angles. When the TunePhos ligands are applied in the Ru-catalyzed enantiomeric hydrogenation of β-keto esters, the ee-values obtained fluctuate with the different dihedral angles of the TunePhos ligands [18]. C4-TunePhos shows comparable or superior enantioselectivity to BINAP in Ru-catalyzed hydrogenation of β-keto esters. More applications of the TunePhos ligands have shown that different asymmetric catalytic reactions may require a different TunePhos ligand with a different dihedral angle. When TunePhos ligands are applied in the Ru-catalyzed hydrogenation of enol acetates, C2-TunePhos is the best ligand in terms of enantioselectivity [19]. However, C3-TunePhos provided the best enantioselectivities for the synthesis of cyclic β-amino acids [20], and the hydrogenation of α-phathalimide ketones [21]. Genêt and Marinetti have developed a non-C_2 symmetric biaryl bisphosphine, MeO-NAPhePHOS, which has shown comparable results to C_2-symmetric biaryl bisphosphine in Ru-catalyzed hydrogenation [22].

Structural variation of BINAP or MeO-BIPHEP can also be made on the aromatic rings of the biaryl backbone (Fig. 26.1). For example, the aromatic rings can be replaced by five- or six-membered heteroaromatic rings. Sannicolò et al. have discovered a series of biheteroaryl bisphosphines such as BITIANP, TetraMe-BITIANP [23], and TetraMe-BITIOP [24]. These ligands have shown comparably good results to BINAP in Ru-catalyzed enantioselective hydrogenation. Chan has reported a dipyridylphosphine ligand P-Phos for Ru-catalyzed enantioselective hydrogenation, and high enantioselectivities and reactivities have been obtained in the hydrogenation of β-keto esters, α-arylacrylic acids, and simple ketones [25].

An *ortho*-substituted BIPHEP ligand, *o*-Ph-HexaMeO-BIPHEP, has been recently developed by Zhang [26]. With two phenyl groups at the *ortho* positions of two diphenylphosphino groups, *o*-Ph-HexaMeO-BIPHEP is specially designed to restrict the rotation of the P-phenyl groups, which is considered to be detrimental for some enantioselective reactions. The design is effective when *o*-Ph-HexaMeO-BIPHEP is employed in the Rh-catalyzed enantioselective hydrogenation of cyclic enamides. While chiral ligands without *ortho*-substituents such as BINAP, BIPHEP, and HexaMeO-BIPHEP provide very poor selectivities, *o*-Ph-HexaMeO-BIPHEP shows excellent enantioselectivity in the hydrogenation of a series of cyclic enamides [26]. Zhang also reported another *ortho*-substituted BIPHEP type ligand – *o*-Ph-MeO-BIPHEP – which afforded excellent enantioselectivities in the hydrogenation of α-dehydroamino acids [27].

Henschke and Casy prepared a biaryl bisphosphine ligand, HexaPhemp, which performed as well as, or better than, the corresponding BINAP ligands [28]. Dellis and Genêt have developed a new electron-deficient atropisomeric ligand based on a SEGPHOS backbone, difluorphos, which has a narrow dihedral angle and electronic-withdrawing substituents. The electron-deficiency was shown to be crucial to reach high levels of enantioselectivity in hydrogenation of some challenging β-keto ester substrates [29].

Chan has discovered a completely atropdiasteroselective synthesis of a biaryl diphosphine based on an enantioselective intramolecular Ullmann coupling or a Fe(III)-promoted oxidative coupling. A chiral atropisomeric biaryl bisphosphine ligand **2** was synthesized through this central-to-axial chirality transfer [30]. Recently, a xylyl-biaryl bisphosphine ligand, Xyl-TetraPHEMP was introduced by Moran, and found to be effective for the Ru-catalyzed hydrogenation of aryl ketones [31].

A family of tunable 4,4′-substituted BINAP was reported by Lin: 4,4-$[SiMe_3]_2$-BINAP **3** and polar 4,4-$[P(O)OH_2]_2$-BINAP **4** have shown high enantioselectivities (up to 99.6% ee) in the hydrogenation of a variety of β-aryl ketoesters [32]. 4,4-$[SiMe_3]_2$-BINAP **3** is also effective for the asymmetric hydrogenation of α-phthalimide ketones and 1,3-diaryl diketones [33]. The 4,4′-bulky groups were shown to be responsible for the enhancement of enantioselectivity and diastereoselectivity in these reactions. Lemaire prepared 4,4′- or 5,5′-diamBINAP, with a 4,4′- or 5,5′-diaminomethyl substituent; the hydrosoluble HBr salt of the Ru complex based on these ligands afforded high enantioselectivity (>97% ee) in the water/organic solvent biphasic hydrogenation of β-keto esters [34]. Lemaire also reported 4,4′- or 5,5′-perfluoroalkylated BINAP, **5** and **6**, which showed the same activities and enantioselectivities as 4,4′- or 5,5′-diamBINAP in the hydrogenation of β-keto esters (Fig. 26.1) [35].

As with most chiral atropisomeric ligands, resolution or enantioselective synthesis is requisite. Mikami developed a novel ligand-accelerated hydrogenation catalyst in which the chirality of an atropos but achiral triphos–Ru complex could be controlled by chiral diamines. Using (*S*)-dm-dabn as controller, a single diastereomeric triphos–Ru complex was obtained through isomerization of the (*R*)-triphos–Ru complex in dichloroethane at 80 °C (Scheme 26.1) [36].

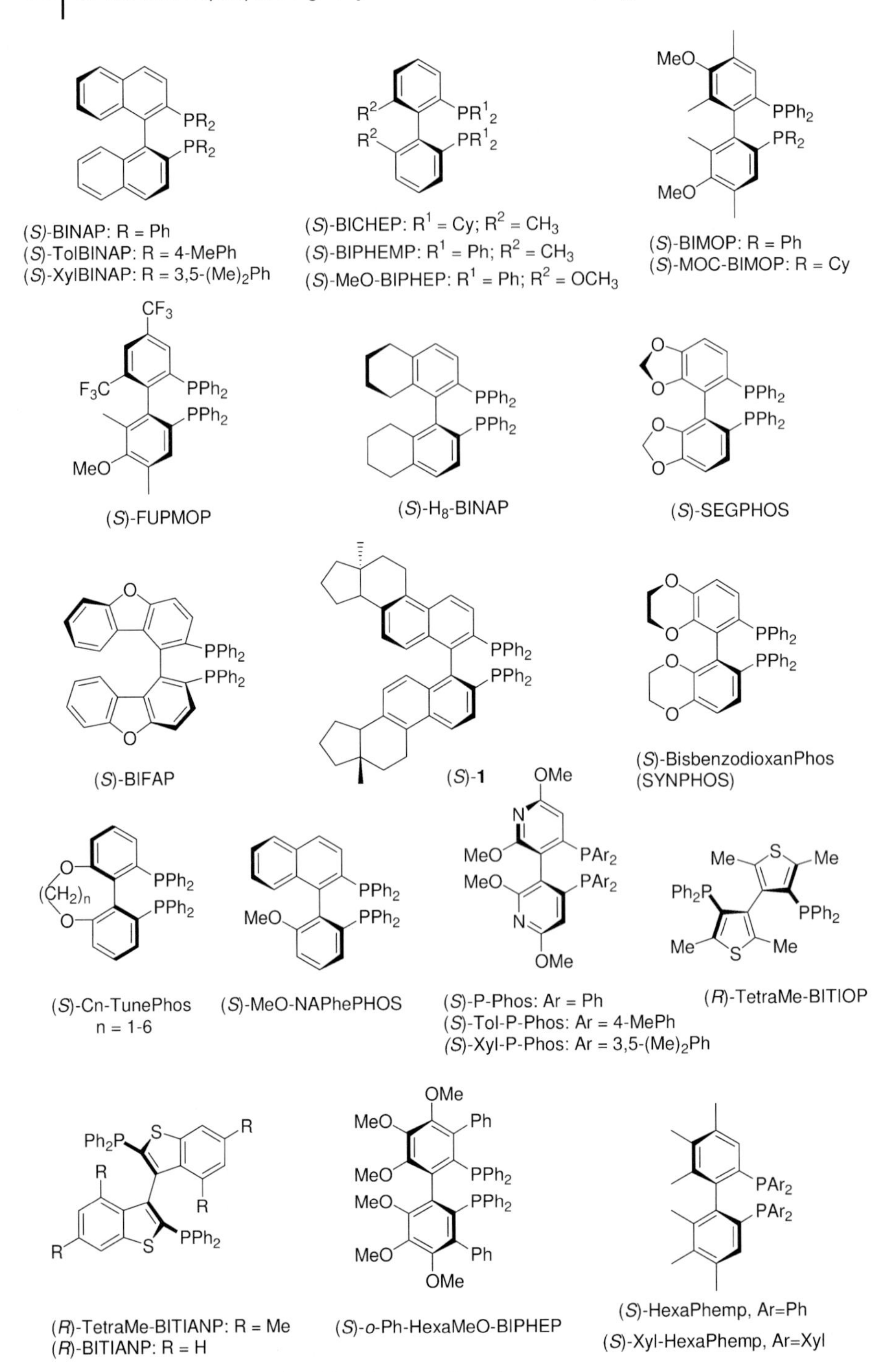

Fig. 26.1 Atropisomeric biaryl bisphosphine ligands.

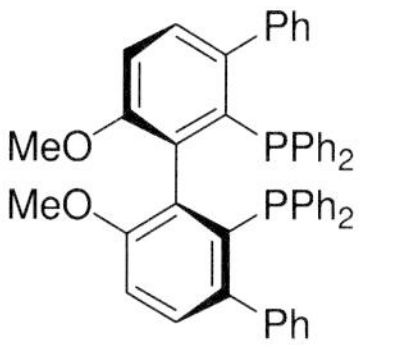

(*S*)-***o*-Ph-MeO-BIPHEP** 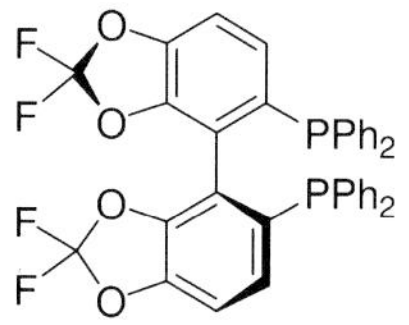 (*S*)-Difluorphos 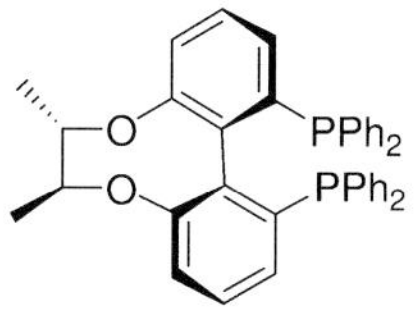 **2**

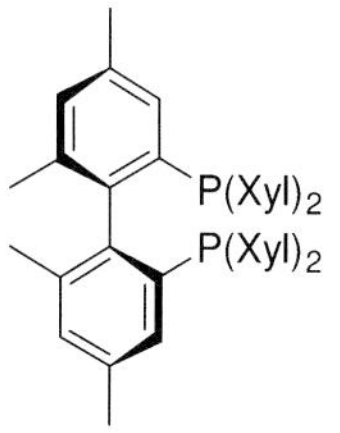

(*S*)-**Xyl-TetraPHEMP** 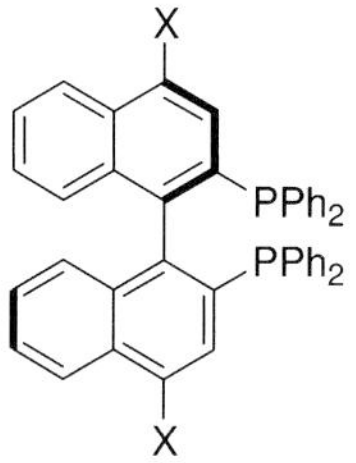 (*R*)-**3**, X=TMS (*R*)-**4**, X=P(O)(OH)$_2$

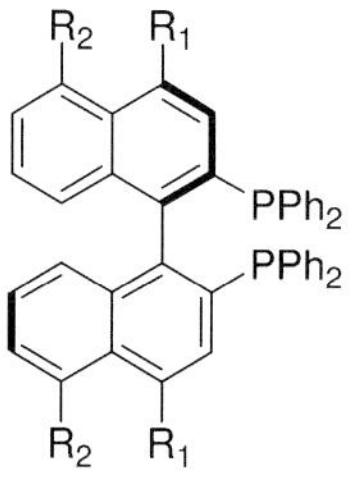

(*R*)-4,4'-diamBINAP, R_1=CH_2NH_2, R_2=H
(*R*)-5,5'-diamBINAP, R_1=H, R_2=CH_2NH_2
(*R*)-**5**, R_1=C_6F_{13}, R_2=H
(*R*)-**6**, R_1=H, R_2=C_6F_{13}

Fig. 26.1 (continued)

Some BINAP or BIPHEP derivatives have also been made in order to make the catalysts water-soluble or recyclable (Fig. 26.2). The literature on supported homogeneous catalysts in the field of asymmetric hydrogenation using BINAP derivatives has recently been reviewed [37]. Davis et al. reported a sulfonated BINAP ligand, BINAP-4-SO_3Na, and found that its water-soluble Ru complex has comparable catalytic properties to the unmodified BINAP-Ru catalyst for hydrogenation of 2-acetamidoacrylic acid [38]. Schmid et al. have developed a water-soluble MeO-BIPHEP type ligand, MeOBIPHEP-S. The ligand has the sulfonato group attached at the *para* position of each P-phenyl group to minimize the possible steric interactions of the sulfonato groups with the inner ligand sphere of a coordinated metal, and thus to retain the high enantioselectivity of the non-sulfonated catalyst. Indeed, MeOBIPHEP-S has shown similarly high enantioselectivity and reactivity to MeO-BIPHEP in the Ru-catalyzed hydrogenation of unsaturated carboxylic acids [39]. Genêt has recently reported some recyclable BINAP ligands such as *Digm*-BINAP and PEG-*Am*-BINAP, which were obtained by tethering BINAP with guanidine and PEG groups, respectively. The Ru catalysts of these ligands maintained high enantioselectivity after three or four recycles [40]. Many polymer-supported BINAP ligands have been developed. For instance, Bayston incorporated the BINAP framework onto an insoluble polymer (polystyrene). The resulting polymer-bound BINAP, after treatment with [Ru(cod)(2-methylallyl)$_2$]$_2$ and HBr, induces high ee-values in the hydrogenation of β-keto esters and acrylic acids [41]. The polymer can be recycled as the cata-

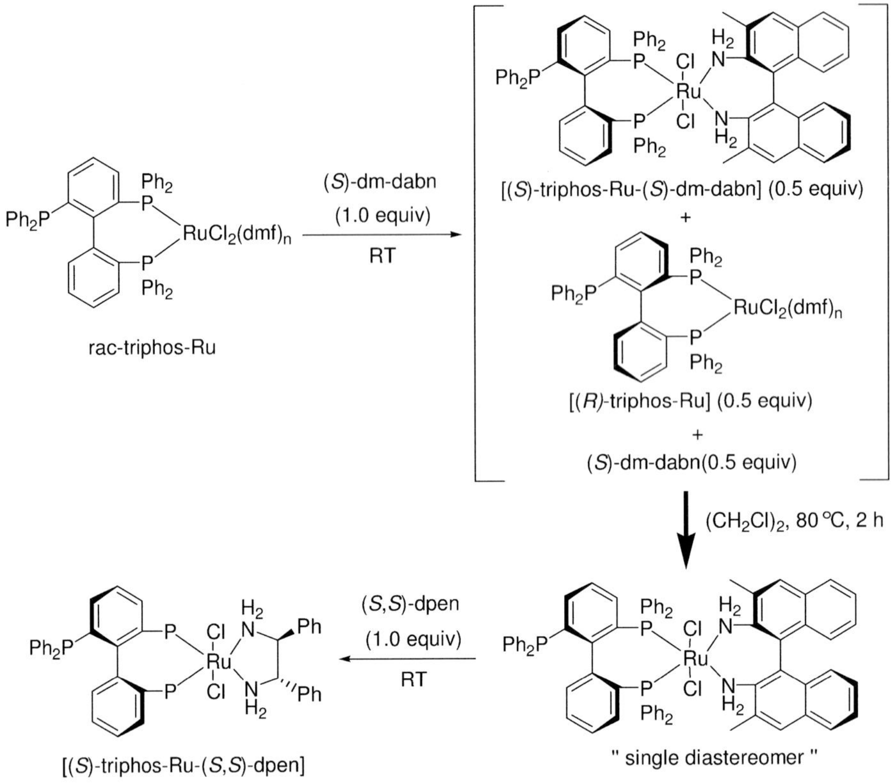

Scheme 26.1

lyst several times, while high ee-values are maintained. Noyori used the same polymer-bound BINAP to create a polymer-bound BINAP/diamine Ru catalyst, which has furnished high ee-values and turnover numbers (TONs) in the hydrogenation of simple ketones [42]. Chan has developed a highly effective polyester-supported BINAP ligand through copolymerization of chiral 5,5′-diaminoBINAP, chiral pentanediol, and terephthaloyl chloride [43]. The ligand has been successfully applied repeatedly in the Ru-catalyzed enantioselective hydrogenation of 2-(6′-methoxy-2-naphthyl)acrylic acid. A dendrimer-supported BINAP ligand has also been reported [44]. Pu has developed several polymer-based chiral ligands such as poly(BINAP) and BINOL-BINAP. These ligands have been applied successfully in the Rh-catalyzed hydrogenation of (*Z*)-methyl α-(benzamido)cinnamate and in the Ru-catalyzed hydrogenation of simple ketones [45]. Lemaire et al. have reported a poly-NAP Ru complex, which provides 99% ee in the hydrogenation of methyl acetoacetate, even after four recycles of the catalyst [46].

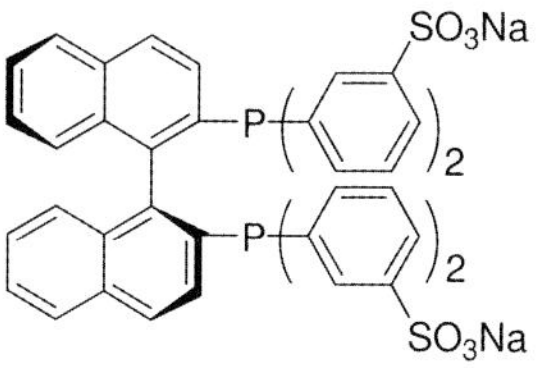

SO_3Na
P
2
P
2
SO_3Na

BINAP-4SO_3Na

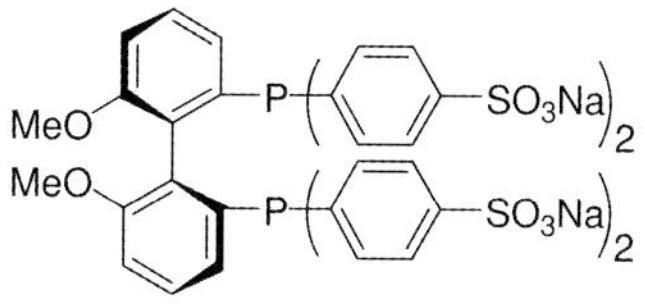

MeO
MeO
P
SO_3Na
2
P
SO_3Na
2

MeOBIPHEP-S

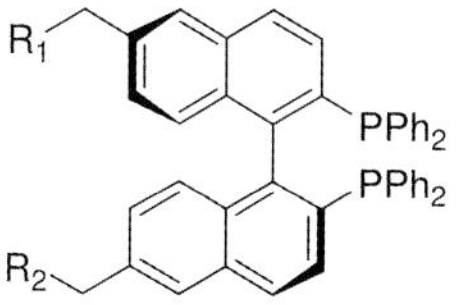

R_1
PPh_2
PPh_2
R_2

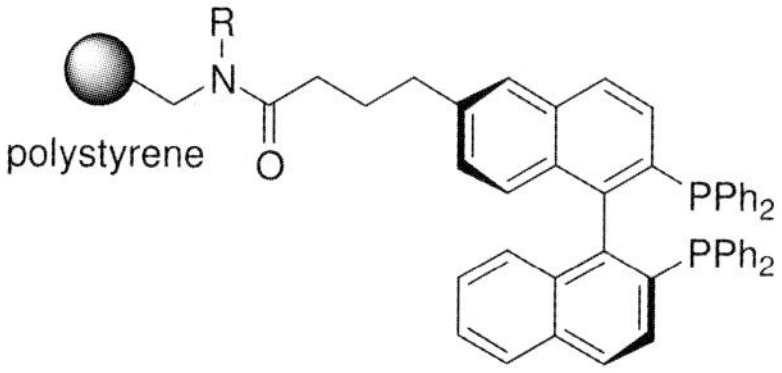

Digm-BINAP: $R_1 = R_2 =$ NH_2^+, Br^- H_2N N H

PEG-*Am*-BINAP: R_1 =
$MeO(CH_2CH_2O)n\text{-}CO(CH_2)_3CONH$, $R_2 = NH_2$

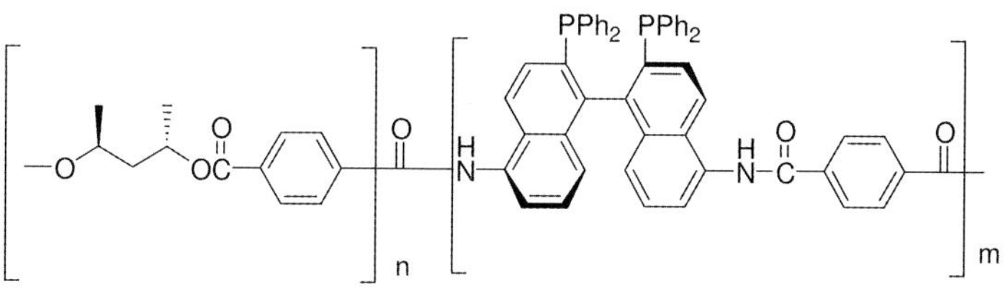

R
N
polystyrene
O
PPh_2
PPh_2

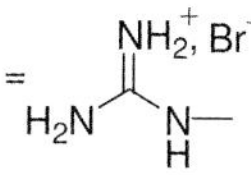

polystyrene-supported BINAP

PPh_2 PPh_2
O
O
O
O
OC
H
N
H
N
C
O
O
n
m

polyester-supported BINAP

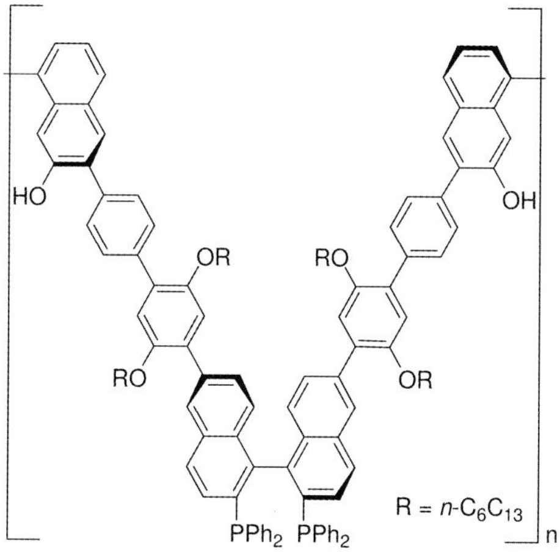

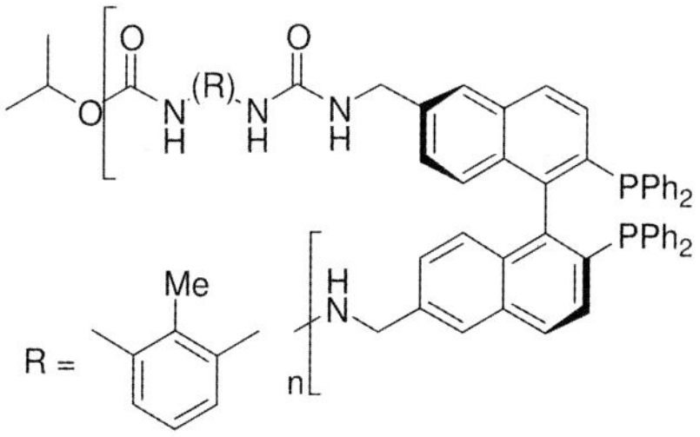

HO
OH
OR
RO
RO
OR
R = *n*-C_6C_{13}
PPh_2 PPh_2
n

BINOL-BINAP

O
O
O
N
H
(R)
N
H
N
H
PPh_2
PPh_2
Me
R =
H
N
n

poly-BINAP

Fig. 26.2 Water-soluble or recyclable BINAP or BIPHEP derivative catalysts.

26.2.2
Chiral Bisphosphine Ligands Based on DIOP Modifications

Kagan's pioneering studies on the development of DIOP has had significant impact on the design of new efficient chiral ligands for enantioselective hydrogenation [47]. However, DIOP itself provides only moderate to good enantioselectivity in the enantioselective hydrogenation of dehydroamino acid derivatives, and its application in highly enantioselective hydrogenation has rarely been disclosed. A possible reason for this is that the seven-membered chelate ring of the DIOP metal complex is conformationally flexible. These conformational ambiguities, as depicted in Figure 26.3, may be responsible for its low efficiency.

Achiwa successfully developed several modified DIOP ligands by varying the electronic and steric properties of DIOP. MOD-DIOP was applied in the rhodium-catalyzed enantioselective hydrogenation of itaconic acid derivatives, and up to 96% ee was obtained [48]. In order to rigidify the conformational flexibility of DIOP, Zhang has introduced a rigid 1,4-diphosphane ligand BICP with two five-membered carbon rings on its backbone (Fig. 26.4). BICP was found to be an efficient ligand for the hydrogenation of α-dehydroamino acids, β-dehydroamino acids, arylenamides, and MOM-protected β-hydroxy enamides [49]. Genov introduced several BICP family ligands, and developed a new catalytic system comprising Ru–**7** or Ru–**8** complexes in combination with a nonchiral 2-(alkylthio)amine or 1,2-diamine and an alkoxide as a base for the highly enantioselective hydrogenation of aryl ketones [50]. Several rigidified DIOP-type ligands have been developed. Zhang [51] and RajanBabu [52] have independently reported the development of DIOP* by introducing two alkyl substituents at the α-positions of the diphenylphosphine groups. The (*S*,*R*,*R*,*S*)-DIOP* was found to provide excellent enantioselectivity in the Rh-catalyzed hydrogenation of arylenamides and MOM-protected β-hydroxy enamides [51]. However, its isomeric ligand (*S*,*S*,*S*,*S*)-DIOP*, which was first synthesized by Kagan [53], provided much lower enantioselectivity. It is believed that the two methyl groups of (*S*,*R*,*R*,*S*)-DIOP* orientate at pseudoequatorial positions in the "effective" conformer of the DIOP* metal complex, thereby stabilizing the "effective" conformer to promote high enantioselectivity. On the other hand, its isomeric ligand (*S*,*S*,*S*,*S*)-DIOP* has two methyl groups at pseudoaxial positions, which destabilize the "effective" conformer and lead to diminished ee-values. Lee has developed 1,4-diphosphane ligands BDPMI, **9** and **10**, with an imidazolidin-2-one backbone [54]. The *gauche* steric interaction between the *N*-substituents and

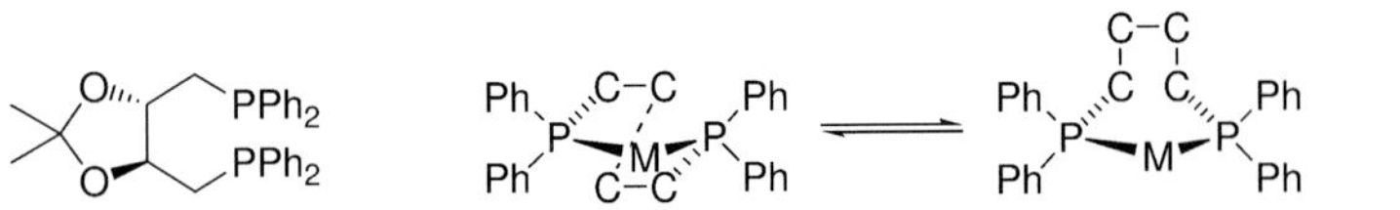

Fig. 26.3 Conformation analysis of DIOP metal complex.

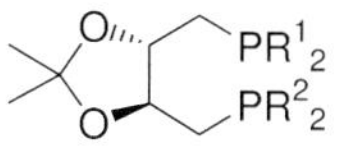

(*S*,*S*)-Cy-DIOP: $R^1 = R^2$ = Cy
(*S*,*S*)-DIOCP: R^1 = Cy; R^2 = Ph
(*S*,*S*)-MOD-DIOP:
$R^1 = R^2$ = 3, 5-$(Me)_2$-4-(MeO)Ph

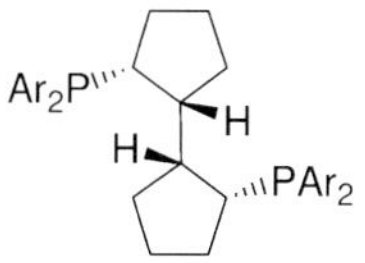

(*R*,*R*)-BICP, Ar = Ph
(*R*,*R*)-**7**, Ar = Xyl
(*R*,*R*)-**8**,
Ar =4-CH_3O-3,5-$(CH_3)_2C_6H_2$

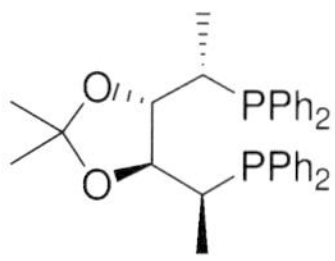

(*R*,*S*,*S*,*R*)-DIOP*

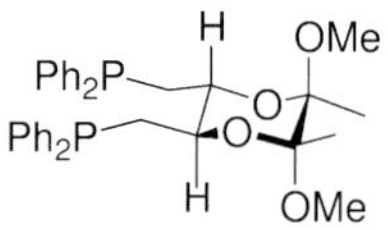

(*R*,*R*,*R*,*R*)-T-Phos

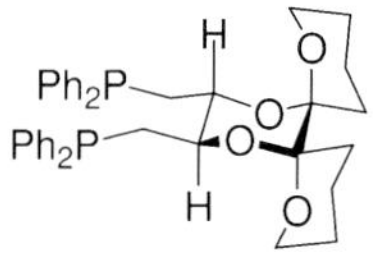

(*R*,*R*,*R*,*R*)-SK-Phos

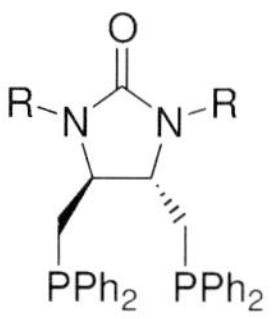

(*S*,*S*)-BDPMI
(*S*,*S*)-**9**: R = Me
(*S*,*S*)-**10**: R = Et

Fig. 26.4 Chiral bisphosphane ligands based on DIOP modifications.

phosphanylmethyl group of the ligands may restrict the conformational flexibility of the seven-membered metal chelate ring. The BDPMI ligands have been successfully applied in the Rh-catalyzed hydrogenation of arylenamides, and up to 99% ee-values have been obtained. A series of 1,4-diphosphane ligands with a conformationally rigid 1,4-dioxane backbone such as T-Phos, and SK-Phos have been developed by Zhang and found to be efficient (up to 99% ee) in the rhodium-catalyzed asymmetric hydrogenation of arylenamides and MOM-protected *β*-hydroxyl enamides [55].

26.2.3
P-Chiral Bisphosphine Ligands

Knowles made the important discovery of the first C_2-symmetric chelating bisphosphine ligand, DIPAMP, which performed much better than the monomeric PAMP [56]. Due to its high catalytic efficiency in the Rh-catalyzed enantioselective hydrogenation of dehydroamino acids, the first P-chiral bisphosphane DIPAMP was quickly employed in the industrial production of L-Dopa [57]. However the development of new efficient P-chiral bisphosphanes was slow, partly because of the difficulties in the ligand synthesis. It was not until Imamoto [58] discovered a series of efficient P-chiral ligands such as BisP* that the development of P-chiral phosphorus ligands regained attention (Fig. 26.5). The BisP* ligands have induced high activity and enantioselectivity in the rhodium-catalyzed hydrogenation of *α*-dehydroamino acids, enamides [59], (*E*)-*β*-(acylamino)-acrylates [60], and *α*,*β*-unsaturated-*α*-acyloxyphosphonates [61]. Mechanistic studies on enantioselective hydrogenation with ^{t}Bu-BisP* as the ligand by Gridnev and Imamoto provided evidence that the Rh-catalyzed hydrogenation can proceed by a different mechanism

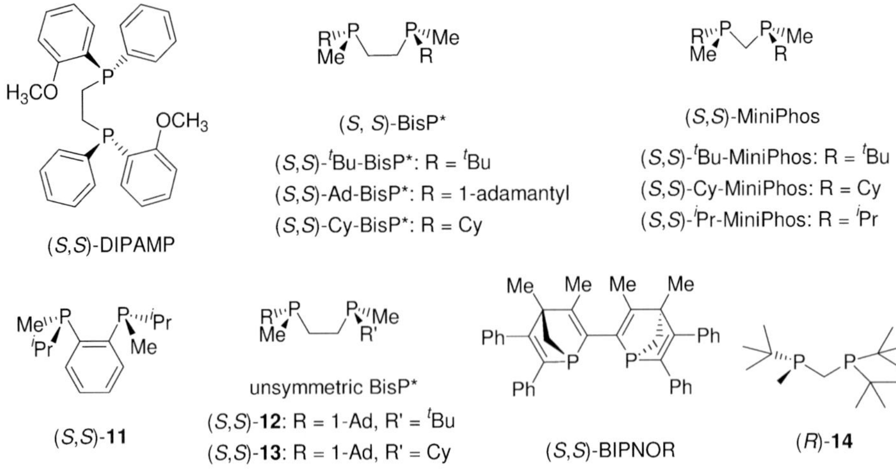

(*S*,*S*)-DIPAMP

(*S*, *S*)-BisP*
(*S*,*S*)-^{t}Bu-BisP*: R = ^{t}Bu
(*S*,*S*)-Ad-BisP*: R = 1-adamantyl
(*S*,*S*)-Cy-BisP*: R = Cy

(*S*,*S*)-MiniPhos
(*S*,*S*)-^{t}Bu-MiniPhos: R = ^{t}Bu
(*S*,*S*)-Cy-MiniPhos: R = Cy
(*S*,*S*)-^{i}Pr-MiniPhos: R = ^{i}Pr

(*S*,*S*)-**11**

unsymmetric BisP*
(*S*,*S*)-**12**: R = 1-Ad, R' = ^{t}Bu
(*S*,*S*)-**13**: R = 1-Ad, R' = Cy

(*S*,*S*)-BIPNOR

(*R*)-**14**

Fig. 26.5 P-chiral ligands.

with an electron-rich phosphorus ligand. A dihydride pathway [62] was suggested, which is different from the classic unsaturated pathway [63, 64] proposed by Halpern and Brown. In addition to Bisp*, several other P-chiral bisphosphanes such as MiniPhos [65], 1,2-bis(isopropylmethylphosphino)benzene (**11**) [66], and unsymmetrical P-chiral BisP* (such as **12** and **13**) [67] have been developed by Imamoto. Imamoto has also developed the P-chirogenic trialkylphosphonium salts derived from Bisp* and MiniPhos. These air-stable salts were conveniently applied in the Rh-catalyzed enantioselective hydrogenation of enamides [68].

Mathey has reported a bisphosphane ligand BIPNOR which contains two chiral bridgehead phosphorus centers [69]. BIPNOR has shown high enantioselectivity in the rhodium-catalyzed hydrogenation of *α*-(acetomido)cinnamic acid and itaconic acid. Recently, a three-hindered quadrant P-chirogenic ligand (*R*)-**14** was also reported by Hoge (Fig. 26.5) [70]. Using (*R*)-**14**–Rh as catalyst, both *E*- and *Z*-(*β*-acylamino) acrylates have been hydrogenated with high enantioselectivities (up to 99% ee) [71].

26.2.4 Other Bisphosphine Ligands

Some other efficient chiral bisphosphane ligands are illustrated in Figure 26.6. These include Bosnich's CHIRAPHOS [72] and PROPHOS [73], Achiwa's BPPM [74], and Rhône-Poulenc's TBPC [75]. A series of modified BPPM ligands such as BCPM and MOD-BPPM were also developed by Achiwa [76]. Some excellent chiral 1,2-bisphosphane ligands such as NORPHOS [77], PYRPHOS (DEGUPHOS) [78], and DPCP [79] for Rh-catalyzed enantioselective hydrogenation were also developed during this period. A few 1,3-bisphosphine ligands such as BDPP (SKEWPHOS) [80] and PPCP [81] were also prepared.

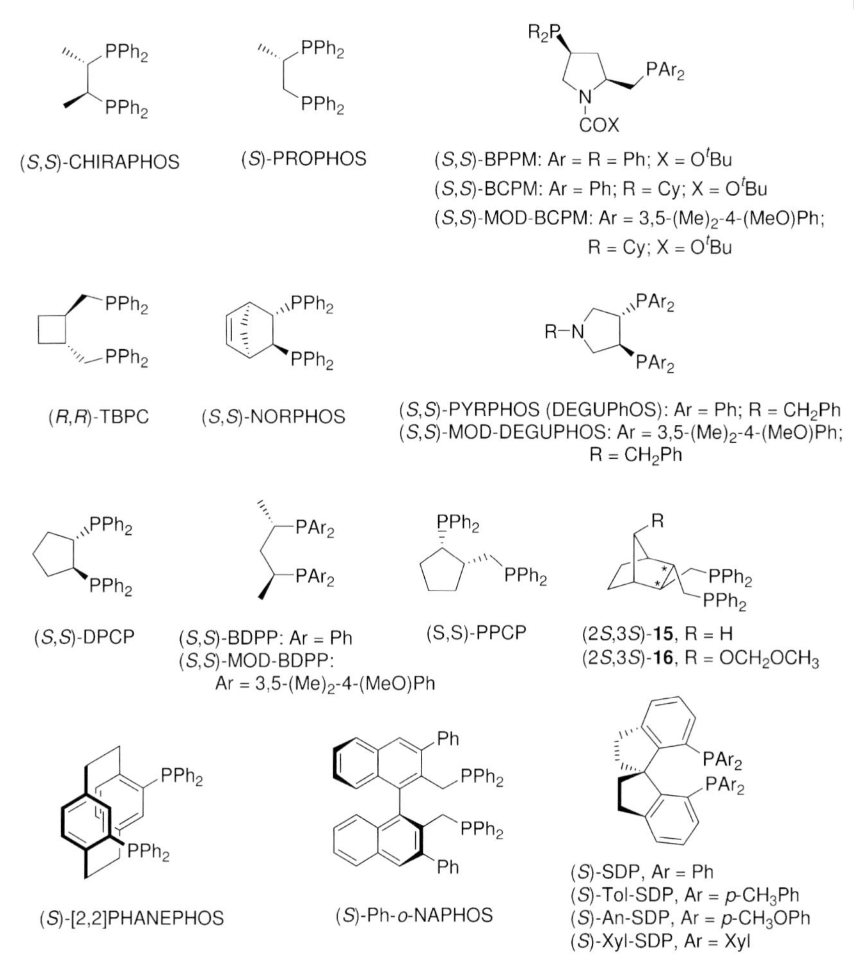

Fig. 26.6 Other efficient chiral bisphosphane ligands.

Pye and Rossen have developed a planar chiral bisphosphine ligand, [2.2] PHANEPHOS, based on a paracyclophane backbone [82]. The ligand has shown excellent enantioselectivity in Rh- or Ru-catalyzed hydrogenations. An *ortho*-phenyl substituted NAPHOS ligand, Ph-*o*-NAPHOS, has been applied successfully in the Rh-catalyzed hydrogenation of α-dehydroamino acid derivatives [83]. Compared to NAPHOS, Ph-*o*-NAPHOS has a more rigid structure and provides higher enantioselectivities. The chiral norbornane diphosphine ligands, **15** and **16**, were reported by Morimoto, and applied in Rh-catalyzed enantioselective hydrogenation [84]. Zhou reported a family of chiral spirodiphosphine ligands such as SDP, containing 1,1′-spirobi-indane as a new scaffold, which are effective for the hydrogenation of simple ketones (Fig. 26.6) [85].

26.3
Applications in Enantioselective Hydrogenation of Alkenes

26.3.1
Enantioselective Hydrogenation of α-Dehydroamino Acid Derivatives

Hydrogenation of α-dehydroamino acid derivatives has been a typical reaction to test the efficiency of new chiral phosphorus ligands. Indeed, a large number of chiral phosphorus ligands with great structural diversity are found to be effective for the Rh-catalyzed hydrogenation of α-dehydroamino acid derivatives. Since (*Z*)-2-(acetamido) cinnamic acid, 2-(acetamido) acrylic acid and their methyl esters are the most frequently applied substrates, and some efficient examples (>95% ee) of hydrogenation of these substrates with different chiral ligands are listed in Table 26.1. Generally, cationic Rh complexes and low hydrogenation pressure are applied in these hydrogenation reactions.

Table 26.1 Enantioselective hydrogenation of α-dehydroamino acid derivatives.

R_1–CH=C(COOR$_2$)(NHCOCH$_3$) $\xrightarrow{\text{Chiral Rh Catalyst}}$ R_1–CH$_2$–CH(COOR$_2$)(NHCOCH$_3$)

A: R_1 = H, R_2 = H   B: R_1 = H, R_2 = CH_3
C: R_1 = Ph, R_2 = H   D: R_1 = Ph, R_2 = CH_3

Ligand	Substrate	SCR	Reaction conditions	% ee of product (config.)	Reference
(*S*)-BINAP	D [a]	100	EtOH, rt, 3 atm H_2	100 (*S*)	1
(*R*)-BICHEP	D [b]	1000	EtOH, rt, 1 atm H_2	95 (*S*)	7c
(*S*)-*o*-Ph-MeO-BIPHEP	A	100	CH_2Cl_2, rt, 1.7 atm H_2	>99 (*S*)	27
(*R,R*)-BICP	A	100	THF, Et_3N, rt, 1 atm H_2	97.5 (*S*)	49a
(*R,R*)-DIPAMP	D	900	MeOH, 50 °C, 3 atm H_2	96 (*S*)	56a
(*S,S*)-^{t}Bu-BisP*	D	500	MeOH, rt, 2 atm H_2	99.9 (*R*)	58a
(*S,S*)-^{t}Bu-MiniPhos	B	500	MeOH, rt, 2 atm H_2	99.9 (*R*)	65
(*S,S*)-**11**	B	500	0 °C, 2 atm H_2	97 (*S*)	66
(*S,S*)-**12**	D	500	MeOH, rt, 2 atm H_2	99.2 (*R*)	67a
(-)-BIPNOR	C	100	EtOH, rt, 3 atm H_2	>98 (*S*)	69a
(*R*)-**14**	A	100	MeOH, rt, 3.4 atm H_2	>99 (*R*)	70
(*R*)-**14**	C	100	MeOH, rt, 3.4 atm H_2	>99 (*R*)	70
(*R,R*)-NORPHOS	C	95	MeOH, rt, 1.1 atm H_2	96 (*R*)	77
(*R,R*)-PYRPHOS	D	50000	MeOH, rt, 61 atm H_2	96.5 (*S*)	78b
(*R*)-PHANEPHOS	B	100	MeOH, rt, 1 atm H_2	99.6 (*R*)	82a
(*S*)-Ph-o-NAPHOS	B	100	MeOH, rt, 3 atm H_2	98.7 (*S*)	83

a) Benzoyl derivative.
b) Ethyl ester.
SCR: substrate:catalyst ratio.

Several chiral ligands, such as PYRPHOS [78b], have been shown to be very efficient ligands for the hydrogenation of α-dehydroamino acid derivatives in terms of both high enantioselectivity and reactivity.

In contrast to the high enantioselectivity achieved for the *Z*-isomeric substrates, hydrogenation of the *E*-isomeric substrates usually proceeds at a much lower rate and gives poor enantioselectivities [86]. With the Rh–BINAP system as the catalyst and tetrahydrofuran (THF) as solvent, hydrogenation of the *Z*- and *E*-isomeric substrates generates products with different configurations [2].

Many synthetic applications of Rh-catalyzed hydrogenation of α-dehydroamino acid derivatives have recently been explored (Scheme 26.2). Takahashi has reported a one-pot sequential enantioselective hydrogenation utilizing a BINAP–Rh and a BINAP–Ru catalyst to synthesize 4-amino-3-hydroxy-5-phenylpentanoic acids in over 95% ee. The process involves a first step in which the dehydroamino acid unit is hydrogenated with the BINAP–Rh catalyst, followed by hydrogenation of the β-keto ester unit with the BINAP–Ru catalyst [87]. A hindered pyridine substituted α-dehydroamino acid derivative has been hydrogenated by a

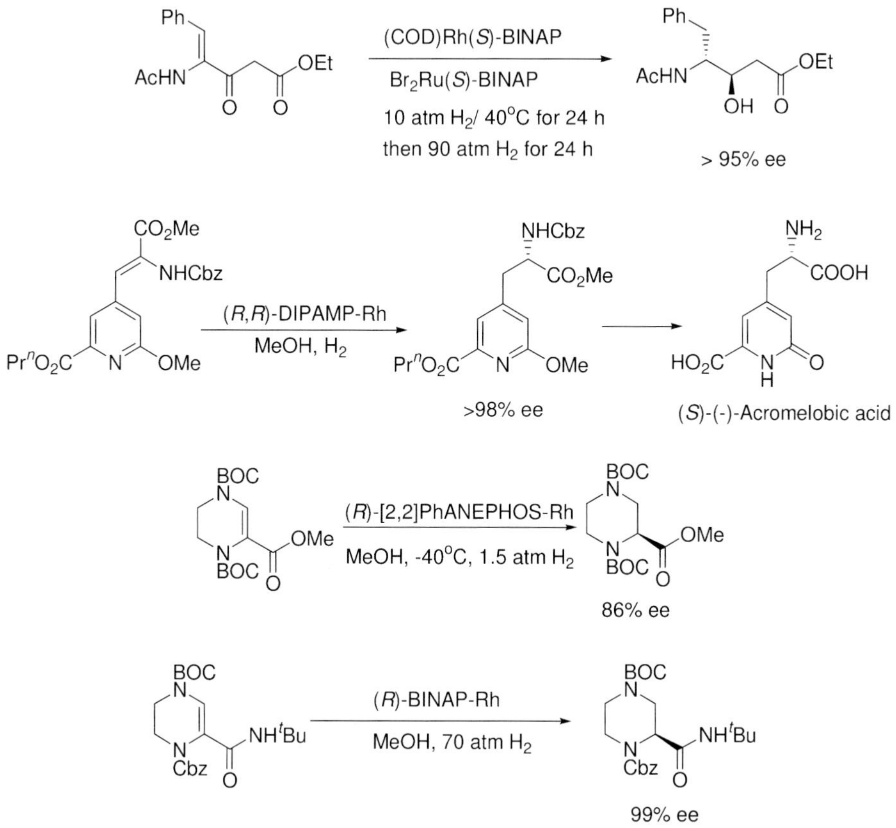

(COD)Rh(*S*)-BINAP
Br_2Ru(*S*)-BINAP
10 atm H_2/ 40°C for 24 h
then 90 atm H_2 for 24 h
> 95% ee

(*R,R*)-DIPAMP-Rh
MeOH, H_2
>98% ee
(*S*)-(-)-Acromelobic acid

(*R*)-[2,2]PhANEPHOS-Rh
MeOH, -40°C, 1.5 atm H_2
86% ee

(*R*)-BINAP-Rh
MeOH, 70 atm H_2
99% ee

Scheme 26.2

Table 26.2 Enantioselective hydrogenation of β,β-dimethyl α-dehydroamino acid esters.

COOMe, NHAc —Chiral Rh Catalyst, H_2→ * COOMe, NHAc

Ligand	SCR	Reaction conditions	% ee of product (config.)	Reference
(*S,S*)-Cy-BisP*	500	MeOH, rt, 6 atm H_2	90.9 (*R*)	58a
(*S,S*)-^{t}Bu-MiniPhos	500	MeOH, rt, 6 atm H_2	87 (*R*)	65
(*S,S*)-**11**	500	rt, 6 atm H_2	87 (*S*)	66
(*S,S*)-**13**	100	MeOH, rt, 20 atm H_2	96.1 (*R*)	67b

SCR: substrate:catalyst ratio.

DIPAMP–Rh complex to give the corresponding chiral α-amino acid derivative in over 98% ee. The chiral product has been used for the synthesis of (*S*)-(–)-acromelobic acid [88]. Hydrogenation of a tetrahydropyrazine derivative catalyzed by a PHANEPHOS–Rh complex at –40 °C gives an intermediate for the synthesis of Crixivan in 86% ee [82a]. Hydrogenation of another tetrahydropyrazine carboxamide derivative catalyzed by an (*R*)-BINAP–Rh catalyst leads to the chiral product in 99% ee [89].

The hydrogenation of β,β-disubstituted α-dehydroamino acids remains a relatively challenging problem. The Rh complexes of chiral ligands such as Cy-BisP* [58a], MiniPhos [65], and unsymmetrical BisP* **13** [67b] have shown high efficiencies for some β,β-disubstituted α-dehydroamino acid substrates. Some efficient examples of hydrogenation of β,β-dimethyl α-dehydroamino acid esters with different chiral phosphorus ligands are listed in Table 26.2.

26.3.2
Enantioselective Hydrogenation of Enamides

Rh-catalyzed hydrogenation of simple enamides has attracted much attention recently. With the development of increasingly efficient chiral phosphorus ligands, extremely high ee-values can be obtained in the Rh-catalyzed hydrogenation of α-aryl enamides. *E*/*Z*-isomeric mixtures of β-substituted enamides can also be hydrogenated, with excellent ee-values. Some efficient examples (>95% ee) of hydrogenation of α-phenylenamide and *E*/*Z*-isomeric mixtures of β-methyl-α-phenylenamide are listed in Table 26.3.

Some alkyl enamides such as *tert*-butylenamide or 1-adamantylenamide can also be hydrogenated with a ^{t}Bu-BisP*–Rh catalyst in 99% ee. Notably, the configurations of the hydrogenation products of these bulky alkyl enamides are opposite to those of aryl enamides. A mechanistic study [90] by Gridnev and Imamoto [59] using nuclear magnetic resonance (NMR) techniques indicates that the hydrogenations of bulky alkyl enamides and aryl enamides involve different

Table 26.3 Enantioselective hydrogenation of enamides.

A: R = H B: R = CH_3 (*E*/*Z*)

Chiral Rh catalysts

Ligand	Substrate	SCR	Reaction conditions	% ee of product (config.)	Reference
(*R,R*)-BICP	B	100	Toluene, rt, 2.7 atm H_2	95.0 (*R*)	49b
(*R,S,S,R*)-DIOP*	A	50	MeOH, rt, 10 atm H_2	98.8 (*R*)	51
	B	50	MeOH, rt, 10 atm H_2	97.3 (*R*)	51
(*S,S*)-**9**	A	100	CH_2Cl_2, rt, 1 atm H_2	98.5 (*R*)	54a
	B	100	CH_2Cl_2, rt, 1 atm H_2	>99 (*R*)	54a
(*R,R,R,R*)-T-Phos	B	100	MeOH, rt, 3.1 atm H_2	98 (*S*)	55
(*R,R,R,R*)-SK-Phos	B	100	MeOH, rt, 3.1 atm H_2	97 (*S*)	55
(*S,S*)-^{t}Bu-BisP*	A	100	MeOH, rt, 3 atm H_2	98 (*R*)	59

SCR: substrate:catalyst ratio.

coordination pathways. *o*-Ph-HexaMeO-BIPHEP [26] has shown high efficiency in the Rh-catalyzed hydrogenation of cyclic enamides. A racemic cyclic enecarbamate has been hydrogenated with an *o*-Ph-HexaMeO-BIPHEP–Rh catalyst to yield the *cis* chiral carbamate in 96% ee [26]. The chiral product can be used directly for the synthesis of sertraline, an anti-depressant. Hydrogenation of some tetra-substituted enamides has also been reported. ^{t}Bu-BisP* and ^{t}Bu-MiniPhos have provided excellent ee-values in the Rh-catalyzed hydrogenation of a β,β-dimethyl-α-phenyl enamide derivatives (Scheme 26.3). Using an *o*-Ph-BIPHEP–Rh catalyst [26], tetra-substituted enamides derived from 1-indanone and 1-tetralone have been hydrogenated with excellent enantioselectivities.

The hydrogenation of a series of *E*/*Z*-isomeric mixtures of α-arylenamides with a MOM-protected β-hydroxyl group catalyzed by a Rh-complex of 1,4-diphosphane T-Phos with a rigid 1,4-dioxane backbone led to chiral β-amino alcohol derivatives in excellent enantioselectivities (Scheme 26.4) [55]. DIOP*–Rh is also effective for this transformation [51 b].

In addition to the Rh chemistry, the Ru–BINAP system has shown excellent enantioselectivity in the hydrogenation of (*Z*)-*N*-acyl-1-alkylidenetetrahydroisoquinolines. Thus, a series of chiral isoquinoline products can be efficiently synthesized [3a, b, 91]. Using Ru–BINAP, the cyclic enamides, 6-bromotetralone-eneacetamide [92] and 7-methyltetralone-eneacetamide [93] are hydrogenated to give the corresponding chiral amide in 97% and 94% ee, respectively (Scheme 26.5). Ru–Biphemp also provided good selectivity (92% ee) in the enantiomeric hydrogenation of a cyclic enamide derived from 3-chromanone (Scheme 26.5) [93].

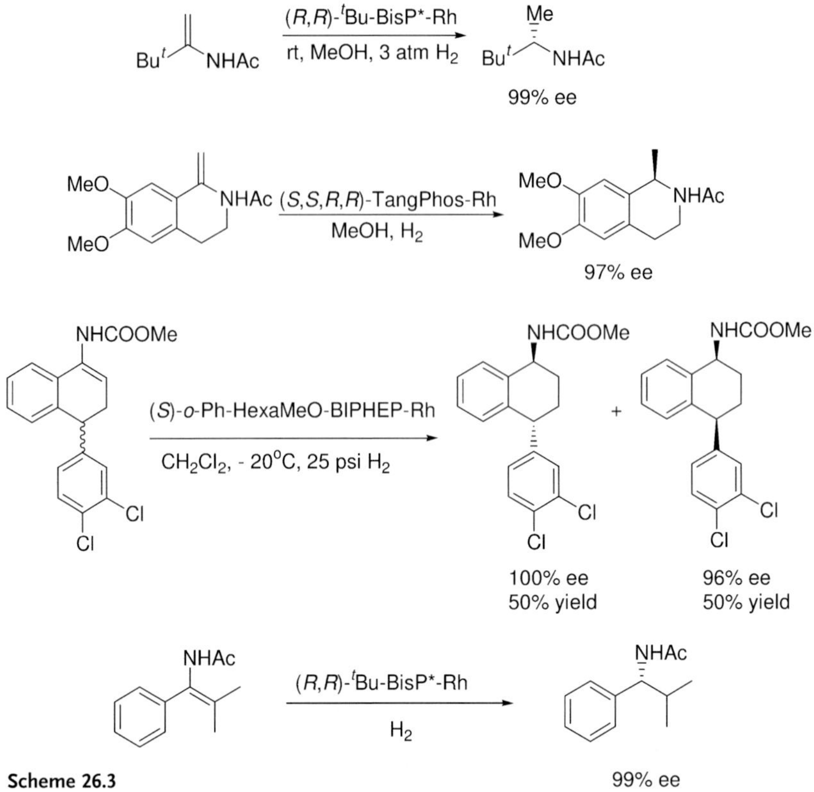

Scheme 26.3

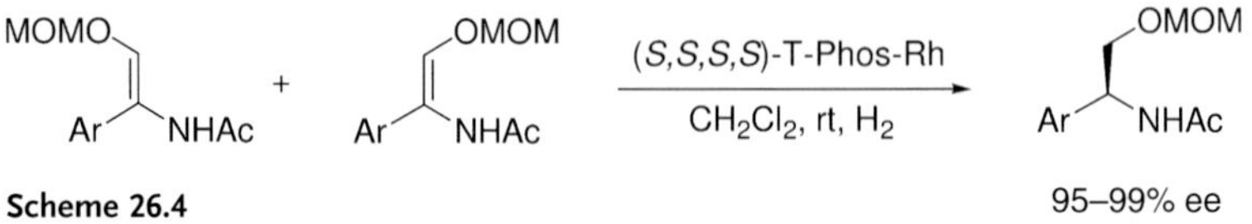

Scheme 26.4

26.3.3
Enantioselective Hydrogenation of (*β*-Acylamino) Acrylates

Enantioselective hydrogenation of (*β*-acylamino) acrylates has gained much attention recently because the *β*-amino acid products are important building blocks for chiral drugs [94]. Since most synthetic methods produce mixtures of *Z*- and *E*-isomeric substrates, it is important that both isomers can be hydrogenated with high enantioselectivity for the practical synthesis of *β*-amino acid derivatives via enantioselective hydrogenation. Some Rh and Ru complexes with chiral phosphorus ligands such as BINAP [95], BICP [49e], BDPMI [54b], ligand **4** [30], ^{t}Bu-BisP* [60], (*R*)-**14** [71], and Xyl-P-Phos [25c] are found to be effective for hydrogenation of (*E*)-alkyl (*β*-acylamino) acrylates. However, only a

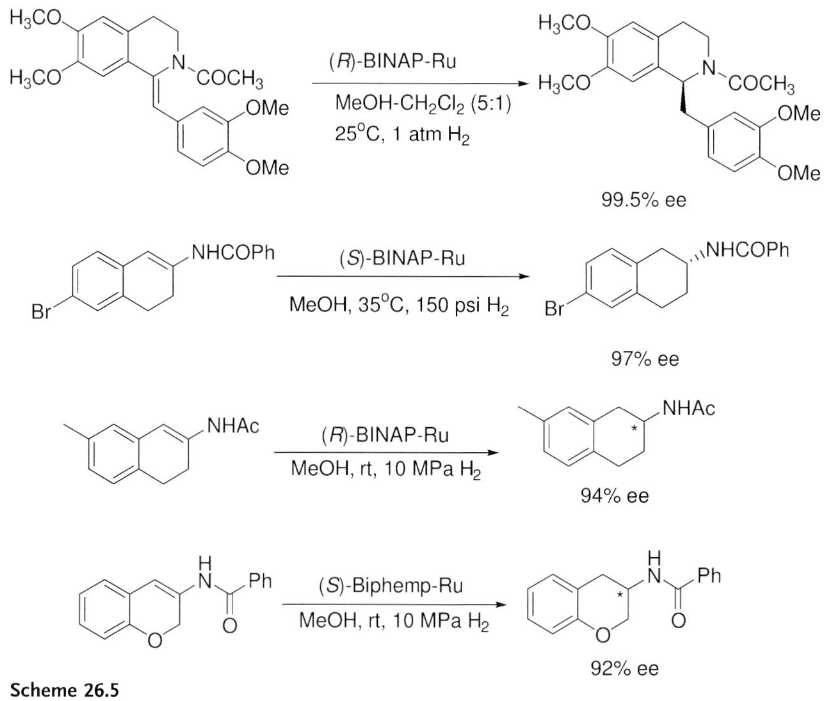

H_3CO, H_3CO, N, $COCH_3$, OMe, OMe

(*R*)-BINAP-Ru
MeOH-CH_2Cl_2 (5:1)
25°C, 1 atm H_2

99.5% ee

NHCOPh, Br

(*S*)-BINAP-Ru
MeOH, 35°C, 150 psi H_2

97% ee

NHAc

(*R*)-BINAP-Ru
MeOH, rt, 10 MPa H_2

94% ee

H, N, Ph, O

(*S*)-Biphemp-Ru
MeOH, rt, 10 MPa H_2

92% ee

Scheme 26.5

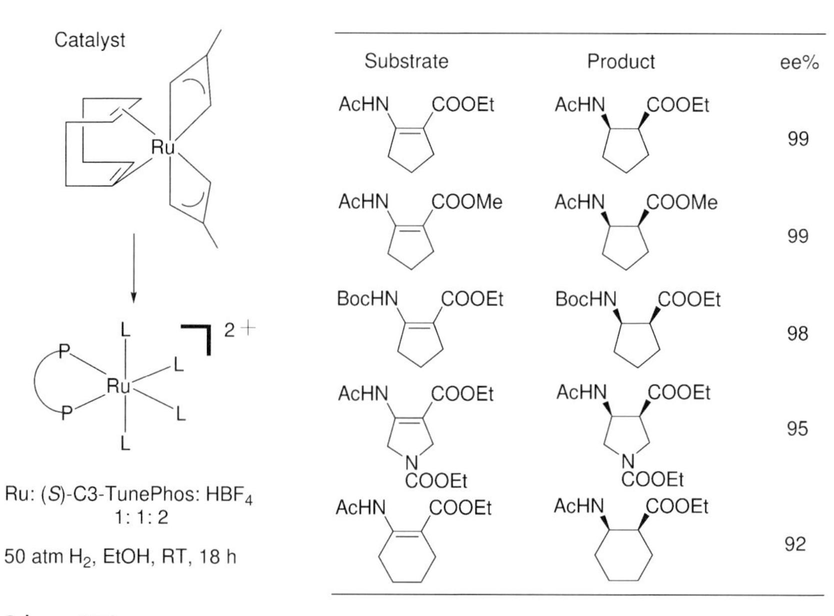

Scheme 26.6

Table 26.4 Enantioselective hydrogenation of (β-acylamino) acrylates.

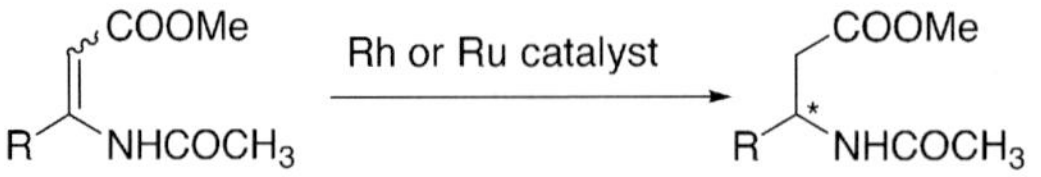

Ligand	R	Geometry	Reaction conditions	% ee of product (config.)	Reference
(*R*)-BINAP–Ru	CH_3	*E*	MeOH, 25 °C, 1 atm H_2	96 (*S*)	95
(*R*)-Xyl-P-Phos–Ru	CH_3	*E*	MeOH, 0 °C, 8 atm H_2	98.1 (*S*)	25c
(*S*)-HexaPHEMP–Rh	CH_3	*E*	MeOH, rt, 9.5 atm H_2	95 (*R*)	28
2–Rh	CH_3	*E*	MeOH, 0 °C, 17 atm H_2	97.7 (*S*)	30
2–Rh	^{i}Pr	*E*	MeOH, 0 °C, 17 atm H_2	98.8 (*S*)	30
(*R,R*)-BICP–Rh	CH_3	*E*	Toluene, rt, 2.7 atm H_2	96.1 (*R*)	49e
(*S,S*)-**9**–Rh	CH_3 [a]	*E*	CH_2Cl_2, rt, 1 atm H_2	94.6 (*R*)	54b
(*S,S*)-^{t}Bu-BisP*–Rh	CH_3	*E*	THF, rt, 3 atm H_2	98.7 (*R*)	60
(*S,S*)-MiniPhos–Rh	CH_3	*E*	THF, rt, 3 atm H_2	96.4 (*R*)	60
(*R*)-**14**–Rh	CH_3	*E*	MeOH, rt, 1.4 atm H_2	99 (*R*)	71
(*S,S*)-**9**–Rh	CH_3 [a]	*Z*	CH_2Cl_2, rt, 6.8 atm H_2	95 (*R*)	54b
(*R*)-**14**–Rh	CH_3	*Z*	EtOAc, rt, 1.4 atm H_2	98 (*R*)	71
(*R*)-**14**–Rh	CH_3	*E/Z* [b]	THF, rt, 1.4 atm H_2	98 (*R*)	71

a) Ethyl ester.
b) *E*:*Z* ratio = 1 : 1.

few chiral ligands such as BDPMI [54b] and (*R*)-**14** [71] can provide over 95% ee hydrogenation for hydrogenation of (*Z*)-alkyl (β-acylamino) acrylates (Table 26.4). With (*R*)-**14**–Rh catalyst, an *E*/*Z*-isomeric mixtures of methyl 3-acetamido-2-butenoate was hydrogenated in THF to give (*R*)-methyl 3-acetamidobutanoate in 98% ee [71].

By employing a Ru catalyst generated *in situ* from Ru(COD)(methallyl)$_2$, (*S*)-C3-TunePhos, and HBF_4, a series of cyclic β-(acylamino) acrylates were hydrogenated with excellent ee-values. As shown in Scheme 26.6, 99% ee was obtained in the hydrogenation of both 2-acetylamino-cyclopent-1-enecarboxylic acid methyl ester and ethyl ester. A heterocyclic β-(acylamino) acrylate is also hydrogenated to give the *cis*-product in excellent enantioselectivity (95% ee). Hydrogenation of a cyclohexenyl substrate provided the corresponding *cis* product in 92% ee [20].

26.3.4 Enantioselective Hydrogenation of Enol Esters

Enol esters have a similar structure as enamides. However, in contrast to many highly enantioselective examples on enantioselective hydrogenation of enamides, only a few successful results have been reported for the hydrogenation of

Table 26.5 Enantioselective hydrogenation of enol esters.

Catalyst	R	R′	Geometry	Reaction conditions	% ee of product (config.)	Reference
(*R*,*R*)-DIPAMP–Rh	CO_2Et	^{i}Pr	*E*/*Z* [a]	MeOH, rt, 3 atm H_2	92 (*S*)	97
(*R*)-BINAP–Ru	CO_2Et	^{i}Pr	*E*/*Z* [a]	MeOH, 50 °C, 50 atm H_2	98 (*S*)	97
(*R*,*R*)-DIPAMP–Rh	CO_2Et	Ph	*Z*	MeOH, rt, 3 atm H_2	88 (*S*)	97
(*S*)-C2-TunePhos–Ru	1-Np	H	N/A	$EtOH/CH_2Cl_2$, rt, 3 atm H_2	97.7 (*S*)	19
2–Rh	1-Np	H	N/A	$EtOH/CH_2Cl_2$, rt, 3.4 atm H_2	96.7 (*R*)	30
2–Rh	*p*-F-C_6H_4	H	N/A	$EtOH/CH_2Cl_2$, rt, 3.4 atm H_2	97.1 (*R*)	30

a) *E*:*Z* ratio = 70 : 30.
N/A: not applicable.

enol esters. One possible reason is that the acyl group of an enol ester has a weaker coordinating ability to the metal catalyst than that of the corresponding enamide substrate. Some Rh and Ru complexes associated with chiral phosphorus ligands such as DIPAMP [96,97] and BINAP [97] are effective for the enantioselective hydrogenation of α-(acyloxy) acrylates. Some chiral phosphorus ligands such as BINAP [98] and TunePhos [19] have been applied to the Rh- or Ru-catalyzed enantioselective hydrogenation of aryl enol acetates without other functionalities (Table 26.5). C2-TunePhos–Ru [19] and **2**-Ru [30] catalysts are found to be equally effective for this transformation.

Enantioselective hydrogenation of a series of enol phosphates with a ^{t}Bu-MiniPhos–Rh or a ^{t}Bu-BisP*–Rh catalyst provides moderate to excellent ee-values (Scheme 26.7) [99].

1 mol% [Rh((*R*,*R*)-^{t}Bu-BisP*)(NBD)](BF_4), MeOH, rt, 4 atm H_2, 18h

96% ee; 89% ee; 93% ee; 98% ee; 99% ee

Scheme 26.7

(S)-SEGPhos-Ru
S/C = 12,270
acetone, 50°C, 40 atm H_2
93.8% ee

(S)-BINAP-Ru
CH_2Cl_2, rt, 100 atm H_2
94% ee

(S)-di-t-Bu-MeO-BINAP-Ru
HBF_4, ^{i}PrOH, 60 atm H_2
n-C_6H_{13} OH
97% ee

Scheme 26.8

Although high hydrogen pressure is required, BINAP and its analogous ligands gave superior results in the Ru-catalyzed hydrogenation of four- and five-membered cyclic lactones or carbonates bearing an exocyclic methylene group (Scheme 26.8) [98]. A (*S*)-SEGPHOS–Ru catalyst provided 93.8% ee in the hydrogenation of a diketene with high turnover numbers (TONs) [100]. With a (*S*)-BINAP–Ru catalyst, 94% ee was obtained in the hydrogenation of 4-methylene-γ-butyrolactone. In the presence of a small amount of HBF_4, a di-*t*-Bu-MeOBIPHEP–Ru catalyst allows the hydrogenation of a 2-pyrone substrate with 97% ee [101].

26.3.5
Enantioselective Hydrogenation of Unsaturated Acids and Esters

26.3.5.1 *α,β*-Unsaturated Carboxylic Acids

Significant advance has been achieved in the enantiomeric hydrogenation of α,β-unsaturated carboxylic acids with chiral Ru catalysts. The Ru–BINAP-dicarboxylate complex has shown excellent enantioselectivities in the hydrogenation of some α,β-unsaturated carboxylic acids, although the catalytic efficiencies are still highly sensitive to the substrates, reaction temperature, and hydrogen pressure [3d]. Other atropisomeric ligands, such as H_8-BINAP [102], MeO-BIPHEP [103], BIPHEMP [103], P-Phos [25], TetraMe-BITIANP [23b], and TetraMe-BITIOP [24] are also effective for this transformation. Ru complexes prepared in different forms may exhibit slightly different efficiencies. Some examples of the hydrogenation of tiglic acid with different metal–ligand complexes are listed in Table 26.6. The H_8-BINAP ligand with a larger dihedral angle gives superior results compared to the BINAP ligand.

With a BINAP–Ru [3d,104], H_8-BINAP–Ru [102], or P-Phos–Ru [25] catalyst, the anti-inflammatory drugs (*S*)-ibuprofen and (*S*)-naproxen could be efficiently synthesized via enantioselective hydrogenation (Scheme 26.9). In these cases, high hydrogenation pressure and low temperature are required to achieve good enantioselectivity. With an (*R*)-BIPHEMP–Ru catalyst, (*S*)-2-(4-fluorophenyl)-3-methylbutanoic acid, a key intermediate for the synthesis of the calcium antago-

Table 26.6 Enantioselective hydrogenation of tiglic acid.

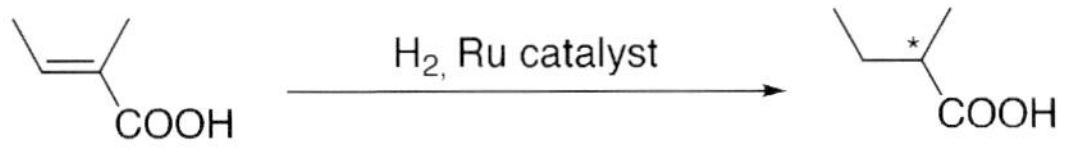

COOH → (H_2, Ru catalyst) → * COOH

Catalyst	SCR	Reaction conditions	% ee of product (config.)	Reference
$Ru(OAc)_2$[(*R*)-BINAP]	100	MeOH, 15–30 °C, 4 atm H_2	91 (*R*)	3d
Ru[(*R*)-BINAP](2-methallyl)$_2$	100	MeOH, 20 °C, 4 atm H_2	90 (*R*)	103
$Ru(OAc)_2$[(*S*)-H8-BINAP]	200	MeOH, 10–25 °C, 1.5 atm H_2	97 (*S*)	102
[(*R*)-MeO-BIPHEP]$RuBr_2$	100	MeOH, 20 °C, 1.4 atm H_2	92 (*R*)	103
[NH_2Et_2][{RuCl[(*S*)-BIPHEMP]}$_2$(μ-Cl)$_3$]	100	MeOH, 20 °C, 4 atm H_2	98 (*S*)	103
Ru(*p*-cymene)[(–)-TetraMe-BITIANP]I_2]	500	MeOH, 25 °C, 10 atm H_2	92 (*S*)	23b
[Ru(–)-TetraMe-BITIOP](2-methally)$_2$]	3000	MeOH, 25 °C, 10 atm H_2	94 (*R*)	24

SCR: substrate:catalyst ratio.

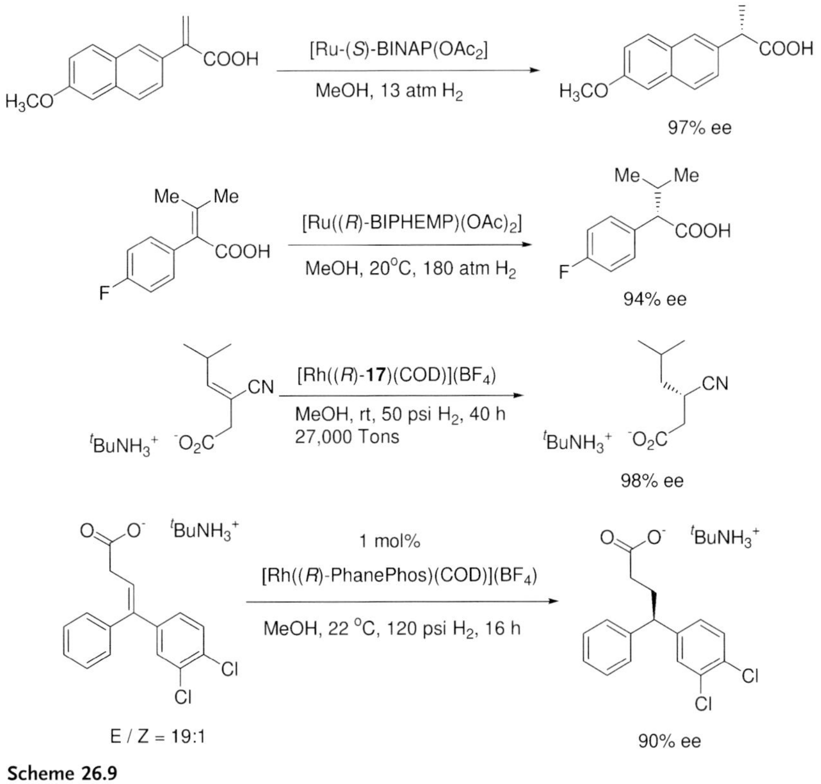

H_3CO … COOH → [Ru-(*S*)-BINAP(OAc$_2$)], MeOH, 13 atm H_2 → H_3CO … COOH, 97% ee

Me Me … COOH, F → [Ru((*R*)-BIPHEMP)(OAc)$_2$], MeOH, 20°C, 180 atm H_2 → Me Me … COOH, F, 94% ee

CN, $^tBuNH_3^+$ ^-O_2C → [Rh((*R*)-**17**)(COD)](BF$_4$), MeOH, rt, 50 psi H_2, 40 h, 27,000 Tons → CN, $^tBuNH_3^+$ ^-O_2C, 98% ee

O O$^-$ $^tBuNH_3^+$, Cl, Cl, E / Z = 19:1 → 1 mol% [Rh((*R*)-PhanePhos)(COD)](BF$_4$), MeOH, 22 °C, 120 psi H_2, 16 h → O O$^-$ $^tBuNH_3^+$, Cl, Cl, 90% ee

Scheme 26.9

nist mibefradil, could be reduced in 94% ee [105]. Using (*R*)-**14**–Rh as catalyst, enantioselective hydrogenation of *tert*-butylammonium (3*Z*)-3-cyano-5-methyl-3-hexenoate produced the precursor to CI-1008 (pregabalin, indicated for psychotic disorder, seizure disorder and pain) with a TON of 27000 and 98% ee (Scheme 26.9) [70]. Using PhanePhos–Rh as catalyst, an isomeric mixture (*E*/*Z*=19:1) of 4,4′-diaryl-3-butenoate was hydrogenated to provide a chiral intermediate for the antidepressant sertraline, with 90% ee (Scheme 26.9) [106].

26.3.5.2 *α*,*β*-Unsaturated Esters, Amides, Lactones, and Ketones

Limited progress has been achieved in the enantioselective hydrogenation of *α*,*β*-unsaturated carboxylic acid esters, amides, lactones, and ketones (Scheme 26.10). The Ru–BINAP system is efficient for the hydrogenation of 2-methylene-*γ*-butyrolactone, and 2-methylene-cyclopentanone [98]. With a dicationic (*S*)-di-*t*-Bu-MeOBIPHEP–Ru complex under a high hydrogen pressure, 3-ethoxy pyrrolidinone could be hydrogenated in isopropanol to give (*R*)-4-ethoxy-*γ*-lactam in 98% ee [39].

26.3.5.3 Itaconic Acids and Their Derivatives

Many chiral phosphorus ligands have shown excellent reactivities and enantioselectivities in the Rh-catalyzed hydrogenation of itaconic acids or esters. Some successful (>95% ee) hydrogenations of itaconic acid or its dimethyl ester with different chiral phosphorus ligands are listed in Table 26.7. High reactivity is observed with electron-rich phosphane ligands such as BICHEP [7c].

In contrast to the many successful examples of hydrogenation of the parent itaconic acid or its dimethyl ester, only a few ligands have been reported to be

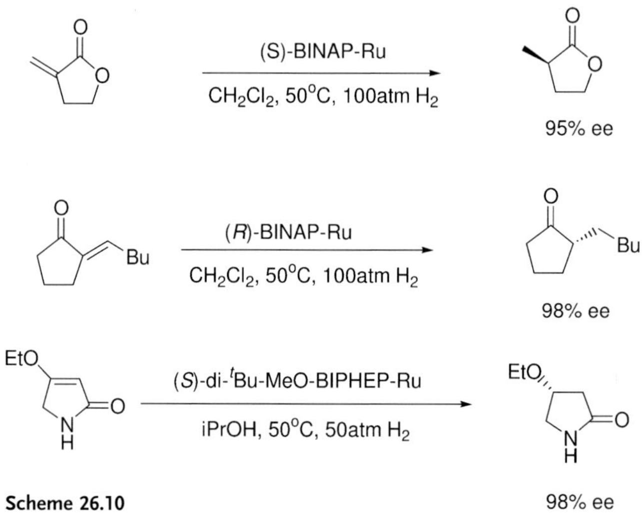

(S)-BINAP-Ru

CH_2Cl_2, 50°C, 100atm H_2

95% ee

O

Bu

(*R*)-BINAP-Ru

CH_2Cl_2, 50°C, 100atm H_2

98% ee

EtO

N H

(*S*)-di-^{t}Bu-MeO-BIPHEP-Ru

iPrOH, 50°C, 50atm H_2

98% ee

Scheme 26.10

Table 26.7 Enantioselective hydrogenation of itaconic acid derivatives.

ROOC–C(=CH2)–CH2–COOR —chiral Rh catalyst→ ROOC–C*H(CH3)–CH2–COOR

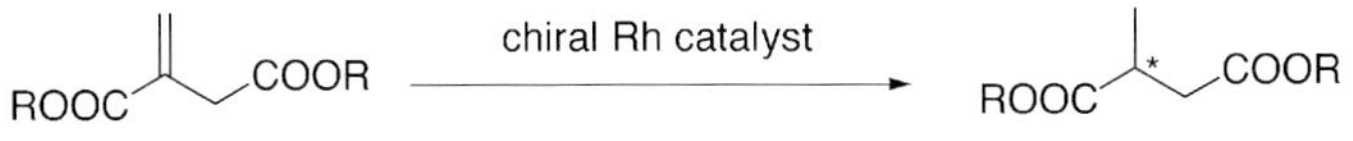

Ligand	R	SCR	Reaction conditions	% ee of product (config.)	Reference
(*R*)-BICHEP	H	1000	EtOH, 25 °C, 1 atm H_2	96 (*R*)	7c
(*S*,*S*)-Ad-BisP*	Me	500	MeOH, rt, 1.6 atm H_2	99.6	58b

SCR: substrate:catalyst ratio.

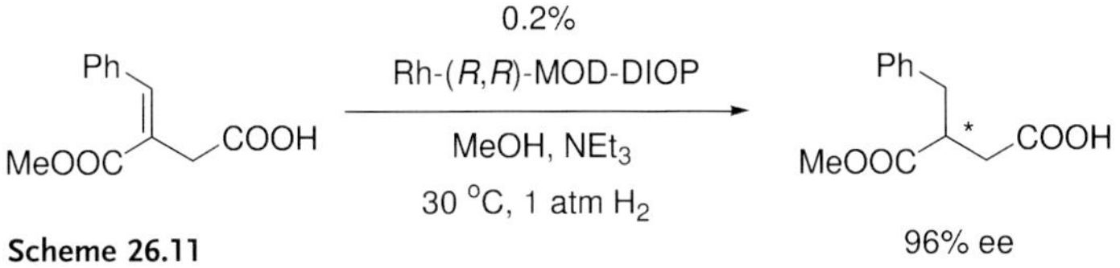

Ph / MeOOC / COOH —0.2% Rh-(*R*,*R*)-MOD-DIOP; MeOH, NEt_3; 30 °C, 1 atm H_2→ Ph / MeOOC / * / COOH, 96% ee

Scheme 26.11

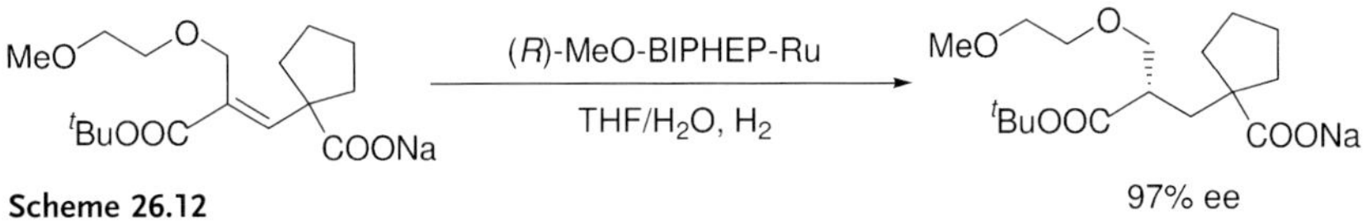

MeO / O / ^{t}BuOOC / COONa —(*R*)-MeO-BIPHEP-Ru; THF/H_2O, H_2→ MeO / O / ^{t}BuOOC / COONa, 97% ee

Scheme 26.12

efficient for the hydrogenation of β-substituted itaconic acid derivatives. Rh complexed with MOD-DIOP [48] is efficient for the hydrogenation of several β-substituted itaconic acid derivatives (Scheme 26.11).

A PYRPHOS ligand was found to be effective for the hydrogenation of a β-aryl- or alkyl-substituted monoamido itaconate [107]. A MeO-BIPHEP–Ru catalyst was successfully applied for the enantioselective hydrogenation of an intermediate for the drug candoxatril in a mixed solvent (THF/H_2O) (Scheme 26.12) [108].

26.3.6
Enantioselective Hydrogenation of Unsaturated Alcohols

Enantioselective hydrogenation of unsaturated alcohols such as allylic and homoallylic alcohols was not very efficient until the discovery of the BINAP–Ru catalyst. With Ru(BINAP)$(OAc)_2$ as the catalyst, geraniol and nerol are successfully hydrogenated to give (*S*)- or (*R*)-citronellol in near-quantitative yield and with 96–99% ee [3c]. A substrate:catalyst ratio (SCR) of up to 48 500 can be applied, and the other double bond at the C6 and C7 positions of the substrate is not reduced. A high hydrogen pressure is required to obtain high enantioselec-

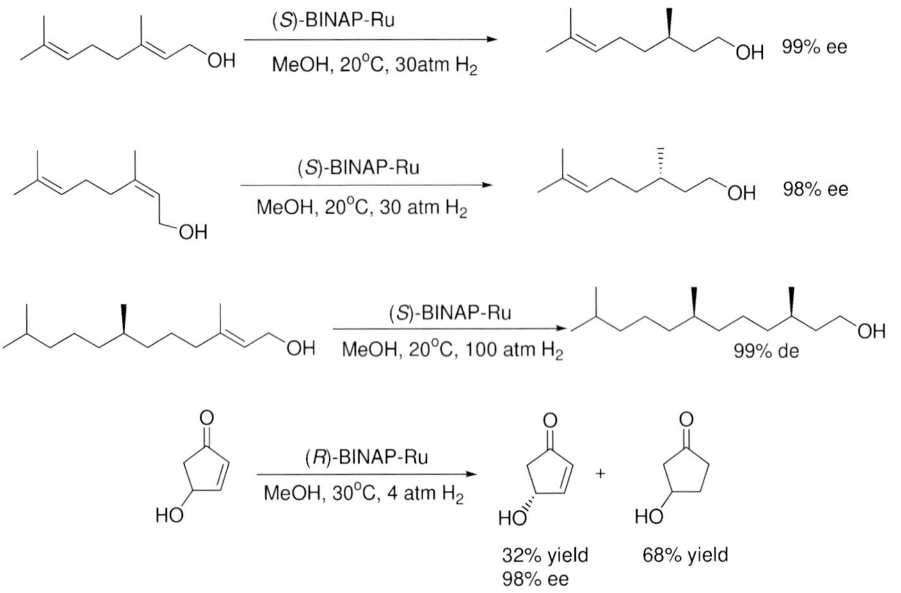

(*S*)-BINAP-Ru, MeOH, 20°C, 30atm H_2 — OH — 99% ee

(*S*)-BINAP-Ru, MeOH, 20°C, 30 atm H_2 — OH — 98% ee

(*S*)-BINAP-Ru, MeOH, 20°C, 100 atm H_2 — OH — 99% de

(*R*)-BINAP-Ru, MeOH, 30°C, 4 atm H_2 — HO — 32% yield, 98% ee + 68% yield

Scheme 26.13

tivity in the hydrogenation of geraniol. Low hydrogen pressure facilitates the isomerization of geraniol to *γ*-geraniol, which leads to the hydrogenation product with the opposite configuration, resulting in a decreased ee-value [109]. In addition to BINAP, other chiral atropisomeric ligands such as MeO-BIPHEP [9], TetraMe-BITIANP [23 b], and TetraMe-BITIOP [24] are also effective for this transformation. The catalytic efficiency of the BINAP–Ru catalyst is strongly sensitive to the substitution patterns of the allylic alcohols. Homoallylic alcohols can also be hydrogenated in high ee-value with the BINAP–Ru catalyst. Its application in the synthesis of (3*R*, 7*R*)-3,7,11-trimethyldodecarol, an intermediate for the synthesis of *α*-tocopherol, is shown in Scheme 26.13. When racemic

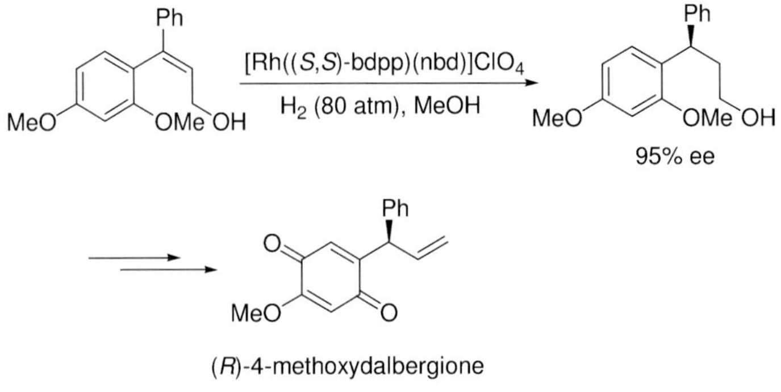

[Rh((*S*,*S*)-bdpp)(nbd)]ClO_4, H_2 (80 atm), MeOH — Ph, MeO, OMe, OH — 95% ee

(*R*)-4-methoxydalbergione

Scheme 26.14

allylic alcohols were subjected to enantioselective hydrogenation with a BINAP–Ru complex, highly efficient kinetic resolutions were achieved [4c]. A racemic 4-hydroxy-2-cyclopentenone was hydrogenated with a (*S*)-BINAP–Ru catalyst to leave unreacted starting material in 98% ee at 68% conversion. The chiral starting material serves as an important building block for three-component coupling prostaglandin synthesis.

A chiral BDPP–Rh complex is an efficient catalyst for the hydrogenation of 3-(2′,4′-dimethoxyphenyl)-3-phenyl-2-propenol. The chiral alcohol product, which was obtained in up to 95% ee, has been used for the synthesis of chiral 4-methoxydalbergione (Scheme 26.14) [110].

26.4 Concluding Remarks

The development of chiral phosphorus ligands has undoubtedly had an enormous impact on the area of enantioselective hydrogenation. Transition-metal catalysts with efficient chiral phosphorus ligands have enabled the synthesis of a variety of chiral products in a very efficient manner, and many practical hydrogenation processes have been exploited in industry for the synthesis of chiral drugs and fine chemicals. However, many challenges remain in the field of enantioselective hydrogenation, and further effort in the quest for new efficient chiral phosphorus ligands, as well as new applications in enantioselective hydrogenation, are needed.

References

1 Miyashita, A., Yasuda, A., Takaya, H., Toriumi, K., Ito, T., Souchi, T., Noyori, R. *J. Am. Chem. Soc.* **1980**, *102*, 7932.

2 Miyashita, A., Takaya, H., Souchi, T., Noyori, R. *Tetrahedron* **1984**, *40*, 1245.

3 (a) Noyori, R., Ohta, M., Hsiao, Y., Kitamura, M., Ohta, T., Takaya, H. *J. Am. Chem. Soc.* **1986**, *108*, 7117; (b) Hitamura, M., Hsiao, Y., Noyori, R., Takaya, H. *Tetrahedron Lett.* **1987**, *28*, 4829; (c) Takaya, H., Ohta, T., Sayo, N., Kumobayashi, H., Akutagawa, S., Inoue, S.-I., Kasahara, I., Noyori, R. *J. Am. Chem. Soc.* **1987**, *109*, 1596; (d) Ohta, T., Takaya, H., Kitamura, M., Nagai, K., Noyori, R. *J. Org. Chem.* **1987**, *52*, 3174.

4 (a) Noyori, R., Ohkuma, T., Kitamura, M., Takaya, H., Sayo, N., Kumobayashi, H., Akutagawa, S. *J. Am. Chem. Soc.* **1987**, *109*, 5856; (b) Kitamura, M., Ohkuma, T., Inoue, S., Sayo, N., Kumobayashi, H., Akutagawa, S., Ohta, T., Takaya, H., Noyori, R. *J. Am. Chem. Soc.* **1988**, *110*, 629; (c) Kitamura, M., Kasahara, I., Manabe, K., Noyori, R., Takaya, H. *J. Org. Chem.* **1988**, *53*, 708; (d) Noyori, R., Ikeda, T., Ohkuma, T., Widhalm, M., Kitamura, M., Takaya, H., Akutagawa, S., Sayo, N., Saito, T., Taketomi, T., Kumobayashi, H. *J. Am. Chem. Soc.* **1989**, *111*, 9134.

5 Ohkuma, T., Ooka, H., Hashiguchi, S., Ikariya, T., Noyori, R. *J. Am. Chem. Soc.* **1995**, *117*, 2675.

6 Ohkuma, T., Ooka, H., Ikariya, T., Noyori, R. *J. Am. Chem. Soc.* **1995**, *117*, 10417.

7 (a) Miyashita, A., Karino, H., Shimamura, J., Chiba, T., Nagano, K., Nohira, H., Takaya, H. *Chem. Lett.* **1989**, 1007;

(b) Miyashita, A., Karino, H., Shimamura, J., Chiba, T., Nagano, K., Nohira, H., Takaya, H. *Chem. Lett.* **1989**, 1849; (c) Chiba, T., Miyashita, A., Nohira, H. *Tetrahedron Lett.* **1991**, *32*, 4745; (d) Chiba, T., Miyashita, A., Nohira, H., Takaya H. *Tetrahedron Lett.* **1993**, *34*, 2351.

8 Schmid, R., Cereghetti, M., Heiser, B., Schönholzer, P., Hansen, H.-J. *Helv. Chim. Acta* **1988**, *71*, 897.

9 Schmid, R., Foricher, J., Cereghetti, M., Schönholzer, P. *Helv. Chim. Acta.* **1991**, *74*, 870.

10 Yamamoto, N., Murata, M., Morimoto, T., Achiwa, K. *Chem. Pharm. Bull.* **1991**, *39*, 1085.

11 Murata, M., Morimoto, T., Achiwa, K. *Synlett* **1991**, 827.

12 Yoshikawa, K., Yamamoto, N., Murata, M., Awano, K., Morimoto, T., Achiwa, K. *Tetrahedron: Asymmetry* **1992**, *3*, 13.

13 (a) Zhang, X., Mashima, K., Koyano, K., Sayo, N., Kumobayashi, H., Akutagawa, S., Takaya, H. *Tetrahedron Lett.* **1991**, *32*, 7283; (b) Zhang, X., Mashima, K., Koyano, K., Sayo, N., Kumobayashi, H., Akutagawa, S., Takaya, H. *J. Chem. Soc., Perkin Trans. 1* **1994**, 2309.

14 Enev, V., Ewers, Ch. L.J., Harre, M., Nickisch, K., Mohr, J.T. *J. Org. Chem.* **1997**, *62*, 7092.

15 Sollewijn Gelpke A.E., Kooijman, H., Spek, A.L., Hiemstra, H. *Chem. Eur. J.* **1999**, *5*, 2472.

16 (a) Saito, T., Yokozawa, T., Zhang, X., Sayo, N. (Takasago International Corporation), U.S. Patent 5872273, **1999**; (b) Saito, T., Yokozawa, T., Ishizaki, T., Moroi, T., Sayo, N., Miura, T., Kumobayashi, H. *Adv. Synth. Catal.* **2001**, *343*, 264.

17 (a) Pai, C.-C., Li, Y.-M., Zhou, Z.-Y., Chan, A.S.C. *Tetrahedron Lett.* **2002**, *43*, 2789; (b) de Paule, S.D., Jeulin, S., Ratovelomanana-Vidal, V., Genêt, J.P., Champion, N., Dellis, P. *Tetrahedron Lett.* **2003**, *44*, 823; (c) de Paule, S.D., Jeulin, S., Ratovelomanana-Vidal, V., Genêt, J.P., Champion, N., Dellis, P. *Eur. J. Org. Chem.* **2003**, 1931.

18 Zhang, Z., Qian, H., Longmire, J., Zhang, X. *J. Org. Chem.* **2000**, *65*, 6223.

19 Wu, S., Wang, W., Tang, W., Lin, M., Zhang, X. *Org. Lett.* **2002**, *4*, 4495.

20 Tang, W., Wu, S., Zhang, X *J. Am. Chem. Soc.* **2003**, *125*, 9570.

21 Lei, A., Wu, S., He, M., Zhang, X. *J. Am. Chem. Soc.* **2004**, *126*, 1626.

22 Michaud, G., Bulliard, M., Ricard, L., Genêt, J.-P., Marinetti, A. *Chem. Eur. J.* **2002**, *8*, 3327.

23 (a) Benincori, T., Brenna, E., Sannicolò, F., Trimarco, L., Antognazza, P., Cesarotti, E. *J. Chem. Soc., Chem. Commun.* **1995**, 685; (b) Benincori, T., Brenna, E., Sannicol, F., Trimarco, L., Antognazza, P., Cesarotti, E. Demartin, F., Pilati, T. *J. Org. Chem.* **1996**, *61*, 6244; (c) Benincori, T., Rizzo, S., Pilati, T., Ponti, A., Sada, M., Pagliarini, E., Ratti, S., Giuseppe, C., de Ferra, L., Sannicolò, F. *Tetrahedron: Asymmetry* **2004**, *15*, 2289.

24 Benincori, T., Cesarotti, E., Piccolo, O., Sannicolò, F. *J. Org. Chem.* **2000**, *65*, 2043.

25 (a) Pai C.-C., Lin, C.-W., Lin, C.-C., Chen, C.-C., Chan, A.S.C. *J. Am. Chem. Soc.* **2000**, *122*, 11513; (b) Wu, J., Chen, H., Kwok, W.H., Lam, K.H., Zhou, Z.Y., Yeung, C.H., Chan, A.S.C. *Tetrahedron Lett.* **2002**, *43*, 1539; (c) Wu, J., Pai C.C., Kwok, W.H., Guo, R.W., Au-Yeung, T.T.L., Yeung, C.H., Chan, A.S.C. *Tetrahedron: Asymmetry* **2003**, *14*, 987; (d) Wu, J., Ji, J.-X., Guo, R., Yeung, C.-H., Chan, A.S.C. *Chem. Eur. J.* **2003**, *9*, 2963.

26 Tang, W., Chi, Y., Zhang, X. *Org. Lett.* **2002**, *4*, 1695.

27 Wu, S., He, M., Zhang, X. *Tetrahedron: Asymmetry* **2004**, *15*, 2177.

28 Henschke, J.P., Burk, M.J., Malan, C.G., Herzberg, D., Peterson, J.A., Wildsmith, A.J., Cobley, C.J., Casy, G. *Adv. Synth. Catal.* **2003**, *345*, 300.

29 Jeulin, S., de Paule, S.D., Ratovelomanana-Vidal, V., Genêt, J.P., Champion, N., Dellis, P. *Angew. Chem. Int. Ed. Engl.* **2004**, *43*, 320.

30 Qiu, L., Wu, J., Chan, S., Au-Yeung, T.T.-L., Ji, J.-X., Guo, R., Pai, C.-C., Zhou, Z., Li, X., Fan, Q., Chan, A.S.C. *Proc. Natl. Acad Sci. USA* **2004**, *101*, 5815.

31 Henschke, J.P., Zanotti-Gerosa, A., Moran, P., Harrison, P., Mullen, B., Casy,

G., Lennon, I.C. *Tetrahedron Lett.* **2003**, *44*, 4379.

32 Hu, A., Ngo, H.L., Lin, W. *Angew. Chem. Int. Ed. Engl.* **2004**, *43*, 2501.

33 Hu, A., Lin, W. *Org. Lett.* **2005**, *7*, 455.

34 Berthod, M., Saluzzo, C., Mignani, G., Lemaire, M. *Tetrahedron: Asymmetry* **2004**, *15*, 639.

35 Berthod, M., Mignani, G., Lemaire, M. *Tetrahedron: Asymmetry* **2004**, *15*, 1121.

36 Aikawa, K., Mikami, K. *Angew. Chem. Int. Ed. Engl.* **2003**, *42*, 5455.

37 Saluzzo, C., Lemaire, M. *Adv. Synth. Catal.* **2002**, *344*, 915.

38 (a) Wan, K.T., Davis, M.E. *J. Chem. Soc., Chem. Commun.* **1993**, 1262; (b) Wan, K.T., Davis, M.E. *Tetrahedron: Asymmetry* **1993**, *4*, 2461; (c) Wan, K.T., Davies, M.E. *J. Catalysis* **1994**, *148*, 1.

39 Schmid, R., Broger, E.A., Cereghetti, M., Crameri, Y., Foricher, J., Lalonde, M., Müller, R.K., Scalone, M., Schoettel, G., Zutter, U. *Pure Appl. Chem.* **1996**, *68*, 131.

40 Guerreiro, P., Ratovelomanana-Vidal, V., Genêt, J.-P., Dellis, P. *Tetrahedron Lett.* **2001**, *42*, 3423.

41 Bayston, D.J., Fraser, J.L., Ashton, M.R., Baxter, A.D., Polywka, M.E.C., Moses, E. *J. Org. Chem.* **1998**, *63*, 3137.

42 Ohkuma, T., Takeno, H., Honda, Y., Noyori, R. *Adv. Synth. Catal.* **2001**, *343*, 369.

43 (a) Fan, Q.-H., Deng, G.-J., Lin, C.-C., Chan, A.S.C. *J. Am. Chem. Soc.* **1999**, *121*, 7407; (b) Fan, Q.-H., Ren, C.-Y., Yeung, C.-H., Hu, W.-H., Chan, A.S.C. *Tetrahedron: Asymmetry* **2001**, *12*, 1241.

44 Fan, Q.-H., Chen, Y.-M., Chen, X.-M., Jiang, D.-Z., Xi, F., Chan, A.S.C. *Chem. Commun.* **2000**, 789.

45 (a) Yu, H.-B., Hu, Q.-S., Pu, L. *Tetrahedron Lett.* **2000**, *41*, 1681; (b) Yu, H.-B., Hu, Q.-S., Pu, L. *J. Am. Chem. Soc.* **2000**, *122*, 6500.

46 a) Lemaire, M., Halle, R. ter, Schulz, E., Colasson, B., Spagnol, M. (CNRS-Rhodia), *Fr. Patent* 99 02119, **1999**; (b) Lemaire, M., Halle, R. ter, Schulz, E., Colasson, B., Spagno, M., Saluzzo, C., Lamouille, T. (CNRS-Rhodia), *Fr. Patent* 99 02510, **1999**; PCT Int. Appl. 2000, WO 0052081.

47 (a) Kagan, H.B., Dang, T.P. *Chem. Commun.* **1971**, 481; (b) Kagan, H.B., Dang, T.P. *J. Am. Chem. Soc.* **1972**, *94*, 6429; (c) Kagan, H.B., Langlois, N., Dang, T.P. *J. Organomet. Chem.* **1975**, *90*, 353.

48 (a) Chiba, M., Takahashi, H., Takahashi, H., Morimoto, T., Achiwa, K. *Tetrahedron Lett.* **1987**, *28*, 3675; (b) Morimoto, T., Chiba, M., Achiwa, K. *Tetrahedron Lett.* **1988**, *29*, 4755; (c) Morimoto, T., Chiba, M., Achiwa, K. *Tetrahedron Lett.* **1989**, *30*, 735; (d) Morimoto, T., Chiba, M., Achiwa, K. *Chem. Pharm. Bull.* **1989**, *37*, 3161; (e) Morimoto, T., Chiba, M., Achiwa, K. *Heterocycles* **1990**, *30*, 363; (f) Morimoto, T., Chiba, M., Achiwa, K. *Tetrahedron Lett.* **1990**, *31*, 261; (g) Yoshikawa, K., Inoguchi, K., Morimoto, T., Achiwa, K. *Heterocycles,* **1990**, *31*, 261; (h) Morimoto, T., Chiba, M., Achiwa, K. *Chem. Pharm. Bull.* **1993**, *41*, 1149.

49 (a) Zhu, G., Cao, P., Jiang, Q., Zhang, X. *J. Am. Chem. Soc.* **1997**, *119*, 1799; (b) Zhu, G., Zhang, X. *J. Org. Chem.* **1998**, *63*, 9590; (c) Zhu, G., Casalnuovo, A.L., Zhang, X. *J. Org. Chem.* **1998**, *63*, 8100; (d) Zhu, G., Zhang, X. *Tetrahedron: Asymmetry* **1998**, *9*, 2415; (e) Zhu, G., Chen, Z., Zhang, X. *J. Org. Chem.* **1999**, *64*, 6907; (f) Cao, P., Zhang, X. *J. Org. Chem.* **1999**, *64*, 2127.

50 Genov, D.G., Ager, D. *Angew. Chem. Int. Ed. Engl.* **2004**, *43*, 2816.

51 (a) Li, W., Zhang, X. *J. Org. Chem.* **2000**, *65*, 5871; (b) Liu, D., Li, W., Zhang, X. *Tetrahedron: Asymmetry* **2004**, *15*, 2181.

52 (a) Yan, Y.-Y., RajanBabu, T.V. *J. Org. Chem.* **2000**, *65*, 900; (b) Yan, Y.-Y., RajanBabu, T.V. *Org. Lett.* **2000**, *2*, 199; (c) Yan, Y.-Y., RajanBabu, T.V. *Org. Lett.* **2000**, *2*, 4137.

53 Kagan, H.B., Fiaud, J.C., Hoornaert, C., Meyer, D., Poulin, J.C. *Bull. Soc. Chim, Belg.* **1979**, *88*, 923.

54 (a) Lee, S.-g., Zhang, Y.J., Song, C.E., Lee, J.K., Choi, J.H. *Angew. Chem. Int. Ed.* **2002**, *41*, 847; (b) Lee, S.-g., Zhang, Y.J. *Org. Lett.* **2002**, *4*, 2429; (c) Lee, S.-g., Zhang, Y.J. *Tetrahedron: Asymmetry* **2002**, *13*, 1039.

55 Li, W., Waldkirch, J.P., Zhang, X. *J. Org. Chem.* **2002**, 7618.

56 (a) Vineyard, B. D., Knowles, W. S., Sabacky, M. J., Bachman, G. L., Weinkauff, O. J. *J. Am. Chem. Soc.* **1977**, *99*, 5946; (b) Knowles, W. S. *Acc. Chem. Res.* **1983**, *16*, 106.

57 Knowles, W. S. *J. Chem. Educ.* **1986**, *63*, 222.

58 (a) Imamoto, T., Watanabe, J., Wada, Y., Masuda, H., Yamada, H., Tsuruta, H., Matsukawa, S., Yamaguchi, K. *J. Am. Chem. Soc.* **1998**, *120*, 1635; (b) Gridnev, I. D., Yamanoi, Y., Higashi, N., Tsuruta, H., Yasutake, M., Imamoto, T. *Adv. Synth. Catal.* **2001**, *343*, 118.

59 Gridnev, I. D., Yasutake, M., Higashi, N., Imamoto, T. *J. Am. Chem. Soc.* **2001**, *123*, 5268.

60 Yasutake, M., Gridnev, I. D., Higashi, N., Imamoto, T. *Org. Lett.* **2001**, *3*, 1701.

61 Gridnev, I. D., Higashi, N., Imamoto, T. *J. Am. Chem. Soc.* **2001**, *123*, 4631.

62. (a) Gridnev, I. D., Higashi, N., Asakura, K., Imamoto, T. *J. Am. Chem. Soc.* **2000**, *122*, 7183; (b) Gridnev, I. D., Higashi, N., Imamoto, T. *J. Am. Chem. Soc.* **2000**, *122*, 10486; (c) Gridnev, I. D., Imamoto, T. *Organometallics* **2001**, *20*, 545.

63 (a) Halpern, J., Riley, D. P., Chan, A. S. C., Pluth, J. J. *J. Am. Chem. Soc.* **1977**, *99*, 8055; (b) Chan, A. S. C., Halpern, J. *J. Am. Chem. Soc.* **1980**, *102*, 838; (c) Chan, A. S. C., Pluth, J. J., Halpern, J. *J. Am. Chem. Soc.* **1980**, *102*, 5952; (d) Halpern, J. *Science* **1982**, *217*, 401; (e) Landis, C. R., Halpern, J. *J. Am. Chem. Soc.* **1987**, *109*, 1746; (f) Ashby, M. T., Halpern, J. *J. Am. Chem. Soc.* **1991**, *113*, 589.

64 (a) Brown, J. M., Chaloner, P. A. *J. Chem. Soc., Chem. Commun.* **1978**, 321; (b) Brown, J. M., Chaloner, P.A. *Tetrahedron Lett.* **1978**, *19*, 1877; (c) Brown, J. M., Chaloner, P. A. *J. Chem. Soc., Chem. Commun.* **1979**, 613; (d) Brown, J. M., Chaloner, P. A. *J. Chem. Soc., Chem. Commun.* **1980**, 344; (e) Brown, J. M., Chaloner, P. A., Glaser, R., Geresh, S. *Tetrahedron* **1980**, *36*, 815; (f) Alcock, N. W., Brown, J. M., Derome, A., Lucy, A. R. *J. Chem. Soc., Chem. Commun.* **1985**, 575; (g) Brown, J. M., Chaloner, P. A., Morris, G. A. *J. Chem. Soc., Perkin Trans.* 2, **1987**, 1583; (h) Brown, J. M. *Chem. Soc. Rev.* **1993**, *22*, 25.

65 Yamanoi, Y., Imamoto, T. *J. Org. Chem.* **1999**, *64*, 2988.

66 Miura, T., Imamoto, T. *Tetrahedron Lett.* **1999**, *40*, 4833.

67 (a) Ohashi, A., Imamoto, T. *Org. Lett.* **2001**, *3*, 373; (b) Ohashi, A., Kikuchi, S.-I., Yasutake, M., Imamoto, T. *Eur. J. Org. Chem.* **2002**, 2535.

68 Danjo, H., Sasaki, W., Miyazaki, T., Imamoto, T. *Tetrahedron Lett.* **2003**, *44*, 3467.

69 (a) Robin, F., Mercier, F., Ricard, L., Mathey, F., Spagnol, M. *Chem. Eur. J.* **1997**, *3*, 1365; (b) Mathey, F., Mercier, F., Robin, F., Ricard, L. *J. Organometallic Chem.* **1998**, *577*, 117.

70 Hoge, G., Wu, H.-P., Kissel, W. S., Pflum, D. A., Greene, D. J., Bao, J. *J. Am. Chem. Soc.* **2004**, *126*, 5966.

71 Wu, H.-P., Hoge, G. *Org. Lett.* **2004**, *6*, 3645.

72. (a) Fryzuk, M. B., Bosnich, B. *J. Am. Chem. Soc.* **1977**, *99*, 6262; (b) Fryzuk, M. B., Bosnich, B. *J. Am. Chem. Soc.* **1979**, *101*, 3043.

73 Fryzuk, M. D., Bosnich, B. *J. Am. Chem. Soc.* **1978**, *100*, 5491.

74 Achima, K. *J. Am. Chem. Soc.* **1976**, *98*, 8265.

75 (a) Phone Poulenc S. A. *Fr. Patent* 2230654, **1974**; (b) Glaser, R., Geresh, S., Twaik, M. *Isr. J. Chem.* **1980**, *20*, 102.

76 (a) Takahashi, H., Hattori, M., Chiba, M., Morimoto, T., Achiwa, K. *Tetrahedron Lett.* **1986**, *27*, 4477; (b) Takahashi, H., Morimoto, T., Achiwa, K. *Chem. Lett.* **1987**, 855; (c) Takahashi, H., Achiwa, K. *Chem. Lett.* **1987**, 1921; (d) Inoguchi, K., Morimoto, T., Achiwa, K. *J. Organomet. Chem.* **1989**, *370*, C9; (e) Takahashi, H., Yamamoto, N., Takeda, H., Achiwa, K. *Chem. Lett.* **1989**, 559; (f) Takahashi, H., Sakuraba, S., Takeda, H., Achiwa, K. *J. Am. Chem. Soc.* **1990**, *112*, 5876; (g) Takeda, H., Hosokawa, S., Aburatani, M., Achiwa, K. *Synlett*, **1991**, 193; (h) Sakuraba, S., Achiwa, K. *Synlett* **1991**, 689; (i) Sakuraba, S., Nakajima, N., Achiwa, K. *Synlett* **1992**, 829; (j) Sakuraba, S., Nakajima, N., Achiwa, K. *Tetrahedron: Asymmetry* **1993**, *7*, 1457; (k) Sakuraba, S., Takahashi, H., Takeda, H., Achiwa, K. *Chem. Pharm. Bull.* **1995**, *43*, 738.

77 Brunner, H., Pieronczyk, W., Schönhammer, B., Streng, K., Bernal, I., Korp, J. *Chem. Ber.* **1981**, *114*, 1137.

78 (a) Nagel, U. *Angew. Chem. Int. Ed.* **1984**, *23*, 435; (b) Nagel, U., Kinzel, E., Andrade, J., Prescher, G. *Chem. Ber.* **1986**, *119*, 3326; (c) Inoguchi, K., Achiwa, K. *Chem. Pharm. Bull.* **1990**. *38*, 818.

79 Allen, D. L., Gibson, V.C., Green, M. L. H., Skinner, J. F., Bashkin, J., Grebenik, P. D. *J. Chem. Soc., Chem. Commun.* **1983**, 895.

80 (a) Bakos, J., Toth, I., Markó, L. *J. Org. Chem.* **1981**, *46*, 5427; (b) NacNeil, P. A., Roberts, N. K., Bosnich, B. *J. Am. Chem. Soc.* **1981**, *103*, 2273; (c) Bakos, J., Tóth, I., Heil, B., Markó, L. *J. Organomet. Chem.* **1985**, *279*, 23.

81 Inoguchi, K., Achiwa, K. *Synlett* **1991**, 49.

82 (a) Pye, P. J., Rossen, K., Reamer, R. A., Tsou, N. N., Volante, R. P., Reider, P. J. *J. Am. Chem. Soc.* **1997**, *119*, 6207; (b) Pye, P. J., Rossen, K., Reamer, R. A., Volante, R. P., Reider, P. J. *Tetrahedron Lett.* **1998**, *39*, 4441; (c) Burk, M. J., Hems, W., Herzberg, D., Malan, C., Zanotti-Gerosa, A. *Org. Lett.* **2000**, *2*, 4173.

83 Zhou, Y.-G., Zhang, X. *Chem. Commun.* **2002**, 1124.

84 Morimoto, T., Yamazaki, A., Achiwa, K. *Chem. Pharm. Bull.* **2004**, *52*, 1367.

85 Xie, J.-H., Wang, L.-X., Fu, Y., Zhu, S.-F., Fan, B.-M., Duan, H.-F., Zhou, Q.-L. *J. Am. Chem. Soc.* **2003**, *125*, 4404.

86 (a) Scott, J. W., Kieth, D. D., Nix, G., Jr., Parrish, D. R., Remington, S., Roth, G. P., Townsend, J. M., Valentine, D., Jr., Yang, R. *J. Org. Chem.* **1981**, *46*, 5086; (b) Vineyard, B. D., Knowles, W. S., Sabacky, M. J., Bachman, G. L., Weinkauff, D. J. *J. Am. Chem. Soc.* **1977**, *99*, 5946.

87 Doi, T., Kokubo, M., Yamamoto, K., Takahashi, T. *J. Org. Chem.* **1998**, *63*, 428.

88 Adamczyk, M., Akireddy, S. R., Reddy, R. E. *Tetrahedron* **2002**, *58*, 6951.

89 Rossen, K., Weissman, S. A., Sager, J., Reamer, R. A., Askin, D., Valante, R. P., Reider, P. J. *Tetrahedron Lett.* **1995**, *36*, 6419.

90 Mechanistic study via a computational approach, see: Feldgus, S., Landis, C. R. *Organometallics* **2001**, *20*, 2374.

91 Kitamura, M., Hsiao, Y., Ohta, M., Tsukamoto, M., Ohta, T., Takaya, H., Noyori, R. *J. Org. Chem.* **1994**, *59*, 297.

92 Tschaen, D. M., Abramson, L., Cai, D., Desmond, R., Dolling, U.-H., Frey, L., Karady, S., Shi, Y.-J., Verhoeven, T. R. *J. Org. Chem.* **1995**, *60*, 4324.

93 Renaud, J. L., Dupau, P., Hay, A.-E., Guingouain , M., Dixneuf, P. H., Bruneau, C. *Adv. Synth. Catal.* **2003**, *345*, 230.

94 (a) Tang, T., Ellman, J. A. *J. Org. Chem.* **1999**, *64*, 12; (b) Sibi, M. P., Shay, H. J., Liu, M., Jasperse, C. P. *J. Am. Chem. Soc.* **1998**, *120*, 6615; (c) Kobayashi, S., Ishitani, H., Ueno, M. *J. Am. Chem. Soc.* **1998**, *120*, 431; (d) Boesch, H., Cesco-Cancian, S., Hecker, L. R., Hoekstra, W. J., Justus, M., Maryanoff, C. A., Scott, L., Shah, R. D., Solms, G., Sorgi, K. L., Stefanick, S. M., Thurnheer, U., Villani, F. J., Jr., Walker, D. G. *Org. Process Res. Dev.* **2001**, *5*, 23; (e) Hoekstra, W. J., Maryanoff, B. E., Damiano, B. P., Andrade-Gordon, P., Cohen, J. H., Costanzo, M. J., Haertlein, B. J., Hecker, L. R., Hulshizer, B. L., Kauffman, J. A., Keane, P., McComsey, D. F., Mitchell, J. A., Scott, L., Shah, R. D., Yabut, S. C. *J. Med. Chem.* **1999**, *42*, 5254; (f) Zhong, H.M., Cohen, J. H., Abdel-Magid, A. F., Kenney, B. D., Maryanoff, C. A., Shah, R. D., Villani, F. J., Jr., Zhang, F., Zhang, X. *Tetrahedron Lett.* **1999**, *40*, 7721; (g) Shankar, B. B., Kirkup, M. P., McCombie, S. W., Clader, J. W., Ganguly, A.K. *Tetrahedron Lett.* **1996**, *37*, 4095; (h) Burnett, D. A., Caplen, M. A., Davis, H. R., Jr., Burrier, R. E., Clader, J. W. *J. Med. Chem.* **1994**, *59*, 1733.

95 Lubell, W. D., Kitamura, M., Noyori, R. *Tetrahedron: Asymmetry* **1991**, *2*, 543.

96 Koenig, K. E., Bachman, G. L., Vineyard, B. D. *J. Org. Chem.* **1980**, *45*, 2362.

97 Schmidt, U., Langner, J., Kirschbaum, B., Braun, C. *Synthesis* **1994**, 1138.

98 Ohta, T., Miyake, T., Seido, N., Kumobayashi, H., Takaya, H. *J. Org. Chem.* **1995**, *60*, 357.

99 Gridnev, I. D., Yasutake, M., Imamoto, T., Beletskaya, I. P. *Proc. Natl. Acad. Sci. USA* **2004**, *101*, 5385.

100 Saito, T., Yokozawa, T., Ishizaki, T., Moroi, T., Sumi, K., Sayo, N. *Chirasource 2000*, Lisbon, Portugal, October 2, **2000**.

101 Fehr, M. J., Consiglio, G., Scalone, M., Schmid, R. *J. Org. Chem.* **1999**, *64*, 5768.

102 (a) Zhang, X., Uemura, T., Matsumura, K., Sayo, N., Kumobayashi, H., Takaya, H. *Synlett* **1994**, 501; (b) Uemura, T., Zhang, X., Matsumura, K., Sayo, N., Kumobayashi, H., Ohta, T., Nozaki, K., Takaya, H. *J. Org. Chem.* **1996**, *61*, 5510.

103 Genêt, J. P., Pinel, C., Ratovelomanana-Vidal, V., Mallart, S., Pfister, X., Bischoff, L., Cano De Andrade, M.C., Darses, S., Galopin, C., Laffitte, J. A. *Tetrahedron: Asymmetry* **1994**, *5*, 675.

104 Manimaran, T., Wu, T.-C., Klobucar, W. D., Kolich, C. H., Stahly, G. P. *Organometallics* **1993**, *12*, 1467.

105 Crameri, Y., Foricher, J., Scalone, M., Schmid, R. *Tetrahedron: Asymmetry* **1997**, *8*, 3617.

106 Boulton, L.T., Lennon, I. C., McCague, R. *Org. Biomol. Chem.* **2003**, *1*, 1094.

107 Jendralla, H., Henning, R., Seuring, B., Herchen, J., Kulitzscher, B., Wunner, J. *Synlett* **1993**, 155.

108 Bulliard, M., Laboue, B., Lastennet, J., Roussiasse, S. *Org. Proc. Res. Dev.* **2001**, *5*, 438.

109 (a) Sun, Y., LeBlond, C., Wang, J., Blackmond, D. G. *J. Am. Chem. Soc.* **1995**, *117*, 12647; (b) Sun, Y., Landau, R. N., Wang, J., LeBlond, C., Blackmond, D. G. *J. Am. Chem. Soc.* **1996**, *118*, 1348; (c) Sun, Y., Wang, J., LeBlond, C., Landau, R. N., Blackmond, D. G. *J. Catal.* **1996**, *161*, 756.

110 (a) Bissel, P., Sablong, R., Lepoittevin, J.-P. *Tetrahedron: Asymmetry* **1995**, *6*, 835; (b) Bissel, P., Nazih, A., Sablong, R., Lepoittevin, J.-P. *Org. Lett.* **1999**, *1*, 1283.

27
Bidentate Ligands Containing a Heteroatom–Phosphorus Bond

Stanton H. L. Kok, Terry T.-L. Au-Yeung, Hong Yee Cheung, Wing Sze Lam, Shu Sun Chan, and Albert S. C. Chan

27.1
Introduction

Bidentate phosphorus ligands containing one or more heteroatom-phosphorus bonds are of high interest because they are relatively easy to prepare, and because a huge multitude of inexpensive, commercially available chiral diols, diamines, amino alcohols and amino acids can serve as the scaffold. Although the heteroatoms in these scaffolds are usually electronegative in nature, the reactivity and enantioselectivity of the metal complexes based on some of these ligands are quite remarkable, and sometimes even surpass those of the complexes based on electron-rich phosphines. This chapter compiles the comprehensive data concerning the asymmetric hydrogenation of various prochiral olefins mediated by the rhodium(I) complexes of this class of chiral ligands.

27.2
Aminophosphine-Phosphinites (AMPPs)

The ease of synthesis from chiral amino alcohols with a wide array of derivatives in one step established its good potential in the field of asymmetric catalysis. The general preparation of "semi-symmetrical" AMPPs involves the nucleophilic attack of two equivalents of chlorophosphine in the presence of a base (Fig. 27.1). A "mixed" AMPP can also be prepared by virtue of the fact that phosphorus-based electrophiles have a strong preference for hydroxy over secondary amine or amide. Clearly, this synthetic method allows the preparation of a large variety of AMPP ligands with adjustable electronic and steric properties. Agbossou recently reviewed the state of the art of AMPPs [1]. In consideration to the modern high-throughput methods, this approach allowed a rapid combinatorial screening of various catalysts and reactions.

In general, applications of AMPP have concentrated on the asymmetric hydrogenation of functionalized olefins, especially dehydroamino acids. Among

The Handbook of Homogeneous Hydrogenation.
Edited by J. G. de Vries and C. J. Elsevier
Copyright © 2007 WILEY-VCH Verlag GmbH & Co. KGaA, Weinheim
ISBN: 978-3-527-31161-3

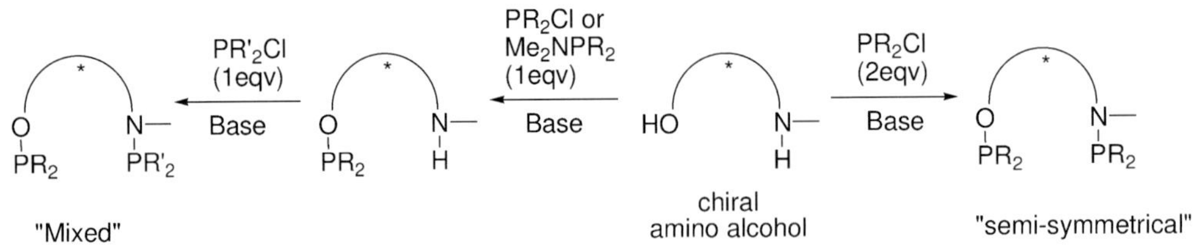

Fig. 27.1 The preparation of AMPPs.

these substrates, (*Z*)-methyl *α*-acetamidocinnamate was the most frequently used benchmark substrate. A strong influence of the solvent on catalytic activity and enantioselectivity was a common phenomenon, and protic solvents were found to be the most effective. However, to avoid the problem of solvolysis of the ligands, polar aprotic solvents were commonly used to obtain the best results. Although the *in-situ* preparation of cationic rhodium complexes was frequently used, no significant dependence of their catalytic performance on the manner of their preparation could be observed. Most Rh complexes allowed the use of atmospheric pressure for hydrogenation with high reaction rate at room temperature. In all cases, the ligands formed a chelation ring with the metal.

Many structurally diverse ephedrine-derived AMPP ligands (Fig. 27.2) have been prepared, and most of these were applied to the asymmetric hydrogenation of olefins. Cesarotti was one of the earliest pioneers in the development of aminophosphine-phosphinite **1** based on (*S*)-2-(ethylamino)butan-1-ol as a starting material. However, the results were only moderate to good [2]. Almost simultaneously with Cesarotti in 1982, Pracejus reported a similar approach [3]. Ephedrine-based Propraphos and its derivatives occupied the major area of this research field. The chiral Rh–Propraphos systems were widely applied in the enantiomeric hydrogenation of *α*-dehydroamino acids, with 31 to 95% e.e. The products included (*S*) and (*R*)-aromatic [4–6] and heteroaromatic alanine derivatives [7–16], and usually have a configuration which is opposite to that of the ligand. 2-Acetamido-cinnamic acid derivatives carrying an electron-withdrawing group at the *para*-position of the phenyl ring could be hydrogenated with relatively high enantioselectivities. In most cases, turnover frequencies (TOFs) could be obtained of up to 3000 h^{-1}, and up to 11 515 h^{-1} for a special case (Table 27.1, entry 392). The use of Rh-**15** in the hydrogenation of dimethyl itaconate gave the product with 80% ee (Table 27.1, entry 432). Structural analogues of Propraphos, Pindophos and Caraphos [7, 17] led to similar ee-values; however, a longer reaction time was required with the Caraphos–Rh complex. Use of (*R*)-Pindophos–Rh in the diastereoselective hydrogenation of dehydrodipetides produced good selectivity (up to 91% ee in the case of *para*-trifluoromethyl-phenylalanyl-phenylalanine [7] (Table 27.2, entry 10). A series of novel ephedrine-based ligands have been shown to be highly effective in the Rh-catalyzed hydrogenation of dehydroamino acids, giving the products with 95–99% ee [18, 19]. The hydrogenation of (*Z*)-acetamidocinnamate with a substrate:catalyst ratio (SCR) of

9*H*-carbazolyl naphthyl 1*H*-indolyl

(*S*)-**1** (Butaphos) R^1 = H; R^2 = Et; R^3 = Et; R^4 = Ph
(*S*)-**2** (NEtAlaNOP) R^1 = H; R^2 = Me; R^3 = Et; R^4 = Ph
(1*R*,2*S*)-**3** R^1 = Ph; R^2 = Me; R^3 = Me; R^4 = Ph
(1*R*)-**4** R^1 = Ph; R^2 = H; R^3 = Me; R^4 = Ph
(1*R*,2*R*)-**5** R^1 = Ph; R^2 = Me; R^3 = Me; R^4 = Ph
(2*R*)-**6** R^1 = H; R^2 = Ph; R^3 = Me; R^4 = Ph
(1*R*,2*S*)-**7** (DPAMPP) R^1 = R^2 = Ph; R^3 = Me; R^4 = Ph
(1*S*,2*S*)-**7** (DPAMPP) R^1 = R^2 = Ph; R^3 = Me; R^4 = Ph
(1*S*,2*R*)-**7** (DPAMPP) R^1 = R^2 = Ph; R^3 = Me; R^4 = Ph
(*R*)-**8** R^1 = naphthyl; R^2 = H; R^3 = *i*Pr; R^4 = Cy
(*S*)-**9** (Propraphos) R^1 = naphthyl; R^2 = H; R^3 = *i*Pr, R^4 = Ph
(*R*)-**9** (Propraphos) R^1 = naphthyl; R^2 = H; R^3 = *i*Pr, R^4 = Ph
(*S*)-**10** (Pindophos) R^1 = 1*H*-indolyl; R^2 = H; R^3 = *i*Pr, R^4 = Ph
(*R*)-**10** (Pindophos) R^1 = 1*H*-indolyl; R^2 = H; R^3 = *i*Pr, R^4 = Ph
(*R*)-**11** (Caraphos) R^1 = 9*H*-carbazolyl; R^2 = H; R^3 = *i*Pr, R^4 = Ph
(*S*)-**12** R^1 = naphthyl; R^2 = H; R^3 = H, R^4 = Ph
(*S*)-**13** R^1 = naphthyl; R^2 = H; R^3 = Me, R^4 = Ph
(*S*)-**14** R^1 = naphthyl; R^2 = H; R^3 = $CHEt_2$, R^4 = Ph
(*S*)-**15** R^1 = naphthyl; R^2 = H; R^3 = Cyclopentyl, R^4 = Ph
(*S*)-**16** R^1 = naphthyl; R^2 = H; R^3 = Cyclohexyl, R^4 = Ph

(1*S*,2*R*)-**17** (EPHOS) R^1 = Ph
(1*S*,3*R*,4S)-**18** R^1 = Me
(1*S*,2*R*,4*R*)-**19** R^1 = *o*-An
(1*S*,2*R*,4*R*)-**20** R^1 = 1-Np
(1*S*,2*R*,4*R*)-**21** R^1 = 2-Np
(1*S*,2*R*,4*S*)-**22** R^1 = *t*Bu

(1*S*,2*R*,4*R*)-**23** R^1 = Ph; R^2 = *o*-An
(1*S*,3*S*,4*R*,6*R*)-**24** R^1 = R^2 = *o*-An

(*R*)-**25** (Ph-tLANOP) R = Ph
(*S*)-**26** (Cy-tLANO) R = Cy
(*S*)-**27** (2-Furyl-tLANOP) R = 2-furyl

(1*R*,3*R*,5*R*)-**28**
(1*S*,3*S*,5*S*)-**28**

29 (ProNOP/prolophos) Ar = Ph
30 (Cy-ProNOP) Ar = Cy
31 (Bu-ProNOP) Ar = Bu

(*S*)-**32** (Ph-oxoProNOP) R = Ph
(*S*)-**33** (Cy-oxoProNOP) R = Cy
(*S*)-**34** ((2-Furyl)-oxoProNOP) R = 2-furyl

(+)-(2*S*,3S)-**35**
(−)-(2*R*,3*R*)-**35**

(1*R*,3*S*,4*S*)-**36** Ar^1 = Ar^2 = Ph
(1*R*,3*S*,4*S*)-**37** Ar^1 = Ar^2 = *p*-Tol
(1*R*,3*S*,4*S*)-**38** Ar^1 = Ph; Ar^2 = *p*-Tol
(1*R*,3*S*,4*S*)-**39** Ar^1 = *p*-Tol; Ar^2 = Ph

exo-**40**

Fig. 27.2 AMPP chiral ligands.

Table 27.1 Enantiomeric hydrogenation using aminophosphine–phosphinites (AMPP).

Entry	Substrate			Catalyst	Conditions				TON	TOF [h^{-1}]	Conv. [%]	ee [%]	Reference(s)
	R^1	R^2	R^3		P[H_2] [bar]	Solvent	Temp. [°C]	Time [h]					
1	H	CO_2H	NHAc	[Rh(COD)(*S*)-**1**]ClO_4	1	EtOH	20	–	200	–	100	55 (*S*)[a]	2a, b
2	H	CO_2H	NHAc	[Rh(COD)(1*S*,2*R*)-**7**]BF_4	50	MeOH	25	1	100	100	100	95.2 (*R*)	18
3	H	CO_2H	NHAc	[Rh(COD)(*S*)-**29**]ClO_4	1	EtOH	20	–	200	–	100	80 (*S*)[a]	2a, b
4	H	CO_2H	NHAc	RhCl(COD)(1*S*,2*S*)-**35**	1	MeOH	25	–	100	–	100	89 (*R*)	37
5	H	CO_2H	NHAc	[Rh(COD) (1*S*,2*S*)-**35**]BF_4	1	MeOH	25	–	100	–	100	86 (*R*)	37
6	H	CO_2H	NHAc	RhCl(COD)(1*R*,2*R*)-**35**	1	MeOH	25	–	100	–	100	89 (*S*)	37
7	H	CO_2H	NHBz	[Rh(COD)(1*S*,2*R*)-**7**]BF_4	50	MeOH	25	1	100	100	100	94.8 (*R*)	18
8	H	CO_2Me	NHAc	[Rh(COD)(*S*)-**2**]ClO_4	1	EtOH/PhH	20	–	200	–	100	66 (*R*)	32
9	H	CO_2Me	NHAc	[Rh(COD)(*S*)- **29**]ClO_4	1	EtOH/PhH	20	–	200	–	100	67 (*S*)	32
10	H	CO_2Me	NHAc	[Rh(COD)(*S*)-**29**]BF_4	1	MeOH	r.t.	–	2500	–	–	79 (*S*)	41
11	H	CO_2Me	NHAc	[Rh(COD)(*S*)-**29**]BF_4	1	DCM	r.t.	–	1500	–	–	69 (*S*)	41
12	H	CO_2Me	NHAc	[Rh(COD)(1*R*,3*S*,4*S*)-**36**]BF_4	1	MeOH	r.t.	–	3500	–	–	85 (*S*)	41
13	H	CO_2Me	NHAc	[Rh(COD)(1*R*,3*S*,4*S*)-**36**]BF_4	1	DCM	r.t.	–	1500	–	–	70 (*S*)	41
14	H	CO_2Me	NHAc	[Rh(COD)(1*R*,3*S*,4*S*)-**37**]BF_4	1	MeOH	r.t.	–	4000	–	–	80 (*S*)	41
15	H	CO_2Me	NHAc	[Rh(COD)(1*R*,3*S*,4*S*)-**37**]BF_4	1	DCM	r.t.	–	2000	–	–	61 (*S*)	41
16	H	CO_2Me	NHAc	[Rh(COD)(1*R*,3*S*,4*S*)-**38**]BF_4	1	MeOH	r.t.	–	3600	–	–	80 (*S*)	41
17	H	CO_2Me	NHAc	[Rh(COD)(1*R*,3*S*,4*S*)-**38**]BF_4	1	DCM	r.t.	–	1200	–	–	63 (*S*)	41
18	H	CO_2Me	NHAc	[Rh(COD)(1*R*,3*S*,4*S*)-**39**]BF_4	1	MeOH	r.t.	–	4500	–	–	77 (*S*)	41
19	H	CO_2Me	NHAc	[Rh(COD)(1*R*,3*S*,4*S*)-**39**]BF_4	1	DCM	r.t.	–	2500	–	–	59 (*S*)	41
20	H	OAc	Ph	[Rh(COD)(1*R*,2*S*)-**3**]BF_4	1	Dioxane	25	–	–	–	–	24 (*R*)	36
21	H	OAc	Ph	[Rh(COD) (1*R*,2*R*)-**5**]BF_4	1	Dioxane	25	–	–	–	–	13 (*R*)	36

22	$(CH_3)_2CH$	CO_2H	NHAc	[Rh(COD)(*S*)-**29**]BF_4	1	EtOH	20	–	200	–	100	96 (*S*)	2b
23	$(CH_3)_2CH$	CO_2H	NHAc	[Rh(COD)(*S*)-**1**]BF_4	1	EtOH	20	–	200	–	100	64 (*S*)	2b
24	$(CH_3)_2CH$	CO_2H	NHBz	[Rh(COD)(*S*)-**1**]ClO_4	1	EtOH	20	–	200	–	100	57 (*S*)	2b
25	$(CH_3)_2CH$	CO_2H	NHBz	[Rh(COD)(*S*)-**29**]ClO_4	1	EtOH	20	–	200	–	100	45 (*S*)	2b
26	Ph	CO_2H	NHAc	[Rh(COD)(*S*)-**1**] ClO_4	1	EtOH	20	–	–	–	–	23 (*S*)[a]	2a, b
27	Ph	CO_2H	NHAc	[Rh(COD)(1*R*,2*S*)-**3**]BF_4	1	Dioxane	25	–	–	–	–	80 (*R*)	36
28	Ph	CO_2H	NHAc	[Rh(COD)(1*R*,2*S*)-**3**]BF_4	1	MeOH	25	–	–	–	–	12 (*R*)	36
29	Ph	CO_2H	NHAc	[Rh(COD)(*R*)-**4**]BF_4	1	Dioxane	25	–	–	–	–	56 (*R*)	36
30	Ph	CO_2H	NHAc	[Rh(COD)(*R*)-**4**]BF_4	1	MeOH	25	–	–	–	–	5 (*R*)	36
31	Ph	CO_2H	NHAc	[Rh(COD) (1*R*,2*R*)-**5**]BF_4	1	Dioxane	25	–	–	–	–	3 (*S*)	36
32	Ph	CO_2H	NHAc	[Rh(COD) (1*R*,2*R*)-**5**]BF_4	1	MeOH	25	–	–	–	–	9 (*R*)	36
33	Ph	CO_2H	NHAc	[Rh(COD)(2*R*)-**6**]BF_4	1	Dioxane	25	–	–	–	–	24 (*R*)	36
34	Ph	CO_2H	NHAc	[Rh(COD)(2*R*)-**6**]BF_4	1	MeOH	25	–	–	–	–	0.5 (*S*)	36
35	Ph	CO_2H	NHAc	[Rh(COD)(1*S*,2*R*)-**7**]BF_4	50	MeOH	25	1	100	100	100	96.5 (*R*)	18
36	Ph	CO_2H	NHAc	[Rh(COD)(*S*)-**9**]BF_4	1	MeOH	25	0.03[b]	50	1667	50	87 (*R*)	8
37	Ph	CO_2H	NHAc	RhCl(COD)(*S*)-**9**	1	MeOH	25	0.12[b]	50	417	50	88 (*R*)	8
38	Ph	CO_2H	NHAc	[Rh(COD)(*R*)-**9**]BF_4	1	MeOH	25	0.092[b]	1000	10870	50	86 (*S*)	9
39	Ph	CO_2H	NHAc	[Rh(COD)(*R*)-**9**]BF_4	1	MeOH	25	0.343[b]	1500	4373	50	85 (*S*)	9
40	Ph	CO_2H	NHAc	Rh(COD)(*S*)-**9**	1	MeOH	25	0.12[b]	50	417	50	88 (*R*)	15
41	Ph	CO_2H	NHAc	Rh(COD)(*S*)-**9**	1	MeOH	25	0.58[b]	500	862	50	85 (*R*)[l]	15
42	Ph	CO_2H	NHAc	[Rh(COD)(*S*)-**9**]BF_4	1	MeOH	25	0.02[b]	50	2500	50	87 (*R*)[m]	15
43	Ph	CO_2H	NHAc	[Rh(COD)(*S*)-**9**]BF_4	1	MeOH	25	0.15[b]	500	3333	50	84 (*R*)[m]	15
44	Ph	CO_2H	NHAc	[Rh(COD)(*S*)-**9**]BF_4	1	MeOH	25	0.22[b]	500	2273	50	89 (*R*)	15
45	Ph	CO_2H	NHAc	[Rh(COD)(*S*)-**9**]BF_4	1	MeOH	25	0.03[b]	50	1667	50	85 (*R*)[m,n]	15
46	Ph	CO_2H	NHAc	[Rh(COD)(*S*)-**9**]BF_4	1	MeOH	25	0.58[b]	50	86	50	82 (*R*)[m,o]	15
47	Ph	CO_2H	NHAc	[Rh(COD)(*R*)-**9**]BF_4	1	PhH	25	1.33[b]	50	38	50	69 (*S*)	15
48	Ph	CO_2H	NHAc	[(*R*)-**9**+CuCl]/[Ru(COD)Cl]$_2$	1	MeOH	25	0.02[b]	50	2500	50	88 (*S*)	15
49	Ph	CO_2H	NHAc	[(*R*)-**9**+CuCl]/[Ru(COD)Cl]$_2$	1	MeOH	25	0.12[b]	500	4167	50	88 (*S*)	15
50	Ph	CO_2H	NHAc	[Rh(COD)(*R*)-**10**]BF_4	1	MeOH	25	0.058[b]	500	8621	50	91 (*S*)	9
51	Ph	CO_2H	NHAc	[Rh(COD)(*R*)-**10**]BF_4	1	MeOH	25	0.117[b]	1000	8547	50	90 (*S*)	9

Table 27.1 (continued)

Entry	Substrate			Catalyst	Conditions				TON	TOF [h^{-1}]	Conv. [%]	ee [%]	Refer-ence(s)
	R[1]	R[2]	R[3]		P[H_2] [bar]	Solvent	Temp. [°C]	Time [h]					
52	Ph	CO_2H	NHAc	[Rh(COD)(*R*)-**10**]BF_4	1	MeOH	25	0.267 [b]	1500	5618	50	90 (*S*)	9
53	Ph	CO_2H	NHAc	[Rh(COD)(*S*)-**12**]BF_4	1	MeOH	25	0.2 [b]	50	250	50	2 (*S*)	8
54	Ph	CO_2H	NHAc	[Rh(COD)(*S*)-**13**]BF_4	1	MeOH	25	0.032 [b]	50	1563	50	45 (*R*)	8
55	Ph	CO_2H	NHAc	[Rh(COD)(*S*)-**13**]BF_4	1	MeOH	25	0.67 [b]	500	746	50	9 (*R*)	8
56	Ph	CO_2H	NHAc	RhCl(COD)(*S*)-**13**	1	MeOH	25	0.25 [b]	50	200	50	20 (*R*)	8
57	Ph	CO_2H	NHAc	[Rh(COD)(*S*)-**14**]BF_4	1	MeOH	25	0.03 [b]	50	1667	50	87 (*R*)	8
58	Ph	CO_2H	NHAc	[Rh(COD)(*S*)-**15**]BF_4	1	MeOH	25	0.03 [b]	50	1667	50	91 (*R*)	8
59	Ph	CO_2H	NHAc	[Rh(COD)(*S*)-**16**]BF_4	1	MeOH	25	0.03 [b]	50	1667	50	89 (*R*)	8
60	Ph	CO_2H	NHAc	[Rh(COD)(1*S*,3*S*,5*S*)-**28**]BF_4	1	MeOH	25	0.017 [b]	50	2941	50	57 (*S*) [m]	40
61	Ph	CO_2H	NHAc	[Rh(COD)(1*R*,3*R*,5*R*)-**28**]BF_4	1	MeOH	25	0.017 [b]	50	2941	50	54 (*R*) [m]	40
62	Ph	CO_2H	NHAc	[Rh(COD)(*S*)-**29**] ClO_4	1	EtOH	20	–	–	–	–	78 (*S*) [a]	2b
63	Ph	CO_2H	NHAc	[Rh(COD)(*S*)-**29**]BF_4	1	MeOH	r.t.	–	–	250	–	80 (*S*) [c]	41
64	Ph	CO_2H	NHAc	[Rh(COD)(*S*)-**29**]BF_4	1	DCM	r.t.	–	–	2000	–	65 (*S*) [c]	41
65	Ph	CO_2H	NHAc	RhCl(COD)(1*S*,2*S*)-**35**	1	MeOH	25	–	100	–	100	85 (*R*)	37
66	Ph	CO_2H	NHAc	[Rh(COD)(1*S*,2*S*)-**35**]BF_4	1	MeOH	25	–	100	–	100	83 (*R*)	37
67	Ph	CO_2H	NHAc	[Rh(COD)(1*R*,2*R*)-**35**]Cl	1	MeOH	25	–	100	–	100	85 (*S*)	37
68	Ph	CO_2H	NHAc	[Rh(COD)(1*R*,3*S*,4*S*)-**36**]BF_4	1	MeOH	r.t.	–	–	3100	–	90 (*S*) [c]	41
69	Ph	CO_2H	NHAc	[Rh(COD)(1*R*,3*S*,4*S*)-**36**]BF_4	1	DCM	r.t.	–	–	600	–	83 (*S*) [c]	41
70	Ph	CO_2H	NHAc	[Rh(COD)(1*R*,3*S*,4*S*)-**37**]BF_4	1	MeOH	r.t.	–	–	2400	–	86 (*S*) [c]	41
71	Ph	CO_2H	NHAc	[Rh(COD)(1*R*,3*S*,4*S*)-**37**]BF_4	1	DCM	r.t.	–	–	300	–	66 (*S*) [c]	41
72	Ph	CO_2H	NHAc	[Rh(COD)(1*R*,3*S*,4*S*)-**38**]BF_4	1	MeOH	r.t.	–	–	2000	–	82 (*S*) [c]	41
73	Ph	CO_2H	NHAc	[Rh(COD)(1*R*,3*S*,4*S*)-**38**]BF_4	1	DCM	r.t.	–	–	300	–	67 (*S*) [c]	41
74	Ph	CO_2H	NHAc	[Rh(COD)(1*R*,3*S*,4*S*)-**39**]BF_4	1	MeOH	r.t.	–	–	2400	–	80 (*S*) [c]	41

75	Ph	CO_2H	NHAc	[Rh(COD)(1*R*,3*S*,4*S*)-**39**]BF_4	1	DCM	r.t.	–	–	300	–	65 (*S*)[c]	41
76	Ph	CO_2H	NHBz	[Rh(COD)(*S*)-**1**]ClO_4	1	EtOH	20	–	200	–	100	49 (*S*)	2b
77	Ph	CO_2H	NHBz	[Rh(COD)(1*S*,2*R*)-**7**]BF_4	50	MeOH	25	1	100	100	100	96.4 (*R*)	18, 19
78	Ph	CO_2H	NHBz	[Rh(COD)(*S*)-**9**]BF_4	1	MeOH	25	0.017[b]	50	2941	50	89 (*R*)	8
79	Ph	CO_2H	NHBz	RhCl(*S*)-**9**	1	MeOH	25	0.25[b]	500	2000	50	89 (*R*)	8
80	Ph	CO_2H	NHBz	[(*R*)-**9**+CuCl]/[Ru(COD)Cl]$_2$	1	MeOH	25	0.5[b]	1500	3000	50	89 (*S*)	15
81	Ph	CO_2H	NHBz	[Rh(COD)(*R*)-**9**]BF_4	1	MeOH	25	0.72[b]	500	694	50	79 (*S*)	15
82	Ph	CO_2H	NHBz	RhCl(*R*)-**9**	1	MeOH	25	0.25[b]	500	200	50	89 (*S*)	15
83	Ph	CO_2H	NHBz	[Rh(COD)(*R*)-**9**]$^+$	1	MeOH	25	0.72[b]	500	694	50	79 (*S*)	42
84	Ph	CO_2H	NHBz	[Rh(COD)(*S*)-**12**]BF_4	1	MeOH	25	0.27[b]	50	185	50	8 (*S*)	8
85	Ph	CO_2H	NHBz	[Rh(COD)(*S*)-**13**]BF_4	1	MeOH	25	0.04[b]	50	1250	50	41 (*R*)	8
86	Ph	CO_2H	NHBz	[Rh(COD)(*S*)-**13**]BF_4	1	MeOH	25	0.67[b]	500	746	50	3 (*R*)	8
87	Ph	CO_2H	NHBz	RhCl(*S*)-**13**	1	MeOH	25	0.2[b]	50	250	50	24 (*R*)	8
88	Ph	CO_2H	NHBz	[Rh(COD)(*S*)-**14**]BF_4	1	MeOH	25	0.07[b]	50	714	50	88 (*R*)	8
89	Ph	CO_2H	NHBz	[Rh(COD)(*S*)-**15**]BF_4	1	MeOH	25	0.03[b]	50	1667	50	94 (*R*)	8
90	Ph	CO_2H	NHBz	[Rh(COD)(*S*)-**16**]BF_4	1	MeOH	25	0.03[b]	50	1667	50	92 (*R*)	8
91	Ph	CO_2H	NHBz	[Rh(COD)(1*S*,3*S*,5*S*)-**28**]BF_4	1	MeOH	25	0.017[b]	50	2941	50	62 (*S*)	40
92	Ph	CO_2H	NHBz	[Rh(COD)(1*R*,3*R*,5*R*)-**28**]BF_4	1	MeOH	25	0.017[b]	50	2941	50	58 (*R*)	40
93	Ph	CO_2H	NHBz	[Rh(COD)(*S*)-**29**]ClO_4	1	EtOH	20	–	200	–	100	62 (*S*)	2b
94	Ph	CO_2Me	NHAc	[Rh(COD)(1*R*,2*S*)-**3**]BF_4	1	Dioxane	25	–	–	–	–	75 (*R*)	36
95	Ph	CO_2Me	NHAc	[Rh(COD)(1*R*,2*S*)-**3**]BF_4	1	MeOH	25	–	–	–	–	12 (*R*)	36
96	Ph	CO_2Me	NHAc	[Rh(COD)(1*R*)-**4**]BF_4	1	Dioxane	25	–	–	–	–	55 (*R*)	36
97	Ph	CO_2Me	NHAc	[Rh(COD)(1*R*)-**4**]BF_4	1	MeOH	25	–	–	–	–	5 (*R*)	36
98	Ph	CO_2Me	NHAc	[Rh(COD)(1*R*,2*R*)-**5**]BF_4	1	Dioxane	25	–	–	–	–	10 (*R*)	36
99	Ph	CO_2Me	NHAc	[Rh(COD)(1*R*,2*R*)-**5**]BF_4	1	MeOH	25	–	–	–	–	2 (S)	36
100	Ph	CO_2Me	NHAc	[Rh(COD)(2*R*)-**6**]BF_4	1	Dioxane	25	–	–	–	–	14 (R)	36
101	Ph	CO_2Me	NHAc	[Rh(COD)(2*R*)-**6**]BF_4	1	MeOH	25	–	–	–	–	6 (*R*)	36
102	Ph	CO_2Me	NHAc	[Rh(COD)(1*R*,2*S*)-**7**]Cl	50	MeOH	r.t.	17	31.3	1.84	31.3	27 (*S*)	18
103	Ph	CO_2Me	NHAc	[Rh(COD)(1*R*,2*S*)-**7**]BF_4	50	MeOH	r.t.	1	100	100	100	96.9 (*S*)	18
104	Ph	CO_2Me	NHAc	[Rh(COD)(1*S*,2*R*)-**7**]BF_4	50	MeOH	r.t.	1	100	100	100	98.3 (*R*)	18

Table 27.1 (continued)

Entry	Substrate			Catalyst	Conditions				TON	TOF [h^{-1}]	Conv. [%]	ee [%]	Reference(s)
	R^1	R^2	R^3		P[H_2] [bar]	Solvent	Temp. [°C]	Time [h]					
105	Ph	CO_2Me	NHAc	[Rh(COD)(1*S*,2*S*)-**7**]BF_4	50	MeOH	r.t.	1	100	100	100	40.6 (*R*)	18
106	Ph	CO_2Me	NHAc	[Rh(COD)(1*R*,2*S*)-**7**]BF_4	10	MeOH	r.t.	1	100	100	100	97.0 (*S*)	18
107	Ph	CO_2Me	NHAc	[Rh(COD)(1*R*,2*S*)-**7**]BF_4	10	Acetone	r.t.	1	–	100	100	95.1 (*S*)	18
108	Ph	CO_2Me	NHAc	[Rh(COD)(1*R*,2*S*)-**7**]BF_4	10	THF	r.t.	1	–	100	100	94.8 (*S*)	18
109	Ph	CO_2Me	NHAc	[Rh(COD)(1*R*,2*S*)-**7**]BF_4	10	IPA	r.t.	1	–	100	100	92.7 (*S*)	18
110	Ph	CO_2Me	NHAc	[Rh(COD)(1*R*,2*S*)-**7**]BF_4	80	MeOH	25	0.5	100	200	100	96.8 (*S*)	18
111	Ph	CO_2Me	NHAc	[Rh(COD)(1*R*,2*S*)-**7**]BF_4	50	MeOH	25	1	100	100	100	96.9 (*S*)	18
112	Ph	CO_2Me	NHAc	[Rh(COD)(1*R*,2*S*)-**7**]BF_4	20	MeOH	25	1	100	100	100	96.4 (*S*)	18
113	Ph	CO_2Me	NHAc	[Rh(COD)(1*R*,2*S*)-**7**]BF_4	1	MeOH	25	4	100	25	100	97.2 (*S*)	18
114	Ph	CO_2Me	NHAc	[Rh(COD)(1*R*,2*S*)-**7**]BF_4	10	Acetone	25	–	100	–	100	96.2 (*S*)	18
115	Ph	CO_2Me	NHAc	[Rh(COD)(1*R*,2*S*)-**7**]BF_4	10	Acetone	25	–	100	–	100	95.8 (*S*)	18
116	Ph	CO_2Me	NHAc	[Rh(COD)(1*R*,2*S*)-**7**]BF_4	10	Acetone	25	–	100	–	100	94.5 (*S*)	18
117	Ph	CO_2Me	NHAc	[Rh(COD)(1*R*,2*S*)-**7**]BF_4	10	Acetone	25	–	100	–	100	93.8 (*S*)	18
118	Ph	CO_2Me	NHAc	[Rh(COD)(1*S*,2*R*)-**7**]BF_4	50	MeOH	25	1	100	100	100	98.3 (*R*)	18
119	Ph	CO_2Me	NHAc	[Rh(COD)(1*S*,2*R*)-**7**]BF_4	50	MeOH	25	4	1000	250	100	97.5 (*R*)	18
120	Ph	CO_2Me	NHAc	[Rh(COD)(1*S*,2*R*)-**7**]BF_4	50	MeOH	25	16	10000	625	100	97.0 (*R*)	18
121	Ph	CO_2Me	NHAc	[Rh(COD)(1*S*,2*R*)-**7**]BF_4	50	MeOH	25	64	41350	646	82.7	97.0 (*R*)	18
122	Ph	CO_2Me	NHAc	[Rh(COD)(1*S*,2*R*)-**7**]BF_4	50	MeOH	25	64	41700	652	41.7	93.0 (*R*)	18
123	Ph	CO_2Me	NHAc	[Rh(COD)(*R*)-**8**]BF_4	1	MeOH	25	1.5 [b]	50	33	50	13 (*R*)	15
124	Ph	CO_2Me	NHAc	Rh(COD)(*R*)-**9**	1	MeOH	25	0.23 [b]	50	217	50	86 (*S*) [m]	15
125	Ph	CO_2Me	NHAc	Rh(COD)(*R*)-**9**	1	MeOH	25	0.33 [b]	50	152	50	85 (*S*) [k]	15
126	Ph	CO_2Me	NHAc	Rh(COD)(*S*)-**9**	1	MeOH	25	0.58 [b]	500	862	50	82 (*R*) [l]	15
127	Ph	CO_2Me	NHAc	[Rh(COD)(*R*)-**9**]BF_4	1	MeOH	25	0.03 [b]	50	1667	50	87 (*S*)	15

128	Ph	CO_2Me	NHAc	[Rh(COD)(*S*)-**9**]BF_4	1	MeOH	25	0.6[b)]	500	833	50	85 (*R*)	15
129	Ph	CO_2Me	NHAc	[Rh(COD)(*R*)-**9**]BF_4	1	MeOH	25	0.03[b)]	50	1667	50	88 (*S*)	15
130	Ph	CO_2Me	NHAc	[Rh(COD)(*R*)-**9**]BF_4	1	MeOH	25	0.03[b)]	50	1667	50	88 (*S*)[q]	15
131	Ph	CO_2Me	NHAc	[(*R*)-**9**+CuCl]/[Ru(COD)Cl]$_2$	1	MeOH	25	0.2[b)]	500	2500	50	85 (*S*)	15
132	Ph	CO_2Me	NHAc	[Rh(COD)(*R*)-**10**]BF_4	1	MeOH	25	0.067[b)]	500	7463	50	90 (*S*)	9
133	Ph	CO_2Me	NHAc	[Rh(COD)(*R*)-**10**]BF_4	1	MeOH	25	0.062[b)]	1000	16129	50	89 (*S*)	9
134	Ph	CO_2Me	NHAc	[Rh(COD)(*R*)-**11**]BF_4	1	MeOH	25	0.017[b)]	50	2941	50	78 (*S*)	17
135	Ph	CO_2Me	NHAc	[Rh(COD)(*S*)-**11**]BF_4	1	MeOH	25	0.017[b)]	50	2941	50	76 (*R*)	17
136	Ph	CO_2Me	NHAc	[Rh(COD)(*S*)-**12**]BF_4	1	MeOH	25	0.18[b)]	50	278	50	35 (*R*)	8
137	Ph	CO_2Me	NHAc	[Rh(COD)(*S*)-**13**]BF_4	1	MeOH	25	0.03[b)]	50	1667	50	47 (*R*)	8
138	Ph	CO_2Me	NHAc	[Rh(COD)(*S*)-**13**]BF_4	1	MeOH	25	0.12[b)]	50	417	50	40 (*R*)	8
139	Ph	CO_2Me	NHAc	RhCl(*S*)-**13**	1	MeOH	25	0.33[b)]	50	152	50	47 (*R*)	8
140	Ph	CO_2Me	NHAc	[Rh(COD)(*S*)-**14**]BF_4	1	MeOH	25	0.03[b)]	50	1667	50	85 (*R*)	8
141	Ph	CO_2Me	NHAc	[Rh(COD)(*S*)-**15**]BF_4	1	MeOH	25	0.03[b)]	50	1667	50	89 (*R*)	8
142	Ph	CO_2Me	NHAc	[Rh(COD)(*S*)-**16**]BF_4	1	MeOH	25	0.03[b)]	50	1667	50	86 (*R*)	8
143	Ph	CO_2Me	NHAc	[Rh(COD)(1*R*,2*S*)-**17**]BF_4	15	DCM	r.t.	18	30	1.7	98	11 (*S*)	31
144	Ph	CO_2Me	NHAc	[Rh(COD)(1*R*,2*S*)-**17**]BF_4	15	PhH	r.t.	22	30	1.4	95	46 (*S*)	31
145	Ph	CO_2Me	NHAc	[Rh(COD)(1*S*,3*R*,4*S*)-**18**]BF_4	15	DCM	r.t.	3	30	10	95	22 (*R*)	31
146	Ph	CO_2Me	NHAc	[Rh(COD)(1*R*,3*R*,4*S*)-**19**]BF_4	15	DCM	r.t.	10.5	30	2.9	99	89 (*S*)	31
147	Ph	CO_2Me	NHAc	[Rh(COD)(1*R*,3*R*,4*S*)-**19**]BF_4	15	PhH	r.t.	20	30	1.5	98	99 (*S*)	31
148	Ph	CO_2Me	NHAc	[Rh(COD)(1*R*,3*R*,4*S*)-**20**]BF_4	15	DCM	r.t.	4	30	7.5	99	88 (*S*)	31
149	Ph	CO_2Me	NHAc	[Rh(COD)(1*R*,3*R*,4*S*)-**20**]BF_4	15	PhH	r.t.	17	30	1.8	98	95 (*S*)	31
150	Ph	CO_2Me	NHAc	[Rh(COD)(1*R*,3*R*,4*S*)-**21**]BF_4	15	DCM	r.t.	4.5	30	6.7	96	16 (*S*)	31
151	Ph	CO_2Me	NHAc	[Rh(COD)(1*S*,3*R*,4*S*)-**22**]BF_4	15	DCM	r.t.	4	30	7.5	95	2 (*S*)	31
152	Ph	CO_2Me	NHAc	[Rh(COD)(1*S*,2*R*,4*R*)-**23**]BF_4	15	DCM	r.t.	13	30	2.3	98	80 (*S*)	31
153	Ph	CO_2Me	NHAc	[Rh(COD)(1*S*,3*S*,4*R*,6*R*)-**24**]BF_4	15	DCM	r.t.	12	30	2.5	94	1 (*S*)	31
154	Ph	CO_2Me	NHAc	[Rh(COD)(1*S*,3*S*,5*S*)-**28**]BF_4	1	MeOH	25	0.017[b)]	50	2941	50	70 (*S*)	40
155	Ph	CO_2Me	NHAc	[Rh(COD)(1*R*,3*R*,5*R*)-**28**]BF_4	1	MeOH	25	0.017[b)]	50	2941	50	69 (*R*)	40
156	Ph	CO_2Me	NHAc	[Rh(COD)(*S*)-**29**]BF_4	1	MeOH	r.t.	–	–	1100	–	80 (*S*)	41
157	Ph	CO_2Me	NHAc	[Rh(COD)(*S*)-**29**]BF_4	1	DCM	r.t.	–	–	2000	–	79 (*S*)	41

Table 27.1 (continued)

Entry	Substrate			Catalyst	Conditions				TON	TOF [h^{-1}]	Conv. [%]	ee [%]	Reference(s)
	R^1	R^2	R^3		P[H_2] [bar]	Solvent	Temp. [°C]	Time [h]					
158	Ph	CO_2Me	NHAc	[Rh(COD)(*S*)-**29**]BF_4	1	EtOAc	r.t.	–	–	500	–	74 (*S*)	41
159	Ph	CO_2Me	NHAc	RhCl(COD)(1*S*,2*S*)-**35**	1	MeOH	25	–	100	–	100	87 (*R*)	37
160	Ph	CO_2Me	NHAc	[Rh(COD)(1*S*,2*S*)-**35**]BF_4	1	MeOH	25	–	100	–	100	87 (*R*)	37
161	Ph	CO_2Me	NHAc	RhCl(COD)(1*R*,2*R*)-**35**	1	MeOH	25	–	100	–	100	87 (*S*)	37
162	Ph	CO_2Me	NHAc	[Rh(COD)(1*R*,3*S*,4*S*)-**36**]BF_4	1	MeOH	r.t.	–	–	3500	–	91 (*S*)	41
163	Ph	CO_2Me	NHAc	[Rh(COD)(1*R*,3*S*,4*S*)-**36**]BF_4	1	DCM	r.t.	–	–	1600	–	88 (*S*)	41
164	Ph	CO_2Me	NHAc	[Rh(COD)(1*R*,3*S*,4*S*)-**36**]BF_4	1	EtOAc	r.t.	–	–	2000	–	83 (*S*)	41
165	Ph	CO_2Me	NHAc	[Rh(COD)(1*R*,3*S*,4*S*)-**36**]BF_4	1	THF	r.t.	–	–	1700	–	84 (*S*)	41
166	Ph	CO_2Me	NHAc	[Rh(COD)(1*R*,3*S*,4*S*)-**37**]BF_4	1	MeOH	r.t.	–	–	2700	–	85 (*S*)	41
167	Ph	CO_2Me	NHAc	[Rh(COD)(1*R*,3*S*,4*S*)-**37**]BF_4	1	DCM	r.t.	–	–	1300	–	82 (*S*)	41
168	Ph	CO_2Me	NHAc	[Rh(COD)(1*R*,3*S*,4*S*)-**37**]BF_4	1	EtOAc	r.t.	–	–	2400	–	78 (*S*)	41
169	Ph	CO_2Me	NHAc	[Rh(COD)(1*R*,3*S*,4*S*)-**37**]BF_4	1	THF	r.t.	–	–	1500	–	78 (*S*)	41
170	Ph	CO_2Me	NHAc	[Rh(COD)(1*R*,3*S*,4*S*)-**38**]BF_4	1	MeOH	r.t.	–	–	2700	–	71 (*S*)	41
171	Ph	CO_2Me	NHAc	[Rh(COD)(1*R*,3*S*,4*S*)-**38**]BF_4	1	DCM	r.t.	–	–	1200	–	78 (*S*)	41
172	Ph	CO_2Me	NHAc	[Rh(COD)(1*R*,3*S*,4*S*)-**38**]BF_4	1	EtOAc	r.t.	–	–	2000	–	76 (*S*)	41
173	Ph	CO_2Me	NHAc	[Rh(COD)(1*R*,3*S*,4*S*)-**39**]BF_4	1	MeOH	r.t.	–	–	3000	–	74 (*S*)	41
174	Ph	CO_2Me	NHAc	[Rh(COD)(1*R*,3*S*,4*S*)-**39**]BF_4	1	DCM	r.t.	–	–	1500	–	78 (*S*)	41
175	Ph	CO_2Me	NHAc	[Rh(COD)(*exo*-**40**)]BF_4	34.5	Acetone	0	7	100	14	100	79.0 (*R*)	39
176	Ph	CO_2Me	NHAc	[Rh(COD)(*exo*-**40**)]BF_4	34.5	Acetone	25	7	100	14	100	77.0 (*R*)	39
177	Ph	CO_2Me	NHAc	[Rh(COD)(*exo*-**40**)]BF_4	17.2	Acetone	25	7	100	14	100	78.0 (*R*)	39
178	Ph	CO_2Me	NHAc	[Rh(COD)(*exo*-**40**)]BF_4	34.5	MeOH	25	7	100	14	100	74.0 (*R*)	39
179	Ph	CO_2Me	NHAc	[Rh(COD)(*exo*-**40**)]BF_4	34.5	THF	25	7	100	14	100	62.0 (*R*)	39
180	Ph	CO_2Me	NHAc	[Rh(COD)(*exo*-**40**)]BF_4	34.5	DCM	25	7	100	14	100	72.0 (*R*)	39
181	Ph	CO_2Me	NHBz	[Rh(COD)(1*S*,2*R*)-**7**]BF_4	50	MeOH	25	1	100	100	100	97.1 (*R*)	18, 19

182	Ph	CO_2Me	NHBz	[Rh(COD)(*S*)-**9**]BF_4	1	MeOH	25	0.67[b)]	50	75	50	81 (*R*)	8
183	Ph	CO_2Me	NHBz	[Rh(COD)(*S*)-**9**]BF_4	1	MeOH	25	0.016[b)]	50	3125	50	89 (*R*)	8, 10
184	Ph	CO_2Me	NHBz	[Rh(COD)(*R*)-**9**]BF_4	1	MeOH	25	0.67[b)]	500	746	50	81 (*S*)	15
185	Ph	CO_2Me	NHBz	Rh(COD)(*R*)-**9**	1	MeOH	25	0.28[b)]	500	1786	50	87 (*S*)	15
186	Ph	CO_2Me	NHBz	[Rh(COD)(*S*)-**12**]BF_4	1	MeOH	25	0.033[b)]	50	1515	50	27 (*R*)	8
187	Ph	CO_2Me	NHBz	[Rh(COD)(*S*)-**13**]BF_4	1	MeOH	25	0.063[b)]	50	794	50	40 (*R*)	8
188	Ph	CO_2Me	NHBz	RhCl(*S*)-**13**	1	MeOH	25	0.35[b)]	50	143	50	41 (*R*)	8
189	Ph	CO_2Me	NHBz	[Rh(COD)(*S*)-**14**]BF_4	1	MeOH	25	0.033[b)]	50	1515	50	87 (*R*)	8
190	Ph	CO_2Me	NHBz	[Rh(COD)(*S*)-**15**]BF_4	1	MeOH	25	0.033[b)]	50	1515	50	92 (*R*)	8
191	Ph	CO_2Me	NHBz	[Rh(COD)(*S*)-**16**]BF_4	1	MeOH	25	0.033[b)]	50	1515	50	88 (*R*)	8
192	Ph	CO_2Me	NHBz	[Rh(COD)(1*S*,3*S*,5*S*)-**28**]BF_4	1	MeOH	25	0.017[b)]	50	2941	50	73 (*S*)	40
193	Ph	CO_2Me	NHBz	[Rh(COD)(1*R*,3*R*,5*R*)-**28**]BF_4	1	MeOH	25	0.017[b)]	50	2941	50	71 (*R*)	40
194	Ph	CO_2Me	NHCbz	[Rh(COD)(*S*)-**9**]BF_4	1	MeOH	25	2[b)]	50	25	50	88 (*R*)	10
195	Ph	CO_2Me	NHBoc	[Rh(COD)(*S*)-**9**]BF_4	1	MeOH	25	0.83[b)]	50	60	50	93 (*R*)	10
196	Ph	NHCOMe	CO_2H	[Rh(COD)(*S*)-**2**]ClO_4	1	EtOH/ PhH (2:1)	20	0.17–0.5	200	400–1176	100	70 (*R*)	32
197	Ph	NHCOMe	CO_2H	[Rh(COD)(*S*)-**29**]ClO_4	1	EtOH/ PhH (2:1)	20	0.17–0.5	200	400–1176	100	86 (*S*)	32
198	Ph	NHCOMe	CO_2Me	[Rh(COD)(*S*)-**2**]ClO_4	1	EtOH/ PhH (2:1)	20	0.17–0.5	200	400–1176	100	48 (*R*)	32
199	Ph	NHBz	CO_2H	[Rh(COD)(*S*)-**2**]ClO_4	1	EtOH/ PhH (2:1)	20	0.17–0.5	200	400–1176	100	53 (*R*)	32
200	Ph	NHBz	CO_2H	[Rh(COD)(*S*)-**29**]ClO_4	1	EtOH/ PhH (2:1)	20	0.17–0.5	200	400–1176	100	61 (*S*)	32
201	Ph	$MePO_2Et$	NHBz	[Rh(COD)(*S*)-**9**]BF_4	1	MeOH	25	0.33[b)]	25	76	50	77 (*Sc*)[g)]	12, 16
202	Ph	$PhPO_2H$	NHBz	[Rh(COD)(*S*)-**9**]BF_4	1	MeOH	25	0.23[b)]	25	109	50	56 (*Sc*)[g)]	12
203	Ph	$PhPO_2H$	NHBz	[Rh(COD)(*S*)-**9**]BF_4	1	MeOH	25	0.33	95	288	95	65 (*Sc*)[g)]	16
204	Ph	$PhPO_2Me$	NHBz	[Rh(COD)(*S*)-**9**]BF_4	1	MeOH	25	0.63[b)]	25	40	50	76 (*Sc*)[g)]	12

Table 27.1 (continued)

Entry	Substrate			Catalyst	Conditions				TON	TOF $[h^{-1}]$	Conv. [%]	ee [%]	Reference(s)
	R^1	R^2	R^3		$P[H_2]$ [bar]	Solvent	Temp. [°C]	Time [h]					
205	Ph	$PhPO_2Me$	NHBz	$[Rh(COD)(S)$-**9**$]BF_4$	1	MeOH	25	0.93	96	103	96	69 (*Sc*)[g]	16
206	Ph	$PhPO_2Et$	NHBz	$[Rh(COD)(S)$-**9**$]BF_4$	1	MeOH	25	1.17[b]	25	21	50	75 (*Sc*)[g]	12
207	Ph	$PhPO_2Et$	NHBz	$[Rh(COD)(S)$-**9**$]BF_4$	1	MeOH	25	2.18[b]	50	23	50	71 (*Sc*)[g]	12
208	Ph	$PhPO_2Et$	NHBz	$[Rh(COD)(S)$-**9**$]BF_4$	1	MeOH	25	0.67[b]	25	37	50	31 (*Sc*)[g]	12
209	Ph	$PhPO_2Et$	NHBz	$[Rh(COD)(S)$-**9**$]BF_4$	1	MeOH	25	0.67[b]	25	37	50	79 (*Sc*)[g]	12
210	Ph	$PhPO_2Et$	NHBz	$[Rh(COD)(S)$-**9**$]BF_4$	1	MeOH	25	2.18	–	–	–	71 (*Sc*)[g]	16
211	Ph	$PO(OMe)_2$	NHBz	$[Rh(COD)(S)$-**9**$]BF_4$	1	MeOH	25	0.23[i]	95	413	95[h]	90 (*S*)[j]	14
212	Ph	$PO(OMe)_2$	NHBz	$[Rh(COD)(R)$-**9**$]BF_4$	1	MeOH	25	0.23[i]	94	409	94[h]	89 (*R*)[j]	14
213	Ph	$PO(OMe)_2$	NHBz	$[Rh(COD)\ (S)$-**9**$]BF_4$	1	MeOH	25	2.33[i]	950	408	95[h]	89 (*S*)[j]	14
214	Ph	$PO(OMe)_2$	NHBz	$[Rh(COD)(S)$-**9**$]BF_4$	1	PhH	25	1.33[i]	95	71	95[h]	82 (*S*)[j]	14
215	Ph	$PO(OMe)_2$	NHBz	$[Rh(COD)(S)$-**9**$]BF_4$	1	THF	25	0.6[i]	94	157	94[h]	83 (*S*)[j]	14
216	Ph	$PO(OMe)_2$	NHBz	$[Rh(COD)(S)$-**15**$]BF_4$	1	MeOH	25	0.2[i]	97	485	97[h]	91 (*S*)[j]	14
217	Ph	$PO(OEt)_2$	NHBz	$[Rh(COD)(S)$-**9**$]BF_4$	1	MeOH	25	0.23[i]	96	417	96[h]	92 (*S*)[j]	14
218	Ph	$PO(O^iPr)_2$	NHBz	$[Rh(COD)(S)$-**9**$]BF_4$	1	MeOH	25	1.03[i]	95	92	95[h]	91 (*S*)[j]	14
219	C_6F_5	CO_2H	NHBz	$[Rh(COD)(R)$-**9**$]^+$	1	MeOH	25	0.17[b]	50	294	50	86 (*S*)	42
220	*o*-Cl-Ph	CO_2Me	NHAc	$[Rh(COD)(1S,2R)$-**7**$]BF_4$	50	MeOH	25	1	100	100	100	92.3 (*R*)	18
221	*o*-Cl-Ph	CO_2Me	NHAc	$[Rh(COD)(1S,2R)$-**7**$]BF_4$	50	Acetone	25	1	100	100	100	98.4 (*R*)	18
222	*o*-Cl-Ph	CO_2Me	NHAc	$[Rh(COD)($*exo*-**40**$)]BF_4$	34.5	Acetone	25	7	95–100	13.6–14.3	95–100	42 (*R*)	39
223	*m*-Cl-Ph	CO_2Me	NHAc	$[Rh(COD)(1S,2R)$-**7**$]BF_4$	50	MeOH	25	1	100	100	100	95.1(*R*)	18
224	*m*-Cl-Ph	CO_2Me	NHAc	$[Rh(COD)($*exo*-**40**$)]BF_4$	34.5	Acetone	25	7	95–100	13.6–14.3	95–100	85 (*R*)	39
225	*p*-Cl-Ph	CO_2H	NHBz	$[Rh(COD)(S)$-**9**$]BF_4$	1	MeOH	25	0.05[b]	100	2000	50	90 (*R*)	11
226	*p*-Cl-Ph	CO_2Me	NHAc	$[Rh(COD)(1S,2R)$-**7**$]BF_4$	50	MeOH	25	1	100	100	100	97.8 (*R*)	18, 19

227	*p*-Cl-Ph	CO_2Me	NHBz	[Rh(COD)(1*S*,2*R*)-**7**]BF_4	50	MeOH	25	1	100	100	100	97.0 (*R*)	18, 19
228	*p*-Cl-Ph	CO_2Me	NHBz	[Rh(COD)(*S*)-**9**]BF_4	1	MeOH	25	0.05 [b]	100	2000	50	89 (*R*)	11
229	*p*-Cl-Ph	CO_2Me	NHBz	[Rh(COD)(*exo*-**40**)]BF_4	34.5	Acetone	25	7	95–100	13.6–14.3	95–100	84 (*R*)	39
230	*p*-Cl-Ph	$PO(OMe)_2$	NHBz	[Rh(COD)(*S*)-**9**]BF_4	1	MeOH	25	0.17 [i]	96	565	96 [h]	90 (*S*) [j]	14
231	*p*-Br-Ph	CO_2Me	NHAc	[Rh(COD)(1*S*,2*R*)-**7**]BF_4	50	MeOH	25	1	100	100	100	98.0 (*R*)	18, 19
232	*p*-Br-Ph	CO_2Me	NHBz	[Rh(COD)(1*S*,2*R*)-**7**]BF_4	50	MeOH	25	1	100	100	100	96.5 (*R*)	18, 19
233	*o*-F-Ph	CO_2H	NHBz	[Rh(COD)(*S*)-**9**]$^+$	1	MeOH	25	0.022 [b]	100	4545	50	91 (*R*)	42
234	*o*-F-Ph	CO_2H	NHBz	[Rh(COD)(*R*)-**9**]$^+$	1	MeOH	25	0.017 [b]	100	5882	50	91 (*S*)	42
235	*o*-F-Ph	CO_2H	NHBz	[(2*S*,3*S*)-**35**-CuCl]/[Ru(-COD)$Cl]_2$ (2:1)	1	MeOH	25	0.17	100	588	50	75 (*R*)	42
236	*o*-F-Ph	CO_2Me	NHBz	[Rh(COD)(*R*)-**9**]$^+$	1	MeOH	25	0.018	50	2778	50	90.4 (*S*)	42
237	*o*-F-Ph	CO_2Me	NHBz	[Rh(COD)(*R*)-**9**]$^+$	1	MeOH	25	0.018	100	5556	50	89 (*S*)	42
238	*o*-F-Ph	CO_2Me	NHBz	[Rh(COD)(*S*)-**9**]$^+$	1	MeOH	25	0.025	50	2000	50	86.4 (*R*)	42
239	*o*-F-Ph	CO_2Me	NHBz	[Rh(COD)(*S*)-**9**]$^+$	1	MeOH	25	0.022	100	4545	50	88 (*R*)	42
240	*o*-F-Ph	$PO(OMe)_2$	NHBz	[Rh(COD)(*S*)-**9**]BF_4	1	MeOH	25	0.23 [i]	97	421	97 [h]	92 (*S*) [j]	14
241	*m*-F-Ph	CO_2H	NHBz	[(1*S*,2*S*)-**35**-CuCl] /[Ru(COD)$Cl]_2$ (2:1)	1	MeOH	25	0.22	100	455	50	71 (*R*)	42
242	*m*-F-Ph	CO_2H	NHBz	[Rh(COD)(*S*)-**9**]$^+$	1	MeOH	25	0.022	100	4545	50	88 (*R*)	42
243	*m*-F-Ph	CO_2H	NHBz	[Rh(COD)(*R*)-**9**]$^+$	1	MeOH	25	0.017	100	5882	50	90 (*S*)	42
244	*m*-F-Ph	CO_2H	NHBz	[Rh(COD)(*R*)-**9**]$^+$	1	MeOH	25	0.33	500	1515	50	89 (*S*)	42
245	*m*-F-Ph	CO_2Me	NHBz	[Rh(COD)(*S*)-**9**]$^+$	1	MeOH	25	0.028	100	3571	50	89 (*R*)	42
246	*m*-F-Ph	CO_2Me	NHBz	[Rh(COD)(*R*)-**9**]$^+$	1	MeOH	25	0.027	100	3703	50	88 (*S*)	42
247	*m*-F-Ph	$PO(OMe)_2$	NHBz	[Rh(COD)(*S*)-**9**]BF_4	1	MeOH	25	0.27 [i]	96	356	96 [h]	90 (*S*) [j]	14
248	*p*-F-Ph	CO_2H	NHBz	[Rh(COD)(*S*)-**9**]$^+$	1	MeOH	25	0.017	100	5882	50	88 (*R*)	42
249	*p*-F-Ph	CO_2H	NHBz	[Rh(COD)(*R*)-**9**]$^+$	1	MeOH	25	0.017	100	5882	50	88 (*S*)	42
250	*p*-F-Ph	CO_2H	NHBz	[Rh(COD)(*S*)-**9**]$^+$	1	MeOH	25	0.28	1000	3571	50	90 (*R*)	42
251	*p*-F-Ph	CO_2H	NHBz	[Rh(COD)(*S*)-**9**]$^+$	1	MeOH	25	2.67	1500	562	50	86 (*R*)	42
252	*p*-F-Ph	CO_2H	NHBz	[Rh(COD)(*S*)-**9**]$^+$ Deuteration	1	MeOH	25	0.017	25	1471	50	90 (*R*)	42
253	*p*-F-Ph	CO_2H	NHBz	[(1*S*,2*S*)-**35**-CuCl]/[Ru(-COD)$Cl]_2$ (2:1)	1	MeOH	25	0.15	100	667	50	75 (*R*)	42

Table 27.1 (continued)

Entry	Substrate			Catalyst	Conditions				TON	TOF [h^{-1}]	Conv. [%]	ee [%]	Reference(s)
	R[1]	R[2]	R[3]		P[H_2] [bar]	Solvent	Temp. [°C]	Time [h]					
254	*p*-F-Ph	CO_2Me	NHAc	[Rh(COD)(1*S*,2*R*)-**7**]BF_4	50	MeOH	25	1	100	100	100	97.2 (*R*)	18, 19
255	*p*-F-Ph	CO_2Me	NHAc	[Rh(COD)(*exo*-**40**)]BF_4	34.5	Acetone	25	7	95–100	13.6–14.3	95–100	80 (*R*)	39
256	*p*-F-Ph	CO_2Me	NHBz	[Rh(COD)(*S*)-**9**]$^+$	1	MeOH	25	0.043 [b]	100	2326	50	89 (*R*)	42
257	*p*-F-Ph	CO_2Me	NHBz	[Rh(COD)(*R*)-**9**]$^+$	1	MeOH	25	0.038 [b]	100	2632	50	90 (*S*)	42
258	*p*-F-Ph	CO_2Me	NHBoc	[Rh(COD)(*S*)-**9**]BF_4	1	MeOH	25	0.15 [b]	50	333	50	92 (*R*)	7, 9
259	*p*-F-Ph	CO_2Me	NHBoc	[Rh(COD)(*S*)-**10**]BF_4	1	MeOH	25	0.07 [b]	50	714	50	94 (*R*)	7, 9
260	*p*-F-Ph	CO_2Me	NHBoc	[Rh(COD)(*R*)-**10**]BF_4	1	MeOH	25	0.07 [b]	50	714	50	94 (*S*)	7, 9
261	*p*-F-Ph	CO_2Me	NHBoc	[Rh(COD)(*R*)-**11**]BF_4	1	MeOH	25	0.13 [b]	50	385	50	87 (*S*)	7, 17
262	*p*-F-Ph	CO_2Me	NHBoc	[Rh(COD)(*S*)-**11**]BF_4	1	MeOH	25	0.13 [b]	50	385	50	86 (*R*)	17
263	*p*-F-Ph	$PhPO_2Et$	NHBz	[Rh(COD)(*S*)-**9**]BF_4	1	MeOH	25	1 [b]	25	25	50	64 (*Sc*) [g]	12
264	*p*-F-Ph	$PhPO_2Et$	NHBz	[Rh(COD)(*S*)-**9**]BF_4	1	MeOH	25	1	95	95	95	64 (*Sc*) [g]	16
265	*p*-F-Ph	$PO(OMe)_2$	NHBz	[Rh(COD)(*S*)-**9**]BF_4	1	MeOH	25	0.23 [i]	96	417	96 [h]	89 (*S*) [j]	14
266	*p*-CF_3-Ph	CO_2H	NHBz	[Rh(COD)(*S*)-**9**]$^+$	1	MeOH	25	0.017 [b]	100	5882	50	90 (*R*)	42
267	*p*-CF_3-Ph	CO_2Me	NHBoc	[Rh(COD)(*S*)-**10**]BF_4	1	MeOH	25	0.1 [b]	50	500	50	93 (*R*)	7, 9
268	*p*-CF_3-Ph	CO_2Me	NHBoc	[Rh(COD)(*S*)-**10**]BF_4	1	MeOH	25	0.07 [b]	50	714	50	95 (*R*)	7, 9
269	*p*-CF_3-Ph	CO_2Me	NHBoc	[Rh(COD)(*R*)-**10**]BF_4	1	MeOH	25	0.08 [b]	50	625	50	94 (*S*)	7, 9, 17
270	*p*-CF_3-Ph	CO_2Me	NHBoc	[Rh(COD)(*R*)-**11**]BF_4	1	MeOH	25	0.17 [b]	50	294	50	86 (*S*)	7
271	*p*-CF_3-Ph	CO_2Me	NHBoc	[Rh(COD)(*S*)-**11**]BF_4	1	MeOH	25	0.17 [b]	50	294	50	85 (*R*)	17
272	*p*-CF_3-Ph	$PO(OMe)_2$	NHBz	[Rh(COD)(*S*)-**9**]BF_4	1	MeOH	25	0.17 [i]	95	559	95 [h]	90 (*S*) [j]	14
273	*p*-CN-Ph	CO_2H	NHBz	[Rh(COD)(*S*)-**9**]BF_4	1	MeOH	25	0.05 [b]	100	2000	50	95 (*R*)	11
274	*p*-NO_2-Ph	CO_2H	NHBz	[Rh(COD)(1*S*,2*R*)-**7**]BF_4	50	MeOH	25	1	100	100	100	97.4 (*R*)	18, 19
275	*p*-NO_2-Ph	CO_2H	NHBz	[Rh(COD)(*S*)-**9**]BF_4	1	MeOH	25	0.05 [b]	100	2000	50	91 (*R*)	11

276	*p*-NO_2-Ph	CO_2H	NHBz	[Rh(COD)(*R*)-**9**]BF_4	1	MeOH	25	0.05[b]	100	2000	50	90 (*S*)	11
277	*p*-NO_2-Ph	CO_2Me	NHAc	[Rh(COD)(1*S*,2*R*)-**7**]BF_4	50	MeOH	25	1	100	100	100	97.5 (*R*)	18, 19
278	*p*-NO_2-Ph	CO_2Me	NHAc	[Rh(COD)(*exo*-**40**)]BF_4	34.5	Acetone	25	7	95–100	13.6–14.3	95–100	90 (*R*)	39
279	*p*-NO_2-Ph	CO_2Me	NHBz	[Rh(COD)(*S*)-**9**]BF_4	1	MeOH	25	0.05[b]	100	2000	50	80 (*R*)	11
280	*p*-NO_2-Ph	CO_2Me	NHBoc	[Rh(COD)(*S*)-**9**]BF_4	1	MeOH	25	0.1[b]	50	500	50	92 (*R*)	7, 9
281	*p*-NO_2-Ph	CO_2Me	NHBoc	[Rh(COD)(*S*)-**10**]BF_4	1	MeOH	25	0.08[b]	50	625	50	93 (*R*)	7, 9
282	*p*-NO_2-Ph	CO_2Me	NHBoc	[Rh(COD)(*R*)-**10**]BF_4	1	MeOH	25	0.1[b]	50	500	50	94 (*S*)	7, 9, 17
283	*p*-NO_2-Ph	CO_2Me	NHBoc	[Rh(COD)(*R*)-**11**]BF_4	1	MeOH	25	0.15[b]	50	385	50	85 (*S*)	7, 17
284	*p*-NO_2-Ph	CO_2Me	NHBoc	[Rh(COD)(*S*)-**11**]BF_4	1	MeOH	25	0.15[b]	50	333	50	85 (*R*)	17
285	*p*-NO_2-Ph	$PO(OMe)_2$	NHBz	[Rh(COD)(*S*)-**9**]BF_4	1	MeOH	25	0.17[i]	95	559	95[h]	91 (*S*)[j]	14
286	*p*-NO_2-Ph	$PhPO_2Et$	NHBz	[Rh(COD)(*S*)-**9**]BF_4	1	MeOH	25	0.75	95	127	95	60 (*Sc*)[g]	16
287	*o*-HO-Ph	CO_2Me	NHBz	[Rh(COD)(1*S*,2*R*)-**7**]BF_4	50	MeOH	25	1	100	100	100	96.3 (*R*)	18, 19
288	*p*-MeO-Ph	CO_2H	NHBz	[Rh(COD)(*S*)-**9**]BF_4	1	MeOH	25	0.05[b]	100	2000	50	90 (*R*)	11
289	*p*-MeO-Ph	CO_2H	NHBz	[Rh(COD)(*R*)-**9**]BF_4	1	MeOH	25	0.05[b]	100	2000	50	92 (*S*)	11
290	*p*-MeO-Ph	CO_2Me	NHBz	[Rh(COD)(*S*)-**9**]BF_4	1	MeOH	25	0.05[b]	100	2000	50	88 (*R*)	11
291	*p*-MeO-Ph	CO_2Me	NHBz	[Rh(COD)(*R*)-**9**]BF_4	1	MeOH	25	0.05[b]	100	2000	50	91 (*S*)	11
292	*p*-MeO-Ph	CO_2Me	NHAc	[Rh(COD)(1*S*,2*R*)-**7**]BF_4	50	MeOH	25	1	100	100	100	97.3 (*R*)	18, 19
293	*p*-MeO-Ph	CO_2Me	NHAc	[Rh(COD)(*exo*-**40**)]BF_4	34.5	Acetone	25	7	95–100	13.6–14.3	95–100	82 (*R*)	39
294	*p*-AcO-Ph	CO_2Me	NHAc	[Rh(COD)(1*S*,2*R*)-**7**]BF_4	50	MeOH	25	1	100	100	100	95.6 (*R*)	18, 19
295	*p*-AcO-Ph	CO_2Me	NHAc	[Rh(COD)(*exo*-**40**)]BF_4	34.5	Acetone	25	7	95–100	13.6–14.3	95–100	80 (*R*)	39
296	*p*-NMe_2-Ph	CO_2H	NHBz	[Rh(COD)(*R*)-**9**]BF_4	1	MeOH	25	0.05[b]	50	1000	50	72 (*S*)	11
297	*p*-NMe_2-Ph	CO_2Me	NHBz	[Rh(COD)(*R*)-**9**]BF_4	1	MeOH	25	0.05[b]	100	2000	50	85 (*S*)	11
298	*o*-Me-Ph	CO_2H	NHBz	[Rh(COD)(*S*)-**9**]BF_4	1	MeOH	25	0.05[b]	100	2000	50	86 (*S*)	11
299	*p*-Me-Ph	CO_2H	NHBz	[Rh(COD)(*S*)-**9**]BF_4	1	MeOH	25	0.05[b]	100	2000	50	89 (*S*)	11
300	*p*-Me-Ph	CO_2Me	NHAc	[Rh(COD)(1*S*,2*R*)-**7**]BF_4	50	MeOH	25	1	100	100	100	97.3 (*R*)	18, 19
301	*p*-Me-Ph	CO_2Me	NHAc	[Rh(COD)(*exo*-**40**)]BF_4	34.5	Acetone	25	7	95–100	13.6–14.3	95–100	62 (*R*)	39

Table 27.1 (continued)

Entry	Substrate R¹	Substrate R²	Substrate R³	Catalyst	Conditions P[H$_2$] [bar]	Conditions Solvent	Conditions Temp. [°C]	Conditions Time [h]	TON	TOF [h^{-1}]	Conv. [%]	ee [%]	Reference(s)
302	*p*-Me-Ph	CO_2Me	NHBoc	[Rh(COD)(*S*)-**10**]BF$_4$	1	MeOH	25	0.12[b]	50	417	50	93 (*R*)	7, 9
303	*p*-Me-Ph	CO_2Me	NHBoc	[Rh(COD)(*R*)-**11**]BF$_4$	1	MeOH	25	0.18	50	278	50	85 (*S*)	7, 17
304	*p*-Me-Ph	CO_2Me	NHBoc	[Rh(COD)(*S*)-**11**]BF$_4$	1	MeOH	25	0.16[b]	50	313	50	84 (*R*)	17
305	4-^{i}Pr-Ph	CO_2H	NHBz	[Rh(COD)(*S*)-**9**]BF$_4$	1	MeOH	25	0.05[b]	100	2000	50	92 (*R*)	11
306	4-^{i}Pr-Ph	CO_2Me	NHBz	[Rh(COD)(*R*)-**9**]BF$_4$	1	MeOH	25	0.05[b]	100	2000	50	89 (*S*)	11
307	4-^{i}Pr-Ph	$PhPO_2Et$	NHBz	[Rh(COD)(*S*)-**9**]BF$_4$	1	MeOH	25	2.35[b]	25	11	50	60 (*Sc*)[g]	12
308	4-^{i}Pr-Ph	$PhPO_2Et$	NHBz	[Rh(COD)(*S*)-**9**]BF$_4$	1	MeOH	25	4.5	95	21	95	60 (*Sc*)[g]	16
309	4-^{i}Pr-Ph	$PO(OMe)_2$	NHBz	[Rh(COD)(*S*)-**9**]BF$_4$	1	MeOH	25	0.4[i]	95	238	95[h]	87 (*S*)[j]	14
310	*p*-^{t}Bu-Ph	CO_2Me	NHBoc	[Rh(COD)(*S*)-**10**]BF$_4$	1	MeOH	25	0.12[b]	50	417	50	92 (*R*)	7, 9
311	*p*-^{t}Bu-Ph	CO_2Me	NHBoc	[Rh(COD)(*R*)-**11**]BF$_4$	1	MeOH	25	0.25[b]	50	200	50	86 (*S*)	7, 17
312	*p*-^{t}Bu-Ph	CO_2Me	NHBoc	[Rh(COD)(*S*)-**11**]BF$_4$	1	MeOH	25	0.25[b]	50	200	50	85 (*R*)	17
313	2,4-dimethyl-Ph	CO_2H	NHBz	[Rh(COD)(*S*)-**9**]BF$_4$	1	MeOH	25	0.05[b]	100	2000	50	79 (*R*)	11
314	2,4-dimethyl-Ph	CO_2Me	NHBz	[Rh(COD)(*R*)-**9**]BF$_4$	1	MeOH	25	0.05[b]	100	2000	50	82 (*S*)	11
315	1-naphthyl	CO_2H	NHBz	[Rh(COD)(*S*)-**9**]BF$_4$	1	MeOH	25	0.05[b]	100	2000	50	86 (*R*)	11
316	1-naphthyl	CO_2H	NHBz	[Rh(COD)(*R*)-**9**]BF$_4$	1	MeOH	25	0.05[b]	100	2000	50	88 (*S*)	11
317	2-naphthyl	CO_2H	NHBz	[Rh(COD)(*S*)-**9**]BF$_4$	1	MeOH	25	0.05[b]	100	2000	50	87 (*R*)	11
318	2-naphthyl	CO_2H	NHBz	[Rh(COD)(*R*)-**9**]BF$_4$	1	MeOH	25	0.05[b]	100	2000	50	92 (*S*)	11
319	2-naphthyl	CO_2Me	NHBz	[Rh(COD)(*S*)-**9**]BF$_4$	1	MeOH	25	0.05[b]	100	2000	50	89 (*R*)	11
320	2-naphthyl	CO_2Me	NHBz	[Rh(COD)(*S*)-**9**]BF$_4$	1	MeOH	25	0.05[b]	100	2000	50	89 (*S*)[e]	11
321	9-phenanthryl	CO_2H	NHBz	[Rh(COD)(*S*)-**9**]BF$_4$	1	MeOH	25	0.05[b]	50	000	50	65 (*R*)	11
322	9-phenanthryl	CO_2H	NHBz	[Rh(COD)(*R*)-**9**]BF$_4$	1	MeOH	25	0.05[b]	25	500	50	63 (*S*)	11
323	3-OMe-4-OAc-Ph	CO_2Me	NHAc	[Rh(COD)(1*S*,2*R*)-**7**]BF$_4$	50	MeOH	25	1	50	100	100	98.1 (*R*)	18, 19
324	3-OAc-4-OMe-Ph	CO_2Me	NHAc	[Rh(COD)(1*R*,2*S*)-**7**]BF$_4$	50	MeOH	25	1	100	100	100	97.4 (*S*)	18

325	3-Me-4-OAc-Ph	CO_2Me	NHAc	[Rh(COD)(*exo*-**40**)]BF_4	34.5	Acetone	25	7	95–100	13.6–14.3	95–100	75 (*R*)	39
326	3-OH-4-OMe-Ph	CO_2H	NHBz	Rh(COD)(*S*)-**9**	1	MeOH	25	0.083 [b]	50	602	50	83 (*R*)	15
327	3,4-$(OMe)_2$Ph	CO_2H	NHAc	[Rh(COD)(*S*)-**9**]BF_4	1	MeOH	25	0.05 [b]	50	1000	50	90 (*R*)	15
328	3,4-$(OMe)_2$Ph	CO_2H	NHBz	Rh(COD)(*S*)-**9**	1	MeOH	25	0.12 [b]	50	417	50	82 (*R*)	15
329	3,4-$(OMe)_2$Ph	CO_2H	NHBz	[(*R*)-**9**+CuCl]/[Ru(COD)Cl]$_2$	1	MeOH	25	0.45 [b]	250	556	50	87 (*S*)	15
330	3,4-$(OMe)_2$Ph	CO_2Me	NHAc	[Rh(COD)(*R*)-**9**]BF_4	1	MeOH	25	0.067 [b]	50	746	50	87 (*S*)	15
331	3,4-$(OMe)_2$Ph	CO_2Me	NHAc	[Rh(COD)(*S*)-**9**]BF_4	1	MeOH	25	0.5 [b]	500	1000	50	81 (*R*) [m,p]	15
332	3,4-$(OMe)_2$Ph	CO_2Me	NHBz	Rh(COD)(*S*)-**9**	1	MeOH	25	0.2 [b]	50	250	50	84 (*R*)	15
333	3,4-$(OMe)_2$Ph	CO_2Me	NHBz	Rh(COD)(*S*)-**9**	1	PhH	25	1.75 [b]	50	29	50	8 (*R*)	15
334	2-Cl-3-OAc-4-OMe-Ph	CO_2Me	NHBz	[Rh(COD)(1*S*,2*R*)-**7**]BF_4	50	MeOH	25	1	100	100	100	98.0 (*R*)	18
335	4-OMe-3-OAc-Ph	NHCOMe	CO_2H	[Rh(COD)(*S*)-**2**]ClO_4	1	EtOH/PhH (2:1)	20	0.17-0.5	200	400-1176	100	83 (*R*)	32
336	4-OMe-3-OAc-Ph	NHCOMe	CO_2H	[Rh(COD)(*S*)-**29**]ClO_4	1	EtOH/PhH (2:1)	20	0.17-0.5	200	400-1176	100	82 (*S*)	32
337	3,4-methylene-dioxyphenyl	CO_2H	NHAc	Rh(COD)(*S*)-**9**	1	PhH	25	0.12 [b]	50	417	50	67 (*R*)	15
338	3,4-methylene-dioxyphenyl	CO_2Me	NHAc	[Rh(COD)(1*S*,2*R*)-**7**]BF_4	50	MeOH	25	1	100	100	100	97.5 (*R*)	18
339	3,4-methylene-dioxyphenyl	CO_2Me	NHAc	[Rh(COD)(*exo*-**40**)]BF_4	34.5	Acetone	25	7	95–100	13.6–14.3	95–100	76 (*R*)	39
340	3,4-methylene-dioxyphenyl	NHCOMe	CO_2H	[Rh(COD)(*S*)-**2**]ClO_4	1	EtOH/PhH (2:1)	20	0.17–0.5	200	400–1176	100	78 (*R*)	32
341	3,4-methylene-dioxyphenyl	NHCOMe	CO_2H	[Rh(COD)(*S*)-**29**]ClO_4	1	EtOH/PhH (2:1)	20	0.17–0.5	200	400–1176	100	81 (*S*)	32
342	2-furyl	CO_2Me	NHAc	[Rh(COD)(1*S*,2*R*)-**7**]BF_4	50	MeOH	25	1	100	100	100	91.1 (*R*)	18
343	2-furyl	CO_2Me	NHAc	[Rh(COD)(*exo*-**40**)]BF_4	34.5	Acetone	25	7	95–100	13.6–14.3	5–100	83 (*R*)	39

Table 27.1 (continued)

Entry	Substrate			Catalyst	Conditions				TON	TOF $[h^{-1}]$	Conv. [%]	ee [%]	Reference(s)
	R^1	R^2	R^3		$P[H_2]$ [bar]	Solvent	Temp. [°C]	Time [h]					
344	Thiophen-2-yl	CO_2H	NHAc	[Rh(COD)(*S*)-**9**]$^+$	1	MeOH	25	0.033 [b]	50	1515	50	90 (*R*)	4
345	Thiophen-2-yl	CO_2H	NHAc	[Rh(COD)(*S*)-**9**]$^+$	1	MeOH	25	0.43 [b]	500	1163	50	89 (*R*)	4
346	Thiophen-2-yl	CO_2H	NHAc	[Rh(COD)(*R*)-**9**]$^+$	1	MeOH	25	0.033 [b]	50	1515	50	90 (*S*)	4
347	Thiophen-2-yl	CO_2H	NHAc	[Rh(COD)(1*S*,2*S*)-**35**]$^+$	1	MeOH	25	0.058 [b]	50	862	50	78 (*R*)	4
348	Thiophen-2-yl	CO_2H	NHBz	[Rh(COD)(1*S*,2*S*)-**35**]$^+$	1	MeOH	25	0.067 [b]	50	746	50	80 (*R*)	4
349	Thiophen-2-yl	CO_2H	NHBz	[Rh(COD)(*R*)-**9**)]$^+$	1	MeOH	25	0.042 [b]	50	1190	50	90 (*S*)	4
350	Thiophen-2-yl	CO_2Me	NHAc	[Rh(COD(*S*)-**9**)]$^+$	1	MeOH	25	0.05 [b]	50	1000	50	88 (*R*)	4
351	Thiophen-2-yl	CO_2Me	NHAc	[Rh(COD(*S*)-**9**)]$^+$	1	MeOH	25	0.33 [b]	250	758	50	86 (*R*)	4
352	Thiophen-2-yl	CO_2Me	NHAc	[Rh(COD)(1*R*,3*R*,5*R*)-**28**]BF_4	1	MeOH	25	0.1 [b]	50	500	50	63 (*R*)	40
353	Thiophen-2-yl	CO_2Me	NHAc	[Rh(COD) (+)(1*S*,2*S*)-**35**]$^+$	1	MeOH	25	0.067 [b]	50	746	50	77 (*R*)	4
354	Thiophen-2-yl	CO_2Me	NHBz	[Rh(COD)(1*S*,2*S*)-**35**]$^+$	1	MeOH	25	0.083	50	602	50	75 (*R*)	4
355	Thiophen-2-yl	CO_2Me	NHBz	[Rh(COD)(*R*)-**9**)]$^+$	1	MeOH	25	0.67 [b]	50	746	50	90 (*S*)	4
356	Thiophen-3-yl	CO_2H	NHAc	[Rh(COD)(*S*)-**9**]$^+$	1	MeOH	25	0.017 [b]	50	2941	50	88 (*R*)	4
357	Thiophen-3-yl	CO_2H	NHAc	[Rh(COD)(*S*)-**9**]$^+$	1	MeOH	25	0.13 [b]	500	3846	50	84 (*R*)	4
358	Thiophen-3-yl	CO_2H	NHAc	[Rh(COD)(1*S*,2*S*)-**35**]$^+$	1	MeOH	25	0.042 [b]	50	1190	50	70 (*R*)	4
359	Thiophen-3-yl	CO_2H	NHBz	[Rh(COD)(*S*)-**9**]$^+$	1	MeOH	25	0.017 [b]	50	2941	50	85 (*R*)	4
360	Thiophen-3-yl	CO_2H	NHBz	[Rh(COD)(*S*)-**9**]$^+$	1	MeOH	25	0.058 [b]	250	4310	50	84 (*R*)	4
361	Thiophen-3-yl	CO_2H	NHBz	[Rh(COD)(+)(1*S*,2*S*)-**35**]$^+$	1	MeOH	25	0.042 [b]	50	1190	50	65 (*R*)	4
362	Thiophen-3-yl	CO_2Me	NHAc	[Rh(COD)(*S*)-**9**)]$^+$	1	MeOH	25	0.02 [b]	50	2500	50	86 (*R*)	4
363	Thiophen-3-yl	CO_2Me	NHAc	[Rh(COD)(*S*)-**9**)]$^+$	1	MeOH	25	0.25 [b]	250	1000	50	83 (*R*)	4
364	Thiophen-3-yl	CO_2Me	NHAc	[Rh(COD)(1*R*,3*R*,5*R*)**28**]BF_4	1	MeOH	25	0.017 [b]	50	2941	50	64 (*R*)	40
365	Thiophen-3-yl	CO_2Me	NHAc	[Rh(COD)(1*S*,2*S*)-**35**]$^+$	1	MeOH	25	0.033 [b]	50	1515	50	72 (*R*)	4

366	Thiophen-3-yl	CO_2Me	NHBz	[Rh(COD)(*S*)-**9**)]$^+$	1	MeOH	25	0.025 [b)]	50	2000	50	85 (*R*)	4
367	Thiophen-3-yl	CO_2Me	NHBz	[Rh(COD)(1*R*,3*R*,5*R*)-**28**]BF_4	1	MeOH	25	0.05 [b)]	50	1000	50	64 (*R*)	40
368	Thiophen-3-yl	CO_2Me	NHBz	[Rh(COD)(+)(1*S*,2*S*)-**35**]$^+$	1	MeOH	25	0.042 [b)]	50	1190	50	70 (*R*)	4
369	Pyridin-3-yl	CO_2H	NHAc	[Rh(COD)(*S*)-**9**]BF_4 [d)]	1	MeOH	25	0.033 [b)]	50	1515	50	89 (*R*)	5
370	Pyridin-3-yl	CO_2H	NHAc	[Rh(COD)(*R*)-**9**]BF_4 [d)]	1	MeOH	25	0.05 [b)]	50	1000	50	90 (*S*)	6
371	Pyridin-3-yl	CO_2H	NHAc	[Rh(COD)(*S*)-**9**]BF_4 [d)]	1	MeOH	25	0.27 [b)]	250	926	50	85 (*R*)	5, 6
372	Pyridin-3-yl	CO_2H	NHAc	[Rh(COD)(*S*)-**9**]BF_4 [d)]	1	MeOH	25	0.45 [b)]	500	1111	50	78 (*R*)	5, 6
373	Pyridin-3-yl	CO_2H	NHAc	[Rh(COD)(+)(1*S*,2*S*)-**35**] BF_4 [d)]	1	MeOH	25	0.083 [b)]	50	602	50	86 (*R*)	5
374	Pyridin-3-yl	CO_2H	NHBz	[Rh(COD)(*S*)-**9**]BF_4 [d)]	1	MeOH	25	0.67 [b)]	50	746	50	86 (*R*) [f)]	5, 6
375	Pyridin-3-yl	CO_2H	NHBz	[Rh(COD)(*R*)-**9**]BF_4 [d)]	1	MeOH	25	0.05 [b)]	50	1000	50	86 (*S*)	6
376	Pyridin-3-yl	CO_2H	NHBz	[Rh(COD)(*S*)-**9**]BF_4 [d)]	1	MeOH	25	0.18 [b)]	50	1389	50	87 (*R*)	6
377	Pyridin-3-yl	CO_2H	NHBz	[Rh(COD)(1*S*,2*S*)-**35**]BF_4 [d)]	1	MeOH	25	0.1 [b)]	50	500	50	60 (*R*)	6
378	Pyridin-3-yl	CO_2Me	NHAc	[Rh(COD)(*S*)-**9**]BF_4 [d)]	1	MeOH	25	0.05 [b)]	50	1000	50	90 (*R*)	6
379	Pyridin-3-yl	CO_2Me	NHAc	[Rh(COD)(*R*)-**9**]BF_4 [d)]	1	MeOH	25	0.033 [b)]	50	1515	50	89 (*S*)	5, 6
380	Pyridin-3-yl	CO_2Me	NHAc	[Rh(COD)(*S*)-**9**]BF_4 [d)]	1	MeOH	25	0.5 [b)]	500	1000	50	83 (*R*)	5, 6
381	Pyridin-3-yl	CO_2Me	NHAc	[Rh(COD)(1*S*,2*S*)-**35**] BF_4 [d)]	1	MeOH	25	0.067 [b)]	50	746	50	84 (*R*)	5
382	Pyridin-3-yl	CO_2Me	NHBz	[Rh(COD)(*S*)-**9**]BF_4 [d)]	1	MeOH	25	0.1 [b)]	50	500	50	88 (*R*)	5, 6
383	Pyridin-3-yl	CO_2Me	NHBz	[Rh(COD)(*S*)-**9**]BF_4 [d)]	1	MeOH	25	0.42 [b)]	250	595	50	84 (*R*)	5, 6
384	Pyridin-3-yl	CO_2Me	NHBz	[Rh(COD)(*S*)-**9**]BF_4 [d)]	1	MeOH	25	0.67 [b)]	500	746	50	81 (*R*)	5, 6
385	Pyridin-3-yl	CO_2Me	NHBz	[Rh(COD)(1*R*,3*R*,5*R*)-**28**]BF_4 [d)]	1	MeOH	25	0.12 [b)]	50	417	50	59 (*R*)	40
386	Pyridin-3-yl	CO_2Me	NHBz	[Rh(COD)(1*S*,2*S*)-**35**] BF_4	1	MeOH	25	0.18 [b)]	50	278	50	70 (*R*)	5
387	Pyridin-4-yl	CO_2H	NHAc	[Rh(COD)(*S*)-**9**]BF_4 [d)]	1	MeOH	25	0.033 [b)]	50	1515	50	89 (*S*)	5, 6
388	Pyridin-4-yl	CO_2H	NHAc	[Rh(COD)(*S*)-**9**]BF_4 [d)]	1	MeOH	25	0.25 [b)]	250	1000	50	84 (*R*)	5, 6
389	Pyridin-4-yl	CO_2H	NHAc	[Rh(COD)(1*S*,2*S*)-**35**] BF_4 [d)]	1	MeOH	25	0.1 [b)]	50	500	50	82 (*R*)	5
390	Pyridin-4-yl	CO_2H	NHBz	[Rh(COD)(*S*)-**9**]BF_4 [d)]	1	MeOH	25	0.05 [b)]	50	1000	50	87 (*R*)	5, 6
391	Pyridin-4-yl	CO_2H	NHBz	Rh(COD)(1*S*,2*S*)-**35**]BF_4 [d)]	1	MeOH	25	0.083 [b)]	50	602	50	74 (*R*)	5, 6
392	Pyridin-4-yl	CO_2Me	NHAc	[Rh(COD)(*S*)-**9**]BF_4 [d)]	1	MeOH	25	0.033 [b)]	50	11515	50	89 (*R*)	5, 6
393	Pyridin-4-yl	CO_2Me	NHAc	[Rh(COD)(*S*)-**9**]BF_4 [d)]	1	MeOH	25	0.42 [b)]	500	1190	50	86 (*R*)	5, 6

Table 27.1 (continued)

Entry	Substrate R1	R2	R3	Catalyst	Conditions P[H_2] [bar]	Solvent	Temp. [°C]	Time [h]	TON	TOF [h^{-1}]	Conv. [%]	ee [%]	Reference(s)
394	Pyridin-4-yl	CO_2Me	NHAc	[Rh(COD)(1*S*,2*S*)-**35**]BF_4 [d]	1	MeOH	25	0.067 [b]	50	746	50	74 (*R*)	5
395	Pyridin-4-yl	CO_2Me	NHBz	[Rh(COD)(*R*)-**9**]BF_4 [d]	1	MeOH	25	0.05 [b]	50	1000	50	90 (*S*)	5, 6
396	Pyridin-4-yl	CO_2Me	NHBz	[Rh(COD)(*S*)-**9**]BF_4 [d]	1	MeOH	25	0.42 [b]	500	1190	50	86 (*R*)	5
397	Pyridin-4-yl	CO_2Me	NHBz	[Rh(COD)(1*R*,3*R*,5*R*)-**28**]BF_4 [d]	1	MeOH	25	0.12 [b]	50	417	50	59 (*R*)	40
398	Pyridin-4-yl	CO_2Me	NHBz	[Rh(COD)(1*S*,2*S*)-**35**] BF_4	1	MeOH	25	0.1 [b]	50	500	50	72 (*R*)	6
399	$PhCH_2$	CO_2H	NHAc	[Rh(COD)(1*R*,2*S*)-7]ClO_4	50	MeOH	r.t.	4	81	20.3	81	2.9 (*S*) [c]	30
400	$PhCH_2$	CO_2Et	NHAc	[Rh(COD)(1*S*,2*R*)-7]BF_4	50	MeOH	25	1	100	100	100	93.1 (*R*)	18
401	$PhCH_2$	CO_2Et	NHAc	[Rh(COD)(1*R*,2*S*)-7]BF_4	50	MeOH	25	1	100	100	100	92.5 (*S*)	18, 30
402	$PhCH_2$	CO_2Et	NHAc	[Rh(COD)(1*R*,2*S*)-7]ClO_4	50	EtOH	r.t.	1	100	100	100	77.1 (*S*)	30
403	$PhCH_2$	CO_2Et	NHAc	[Rh(COD)(1*R*,2*S*)-7]ClO_4	50	IPA	r.t.	1	100	100	100	83.9 (*S*)	30
404	$PhCH_2$	CO_2Et	NHAc	[Rh(COD)(1*R*,2*S*)-7]ClO_4	50	THF	r.t.	1	100	100	100	88.3 (*S*)	30
405	$PhCH_2$	CO_2Et	NHAc	[Rh(COD)(1*R*,2*S*)-7]ClO_4	50	CH_2Cl_2	r.t.	1	76.1	76.1	76.1	52.4 (*S*)	30
406	$PhCH_2$	CO_2Et	NHAc	[Rh(COD)(1*R*,2*S*)-7]ClO_4	50	Acetone	r.t.	1	89.3	89.3	89.3	79.0 (*S*)	30
407	$PhCH_2$	CO_2Et	NHAc	[Rh(COD)(1*R*,2*S*)-7]ClO_4	50	PhH	r.t.	1	100	100	100	80.2 (*S*)	30
408	$PhCH_2$	CO_2Et	NHAc	[Rh(COD)(1*R*,2*S*)-7]ClO_4	3	MeOH	r.t.	5	100	20	100	94.6 (*S*)	30
409	$PhCH_2$	CO_2Et	NHAc	[Rh(COD)(1*R*,2*S*)-7]ClO_4	20	MeOH	r.t.	3	100	33	100	95.2 (*S*)	30
410	$PhCH_2$	CO_2Et	NHAc	[Rh(COD)(1*R*,2*S*)-7]ClO_4	50	MeOH	r.t.	1	100	100	100	95.7 (*S*)	30
411	$PhCH_2$	CO_2Et	NHAc	[Rh(COD)(1*R*,2*S*)-7]ClO_4	50	MeOH	–10	4	100	25	100	93.6 (*S*)	30
412	$PhCH_2$	CO_2Et	NHAc	[Rh(COD)(1*R*,2*S*)-7]ClO_4	50	MeOH	10	1	100	100	100	95.7 (*S*)	30
413	$PhCH_2$	CO_2Et	NHAc	[Rh(COD)(1*R*,2*S*)-7]ClO_4	50	MeOH	30	1	100	100	100	93.3 (*S*)	30
414	$PhCH_2$	CO_2Et	NHAc	[Rh(COD)(1*R*,2*S*)-7]ClO_4	50	MeOH	50	0.5	100	200	100	70.8 (*S*)	30
415	$PhCH_2$	CO_2Et	NHAc	[Rh(COD)(1*R*,2*S*)-7]ClO_4	50	MeOH	r.t.	1	19	19	19	85.0 (*S*)	30
416	$PhCH_2$	CO_2Et	NHBz	[Rh(COD)(1*R*,2*S*)-7]ClO_4	50	MeOH	r.t.	1	100	100	100	94.6 (*S*)	30
417	$PhCH_2$	CO_2Et	NHCbz	[Rh(COD)(1*R*,2*S*)-7]ClO_4	50	MeOH	r.t.	1	20	20	20	58.4 (*S*)	30
418	$PhCH_2$	CO_2Et	$NHCO_2CH_2$-$CH(CH_3)_2$	[Rh(COD)(1*R*,2*S*)-7]ClO_4	50	MeOH	r.t.	1	94	94	94	93.1 (*S*)	30

419	$PhCH_2$	CO_2Et	NHBoc	[Rh(COD)(1*R*,2*S*)-**7**]ClO_4	50	MeOH	r.t.	4	43	10.8	43	78.1 (*S*)	30
420	H	CO_2H	CH_2CO_2H	[Rh(COD)(*S*)-**1**]ClO_4	1	EtOH	20	–	–	–	–	10 (*R*)[a)]	2a, b
421	H	CO_2H	CH_2CO_2H	[Rh(COD)(1*R*,2*S*)-**3**]BF_4	1	Dioxane	25	–	–	–	–	64 (*S*)	36
422	H	CO_2H	CH_2CO_2H	[Rh(COD)(1*R*,2*S*)-**3**]BF_4	1	MeOH	25	–	–	–	–	59 (*S*)	36
423	H	CO_2H	CH_2CO_2H	[Rh(COD)(1*R*)-**4**]BF_4	1	Dioxane	25	–	–	–	–	3 (*S*)	36
424	H	CO_2H	CH_2CO_2H	[Rh(COD)(1*R*)-**4**]BF_4	1	MeOH	25	–	–	–	–	0	36
425	H	CO_2H	CH_2CO_2H	[Rh(COD)(1*R*,2*R*)-**5**]BF_4	1	Dioxane	25	–	–	–	–	12 (*R*)	36
426	H	CO_2H	CH_2CO_2H	[Rh(COD)(1*R*,2*R*)-**5**]BF_4	1	MeOH	25	–	–	–	–	8 (*R*)	36
427	H	CO_2H	CH_2CO_2H	[Rh(COD)(2*R*)-**6**]BF_4	1	Dioxane	25	–	–	–	–	31 (*R*)	36
428	H	CO_2H	CH_2CO_2H	[Rh(COD)(2*R*)-**6**]BF_4	1	MeOH	25	–	–	–	–	14 (*R*)	36
429	H	CO_2Me	CH_2CO_2Me	[Rh(COD)(*R*)-**9**]BF_4	1	CD_3OD	25	–	–	–	–	70 (*S*)	13
430	H	CO_2Me	CH_2CO_2Me	[Rh(COD)(*R*)-**13**]BF_4	1	CD_3OD	25	–	–	–	–	40 (*S*)	13
431	H	CO_2Me	CH_2CO_2Me	[Rh(COD)(*R*)-**14**]BF_4	1	CD_3OD	25	–	–	–	–	78 (*S*)	13
432	H	CO_2Me	CH_2CO_2Me	[Rh(COD)(*R*)-**15**]BF_4	1	CD_3OD	25	–	–	–	–	80 (*S*)	13
433	H	CO_2Me	CH_2CO_2Me	[Rh(COD)(*R*)-**16**]BF_4	1	CD_3OD	25	–	–	–	–	25 (*S*)	13
434	H	CO_2H	CH_2CO_2H	[Rh(COD)(*S*)-**29**]ClO_4	1	EtOH	20	–	–	–	–	20 (*R*)[a)]	2a, b

a) Optical yield.
b) t/2 for uptake 50% of theoretical hydrogen volume.
c) ee determination of the corresponding ester using diazomethane.
d) Addition of 1.5 equiv. HBF_4 .
e) ee determination after recrystallization.
f) Partial reaction with methanol to produce the corresponding ester.
g) The ee-values with respect to the α-carbon atom can be determined from the enantiomeric excesses of the diastereomer pairs of the ester.
h) Crude yield after evaporation of solvent.
i) Approx. reaction time = approx. time for uptake of half of the H_2 volume $\times 2$.
j) Configuration (*S*) corresponds to the D-configuration of amino carboxylic acids.
k) Catalyst [RhCOD(*R*)-**9**]$^+$ – half-year exposure to air.
l) Preformed ligand-RhCl(Benzene).
m) Preformed complex.
n) Exposed to air.
o) Stirred for 1 h in 50% methanol/water.
p) Stock solution in benzene after 10 days (29 mL methanol + 1 mL stock solution).
q) Ligand stored for one month on air.
r) Catalyst solution agitated for 30 min with air.

Table 27.2 Asymmetric hydrogenation of other prochiral olefins.

AcO 11

BocHN … N H … CO_2Me, Ph, R — 12

12a: R = Ph
12b: R = p-CH_3-Ph
12c: R = p-F-Ph
12d: R = p-CF_3-Ph

Entry	Catalyst	Substrate	Conditions				TON	TOF [h^{-1}]	Conv. [%]	ee [%]	Reference
			P[H_2] [bar]	Solvent	Temp. [°C]	Time h]					
1	[Rh(*R*)-**25**]BF_4	11	10	EtOAc	r.t.	18	100	5.6	100	71 (*S*)	34
2	[Rh(*S*)-**26**]BF_4	11	10	EtOAc	r.t.	18	100	5.6	100	71 (*R*)	34
3	[Rh(*S*)-**27**]BF_4	11	10	EtOAc	r.t.	18	100	5.6	100	5 (*S*)	34
4	[Rh(*S*)-**32**]BF_4	11	10	EtOAc	r.t.	18	100	5.6	100	30 (*R*)	34
5	[Rh(*S*)-**33**]BF_4	11	10	EtOAc	r.t.	18	100	5.6	100	95 (*R*)	34
6	[Rh(*S*)-**34**]BF_4	11	10	EtOAc	r.t.	18	100	5.6	100	31 (*R*)	34
7	[Rh(COD)(*R*)-**10**]BF_4	12a	1	MeOH	25	1	–	–	–	77[a)]	7
8	[Rh(COD)(*R*)-**10**]BF_4	12b	1	MeOH	25	1.17	–	–	–	76[a)]	7
9	[Rh(COD)(*R*)-**10**]BF_4	12c	1	MeOH	25	1.3	–	–	–	78[a)]	7
10	[Rh(COD)(*R*)-**10**]BF_4	12d	1	MeOH	25	1.17	–	–	–	91[a)]	7

a) The ee-values with respect to the α-carbon atom can be determined from the enantiomeric excesses of the diastereomer pairs of the ester (see [12]).

10000 was completed within 16 h, giving the desired product in 97% ee (Table 27.1, entry 120; 98.3% ee at SCR 100, Table 27.1, entry 118). These results were comparable to those using phosphine and phosphinite ligands (e.g., DuPhos, 99% ee [20]; DIPAMP, 96% ee [21]; TRAP, 92% ee [22]; DIOP, 55% ee [23]; CAPP, 95.6% ee [24]; BPPFA, 21% ee [25]; Ph-*β*-Glup, 91.5% ee [26]; SpirOP, 95.7% ee [27]). Further application in the hydrogenation of methyl 2-acetamido-3-(3-methoxy-4-acetoxyphenyl)-acrylate (a crucial intermediate in the synthesis of L-dopa [28]) was successfully achieved in 97.4% ee (Table 27.1, entry 324). Similarly, the enantioselective hydrogenation of ethyl (*Z*)-2-acetamido-4-phenylcrotonate gave the homophenylalanine derivative in 92.5% ee (Table 27.1, entry 401). This product is a key component of (*S*,*S*)-benazepril, an angiotensin-converting enzyme inhibitor widely used as an antihypertensive agent [29]. Jiang studied the hydrogenation of *N*-protected (*Z*)-2-aminocrotonates and found that the enantioselectivity and activity were strongly dependent on the type of *N*-protecting group used (NHAc, 95.7% ee with 100% conv.; $NHCO_2Me$, 85% ee with 19% conv.; Table 27.1, entry 412 versus 415) [30]. The results from using Rh–DPAMPP compared favorably with many commonly used chiral Rh–diphosphine catalysts (e.g., Rh–BINAP, 21.8% ee with 100% conv.; Rh–DIPAMP, 50.8% ee with 100% conv.; Rh–BDPP, 69.4% ee with 100% conv.; Rh–PPM, 14.4% ee with 7.9% conv. under the same reaction conditions).

The introduction of extra stereogenicity at the phosphorus centers is one of the methods used to increase chiral induction. Indeed, replacement of the pro-*R* phenyl with an *o*-anisyl group on the ephedrine backbone of EPHOS **17** gave **19** which was highly effective in the hydrogenation of methyl *α*-acetamidocinnamate, giving the product in 99% ee (Table 27.1, entry 147). In contrast, the use of EPHOS **17** induced only 46% ee (Table 27.1, entry 144) under the same conditions [31]. Similarly, replacement of the phenyl group with 1-naphthyl gave ligand **20** which led to 95% ee in the same reaction (Table 27.1, entry 149). It is interesting to note that the structurally similar *o*-anisyl ligands **19** and **23** derived from (+) and (–)-ephedrine, respectively, both induce high ee-values with the same (*S*) configuration of the product amino ester. This clearly shows the predominance of the chiral P center over the carbon backbone effect. However, the poor result obtained in the case of **24** bearing *o*-anisyl (*S*p) aminophosphine and (*R*p) phosphinite groups might be due to the quasi-meso structure that did not give any asymmetric induction (Fig. 27.3c) [1, 31].

Petit reported a close analogue to ProNOP lacking the rigid pyrrolidine ring, yet, the results with both ligands in the hydrogenation of (*E*)-acetamidocinnamic acid derivatives were similar (ProNOP, 61–86% ee; NETAlaNOP, 53–83% ee) [32]. With these substrates, it is important that the problem of *E*/*Z* isomerization (as highlighted by Noyori) should be considered [33]. Ligands **25** to **27** are the only type of amidophosphine-phosphinites applied in asymmetric hydrogenation [34]. Ligand **33** was found to be highly effective in the hydrogenation of 4-oxoisophorone enol acetate (100% conversion, 95% ee (*R*): Table 27.2, entry 5; Scheme 27.1). The product, (*S*)-phorenol acetate, is an intermediate in the synthesis of the natural pigment zeaxanthin [35]. The Rh complexes with **25** or **26**

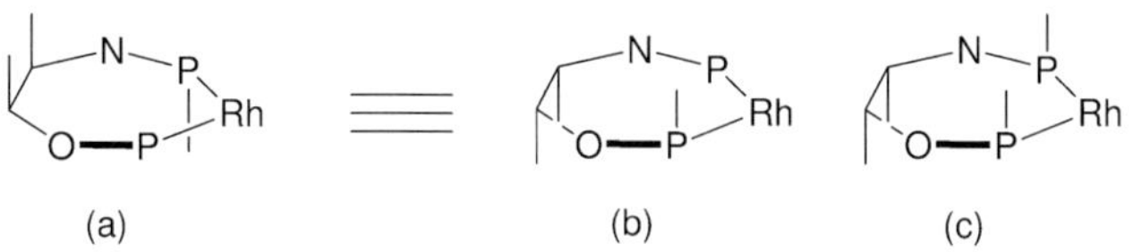

Fig. 27.3 Quasi-meso effect in asymmetric induction.

gave 71% ee in the same reaction (Table 27.2, entries 1 and 2). The other ephedrine-based ligands, including **1** and **3** to **6** [2, 36], afforded poor to moderate enantioselectivity in the rhodium-catalyzed hydrogenation of 2-acetamido-cinnamic acid derivatives. Structural variation leading to increased rigidity of the ligand backbone is one of the promising methods to enhance enantioselectivity. Cesarotti developed an aminophosphine-phosphinite based on the rigid pyrrolidine structure of prolinol [2]. However, results with this ligand in the enantiomeric hydrogenation of dehydroamino acid derivatives were only poor to moderate. Petit [32] improved the results and obtained up to 86% ee. Interestingly, when using $[Rh(COD)(L)]ClO_4$ as catalyst, both the *Z* and *E* isomeric substrates were converted to products with the same configuration (Table 27.1, entries 93 versus 200).

During the late 1980s, Döbler and Pracejus introduced the bicyclic [2.2.1] system to provide extra conformational rigidity to the ligand backbone [37]. The synthesis of these new chiral ligands was based on the resolution of the amino alcohol obtained from the aminolysis of *exo*-norbornane epoxide [38], followed by reaction with the corresponding chlorophosphine. Indeed, the *in-situ*-prepared cationic or neutral Rh catalysts based on ligand **35** resulted in better enantioselectivity in hydrogenation of 2-acetamido-acrylic acid (up to 89% ee: Table 27.1, entry 6).

The ease of preparation of (1*S*,2*R*)-1-hydroxylmethyl-2-amino-7,7-dimethylbicyclo [2.2.1] heptane from ketopinic acid prompted us to synthesize a new AMPP ligand (i.e., *exo*-**40**) [39]. The rhodium-catalyzed enantioselective hydrogenation of 2-acetamido-acrylic acid using this ligand gave the product in 77% ee with 95–100% conversion. Electron-withdrawing groups on the β-substituted phenyl ring of the substrate resulted in significant enantioselectivity enhancement (Ph, 77% ee; 4-MePh, 62% ee; 4-NO_2Ph, 90% ee). The effect of solvents on the enantioselectivity of the reaction was also quite significant, with acetone being found the best. Döbler performed the enantiomeric hydrogenation of standard dehydroamino acid and other heteroaryl derivatives using the rhodium complex based on bicyclo [3.3.0]-octane (i.e., **28**), resulting in moderate enantioselectivities (58–73% ee) [40]. More recently, ligands **36** to **39** based on the bicyclic

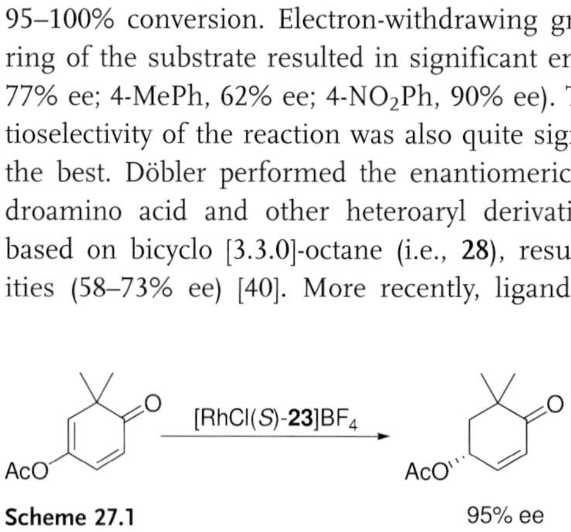

Scheme 27.1

[2.2.1] system were prepared [41]. The effect of the additional *P*-stereogenic center(s) was also explored with these ligands. The application of these ligands in enantiomeric hydrogenation resulted in products with up to 91% ee and TOFs ranging from 600 to 4000 h^{-1}. The substitution of one or both P-phenyl groups by a *p*-tolyl group resulted in a slight decrease of the enantioselectivity, regardless of the heteroatom linker (N or O) [42].

27.3
Bisphosphinamidite Ligands

The early development of this type of ligand was concentrated during the late 1970s and early 1980s. In 1976, the first article published on this topic was written by Giongo and co-workers, who described the initial synthesis of a chiral bisphosphinamidite **41** and its application in the enantiomeric hydrogenation of a number of dehydroamino acid derivatives [43]. The resulting enantioselectivities were comparable to the state-of-the-art ligand DIOP. In pursuing the same line of research, the Pracejus group also prepared (*S*,*S*)-**41** and achieved similar results [44]. Subsequently, the Giongo group further introduced other C_2-symmetric, 1,2-diamine-tethered bisphosphinites **42–49** with ee-values reaching 94% in the hydrogenation of (*Z*)-2-acetamidoacrylic acid (Table 27.3) [45]. Interestingly, both Giongo and Onuma noticed that when the hydrogen atoms of the amino groups were replaced with methyl groups whilst keeping the backbone chirality unchanged (as in the cases of **45** versus **46** and **47** versus **48**), a reversal of product configuration was observed. The Onuma group rationalized this by proposing a model wherein the helicity of the edge-phenyl groups on the phosphorus atoms were of opposite sense in the presence and absence of the methyl groups, respectively, as a result of a change of chirality on the nitrogen atoms. Non-C_2-symmetric pyrrolidine-based ligands **50–52** were also tested in asymmetric hydrogenation [47], though the results obtained were unsatisfactory. The use of a 1,4-diamino bridged bisphosphinamidite **53** was described in a recent publication in which excellent selectivity was recorded for the hydrogenation of α-acylaminocinnamic acid [48].

Surprisingly, given that many *P*-chiral ligands are efficient chiral inducers, only one example has been reported of a C_2-symmetric, *P*-chiral bisphosphinamidite. Ligand **54** was prepared by Wills et al. and tested in the Rh-catalyzed enantioselective hydrogenation of α-acylaminoacrylate to give disappointingly low selectivity (33% ee) and low efficiency (TON = 20, TOF = 0.4) [49].

Bisphosphinamidites which are supported by an axially chiral framework are another important class of ligands. Although reported as early as 1980 [50], no reports on the use of binaphthyl-based bisphosphinamidite in asymmetric catalysis were published during the decade thereafter. As described above, the selectivity and substrate generality in these early attempts were very limited in scope. In 1998, we unveiled that by partially hydrogenating BINAM to H_8-BINAM and

Table 27.3 Bisphosphinamidite ligands.

$R^1CH=C(R^2)R^3 \xrightarrow[\text{solvent}]{Rh(I)/\mathbf{L}^*,\ H_2} R^1CH_2-CH(R^2)R^3$

Entry	Substrate R¹	Substrate R²	Substrate R³	Ligand (L)	$P[H_2]$ [atm]	Solvent	Temp. [°C]	Time [h]	TON Sub:Rh	TOF [h^{-1}]	Conv. [%]	ee [%]	Reference(s)
1	H	Ph	NHAc	**55a**	1.0	THF	r.t.	0.5	200	100	100	93	51a, 52
2	H	Ph	NHAc	**55b**	2.0	THF	r.t.	0.17	200	1200	100	98	54
3	H	Ph	NHAc	**56a**	1.0	THF	r.t.	0.5	200	400	100	97	51b
4	H	Ph	NHAc	**56b**	2.0	THF	r.t.	0.17	200	1200	100	96	54
5	H	3-Cl-Ph	NHAc	**55b**	2.0	THF	r.t.	0.17	200	1200	100	98	54
6	H	3-Cl-Ph	NHAc	**56b**	2.0	THF	r.t.	0.17	200	1200	100	98	54
7	H	4-Cl-Ph	NHAc	**55a**	1.0	THF	r.t.	0.5	200	400	100	95	51a
8	H	4-Cl-Ph	NHAc	**56a**	1.0	THF	0	0.5	200	400	100	97	51a, b
9	H	3-Br-Ph	NHAc	**55b**	2.0	THF	r.t.	0.17	200	1200	100	98	54
10	H	3-Br-Ph	NHAc	**56b**	2.0	THF	r.t.	0.17	200	1200	100	97	54
11	H	4-Br-Ph	NHAc	**55b**	2.0	THF	r.t.	0.17	200	1200	100	98	54
12	H	4-Br-Ph	NHAc	**56b**	2.0	THF	r.t.	0.17	200	1200	100	95	54
13	H	4-F-Ph	NHAc	**55a**	1.0	THF	r.t.	0.5	172	344	86	90	51a
14	H	4-F-Ph	NHAc	**56a**	1.0	THF	0	0.5	200	400	100	96	51a, b
15	H	4-CF_3-Ph	NHAc	**55a**	1.0	THF	r.t.	0.5	500	400	100	95	51a, 52
16	H	4-CF_3-Ph	NHAc	**55b**	2.0	THF	r.t.	0.17	200	1200	100	99	54
17	H	4-CF_3-Ph	NHAc	**56a**	1.0	THF	0	0.5	200	400	100	99	51a, b
18	H	4-CF_3-Ph	NHAc	**56a**	1.0	THF	5	0.5	1000	2000	100	99	51a
19	H	4-CF_3-Ph	NHAc	**56b**	2.0	THF	r.t.	0.17	200	1200	100	97	54
20	H	3-CH_3O-Ph	NHAc	**55b**	2.0	THF	r.t.	0.17	200	1200	100	97	54

21	H	3-CH_3O-Ph	NHAc	**56b**	2.0	THF	r.t.	0.17	200	1200	100	97	54
22	H	3-CH_3-Ph	NHAc	**55a**	1.0	THF	r.t.	0.5	186	372	93	95	51b
23	H	3-CH_3-Ph	NHAc	**55b**	2.0	THF	r.t.	0.17	200	1200	100	98	54
24	H	3-CH_3-Ph	NHAc	**56a**	1.0	THF	0	0.5	200	400	100	98	51a, b
25	H	3-CH_3-Ph	NHAc	**56b**	2.0	THF	r.t.	0.17	200	1200	100	98	54
26	H	4-CH_3-Ph	NHAc	**55a**	1.0	THF	r.t.	0.5	500	1000	100	95	51a
27	H	4-CH_3-Ph	NHAc	**55b**	2.0	THF	r.t.	0.17	200	1200	100	96	54
28	H	4-CH_3-Ph	NHAc	**55b**	2.0	THF	r.t.	0.17	200	1200	100	94	54
29	H	4-CH_3-Ph	NHAc	**56a**	1.0	THF	0	0.5	200	400	100	97	51a, b
30	H	4-CH_3-Ph	NHAc	**56b**	2.0	THF	r.t.	0.17	200	1200	100	96	54
31	H	4-CH_3-Ph	NHAc	**56b**	2.0	THF	r.t.	0.17	200	1200	100	94	54
32	H	4-Et-Ph	NHAc	**55b**	2.0	THF	r.t.	0.17	200	1200	100	97	54
33	H	4-Et-Ph	NHAc	**56b**	2.0	THF	r.t.	0.17	200	1200	100	97	54
34	H	2-furyl	NHAc	**55a**	1.0	THF	r.t.	0.5	200	400	100	96	51a
35	H	2-furyl	NHAc	**55b**	2.0	THF	r.t.	0.17	200	1200	100	98	54
36	H	2-furyl	NHAc	**56a**	1.0	THF	0	0.5	200	400	100	98	51a, b
37	H	2-furyl	NHAc	**56b**	2.0	THF	r.t.	0.17	200	1200	100	96	54
38	Me	Ph	NHAc	**55b**	2.0	THF	r.t.	0.17	200	1200	100	93	54
39	Me	Ph	NHAc	**56b**	2.0	THF	r.t.	0.17	200	1200	100	94	54
40	Me	4-Cl-Ph	NHAc	**56a**	1	THF	0	2	200	100	100	80.3	51b
41	Me	4-CH_3-Ph	NHAc	**56a**	1	THF	0	2	193	97	97	77	51b
42	H	CO_2H	NHAc	**41a**	1	MeOH	25	–	570	–	95	73 [a]	43
43	H	CO_2H	NHAc	**41a**	1	EtOH	25	–	–	–	90–100	76.7	45b (*R*)
44	H	CO_2H	NHAc	**41b**	1	EtOH	25	–	125	–	90–100	25.1	45b (*S*)
45	H	CO_2H	NHAc	**42**	25	EtOH	25	–	125	–	90–100	83 [a]	45a
46	H	CO_2H	NHAc	**42**	5	EtOH	25	–	125	–	90–100	83.9 [a]	45b
47	H	CO_2H	NHAc	**43**	1	EtOH	25	–	125	–	90–100	78.9 [a]	45b
48	H	CO_2H	NHAc	**44**	1	EtOH	25	–	125	–	90–100	88.1	45b
49	H	CO_2H	NHAc	**45**	1	EtOH	25	–	125	–	90–100	89.5	45b

Table 27.3 (continued)

Entry	Substrate R[1]	Substrate R[2]	Substrate R[3]	Ligand (L)	P[H_2] [atm]	Solvent	Temp. [°C]	Time [h]	TON Sub:Rh	TOF [h^{-1}]	Conv. [%]	ee [%]	Reference(s)
50	H	CO_2H	NHAc	**46**	5	EtOH	25	–	125	–	90–100	86.2	45b
51	H	CO_2H	NHAc	**47a**	1	EtOH	25	–	125	–	90–100	24.0	45b
52	H	CO_2H	NHAc	**48**	1	EtOH	25	–	125	–	90–100	90.9	45b
53	H	CO_2H	NHAc	**49**	5	EtOH	25	–	125	–	90–100	12.0	45b
54	H	CO_2H	NHAc	(*S,S*)-**50**	1	EtOH	25	–	125	–	–	33	47a
55	H	CO_2H	NHAc	(*S,R*)-**50**	1	EtOH	25	–	125	–	–	61	47a
56	H	CO_2H	NHAc	**51**	1	EtOH	25	–	125	–	–	68	47a
57	H	CO_2H	NHAc	**53**	24	*i*PrOH	r.t.	24	77	32	77	68	48
58	H	CO_2H	NHAc	**55a**	2.0	EtOH	r.t.	0.17	100	600	100	93.5	52
59	H	CO_2H	NHAc	**55b**	3.4	MeOH	r.t.	0.17	500	3000	100	98	53, 54
60	H	CO_2H	NHAc	**56a**	2.0	EtOH	r.t.	0.17	100	600	100	99	51b, 52
61	H	CO_2H	NHAc	**58**	1	EtOH	r.t.	1.5	74.5	50	15	78	55
62	Ph	CO_2H	NHAc	**41a**	1	MeOH	25	–	300	–	95	84[a)]	43, 45a
63	Ph	CO_2H	NHBz	**41a**	1	MeOH	25	–	50	–	70	68[a)]	43
64	Ph	CO_2H	NHBz	**41a**	1	EtOH	25	0.08–0.67	–	1008	90–100	75	45a
65	Ph	CO_2H	NHAc	**41a**	1	MeOH	25	0.03	48	1600	45	81.7[a)]	44
66	Ph	CO_2H	NHAc	**41a**	1	EtOH	25	–	125	–	90–100	77.3	45b, 47a
67	Ph	CO_2H	NHAc	**41b**	1	EtOH	25	–	125	–	90–100	40.8	45b, 47a
68	Ph	CO_2H	NHAc	**42**	10	EtOH	0	–	125	–	90–100	93[a)]	45a
69	Ph	CO_2H	NHAc	**42**	5	EtOH	25	–	125	–	90–100	80.6[a)]	45b
70	Ph	CO_2H	NHAc	**43**	1	EtOH	25	–	125	–	90–100	74.8[a)]	45b
71	Ph	CO_2H	NHAc	**44**	1	EtOH	25	–	125	–	90–100	91.9	45b
72	Ph	CO_2H	NHAc	**45**	2	EtOH	25	–	125	–	90–100	94.4	45b,c, 47a
73	Ph	CO_2H	NHAc	**46**	5	EtOH	25	–	125	–	90–100	68.4	45b,c, 47a

74	Ph	CO_2H	NHAc	**47a**	1	EtOH	25	–	125	–	90–100	47.0	45b, 47a
75	Ph	CO_2H	NHAc	**47a**	7.8	EtOH : PhH 1 : 1	r.t.	–	–	–	–	70	46
76	Ph	CO_2H	NHAc	**47b**	7.8	EtOH : PhH 1 : 1	r.t.	–	–	–	–	72	46
77	Ph	CO_2H	NHBz	**47a**	7.8	EtOH : PhH 1 : 1	r.t.	–	–	–	–	62	46
78	Ph	CO_2H	NHBz	**47b**	7.8	EtOH : PhH 1 : 1	r.t.	–	–	–	–	60	46
79	Ph	CO_2H	NHAc	**48**	1	EtOH	25	–	125	–	90–100	92.1	45b, 46, 47a
80	Ph	CO_2H	NHBz	**48**	7.8	EtOH	r.t.	–	–	–	–	92	46
81	Ph	CO_2H	NHAc	**49**	5	EtOH	25	–	125	–	90–100	rac	45b
82	Ph	CO_2H	NHAc	(*S,S*)-**50**	4.5	EtOH	25	–	125	–	–	35	47a
83	Ph	CO_2H	NHAc	(*S,R*)-**50**	4.5	EtOH	25	–	125	–	–	59	47a
84	Ph	CO_2H	NHAc	**51**	4.5	EtOH	25	–	125	–	–	69	47a
85	Ph	CO_2H	NHAc	**53**	1	*i*PrOH	r.t.	24	100	4.2	100	98	48
86	Ph	CO_2H	NHAc	**55a**	2.0	EtOH	r.t.	0.17	100	600	100	90.3	52
87	Ph	CO_2H	NHAc	**55b**	3.4	MeOH	r.t.	0.17	500	3000	100	98	53, 54
88	Ph	CO_2H	NHAc	**56a**	2.0	EtOH	r.t.	0.17	100	600	100	94.2	52
89	Ph	CO_2H	NHAc	**57**	6.7	$(CH_3)_2CO$	25	5	100	20	100	79.6	56
90	Ph	CO_2H	NHAc	**58**	1	EtOH	r.t.	1.5	500	333	100	>98	55
91	Ph	CO_2Me	NHAc	**41a**	1	MeOH	25	–	450	–	100	49 [a)]	43
92	Ph	CO_2Me	NHAc	**41a**	–	C_6H_6	25	0.03*	50	313	45	82.5 [a)]	44
93	Ph	CO_2Me	NHAc	**41a**	1	EtOH	25	0.08–0.67	–	828	90–100	55	45a
94	Ph	CO_2Me	NHAc	**54**	1	MeOH	–	48	20	0.4	95	33	49
95	Ph	CO_2Me	NHAc	**55a**	2.0	THF	r.t.	0.17	100	600	100	90	52
96	Ph	CO_2Me	NHAc	**55a**	3.4	MeOH	r.t.	0.17	500	3000	100	91	53, 54
97	Ph	CO_2Me	NHAc	**55b**	3.4	MeOH	r.t.	0.5	5000	10000	100	98.6	53, 54
98	Ph	CO_2Me	NHAc	**55c**	3.4	MeOH	r.t.	0.5	500	1000	100	13	53, 54
99	Ph	CO_2Me	NHAc	**56a**	2.0	THF	r.t.	0.17	100	600	100	96	51b, 52
100	Ph	CO_2Me	NHAc	**57**	6.7	$(CH_3)_2CO$	25	5	200	20	100	73.7	56

Table 27.3 (continued)

Entry	Substrate			Ligand (L)	$P[H_2]$ [atm]	Solvent	Temp. [°C]	Time [h]	TON Sub:Rh	TOF [h^{-1}]	Conv. [%]	ee [%]	Refer-ence(s)
	R^1	R^2	R^3										
101	Ph	CO_2Me	NHAc	**58**	1	$(CH_3)_2CO$	r.t.	1	500	500	100	>99	55
102	Ph	CO_2NH_2	NHAc	**47a**	7.8	EtOH : PhH 1 : 1	r.t.	–	–	–	–	92	46
103	Ph	CO_2NH_2	NHAc	**47b**	–	EtOH : PhH 1 : 1	r.t.	–	–	–	–	92	46
104	Ph	CO_2NH_2	NHAc	**47b**	–	EtOH : PhH 1 : 1	r.t.	–	–	–	–	70	46
105	2-Cl-Ph	CO_2H	NHAc	**55a**	2.0	EtOH	r.t.	0.17	100	600	100	90	52
106	2-Cl-Ph	CO_2H	NHAc	**56a**	2.0	THF	r.t.	0.17	100	600	100	94	51b, 52
107	2-Cl-Ph	CO_2H	NHAc	**57**	6.7	$(CH_3)_2CO$	25	5	100	20	100	78.1	56
108	2-Cl-Ph	CO_2H	NHAc	**58**	1	EtOH	r.t.	1.5	500	333	100	96	55
109	2-Cl-Ph	CO_2Me	NHAc	**55a**	2.0	THF	r.t.	0.17	100	600	100	90	52
110	2-Cl-Ph	CO_2Me	NHAc	**55b**	3.4	MeOH	r.t.	0.17	500	3000	100	97	53, 54
111	2-Cl-Ph	CO_2Me	NHAc	**56a**	2.0	THF	r.t.	0.17	100	600	100	97	52
112	2-Cl-Ph	CO_2Me	NHAc	**57**	6.7	$(CH_3)_2CO$	25	5	100	20	100	72.0	56
113	2-Cl-Ph	CO_2Me	NHAc	**58**	1	$(CH_3)_2CO$	r.t.	1	500	500	100	96	55
114	3-Cl-Ph	CO_2H	NHAc	**55a**	2.0	EtOH	r.t.	0.17	100	600	100	88	52
115	3-Cl-Ph	CO_2H	NHAc	**56a**	2.0	THF	r.t.	0.17	100	600	100	93	51b, 52
116	3-Cl-Ph	CO_2H	NHAc	**57**	6.7	$(CH_3)_2CO$	25	5	100	20	100	76.3	56
117	3-Cl-Ph	CO_2H	NHAc	**58**	1	EtOH	r.t.	1.5	500	333	100	95	55
118	3-Cl-Ph	CO_2Me	NHAc	**55a**	2.0	THF	r.t.	0.17	100	600	100	90	52
119	3-Cl-Ph	CO_2Me	NHAc	**55b**	3.4	MeOH	r.t.	0.17	500	3000	100	97	53, 54
120	3-Cl-Ph	CO_2Me	NHAc	**56a**	2.0	THF	r.t.	0.17	100	600	100	94	51b, 52
121	3-Cl-Ph	CO_2Me	NHAc	**57**	6.7	$(CH_3)_2CO$	25	5	100	20	100	72.1	56
122	3-Cl-Ph	CO_2Me	NHAc	**58**	1	$(CH_3)_2CO$	r.t.	1	500	500	100	>99	55
123	4-Cl-Ph	CO_2H	NHAc	**55a**	2.0	EtOH	r.t.	0.17	100	600	100	86	52
124	4-Cl-Ph	CO_2H	NHAc	**56a**	2.0	EtOH	r.t.	0.17	100	600	100	93	51b, 52

125	4-Cl-Ph	CO_2H	NHAc	**57**	6.7	$(CH_3)_2CO$	25	5	100	20	100	79.1	56
126	4-Cl-Ph	CO_2Me	NHAc	**55a**	2.0	THF	r.t.	0.17	100	600	100	88	52
127	4-Cl-Ph	CO_2Me	NHAc	**55b**	3.4	MeOH	r.t.	0.17	500	3000	100	98	53, 54
128	4-Cl-Ph	CO_2Me	NHAc	**56a**	2.0	THF	r.t.	0.17	100	600	100	94	51b, 52
129	4-Cl-Ph	CO_2Me	NHAc	**57**	6.7	$(CH_3)_2CO$	25	5	100	20	100	74.0	56
130	4-Cl-Ph	CO_2Me	NHAc	**58**	1	$(CH_3)_2CO$	r.t.	1	500	500	100	>99	55
131	4-Cl-Ph	CO_2Me	NHBz	**55b**	3.4	MeOH	r.t.	0.17	500	3000	100	99	53
132	4-Br-Ph	CO_2Me	NHAc	**55b**	3.4	MeOH	r.t.	0.17	500	3000	100	98	53
133	4-Br-Ph	CO_2Me	NHAc	**56a**	2.0	THF	r.t.	0.17	100	600	100	96	51b, 52
134	4-Br-Ph	CO_2Me	NHAc	**58**	1	$(CH_3)_2CO$	r.t.	1	500	500	100	>99	55
135	4-Br-Ph	CO_2Me	NHBz	**56a**	2.0	THF	r.t.	0.17	100	600	100	96	51b, 52
136	4-F-Ph	CO_2Me	NHAc	**55b**	3.4	MeOH	r.t.	0.17	500	3000	100	98	53, 54
137	4-F-Ph	CO_2Me	NHAc	**56a**	2.0	THF	r.t.	0.17	100	600	100	93	51b, 52
138	4-F-Ph	CO_2Me	NHAc	**57**	6.7	$(CH_3)_2CO$	25	5	100	20	100	73.0	56
139	4-F-Ph	CO_2Me	NHAc	**58**	1	$(CH_3)_2CO$	r.t.	1	500	500	100	>99	55
140	4-F-Ph	CO_2Me	NHBz	**56a**	2.0	THF	r.t.	0.17	100	600	100	94	51b, 52
141	4-F-Ph	CO_2Me	NHBz	**55b**	3.4	MeOH	r.t.	0.17	500	3000	100	99	53, 54
142	4-NO_2-Ph	CO_2H	NHAc	**56a**	2.0	EtOH	r.t.	0.17	100	600	100	90	51b, 52
143	4-NO_2-Ph	CO_2H	NHAc	**57**	6.7	$(CH_3)_2CO$	25	5	100	20	100	76.4	56
144	4-NO_2-Ph	CO_2Me	NHAc	**55a**	3.4	MeOH	r.t.	0.17	500	3000	100	82	53, 54
145	4-NO_2-Ph	CO_2Me	NHAc	**55b**	3.4	MeOH	r.t.	0.17	500	3000	100	96	53, 54
146	4-NO_2-Ph	CO_2Me	NHAc	**56a**	2.0	THF	r.t.	0.17	100	600	100	91	51b, 52
147	4-NO_2-Ph	CO_2Me	NHAc	**57**	6.7	$(CH_3)_2CO$	25	5	100	20	100	73.4	56
148	4-NO_2-Ph	CO_2Me	NHAc	**58**	1	$(CH_3)_2CO$	r.t.	1	500	500	100	94	55
149	4-CH_3-Ph	CO_2Me	NHAc	**55a**	3.4	MeOH	r.t.	0.17	500	3000	100	89	53, 54
150	4-CH_3-Ph	CO_2Me	NHAc	**55b**	3.4	MeOH	r.t.	0.17	500	3000	100	98	53, 54
151	4-CH_3-Ph	CO_2Me	NHAc	**56a**	2.0	THF	r.t.	0.17	100	600	100	94	51b, 52
152	4-CH_3-Ph	CO_2Me	NHAc	**57**	6.7	$(CH_3)_2CO$	25	5	200	20	100	71.2	56
153	4-CH_3-Ph	CO_2Me	NHAc	**58**	1	$(CH_3)_2CO$	r.t.	1	500	500	100	>98	55

Table 27.3 (continued)

Entry	Substrate R^1	R^2	R^3	Ligand (L)	$P[H_2]$ [atm]	Solvent	Temp. [°C]	Time [h]	TON Sub:Rh	TOF [h^{-1}]	Conv. [%]	ee [%]	Refer-ence(s)
154	4-CH_3-Ph	CO_2Me	NHBz	**56a**	2.0	THF	r.t.	0.17	100	600	100	95	51b, 52
155	2-CH_3O-Ph	CO_2H	NHAc	**56a**	2.0	EtOH	r.t.	0.17	100	600	100	93	51b, 52
156	2-CH_3O-Ph	CO_2H	NHAc	**57**	6.7	$(CH_3)_2CO$	25	5	100	20	100	79.0	56
157	2-CH_3O-Ph	CO_2H	NHAc	**58**	1	EtOH	r.t.	1.5	500	333	100	84	55
158	4-CH_3O-Ph	CO_2Me	NHAc	**53**	1	*i*PrOH	r.t.	24	100	4.2	100	91	48
159	4-CH_3O-Ph	CO_2Me	NHAc	**55a**	2.0	THF	r.t.	0.17	100	600	100	93	52
160	4-CH_3O-Ph	CO_2Me	NHAc	**55a**	3.4	MeOH	r.t.	0.17	500	3000	100	87	53, 54
161	4-CH_3O-Ph	CO_2Me	NHAc	**55b**	3.4	MeOH	r.t.	0.17	500	3000	100	98	53, 54
162	4-CH_3O-Ph	CO_2Me	NHAc	**56a**	2.0	THF	r.t.	0.17	100	600	100	93	51b, 52
163	4-CH_3O-Ph	CO_2Me	NHAc	**57**	6.7	$(CH_3)_2CO$	25	5	100	20	100	72.7	56
164	4-CH_3O-Ph	CO_2Me	NHAc	**58**	1	$(CH_3)_2CO$	r.t.	1	500	500	100	>99	55
165	4-CH_3O-Ph	CO_2Me	NHBz	**56a**	2.0	THF	r.t.	0.17	100	600	100	95	51b, 52
166	4-AcO-Ph	CO_2Me	NHAc	**55a**	3.4	MeOH	r.t.	0.17	500	3000	100	87	53,54
167	4-AcO-Ph	CO_2Me	NHAc	**55b**	3.4	MeOH	r.t.	0.17	500	3000	100	98	53, 54
168	4-HO-Ph	CO_2Me	NHAc	**57**	6.7	$(CH_3)_2CO$	25	5	100	20	100	74.6	56
169	3,4-(CH_2O_2)-Ph	CO_2H	NHAc	**41a**	1	MeOH	25		50		90	75[a]	43
170	3,4-(CH_2O_2)-Ph	CO_2H	NHAc	**41a**	1	EtOH	25		–	612	90–100	77	45a
171	3,4-(CH_2O_2)-Ph	CO_2H	NHAc	**55a**	2.0	EtOH	r.t.	0.17	100	600	100	77	52
172	3,4-(CH_2O_2)-Ph	CO_2H	NHAc	**56a**	2.0	EtOH	r.t.	0.17	100	600	100	91	51b, 52
173	3,4-(CH_2O_2)-Ph	CO_2H	NHAc	**57**	6.7	$(CH_3)_2CO$	25	5	100	20	100	80.3	56
174	3,4-(CH_2O_2)-Ph	CO_2H	NHAc	**58**	1	EtOH	r.t.	1.5	500	333	100	92	55
175	3,4-(CH_2O_2)-Ph	CO_2Me	NHAc	**55a**	3.4	MeOH	r.t.	0.17	500	3000	100	81	53
176	3,4-(CH_2O_2)-Ph	CO_2Me	NHAc	**55b**	3.4	MeOH	r.t.	0.17	500	3000	100	98	53, 54
177	3,4-(CH_2O_2)-Ph	CO_2Me	NHAc	**56a**	2.0	THF	r.t.	0.17	100	600	100	93	51b, 52
178	3,4-(CH_2O_2)-Ph	CO_2Me	NHAc	**57**	6.7	$(CH_3)_2CO$	25	5	100	20	100	76.2	56

179	3,4-(CH_2O_2)-Ph	CO_2Me	NHAc	**58**	1	$(CH_3)_2CO$	r.t.	1	500	500	100	>98	55
180	3-CH_3O,4-Ac-Ph	CO_2H	NHAc	**58**	1	EtOH	r.t.	1.5	500	333	100	94	55
181	3-CH_3O,4-AcO-Ph	CO_2H	NHAc	**41 a**	1	EtOH	25	–	396	–	90–100	87	45 a
182	2-furyl	CO_2Me	NHAc	**55 a**	3.4	MeOH	r.t.	0.17	500	3000	100	84	53, 54
183	2-furyl	CO_2Me	NHAc	**55 b**	3.4	MeOH	r.t.	0.17	500	3000	100	98	53, 54
184	2-furyl	CO_2Me	NHAc	**56 a**	2.0	THF	r.t.	0.17	100	600	100	91	51 b, 52
185	2-furyl	CO_2Me	NHAc	**58**	1	$(CH_3)_2CO$	r.t.	1.5	500	333	100	>99	55
186	2-furyl	CO_2H	NHBz	**56 a**	2.0	THF	r.t.	0.17	100	600	100	94	51 b, 52
187	2-furyl	CO_2Me	NHBz	**56 a**	2.0	THF	r.t.	0.17	100	600	100	93	51 b, 52
188	(*E*)-PhCH=CH	CO_2Me	NHAc	**56 a**	2.0	THF	r.t.	0.17	100	600	100	90	51 b, 52
189	H	CO_2Et	OC(O)Me	**58**	1	THF	r.t.	1.5	250	167	100	97	55
190	H	CO_2Me	NHAc	**53**	1	*i*PrOH	r.t.	24	77	3.2	77	68	48
191	H	CO_2Me	NHAc	**55 a**	2.0	THF	r.t.	0.17	100	600	100	93	51, 52
192	H	CO_2Me	NHAc	**55 b**	3.4	MeOH	r.t.	0.17	500	3000	100	97	53, 54
193	H	CO_2Me	NHAc	**56 a**	2.0	THF	r.t.	0.17	100	600	100	97	52
194	H	CO_2Me	NHAc	**58**	1	$(CH_3)_2CO$	r.t.	1	500	500	100	95	55
195	*N*-Ac-3-indole	CO_2Me	NHBz	**56 a**	2.0	THF	r.t.	0.17	100	600	100	93	52
196	H	CO_2H	CH_2CO_2H	**41 a**	5	EtOH	25	–	50	–	90–100	12.3	45 b
197	H	CO_2H	CH_2CO_2H	**41 b**	5	EtOH	25	–	50	–	90–100	35.1	45 b
198	H	CO_2H	CH_2CO_2H	**42**	5	EtOH	25	–	50	–	90–100	7.9	45 b
199	H	CO_2H	CH_2CO_2H	**43**	5	EtOH	25	–	50	–	90–100	36.6	45 b
200	H	CO_2H	CH_2CO_2H	**44**	5	EtOH	25	–	50	–	90–100	25.3	45 b

Table 27.3 (continued)

Entry	Substrate			Ligand (L)	$P[H_2]$ [atm]	Solvent	Temp. [°C]	Time [h]	TON Sub : Rh	TOF [h^{-1}]	Conv. [%]	ee [%]	Reference(s)
	R^1	R^2	R^3										
201	H	CO_2H	CH_2CO_2H	**45**	5	EtOH	25	–	50	–	90–100	71.4	45b,c
202	H	CO_2H	CH_2CO_2H	**46**	5	EtOH	25	–	50	–	90–100	5.8	45b,c
203	H	CO_2H	CH_2CO_2H	**47a**	5	EtOH	25	–	50	–	90–100	60.3	45b
204	H	CO_2H	CH_2CO_2H	**48**	5	EtOH	25	–	50	–	90–100	8.2	45b
205	H	CO_2H	CH_2CO_2H	**49**	5	EtOH	25	–	50	–	90–100	5.8	45b
206	H	CO_2H	CH_2CO_2H	**58**	1	THF	r.t.	1.5	250	167	100	68	55
207	H	CO_2Me	CH_2CO_2Me	**58**	1	THF	r.t.	1.5	250	167	100	93	55
208	H	CO_2Me	OC(O)Me	**58**	1	THF	r.t.	1.5	250	167	100	96	55
209	CH_2OH	Me	$Me_2C{=}CH(CH_2)_2$	**52**	10	C_6H_6	r.t.	8	48	6.0	95	68	47b
210	CH_2OH	$Me_2C{=}CH(CH_2)_2$	Me	**52**	10	C_6H_6	r.t.	8	49	6.1	97	61	47b

a) Optical yield.

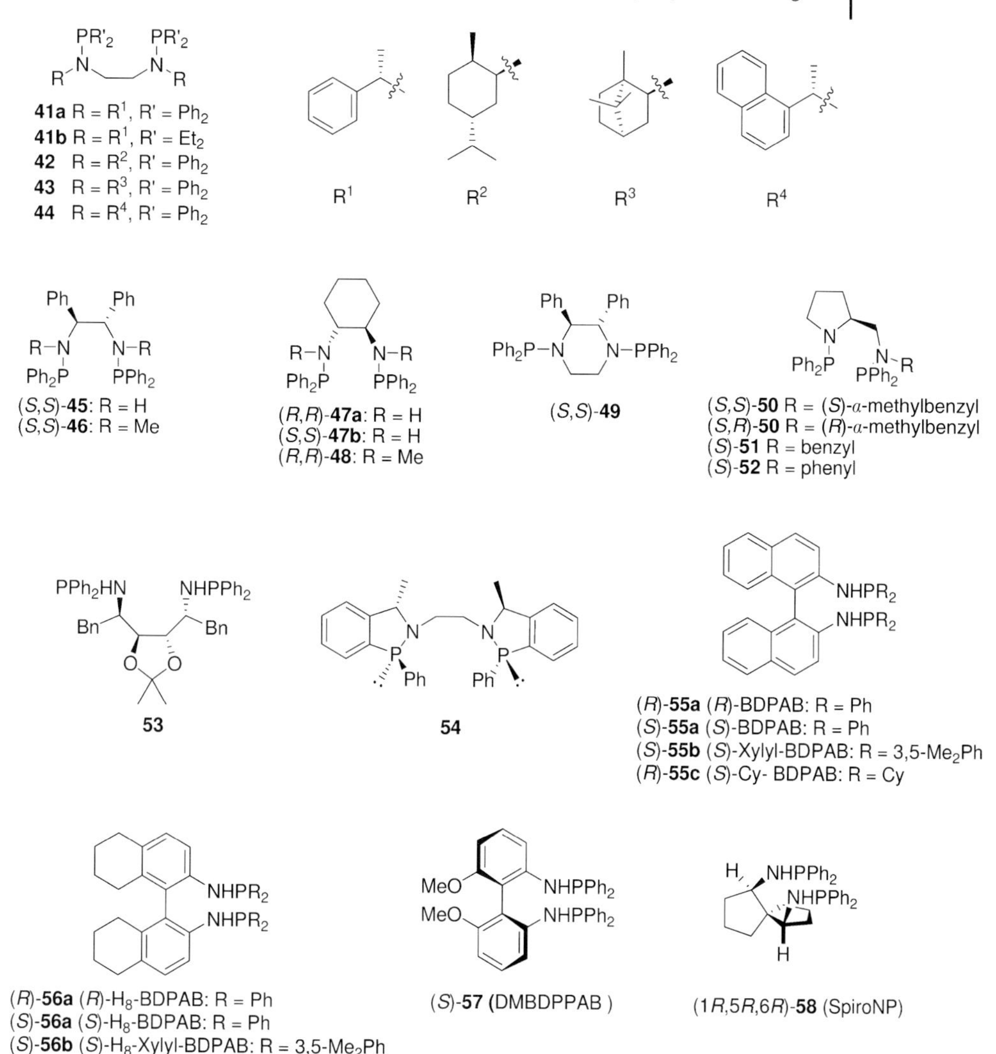

Fig. 27.4 Bisphosphinamidite ligands.

subsequently preparing the corresponding 2,2′-bis(diphenylphosphinoamino)-1,1′-binaphthyl (BDPABs), the enantioselectivities in the hydrogenation of enamides were significantly improved in the case of (*S*)-**55a** versus (*S*)-**56a** [51]. Similarly, in the asymmetric hydrogenation of (*Z*)-2-acetamido-3-arylacrylic acids, the same observation was noted [52]. A boost in ee-value was also induced by replacing Ph with 3,5-Me_2Ph (**55a** versus **55b**) [53, 54]. A TOF as high as 3000 and a selectivity of up to 99% ee with the use of **55b** were observed, indicative of its high efficiency and effectiveness. In our recent findings, the conformationally rigid SpiroNP **58** also led to high enantioselectivities in the asymmetric hydrogenation of dehydroamino acid derivatives [55]. An analogous biphenyl-based ligand **57**, however, was much less efficient than the binaphthyl-based or spiro-based counterparts [56].

27.4
Mixed Phosphine-Phosphoramidites and Phosphine-Aminophosphine Ligands

In contrast to the remarkable development of C_2-symmetrical ligands and C_1-nonsymmetrical ligands, the mixed bidentate ligands mentioned in the title were rather underdeveloped. The use of a ferrocene-based chiral backbone led to a promising class of new ligands having a wide scope and inducing good activity. Bophoz [57, 58] (Fig. 27.5, **68**) represented the first mixed phosphine-aminophosphine ligands for asymmetric catalysis with a wide scope of alkene substrates, including α,β-unsaturated acids, enamides, and acetamidocinnamic acid derivatives. The TON of these catalytic asymmetric hydrogenations was gener-

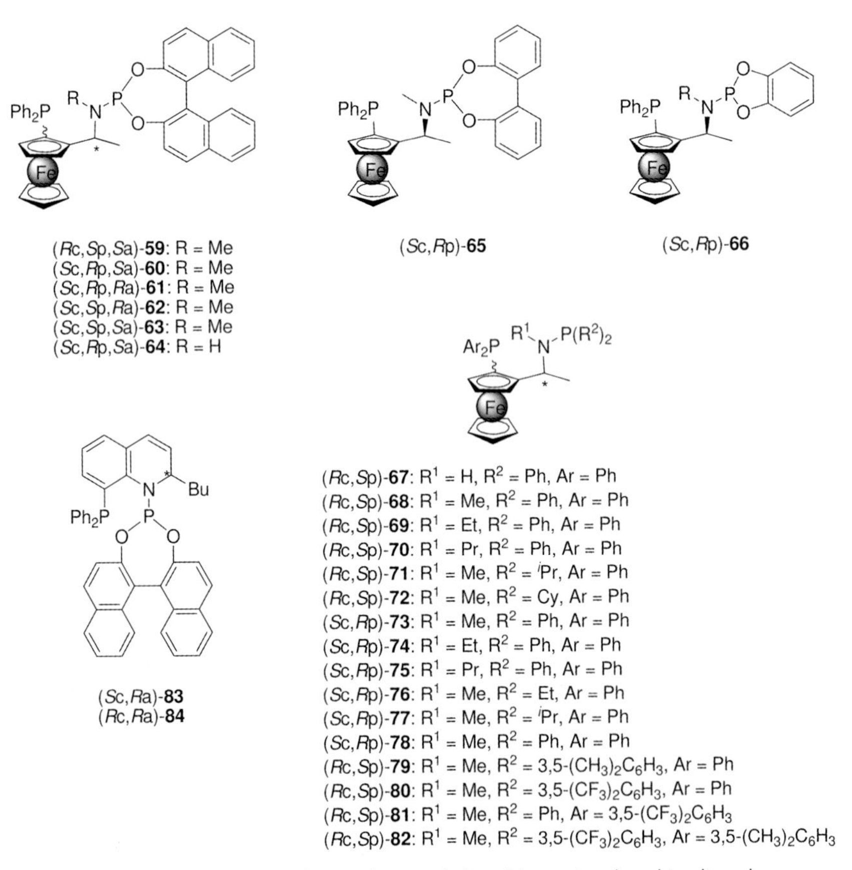

Fig. 27.5 Mixed phosphine-phosphoramidites and phosphine-aminophosphine ligands.

ally in the range of 15.8 to 100. The potential of industrial usage was increased by improving its SCR to 10000 (Table 27.4, entry 53). The shelf stability is an attractive point of this type of ligand. Somewhat surprisingly, Maligres and Krska found that **73** was unable to induce enantioselectivity in the Ru(II)-catalyzed hydrogenation of (*Z*)-α-phenoxybutenoic acid (Table 27.4, entry 96) [59].

The introduction of a third chiral element onto the chiral backbone was of interest, and we constructed a modified PPFA [60] with extra axial chirality from BINOL (Fig. 27.5, **59**) [61]. This type of ligand contains three chirality elements. Indeed, the enantioselectivity and activity remained excellent in the enantiomeric hydrogenation of α-dehydroamino acid derivatives and enamides using these ligands (Table 27.4, entries 1, 52, 67, 69–71, 79), regardless of the electronic properties of the *para*-substituting group. A similar approach was taken by Zheng's group using different diastereomers [62]. The scope was further extended to dimethyl itaconate, which was hydrogenated with a higher TON and with excellent ee and activity (Table 27.4, entry 91).

Recently, we developed three new fluorinated ferrocenyl phosphine-aminophosphine ligands derived from *N*,*N*-dimethyl-1-ferrocenylethylamine (Ugi's amine) [63]. These ligands were efficiently applied in the Rh-catalyzed hydrogenation of various aryl enamides (92.1 to 99.7% ee) and α-dehydroamino acid derivatives (98.5 to 99.7% ee), with complete conversion. The Rh–complex based on **80** led to somewhat lower enantioselectivities in the hydrogenation of arylenamides with *para*-EDG; however, the enantioselectivities were almost equally high for substrates containing *para*-EDG or *para*-EWG (98.5 to 99.7% ee) at 5 °C. These ligands also showed a remarkable air- and water-stability.

The ferrocene-based ligands have proven to be promising in most aspects of asymmetric catalysis. The only drawback was, however, the laborious resolution of Ugi's amine [64] which is used as a starting material, although this problem was solved by the facile asymmetric hydrogenation of ferrocenyl ketones using (XylylP-Phos-Ru–DPEN)Cl_2 (with a nonoptimized SCR of up to 100000 on a 150-g scale; Scheme 27.2) [65]. With this method in hand, it became more flexible and almost effortless to generate a large structural diversity of ferrocene-based chiral ligands.

Mixed phosphine-phosphoramidite ligands QUINAPHOS **83** and **84**, as developed by Leitner, worked well for the Rh-catalyzed hydrogenation of itaconic acid and α-dehydroamino acid derivatives. The ligand **84** also exerted extra reactivity, leading to an average TOF of 36000 h^{-1} in the hydrogenation of dimethyl itaconate after the addition of a second batch of substrate with SCR 6000:1 [66]. In contrast to BINAPHOS-type ligands, the major asymmetric induction relied on the 2-position of the alkyl groups embedded in the fairly rigid heterocyclic skeleton.

Table 27.4 Mixed phosphine–phosphoramidites and phosphine–aminophosphine ligands.

$R^1CH=C(R^2)R^3 \longrightarrow R^1CH_2\text{–}C^*H(R^2)R^3$

Entry	Substrate			Catalyst	Conditions				TON	TOF [h^{-1}]	Conv. [%]	ee [%]	Refer-ence(s)
	R^1	R^2	R^3		P[H$_2$] [bar]	Solvent	Temp. [°C]	Time [h]					
1	H	Ph	NHAc	Rh-(*Rc*,*Sp*,*Sa*)-**59**	20.7	THF	r.t.	7 [a]	99	14.1	>99	87.5 (*S*)	61
2	H	Ph	NHAc	Rh-(*Sc*,*Rp*,*Sa*)-**60**	10	DCM	r.t.	1	5000	5000	100	99.3 (*R*)	62
3	H	Ph	NHAc	Rh-(*Sc*,*Rp*,*Ra*)-**61**	10	DCM	r.t.	1	100	100	100	10.6 (*S*)	62
4	H	Ph	NHAc	Rh-(*Sc*,*Sp*,*Ra*)-**62**	10	DCM	r.t.	1	100	100	100	99.6 (*S*)	62
5	H	Ph	NHAc	Rh-(*Sc*,*Sp*,*Sa*)-**63**	10	DCM	r.t.	1	100	100	100	82.6 (*R*)	62
6	H	Ph	NHAc	Rh-(*Sc*,*Rp*)-**65**	10	DCM	r.t.	1	100	100	100	81.5 (*S*)	62
7	H	Ph	NHAc	Rh-(*Sc*,*Rp*)-**66**	10	DCM	r.t.	1	100	100	100	78.1 (*R*)	62
8	H	Ph	NHAc	Rh-(*Sc*,*Rp*)-**73**	10	DCM	r.t.	1	100	100	100	61.8 (*R*)	62
9	H	Ph	NHAc	Rh-(*Rc*,*Sp*)-**67**	20.7	DCM	r.t.	10	100	10	100	70.0 (*S*)	63
10	H	Ph	NHAc	Rh-(*Rc*,*Sp*)-**68**	20.7	DCM	r.t.	8	100	12.5	100	80.6 (*S*)	63
11	H	Ph	NHAc	Rh-(*Rc*,*Sp*)-**80**	20.7	DCM	r.t.	16	100	6.25	100	94.6 (*S*)	63
12	H	Ph	NHAc	Rh-(*Rc*,*Sp*)-**81**	20.7	DCM	r.t.	10	100	10	100	35.0 (*S*)	63
13	H	Ph	NHAc	Rh-(*Rc*,*Sp*)-**82**	20.7	*i*-PrOH	r.t.	16	100	6.25	100	94.4 (*S*)	63
14	H	Ph	NHAc	Rh-(*Rc*,*Sp*)-**80**	20.7	Toluene	r.t.	16	100	6.25	100	93.5 (*S*)	63
15	H	Ph	NHAc	Rh-(*Rc*,*Sp*)-**80**	20.7	THF	r.t.	16	100	6.25	100	96.5 (*S*)	63
16	H	Ph	NHAc	Rh-(*Rc*,*Sp*)-**82**	20.7	THF	r.t.	8	100	12.5	100	96.2 (*S*)	63
17	H	Ph	NHAc	Rh-(*Rc*,*Sp*)-**80**	20.7	THF	r.t.	–	100	–	100	96.1 (*S*) [d]	63
18	H	Ph	NHAc	Rh-(*Rc*,*Sp*)-**80**	20.7	THF	r.t.	–	500	–	100	95.8 (*S*) [d]	63
19	H	Ph	NHAc	Rh-(*Rc*,*Sp*)-**80**	20.7	THF	r.t.	–	100	–	100	95.5 (*S*) [d]	63
20	H	Ph	NHAc	Rh-(*Rc*,*Sp*)-**80**	20.7	THF	r.t.	–	100	–	100	95.2 (*S*) [d]	63

21	H	Ph	NHAc	Rh-(*Rc*,*Sp*)-**80**	20.7	THF/H_2O 95/5	r.t.	–	100	–	100	95.1 (*S*) [d]	63
22	H	Ph	NHAc	Rh-(*Rc*,*Sp*)-**80**	20.7	THF/H_2O 70/30	r.t.	–	100	–	100	77.2 (*S*) [d]	63
23	H	Ph	NHAc	Rh-(*Rc*,*Sp*)-**80**	20.7	THF	r.t.	16	100	6.25	100	96.5 (*S*)	63
24	H	Ph	NHAc	Rh-(*Rc*,*Sp*)-**80**	20.7	THF	r.t.	16	200	12.5	100	95.8 (*S*)	63
25	H	Ph	NHAc	Rh-(*Rc*,*Sp*)-**80**	20.7	THF	r.t.	16	500	31.25	100	96.4 (*S*)	63
26	H	Ph	NHAc	Rh-(*Rc*,*Sp*)-**80**	20.7	THF	r.t.	16	1000	62.5	100	95.8 (*S*)	63
27	H	Ph	NHAc	Rh-(*Rc*,*Sp*)-**80**	20.7	THF	5	30	500	16.67	100	98.3 (*S*)	63
28	H	*p*-Cl-Ph	NHAc	Rh-(*Sc*,*Rp*,*Sa*)-**60**	10	DCM	r.t.	1	1000	1000	100	98.8 (*R*)	62
29	H	*p*-Br-Ph	NHAc	Rh-(*Sc*,*Rp*,*Sa*)-**60**	10	DCM	r.t.	1	1000	1000	100	99.0 (*R*)	62
30	H	*p*-Br-Ph	NHAc	Rh-(*Rc*,*Sp*)-**80**	20.7	THF	5	30	500	16.67	100	99.7 (*S*)	63
31	H	*p*-Br-Ph	NHAc	Rh-(*Rc*,*Sp*)-**80**	20.7	THF	r.t.	16	500	31.25	100	99.3 (*S*)	63
32	H	*p*-F-Ph	NHAc	Rh-(*Sc*,*Rp*,*Sa*)-**60**	10	DCM	r.t.	1	1000	1000	100	98.7 (*R*)	62
33	H	*p*-CF_3-Ph	NHAc	Rh-(*Rc*,*Sp*)-**67**	20.7	DCM	r.t.	10	100	10	100	73.1 (*S*)	63
34	H	*p*-CF_3-Ph	NHAc	Rh-(*Sc*,*Rp*,*Sa*)-**60**	10	DCM	r.t.	1	1000	1000	100	99.2 (*R*)	62
35	H	*p*-CF_3-Ph	NHAc	Rh-(*Rc*,*Sp*)-**68**	20.7	DCM	r.t.	8	100	12.5	100	79.6 (*S*)	63
36	H	*p*-CF_3-Ph	NHAc	Rh-(*Rc*,*Sp*)-**80**	20.7	THF	r.t.	16	500	31.25	100	97.1 (*S*)	63
37	H	*p*-CF_3-Ph	NHAc	Rh-(*Rc*,*Sp*)-**80**	20.7	THF	5	30	500	16.67	100	98.6 (*S*)	63
38	H	*m*-CH_3Ph	NHAc	Rh-(*Rc*,*Sp*)-**80**	20.7	THF	5	30	500	16.67	100	98.5 (*S*)	63
39	H	*p*-CH_3Ph	NHAc	Rh-(*Rc*,*Sp*)-**80**	20.7	THF	r.t.	16	500	31.25	100	92.1 (*S*)	63
40	H	*p*-CH_3Ph	NHAc	Rh-(*Rc*,*Sp*)-**80**	20.7	THF	5	30	500	16.67	100	99.4 (*S*)	63
41	H	*m*-CH_3OPh	NHAc	Rh-(*Rc*,*Sp*)-**80**	20.7	THF	5	30	500	16.67	100	99.0 (*S*)	63
42	H	*p*-CH_3OPh	NHAc	Rh-(*Rc*,*Sp*)-**80**	20.7	THF	r.t.	16	500	31.25	100	93.5 (*S*)	63
43	H	*p*-CH_3OPh	NHAc	Rh-(*Rc*,*Sp*)-**80**	20.7	THF	5	30	500	16.67	100	99.3 (*S*)	63
44	H	CO_2H	NHAc	Rh-(*Rc*,*Sp*)-**68**	0.7	THF	r.t.	1	95	95	>95	96 (*S*)	57
45	H	CO_2H	NHAc	Rh-(Sc,*Rp*)-**73**	0.69–1.38	THF	25	24	100	4.2	100	96.1 (*R*) [b]	58
46	H	CO_2Me	NHAc	Rh-(*Rc*,*Sp*)-**68**	0.7	THF	r.t.	1	95	95	>95	98.5 (*S*)	57
47	H	CO_2Me	NHAc	Rh-(Sc,*Rp*)-**73**	0.69–1.38	THF	25	1	40	40	100	98.4 (*R*)	58
48	H	CO_2Me	NHAc	Rh-(*Rc*,*Ra*)-**84**	30	DCM	r.t.	24	990	41.3	>99	97.8 (*S*)	62

Table 27.4 (continued)

Entry	Substrate			Catalyst	Conditions				TON	TOF [h^{-1}]	Conv. [%]	ee [%]	Reference(s)
	R^1	R^2	R^3		$P[H_2]$ [bar]	Solvent	Temp. [°C]	Time [h]					
49	H	CO_2Me	$NHCO_2Bn$	Rh-(*Rc*,*Sp*)-**68**	0.7	THF	r.t.	1	95	95	>95	98 (*S*)	57, 58
50	Me[c]	CO_2Me	NHCO (2-oxopyrrolidin-1-yl)	Rh-(*Rc*,*Sp*)-**68**	2.8	THF	25	18	20.9	1.2	99	96.2 (*S*)	58
51	Ph	CO_2H	NHAc	Rh- (*Rc*,*Sp*)-**68**	0.69–1.38	THF	25	1	100	100	100	99.4 (*S*)[b]	57, 58
52	Ph	CO_2Me	NHAc	Rh-(*Rc*,*Rp*,*Sa*)-**59**	20.7	THF	r.t.	7[a]	99	14.1	>99	99.0 (*S*)	61
53	Ph	CO_2Me	NHAc	Rh-(*Sc*,*Rp*,*Sa*)-**60**	10	DCM	r.t.	1	10000	10000	100	99.0 (*R*)	62
54	Ph	CO_2Me	NHAc	Rh-(*Rc*,*Sp*)-**67**	0.7	THF	r.t.	1	95	95	95	97.2 (*S*)	57, 58
55	Ph	CO_2Me	NHAc	Rh- (*Rc*,*Sp*)-**68**	0.7	THF	r.t.	1	95	95	>95	99.1 (*S*)	57, 58
56	Ph	CO_2Me	NHAc	Rh-(*Rc*,*Sp*)-**68**	20.7	THF	r.t.	–	200	–	100	99.0 (*S*)	63
57	Ph	CO_2Me	NHAc	Rh- (*Rc*,*Sp*)-**68**	3.1	THF	r.t.	1.2	9630	8025	96.3	96.8 (*S*)	57, 58
58	Ph	CO_2Me	NHAc	Rh-(*Rc*,*Sp*)-**69**	0.7	THF	r.t.	1	95	95	95	94.3 (*S*)	57, 58
59	Ph	CO_2Me	NHAc	Rh-(*Rc*,*Sp*)-**70**	0.7	THF	r.t.	1	95	95	>95	93.3 (*S*)	57
60	Ph	CO_2Me	NHAc	Rh-(*Sc*,*Rp*)-**75**	0.69–1.38	THF	25	1	100	100	100	93.3 (*R*)	58
61	Ph	CO_2Me	NHAc	Rh-(*Rc*,*Sp*)-**80**	20.7	THF	r.t.	–	200	–	100	99.2 (*S*)	63
62	Ph	CO_2Me	NHAc	Rh-(*Rc*,*Sp*)-**81**	20.7	THF	r.t.	–	200	–	100	96.1 (*S*)	63
63	Ph	CO_2Me	NHAc	Rh-(*Rc*,*Sp*)-**82**	20.7	THF	r.t.	–	200	–	100	98.5 (*S*)	63
64	Ph	CO_2Me	NHBz	Rh-(*Rc*,*Sp*)-**68**	0.69–1.38	THF	25	6	100	16.7	100	98.4 (*S*)	58
65	Ph	CO_2Me	$NHCO_2$*t*-Bu	Rh-(*Rc*,*Sp*)-**68**	0.7	THF	r.t.	1	95	95	>95	99.5 (*S*)	57, 58
66	Bn	CO_2Et	$NHCO_2Bn$	Rh-(*Rc*,*Sp*)-**68**	0.69–1.38	THF	25	6	99	16.5	99	98.4 (*S*)	58
67	*p*-Cl-Ph	CO_2Me	NHAc	Rh-(*Rc*,*Sp*,*Sa*)-**59**	20.7	THF	r.t.	7[a]	99	14.1	>99	99.0 (*S*)	61
68	*p*-Cl-Ph	CO_2Me	NHAc	Rh-(*Rc*,*Sp*)-**68**	0.69–1.38	THF	25	2	50	25	100	98.8 (*S*)	58
69	*p*-Br-Ph	CO_2Me	NHAc	Rh-(*Rc*,*Sp*,*Sa*)-**59**	20.7	THF	r.t.	7[a]	99	14.1	>99	99.0 (*S*)	61
70	*p*-F-Ph	CO_2Me	NHAc	Rh-(*Rc*,*Sp*,*Sa*)-**59**	20.7	THF	r.t.	7[a]	99	14.1	>99	99.0 (*S*)	61

71	*p*-NO_2-Ph	CO_2Me	NHAc	Rh-(*Rc*,*Sp*,*Sa*)-**59**	20.7	THF	r.t.	7[a]	99	14.1	>99	99.6 (*S*)	61
72	*p*-NO_2-Ph	CO_2Me	NHAc	Rh-(*Rc*,*Sp*)-**68**	0.69–1.38	THF	25	0.5	100	200	100	97.7 (*S*)	58
73	*p*-NO_2Ph	CO_2Me	NHAc	Rh-(*Rc*,*Sp*)-**68**	20.7	THF	r.t.	–	200	–	100	99.5 (*S*)	63
74	*p*-NO_2Ph	CO_2Me	NHAc	Rh-(*Rc*,*Sp*)-**80**	20.7	THF	r.t.	–	200	–	100	99.7 (*S*)	63
75	*p*-CN-Ph	CO_2Me	NHAc	Rh-(*Rc*,*Sp*)-**68**	0.69–1.38	THF	25	1	100	100	100	99.0 (*S*)	58
76	*o*-MeO-Ph	CO_2Me	NHAc	Rh-(*Rc*,*Sp*)-**68**	0.69–1.38	THF	25	2	49.8	24.9	99.5	97.7 (*S*)	58
77	*m*-MeO-Ph	CO_2Me	NHAc	Rh-(*Rc*,*Sp*)-**68**	0.69–1.38	THF	25	0.5	50	100	100	98.0 (*S*)	58
78	*p*-MeO-Ph	CO_2Me	NHAc	Rh-(*Rc*,*Sp*)-**68**	0.69–1.38	THF	25	0.5	45	90	90	97.9 (*S*)	58
79	*p*-Me-Ph	CO_2Me	NHAc	Rh-(*Rc*,*Sp*,*Sa*)-**59**	20.7	THF	r.t.	7[a]	99	14.1	>99	97.4 (*S*)	61
80	Cyclopropyl	CO_2Me	NHBz	Rh-(*Sc*,*Rp*)-**73**	0.69–1.38	THF	25	24	100	4.2	100	91.6 (*R*)	58
81	Cyclopropyl	CO_2Me	$NHCO_2t$-Bu	Rh-(*Rc*,*Sp*)-**68**	0.69–1.38	THF	25	6	90	15	90	98.6 (*S*)	58
82	Cyclopropyl	Bn	$NHCO_2t$-Bu	Rh-(*Rc*,*Sp*)-**68**	0.69–1.38	Acetone	25	1	94	94	94	>99 (*S*)	57, 58
83	1-Naphthyl	CO_2Me	NHAc	Rh-(*Rc*,*Sp*)-**68**	0.69–1.38	THF	25	6	95	15.8	>95	99.3 (*S*)	58
84	1-Naphthyl	CO_2Me	$NHCO_2t$-Bu	Rh-(*Rc*,*Sp*)-**68**	0.69–1.38	THF	25	6	95	15.8	>95	98.2 (*S*)	58
85	2-Naphthyl	CO_2Me	NHAc	Rh-(*Rc*,*Sp*)-**68**	0.69–1.38	THF	25	1	95	95	>95	98.1 (*S*)	58
86	2-Naphthyl	CO_2Me	$NHCO_2t$-Bu	Rh-(*Rc*,*Sp*)-**68**	0.69–1.38	THF	25	6	97	16.2	97	97.4 (*S*)	58
87	3-furyl	CO_2Me	NHCOPh	Rh-(*Sc*,*Rp*)-**73**	0.69–1.38	THF	25	6	100	16.7	100	96.6 (*R*)	58
88	3-furyl	CO_2Me	$NHCO_2t$-Bu	Rh-(*Rc*,*Sp*)-**68**	0.69–1.38	THF	25	6	98	16.3	98	97.2 (*S*)	58
89	H	CO_2H	CH_2CO_2H	Rh-(*Rc*,*Sp*)-**67**	20.7	MeOH	r.t.	6	95	15.8	>95	94.0 (*R*)	57
90	H	CO_2H	CH_2CO_2H	Rh-(*Rc*,*Sp*)-**68**	20.7	MeOH	r.t.	6	95	15.8	>95	97.4 (*R*)	57
91	H	CO_2Me	CH_2CO_2Me	Rh-(*Sc*,*Rp*,*Sa*)-**60**	10	DCM	r.t.	0.5	10000	20000	100	99.1 (*S*)	62
92	H	CO_2Me	CH_2CO_2Me	Rh-(*Rc*,*Sp*)-**67**	20.7	MeOH	r.t.	6	95	15.8	>95	91.6 (*R*)	57, 58
93	H	CO_2Me	CH_2CO_2Me	Rh-(*Rc*,*Sp*)-**68**	10	MeOH	r.t.	6	95	15.8	>95	94.0 (*R*)	57, 58
94	H	CO_2Me	CH_2CO_2Me	Rh-(*Sc*,*Ra*)-**83**	30	DCM	r.t.	24	990	41.3	>99	78.8 (*R*)	66
95	H	CO_2Me	CH_2CO_2Me	Rh-(*Rc*,*Ra*)-**84**	30	DCM	r.t.	24	990	41.3	>99	98.8 (*R*)	66
96	Me	CO_2H	OPh	Ru-(*Sc*,*Rp*)-**73**	6.2	MeOH/ EtOH/DCM 80/13/7	20–25	20	17.4	0.9	100	rac	59
97	Ph	CO_2H	CH_2CO_2H	Rh-(*Rc*,*Sp*)-**67**	20.7	MeOH	r.t.	6	95	15.8	>95	99.0 (*R*)	57
98	Ph	CO_2H	CH_2CO_2H	Rh-(*Rc*,*Sp*)-**68**	20.7	MeOH	r.t.	6	95	15.8	>95	89.0 (*R*)	57
99	Ph	CO_2Me	CH_2CO_2Me	Rh-(*Rc*,*Sp*)-**67**	20.7	MeOH	r.t.	6	95	15.8	>95	80.0 (*R*)	57

a) Average value.
b) The ee-value was determined by the corresponding methyl ester.
c) The ratio of *Z*/*E* is not provided.
d) The catalyst was prepared *in situ* in air.

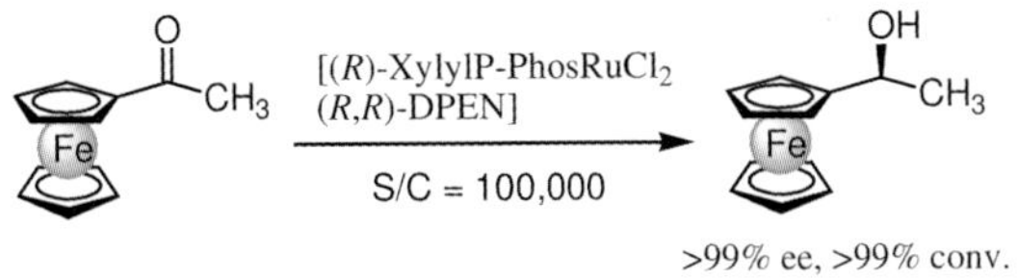

$[(R)\text{-XylylP-PhosRuCl}_2$ $(R,R)\text{-DPEN}]$

S/C = 100,000

>99% ee, >99% conv.

Scheme 27.2

27.5
Bisphosphinite Ligands (One P–O Bond)

A large number of bidentate phosphinites have been reported, with sugars being the most abundantly used backbone. *trans*-BDPCH **85** (Fig. 27.6) is the earliest example of a bisphosphinite used in the rhodium-catalyzed asymmetric hydrogenation of functionalized olefins inducing moderate ee-values (48.5–78.9%) [67]. A similar approach using a more rigid pentacyclic system as backbone (*trans*-BDPCP **86**) induced only poor to moderate ee-values. The best ee-value (78.9%) was obtained in the enantiomeric hydrogenation of *α*-acetamidoacrylic acid [68]. In 2000, Leitner developed a perfluorinated analogue **87** which induced 72% ee in the Rh-catalyzed hydrogenation of dimethyl methylsuccinate (Table 27.5, entry 934) in a supercritical CO_2 ($scCO_2$) and perfluorinated alcohol solvent [69]. An average TOF up to 40 000 h^{-1} was obtained with this system.

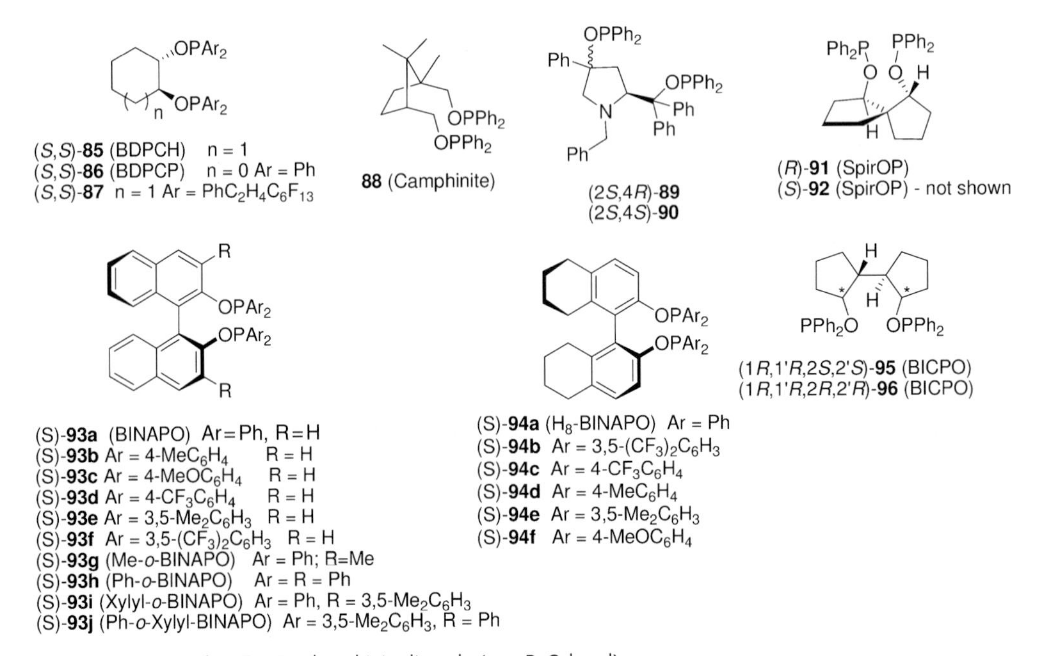

Fig. 27.6 Bisphosphinite ligands (one P–O bond).

The binaphthyl system has served as the basis of a several classes of ligands. It has been suggested that the highly skewed position of the naphthyl rings in BINAP is the determining factor in its effectiveness in asymmetric catalytic reactions [70]. In an early study by Grubbs, the use of atropisomeric BINAPO **93a** based on the binaphthol skeleton induced 6 to 76% ee in the Rh-catalyzed hydrogenation of α-dehydroamino acids and enamides [71]. Interestingly, we found that the partially hydrogenated H_8-BINAPO was more effective than BINAPO in the Rh-catalyzed hydrogenation of (*Z*)-acetamido-3-arylacrylic acids and their methyl esters (63.9–84% ee) [52]. In fact, recent research showed that chiral catalysts derived from the 5,5′,6,6′,7,7′,8,8′-octahydro-1,1′-bi-2,2′-naphthyl backbone (e.g., H_8-BINAP [72], H_8-BINOL [73, 74], H_8-BINAM [9], H_8-BDPAB [51], H_8-binaphthoxy [75], H_8-MAPs [76]) exhibited higher efficiency and enantioselectivity in asymmetric catalytic reactions than those prepared from the parent binaphthyl backbone, probably due to the steric and electronic modulation in the H_8-binaphthyl backbone [77]. A systematic quantification of the electronic and steric influences of these ligands were carried out by Bakos and Gergely [78]. A detrimental effect of *para*-electron-withdrawing substituents on the phenyl rings of this class of ligands was observed on enantioselectivity and activity in the hydrogenation of dimethyl itaconate (**94b** with 51.6% ee, Table 27.5, entry 939). In contrast, *para*-electron-donating groups (i.e., *p*-OMe group) enhanced the enantioselectivity (93.9% ee, Table 27.5, entry 944). Similarly, the use of **94f** (i.e., *p*-OMe group) in the hydrogenation of methyl (*Z*)-α-acetamido-cinnamate gave 98.6% ee. The 3,3′-disubstituted bisphosphinite ligand *o*-BINAPO (**93a, b**) reported by Zhang was successfully applied in the hydrogenation of enamides (67.2–96.3% ee) and α-dehydroamino acid derivatives (81.5–99.9% ee) [79]. The further demonstration of its application in the hydrogenation of β-aryl-substituted β-(acylamino)acrylates was also successful, leading to formation of the products with 80 to 99% ee (with **93i**) [80].

Chiral ligands **88** [81], **89** and **90** [82] with rigid backbones were found to be less effective in Rh-catalyzed hydrogenation reactions. In 1997, we introduced the novel ligand SpirOP (**91** and **92**) based on a rigid spiro backbone which mimics the binaphthyl rings in BINAP in its most effective state (skewed position), giving rise to an eight-membered chelate ring [27, 83]. Indeed, the desired hydrogenation product 2-acetamidopropionic acid was obtained in >99.9% ee, with complete conversion in 10 min using Rh–SpirOP. Similarly remarkable activity and enantioselectivity was found upon hydrogenation of the corresponding methyl ester using the same catalyst (99% ee, 99.9% conv.). The TOF of the hydrogenation of 2-acetamido-acrylic acid could be further increased to 10 000 h^{-1} at ambient temperature whilst retaining 96.8% ee. The substrate's scope was also excellent (>97% ee for (*Z*)-2-acetamido-3-arylacrylic acids and 94.2–97.2% ee for the corresponding methyl esters). It is of interest to note that the Rh–SpirOP complexes in methanol showed unexpected stability based on a ^{31}P-NMR study at ambient temperature for two days. It was further demonstrated that the use of SpirOP in the hydrogenation of α-phenylenamide gave rise to good to excellent enantioselectivities (85.6–97.4% ee) [84].

Table 27.5 Enantiomeric hydrogenation using bisphosphinite, bisphosphonite, or bisphosphite.

Entry	Substrate			Catalyst	Conditions				TON	TOF [h^{-1}]	Conv. [%]	ee [%]	Reference(s)
	R^1	R^2	R^3		P[H_2] [bar]	Solvent	Temp. [°C]	Time [h]					
1	H	CO_2H	NHAc	[Rh(1,5-hexadiene) (+)-*trans*-**85**]Cl	50	–	–20	24	–	–	–	78.9 (S)	67
2	H	CO_2H	NHAc	[Rh(1,5-hexadiene) (+)-*trans*-**86**]Cl	50	–	–20	24	–	–	–	0	68 a
3	H	CO_2H	NHAc	[Rh(COD)**91**]BF_4	1	MeOH	25	0.167	100	600	>99.9	>99.9 (*R*)	27
4	H	CO_2H	NHAc	[Rh(COD)**91**]BF_4	13.8	MeOH	25	1	10000	10000	>99.9	96.8 (*R*)	27
5	H	CO_2H	NHAc	[Rh(COD)**95**]BF_4	1	IPA	r.t.	24	100	4	100	94.8 (*S*)	85
6	H	CO_2H	NHAc	[Rh(COD)**97a**]BF_4	1	MeOH	25	0.0317 [f]	50	1579	50	72.6	26
7	H	CO_2H	NHAc	[Rh(COD)**97a**]BF_4	1	MeOH	25	0.0167 [f]	50	3000	50	97.7 (*S*)	91 b
8	H	CO_2H	NHAc	[Rh(COD)**97c**]BF_4	1	MeOH	25	0.0167 [f]	50	3000	50	96.5 (*S*)	91 b
9	H	CO_2H	NHAc	[Rh(COD)**98a**]BF_4	1	MeOH	25	0.047 [f]	50	1071	50	97.7 (*S*)	26
10	H	CO_2H	NHAc	[Rh(COD)**98a**]BF_4	1	MeOH	25	0.25	50	200	50	97.7 (*S*)	26
11	H	CO_2H	NHAc	[Rh(COD)**98a**]BF_4	100	MeOH	25	–	100 [c]	–	–	93.9 (*S*)	26
12	H	CO_2H	NHAc	[Rh(COD)**98b**]BF_4	2.8	THF	r.t.	3	100	33	100	97 (*S*)	98
13	H	CO_2H	NHAc	[Rh(COD)**98b**]SbF_6	2–2.8	THF	r.t.	2–3	–	–	–	96.9	95 b
14	H	CO_2H	NHAc	[Rh(COE)**93a**]Cl	102	Tol	25	24	25	1	50 [b]	9 [a]	71
15	H	CO_2H	NHAc	[Rh(COE)**93a**]Cl	95	Tol/acetone 1:1	0	24	50	2	100 [b]	6 [a]	71
16	H	CO_2H	NHAc	[Rh(NBD)**97a**]PF_6	1	EtOH	30	0.33	40	120	100 [b]	67 (*S*)	88
17	H	CO_2H	NHAc	[Rh(NBD)**97a**]PF_6	1	EtOH	30	0.33	100	300	100 [b]	68 (*S*)	88
18	H	CO_2H	NHAc	[Rh(NBD)**97a**]PF_6	1	EtOH	0	0.5	100	200	100 [b]	74 (*S*)	88
19	H	CO_2H	NHAc	[Rh(NBD)**97a**]PF_6	1	EtOH	–20	1	100	100	100 [b]	80 (*S*)	88

20	H	CO_2H	NHAc	[Rh(COD)**102a**]BF_4	2–2.8	THF	r.t.	2–3	–	–	–	95.0 (*S*)	95b
21	H	CO_2H	NHAc	[Rh(COD)**103a**]SbF_6	2–2.8	THF	r.t.	2–3	–	–	–	90.8 (*R*)	95b
22	H	CO_2H	NHAc	[Rh(COD)**114a**]BF_4	1	MeOH	25	0.017 [f]	50	2941	50	59 (*S*)	99
23	H	CO_2H	NHAc	[Rh(COD)**114a**]BF_4	1	PhH	25	6.67 [f]	50	8	50	48 (*S*)	99
24	H	CO_2H	NHAc	[Rh(COD)**114a**]BF_4	1	H_2O	25	7.58 [f]	50	7	50	14 (*S*)	99
25	H	CO_2H	NHAc	[Rh(COD)**114b**]BF_4	1	MeOH	25	0.017 [f]	50	2941	50	56 (*S*)	99
26	H	CO_2H	NHAc	[Rh(COD)**114b**]BF_4	1	PhH	25	8.08 [f]	50	6	50	71 (*S*)	99
27	H	CO_2H	NHAc	[Rh(COD)**114b**]BF_4	1	H_2O+Triton X-100 (0.1 mmol)	25	0.01 [f]	50	500	50	42 (*S*)	99
28	H	CO_2H	NHAc	[Rh(COD)**114b**]BF_4	1	H_2O+Triton X-100 (0.5 mmol)	25	0.05 [f]	50	1000	50	42 (*S*)	99
29	H	CO_2H	NHAc	[Rh(COD)**115a**]SbF_6	2.8	THF	r.t.	3	100	33	100	86 (*S*)	98
30	H	CO_2H	NHAc	[Rh(COD)**115a**]SbF_6	2.8	H_2O	r.t.	19	100	5	100	14 (*S*)	98
31	H	CO_2H	NHAc	[Rh(COD)**115b**]SbF_6	2.8	THF	r.t.	3	100	33	100	90 (*S*)	98
32	H	CO_2H	NHAc	[Rh(COD)**115c**]BF_4	2.8	THF	r.t.	–	150	–	100	87 (*S*)	98
33	H	CO_2H	NHAc	[Rh(COD)**115c**]BF_4	2.8	MeOH	r.t.	–	150	–	100	54 (*S*)	98
34	H	CO_2H	NHAc	[Rh(COD)**115c**]BF_4	2.8	H_2O	r.t.	–	150	–	100	53 (*S*)	98
35	H	CO_2H	NHAc	[Rh(COD)**115c**]BF_4	2.8	H_2O/EtOAc (1:1)	r.t.	–	100	–	100	6 (*S*)	98
36	H	CO_2H	NHAc	[Rh(COD)**115d**]BF_4	2.8	THF	r.t.	–	100	–	100	93 (*S*)	98
37	H	CO_2H	NHAc	[Rh(COD)**115d**]BF_4	2.8	MeOH	r.t.	–	100	–	100	37 (*S*)	98
38	H	CO_2H	NHAc	[Rh(COD)**115d**]BF_4	2.8	EtOH	r.t.	–	100	–	100	89 (*S*)	98
39	H	CO_2H	NHAc	[Rh(COD)**115d**]BF_4	2.8	H_2O	r.t.	–	100	–	100	2 (*S*)	98
40	H	CO_2H	NHAc	[Rh(COD)**115d**]BF_4	2.8	H_2O/EtOAc (1:1)	r.t.	–	100	–	100	2 (*S*)	98
41	H	CO_2H	NHAc	[Rh(COD)**115e**]BF_4	2.8	THF	r.t.	–	150	–	100	0	98
42	H	CO_2H	NHAc	[Rh(COD)**119d**]SbF_6	2.8	H_2O	r.t.	2	125	63	100	59 (*S*)	102
43	H	CO_2H	NHAc	[Rh(COD)**119e**]SbF_6	2.8	H_2O/EtOAc	r.t.	20	125	6	100	65 (*S*) [a]	102
44	H	CO_2H	NHAc	[Rh(COD)**128**]Cl	10	MeOH	r.t.	0.25	230	920	100	26 (*S*)	106
45	H	CO_2H	NHAc	[Rh(COD)**129**]Cl	56	MeOH	r.t.	1	260	260	100	14 (*R*)	106
46	H	CO_2H	NHAc	[Rh(COD)**132**]BF_4	34.5	Acetone	25	0.25	100	400	100	96.7 (*R*)	107
47	H	CO_2H	NHAc	[Rh(COD)**141b**]BF_4	3.5	MeOH	r.t.	0.5	1000	2000	>99	97 (*S*)	111
48	H	CO_2H	NHAc	[Rh(COD)**166**]BF_4	51	H_2O	r.t.	24	50	2	100	10 (*R*) [j]	129

Table 27.5 (continued)

Entry	Substrate			Catalyst	Conditions				TON	TOF $[h^{-1}]$	Conv. [%]	ee [%]	Reference(s)
	R^1	R^2	R^3		$P[H_2]$ [bar]	Solvent	Temp. [°C]	Time [h]					
49	H	CO_2Me	NHAc	[Rh(COD)**91**]BF_4	1	MeOH	25	0.167	100	600	>99.9	99.0 (*R*)	27
50	H	CO_2Me	NHAc	[Rh(COE)**93a**]Cl	99	Tol	0	53.5	30	0.6	60[b)]	44[a)]	71
51	H	CO_2Me	NHAc	[Rh(COE)**93a**]Cl	91	Tol/acetone 1:1	0	68.5	50	0.7	100[b)]	76[a)]	71
52	H	CO_2Me	NHAc	[Rh(COD)**93a**]PF_6	3	Tol	r.t.	12	100	8.3	100	73.2 (*S*)	79
53	H	CO_2Me	NHAc	[Rh(COD)**93g**]PF_6	3	Tol	r.t.	12	100	8.3	100	94.8 (*S*)	79
54	H	CO_2Me	NHAc	[Rh(COD)**93h**]PF_6	3	Tol	r.t.	12	100	8.3	100	99.9 (*S*)	79
55	H	CO_2Me	NHAc	[Rh(COD)**93i**]PF_6	3	Tol	r.t.	12	100	8.3	100	95.4 (*S*)	79
56	H	CO_2Me	NHAc	[Rh(COD)**93j**]PF_6	3	Tol	r.t.	12	100	8.3	100	93 (*S*)	79
57	H	CO_2Me	NHAc	[Rh(COD)**94a**]BF_4	6.9	DCM	r.t.	0.17	245	1441	49	81 (*S*)	52
58	H	CO_2Et	NHAc	[Rh(COD)**97a**]BF_4	1	MeOH	25	0.085[f)]	50	588	50	58.2	26
59	H	CO_2Me	NHAc	[Rh(COD)**97a**]BF_4	1	MeOH	25	0.0267[f)]	50	1875	50	73.4	26
60	H	CO_2Me	NHAc	[Rh(COD)**97a**]BF_4	1	MeOH	25	0.0167[f)]	50	3000	50	90.9 (*S*)	91b
61	H	CO_2Me	NHAc	[Rh(NBD)**97a**]PF_6	1	EtOH	30	0.167	100	600	100[b)]	53 (*S*)	88
62	H	CO_2Me	NHAc	[Rh(NBD)**97a**]PF_6	1	EtOH	0	0.5	100	200	100[b)]	78 (*S*)	88
63	H	CO_2Me	NHAc	[Rh(COD)**97b**]BF_4	1	H_2O	25	0.33	50	150	50	44 (S)	99
64	H	CO_2Me	NHAc	[Rh(COD)**97b**]BF_4	1	H_2O + 0.1 mmol $LiBF_4$	25	0.52	50	97	50	43 (S)	99
65	H	CO_2Me	NHAc	[Rh(COD)**97b**]BF_4	1	H_2O + 0.1 mmol $NaBF_4$	25	0.55	50	91	50	40 (S)	99
66	H	CO_2Me	NHAc	[Rh(COD)**97b**]BF_4	1	H_2O + 0.1 mmol KBF_4	25	0.53	50	94	50	42 (S)	99
67	H	CO_2Me	NHAc	[Rh(COD)**97b**]BF_4	1	H_2O + 0.1 mmol $RbBF_4$	25	0.62	50	81	50	42 (S)	99

68	H	CO_2Me	NHAc	[Rh(COD)**97b**]BF_4	1	H_2O + 0.1 mmol $CsBF_4$	25	0.62	50	81	50	41 (S)	99
69	H	CO_2Me	NHAc	[Rh(COD)**97b**]BF_4	1	H_2O + Triton X100 (0.1 mmol)	25	0.067	50	750	50	69 (S)	99
70	H	CO_2Me	NHAc	[Rh(COD)**97b**]BF_4	1	H_2O + Triton X100 + 0.1 mmol $LiBF_4$	25	0.12	50	429	50	68 (S)	99
71	H	CO_2Me	NHAc	[Rh(COD)**97b**]BF_4	1	H_2O + Triton X100 + 0.1 mmol $NaBF_4$	25	0.1	50	500	50	68 (S)	99
72	H	CO_2Me	NHAc	[Rh(COD)**97b**]BF_4	1	H_2O + Triton X100 + 0.1 mmol KBF_4	25	0.1	50	500	50	68 (S)	99
73	H	CO_2Me	NHAc	[Rh(COD)**97b**]BF_4	1	H_2O+ Triton X100 + 0.1 mmol $RbBF_4$	25	0.15	50	333	50	68 (S)	99
74	H	CO_2Me	NHAc	[Rh(COD)**97b**]BF_4	1	H_2O + Triton X100 + 0.1 mmol $CsBF_4$	25	0.13	50	375	50	68 (S)	99
75	H	CO_2Me	NHAc	[Rh(COD)**97c**]BF_4	1	MeOH	25	0.0333 [f]	50	1500	50	95.2 (*S*)	91b
76	H	CO_2Me	NHAc	[Rh(COD)**98a**]BF_4	1	MeOH	25	0.0217 [f]	50	2308	50	90.9	26
77	H	CO_2Et	NHAc	[Rh(COD)**98a**]BF_4	1	MeOH	25	0.0233 [f]	50	2143	50	83.0	26
78	H	CO_2Me	NHAc	[Rh(COD)**98a**]BF_4	1	MeOH	25	0.016 [f]	50	3333	50	90.6 (*S*)	91e
79	H	CO_2Me	NHAc	[Rh(COD)**98i**]BF_4	1	MeOH	25	0.0333 [f]	50	1500	50	95 (*S*)	97
80	H	CO_2Me	NHAc	[Rh(COD)**98i**]BF_4	1	H_2O	25	0.467 [f]	50	107	50	79 (*S*)	97
81	H	CO_2Me	NHAc	[Rh(COD)**98i**]BF_4	1	H_2O+SDS, 0.035 [g]	25	0.133 [f]	50	375	50	93 (*S*)	97
82	H	CO_2Me	NHAc	[Rh(COD)**98i**]BF_4	1	H_2O+SDS, 0.173 [g]	25	0.0417 [f]	50	1200	50	97 (*S*)	97
83	H	CO_2Me	NHAc	[Rh(COD)**114a**]BF_4	1	PhH	25	0.017 [f]	50	2941	50	41 (*S*)	99
84	H	CO_2Me	NHAc	[Rh(COD)**114a**]BF_4	1	H_2O	25	1.22 [f]	50	41	50	34 (*S*)	99
85	H	CO_2Me	NHAc	[Rh(COD)**114a**]BF_4	1	MeOH	25	0.017 [f]	50	2941	50	75 (*S*)	99
86	H	CO_2Me	NHAc	[Rh(COD)**114b**]BF_4	1	MeOH	25	0.017 [f]	50	2941	50	71 (*S*)	99
87	H	CO_2Me	NHAc	[Rh(COD)**114b**]BF_4	1	PhH	25	0.017 [f]	50	2941	50	36 (*S*)	99
88	H	CO_2Me	NHAc	[Rh(COD)**114b**]BF_4	1	H_2O	25	0.33 [f]	50	152	50	44 (*S*)	99
89	H	CO_2Me	NHAc	[Rh(COD)**114b**]BF_4	1	H_2O+Triton X-100 (0.1 mmol)	25	0.067 [f]	50	746	50	69 (*S*)	99

Table 27.5 (continued)

Entry	Substrate			Catalyst	Conditions				TON	TOF [h^{-1}]	Conv. [%]	ee [%]	Reference(s)
	R^1	R^2	R^3		P[H_2] [bar]	Solvent	Temp. [°C]	Time [h]					
90	H	CO_2Me	NHAc	[Rh(COD)**114b**]BF_4	1	H_2O + Triton X-100 (0.5 mmol)	25	0.05 [f]	50	1000	50	70 (*S*)	99
91	H	CO_2Me	NHAc	[Rh(COD)**115d**]BF_4	2.8	THF	r.t.	3	100	33	100	93 (*S*)	98
92	H	CO_2Me	NHAc	[Rh(COD)**115d**]BF_4	2.8	EtOH	r.t.	1	150	150	100	89 (*S*)	98
93	H	CO_2Me	NHAc	[Rh(COD)**115d**]BF_4	2.8	H_2O/THF (3:1)	r.t.	2	100	50	100	87 (*S*)	98
94	H	CO_2Me	NHAc	[Rh(COD)**115d**]BF_4	2.8	MeOH	r.t.	3	100	33	100	37 (*S*)	98
95	H	CO_2Me	NHAc	[Rh(COD)**115d**]BF_4	2.8	MeOH/H_2O (1:1)	r.t.	7	100	14	100	90 (*S*)	98
96	H	CO_2Me	NHAc	[Rh(COD)**115d**]BF_4	2.8	MeOH/H_2O (1:3)	r.t.	3	100	33	100	74 (*S*)	98
97	H	CO_2Me	NHAc	[Rh(COD)**115d**]BF_4	2.8	MeOH/H_2O (1:20)	r.t.	3	150	50	100	57 (*S*)	98
98	H	CO_2Me	NHAc	[Rh(COD)**115d**]BF_4	2.8	H_2O	r.t.	21	100	5	100	2 (*S*)	98
99	H	CO_2Me	NHAc	[Rh(COD)**115d**]BF_4	2.8	H_2O	r.t.	12	52.5	4	35	58 (*S*)	98
100	H	CO_2Me	NHAc	[Rh(COD)**115d**]BF_4	2.8	EtOH/H_2O (1:1)	r.t.	1	150	150	100	85 (*S*)	98
101	H	CO_2Me	NHAc	[Rh(COD)**115g**]SbF_6	2.8	H_2O	r.t.	1	150	150	100	61 (*S*)	98
102	H	CO_2Me	NHAc	[Rh(COD)**115g**]SbF_6	2.8	H_2O	r.t.	3	100	33	97	65 (*S*)	98
103	H	CO_2Me	NHAc	[Rh(COD)**115g**]SbF_6	2.8	THF	r.t.	1	150	150	100	86 (*S*)	98
104	H	CO_2Me	NHAc	[Rh(COD)**117**]BF_4	5	H_2O	r.t.	1.5	50	33	100	80 (*S*)	101
105	H	CO_2Me	NHAc	[Rh(COD)**119a**]SbF_6	2.07	THF	r.t.	–	–	–	~100	65 (*S*)	102
106	H	CO_2Me	NHAc	[Rh(COD)**119b**]SbF_6	2.07	THF	r.t.	–	–	–	~100	83 (*S*)	102
107	H	CO_2Me	NHAc	[Rh(COD)**119d**]SbF_6	2.8	H_2O	r.t.	2	125	63	100	55 (*S*)	102
108	H	CO_2Me	NHAc	[Rh(COD)**119e**]SbF_6	2.8	H_2O	r.t.	2	125	63	100	49 (*S*)	102
109	H	CO_2Me	NHAc	[Rh(COD)ent-**120**]BF_4	1	Acetone	r.t.	–	100	–	100	18 (*R*)	105
110	H	CO_2Me	NHAc	[Rh(COD)**121a**]BF_4	1	Acetone	r.t.	–	100	–	100	5 (*R*)	105
111	H	CO_2Me	NHAc	[Rh(COD)**121b**]BF_4	1	Acetone	r.t.	–	100	–	100	59 (*R*)	105

112	H	CO_2Me	NHAc	[Rh(COD)**121d**]BF_4	1	Acetone	r.t.	–	100	–	100	26 (*R*)	105
113	H	CO_2Me	NHAc	[Rh(COD)**122a**]BF_4	1	Acetone	r.t.	0.25	500	2000	100	76 (*R*)	105
114	H	CO_2Me	NHAc	[Rh(COD)**122b**]BF_4	1	Acetone/DCM 13:2	r.t.	–	100	–	100	85 (*R*)	105
115	H	CO_2Me	NHAc	[Rh(COD)**122b**]BF_4	1	Acetone/DCM 13:2	–25	0.33	100	303	100	91 (*R*)	105
116	H	CO_2Me	NHAc	[Rh(COD)**122c**]BF_4	1	Acetone	r.t.	–	100	–	100	80 (*R*)	105
117	H	CO_2Me	NHAc	[Rh(COD)**122d**]BF_4	1	Acetone/DCM 13:2	r.t.	–	100	–	100	78 (*R*)	105
118	H	CO_2Me	NHAc	[Rh(COD)**122e**]BF_4	1	Acetone	r.t.	–	100	–	100	87 (*R*)	105
119	H	CO_2Me	NHAc	[Rh(COD)**122e**]BF_4	1	Acetone	–25	0.5	100	200	100	93 (*R*)	105
120	H	CO_2Me	NHAc	[Ir(COD)**126**]BF_4	1	DCM	25	0.42	100	238	100	78 (*R*)	104
121	H	CO_2Me	NHAc	[Rh(COD)**126**]BF_4	1	DCM	25	0.25	100	400	100	8 (*R*)	104
122	H	CO_2Me	NHAc	[Ir(COD)**127**]BF_4	1	DCM	25	0.75	76	101	76	15 (*R*)	104
123	H	CO_2Me	NHAc	[Rh(COD)**127**]BF_4	1	DCM	25	0.02	100	5000	100	76 (*R*)	104
124	H	CO_2Me	NHAc	[Rh(COD)**127**]BF_4	1	DCM	25	0.75	96	128	96	81 (*R*)	104
125	H	CO_2Me	NHAc	[Rh(COD)**133a**]BF_4	1	IPA	r.t.	24	72	3	72	41	48
126	H	CO_2Me	NHAc	[Rh(COD)**133b**]BF_4	1	IPA	r.t.	24	77	2	77	46	48
127	H	CO_2Me	NHAc	[Rh(COD)**133c**]BF_4	1	IPA	r.t.	24	80	3	80	48	48
128	H	CO_2Me	NHAc	[Rh(COD)**136**]BF_4	1.3	DCM	r.t.	20	1000	50	100	90 (*R*)	109
129	H	CO_2Me	NHAc	[Rh(COD)**136**]BF_4	1.5	DCM	25	3	479	160	100	90 (*R*)	110
130	H	CO_2Me	NHAc	[Rh(COD)**138**]BF_4	1.3	DCM	r.t.	20	1000	50	100	99.5 (*R*)	109
131	H	CO_2Me	NHAc	[Rh(COD)**140**]BF_4	1.5	DCM	25	3	474	158	99	23 (*R*)	110
132	H	CO_2Me	NHAc	[Rh(COD)**141a**]BF_4	3.5	MeOH	r.t.	0.5	1000	2000	>99	96 (*S*)	111
133	H	CO_2Me	NHAc	[Rh(COD)**141a**]BF_4	3.5	MeOH/H_2O 9:1	r.t.	3	1000	333	>99	96 (*S*)	111
134	H	CO_2Me	NHAc	[Rh(COD)**141a**]BF_4	3.5	DCM	r.t.	1	1000	1000	>99	98 (*S*)	111
135	H	CO_2Me	NHAc	[Rh(COD)**141a**]BF_4	3.5	Tol	r.t.	0.5	1000	2000	>99	99 (*S*)	111
136	H	CO_2Me	NHAc	[Rh(COD)**141b**]BF_4	3.5	MeOH	r.t.	0.5	1000	2000	>99	99 (*S*)	111
137	H	CO_2Me	NHAc	[Rh(COD)**141c**]BF_4	3.5	MeOH	r.t.	1	50	50	5	–	111
138	H	CO_2Me	NHAc	[Rh(COD)**141c**]BF_4	3.5	MeOH	r.t.	16	980	61	98	74 (*S*)	111
139	H	CO_2Me	NHAc	[Rh(COD)**141d**]BF_4	3.5	MeOH	r.t.	21	250	12	25	46 (*S*)	111
140	H	CO_2Me	NHAc	[Ir(COD)**142a**]BF_4	5	DCM : MeOH 2 : 1	40	20	20	1	20	19 (*S*)	117d

Table 27.5 (continued)

Entry	Substrate			Catalyst	Conditions				TON	TOF [h^{-1}]	Conv. [%]	ee [%]	Reference(s)
	R^1	R^2	R^3		$P[H_2]$ [bar]	Solvent	Temp. [°C]	Time [h]					
141	H	CO_2Me	NHAc	[Ir(COD)**142a**]BF_4	5	DCM : MeOH 2 : 1	40	20	9	0.45	9	7 (*S*)	117d
142	H	CO_2Me	NHAc	[Ir(COD)**142a**]BF_4	1	DCM : MeOH 2 : 1	40	20	31	1.6	31	35 (*S*)	117d
143	H	CO_2Me	NHAc	[Rh(COD)**142a**]BF_4	5	Tol : MeOH 2 : 1	40	20	94	4.7	94	33 (*S*)	117d
144	H	CO_2Me	NHAc	[Ir(COD)**142b**]BF_4	5	DCM : MeOH 2 : 1	40	20	10	0.5	10	6 (*S*)	117d
145	H	CO_2Me	NHAc	[Ir(COD)**142b**]BF_4	1	DCM : MeOH 2 : 1	40	20	22	1.1	22	24 (*S*)	117d
146	H	CO_2Me	NHAc	[Rh(COD)**142b**]BF_4	5	Tol : MeOH 2 : 1	40	20	99	5	99	35 (*S*)	117d
147	H	CO_2Me	NHAc	[Rh(COD)**142b**]BF_4	2	Tol : MeOH 2 : 1	40	20	100	5	100	21 (*S*)	117d
148	H	CO_2Me	NHAc	[Rh(COD)**142b**]BF_4	5	DCM	40	20	100	5	100	10 (*S*)	117d
149	H	CO_2Me	NHAc	[Ir(COD)**143a**]BF_4	1	DCM : MeOH 2 : 1	40	20	30	1.5	30	37 (*R*)	117c
150	H	CO_2Me	NHAc	[Rh(COD)**143a**]BF_4	5	Tol : MeOH 2 : 1	40	20	56	2.8	56	4 (*R*)	117c
151	H	CO_2Me	NHAc	[Ir(COD)**143b**]BF_4	1	DCM : MeOH 2 : 1	40	20	24	1.2	24	28 (*R*)	117c
152	H	CO_2Me	NHAc	[Rh(COD)**143b**]BF_4	5	Tol : MeOH 2 : 1	40	20	88	4.4	88	6 (*R*)	117c
153	H	CO_2Me	NHAc	[Rh(COD)**144a**]BF_4	5	DCM	25	8	98	12.3	98	92 (*S*)	117b
154	H	CO_2Me	NHAc	[Rh(COD)**144c**]BF_4	5	DCM	25	6	100	16.7	100	97 (*S*)	117b
155	H	CO_2Me	NHAc	[Rh(COD)**144c**]BF_4	30	DCM	5	4	1000	250	100	>99 (*S*)	117b
156	H	CO_2Me	NHAc	[Rh(COD)**145a**]BF_4	5	DCM	25	8	100	12.5	100	3 (*S*)	117b
157	H	CO_2Me	NHAc	[Rh(COD)**146a**]BF_4	5	DCM	25	8	97	12.1	97	71 (*S*)	117b
158	H	CO_2Me	NHAc	[Rh(COD)**146c**]BF_4	5	DCM	25	8	92	11.5	92	29 (*S*)	117b
159	H	CO_2Me	NHAc	[Rh(COD)**149a**]BF_4	0.3	DCM	20	20	660	33	66	43.8 (*S*)	116
160	H	CO_2Me	NHAc	[Rh(COD)**149b**]BF_4	0.3	DCM	20	20	770	39	77	23.2 (*S*)	116
161	H	CO_2Me	NHAc	[Rh(COD)**149c**]BF_4	0.3	DCM	20	20	1000	50	>99	88.8 (*R*)	116
162	H	CO_2Me	NHAc	[Rh(COD)**149e**]BF_4	0.3	DCM	20	20	1000	50	>99	80.7 (*R*)	116
163	H	CO_2Me	NHAc	[Rh(COD)**159a**]BF_4	1	MeOH	r.t.	4	97	24.3	97	93 (*R*)	123
164	H	CO_2Me	NHAc	[Rh(COD)**159a**]BF_4	1	THF	r.t.	4	84	21	84	94 (*R*)	123

165	H	CO_2Me	NHAc	[Rh(COD)**159a**]BF_4	1	DCM/MeOH 9:1	r.t.	14	78	5.6	78	98 (*R*)	123
166	H	CO_2Me	NHAc	[Rh(COD)**159a**]BF_4	1	DCM	r.t.	14	77	5.5	77	99 (*R*)	123
167	H	CO_2Me	NHAc	[Rh(COD)**159b**]BF_4	1	DCM	r.t.	2.5	39	15.6	39	96 (*S*)	123
168	H	CO_2Me	NHAc	[Rh(COD)**159c**]BF_4	1	DCM	r.t.	14	100	7.1	100	4 (*S*)	123
169	H	CO_2Me	NHAc	[Rh(COD)**159d**]BF_4	1	DCM	r.t.	2.5	100	40	100	96 (*S*)	123
170	H	CO_2Me	NHAc	[Rh(COD)**159e**]BF_4	1	DCM	r.t.	2.5	100	40	100	95 (*R*)	123
171	H	CO_2Me	NHAc	[Rh(COD)**159f**]BF_4	1	DCM	r.t.	2.5	100	40	100	6 (*R*)	123
172	H	CO_2Me	NHAc	[Rh(COD)**159g**]BF_4	1	DCM	r.t.	2.5	100	40	100	95 (*S*)	123
173	H	CO_2Me	NHAc	[Rh(COD)**159h**]BF_4	1	DCM	r.t.	5	21	4.2	21	58 (*R*)	123
174	H	CO_2Me	NHAc	[Rh(COD)**160a**]BF_4	1	DCM	r.t.	2.5	32	12.8	32	61 (*S*)	123
175	H	CO_2Me	NHAc	[Rh(COD)**160b**]BF_4	1	DCM	r.t.	1	100	100	100	96 (*R*)	123
176	H	CO_2Me	NHAc	[Rh(COD)**163a**]PF_6	1	THF	r.t.	12	100	8.3	100	>99 (*S*)	128
177	H	CO_2Me	NHAc	[Rh(COD)**163b**]PF_6	1	THF	r.t.	12	100	8.3	100	96 (*S*)	128
178	H	CO_2Me	NHAc	[Rh(COD)**164a**]PF_6	1	THF	r.t.	12	100	8.3	100	>99 (*S*)	128
179	H	CO_2Me	NHAc	[Rh(COD)**164b**]PF_6	1	THF	r.t.	12	100	8.3	100	77 (*S*)	128
180	H	CO_2Me	NHAc	[Rh(COD)**166**]BF_4	51	H_2O/ EtOAc (1:1)	r.t.	24	50	2	100	18 (*R*)	129
181	H	CO_2Me	NHAc	[Rh(COD)**167a**]BF_4	1	DCM	r.t.	0.08	100	>1200	100	88.2 (*S*)	121 a, b
182	H	CO_2Me	NHAc	[Rh(COD)**167b**]BF_4	1	MeOH	25	1.3	100	77	100	91 (*R*)	121 b
183	H	CO_2Me	NHAc	[Rh(COD)**167b**]BF_4	1	DCM	25	2.5	100	40	100	>99 (*R*)	121 b
184	H	CO_2Me	NHAc	[Rh(COD)**167b**]BF_4	1	Tol	25	10	100	10	100	97 (*R*)	121 b
185	H	CO_2Me	NHAc	[Rh(COD)**167b**]BF_4	1	THF	25	2	100	50	100	92 (*R*)	121 b
186	H	CO_2Me	NHAc	[Rh(COD)**167b**]BF_4	1	DCM	25	2.5	100	40	100	>99 (*R*) [k)]	121 b
187	H	CO_2Me	NHAc	[Rh(COD)**167b**]BF_4	1	DCM	25	2.5	100	40	100	>99 (*R*) [l)]	121 a, b
188	H	CO_2Me	NHAc	[Rh(nbd)**167b**]BF_4	1	DCM	25	2.5	100	40	100	>99 (*R*) [l)]	121 b
189	H	CO_2Me	NHAc	[Rh(COD)**167c**]BF_4	1	DCM	r.t.	0.33	100	303	100	98.3 (*S*)	121 a, b
190	H	CO_2Me	NHAc	[Rh(COD)**167d**]BF_4	1	DCM	r.t.	0.33	100	303	100	97.6 (*R*)	121 a, b
191	H	CO_2Me	NHAc	[Rh(COD)**168a**]BF_4	5	DCM	25	8	100	13	100	92 (*S*)	126
192	H	CO_2Me	NHAc	[Rh(COD)**168a**]BF_4	30	DCM	25	12	100	8	100	98 (*S*)	126
193	H	CO_2Me	NHAc	[Rh(COD)**168b**]BF_4	5	DCM	25	8	71	9	71	82 (*S*)	126
194	H	CO_2Me	NHAc	[Rh(COD)**168c**]BF_4	5	DCM	25	8	46	6	46	15 (*S*)	126
195	H	CO_2Me	NHAc	[Rh(COD)**168d**]BF_4	5	DCM	25	8	33	4	33	12 (*S*)	126
196	H	Ph	NHAc	[Rh(COD)**91**]BF_4	6.9	MeOH	r.t.	0.167	100	600	100	83.3 (*R*)	84

Table 27.5 (continued)

Entry	Substrate			Catalyst	Conditions				TON	TOF $[h^{-1}]$	Conv. [%]	ee [%]	Reference(s)
	R^1	R^2	R^3		$P[H_2]$ [bar]	Solvent	Temp. [°C]	Time [h]					
197	H	Ph	NHAc	[Rh(COD)**91**]BF_4	6.9	IPA	r.t.	0.167	100	600	100	83.5 (*R*)	84
198	H	Ph	NHAc	[Rh(COD)**91**]BF_4	6.9	Acetone	r.t.	0.167	100	600	100	83.1 (*R*)	84
199	H	Ph	NHAc	[Rh(COD)**91**]BF_4	6.9	THF	r.t.	0.167	100	600	100	81.9 (*R*)	84
200	H	Ph	NHAc	[Rh(COD)**91**]BF_4	6.9	DCM	r.t.	0.167	100	600	100	82.5 (*R*)	84
201	H	Ph	NHAc	[Rh(COD)**91**]BF_4	6.9	Tol	r.t.	0.167	100	600	100	79.3 (*R*)	84
202	H	Ph	NHAc	[Rh(COD)**91**]ClO_4	1	IPA	0	0.167	100	600	100	89.0 (*R*)	84
203	H	4-Cl-Ph	NHAc	[Rh(COD)**91**]ClO_4	1	IPA	0	0.167	100	600	100	86.1 (*R*)	84
204	H	4-F-Ph	NHAc	[Rh(COD)**91**]ClO_4	1	IPA	0	0.167	100	600	100	87.9 (*R*)	84
205	H	4-CF_3-Ph	NHAc	[Rh(COD)**91**]ClO_4	1	IPA	0	0.167	100	600	100	90.0 (*R*)	84
206	H	3-Me-Ph	NHAc	[Rh(COD)**91**]ClO_4	1	IPA	0	0.167	100	600	100	85.6 (*R*)	84
207	H	4-Me-Ph	NHAc	[Rh(COD)**91**]ClO_4	1	IPA	0	0.167	100	600	100	86.5 (*R*)	84
208	*i*-Pr	CO_2H	NHAc	[Rh(COD)**95**]BF_4	1	IPA	r.t.	24	100	4	100	45.7 (*S*)	85
209	*i*-Pr	CO_2H	NHAc	[Rh(COD)**97a**]BF_4	1	MeOH	25	0.0316 [f]	50	1579	50	57.2	26
210	*i*-Pr	CO_2H	NHAc	[Rh(COD)**98a**]BF_4	1	MeOH	25	0.0133 [f]	50	3750	50	95.3	26
211	*i*-Pr	CO_2H	NHAc	[Rh(COD)**98a**]SbF_6	2–2.8	THF	r.t.	2–3	–	–	–	90.0 (*S*)	95 b
212	*i*-Pr	CO_2H	NHAc	[Rh(COD)**98b**]SbF_6	2–2.8	THF	r.t.	2–3	–	–	–	91.0 (*S*)	95 b
213	*i*-Pr	CO_2H	NHAc	[Rh(COD)**98c**]SbF_6	2–2.8	THF	r.t.	2–3	–	–	–	64.4	95 b
214	*i*-Pr	CO_2H	NHAc	[Rh(COD)**98d**]SbF_6	2–2.8	THF	r.t.	2–3	–	–	–	26.0	95 b
215	*i*-Pr	CO_2H	NHAc	[Rh(COD)**98g**]SbF_6	2–2.8	THF	r.t.	2–3	–	–	–	83.6	95 b
216	*i*-Pr	CO_2H	NHAc	[Rh(COD)**103a**]SbF_6	2–2.8	THF	r.t.	2–3	–	–	–	89.2 (*R*)	95 b
217	*i*-Pr	CO_2Me	NHAc	[Rh(COD)**98a**]BF_4	1	MeOH	25	0.02 [f]	50	2500	50	86.1	26
218	*i*-Pr (*Z*)	CO_2Me	NHAc	[Rh(COD)**98b**]SbF_6	2–2.8	THF	r.t.	2–3	–	–	–	92.0	95 b
219	*i*-Pr (*Z*/*E*)	CO_2Me	NHAc	[Rh(COD)**98b**]SbF_6	2–2.8	THF	r.t.	2–3	–	–	–	86.5	95 b
220	*i*-Pr	CO_2Me	NHAc	[[Rh(COD)**98d**]SbF_6	2–2.8	THF	r.t.	2–3	–	–	–	5.6 (*R*)	95 b

221	*i*-Pr	CO_2Me	NHAc	[Rh(COD)**98g**]SbF_6	2–2.8	THF	r.t.	2–3	–	–	–	87.2 (*S*)	95b
222	*i*-Pr	CO_2Me	NHAc	[Rh(COD)**103a**]SbF_6	2–2.8	THF	r.t.	2–3	–	–	–	86.9 (*R*)	95b
223	*i*-Pr	CO_2Me	NHAc	[Rh(COD)**103g**]SbF_6	2–2.8	THF	r.t.	2–3	–	–	–	86.5 (*R*)	95b
224	*i*-Pr (*E*)	NHAc	CO_2Me	[Rh(COD)**98b**]SbF_6	2–2.8	THF	r.t.	2–3	–	–	–	73.3	95b
225	Bn	CO_2Me	NHAc	[Rh(COD)**98b**]SbF_6	2–2.8	THF	r.t.	2–3	–	–	–	40.6 (*S*)	95b
226	Bn	CO_2Me	NHAc	[Rh(COD)**103a**]SbF_6	2–2.8	THF	r.t.	2–3	–	–	–	67.0 (*R*)	95b
227	Ph	CO_2H	NHAc	[Rh(1,5-hexadiene) (+)-*trans*-**85**]Cl	50	–	0	24	–	–	–	68.5 (*S*) [a]	67
228	Ph	CO_2H	NHAc	[Rh(1,5-hexadiene)d-*trans*-**86**]Cl	50	–	0	–	–	–	–	12 (*S*) [a]	68a
229	Ph	CO_2H	NHAc	[Rh(COD)**88**]Cl	20.7	PhH/EtOH 1:1	60	24	50 [c]	2.1	100 [d]	4.3	81
230	Ph	CO_2H	NHAc	[Rh(COD)**91**]BF_4	1	MeOH	25	0.167	100	600	>99.9	97.9 (*R*)	27
231	Ph	CO_2H	NHAc	[Rh(COD)**93a**]BF_4	6.9	MeOH	r.t.	0.5	356	712	71.2	18 (*S*)	52
232	Ph	CO_2H	NHAc	[Rh(COD)**94a**]BF_4	6.9	MeOH	r.t.	0.5	410	820	81.9	74.2 (*S*)	52
233	Ph	CO_2H	NHAc	[Rh(COD)**95**]BF_4	1	DCE	r.t.	24	100	4	100	88.2 (*S*)	85
234	Ph	CO_2H	NHAc	[Rh(COD)**95**]BF_4	1	THF	r.t.	24	100	4	100	89.1 (*S*)	85
235	Ph	CO_2H	NHAc	[Rh(COD)**95**]BF_4	1	THF:Et_3N = 1:1	r.t.	24	30	1	30	30.9 (*S*)	85
236	Ph	CO_2H	NHAc	[Rh(COD)**95**]BF_4	1	MeOH	r.t.	24	100	4	100	92.4 (*S*)	85
237	Ph	CO_2H	NHAc	[Rh(COD)**95**]BF_4	1	MeOH : Et_3N = 1 : 1	r.t.	24	100	4	100	67.9 (*S*)	85
238	Ph	CO_2H	NHAc	[Rh(COD)**95**]BF_4	1	EtOH	r.t.	24	100	4	100	92.0 (*S*)	85
239	Ph	CO_2H	NHAc	[Rh(COD)**95**]BF_4	1	CF_3CH_2OH	r.t.	24	100	4	100	80.3 (*S*)	85
240	Ph	CO_2H	NHAc	[Rh(COD)**95**]BF_4	1	*t*-BuOH	r.t.	24	100	4	100	91.1 (*S*)	85
241	Ph	CO_2H	NHAc	[Rh(COD)**95**]BF_4	1	IPA	r.t.	24	100	4	100	94.7 (*S*)	85
242	Ph	CO_2H	NHAc	[Rh(COD)**95**]BF_4	1	IPA	0	24	100	4	100	96.1 (*S*)	85
243	Ph	CO_2H	NHAc	[Rh(COD)**95**]BF_4	1	IPA	r.t.	24	86.6	3.6	86.6	63.9 (*S*)	85
244	Ph	CO_2H	NHAc	[Rh(COD)**96**]BF_4	1	IPA	r.t.	24	100	4.2	100	83.5 (*R*)	85
245	Ph	CO_2H	NHAc	[Rh(COD)**97a**]BF_4	1	MeOH	25	0.11 [f]	50	455	50	73.2 (*S*)	26
246	Ph	CO_2H	NHAc	[Rh(COD)**97a**]BF_4	1	EtOH	25	0.1 [f]	50	500	50	71 (*S*) [a]	26
247	Ph	CO_2H	NHAc	$[Rh(COD)Cl]_2$ + **97a** (neutral)	1	EtOH	25	3	50	16.7	50	69 (*S*)	26

Table 27.5 (continued)

Entry	Substrate			Catalyst	Conditions				TON	TOF [h^{-1}]	Conv. [%]	ee [%]	Reference(s)
	R^1	R^2	R^3		$P[H_2]$ [bar]	Solvent	Temp. [°C]	Time [h]					
248	Ph	CO_2H	NHAc	[Rh(NBD)**97a**]PF_6	1	EtOH	30	1	100	100	100 [b]	61 (*S*)	88
249	Ph	CO_2H	NHAc	[Rh(NBD)**97a**]PF_6	1	EtOH	0	1.5	100	67	100 [b]	75 (*S*)	88
250	Ph	CO_2H	NHAc	[Rh(COD)**97a**]BF_4	1	MeOH	25	0.0333 [f]	50	1500	50	96.6 (*S*)	91b
251	Ph	CO_2H	NHAc	[Rh(COD)**97a**]BF_4	1	PhH	25	0.183 [f]	50	273	50	98.6 (*S*)	91b
252	Ph	CO_2H	NHAc	[Rh(COD)**97a**]BF_4	1	Tol	25	0.317 [f]	50	158	50	98.9 (*S*)	91b
253	Ph	CO_2H	NHAc	[Rh(COD)**97a**]BF_4	1	Tol	25	0.32	50	156	50	98.9 (*S*)	91b
254	Ph	CO_2H	NHAc	[Rh(COD)**97a**]Cl	50	–	25	8	100 [c]	–	–	46 (*S*) [a]	89
255	Ph	CO_2H	NHAc	[Rh(COD)**97a**]ClO_4	1	EtOH	25	–	50	–	100	61 (*S*)	90
256	Ph	CO_2H	NHAc	[Rh(COD)**97b**]Cl	50	–	25	8	100 [c]	–	–	36 (*S*) [a]	89
257	Ph	CO_2H	NHAc	[Rh(COD)**97b**]BF_4	1	H_2O	25	0.52	50	91	50	80 (S)	99
258	Ph	CO_2H	NHAc	[Rh(COD)**97b**]BF_4	1	H_2O + 0.1 mmol $LiBF_4$	25	0.5	50	100	50	64 (S)	99
259	Ph	CO_2H	NHAc	[Rh(COD)**97b**]BF_4	1	H_2O + 0.1 mmol $NaBF_4$	25	0.3	50	167	50	83 (S)	99
260	Ph	CO_2H	NHAc	[Rh(COD)**97b**]BF_4	1	H_2O + 0.1 mmol KBF_4	25	0.45	50	111	50	82 (S)	99
261	Ph	CO_2H	NHAc	[Rh(COD)**97b**]BF_4	1	H_2O + 0.1 mmol $RbBF_4$	25	0.33	50	150	50	82 (S)	99
262	Ph	CO_2H	NHAc	[Rh(COD)**97b**]BF_4	1	H_2O + 0.1 mmol $CsBF_4$	25	0.47	50	120	50	83 (S)	99
263	Ph	CO_2H	NHAc	[Rh(COD)**97c**]BF_4	1	MeOH	25	0.0667 [f]	50	750	50	95.1 (*S*)	91b
264	Ph	CO_2H	NHAc	[Rh(COD)**97c**]BF_4	1	PhH	25	0.467 [f]	50	107	50	85.5 (*S*)	91b
265	Ph	CO_2H	NHAc	[Rh(COD)**97c**]BF_4	1	Tol	25	8 [f]	50	6.25	50	82.3 (*S*)	91b
266	Ph	CO_2H	NHAc	[Rh(COD)**97d**]BF_4	1	MeOH	25	–	–	–	–	53.7 (*S*)	26

267	Ph	CO_2H	NHAc	[Rh(COD)**98a**]BF_4	1	MeOH	25	0.0233 [f]	50	2143	50	96.6 (*S*)	26
268	Ph	CO_2H	NHAc	[Rh(COD)**98a**]BF_4	1	MeOH	–27	–	–	–	–	99.3 (*S*)	26
269	Ph	CO_2H	NHAc	[Rh(COD)**98a**]BF_4	1	MeOH	–22.2	–	–	–	–	98.3 (*S*)	26
270	Ph	CO_2H	NHAc	[Rh(COD)**98a**]BF_4	1	MeOH	0.4	–	–	–	–	97.7 (*S*)	26
271	Ph	CO_2H	NHAc	[Rh(COD)**98a**]BF_4	1	MeOH	25	–	–	–	–	97.1 (*S*)	26
272	Ph	CO_2H	NHAc	[Rh(COD)**98a**]BF_4	1	MeOH	55.2	–	–	–	–	92.7 (*S*)	26
273	Ph	CO_2H	NHAc	[Rh(COD)**98a**]BF_4	1	EtOH	25	0.117 [f]	50	429	50	96 (*S*) [a]	26
274	Ph	CO_2H	NHAc	$[Rh(COD)Cl]_2$ + **98a** (neutral)	1	EtOH	25	1.67	50	30	50	87 (*S*) [a]	26
275	Ph	CO_2H	NHAc	[Rh(COD)**98a**]BF_4	1	MeOH	25	0.03	50	1667	50	96.5 (*S*)	91 e
276	Ph	CO_2H	NHAc	[Rh(COD)**98a**]SbF_6	2–2.8	THF	r.t.	2–3	–	–	–	94.0	95 a, b
277	Ph	CO_2H	NHAc	[Rh(COD)**98b**]SbF_6	2–2.8	THF	r.t.	2–3	–	–	–	99.0	95 a, b
278	Ph	CO_2H	NHAc	[Rh(COD)**98c**]SbF_6	2–2.8	THF	r.t.	2–3	–	–	–	60.0	95 a, b
279	Ph	CO_2H	NHAc	[Rh(COD)**98d**]SbF_6	2–2.8	THF	r.t.	–	–	–	–	71	95 a
280	Ph	CO_2H	NHAc	[Rh(COD)**98 f**]OTf	2–2.8	THF	r.t.	2–3	–	–	–	96.0	95 b
281	Ph	CO_2H	NHAc	[Rh(COD)**98 f**]SbF_6	2–2.8	THF	r.t.	2–3	–	–	–	93.0	95 b
282	Ph	CO_2H	NHAc	[Rh(COD)**98g**]SbF_6	2–2.8	THF	r.t.	2–3	–	–	–	97.6	95 b
283	Ph	CO_2H	NHAc	[Rh(COD)**98h**]SbF_6	2–2.8	THF	r.t.	2–3	–	–	–	91.0	95 b
284	Ph	CO_2H	NHAc	[Rh(COD)**100**]Cl	50	–	25	8	100 [c]	–	–	80 (*S*) [a]	89
285	Ph	CO_2H	NHAc	[Rh(COD)**100b**]BF_4	1	MeOH	25	–	50	–	50	90 (*S*)	91 e
286	Ph	CO_2H	NHAc	[Rh(COD)**101a**]BF_4	1	MeOH	25	0.0617 [f]	50	810	50	94.5 (*S*)	91 c
287	Ph	CO_2H	NHAc	[Rh(COD)**101b**]BF_4	1	MeOH	25	0.0633 [f]	50	789	50	96.2 (*S*)	91 c
288	Ph	CO_2H	NHAc	[Rh(COD)**101c**]BF_4	1	MeOH	25	0.0633 [f]	50	789	50	91.4 (*S*)	91 c
289	Ph	CO_2H	NHAc	[Rh(COD)**101d**]BF_4	1	MeOH	25	0.0683 [f]	50	732	50	94.9 (*S*)	91 c
290	Ph	CO_2H	NHAc	[Rh(COD)**101e**]BF_4	1	MeOH	25	0.0817 [f]	50	612	50	90.4 (*S*)	91 c
291	Ph	CO_2H	NHAc	[Rh(COD)**101 f**]BF_4	1	MeOH	25	0.142 [f]	50	353	50	93.6 (*S*)	91 c
292	Ph	CO_2H	NHAc	[Rh(COD)**102a**]BF_4	2–2.8	THF	r.t.	2–3	–	–	–	94.5	95 b
293	Ph	CO_2H	NHAc	[Rh(COD)**102a**]BF_4	2–2.8	THF	r.t.	–	–	–	–	98.3	95 a
294	Ph	CO_2H	NHAc	[Rh(COD)**102b**]SbF_6	2–2.8	THF	r.t.	2–3	–	–	–	94.5	95 b
295	Ph	CO_2H	NHAc	[Rh(COD)**103a**]BF_4	2–2.8	THF	r.t.	2–3	–	–	–	95.8	95 b
296	Ph	CO_2H	NHAc	[Rh(COD)**103a**]SbF_6	2–2.8	THF	r.t.	2–3	–	–	–	97.0	95 b

Table 27.5 (continued)

Entry	Substrate			Catalyst	Conditions				TON	TOF [h^{-1}]	Conv. [%]	ee [%]	Reference(s)
	R[1]	R[2]	R[3]		P[H_2] [bar]	Solvent	Temp. [°C]	Time [h]					
297	Ph	CO_2H	NHAc	[Rh(COD)**103a**]SbF_6	2–2.8	THF	r.t.	–	–	–	–	93	95 a
298	Ph	CO_2H	NHAc	[Rh(COD)**111**]ClO_4	1	EtOH	25	0.3	50	167	100	55 (*R*)	90
299	Ph	CO_2H	NHAc	[Rh(COD)**111**]ClO_4	1	EtOH	0	1	50	50	100	70 (*R*)	90
300	Ph	CO_2H	NHAc	[Rh(COD)**112b**]BF_4	1	MeOH	25	0.12	50	417	50	1 (*S*)	91 e
301	Ph	CO_2H	NHAc	[Rh(COD)**113a**]Cl	50	–	25	8	100 [c]	–	–	0	89
302	Ph	CO_2H	NHAc	[Rh(COD)**113b**]Cl	50	–	25	8	100 [c]	–	–	0	89
303	Ph	CO_2H	NHAc	[Rh(COD)**113c**]BF_4	1	MeOH	25	0.17	50	294	50	2 (*S*)	91 e
304	Ph	CO_2H	NHAc	[Rh(COD)**113d**]BF_4	1	MeOH	25	0.12	50	429	50	46 (*S*)	91 e
305	Ph	CO_2H	NHAc	[Rh(COD)**113e**]BF_4	1	MeOH	25	0.12	50	417	50	46 (*S*)	91 e
306	Ph	CO_2H	NHAc	[Rh(COD)**114a**]BF_4	1	MeOH	25	0.083 [f]	50	602	50	55 (*S*)	99
307	Ph	CO_2H	NHAc	[Rh(COD)**114a**]BF_4	1	PhH	25	0.68 [f]	50	74	50	58 (*S*)	99
308	Ph	CO_2H	NHAc	[Rh(COD)**114b**]BF_4	1	MeOH	25	0.05 [f]	50	1000	50	52 (*S*)	99
309	Ph	CO_2H	NHAc	[Rh(COD)**114b**]BF_4	1	PhH	25	0.052 [f]	50	97	50	80 (*S*)	99
310	Ph	CO_2H	NHAc	[Rh(COD)**115d**]BF_4	2.8	THF	r.t.	–	150	–	100	97 (*S*)	98
311	Ph	CO_2H	NHAc	[Rh(COD)**115d**]BF_4	2.8	H_2O/EtOAc (1:1)	r.t.	–	100	–	100	7 (*S*)	98
312	Ph	CO_2H	NHAc	[Rh(COD)**117**]BF_4	5	H_2O/MeOH/ EtOAc (0.6:0.4:2)	r.t.	3	100	33	100	96 (*S*)	101
313	Ph	CO_2H	NHAc	[Rh(COD)**117**]BF_4	5	H_2O/MeOH(3:2)	r.t.	1.5	100	67	100	95 (*S*)	101
314	Ph	CO_2H	NHAc	[Rh(COD)**119d**]SbF_6	2.8	H_2O/THF(1:1)	r.t.	24	15	0.6	12	65 (*S*)	102
315	Ph	CO_2H	NHAc	[Rh(COD)**119d**]SbF_6	2.8	THF	r.t.	2	125	63	100	70 (*S*)	102
316	Ph	CO_2H	NHAc	[Rh(COD)**124**]BF_4	1	EtOH	25	1	100	100	100	30 (*R*)	103
317	Ph	CO_2H	NHAc	[Rh(COD)**124**]BF_4	1	THF	25	1	100	100	100	40 (*R*)	103
318	Ph	CO_2H	NHAc	[Rh(COD)**124**]BF_4	1	THF	0	1	100	100	100	52 (*R*)	103

319	Ph	CO_2H	NHAc	[Rh(COD)**124**]BF_4	1	THF	–78	3	100	33	100	10 (*R*)	103
320	Ph	CO_2H	NHAc	[Rh(COD)**124**]BPh_4	1	THF	25	1	100	100	100	30 (*R*)	103
321	Ph	CO_2H	NHAc	[Rh(COD)**124**]BPh_4	1	EtOH	25	1	100	100	100	35 (*R*)	103
322	Ph	CO_2H	NHAc	[Rh(COD)**124**]Cl	20.4	PhH/EtOH 1:1	60	24	100	2	100	8.2 (*R*)	103
323	Ph	CO_2H	NHAc	[Rh(COD)**124**]ClO_4	1	THF	25	1	100	100	100	36 (*R*)	103
324	Ph	CO_2H	NHAc	[Rh(COD)**124**]ClO_4	1	EtOH	25	1	100	100	100	28 (*R*)	103
325	Ph	CO_2H	NHAc	[Rh(COD)**124**]PF_6	1	THF	25	1	100	100	100	32 (*R*)	103
326	Ph	CO_2H	NHAc	[Rh(COD)**124**]PF_6	1	EtOH	25	1	100	100	100	30 (*R*)	103
327	Ph	CO_2H	NHAc	[Rh(COD)**125**]BF_4	1	THF	25	1	100	100	100	54 (*S*)	103
328	Ph	CO_2H	NHAc	[Rh(COD)**126**]Cl	50	–	25	1	100 [c)]	100	100	62 (*R*) [a)]	89
329	Ph	CO_2H	NHAc	[Rh(COD)**128**]Cl	51	MeOH	r.t.	0.25	230	920	100	36 (*S*)	106
330	Ph	CO_2H	NHAc	[Rh(COD)**129**]Cl	51	MeOH	r.t.	1	270	270	100	15 (*R*)	106
331	Ph	CO_2H	NHAc	[Rh(COD)**131**]BF_4	1	PhH	25	6	89.5	15	89.5	40.8 (*R*)	108
332	Ph	CO_2H	NHAc	[Rh(COD)**132**]BF_4	1	Acetone	25	0.25–1	100	100–400	100	90.1 (*R*)	107
333	Ph	CO_2H	NHAc	[Rh(COD)**132**]BF_4	6.9	Acetone	25	0.25–1	100	100–400	100	92.8 (*R*)	107
334	Ph	CO_2H	NHAc	[Rh(COD)**132**]BF_4	34.5	Acetone	25	0.25–1	100	100–400	100	94.4 (*R*)	107
335	Ph	CO_2H	NHAc	[Rh(COD)**132**]BF_4	34.5	Acetone	–15	0.25–1	100	100–400	100	97.1 (*R*)	107
336	Ph	CO_2H	NHAc	[Rh(COD)**133a**]BF_4	1	IPA	r.t.	24	98	4	98	94 [j)]	48
337	Ph	CO_2H	NHAc	[Rh(COD)**133b**]BF_4	1	IPA	r.t.	24	94	4	94	89 [j)]	48
338	Ph	CO_2H	NHAc	[Rh(COD)**133c**]BF_4	1	IPA	r.t.	24	94	4	98	97 [j)]	48
339	Ph	CO_2H	NHAc	[Rh(COD)**134**]BF_4	1	PhH	25	5.0	85.6	17	85.6	25.6 (*R*)	108
340	Ph	CO_2H	NHAc	[Rh(COD)**135**]BF_4	1	PhH	25	7.5	93.4	12	93.4	57.1 (*R*)	108
341	Ph	CO_2H	NHAc	[Rh(COD)**141a**]BF_4	3.5	MeOH	r.t.	0.5	1000	2000	>99	93 (*S*)	111
342	Ph	CO_2H	NHAc	[Rh(COD)**141b**]BF_4	3.5	MeOH	r.t.	0.5	1000	2000	>99	99 (*S*)	111
343	Ph	CO_2H	NHAc	[Rh(COD)**142a**]BF_4	5	Tol:MeOH 2:1	40	20	100	5	100	31 (*S*)	117d
344	Ph	CO_2H	NHAc	[Rh(COD)**142b**]BF_4	5	Tol:MeOH 2:1	40	20	100	5	100	30 (*S*)	117d
345	Ph	CO_2H	NHAc	[Ir(COD)**142b**]BF_4	5	DCM:MeOH 2:1	40	20	18	0.9	18	15 (*S*)	117d

Table 27.5 (continued)

Entry	Substrate			Catalyst	Conditions				TON	TOF [h^{-1}]	Conv. [%]	ee [%]	Reference(s)
	R^1	R^2	R^3		$P[H_2]$ [bar]	Solvent	Temp. [°C]	Time [h]					
346	Ph	CO_2H	NHAc	[Rh(COD)**163a**]PF_6	1	THF	r.t.	12	99	8.3	100	99 (*S*)	128
347	Ph	CO_2H	NHAc	[Rh(COD)**164a**]PF_6	1	THF	r.t.	12	>99	8.3	100	>99 (*S*)	128
348	Ph	CO_2H	NHAc	[Rh(COD)**166**]BF_4	51	H_2O/EtOAc (1:1)	r.t.	24	50	2.1	100	50 (*R*)[j]	129
349	Ph	CO_2H	NHAc	[Rh(NBD)**107**]ClO_4	1.48	PhH : EtOH = 1 : 1	25	24	100	4	100	24.8 (*R*)	93
350	Ph	CO_2H	NHAc	[Rh(NBD)**107**]ClO_4	1.48	PhH : EtOH = 1 : 1	25	24	47.5	2	95	31.0 (*R*)	93
351	Ph	CO_2H	NHAc	[Rh(NBD)**107**]ClO_4	1.48	PhH : EtOH = 1 : 1	25	24	18.4	0.8	92	16.2 (*R*)	93
352	Ph	CO_2H	NHAc	[Rh(NBD)**107**]ClO_4	1.48	PhH : EtOH = 1 : 1	40	24	95	4	95	14.9 (*R*)	93
353	Ph	CO_2H	NHAc	[Rh(NBD)**107**]ClO_4	1.48	PhH : EtOH = 1 : 1	60	24	100	4	100	12.4 (*R*)	93
354	Ph	CO_2H	NHAc	[Rh(NBD)**107**]ClO_4	1.48	PhH : EtOH = 1 : 1	80	24	96	4	96	5.3 (*R*)	93
355	Ph	CO_2H	NHAc	[Rh(NBD)**107**]ClO_4	1.48	PhH : EtOH = 1 : 1	–15 to –20	7	53	8	53	62.7 (*R*)	93
356	Ph	CO_2H	NHAc	[Rh(NBD)**107**]ClO_4	1.48	PhH : EtOH = 1 : 1	–15 to –20	7	18.8	3	94	27.7 (*R*)	93
357	Ph	CO_2H	NHAc	[Rh(NBD)**107**]ClO_4	1.97	PhH : EtOH = 1 : 1	25	24	90	4	90	20.9 (*R*)	93
358	Ph	CO_2H	NHAc	[Rh(NBD)**108**]ClO_4	1.48	PhH : EtOH = 1 : 1	25	24	100	4	100	63.4 (*S*)	93
359	Ph	CO_2H	NHAc	[Rh(NBD)**108**]ClO_4	1.48	PhH : EtOH = 1 : 1	25	24	50	2	100	68.2 (*S*)	93
360	Ph	CO_2H	NHAc	[Rh(NBD)**108**]ClO_4	1.48	PhH : EtOH = 1 : 1	25	24	20	0.9	100	44.1 (*S*)	93
361	Ph	CO_2H	NHAc	[Rh(NBD)**108**]ClO_4	1.48	PhH : EtOH = 1 : 1	25	1	96	96	96	60.4 (*S*)	93
362	Ph	CO_2H	NHAc	[Rh(NBD)**108**]ClO_4	1.48	PhH : EtOH = 1 : 1	25	2	100	50	100	66.9 (*S*)	93
363	Ph	CO_2H	NHAc	[Rh(NBD)**108**]ClO_4	1.48	PhH : EtOH = 1 : 1	25	4	94	24	94	59.6 (*S*)	93
364	Ph	CO_2H	NHAc	[Rh(NBD)**108**]ClO_4	1.48	PhH : EtOH = 1 : 1	40	24	100	4	100	45.9 (*S*)	93
365	Ph	CO_2H	NHAc	[Rh(NBD)**108**]ClO_4	1.48	PhH : EtOH = 1 : 1	60	24	100	4	100	26.3 (*S*)	93
366	Ph	CO_2H	NHAc	[Rh(NBD)**108**]ClO_4	1.48	PhH : EtOH = 1 : 1	80	24	100	4	100	12.9 (*S*)	93
367	Ph	CO_2H	NHAc	[Rh(NBD)**108**]ClO_4	1.48	PhH : EtOH = 1 : 1	–15 to –20	6	20	3	20	80.1 (*S*)	93

368	Ph	CO_2H	NHAc	[Rh(NBD)**108**]ClO_4	1.48	PhH : EtOH = 1 : 1	–15 to –20	8	20	3	100	74.1 (*S*) 93
369	Ph	CO_2H	NHAc	[Rh(NBD)**108**]ClO_4	1.48	PhH : EtOH = 1 : 1	0	24	100	4	100	72.1 (*S*) 93
370	Ph	CO_2H	NHAc	[Rh(NBD)**108**]ClO_4	1.48	PhH : EtOH = 1 : 1	–5	24	93	4	93	78.4 (*S*) 93
371	Ph	CO_2H	NHAc	[Rh(NBD)**108**]ClO_4	1.48	PhH : EtOH = 1 : 1	–15	24	79	3	79	90.4 (*S*) 93
372	Ph	CO_2H	NHAc	[Rh(NBD)**108**]ClO_4	1.09	PhH : EtOH = 1 : 1	25	24	70	3	70	21.9 (*S*) 93
373	Ph	CO_2H	NHAc	[Rh(NBD)**108**]ClO_4	19.70	PhH : EtOH = 1 : 1	25	50	100	2	100	22.9 (*S*) 93
374	Ph	CO_2H	NHAc	[Rh(NBD)**109**]ClO_4 dimer	1.48	–	25	24	63	3	63 [b]	13.8 (*R*) 94
375	Ph	CO_2H	NHAc	[Rh(NBD)**109**]ClO_4 dimer	1.48	–	25	24	50	2	100 [b]	11.6 (*R*) 94
376	Ph	CO_2H	NHAc	[Rh(NBD)**109**]ClO_4 dimer	1.48	–	25	24	20	0.8	100 [b]	14.9 (*R*) 94
377	Ph	CO_2H	NHAc	[Rh(NBD)**109**]ClO_4 dimer	1.48	–	40	24	100	4	100 [b]	8.2 (*R*) 94
378	Ph	CO_2H	NHAc	[Rh(NBD)**109**]ClO_4 dimer	1.48	–	60	24	100	4	100 [b]	1.6 (*S*) 94
379	Ph	CO_2H	NHAc	[Rh(NBD)**109**]ClO_4 dimer	1.48	–	80	24	100	4	100 [b]	2.9 (*S*) 94
380	Ph	CO_2H	NHAc	[Rh(NBD)**109**]ClO_4 dimer	1.48	–	–15 to –20	7	93	13	93 [b]	26.3 (*R*) 94
381	Ph	CO_2H	NHAc	[Rh(NBD)**109**]ClO_4 dimer	1.48	–	–15 to –20	7	20	3	100 [b]	29.3 (*R*) 94
382	Ph	CO_2H	NHAc	[Rh(NBD)**110**]ClO_4 dimer	1.48	–	25	24	100	4	100 [b]	4.8 (*R*) 94
383	Ph	CO_2H	NHAc	[Rh(NBD)**110**]ClO_4 dimer	1.48	–	25	24	50	2	100 [b]	6.9 (*R*) 94

Table 27.5 (continued)

Entry	Substrate			Catalyst	Conditions				TON	TOF [h^{-1}]	Conv. [%]	ee [%]	Reference(s)
	R^1	R^2	R^3		P[H_2] [bar]	Solvent	Temp. [°C]	Time [h]					
384	Ph	CO_2H	NHAc	[Rh(NBD)**110**]ClO_4 dimer	1.48	–	25	24	19.4	0.8	97 [b]	2.5 (*R*)	94
385	Ph	CO_2H	NHAc	[Rh(NBD)**110**]ClO_4 dimer	1.48	–	40	24	100	4	100 [b]	3.1 (*R*)	94
386	Ph	CO_2H	NHAc	[Rh(NBD)**110**]ClO_4 dimer	1.48	–	60	24	100	4	100 [b]	1.5 (*R*)	94
387	Ph	CO_2H	NHAc	[Rh(NBD)**110**]ClO_4 dimer	1.48	–	80	24	100	4	100 [b]	2.1 (*R*)	94
388	Ph	CO_2H	NHAc	[Rh(NBD)**110**]ClO_4 dimer	1.48	–	–15 to –20	7	92	13	92 [b]	12.1 (*R*)	94
389	Ph	CO_2H	NHAc	[Rh(NBD)**110**]ClO_4 dimer	1.48	–	–15 to –20	7	18.6	3	93 [b]	9.5 (*R*)	94
390	Ph	CO_2H	NHAc	$[RhCl(COD)]_2$ + **165** + Et_3N	1	EtOH : PhH =1 : 1	r.t.	48	100	2	100	4.7 (*R*) [a]	119
391	Ph	CO_2H	NHBz	[Rh(COD)**95**]BF_4	1	IPA	r.t.	24	100	4	100	89.2 (*S*)	85
392	Ph	CO_2H	NHBz	[Rh(COD)**97a**]BF_4	1	MeOH	25	0.0333	50	1500	50	95.0 (*S*)	91 b
393	Ph	CO_2H	NHBz	[Rh(COD)**97c**]BF_4	1	MeOH	25	0.05 [f]	50	1000	50	93.7 (*S*)	91 b
394	Ph	CO_2H	NHBz	[Rh(COD)**98a**]BF_4	1	MeOH	25	0.117 [f]	50	429	50	96 (*S*)	26
395	Ph	CO_2H	NHBz	[Rh(COD)**98a**]BF_4	50	MeOH	25		100 [b]			95 (*S*)	26
396	Ph	CO_2H	NHBz	[Rh(COD)**128**]Cl	10	MeOH	r.t.	1	110	110	100	44 (*S*)	106
397	Ph	CO_2H	NHBz	[Rh(COD)**129**]Cl	51	MeOH	r.t.	1	100	100	100	17 (*R*)	106
398	Ph	CO_2H	NHBz	[Rh(COD)**163a**]PF_6	1	THF	r.t.	12	>99	8.3	100	>99 (*S*)	128
399	Ph	CO_2H	NHBz	[Rh(COD)**164a**]PF_6	1	THF	r.t.	12	>99	8.3	100	>99 (*S*)	128
400	Ph	CO_2Me	NHAc	[Rh(1,5-hexadiene)d-*trans*-**86**]Cl	50	–	50	–	–	–	–	43 (*S*) [a]	68 a
401	Ph	CO_2Me	NHAc	[Rh(COD)**88**]Cl	69	PhH : EtOH =1 : 1	100	48	50 [c]	–	[d]	10.3	81
402	Ph	CO_2Me	NHAc	[Rh(COD)**91**]BF_4	1	MeOH	25	0.167	100	600	>99.9	95.7 (*R*)	27

403	Ph	CO_2Me	NHAc	[Rh(COD)**93a**]BF_4	6.9	DCM	r.t.	0.17	428	2518	85.5	64 (*S*)	52
404	Ph	CO_2Me	NHAc	[Rh(COE)**93a**]Cl	97	Tol/acetone 1:1	0	24	20.5	0.9	41 [b]	76 [a]	71
405	Ph	CO_2Me	NHAc	[Rh(COD)**94a**]BF_4	6.9	DCM	r.t.	0.17	500	2941	100	84 (*S*)	52
406	Ph	CO_2Me	NHAc	$[Rh(COD)Cl]_2$ + **97a** (cationic)	1	PhH	25	0.117 [f]	50	429	50	6 (*S*) [a]	26
407	Ph	CO_2Me	NHAc	$[Rh(COD)Cl]_2$ + **97a** (neutral)	1	EtOH	25	5.17 [f]	50	10	50	63 (*S*) [a]	26
408	Ph	CO_2Me	NHAc	$[Rh(COD)Cl]_2$ + **97a** (neutral)	1	PhH	25	>83.3 [f]	50	0.6	50	14 (*S*) [a]	26
409	Ph	CO_2Me	NHAc	[Rh(COD)**97a**]BF_4	1	MeOH	25	0.113 [f]	50	441	50	72.2 (*S*)	26
410	Ph	CO_2Me	NHAc	[Rh(COD)**97a**]BF_4	1	MeOH	–21.3	–	–	–	–	82.3 (*S*)	26
411	Ph	CO_2Me	NHAc	[Rh(COD)**97a**]BF_4	1	MeOH	0.5	–	–	–	–	77.8 (*S*)	26
412	Ph	CO_2Me	NHAc	[Rh(COD)**97a**]BF_4	1	EtOH	25	0.1 [f]	50	500	50	73 (*S*) [a]	26, 91e
413	Ph	CO_2Me	NHAc	[Rh(COD)**97a**]BF_4	1	MeOH	25	–	–	–	–	73 (*S*)	91a
414	Ph	CO_2Me	NHAc	[Rh(COD)**97a**]BF_4	1	MeOH	25	0.1 [f]	50	500	50	91.5 (*S*)	91b
415	Ph	CO_2Me	NHAc	[Rh(COD)**97a**]BF_4	1	MeOH	25	0.12	50	417	50	72 (*S*)	91e
416	Ph	CO_2Me	NHAc	[Rh(COD)**97a**]BF_4	1	PhH	25	0.08	50	625	50	6 (*R*)	91e
417	Ph	CO_2Me	NHAc	[Rh(COD)**97a**]Cl	50	–	25	8	100 [c]	–	–	8 (*S*) [a]	89
418	Ph	CO_2Me	NHAc	[Rh(COD)**97a**]ClO_4	1	EtOH	25	–	50	–	100	60(*S*)	90
419	Ph	CO_2Me	NHAc	[Rh(NBD)**97a**]PF_6	1	EtOH	30	0.5	100	200	100 [b]	60 (*S*)	88
420	Ph	CO_2Me	NHAc	[Rh(NBD)**97a**]PF_6	1	EtOH	0	3	100	33	100	65 (*S*)	88
421	Ph	CO_2Me	NHAc	[Rh(COD)**97b**]BF_4	1	H_2O + 0.1 mmol $LiBF_4$	25	0.033	50	1500	50	45 (S)	99
422	Ph	CO_2Me	NHAc	[Rh(COD)**97b**]BF_4	1	H_2O + 0.1 mmol $NaBF_4$	25	0.017	50	3000	50	41 (S)	99
423	Ph	CO_2Me	NHAc	[Rh(COD)**97b**]BF_4	1	H_2O + 0.1 mmol KBF_4	25	0.017	50	3000	50	41 (S)	99
424	Ph	CO_2Me	NHAc	[Rh(COD)**97b**]BF_4	1	H_2O + 0.1 mmol $RbBF_4$	25	0.033	50	1500	50	41 (S)	99
425	Ph	CO_2Me	NHAc	[Rh(COD)**97b**]BF_4	1	H_2O + 0.1 mmol $CsBF_4$	25	0.033	50	1500	50	41 (S)	99

Table 27.5 (continued)

Entry	Substrate			Catalyst	Conditions				TON	TOF $[h^{-1}]$	Conv. [%]	ee [%]	Refer-ence(s)
	R^1	R^2	R^3		$P[H_2]$ [bar]	Solvent	Temp. [°C]	Time [h]					
426	Ph	CO_2Me	NHAc	[Rh(COD)**97b**]BF_4	1	H_2O	25	0.033	50	1500	50	41 (*S*)	99
427	Ph	CO_2Me	NHAc	[Rh(COD)**97b**]Cl	50	–	25	8	100 [c]	–	–	10 (*S*) [a]	89
428	Ph	CO_2Me	NHAc	[Rh(COD)**97c**]BF_4	1	MeOH	25	0.58	50	86	50	61 (*S*)	91a
429	Ph	CO_2Me	NHAc	[Rh(COD)**97c**]BF_4	1	MeOH + Triton X-100	25	0.07	50	714	50	87 (*S*)	91a
430	Ph	CO_2Me	NHAc	[Rh(COD)**97c**]BF_4	1	MeOH + Tween 20	25	0.12	50	417	50	86 (*S*)	91a
431	Ph	CO_2Me	NHAc	[Rh(COD)**97c**]BF_4	1	MeOH + Tween 40	25	0.1	50	500	50	86 (*S*)	91a
432	Ph	CO_2Me	NHAc	[Rh(COD)**97c**]BF_4	1	MeOH + Tween 60	25	0.12	50	417	50	85 (*S*)	91a
433	Ph	CO_2Me	NHAc	[Rh(COD)**97c**]BF_4	1	MeOH + Tween 80	25	0.13	50	385	50	87 (*S*)	91a
434	Ph	CO_2Me	NHAc	[Rh(COD)**97c**]BF_4	1	MeOH + Brij 56	25	0.1	50	500	50	83 (*S*)	91a
435	Ph	CO_2Me	NHAc	[Rh(COD)**97c**]BF_4	1	MeOH + Brij 58	25	0.08	50	625	50	85 (*S*)	91a
436	Ph	CO_2Me	NHAc	[Rh(COD)**97c**]BF_4	1	MeOH + Brij 76	25	0.12	50	417	50	83 (*S*)	91a
437	Ph	CO_2Me	NHAc	[Rh(COD)**97c**]BF_4	1	MeOH + Brij 78	25	0.1	50	500	50	82 (*S*)	91a
438	Ph	CO_2Me	NHAc	[Rh(COD)**97c**]BF_4	1	MeOH	25	0.05 [f]	50	1000	50	94.8 (*S*)	91b
439	Ph	CO_2Me	NHAc	[Rh(COD)**97d**]BF_4	1	MeOH	25	–	–	–	–	17.2 (*S*)	26
440	Ph	CO_2Me	NHAc	[Rh(COD)**97d**]BF_4	1	MeOH	25	–	–	–	–	63 (*S*)	91a
441	Ph	CO_2Me	NHAc	$[Rh(COD)Cl]_2$ + **98a** (cationic)	1	PhH	25	0.117 [f]	50	429	50	69 (*S*) [a]	26
442	Ph	CO_2Me	NHAc	$[Rh(COD)Cl]_2$ + **98a** (neutral)	1	EtOH	25	10.7 [f]	50	4.69	50	79 (*S*) [a]	26
443	Ph	CO_2Me	NHAc	[Rh(COD)**98a**]BF_4	1	EtOH	25	0.117 [f]	50	429	50	89 (*S*) [a]	26
444	Ph	CO_2Me	NHAc	[Rh(COD)**98a**]BF_4	1	MeOH	25	0.103 [f]	50	484	50	91.1 (*S*)	26
445	Ph	CO_2Me	NHAc	[Rh(COD)**98a**]BF_4	100	MeOH	25	0.00083 [f]	50	60241	50	91.5 (*S*)	26
446	Ph	CO_2Me	NHAc	[Rh(COD)**98a**]BF_4	1	MeOH	–20	–	–	–	–	95.4 (*S*)	26

447	Ph	CO_2Me	NHAc	[Rh(COD)**98a**]BF_4	1	MeOH	–5.2	–	–	–	–	93.9 (*S*)	26
448	Ph	CO_2Me	NHAc	[Rh(COD)**98a**]BF_4	1	MeOH	10.1	–	–	–	–	93.2 (*S*)	26
449	Ph	CO_2Me	NHAc	[Rh(COD)**98a**]BF_4	1	MeOH	25	–	–	–	–	90.5 (*S*)	26
450	Ph	CO_2Me	NHAc	[Rh(COD)**98a**]BF_4	1	MeOH	40.6	–	–	–	–	88.0 (*S*)	26
451	Ph	CO_2Me	NHAc	[Rh(COD)**98a**]BF_4	1	MeOH	54.6	–	–	–	–	86.2 (*S*)	26
452	Ph	CO_2Me	NHAc	[Rh(COD)**98a**]BF_4	1	MeOH	25	–	–	–	–	91 (*S*)	91a
453	Ph	CO_2Me	NHAc	[Rh(COD)**98a**]BF_4	1	MeOH	25	0.1 [f]	50	500	50	91.5 (*S*)	91e
454	Ph	CO_2Me	NHAc	[Rh(COD)**98a**]BF_4	1	$ClCH_2CH_2Cl$	25	0.08	50	625	50	90 (*S*)	91e
455	Ph	CO_2Me	NHAc	[Rh(COD)**98a**]BF_4	1	CH_2Cl_2	25	0.17	50	294	50	89 (*S*)	91e
456	Ph	CO_2Me	NHAc	[Rh(COD)**98a**]BF_4	1	*o*-Xylene	25	0.03	50	1667	50	83 (*S*)	91e
457	Ph	CO_2Me	NHAc	[Rh(COD)**98a**]BF_4	1	*m*-Xylene	25	0.08	50	625	50	85 (*S*)	91e
458	Ph	CO_2Me	NHAc	[Rh(COD)**98a**]BF_4	1	*p*-Xylene	25	0.13	50	385	50	81 (*S*)	91e
459	Ph	CO_2Me	NHAc	[Rh(COD)**98a**]BF_4	1	EtOH	25	0.083 [f]	50	600	50	89 (*S*)	91e
460	Ph	CO_2Me	NHAc	[Rh(COD)**98a**]BF_4	1	THF	25	0.067 [f]	50	750	50	86.1 (*S*)	91e
461	Ph	CO_2Me	NHAc	[Rh(COD)**98a**]BF_4	1	PhH	25	0.117 [f]	50	429	50	81.0 (*S*)	91e
462	Ph	CO_2Me	NHAc	[Rh(COD)**98a**]BF_4	1	Tol	25	0.1 [f]	50	500	50	81.0 (*S*)	91e
463	Ph	CO_2Me	NHAc	[Rh(COD)**98a**]BF_4	2–2.8	THF	r.t.	2–3	–	–	–	84.7	95b
464	Ph	CO_2Me	NHAc	[Rh(COD)**98a**]SbF_6	2–2.8	THF	r.t.	2–3	–	–	–	90.2	95b
465	Ph	CO_2Me	NHAc	[Rh(COD)**98b**]BF_4	2–2.8	THF	r.t.	2–3	–	–	–	94.4	95b
466	Ph	CO_2Me	NHAc	[Rh(COD)**98b**]SbF_6	2–2.8	THF	r.t.	2–3	–	–	–	97.4	95b
467	Ph	CO_2Me	NHAc	[Rh(COD)**98c**]BF_4	2–2.8	THF	r.t.	2–3	–	–	–	6.2	95b
468	Ph	CO_2Me	NHAc	[Rh(COD)**98c**]SbF_6	2–2.8	THF	r.t.	2–3	–	–	–	2.0	95b
469	Ph	CO_2Me	NHAc	[Rh(COD)**98d**]BF_4	2–2.8	THF	r.t.	2–3	–	–	–	7.2	95b
470	Ph	CO_2Me	NHAc	[Rh(COD)**98e**]BF_4	2–2.8	THF	r.t.	2–3	–	–	–	9.8	95b
471	Ph	CO_2Me	NHAc	[Rh(COD)**98e**]SbF_6	2–2.8	THF	r.t.	2–3	–	–	–	2.0	95b
472	Ph	CO_2Me	NHAc	[Rh(COD)**98g**]BF_4	2–2.8	THF	r.t.	2–3	–	–	–	98.2	95b
473	Ph	CO_2Me	NHAc	[Rh(COD)**98g**]SbF_6	2–2.8	THF	r.t.	2–3	–	–	–	99.0	95b
474	Ph	CO_2Me	NHAc	[Rh(COD)**98h**]SbF_6	2–2.8	THF	r.t.	2–3	–	–	–	81.0	95b
475	Ph	CO_2Me	NHAc	[Rh(COD)**98i**]BF_4	1	MeOH	25	0.0567 [f]	50	882	50	95 (*S*)	97
476	Ph	CO_2Me	NHAc	[Rh(COD)**98i**]BF_4	1	H_2O	25	6 [f]	50	8	50	84 (*S*)	97
477	Ph	CO_2Me	NHAc	[Rh(COD)**98i**]BF_4	1	H_2O+SDS, 0.035 [g]	25	1.1 [f]	50	45	50	94 (*S*)	97

Table 27.5 (continued)

Entry	Substrate			Catalyst	Conditions				TON	TOF $[h^{-1}]$	Conv. [%]	ee [%]	Refer-ence(s)
	R^1	R^2	R^3		$P[H_2]$ [bar]	Solvent	Temp. [°C]	Time [h]					
478	Ph	CO_2Me	NHAc	[Rh(COD)**98i**]BF_4	1	H_2O+SDS, 0.173 [g]	25	0.11 [f]	50	469	50	97 (*S*)	97
479	Ph	CO_2Me	NHAc	[Rh(COD)**98i**]BF_4	1	H_2O+Triton X-100, 0.03 [g]	25	1.083 [f]	50	46	50	95 (*S*)	97
480	Ph	CO_2Me	NHAc	[Rh(COD)**98i**]BF_4	1	H_2O+Triton X-100, 0.1 [g]	25	0.33 [f]	50	150	50	95 (*S*)	97
481	Ph	CO_2Me	NHAc	[Rh(COD)**100**]Cl	50	–	25	8	100 [c]	–	–	10 (*S*) [a]	89
482	Ph	CO_2Me	NHAc	[Rh(COD)**100b**BF_4	1	MeOH	25	–	–	–	–	89 (*S*)	91 a
483	Ph	CO_2Me	NHAc	[Rh(COD)**100b**]BF_4	1	MeOH	25	–	50	–	50	80 (*S*)	91 e
484	Ph	CO_2Me	NHAc	[Rh(COD)**101a**]BF_4	1	MeOH	25	0.09 [f]	50	555	50	91.1 (*S*)	91 e
485	Ph	CO_2Me	NHAc	[Rh(COD)**101b**]BF_4	1	MeOH	25	0.093 [f]	50	536	50	91.8 (*S*)	91 c
486	Ph	CO_2Me	NHAc	[Rh(COD)**101c**]BF_4	1	MeOH	25	0.11 [f]	50	469	50	89.2 (*S*)	91 c
487	Ph	CO_2Me	NHAc	[Rh(COD)**101d**]BF_4	1	MeOH	25	0.15 [f]	50	345	50	90.8 (*S*)	91 c
488	Ph	CO_2Me	NHAc	[Rh(COD)**101e**]BF_4	1	MeOH	25	0.12 [f]	50	407	50	89.3 (*S*)	91 c
489	Ph	CO_2Me	NHAc	[Rh(COD)**101 f**]BF_4	1	MeOH	25	0.14 [f]	50	361	50	83.8 (*S*)	91 c
490	Ph	CO_2Me	NHAc	[Rh(COD)**102a**]BF_4	2–2.8	THF	r.t.	2–3	–	–	–	98.3	95 b
491	Ph	CO_2Me	NHAc	[Rh(COD)**102a**]SbF_6	2–2.8	THF	r.t.	2–3	–	–	–	98.4	95 b
492	Ph	CO_2Me	NHAc	[Rh(COD)**102b**]SbF_6	2–2.8	THF	r.t.	2–3	–	–	–	94.9	95 b
493	Ph	CO_2Me	NHAc	[Rh(COD)**103a**]BF_4	2–2.8	THF	r.t.	2–3	–	–	–	93.0	95 b
494	Ph	CO_2Me	NHAc	[Rh(COD)**103a**]SbF_6	2–2.8	THF	r.t.	2–3	–	–	–	96.3	95 b
495	Ph	CO_2Me	NHAc	[Rh(COD)**103b**]BF_4	2–2.8	THF	r.t.	–	–	–	–	71.1	95 a
496	Ph	CO_2Me	NHAc	[Rh(COD)**103b**]BF_4	2–2.8	THF	r.t.	2–3	–	–	–	87.4	95 b
497	Ph	CO_2Me	NHAc	[Rh(COD)**103c**]BF_4	2–2.8	THF	r.t.	2–3	–	–	–	1.0	95 a, b
498	Ph	CO_2Me	NHAc	[Rh(COD)**103d**]BF_4	2–2.8	THF	r.t.	–	–	–	–	2.3	95 a, b
499	Ph	CO_2Me	NHAc	[Rh(COD)**103e**]BF_4	2–2.8	THF	r.t.	2–3	–	–	–	2.0	95 a, b

500	Ph	CO_2Me	NHAc	[Rh(COD)**103f**]BF_4	2–2.8	THF	r.t.	2–3	–	–	–	84.7	95a, b
501	Ph	CO_2Me	NHAc	[Rh(COD)**104a**]BF_4	2–2.8	THF	r.t.	2–3	–	–	–	92.4	95b
502	Ph	CO_2Me	NHAc	[Rh(COD)**104b**]BF_4	2–2.8	THF	r.t.	2–3	–	–	–	84.0	95b
503	Ph	CO_2Me	NHAc	[Rh(COD)**104d**]BF_4	2–2.8	THF	r.t.	2–3	–	–	–	11.0	95b
504	Ph	CO_2Me	NHAc	[Rh(COD)**105a**]BF_4	2–2.8	THF	r.t.	2–3	–	–	–	65.1	95a, b
505	Ph	CO_2Me	NHAc	[Rh(COD)**106a**]SbF_6	2–2.8	THF	r.t.	2–3	–	–	–	83.2	95b
506	Ph	CO_2Me	NHAc	[Rh(COD)**111**]ClO_4	1	EtOH	25	24	50	2	100	52 (*R*)	90
507	Ph	CO_2Me	NHAc	[Rh(COD)**112a**]BF_4	2–2.8	THF	r.t.	2–3	–	–	–	72.2	95b
508	Ph	CO_2Me	NHAc	[Rh(COD)**112b**]BF_4	1	MeOH	25	0.07	50	714	50	1.5 (*S*)	91e
509	Ph	CO_2Me	NHAc	[Rh(COD)**113a**]Cl	50	–	25	8	100 [c]	–	–	46 (*S*) [a]	89
510	Ph	CO_2Me	NHAc	[Rh(COD)**113b**]Cl	50	–	25	8	100 [c]	–	–	20 (*S*) [a]	89
511	Ph	CO_2Me	NHAc	[Rh(COD)**113c**]BF_4	1	MeOH	25	0.07	50	714	50	66 (*S*)	91a, e
512	Ph	CO_2Me	NHAc	[Rh(COD)**113c**]BF_4	1	MeOH	25	0.067	50	750	50	66 (*S*)	91e
513	Ph	CO_2Me	NHAc	[Rh(COD)**113d**]BF_4	1	MeOH	25	–	–	–	–	77 (*S*)	91a
514	Ph	CO_2Me	NHAc	[Rh(COD)**113d**]BF_4	1	MeOH	25	0.13	50	750	50	83 (*S*)	91e
515	Ph	CO_2Me	NHAc	[Rh(COD)**113e**]BF_4	1	MeOH	25	0.13	50	385	50	83 (*S*)	91a, e
516	Ph	CO_2Me	NHAc	[Rh(COD)**113f**]BF_4	1	MeOH	25	–	–	–	–	59 (*S*)	91a
517	Ph	CO_2Me	NHAc	[Rh(COD)**114a**]BF_4	1	MeOH	25	0.083 [f]	50	602	50	57 (*S*)	99
518	Ph	CO_2Me	NHAc	[Rh(COD)**114a**]BF_4	1	PhH	25	0.05 [f]	50	1000	50	43 (*S*)	99
519	Ph	CO_2Me	NHAc	[Rh(COD)**114b**]BF_4	1	MeOH	25	0.05 [f]	50	1000	50	53 (*S*)	99
520	Ph	CO_2Me	NHAc	[Rh(COD)**114b**]BF_4	1	PhH	25	0.033 [f]	50	1515	50	41 (*S*)	99
521	Ph	CO_2Me	NHAc	[Rh(COD)**116a**]SbF_6	2.07	THF	r.t.	–	–	–	–	35 (*R*)	102
522	Ph	CO_2Me	NHAc	[Rh(COD)**116b**]SbF_6	2.07	THF	r.t.	–	–	–	–	30 (*R*)	102
523	Ph	CO_2Me	NHAc	[Rh(COD)**117**]BF_4	5	H_2O	r.t.	6	20	3	100	88 (*S*)	101
524	Ph	CO_2Me	NHAc	[Rh(COD)**117**]BF_4	5	H_2O + 10 wt% SDS	r.t.	1	100	100	100	99.9 (*S*) [b]	101
525	Ph	CO_2Me	NHAc	[Rh(COD)**117**]BF_4	5	H_2O/EtOAc (1:1)	r.t.	1.5	50	33	100	87 (*S*)	101
526	Ph	CO_2Me	NHAc	[Rh(COD)**117**]BF_4	5	H_2O/MeOH/ EtOAc (0.6:0.4:1)	r.t.	3	100	33	100	98 (*S*)	101
527	Ph	CO_2Me	NHAc	[Rh(COD)**117**]BF_4	5	H_2O/MeOH (3:2)	r.t.	1.5	100	67	100	94 (*S*)	101
528	Ph	CO_2Me	NHAc	[Rh(COD)**118**]SbF_6	2.07	THF	r.t.	–	–	–	–	25 (*R*)	102
529	Ph	CO_2Me	NHAc	[Rh(COD)**119a**]SbF_6	2.07	THF	r.t.	–	–	–	~100	69 (*S*)	102

Table 27.5 (continued)

Entry	Substrate			Catalyst	Conditions				TON	TOF $[h^{-1}]$	Conv. [%]	ee [%]	Refer-ence(s)
	R^1	R^2	R^3		$P[H_2]$ [bar]	Solvent	Temp. [°C]	Time [h]					
530	Ph	CO_2Me	NHAc	[Rh(COD)**119b**]SbF_6	2.07	THF	r.t.	–	–	–	~100	87 (*S*)	102
531	Ph	CO_2Me	NHAc	[Rh(COD)**119c**]BF_4	5	H_2O/EtOAc (1:1)	r.t.	1.5	50	33	100	68 (*S*)	101
532	Ph	CO_2Me	NHAc	[Rh(COD)**119c**]BF_4	5	H_2O/MeOH/EtOAc (0.6:0.4:1)	r.t.	3	100	33	100	76 (*S*)	101
533	Ph	CO_2Me	NHAc	[Rh(COD)**119c**]BF_4	5	H_2O/MeOH (3:2)	r.t.	1.5	100	67	100	75 (*S*)	101
534	Ph	CO_2Me	NHAc	[Rh(COD)**119d**]BF_4	5	H_2O	r.t.	6	20	3	100	55 (*S*)	101
535	Ph	CO_2Me	NHAc	[Rh(COD)**119d**]BF_4	5	H_2O + 10 wt% SDS	r.t.	1	100	100	100	90 (*S*) [b]	101
536	Ph	CO_2Me	NHAc	[Rh(COD)ent-**120**]BF_4	1	Acetone	r.t.	0.08	100	1250	100	27 (*R*)	105
537	Ph	CO_2Me	NHAc	[Rh(COD)**121a**]BF_4	1	Acetone	r.t.	0.08	100	1250	100	18 (*R*)	105
538	Ph	CO_2Me	NHAc	[Rh(COD)**121b**]BF_4	1	Acetone	r.t.	0.08	100	1250	100	59 (*R*)	105
539	Ph	CO_2Me	NHAc	[Rh(COD)**121d**]BF_4	1	Acetone	r.t.	0.08	100	1250	100	32 (*R*)	105
540	Ph	CO_2Me	NHAc	[Rh(COD)**122a**]BF_4	1	Acetone	r.t.	0.08	95	1188	95	73 (*R*)	105
541	Ph	CO_2Me	NHAc	[Rh(COD)**122b**]BF_4	1	Acetone/DCM 13:2	r.t.	0.08	96	1200	96	81 (*R*)	105
542	Ph	CO_2Me	NHAc	[Rh(COD)**122c**]BF_4	1	Acetone	r.t.	0.08	100	1250	100	77 (*R*)	105
543	Ph	CO_2Me	NHAc	[Rh(COD)**122d**]BF_4	1	Acetone/DCM 13:2	r.t.	0.08	100	1250	100	75 (*R*)	105
544	Ph	CO_2Me	NHAc	[Rh(COD)**122e**]BF_4	1	Acetone	r.t.	0.08	100	1250	100	86 (*R*)	105
545	Ph	CO_2Me	NHAc	[Rh(COD)**123b**]SbF_6	2–2.8	THF	r.t.	2–3	–	–	–	49.0	95b
546	Ph	CO_2Me	NHAc	[Rh(COD)**124**]BF_4	1	THF	25	1	100	100	100	24 (*R*)	103
547	Ph	CO_2Me	NHAc	[Rh(COD)**124**]Cl	68	PhH : EtOH = 1 : 1	100	48	100	2.1	100	3.4 (*R*)	103
548	Ph	CO_2Me	NHAc	[Rh(COD)**125**]BF_4	1	THF	25	1	100	100	100	35 (*S*)	103
549	Ph	CO_2Me	NHAc	[Rh(COD)**126**]BF_4	1	DCM	25	0.02	100	5000	100	10 (*R*)	104
550	Ph	CO_2Me	NHAc	[Rh(COD)**126**]Cl	50	–	25	1	100 [c]	100	100	48 (*R*) [a]	89
551	Ph	CO_2Me	NHAc	[Ir(COD)**126**]BF_4	1	DCM	25	0.42	100	238	100	20 (*R*)	104
552	Ph	CO_2Me	NHAc	[Rh(COD)**127**]BF_4	1	DCM	25	0.02	100	5000	100	35 (*R*)	104

553	Ph	CO_2Me	NHAc	[Ir(COD)**127**]BF_4	1	DCM	25	0.75	100	133	100	10 (*R*)	104
554	Ph	CO_2Me	NHAc	[Rh(COD)**131**]BF_4	1	PhH	25	5.5	99.9	18	99.9	31.5 (*R*)	108
555	Ph	CO_2Me	NHAc	[Rh(COD)**132**]BF_4	34.5	Acetone	25	0.25	100	400	100	91.6 (*R*)	107
556	Ph	CO_2Me	NHAc	[Rh(COD)**132**]BF_4	34.5	MeOH	25	0.25	100	400	100	84 (*R*)	107
557	Ph	CO_2Me	NHAc	[Rh(COD)**132**]BF_4	34.5	IPA	25	0.25	100	400	100	89.4 (*R*)	107
558	Ph	CO_2Me	NHAc	[Rh(COD)**132**]BF_4	34.5	THF	25	0.25	100	400	100	86.3 (*R*)	107
559	Ph	CO_2Me	NHAc	[Rh(COD)**132**]BF_4	34.5	DCM	25	0.25	100	400	100	86.2 (*R*)	107
560	Ph	CO_2Me	NHAc	[Rh(COD)**132**]BF_4	34.5	PhH	25	0.25	100	400	100	82.9 (*R*)	107
561	Ph	CO_2Me	NHAc	[Rh(COD)**134**]BF_4	1	PhH	25	4.5	92.5	21	92.5	24.6 (*R*)	108
562	Ph	CO_2Me	NHAc	[Rh(COD)**135**]BF_4	1	PhH	25	7.5	96.6	13	96.6	46.2 (*R*)	108
563	Ph	CO_2Me	NHAc	[Rh(COD)**136**]BF_4	1.5	MeOH	25	20	388	19	81	19 (*R*)	110
564	Ph	CO_2Me	NHAc	[Rh(COD)**139**]BF_4	1.2	MeOH	–		100	–	100	54	113
565	Ph	CO_2Me	NHAc	[Rh(COD)**139**]BF_4	5	MeOH	–	2	94	47	94	47	113
566	Ph	CO_2Me	NHAc	[Rh(COD)**140**]BF_4	1.5	DCM	25	20	479	24	100	14 (*R*)	110
567	Ph	CO_2Me	NHAc	[Rh(COD)**141a**]BF_4	3.5	MeOH	r.t.	0.5	1000	2000	>99	95 (*S*)	111
568	Ph	CO_2Me	NHAc	[Rh(COD)**141a**]BF_4	5	MeOH	r.t.	2	5000	2500	>99	95 (*S*)	111
569	Ph	CO_2Me	NHAc	[Rh(COD)**141b**]BF_4	3.5	MeOH	r.t.	0.5	980	1960	98	97 (*S*)	111
570	Ph	CO_2Me	NHAc	[Rh(COD)**141b**]BF_4	3.5	Tol	r.t.	2	1000	500	>99	99 (*S*)	111
571	Ph	CO_2Me	NHAc	[Rh(COD)**141b**]BF_4	5	MeOH	r.t.	6	5000	833	>99	98.5 (*S*)	111
572	Ph	CO_2Me	NHAc	[Rh(COD)**144a**]BF_4	5	DCM	25	8	96	12	96	91 (*S*)	117b
573	Ph	CO_2Me	NHAc	[Rh(COD)**144c**]BF_4	5	DCM	25	6	100	16.7	100	98 (*S*)	117b
574	Ph	CO_2Me	NHAc	[Rh(COD)**144c**]BF_4	30	DCM	5	4	1000	250	100	>99 (*S*)	117b
575	Ph	CO_2Me	NHAc	[Rh(COD)**145a**]BF_4	5	DCM	25	8	100	12.5	100	2 (*S*)	117b
576	Ph	CO_2Me	NHAc	[Rh(COD)**146a**]BF_4	5	DCM	25	8	98	12.3	98	70 (*S*)	117b
577	Ph	CO_2Me	NHAc	[Rh(COD)**146c**]BF_4	5	DCM	25	8	96	12	96	32 (*S*)	117b
578	Ph	CO_2Me	NHAc	[Rh(COD)**147**]BF_4	1	THF	25	–	–	–	–	13	115
579	Ph	CO_2Me	NHAc	[Rh(COD)**150a**]BF_4	1	DCM	25	2.2	100	45.5	100	30 (*S*)	118
580	Ph	CO_2Me	NHAc	[Rh(COD)**150b**]BF_4	1	DCM	25	3.3	100	30.3	100	18 (*R*)	118
581	Ph	CO_2Me	NHAc	[Rh(COD)**150c**]BF_4	1	DCM	25	4.3	100	23.3	100	30 (*R*)	118
582	Ph	CO_2Me	NHAc	[Rh(COD)**150d**]BF_4	1	DCM	25	3	100	33.3	100	48 (*S*)	118

Table 27.5 (continued)

Entry	Substrate			Catalyst	Conditions				TON	TOF [h^{-1}]	Conv. [%]	ee [%]	Reference(s)
	R^1	R^2	R^3		$P[H_2]$ [bar]	Solvent	Temp. [°C]	Time [h]					
583	Ph	CO_2Me	NHAc	[Rh(COD)**154a**]OTf	5	MeOH	r.t.	18	200	11.1	100	81 (*S*)	127
584	Ph	CO_2Me	NHAc	[Rh(COD)**154a**]OTf	5	Tol	r.t.	18	200	11.1	100	85 (*S*)	127
585	Ph	CO_2Me	NHAc	[Rh(COD)**154b**]OTf	5	Tol	r.t.	18	200	11.1	100	89 (*S*)	127
586	Ph	CO_2Me	NHAc	[Rh(COD)**154c**]OTf	5	Tol	r.t.	18	200	11.1	100	50 (*R*)	127
587	Ph	CO_2Me	NHAc	[Rh(COD)(*S*,*S*ax)-**155b**] BF_4	4.1	DCM	r.t.	24	450	18.8	90	70.3 (*R*)	124 a
588	Ph	CO_2Me	NHAc	[Rh(COD) (*S*,*R*ax)-**155b**]B F_4	4.1	DCM	r.t.	24	500	18.8	100	99.0 (*S*)	124 a
589	Ph	CO_2Me	NHAc	[Rh(COD)**155c**]BF_4	4.1	DCM	r.t.	24	225	9.4	45	22.0 (*S*)	124 a
590	Ph	CO_2Me	NHAc	[Rh(COD)**156a**]BF_4	4.1	DCM	r.t.	16	500	31.3	100	99.5 (*R*)	124 a
591	Ph	CO_2Me	NHAc	[Rh(COD)**156b**]BF_4	4.1	DCM	r.t.	16	500	31.3	100	56.1 (*R*)	124 a
592	Ph	CO_2Me	NHAc	[Rh(COD)**156c**]BF_4	4.1	DCM	r.t.	16	100	6.3	20	90.6 (*R*)	124 a
593	Ph	CO_2Me	NHAc	[Rh(COD)**159a**]BF_4	1	DCM	r.t.	24	100	4.2	100	97 (*R*)	123
594	Ph	CO_2Me	NHAc	[Rh(COD)**159b**]BF_4	1	DCM	r.t.	24	100	4.2	100	92 (*S*)	123
595	Ph	CO_2Me	NHAc	[Rh(COD)**159c**]BF_4	1	DCM	r.t.	24	50	2.1	100	6 (*S*)	123
596	Ph	CO_2Me	NHAc	[Rh(COD)**159d**]BF_4	1	DCM	r.t.	12	100	8.3	100	95 (*S*)	123
597	Ph	CO_2Me	NHAc	[Rh(COD)**159e**]BF_4	1	DCM	r.t.	12	100	8.3	100	95 (*R*)	123
598	Ph	CO_2Me	NHAc	[Rh(COD)**159 f**]BF_4	1	DCM	r.t.	12	50	4.2	100	3 (*R*)	123
599	Ph	CO_2Me	NHAc	[Rh(COD)**159g**]BF_4	1	DCM	r.t.	12	50	4.2	100	89 (*S*)	123
600	Ph	CO_2Me	NHAc	[Rh(COD)**159h**]BF_4	1	DCM	r.t.	24	85	3.5	85	63 (*R*)	123
601	Ph	CO_2Me	NHAc	[Rh(COD)**160a**]BF_4	1	DCM	r.t.	12	50	4.2	100	65 (*S*)	123
602	Ph	CO_2Me	NHAc	[Rh(COD)**160b**]BF_4	1	DCM	r.t.	1	50	50	100	95 (*R*)	123
603	Ph	CO_2Me	NHAc	[Rh(COD)**163a**]PF_6	1	THF	r.t.	12	>99	8.3	100	>99 (*S*)	128
604	Ph	CO_2Me	NHAc	[Rh(COD)**164a**]PF_6	1	THF	r.t.	12	>99	8.3	100	>99 (*S*)	128

605	Ph	CO_2Me	NHAc	[Rh(COD)**166**]BF_4	1	DCM	r.t.	1	100	100	100	37 (*R*)	129
606	Ph	CO_2Me	NHAc	[Rh(COD)**166**]BF_4	1	DCE	r.t.	0.5	100	200	100	69 (*R*)	129
607	Ph	CO_2Me	NHAc	[Rh(COD)**166**]BF_4	50	DCE	r.t.	2	88	44	88	33 (*R*)	129
608	Ph	CO_2Me	NHAc	[Rh(COD)**166**]BF_4	30	H_2O	r.t.	24	40	1.7	100	13 (*R*)	129
609	Ph	CO_2Me	NHAc	[Rh(COD)**166**]BF_4	50	H_2O	r.t.	24	40	1.7	100	72 (*R*)	129
610	Ph	CO_2Me	NHAc	[Rh(COD)**166**]BF_4	70	H_2O	r.t.	24	40	1.7	100	62 (*R*)	129
611	Ph	CO_2Me	NHAc	[Rh(COD)**166**]BF_4	50	MeOH	r.t.	9	50	5.6	100	50 (*R*)	129
612	Ph	CO_2Me	NHAc	[Rh(COD)**166**]BF_4	50	H_2O/EtOAc (1:1)	r.t.	12(24)[h]	50	4.2	100 (100)[i]	73 (70)[i](*R*)	129
613	Ph	CO_2Me	NHAc	[Rh(COD)**167a**]BF_4	1	DCM	25	0.17	100	588	100	84.1 (*S*)	121a, b
614	Ph	CO_2Me	NHAc	[Rh(COD)**167b**]BF_4	1	DCM	25	3	100	33	100	98.8 (*R*)	121a, b
615	Ph	CO_2Me	NHAc	[Rh(COD)**167c**]BF_4	1	DCM	r.t.	0.5	100	200	100	98.0 (*S*)	121a
616	Ph	CO_2Me	NHAc	[Rh(COD)**167c**]BF_4	1	DCM	25	0.5	100	200	100	91 (*S*)	121b
617	Ph	CO_2Me	NHAc	[Rh(COD)**167d**]BF_4	1	DCM	r.t.	0.5	100	200	100	94.3 (*R*)	121a, b
618	Ph	CO_2Me	NHAc	[Rh(COD)**168a**]BF_4	5	DCM	25	8	77	6	77	94 (*S*)	126
619	Ph	CO_2Me	NHAc	[Rh(COD)**168a**]BF_4	30	DCM	25	12	72	6	72	98 (*S*)	126
620	Ph	CO_2Me	NHAc	[Rh(COD)**168b**]BF_4	5	DCM	25	8	53	7	53	85 (*S*)	126
621	Ph	CO_2Me	NHAc	[Rh(COD)**168c**]BF_4	5	DCM	25	8	29	4	29	18 (*S*)	126
622	Ph	CO_2Me	NHAc	[Rh(COD)**168d**]BF_4	5	DCM	25	8	35	4	35	17 (*S*)	126
623	Ph	CO_2Me	NHBz	[Rh(COD)**97a**]BF_4	1	MeOH	25	0.1[f]	50	500	50	87.3 (*S*)	91b
624	Ph	CO_2Me	NHBz	[Rh(COD)**97c**]BF_4	1	MeOH	25	0.05[f]	50	1000	50	91.6 (*S*)	91b
625	Ph	CO_2Me	NHBz	[Rh(COD)**98a**]BF_4	1	MeOH	25	0.117[f]	50	429	50	77 (*S*)	26
626	Ph	CO_2Me	NHBz	[Rh(COD)**98a**]BF_4	50	MeOH	25	–	100[b]	–	–	77 (*S*)	26
627	Ph	CO_2Me	NHBz	[Rh(COD)**117**]BF_4	5	H_2O/MeOH/ EtOAc (0.6:0.4:1)	r.t.	3	100	33	100	92 (*S*)	101
628	Ph	CO_2Me	NHBz	[Rh(COD)**117**]BF_4	5	H_2O/MeOH(3:2)	r.t.	1.5	100	67	100	90 (*S*)	101
629	Ph	CO_2Me	NHBz	[Rh(COD)**163a**]PF_6	1	THF	r.t.	12	>99	8.3	100	>99 (*S*)	128
630	Ph	CO_2Me	NHBz	[Rh(COD)**164a**]PF_6	1	THF	r.t.	12	>99	8.3	100	>99 (*S*)	128
631	Ph	CO_2Me	NHCbz	[Rh(COD)**98a**]BF_4	1	MeOH	25	–	–	–	–	57 (*S*)[a]	10
632	Ph	CO_2Et	NHAc	[Rh(COD)**89**]PF_6	5	EtOH	60	7	90	13	90	7 (*S*)	82

Table 27.5 (continued)

Entry	Substrate			Catalyst	Conditions				TON	TOF $[h^{-1}]$	Conv. [%]	ee [%]	Reference(s)
	R^1	R^2	R^3		$P[H_2]$ [bar]	Solvent	Temp. [°C]	Time [h]					
633	Ph	CO_2Et	NHAc	[Rh(COD)**90**]PF_6	5	EtOH	60	3	85	28	85	12 (*S*)	82
634	Ph	CO_2Et	NHAc	[Rh(COD)**97a**]BF_4	1	MeOH	25	0.27 [f]	50	187	50	58.3	26
635	Ph	CO_2Et	NHAc	[Rh(COD)**97a**]BF_4	1	MeOH	25	0.1 [f]	50	500	50	90.6 (*S*)	91b
636	Ph	CO_2Et	NHAc	[Rh(COD)**97c**]BF_4	1	MeOH	25	0.05 [f]	50	1000	50	94.4 (*S*)	91b
637	Ph	CO_2Et	NHAc	[Rh(COD)**98a**]BF_4	1	MeOH	25	0.08 [f]	50	625	50	90.2	26
638	2-Cl-Ph	CO_2H	NHAc	[Rh(COD)**91**]BF_4	1	MeOH	25	0.167	100	600	>99.9	97.3 (*R*)	27
639	2-Cl-Ph	CO_2H	NHAc	[Rh(COD)**95**] BF_4	1	IPA	r.t.	24	100	4	100	92.9 (*S*)	85
640	2-Cl-Ph	CO_2H	NHAc	[Rh(COD)**132**]BF_4	34.5	Acetone	25	0.25–1	100	100–400	100	92.3 (*R*)	107
641	2-Cl-Ph	CO_2H	NHAc	[Rh(COD)**163a**]PF_6	1	THF	r.t.	12	>99	8.3	100	>99 (*S*)	128
642	2-Cl-Ph	CO_2H	NHAc	[Rh(COD)**164a**]PF_6	1	THF	r.t.	12	>99	8.3	100	>99 (*S*)	128
643	2-Cl-Ph	CO_2Me	NHAc	[Rh(COD)**94a**]BF_4	6.9	DCM	r.t.	0.17	500	2941	100	85 (*S*)	52
644	2-Cl-Ph	CO_2Me	NHAc	[Rh(COD)**93h**]PF_6	3	Tol	r.t.	12	100	8.3	100	81.5 (*S*)	79
645	2-Cl-Ph	CO_2Me	NHAc	[Rh(COD)**163a**]PF_6	1	THF	r.t.	12	>99	8.3	100	>99 (*S*)	128
646	2-Cl-Ph	CO_2Me	NHAc	[Rh(COD)**164a**]PF_6	1	THF	r.t.	12	>99	8.3	100	>99 (*S*)	128
647	3-Cl-Ph	CO_2H	NHAc	[Rh(COD)**91**]BF_4	1	MeOH	25	0.167	100	600	>99.9	97.4 (*R*)	27
648	3-Cl-Ph	CO_2H	NHAc	[Rh(COD)**93a**]BF_4	6.9	MeOH	r.t.	0.5	282	564	56.3	10.5 (*S*)	52
649	3-Cl-Ph	CO_2H	NHAc	[Rh(COD)**94a**]BF_4	6.9	MeOH	r.t.	0.5	440	880	87.9	37 (*S*)	52
650	3-Cl-Ph	CO_2H	NHAc	[Rh(COD)**132**]BF_4	34.5	Acetone	25	0.25–1	100	100–400	100	90.3 (*R*)	107
651	3-Cl-Ph	CO_2Me	NHAc	[Rh(COD)**93a**]BF_4	6.9	DCM	r.t.	0.17	352	2071	70.3	54.7 (*S*)	52
652	3-Cl-Ph	CO_2Me	NHAc	[Rh(COD)**94a**]BF_4	6.9	DCM	r.t.	0.17	500	2941	100	78.3 (*S*)	52
653	3-Cl-Ph	CO_2Me	NHAc	[Rh(COD)**166**]BF_4	51	H_2O/EtOAc (1:1)	r.t.	24	50	2	100	67 (*R*)	129
654	4-Cl-Ph	CO_2H	NHAc	[Rh(COD)**91**]BF_4	1	MeOH	25	0.167	100	600	>99.9	97.3 (*R*)	27
655	4-Cl-Ph	CO_2H	NHAc	[Rh(COD)**132**]BF_4	34.5	Acetone	25	0.25–1	100	100–400	100	93.3 (*R*)	107
656	4-Cl-Ph	CO_2H	NHAc	[Rh(COD)**132**]BF_4	34.5	Acetone	0–5	0.25–1	100	100–400	100	94.6 (*R*)	107

657	4-Cl-Ph	CO_2H	NHAc	[Rh(COD)**132**]BF_4	34.5	Acetone	–15	0.25–1	100	100–400	100	96.3 (*R*)	107
658	4-Cl-Ph	CO_2Me	NHAc	[Rh(COD)**91**]BF_4	1	MeOH	25	0.167	100	600	>99.9	94.2 (*R*)	27
659	4-Cl-Ph	CO_2Me	NHAc	[Rh(COD)**94a**]BF_4	6.9	DCM	r.t.	0.17	500	2941	100	80.8 (*S*)	52
660	4-Cl-Ph	CO_2Me	NHAc	[Rh(COD)**132**]BF_4	34.5	Acetone	25	0.25	100	400	100	91.3 (*R*)	107
661	3-Br-Ph	CO_2H	NHAc	[Rh(COD)**95**] BF_4	1	IPA	r.t.	24	100	4	100	93.5 (*S*)	85
662	3-Br-Ph	CO_2H	NHAc	[Rh(COD)**98a**]BF_4	2.8	THF	r.t.	3	100	33	100	89 (*S*)	98
663	3-Br-Ph	CO_2H	NHAc	[Rh(COD)**98b**]BF_4	2.8	THF	r.t.	3	100	33	100	97 (*S*)	98
664	3-Br-Ph	CO_2H	NHAc	[Rh(COD)**103a**]SbF_6	2–2.8	THF	r.t.	–	–	–	–	96.4	95 a
665	3-Br-Ph	CO_2H	NHAc	[Rh(COD)**115a**]SbF_6	2.8	THF	r.t.	3	100	33	100	74 (*S*)	98
666	3-Br-Ph	CO_2H	NHAc	[Rh(COD)**115b**]SbF_6	2.8	THF	r.t.	3	100	33	100	95 (*S*)	98
667	3-Br-Ph	CO_2H	NHAc	[Rh(COD)**115d**]BF_4	2.8	THF	r.t.	–	100	–	100	96 (*S*)	98
668	3-Br-Ph	CO_2H	NHAc	[Rh(COD)**163a**]PF_6	1	THF	r.t.	12	>99	8.3	100	>99 (*S*)	128
669	3-Br-Ph	CO_2H	NHAc	[Rh(COD)**164a**]PF_6	1	THF	r.t.	12	>99	8.3	100	>99 (*S*)	128
670	3-Br-Ph	CO_2Me	NHAc	[Rh(COD)**93h**]PF_6	3	Tol	r.t.	12	100	8.3	100	92.6 (*S*)	79
671	3-Br-Ph	CO_2Me	NHAc	[Rh(COD)**98a**]SbF_6	2–2.8	THF	r.t.	2–3	–	–	–	89.2	95 b
672	3-Br-Ph	CO_2Me	NHAc	[Rh(COD)**98b**]SbF_6	2–2.8	THF	r.t.	2–3	–	–	–	96.8	95 b
673	3-Br-Ph	CO_2Me	NHAc	[Rh(COD)**103a**]SbF_6	2–2.8	THF	r.t.	2–3	–	–	–	96.4	95 b
674	3-Br-Ph	CO_2Me	NHAc	[Rh(COD)**163a**]PF_6	1	THF	r.t.	12	>99	8.3	100	>99 (*S*)	128
675	3-Br-Ph	CO_2Me	NHAc	[Rh(COD)**164a**]PF_6	1	THF	r.t.	12	>99	8.3	100	>99 (*S*)	128
676	4-Br-Ph	CO_2H	NHAc	[Rh(COD)**98b**]SbF_6	2–2.8	THF	r.t.	2–3	–	–	–	98.0	95 b
677	4-Br-Ph	CO_2H	NHAc	[Rh(COD)**98c**]SbF_6	2–2.8	THF	r.t.	2–3				47.0	95 b
678	4-Br-Ph	CO_2H	NHAc	[Rh(COD)**103a**]SbF_6	2–2.8	THF	r.t.	2–3	–	–	–	96.4	95 b
679	4-Br-Ph	CO_2Me	NHAc	[Rh(COD)**91**]BF_4	1	MeOH	25	0.167	100	600	>99.9	96.3 (*R*)	27
680	4-Br-Ph	CO_2Me	NHAc	[Rh(COD)**132**]BF_4	34.5	Acetone	25	0.25	100	400	100	91.2 (*R*)	107
681	4-Br-Ph	CO_2Me	NHAc	[Rh(COD)**154a**]OTf	5	MeOH	r.t.	18	200	11.1	100	82 (*S*)	127
682	4-Br-Ph	CO_2Me	NHAc	[Rh(COD)**154a**]OTf	5	Tol	r.t.	18	200	11.1	100	87 (*S*)	127
683	4-Br-Ph	CO_2Me	NHAc	[Rh(COD)**154b**]OTf	5	Tol	r.t.	18	200	11.1	100	87 (*S*)	127
684	4-Br-Ph	CO_2Me	NHAc	[Rh(COD)**154c**]OTf	5	Tol	r.t.	18	196	10.9	98	29 (*R*)	127
685	2-F-Ph	CO_2H	NHAc	[Rh(COD)**98a**]BF_4	2.8	THF	r.t.	3	100	33	100	89 (*S*)	98
686	2-F-Ph	CO_2H	NHAc	[Rh(COD)**98b**]BF_4	2.8	THF	r.t.	3	100	33	100	97 (*S*)	98
687	2-F-Ph	CO_2H	NHAc	[Rh(COD)**115a**]SbF_6	2.8	THF	r.t.	3	100	33	100	63 (*S*)	98

Table 27.5 (continued)

Entry	Substrate			Catalyst	Conditions				TON	TOF [h^{-1}]	Conv. [%]	ee [%]	Refer-ence(s)
	R^1	R^2	R^3		$P[H_2]$ [bar]	Solvent	Temp. [°C]	Time [h]					
688	2-F-Ph	CO_2H	NHAc	[Rh(COD)**115b**]SbF_6	2.8	THF	r.t.	3	100	33	100	96 (*S*)	98
689	2-F-Ph	CO_2Me	NHAc	[Rh(COD)**98a**]SbF_6	2–2.8	THF	r.t.	2–3	–	–	–	89.1	95 b
690	2-F-Ph	CO_2Me	NHAc	[Rh(COD)**98b**]SbF_6	2–2.8	THF	r.t.	2–3	–	–	–	96.8	95 b
691	2-F-Ph	CO_2Me	NHAc	[Rh(COD)**98g**]SbF_6	2–2.8	THF	r.t.	2–3	–	–	–	97.8	95 b
692	2-F-Ph	CO_2Me	NHAc	[Rh(COD)**103a**]SbF_6	2–2.8	THF	r.t.	2–3	–	–	–	95.6	95 b
693	2-F-Ph	CO_2Me	NHAc	[Rh(COD)**115a**]SbF_6	2.8	THF	r.t.	3	100	33	100	66 (*S*)	98
694	3-F-Ph	CO_2H	NHAc	[Rh(COD)**115d**]BF_4	2.8	THF	r.t.	–	100	–	100	95 (*S*)	98
695	3-F-Ph	CO_2H	NHAc	[Rh(COD)**115e**]BF_4	2.8	THF	r.t.	–	100	–	100	2 (*S*)	98
696	3-F-Ph	CO_2Me	NHAc	[Rh(COD)**98a**]SbF_6	2–2.8	THF	r.t.	2–3	–	–	–	88.9	95 b
697	3-F-Ph	CO_2Me	NHAc	[Rh(COD)**98b**]SbF_6	2–2.8	THF	r.t.	2–3	–	–	–	97.1	95 b
698	3-F-Ph	CO_2Me	NHAc	[Rh(COD)**103a**]SbF_6	2–2.8	THF	r.t.	2–3	–	–	–	96.3	95 b
699	4-F-Ph	CO_2H	NHAc	[Rh(COD)**95**] BF_4	1	IPA	r.t.	24	100	4	100	91.1 (*S*)	85
700	4-F-Ph	CO_2H	NHAc	[Rh(COD)**103a**]SbF_6	2–2.8	THF	r.t.	2–3	–	–	–	96.4	95 b
701	4-F-Ph	CO_2H	NHAc	[Rh(COD)**163a**]PF_6	1	THF	r.t.	12	99	8.3	100	99 (*S*)	128
702	4-F-Ph	CO_2H	NHAc	[Rh(COD)**164a**]PF_6	1	THF	r.t.	12	>99	8.3	100	>99 (*S*)	128
703	4-F-Ph	CO_2Me	NHAc	[Rh(COD)**91**]BF_4	1	MeOH	25	0.167	100	600	>99.9	95.5 (*R*)	27
704	4-F-Ph	CO_2Me	NHAc	[Rh(COD)**93h**]PF_6	3	Tol	r.t.	12	100	8.3	100	93.4 (*S*)	79
705	4-F-Ph	CO_2Me	NHAc	[Rh(COD)**98a**]BF_4	2–2.8	THF	r.t.	2–3	–	–	–	84.0	95b
706	4-F-Ph	CO_2Me	NHAc	[Rh(COD)**98a**]SbF_6	2–2.8	THF	r.t.	2–3	–	–	–	85.0	95 a, b
707	4-F-Ph	CO_2Me	NHAc	[Rh(COD)**98b**]SbF_6	2–2.8	THF	r.t.	2–3	–	–	–	97.2	95 a, b
708	4-F-Ph	CO_2Me	NHAc	[Rh(COD)**98c**]SbF_6	2–2.8	THF	r.t.	2–3	–	–	–	13.0	95 a, b
709	4-F-Ph	CO_2Me	NHAc	[Rh(COD)**98d**]SbF_6	2–2.8	THF	r.t.	2–3	–	–	–	9.0	95 a, b
710	4-F-Ph	CO_2Me	NHAc	[Rh(COD)**98f**]SbF_6	2–2.8	THF	r.t.	2–3	–	–	–	89.0	95 b
711	4-F-Ph	CO_2Me	NHAc	[Rh(COD)**98g**]SbF_6	2–2.8	THF	r.t.	2–3	–	–	–	98.7	95 b

712	4-F-Ph	CO_2Me	NHAc	[Rh(COD)**98h**]SbF_6	2–2.8	THF	r.t.	2–3	–	–	–	81.0	95b
713	4-F-Ph	CO_2Me	NHAc	[Rh(COD)**102a**]BF_4	2–2.8	THF	r.t.	2–3	–	–	–	97.8	95b
714	4-F-Ph	CO_2Me	NHAc	[Rh(COD)**103a**]SbF_6	2–2.8	THF	r.t.	2–3	–	–	–	96.2	95b
715	4-F-Ph	CO_2Me	NHAc	[Rh(COD)**103b**]SbF_6	2–2.8	THF	r.t.	2–3	–	–	–	73.5	95b
716	4-F-Ph	CO_2Me	NHAc	[Rh(COD)**103c**]SbF_6	2–2.8	THF	r.t.	2–3	–	–	–	<1	95b
717	4-F-Ph	CO_2Me	NHAc	[Rh(COD)**103d**]SbF_6	2–2.8	THF	r.t.	2–3	–	–	–	11.0	95b
718	4-F-Ph	CO_2Me	NHAc	[Rh(COD)**103e**]SbF_6	2–2.8	THF	r.t.	2–3	–	–	–	<1	95b
719	4-F-Ph	CO_2Me	NHAc	[Rh(COD)**103f**]SbF_6	2–2.8	THF	r.t.	2–3	–	–	–	87.0	95b
720	4-F-Ph	CO_2Me	NHAc	[Rh(COD)**104a**]BF_4	2–2.8	THF	r.t.	2–3	–	–	–	92.0	95b
721	4-F-Ph	CO_2Me	NHAc	[Rh(COD)**132**]BF_4	34.5	Acetone	25	0.25	100	400	100	91.2 (*R*)	107
722	4-F-Ph	CO_2Me	NHAc	[Rh(COD)**163a**]PF_6	1	THF	r.t.	12	99	8.3	100	99 (*S*)	128
723	4-F-Ph	CO_2Me	NHAc	[Rh(COD)**164a**]PF_6	1	THF	r.t.	12	>99	8.3	100	>99 (*S*)	128
724	4-F-Ph	CO_2Me	NHCbz	[Rh(COD)**98a**]SbF_6	2–2.8	THF	r.t.	2–3	–	–	–	62.0	95a, b
725	4-F-Ph	CO_2Me	NHCbz	[Rh(COD)**98b**]SbF_6	2–2.8	THF	r.t.	–	–	–	–	97	95a
726	4-F-Ph	CO_2Me	NHCbz	[Rh(COD)**98b**]SbF_6	2–2.8	THF	r.t.	2–3	–	–	–	95.7	95b
727	4-F-Ph	CO_2Me	NHCbz	[Rh(COD)**98c**]SbF_6	2–2.8	THF	r.t.	–	–	–	–	<1	95a
728	4-F-Ph	CO_2Me	NHCbz	[Rh(COD)**98c**]SbF_6	2–2.8	THF	r.t.	2–3	–	–	–	<3	95b
729	4-F-Ph	CO_2Me	NHCbz	[Rh(COD)**98d**]SbF_6	2–2.8	THF	r.t.	–	–	–	–	54	95a
730	4-F-Ph	CO_2Me	NHCbz	[Rh(COD)**98d**]SbF_6	2–2.8	THF	r.t.	2–3	–	–	–	<5	95b
731	4-F-Ph	CO_2Me	NHCbz	[Rh(COD)**98f**]SbF_6	2–2.8	THF	r.t.	2–3	–	–	–	85.0	95b
732	4-F-Ph	CO_2Me	NHCbz	[Rh(COD)**102a**]BF_4	2–2.8	THF	r.t.	2–3	–	–	–	96.0	95b
733	4-F-Ph	CO_2Me	NHCbz	[Rh(COD)**103a**]SbF_6	2–2.8	THF	r.t.	2–3	–	–	–	90.0	95b
734	4-F-Ph	CO_2Me	NHCbz	[Rh(COD)**123a**]SbF_6	2–2.8	THF	r.t.	2–3	–	–	–	56.8	95b
735	4-F-Ph	CO_2Me	NHCbz	[Rh(COD)**123b**]SbF_6	2–2.8	THF	r.t.	2–3	–	–	–	53.0	95b
736	4-F-Ph	CO_2Me	NHCbz	[Rh(COD)**123f**]SbF_6	2–2.8	THF	r.t.	2–3	–	–	–	57.0	95b
737	4-NO_2-Ph	CO_2H	NHAc	[Rh(COD)**91**]BF_4	1	MeOH	25	0.167	100	600	>99.9	97.0 (*R*)	27
738	4-NO_2-Ph	CO_2Me	NHAc	[Rh(COD)**132**]BF_4	34.5	Acetone	25	0.25	100	400	100	90.5 (*R*)	107
739	4-HO-Ph	CO_2H	NHAc	[Rh(1,5-hexadiene) (+)-*trans*-**85**]Cl	50	–	15	24	–	–	–	48.5 (S) [a]	67
740	4-HO-Ph	CO_2Me	NHAc	[Rh(COD)**132**]BF_4	34.5	Acetone	25	0.25	100	400	100	91.5 (*R*)	107
741	3-MeO-Ph	CO_2H	NHAc	[Rh(COD)**98a**]SbF_6	2–2.8	THF	r.t.	2–3	–	–	–	91.0	95a, b

Table 27.5 (continued)

Entry	Substrate			Catalyst	Conditions				TON	TOF $[h^{-1}]$	Conv. [%]	ee [%]	Reference(s)
	R^1	R^2	R^3		$P[H_2]$ [bar]	Solvent	Temp. [°C]	Time [h]					
742	3-MeO-Ph	CO_2H	NHAc	[Rh(COD)**98b**]SbF_6	2–2.8	THF	r.t.	2–3	–	–	–	97.0	95a, b
743	3-MeO-Ph	CO_2H	NHAc	[Rh(COD)**98c**]SbF_6	2–2.8	THF	r.t.	2–3	–	–	–	53.0	95a, b
744	3-MeO-Ph	CO_2H	NHAc	[Rh(COD)**98d**]SbF_6	2–2.8	THF	r.t.	2–3	–	–	–	5.0	95a, b
745	3-MeO-Ph	CO_2H	NHAc	[Rh(COD)**103a**]SbF_6	2–2.8	THF	r.t.	2–3	–	–	–	95.9	95b
746	3-MeO-Ph	CO_2H	NHAc	[Rh(COD)**103b**]SbF_6	2–2.8	THF	r.t.	2–3	–	–	–	73.4	95b
747	3-MeO-Ph	CO_2H	NHAc	[Rh(COD)**103c**]SbF_6	2–2.8	THF	r.t.	2–3	–	–	–	<1	95b
748	3-MeO-Ph	CO_2H	NHAc	[Rh(COD)**103d**]SbF_6	2–2.8	THF	r.t.	2–3	–	–	–	2.3	95b
749	3-MeO-Ph	CO_2H	NHAc	[Rh(COD)**103e**]SbF_6	2–2.8	THF	r.t.	2–3	–	–	–	2.1	95b
750	3-MeO-Ph	CO_2H	NHAc	[Rh(COD)**103f**]SbF_6	2–2.8	THF	r.t.	2–3	–	–	–	85.3	95b
751	3-MeO-Ph	CO_2H	NHAc	[Rh(COD)**104a**]BF_4	2–2.8	THF	r.t.	2–3	–	–	–	93.1	95b
752	3-MeO-Ph	CO_2H	NHAc	[Rh(COD)**132**]BF_4	34.5	Acetone	25	0.25–1	100	–	100	93.2 (*R*)	107
753	3-MeO-Ph	CO_2Me	NHAc	[Rh(COD)**98a**]SbF_6	2–2.8	THF	r.t.	2–3	–	–	–	88.0	95b
754	3-MeO-Ph	CO_2Me	NHAc	[Rh(COD)**98b**]SbF_6	2–2.8	THF	r.t.	2–3	–	–	–	96.8	95b
755	3-MeO-Ph	CO_2Me	NHAc	[Rh(COD)**98c**]SbF_6	2–2.8	THF	r.t.	2–3	–	–	–	21.0	95b
756	3-MeO-Ph	CO_2Me	NHAc	[Rh(COD)**98g**]SbF_6	2–2.8	THF	r.t.	2–3	–	–	–	98.8	95b
757	3-MeO-Ph	CO_2Me	NHAc	[Rh(COD)**116a**]SbF_6	2.07	THF	r.t.	–	–	–	–	70 (*R*)	1, 2
758	3-MeO-Ph	CO_2Me	NHAc	[Rh(COD)**116b**]SbF_6	2.07	THF	r.t.	–	–	–	–	40 (*R*)	102
759	3-MeO-Ph	CO_2Me	NHAc	[Rh(COD)**119a**]SbF_6	2.07	THF	r.t.	–	–	–	~100	70 (*S*)	102
760	3-MeO-Ph	CO_2Me	NHAc	[Rh(COD)**119b**]SbF_6	2.07	THF	r.t.	–	–	–	~100	92 (*S*)	102
761	4-MeO-Ph	CO_2H	NHAc	[Rh(COD)**95**]BF_4	1	IPA	r.t.	24	100	4	100	93.2 (*S*)	85
762	4-MeO-Ph	CO_2Me	NHAc	[Rh(COD)**91**]BF_4	1	MeOH	25	0.167	100	600	>99.9	96.2 (*R*)	27
763	4-MeO-Ph	CO_2Me	NHAc	[Rh(COD)**93h**]PF_6	3	Tol	r.t.	12	100	8.3	100	87.2 (*S*)	79
764	4-MeO-Ph	CO_2Me	NHAc	[Rh(COD)**94f**]BF_4	7	DCM	r.t.	0.1	500	5000	100	96.8 (*S*)	78
765	4-MeO-Ph	CO_2Me	NHAc	[Rh(COD)**94f**]BF_4	1	DCM	r.t.	0.42	488	1162	97.5	98.6 (*S*)	78
766	4-MeO-Ph	CO_2Me	NHAc	[Rh(COD)**117**]BF_4	5	H_2O/MeOH/ EtOAc(0. 6:0.4:1)	r.t.	3	100	33	100	98 (*S*)	101

767	4-MeO-Ph	CO_2Me	NHAc	[Rh(COD)**117**]BF_4	5	H_2O/MeOH (3:2)	r.t.	3	100	33	100	98 (*S*)	101
768	4-MeO-Ph	CO_2Me	NHAc	[Rh(COD)**132**]BF_4	34.5	Acetone	25	0.25	100	400	100	91.4 (*R*)	107
769	4-MeO-Ph	CO_2Me	NHAc	[Rh(COD)**133a**]BF_4	1	IPA	r.t.	24	97	4	97	90	48
770	3-Me-Ph	CO_2Me	NHAc	[Rh(COD)**93h**]SbF_6	3	THF	r.t.	12	100	8.3	100	96.3 (*S*)	79
771	4-Me-Ph	CO_2Me	NHAc	[Rh(COD)**91**]BF_4	1	MeOH	25	0.167	100	600	>99.9	95.6 (*R*)	27
772	4-Me-Ph	CO_2Me	NHAc	[Rh(COD)**94a**]BF_4	6.9	DCM	r.t.	0.17	500	2941	100	83.5 (*S*)	52
773	4-Me-Ph	CO_2Me	NHAc	[Rh(COD)**94d**]BF_4	7	DCM	r.t.	0.42	500	1190	100	92.5 (*S*)	78
774	4-Me-Ph	CO_2Me	NHAc	[Rh(COD)**132**]BF_4	34.5	Acetone	25	0.25	100	400	100	90.6 (*R*)	107
775	4-Me-Ph	CO_2Me	NHAc	[Rh(COD)**154a**]OTf	5	MeOH	r.t.	18	40	2.2	20	73 (*S*)	127
776	4-Me-Ph	CO_2Me	NHAc	[Rh(COD)**154a**]OTf	5	Tol	r.t.	18	120	6.7	60	85 (*S*)	127
777	4-Me-Ph	CO_2Me	NHAc	[Rh(COD)**154b**]OTf	5	Tol	r.t.	18	200	11.1	100	85 (*S*)	127
778	4-Me-Ph	CO_2Me	NHAc	[Rh(COD)**154c**]OTf	5	Tol	r.t.	18	186	10.3	93	34 (*R*)	127
779	4-Me-Ph	CO_2Me	NHBz	[Rh(COD)**94a**]BF_4	6.9	DCM	r.t.	0.17	479	2818	95.7	80 (*S*)	52
780	4-CF_3-Ph	CO_2Me	NHAc	[Rh(COD)**93h**]SbF_6	3	THF	r.t.	12	100	8.3	100	95.7 (*S*)	79
781	4-CF_3-Ph	CO_2Me	NHAc	[Rh(COD)**94c**]BF_4	7	DCM	r.t.	0.83	409	493	81.8	48.7 (*S*)	78
782	2-Naphthyl	CO_2H	NHAc	[Rh(COD)**95**]BF_4	1	IPA	r.t.	24	100	4	100	91.4 (*S*)	85
783	2-Naphthyl	CO_2H	NHAc	[Rh(COD)**98a**]SbF_6	2–2.8	THF	r.t.	2–3	–	–	–	94.0	95b
784	2-Naphthyl	CO_2H	NHAc	[Rh(COD)**98b**]SbF_6	2–2.8	THF	r.t.	2–3	–	–	–	98.0	95b
785	2-Naphthyl	CO_2H	NHAc	[Rh(COD)**98c**]SbF_6	2–2.8	THF	r.t.	2–3	–	–	–	22.0	95b
786	2-Naphthyl	CO_2H	NHAc	[Rh(COD)**98d**]SbF_6	2–2.8	THF	r.t.	2–3	–	–	–	26.6	95b
787	2-Naphthyl	CO_2H	NHAc	[Rh(COD)**103a**]SbF_6	2–2.8	THF	r.t.	–	–	–	–	96	95a
788	2-Naphthyl	CO_2H	NHAc	[Rh(COD)**163a**]PF_6	1	THF	r.t.	12	>99	8.3	100	>99 (*S*)	128
789	2-Naphthyl	CO_2H	NHAc	[Rh(COD)**164a**]PF_6	1	THF	r.t.	12	>99	8.3	100	>99 (*S*)	128
790	2-Naphthyl	CO_2Me	NHAc	[Rh(COD)**93h**]PF_6	3	Tol	r.t.	12	100	8.3	100	97.3 (*S*)	79
791	2-Naphthyl	CO_2Me	NHAc	[Rh(COD)**93h**]SbF_6	3	THF	r.t.	12	100	8.3	100	94.1 (*S*)	79
792	2-Naphthyl	CO_2Me	NHAc	[Rh(COD)**98a**]SbF_6	2–2.8	THF	r.t.	2–3	–	–	–	86.5	95b
793	2-Naphthyl	CO_2Me	NHAc	[Rh(COD)**98b**]SbF_6	2–2.8	THF	r.t.	2–3	–	–	–	97.1	95b
794	2-Naphthyl	CO_2Me	NHAc	[Rh(COD)**98d**]SbF_6	2–2.8	THF	r.t.	2–3	–	–	–	10.8	95b
795	2-Naphthyl	CO_2Me	NHAc	[Rh(COD)**103a**]SbF_6	2–2.8	THF	r.t.	2–3	–	–	–	96.0	95b
796	2-Naphthyl	CO_2Me	NHAc	[Rh(COD)**104a**]BF_4	2–2.8	THF	r.t.	2–3	–	–	–	93.0	95b

Table 27.5 (continued)

Entry	Substrate			Catalyst	Conditions				TON	TOF [h^{-1}]	Conv. [%]	ee [%]	Reference(s)
	R^1	R^2	R^3		$P[H_2]$ [bar]	Solvent	Temp. [°C]	Time [h]					
797	2-Naphthyl	CO_2Me	NHAc	[Rh(COD)**117**]BF_4	5	H_2O/MeOH/EtOAc(0.6:0.4:1)	r.t.	3	100	33	100	96 (*S*)	101
798	2-Naphthyl	CO_2Me	NHAc	[Rh(COD)**117**]BF_4	5	H_2O/MeOH (3:2)	r.t.	3	100	33	100	95 (*S*)	101
799	2-Naphthyl	CO_2Me	NHAc	[Rh(COD)**163a**]PF_6	1	THF	r.t.	12	>99	8.3	100	>99 (*S*)	128
800	2-Naphthyl	CO_2Me	NHAc	[Rh(COD)**164a**]PF_6	1	THF	r.t.	12	>99	8.3	100	>99 (*S*)	128
801	3,5-F_2-Ph	CO_2H	NHAc	[Rh(COD)**103a**]SbF_6	2–2.8	THF	r.t.	–	–	–	–	96.2	95 a, b
802	3,5-F_2-Ph	CO_2Me	NHAc	[Rh(COD)**98a**]SbF_6	2–2.8	THF	r.t.	2–3	–	–	–	88.3	95 b
803	3,5-F_2-Ph	CO_2Me	NHAc	[Rh(COD)**98b**]SbF_6	2–2.8	THF	r.t.	2–3	–	–	–	97.0	95 b
804	3,5-F_2-Ph	CO_2Me	NHAc	[Rh(COD)**98g**]SbF_6	2–2.8	THF	r.t.	2–3	–	–	–	98.4	95 b
805	3,5-F_2-Ph	CO_2Me	NHAc	[Rh(COD)**103b**]SbF_6	2–2.8	THF	r.t.	2–3	–	–	–	73.0	95 b
806	3,5-F_2-Ph	CO_2Me	NHAc	[Rh(COD)**103c**]SbF_6	2–2.8	THF	r.t.	2–3	–	–	–	3.0	95 b
807	3,5-F_2-Ph	CO_2Me	NHAc	[Rh(COD)**103d**]SbF_6	2–2.8	THF	r.t.	2–3	–	–	–	5.6	95 b
808	3,5-F_2-Ph	CO_2Me	NHAc	[Rh(COD)**103e**]SbF_6	2–2.8	THF	r.t.	2–3	–	–	–	2.7	95 b
809	3,5-F_2-Ph	CO_2Me	NHAc	[Rh(COD)**103 f**]SbF_6	2–2.8	THF	r.t.	2–3	–	–	–	85.1	95 b
810	3,5-Me_2-P	CO_2Me	NHAc	[Rh(COD)**93e**BF_4	7	DCM	r.t.	0.42	500	1190	100	93.9 (*S*)	78
811	3,5-Me_2-P	CO_2Me	NHAc	[Rh(COD)**94e**]BF_4	7	DCM	r.t.	0.22	500	2273	100	95.4 (*S*)	78
812	3,5-$(CF_3)_2$-Ph	CO_2Me	NHAc	[Rh(COD)**94b**]BF_4	7	DCM	r.t.	1.75	147	84	29.3	30.9 (*S*)	78
813	3,5-$(CF_3)_2$-Ph	CO_2Me	NHAc	[Rh(COD)**98b**]SbF_6	2–2.8	THF	r.t.	2–3	–	–	–	95.8	95b
814	3,5-$(CF_3)_2$-Ph	CO_2Me	NHAc	[Rh(NBD)**98b**]SbF_6	2.8	THF	r.t.	0.25	1000	4000	100	96.1 (*S*)	95 b
815	3,5-$(CF_3)_2$-Ph	CO_2Me	NHAc	[Rh(COD)**98a**]SbF_6	2–2.8	THF	r.t.	2–3	–	–	–	85.2	95 b
816	3,5-$(CF_3)_2$-Ph	CO_2Me	NHAc	[Rh(COD)**98g**]SbF_6	2–2.8	THF	r.t.	2–3	–	–	–	97.1	95 b
817	3,5-$(CF_3)_2$-Ph	CO_2Me	NHAc	[Rh(COD)**98g**]BF_4	2–2.8	THF	r.t.	2–3	–	–	–	96.9	95 b

818	3,5-$(CF_3)_2$-Ph	CO_2Me	NHAc	[Rh(COD)**102a**]BF_4	2–2.8	THF	r.t.	2–3	–	–	–	93.7	95 b
819	3,5-$(CF_3)_2$-Ph	CO_2Me	NHAc	[Rh(COD)**103a**]SbF_6	2–2.8	THF	r.t.	2–3	–	–	–	97.4	95 b
820	3,5-$(CF_3)_2$-Ph	CO_2Me	NHAc	[Rh(COD)**103b**]SbF_6	2–2.8	THF	r.t.	2–3	–	–	–	77.9	95 b
821	3,5-$(CF_3)_2$-Ph	CO_2Me	NHAc	[Rh(COD)**103c**]SbF_6	2–2.8	THF	r.t.	2–3	–	–	–	3.2	95 b
822	3,5-$(CF_3)_2$-Ph	CO_2Me	NHAc	[Rh(COD)**103d**]SbF_6	2–2.8	THF	r.t.	2–3	–	–	–	<1	95 b
823	3,5-$(CF_3)_2$-Ph	CO_2Me	NHAc	[Rh(COD)**103e**]SbF_6	2–2.8	THF	r.t.	2–3	–	–	–	<1	95 b
824	3,5-$(CF_3)_2$-Ph	CO_2Me	NHAc	[Rh(COD)**103 f**]SbF_6	2–2.8	THF	r.t.	2–3	–	–	–	83.9	95 b
825	3,4-$(MeO)_2$-Ph	CO_2H	NHAc	[Rh(COD)**97a**]BF_4	1	MeOH	25	0.1 [f]	50	500	50	96.7 (*S*)	91 b
826	3,4-$(MeO)_2$-Ph	CO_2H	NHAc	[Rh(COD)**97c**]BF_4	1	MeOH	25	0.083 [f]	50	600	50	94.8 (*S*)	91 b
827	3,4-$(MeO)_2$-Ph	CO_2H	NHBz	[Rh(COD)**97a**]BF_4	1	MeOH	25	0.083 [f]	50	600	50	95.1 (*S*)	91 b
828	3,4-$(MeO)_2$-Ph	CO_2H	NHBz	[Rh(COD)**97c**]BF_4	1	MeOH	25	0.067 [f]	50	750	50	92.0 (*S*)	91 b
829	3,4-$(MeO)_2$-Ph	CO_2Me	NHAc	[Rh(COD)**97a**]BF_4	1	MeOH	25	0.37 [f]	50	136	50	92.4 (*S*)	91 b
830	3,4-$(MeO)_2$-Ph	CO_2Me	NHAc	[Rh(COD)**97c**]BF_4	1	MeOH	25	0.2 [f]	50	250	50	95.7 (*S*)	91 b
831	3,4-$(MeO)_2$-Ph	CO_2Me	NHBz	[Rh(COD)**97a**]BF_4	1	MeOH	25	0.13 [f]	50	375	50	87.7 (*S*)	91 b
832	3,4-$(MeO)_2$-Ph	CO_2Me	NHBz	[Rh(COD)**97c**]BF_4	1	MeOH	25	0.12 [f]	50	429	50	91.2 (*S*)	91 b
833	3,4-$(MeO)_2$-Ph	CO_2Et	NHAc	[Rh(COD)**97a**]BF_4	1	MeOH	25	0.13 [f]	50	375	50	90.6 (*S*)	91 b
834	3,4-$(MeO)_2$-Ph	CO_2Et	NHAc	[Rh(COD)**97c**]BF_4	1	MeOH	25	0.083 [f]	50	600	50	95.2 (*S*)	91 b
835	3,4-$(MeO)_2$-Ph	CO_2Et	NHBz	[Rh(COD)**97a**]BF_4	1	MeOH	25	0.17 [f]	50	300	50	88.9 (*S*)	91 b
836	3,4-$(MeO)_2$-Ph	CO_2Et	NHBz	[Rh(COD)**97c**]BF_4	1	MeOH	25	0.13 [f]	50	375	50	90.5 (*S*)	91 b
837	3,4-$(MeO)_2$-Ph	CO_2*i*-Pr	NHAc	[Rh(COD)**97a**]BF_4	1	MeOH	25	0.37 [f]	50	136	50	91.3 (*S*)	91 b
838	3,4-$(MeO)_2$-Ph	CO_2*i*-Pr	NHAc	[Rh(COD)**97c**]BF_4	1	MeOH	25	0.18 [f]	50	273	50	94.7 (*S*)	91 b
839	3,4-$(MeO)_2$-Ph	CO_2*i*-Pr	NHBz	[Rh(COD)**97a**]BF_4	1	MeOH	25	0.33 [f]	50	150	50	89.1 (*S*)	91 b
840	3,4-$(MeO)_2$-Ph	CO_2*i*-Pr	NHBz	[Rh(COD)**97c**]BF_4	1	MeOH	25	0.23 [f]	50	214	50	92.7 (*S*)	91 b
841	3,4-$(MeO)_2$-Ph	CO_2-C_2H_4OH	NHBz	[Rh(COD)**97a**]BF_4	1	MeOH	25	0.17 [f]	50	300	50	87.3 (*S*)	91 b
842	3,4-$(MeO)_2$-Ph	CO_2-C_2H_4OH	NHBz	[Rh(COD)**97c**]BF_4	1	MeOH	25	0.13 [f]	50	375	50	89.9 (*S*)	91 b
843	3-MeO-4-HO-Ph	CO_2H	NHBz	[Rh(COD)**97a**]BF_4	1	MeOH	25	0.067 [f]	50	750	50	96.9 (*S*)	91 b
844	3-MeO-4-HO-Ph	CO_2H	NHBz	[Rh(COD)**97c**]BF_4	1	MeOH	25	0.067 [f]	50	750	50	94.1 (*S*)	91 b

Table 27.5 (continued)

Entry	Substrate			Catalyst	Conditions				TON	TOF [h^{-1}]	Conv. [%]	ee [%]	Reference(s)
	R^1	R^2	R^3		$P[H_2]$ [bar]	Solvent	Temp. [°C]	Time [h]					
845	3-MeO-4-HO-Ph	CO_2Me	NHAc	[Rh(COD)**97a**]BF_4	1	MeOH	25	0.2 [f]	50	250	50	91.7 (*S*)	91b
846	3-MeO-4-HO-Ph	CO_2Me	NHAc	[Rh(COD)**97c**]BF_4	1	MeOH	25	0.083 [f]	50	600	50	95.0 (*S*)	91b
847	3-MeO-4-HO-Ph	CO_2Me	NHAc	[Rh(COD)**132**]BF_4	34.5	Acetone	25	0.25	100	400	100	90.2 (*R*)	107
848	3-MeO-4-HO-Ph	CO_2Me	NHBz	[Rh(COD)**97a**]BF_4	1	MeOH	25	0.18 [f]	50	273	50	89.0 (*S*)	91b
849	3-MeO-4-HO-Ph	CO_2Me	NHBz	[Rh(COD)**97c**]BF_4	1	MeOH	25	0.1 [f]	50	500	50	92.1 (*S*)	91b
850	3-MeO-4-HO-Ph	$CO_2C_2H_4OH$	NHAc	[Rh(COD)**97a**]BF_4	1	MeOH	25	0.22 [f]	50	231	50	91.7 (*S*)	91b
851	3-MeO-4-HO-Ph	$CO_2C_2H_4OH$	NHAc	[Rh(COD)**97c**]BF_4	1	MeOH	25	0.12 [f]	50	429	50	95.5 (*S*)	91b
852	3-MeO-4-HO-Ph	$CO_2C_2H_4OH$	NHBz	[Rh(COD)**97a**]BF_4	1	MeOH	25	0.27 [f]	50	188	50	88.4 (*S*)	91b
853	3-MeO-4-HO-Ph	$CO_2C_2H_4OH$	NHBz	[Rh(COD)**97c**]BF_4	1	MeOH	25	0.2 [f]	50	250	50	90.2 (*S*)	91b
854	3-MeO-4-AcOPh	CO_2H	NHAc	[Rh(COD)**98a**]BF_4	1	MeOH	25	0.095 [f]	50	526	50	94 (*S*)	26
855	3-MeO-4-AcOPh	CO_2H	NHAc	[Rh(COD)**97a**]BF_4	1	MeOH	25	0.08 [f]	50	625	50	71 (*S*)	26
856	3-MeO-4-AcOPh	CO_2H	NHAc	[Rh(COD)**97a**]BF_4	1	MeOH	25	0.17 [f]	50	300	50	96.0 (*S*)	91b
857	3-MeO-4-AcOPh	CO_2H	NHAc	[Rh(COD)**97c**]BF_4	1	MeOH	25	0.15 [f]	50	333	50	95.2 (*S*)	91b
858	3-MeO-4-OAcPh	CO_2H	NHAc	[Rh(COD)**95**]BF_4	1	IPA	r.t.	24	100	4	100	95.0 (*S*)	85
859	3-MeO-4-AcOPh	CO_2H	NHAc	[Rh(COD)**124**]BF_4	1	THF	25	1	100	100	100	36 (*R*)	103
860	3-MeO-4-AcOPh	CO_2H	NHAc	[Rh(COD)**125**]BF_4	1	THF	25	1	100	100	100	65 (*S*)	103
861	3-MeO-4-AcOPh	CO_2Me	NHAc	[Rh(COD)**97a**]BF_4	1	MeOH	25	0.083 [f]	50	600	50	92.4 (*S*)	91b
862	3-MeO-4-AcOPh	CO_2Me	NHAc	[Rh(COD)**97c**]BF_4	1	MeOH	25	0.05 [f]	50	1000	50	95.6 (*S*)	91b
863	3-MeO-4-AcOPh	CO_2Me	NHAc	[Rh(COD)**98a**]BF_4	1	MeOH	25	0.15 [f]	50	333	50	91 (*S*)	26
864	3-MeO-4-AcOPh	CO_2Me	NHBz	[Rh(COD)**97a**]BF_4	1	MeOH	25	0.083 [f]	50	600	50	87.2 (*S*)	91b
865	3-MeO-4-AcOPh	CO_2Me	NHBz	[Rh(COD)**97c**]BF_4	1	MeOH	25	0.067 [f]	50	750	50	91.3 (*S*)	91b
866	3-MeO-4-AcOPh	CO_2Et	NHAc	[Rh(COD)**98a**]BF_4	1	MeOH	25	0.14 [f]	50	353	50	87 (*S*)	26
867	3-MeO-4-AcOPh	CO_2Et	NHBz	[Rh(COD)**97c**]BF_4	1	MeOH	25	0.083 [f]	50	600	50	90.5 (*S*)	91b
868	3,4-(OCH_2O)-Ph	CO_2H	NHAc	[Rh(COD)**132**]BF_4	34.5	Acetone	25	0.25–1	100	100–400	100	94.2 (*R*)	107

869	3,4-(OCH_2O)-Ph	CO_2Me	NHAc	[Rh(COD)**91**]BF_4	1	MeOH	25	0.167	100	600	>99.9	94.9 (*R*)	27
870	3,4-(OCH_2O)-Ph	CO_2Me	NHAc	[Rh(COD)**132**]BF_4	34.5	Acetone	25	0.25	100	400	100	93.2 (*R*)	107
871	4-Ph-Ph	CO_2Me	NHAc	[Rh(COD)**93h**]SbF_6	3	THF	r.t.	12	100	8.3	100	94.2 (*S*)	79
872	2-Furyl	CO_2Me	NHAc	[Rh(COD)**91**]BF_4	1	MeOH	25	0.167	100	600	>99.9	97.2 (*R*)	27
873	2-Furyl	CO_2Me	NHBz	[Rh(COD)**94a**]BF_4	6.9	DCM	r.t.	0.17	500	2941	100	63.9 (*S*)	52
874	Thiophen-2-yl	CO_2H	NHAc	[Rh(COD)**95**]BF_4	1	IPA	r.t.	24	100	4	100	90.1 (*S*)	85
875	Thiophen-2-yl	CO_2H	NHAc	[Rh(COD)**98a**]BF_4	2.8	THF	r.t.	3	100	33	100	85 (*S*)	98
876	Thiophen-2-yl	CO_2H	NHAc	[Rh(COD)**98b**]BF_4	2.8	THF	r.t.	3	100	33	100	96 (*S*)	98
877	Thiophen-2-yl	CO_2H	NHAc	[Rh(COD)**115a**]SbF_6	2.8	THF	r.t.	3	91	30	91	26 (*S*)	98
878	Thiophen-2-yl	CO_2H	NHAc	[Rh(COD)**115b**]SbF_6	2.8	THF	r.t.	3	28	9	28	80 (*S*)	98
879	Thiophen-2-yl	CO_2Me	NHAc	[Rh(COD)**98a**]SbF_6	2–2.8	THF	r.t.	2–3	–	–	–	85.2	95b
880	Thiophen-2-yl	CO_2Me	NHAc	[Rh(COD)**98b**]SbF_6	2–2.8	THF	r.t.	2–3	–	–	–	95.6	95b
881	Thiophen-2-yl	CO_2Me	NHAc	[Rh(COD)**98g**]SbF_6	2–2.8	THF	r.t.	2–3	–	–	–	97.2	95b
882	Thiophen-2-yl	CO_2Me	NHAc	[Rh(COD)**103a**]SbF_6	2–2.8	THF	r.t.	2–3	–	–	–	96	95b
883	Thiophen-2-yl	CO_2Me	NHAc	[Rh(COD)**163a**]PF_6	1	THF	r.t.	12	95	7.9	100	95 (*S*)	128
884	Thiophen-2-yl	CO_2Me	NHAc	[Rh(COD)**164a**]PF_6	1	THF	r.t.	12	95	7.9	100	95 (*S*)	128
885	Thiophen-3-yl	CO_2H	NHAc	[Rh(COD)**98a**]BF_4	2.8	THF	r.t.	3	100	33	100	87 (*S*)	98
886	Thiophen-3-yl	CO_2H	NHAc	[Rh(COD)**98b**]BF_4	2.8	THF	r.t.	3	100	33	100	97 (*S*)	98
887	Thiophen-3-yl	CO_2H	NHAc	[Rh(COD)**115a**]SbF_6	2.8	THF	r.t.	3	86	29	86	28 (*S*)	98
888	Thiophen-3-yl	CO_2H	NHAc	[Rh(COD)**115b**]SbF_6	2.8	THF	r.t.	3	68	23	68	92 (*S*)	98
889	Thiophen-3-yl	CO_2Me	NHAc	[Rh(COD)**98a**]SbF_6	2–2.8	THF	r.t.	2–3	–	–	–	86.6	95b
890	Thiophen-3-yl	CO_2Me	NHAc	[Rh(COD)**98b**]SbF_6	2–2.8	THF	r.t.	2–3	–	–	–	96.7	95b
891	Thiophen-3-yl	CO_2Me	NHAc	[Rh(COD)**98g**]SbF_6	2–2.8	THF	r.t.	2–3	–	–	–	98.8	95b
892	Thiophen-3-yl	CO_2Me	NHAc	[Rh(COD)**103a**]SbF_6	2–2.8	THF	r.t.	2–3	–	–	–	97.0	95a, b
893	H	Ph	NHAc	[Rh(COD)**93a**]SbF_6	3	THF	r.t.	12	100	8.3	100	28.3 (*S*)	79
894	H	Ph	NHAc	[Rh(COD)**93g**]SbF_6	3	THF	r.t.	12	100	8.3	100	67.2 (*S*)	79
895	H	Ph	NHAc	[Rh(COD)**93h**]SbF_6	3	THF	r.t.	12	100	8.3	100	94.3 (*S*)	79
896	H	Ph	NHAc	[Rh(COD)**93i**]SbF_6	3	THF	r.t.	12	100	8.3	100	89.4 (*S*)	79
897	H	Ph	NHAc	[Rh(COD)**93j**]SbF_6	3	THF	r.t.	12	100	8.3	100	90.3 (*S*)	79
898	CO_2Me	Ph	NHAc	[Ru(*p*-cymene)**93a**]Cl	5.5	EtOH	50	20	25	1.25	100	2 (*S*)	80
899	CO_2Me	Ph	NHAc	[Ru(*p*-cymene)**93g**]Cl	5.5	EtOH	50	20	25	1.25	100	22 (*S*)	80

Table 27.5 (continued)

Entry	Substrate			Catalyst	Conditions				TON	TOF $[h^{-1}]$	Conv. [%]	ee [%]	Reference(s)
	R^1	R^2	R^3		$P[H_2]$ [bar]	Solvent	Temp. [°C]	Time [h]					
900	CO_2Me	Ph	NHAc	[Ru(*p*-cymene)**93h**]Cl	5.5	EtOH	50	20	25	1.25	100	98 (*S*)	80
901	CO_2Me	Ph	NHAc	[Ru(*p*-cymene)**93i**]Cl	5.5	EtOH	50	20	25	1.25	100	99 (*S*)	80
902	CO_2Me	Ph	NHAc	[Ru(*p*-cymene)**93j**]Cl	5.5	EtOH	50	20	25	1.25	100	97 (*S*)	80
903	CO_2Et	Ph	NHAc	[Ru(*p*-cymene)**93i**]Cl	5.5	EtOH	50	20	25	1.25	100	98 (*S*)	80
904	CO_2Me	2-MeO-Ph	NHAc	[Ru(*p*-cymene)**93i**]Cl	5.5	EtOH	50	20	25	1.25	100	80 (*S*)	80
905	CO_2Me	4-MeO-Ph	NHAc	[Ru(*p*-cymene)**93i**]Cl	5.5	EtOH	50	20	25	1.25	100	99 (*S*)	80
906	CO_2Me	2-Me-Ph	NHAc	[Ru(*p*-cymene)**93i**]Cl	5.5	EtOH	50	20	25	1.25	100	96 (*S*)	80
907	CO_2Me	4-Me-Ph	NHAc	[Ru(*p*-cymene)**93i**]Cl	5.5	EtOH	50	20	25	1.25	100	99 (*S*)	80
908	CO_2Me	4-Br-Ph	NHAc	[Ru(*p*-cymene)**93i**]Cl	5.5	EtOH	50	20	25	1.25	100	97 (*S*)	80
909	CO_2Me	4-Cl-Ph	NHAc	[Ru(*p*-cymene)**93i**]Cl	5.5	EtOH	50	20	25	1.25	100	97 (*S*)	80
910	CO_2Me	4-F-Ph	NHAc	[Ru(*p*-cymene)**93i**]Cl	5.5	EtOH	50	20	25	1.25	100	99 (*S*)	80
911	CO_2Et	4-Br-Ph	NHAc	[Ru(*p*-cymene)**93i**]Cl	5.5	EtOH	50	20	25	1.25	100	93 (*S*)	80
912	CO_2Et	4-Cl-Ph	NHAc	[Ru(*p*-cymene)**93i**]Cl	5.5	EtOH	50	20	25	1.25	100	95 (*S*)	80
913	CO_2Et	4-F-Ph	NHAc	[Ru(*p*-cymene)**93i**]Cl	5.5	EtOH	50	20	25	1.25	100	98 (*S*)	80
914	H	CO_2H	CH_2CO_2H	[Ir(COD)**142a**]BF_4	5	DCM : MeOH 2 : 1	40	4	100	25	100	35 (*R*)	117d
915	H	CO_2H	CH_2CO_2H	[Ir(COD)**142a**]BF_4	5	DCM : MeOH 2 : 1	25	12	70	5.8	70	34 (*R*)	117d
916	H	CO_2H	CH_2CO_2H	[Ir(COD)**142a**]BF_4	1	DCM : MeOH 2 : 1	40	4	100	25	100	54 (*R*)	117d
917	H	CO_2H	CH_2CO_2H	[Ir(COD)**142a**]BF_4	1	DCM : MeOH 2 : 1	25	8	50	6.3	50	40 (*R*)	117d
918	H	CO_2H	CH_2CO_2H	[Rh(COD)**142a**]BF_4	5	Tol : MeOH 2 : 1	40	6	70	11.7	70	45 (*R*)	117d

919	H	CO_2H	CH_2CO_2H	[Rh(COD)**142a**]BF_4	5	DCM : MeOH 2 : 1	40	6	13	2.2	13	–	117d
920	H	CO_2H	CH_2CO_2H	[Ir(COD)**142b**]BF_4	5	DCM : MeOH 2 : 1	40	4	100	25	100	29 (*R*)	117d
921	H	CO_2H	CH_2CO_2H	[Ir(COD)**142b**]BF_4	5	DCM : MeOH 2 : 1	25	12	68	5.7	68	32 (*R*)	117d
922	H	CO_2H	CH_2CO_2H	[Ir(COD)**142b**]BF_4	1	DCM : MeOH 2 : 1	40	4	87	21.8	87	47 (*R*)	117d
923	H	CO_2H	CH_2CO_2H	[Ir(COD)**142b**]BF_4	1	DCM : MeOH 2 : 1	25	8	70	8.8	70	26 (*R*)	117d
924	H	CO_2H	CH_2CO_2H	[Rh(COD)**142b**]BF_4	5	Tol : MeOH 2 : 1	40	6	99	16.5	99	49 (*R*)	117d
925	H	CO_2H	CH_2CO_2H	[Ir(COD)**142c**]BF_4	5	DCM : MeOH 2 : 1	40	20	100	5	100	13 (*R*)	117d
926	H	CO_2H	CH_2CO_2H	[Ir(COD)**142c**]BF_4	5	DCM : MeOH 2 : 1	25	12	44	3.7	44	11 (*R*)	117d
927	H	CO_2H	CH_2CO_2H	[Rh(COD)**142c**]BF_4	5	Tol : MeOH 2 : 1	40	20	20	1	20	20 (*R*)	117d
928	H	CO_2H	CH_2CO_2H	[Ir(COD)**143a**]BF_4	1	DCM : MeOH 2 : 1	40	6	100	16.7	100	15 (*S*)	117c
929	H	CO_2H	CH_2CO_2H	[Rh(COD)**143a**]BF_4	5	Tol : MeOH 2 : 1	40	20	100	5	100	11 (*R*)	117c
930	H	CO_2H	CH_2CO_2H	[Ir(COD)**143b**]BF_4	1	DCM : MeOH 2 : 1	40	6	100	16.7	100	13 (*S*)	117c
931	H	CO_2H	CH_2CO_2H	[Rh(COD)**143b**]BF_4	5	Tol : MeOH 2 : 1	40	20	100	5	100	10 (*R*)	117c
932	H	CO_2H	CH_2CO_2H	[Ir(COD)**143c**]BF_4	5	DCM : MeOH 2 : 1	40	20	47	2.4	47	8 (*R*)	117c
933	H	CO_2H	CH_2CO_2H	[Rh(COD)**143c**]BF_4	5	Tol : MeOH 2 : 1	40	20	100	5	100	50 (*R*)	117c
934	H	CO_2Me	CH_2CO_2Me	[Rh(COD)**87**]BARF	30–45	$scCO_2$	40–45	20	1000	50	100	73 (*R*)	69

Table 27.5 (continued)

Entry	Substrate			Catalyst	Conditions				TON	TOF [h^{-1}]	Conv. [%]	ee [%]	Reference(s)
	R^1	R^2	R^3		P[H$_2$] [bar]	Solvent	Temp. [°C]	Time [h]					
935	H	CO_2Me	CH_2CO_2Me	[Rh(COD)**93a**]BF_4	20	DCM	r.t.	0.28	500	1786	100	81.3 (*R*)	78
936	H	CO_2Me	CH_2CO_2Me	[Rh(COD)**93b**]BF_4	20	DCM	r.t.	0.17	500	2941	100	81.0 (*R*)	78
937	H	CO_2Me	CH_2CO_2Me	[Rh(COD)**93d**]BF_4	20	DCM	r.t.	0.5	500	1000	100	65.6 (*R*)	78
938	H	CO_2Me	CH_2CO_2Me	[Rh(COD)**93f**]BF_4	20	DCM	r.t.	1.5	412	275	82.3	50.9 (*R*)	78
939	H	CO_2Me	CH_2CO_2Me	[Rh(COD)**94b**]BF_4	20	DCM	r.t.	1.5	439	293	87.7	51.6 (*R*)	78
940	H	CO_2Me	CH_2CO_2Me	[Rh(COD)**94c**]BF_4	20	DCM	r.t.	0.33	500	1515	100	72.9 (*R*)	78
941	H	CO_2Me	CH_2CO_2Me	[Rh(COD)**94d**]BF_4	20	DCM	r.t.	0.12	500	4167	100	89.5 (*R*)	78
942	H	CO_2Me	CH_2CO_2Me	[Rh(COD)**94e**]BF_4	20	DCM	r.t.	0.15	500	3333	100	91.5 (*R*)	78
943	H	CO_2Me	CH_2CO_2Me	[Rh(COD)**94f**]BF_4	20	DCM	r.t.	0.083	500	6024	100	92.2 (*R*)	78
944	H	CO_2Me	CH_2CO_2Me	[Rh(COD)**94f**]BF_4	1	DCM	r.t.	0.42	500	1190	100	93.9 (*R*)	78
945	H	CO_2Me	CH_2CO_2Me	[Rh(COD)**97a**]ClO_4 +0.1 Et_3N	1	EtOH	25	0.5	50	100	100	29 (*R*)	90
946	H	CO_2Me	CH_2CO_2Me	[Rh(COD)**97a**]ClO_4 +0.2 Et_3N	1	EtOH	25	0.5	50	100	100	16 (*R*)	90
947	H	CO_2Me	CH_2CO_2Me	[Rh(COD)**111**]ClO_4	1	EtOH	25	0.5	50	100	100	45 (*R*)	90
948	H	CO_2Me	CH_2CO_2Me	[Rh(COD)**111**]ClO_4	1	EtOH	0	1	50	50	100	33 (*R*)	90
949	H	CO_2Me	CH_2CO_2Me	[Rh(COD)**111**]ClO_4	1	EtOH	25	0.3	50	167	100	31 (*S*)	90
950	H	CO_2Me	CH_2CO_2Me	[Rh(COD)**111**]ClO_4	1	EtOH	0	0.5	50	100	100	54 (*S*)	90
951	H	CO_2Me	CH_2CO_2Me	[Rh(COD)**111**]ClO_4 +0.1 Et_3N	1	EtOH	25	0.1	50	500	100	51 (*S*)	90
952	H	CO_2Me	CH_2CO_2Me	[Rh(COD)**111**]ClO_4 +0.2 Et_3N	1	EtOH	25	0.3	50	167	100	28 (*S*)	90
953	H	CO_2Me	CH_2CO_2Me	[Rh(COD)ent-**120**]BF_4	1	DCM	r.t.	0.08	100	1250	100	4 (*R*)	105

954	H	CO_2Me	CH_2CO_2Me	[Rh(COD)**121a**]BF_4	1	DCM	r.t.	0.08	100	1250	100	53 (*R*)	105
955	H	CO_2Me	CH_2CO_2Me	[Rh(COD)**121b**]BF_4	1	DCM	r.t.	0.08	100	1250	100	9 (*R*)	105
956	H	CO_2Me	CH_2CO_2Me	[Rh(COD)**121d**]BF_4	1	DCM	r.t.	0.08	100	1250	100	19 (*R*)	105
957	H	CO_2Me	CH_2CO_2Me	[Rh(COD)**122a**]BF_4	1	DCM	r.t.	0.08	100	1250	100	48 (*S*)	105
958	H	CO_2Me	CH_2CO_2Me	[Rh(COD)**122b**]BF_4	1	DCM	r.t.	0.08	100	1250	100	48 (*S*)	105
959	H	CO_2Me	CH_2CO_2Me	[Rh(COD)**122b**]BF_4	1	Acetone/DCM 13:2	r.t.	1.25	100	80	100	29 (*S*)	105
960	H	CO_2Me	CH_2CO_2Me	[Rh(COD)**122c**]BF_4	1	DCM	r.t.	0.08	100	1250	100	54 (*S*)	105
961	H	CO_2Me	CH_2CO_2Me	[Rh(COD)**122d**]BF_4	1	DCM	r.t.	0.08	100	1250	100	53 (*S*)	105
962	H	CO_2Me	CH_2CO_2Me	[Rh(COD)**122d**]BF_4	1	Acetone/DCM 13:2	r.t.	1	100	100	100	63 (*S*)	105
963	H	CO_2Me	CH_2CO_2Me	[Rh(COD)**122e**]BF_4	1	DCM	r.t.	0.08	100	1250	100	51 (*S*)	105
964	H	CO_2Me	CH_2CO_2Me	[Ir(COD)**126**]BF_4	1	DCM	25	5	2	0.4	2	3 (*R*)	104
965	H	CO_2Me	CH_2CO_2Me	[Rh(COD)**126**]BF_4	1	DCM	25	0.83	100	120	100	9 (*S*)	104
966	H	CO_2Me	CH_2CO_2Me	[Ir(COD)**127**]BF_4	1	DCM	25	28	100	3.6	100	24 (*S*)	104
967	H	CO_2Me	CH_2CO_2Me	[Rh(COD)**127**]BF_4	1	DCM	25	0.42	99	236	99	15 (*R*)	104
968	H	CO_2Me	CH_2CO_2Me	[Rh(COD)**136**]BF_4	1.3	DCM	r.t.	20	2000	100	100	97–99 (*R*)	109
969	H	CO_2Me	CH_2CO_2Me	[Rh(COD)**136**]BF_4	1.3	DCM	r.t.	20	1000	50	100	97–99 (*R*)	109
970	H	CO_2Me	CH_2CO_2Me	[Rh(COD)**137a**]BF_4	20	DCM	23	0.5	1000	2000	100	59.7 (*R*)	112
971	H	CO_2Me	CH_2CO_2Me	[Rh(COD)**137b**]BF_4	20	DCM	23	0.17	1000	5882	100	88.5 (*R*)	112
972	H	CO_2Me	CH_2CO_2Me	[Rh(COD)**138**]BF_4	1.3	DCM	r.t.	20	2000	100	100	>99.5 (*R*)	109
973	H	CO_2Me	CH_2CO_2Me	[Rh(COD)**138**]BF_4	1.3	DCM	r.t.	20	5380	269	100	>99.5 (*R*)	109
974	H	CO_2Me	CH_2CO_2Me	[Rh(COD)**138**]BF_4	1.3	DCM	r.t.	20	1000	500	100	>99.5 (*R*)	109
975	H	CO_2Me	CH_2CO_2Me	[Rh(COD)**142a**]BF_4	5	DCM	25	8	12	1.5	12	22 (*R*)	117b
976	H	CO_2Me	CH_2CO_2Me	[Rh(COD)**143a**]BF_4	5	DCM	25	8	28	3.5	28	64 (*R*)	117b
977	H	CO_2Me	CH_2CO_2Me	[Rh(COD)**144a**]BF_4	5	DCM	25	8	90	11.3	90	90 (*R*)	117a
978	H	CO_2Me	CH_2CO_2Me	[Rh(COD)**144a**]BF_4	5	Tol	25	8	16	2	16	2 (*S*)	117b
979	H	CO_2Me	CH_2CO_2Me	[Rh(COD)**144a**]BF_4	5	DCM	25	8	90	11.3	90	90 (*R*)	117b
980	H	CO_2Me	CH_2CO_2Me	[Rh(COD)**144a**]BF_4	5	AcOEt	25	8	8	1	8	2 (*R*)	117b
981	H	CO_2Me	CH_2CO_2Me	[Rh(COD)**144a**]BF_4	5	THF	25	8	99	12.4	99	12 (*R*)	117b
982	H	CO_2Me	CH_2CO_2Me	[Rh(COD)**144a**]BF_4	1	DCM	25	20	100	5	100	10 (*R*)	117b

Table 27.5 (continued)

Entry	Substrate			Catalyst	Conditions				TON	TOF [h^{-1}]	Conv. [%]	ee [%]	Reference(s)
	R^1	R^2	R^3		P[H_2] [bar]	Solvent	Temp. [°C]	Time [h]					
983	H	CO_2Me	CH_2CO_2Me	[Rh(COD)**144a**]BF_4	2	DCM	25	8	66	8.3	66	90 (*R*)	117b
984	H	CO_2Me	CH_2CO_2Me	[Rh(COD)**144a**]BF_4	10	DCM	25	3	90	30	90	90 (*R*)	117b
985	H	CO_2Me	CH_2CO_2Me	[Rh(COD)**144a**]BF_4	30	DCM	25	0.8	100	125	100	91 (*R*)	117b
986	H	CO_2Me	CH_2CO_2Me	[Rh(COD)**144a**]BF_4	5	DCM	25	8	90	11.3	90	90 (*R*)	117b
987	H	CO_2Me	CH_2CO_2Me	[Rh(COD)**144b**]BF_4	5	DCM	25	8	82	10.3	82	85 (*R*)	117a, b
988	H	CO_2Me	CH_2CO_2Me	[Rh(COD)**144c**]BF_4	5	DCM	25	6	100	16.7	100	97 (*R*)	117a, b
989	H	CO_2Me	CH_2CO_2Me	[Rh(COD)**144c**]BF_4	30	DCM	5	4	1000	250	100	>99 (*R*)	117b
990	H	CO_2Me	CH_2CO_2Me	[Rh(COD)**144d**]BF_4	5	DCM	25	8	50	6.3	50	50 (*S*)	117a, b
991	H	CO_2Me	CH_2CO_2Me	[Rh(COD)**144e**]BF_4	5	DCM	25	8	46	5.8	46	52 (*R*)	117a
992	H	CO_2Me	CH_2CO_2Me	[Rh(COD)**144e**]BF_4	5	DCM	25	8	46	5.8	46	52 (*R*)	117b
993	H	CO_2Me	CH_2CO_2Me	[Rh(COD)**144f**]BF_4	5	DCM	25	8	100	12.5	100	90 (*S*)	117b
994	H	CO_2Me	CH_2CO_2Me	[Rh(COD)**144g**]BF_4	5	DCM	25	8	100	12.5	100	92 (*R*)	117b
995	H	CO_2Me	CH_2CO_2Me	[Rh(COD)**145a**]BF_4	5	DCM	25	8	100	12.5	100	2 (*R*)	117a, b
996	H	CO_2Me	CH_2CO_2Me	[Rh(COD)**145b**]BF_4	5	DCM	25	8	98	12.3	98	2 (*R*)	117a, b
997	H	CO_2Me	CH_2CO_2Me	[Rh(COD)**145c**]BF_4	5	DCM	25	8	100	12.5	100	3 (*R*)	117a, b
998	H	CO_2Me	CH_2CO_2Me	[Rh(COD)**146a**]BF_4	5	DCM	25	8	87	10.9	87	67 (*R*)	117a, b
999	H	CO_2Me	CH_2CO_2Me	[Rh(COD)**146b**]BF_4	5	DCM	25	8	80	10	80	63 (*R*)	117a, b
1000	H	CO_2Me	CH_2CO_2Me	[Rh(COD)**146c**]BF_4	5	DCM	25	8	73	9.1	73	29 (*R*)	117a, b
1001	H	CO_2Me	CH_2CO_2Me	[Rh(COD)**146d**]BF_4	5	DCM	25	8	69	8.6	69	27 (*R*)	117a, b
1002	H	CO_2Me	CH_2CO_2Me	[Rh(COD)**149a**]BF_4	0.3	DCM	20	20	325	16	65	21.0 (*S*)	116
1003	H	CO_2Me	CH_2CO_2Me	[Rh(COD)**149b**]BF_4 [k]	0.3	DCM	20	20	1000	50	>99	87.8 (*S*)	116
1004	H	CO_2Me	CH_2CO_2Me	[Rh(COD)**149c**]BF_4	0.3	DCM	–10	20	1000	50	>99	96.2 (*R*)	116
1005	H	CO_2Me	CH_2CO_2Me	[Rh(COD)**149c**]BF_4 [k]	0.3	DCM	20	20	1000	50	>99	94.5 (*R*)	116
1006	H	CO_2Me	CH_2CO_2Me	[Rh(COD)**149d**]BF_4 [k]	0.3	DCM	20	20	740	37	74	38.9 (*S*)	116

1007	H	CO_2Me	CH_2CO_2Me	[Rh(COD)**149e**]BF_4	0.3	DCM	20	20	1000	50	>99	96.8 (*R*)	116
1008	H	CO_2Me	CH_2CO_2Me	[Rh(COD)**149e**]BF_4	0.3	DCM	–10	20	1000	50	>99	98.2 (*R*)	116
1009	H	CO_2Me	CH_2CO_2Me	[Rh(COD)**149f**]BF_4	0.3	DCM	20	20	60	3	24	5.2 (*R*)	116
1010	H	CO_2Me	CH_2CO_2Me	[Rh(COD)**149g**]BF_4	0.3	DCM	20	20	1000	50	>99	49.3 (*R*)	116
1011	H	CO_2Me	CH_2CO_2Me	[Rh(COD)**150a**]BF_4	1	DCM	25	0.2	100	500	100	63 (*R*)	118
1012	H	CO_2Me	CH_2CO_2Me	[Rh(COD)**150b**]BF_4	1	DCM	25	3.3	100	30.3	100	66 (*R*)	118
1013	H	CO_2Me	CH_2CO_2Me	[Rh(COD)**150c**]BF_4	1	DCM	25	3.3	100	30.3	100	14 (*R*)	118
1014	H	CO_2Me	CH_2CO_2Me	[Rh(COD)**150d**]BF_4	1	DCM	25	0.25	100	400	100	70 (*R*)	118
1015	H	CO_2Me	CH_2CO_2Me	[Rh(COD)**151**]BF_4	1.3	DCM	22	2.5	1000	400	100	88 (*R*)	120
1016	H	CO_2Me	CH_2CO_2Me	[Rh(COD)**151**]BF_4	1.3	DCM	22	3.2	1860	581	93	87 (*R*)	120
1017	H	CO_2Me	CH_2CO_2Me	[Rh(COD)**152**]BF_4	1.3	DCM	22	1.5	1000	667	100	77 (*S*)	120
1018	H	CO_2Me	CH_2CO_2Me	[Rh(COD)**152**]BF_4	1.3	DCM	22	1.5	2000	1333	100	79 (*S*)	120
1019	H	CO_2Me	CH_2CO_2Me	[Rh(COD)**153**]BF_4	1.3	DCM	22	2.5	1000	400	100	52 (*S*)	120
1020	H	CO_2Me	CH_2CO_2Me	[Rh(COD)**153**]BF_4	1.3	DCM	22	2.8	1800	643	90	60 (*S*)	120
1021	H	CO_2Me	CH_2CO_2Me	[Rh(COD)**155c**]BF_4	4.1	DCM	r.t.	17	500	29.4	100	49.2 (*S*)	124b
1022	H	CO_2Me	CH_2CO_2Me	[Rh(COD)**156a**]BF_4	4.1	DCM	r.t.	17	500	29.4	100	99.3 (*S*)	124b
1023	H	CO_2Me	CH_2CO_2Me	[Rh(COD)**156a**]BF_4	5.1	DCM	r.t.	17	3000	176.5	100	99.8 (*S*)	124b
1024	H	CO_2Me	CH_2CO_2Me	[Rh(COD)**156a**]BF_4	4.1	DCM	r.t.	24	10000	416.7	100	99.6 (*S*)	124b
1025	H	CO_2Me	CH_2CO_2Me	[Rh(COD)**156b**]BF_4	4.1	DCM	r.t.	17	500	29.4	100	30.8 (*R*)	124b
1026	H	CO_2Me	CH_2CO_2Me	[Rh(COD)**156c**]BF_4	4.1	DCM	r.t.	17	500	29.4	100	1.8 (*R*)	124b
1027	H	CO_2H	Ph	[Rh(1,5-hexadiene) (+)-*trans*-**85**]Cl	50	–	50	–	–	–	–	0.7 (S) [a]	68a
1028	H	CO_2H	Ph	[Rh(1,5-hexadiene)d-*trans*-**86**]Cl	50	–	50	–	–	–	–	0	68a
1029	H	CO_2H	Ph	[Rh(COD)**88**]Cl	20.7	PhH : EtOH = 1 : 1	60	24	50 [c]	–	100 [d]	2	81
1030	H	CO_2H	Ph	[Rh(COD)**97a**]ClO_4	1	EtOH	25	1	50	50	100	2 (*S*)	90
1031	H	CO_2H	Ph	[Rh(COD)**97a**]ClO_4 + 0.1 Et_3N	1	EtOH	25	24	50	2	100	2 (*S*)	90
1032	H	CO_2H	Ph	[Rh(COD)**124**]BF_4	1	THF	25	1	100	100	100	27 (*R*)	103

Table 27.5 (continued)

Entry	Substrate			Catalyst	Conditions				TON	TOF $[h^{-1}]$	Conv. [%]	ee [%]	Reference(s)
	R^1	R^2	R^3		$P[H_2]$ [bar]	Solvent	Temp. [°C]	Time [h]					
1033	H	CO_2H	Ph	[Rh(COD)**124**]Cl	20.4	PhH : EtOH = 1 : 1	60	24	100	4	100	2.1 (*R*)	103
1034	H	CO_2H	Ph	[Rh(COD)**125**]BF_4	1	THF	25	1	100	100	100	17 (*S*)	103
1035	H	CO_2H	Ph	[Rh(COD)**148**]BF_4	1	Acetone	r.t.	2	–	–	>90	2–10	114
1036	H	CO_2Me	Ph	[Rh(1,5-hexadiene) (+)-*trans*-**85**]Cl	50	–	50	–	–	–	–	4.5 (*S*) [a]	68a
1037	H	CO_2Me	Ph	[Rh(1,5-hexadiene)d-*trans*-**86**]Cl	50	–	50	–	–	–	–	20 (*S*) [a]	68a
1038	H	CO_2Me	Ph	[Rh(COD)**88**]Cl	69	PhH : EtOH = 1 : 1	100	48	50 [c]	–	[d]	1.0	81
1039	H	CO_2Me	Ph	[Rh(COD)**98a**]BF_4	1	MeOH	25	9 [f]	50	6	50	64 (*S*)	26
1040	H	CO_2Me	Ph	[Rh(COD)**98a**]BF_4	50	MeOH	25	–	100 [b]			64 (*S*)	26
1041	H	CO_2Me	Ph	[Rh(COD)**124**]BF_4	1	THF	25	1	100	100	100	12 (*R*)	103
1042	H	CO_2Me	Ph	[Rh(COD)**124**]Cl	68	PhH : EtOH = 1 : 1	100	48	100	2	100	0	103
1043	H	CO_2Me	Ph	[Rh(COD)**125**]BF_4	1	THF	25	1	100	100	100	10 (*S*)	103
1044	Ph	CO_2H	Me	[Rh(COD)**88**]Cl	20.7	PhH : EtOH = 1 : 1	60	24	50 [c]	–	100 [d]	14.3	81
1045	Ph	CO_2H	Me	[Rh(COD)**124**]BF_4	1	THF	25	1	100	100	100	54 (*R*)	103
1046	Ph	CO_2H	Me	[Rh(COD)**124**]Cl	20.4	PhH : EtOH = 1 : 1	60	24	100	4	100	7.1 (*R*)	103
1047	Ph	CO_2H	Me	[Rh(COD)**125**]BF_4	1	THF	25	1	100	100	100	48 (*S*)	103
1048	CO_2H	Me	Ph	[Rh(COD)**97a**]ClO_4	1	EtOH	25	24	50	2	100	5 (*S*)	90
1049	CO_2H	Me	Ph	[Rh(COD)**97a**]ClO_4 +0.1 Et_3N	1	EtOH	25	24	50	2	100	5 (*S*)	90
1050	CO_2H	Ph	Me	[Rh(COD)**97a**]ClO_4	1	EtOH	25	24	50	2	100	5 (*S*)	90
1051	CO_2H	Ph	Me	[Rh(COD)**97a**]ClO_4 +0.1 Et_3N	1	EtOH	25	24	50	2	100	17 (*S*)	90

1052	Ph	CO_2Me	Me	[Rh(COD)**88**]Cl	20.7	PhH : EtOH = 1 : 1	100	48	50 [c]	–	[d]	4.3	81
1053	Ph	CO_2Me	Me	[Rh(COD)**124**]Cl	68	PhH : EtOH = 1 : 1	100	48	100	2	100	2.3 (*R*)	103
1054	Ph	CO_2H	Ph	[Rh(COD)**88**]Cl	20.7	PhH : EtOH = 1 : 1	60	24	50 [c]	–	[d]	12.0	81
1055	Ph	CO_2Me	Ph	[Rh(COD)**88**]Cl	20.7	PhH : EtOH = 1 : 1	100	48	50 [c]	–	[d]	4.6	81
1056	CO_2H	Me	CO_2H	[Rh(COD)**97a**]ClO_4	1	EtOH	25	24	50	2	100	23 (*S*)	90
1057	CO_2H	Me	CO_2H	[Rh(COD)**111**]ClO_4	1	EtOH	25	24	50	2	100	7 (*S*)	90
1058	CO_2H	Me	CO_2H	$[Rh($**128**$)Cl]_2$	56	MeOH	r.t.	17	100	6	100	24 (*S*) [a]	106
1059	CO_2H	Me	CO_2H	[Rh(COD)**128**]Cl	56	PhH	75	17	100	5.9	100	24 (*S*)	106
1060	CO_2H	Me	CO_2H	[Rh(COD)**129**]Cl	56	PhH	75	–	–	–	0	0	106
1076	H	Ph	C_2H_5	$[Rh($**129**$)Cl]_2$	50	MeOH	r.t.	15	140	9	100	37 (*R*) [a]	106
1077	Ph	H	Me	[Rh(COD)**97a**]BF_4	1	MeOH	25	0.03	50	1667	50	96.6 (*S*)	91b
1078	Ph	H	Me	[Rh(COD)**97c**]BF_4	1	MeOH	25	0.07	50	714	50	95.1 (*S*)	91b
1079	Ph	Me	Me	[Rh(COD)**97a**]BF_4	1	MeOH	25	0.1	50	500	50	91.5 (*S*)	91b
1080	Ph	Me	Me	[Rh(COD)**97c**]BF_4	1	MeOH	25	0.05	50	1000	50	94.8 (*S*)	91b
1081	Ph	Et	Me	[Rh(COD)**97a**]BF_4	1	MeOH	25	0.1	50	500	50	90.6 (*S*)	91b
1082	Ph	Et	Me	[Rh(COD)**97c**]BF_4	1	MeOH	25	0.05	50	1000	50	94.4 (*S*)	91b
1083	Ph	Et	Me	[Rh(COD)**97a**]BF_4	1	MeOH	25	0.1	50	500	50	90.6 (*S*)	91b
1084	Ph	Et	Me	[Rh(COD)**97c**]BF_4	1	MeOH	25	0.05	50	1000	50	94.4 (*S*)	91b
1085	Ph	H	Ph	[Rh(COD)**97a**]BF_4	1	MeOH	25	0.03	50	1667	50	95.0 (*S*)	91b
1086	Ph	H	Ph	[Rh(COD)**97c**]BF_4	1	MeOH	25	0.05	50	1000	50	93.7 (*S*)	91b
1087	Ph	Me	Ph	[Rh(COD)**97a**]BF_4	1	MeOH	25	0.1	50	500	50	87.3 (*S*)	91b
1088	Ph	Me	Ph	[Rh(COD)**97c**]BF_4	1	MeOH	25	0.05	50	1000	50	91.6 (*S*)	91b
1089													
1090	3,4-$(MeO)_2$-C_6H_3	H	Me	[Rh(COD)**97a**]BF_4	1	MeOH	25	0.1	50	500	50	96.7 (*S*)	91b
1091	3,4-$(MeO)_2$-C_6H_3	H	Me	[Rh(COD)**97c**]BF_4	1	MeOH	25	0.08	50	625	50	94.8 (*S*)	91b
1092	3-MeO-4-AcO-C_6H_3	H	Me	[Rh(COD)**97a**]BF_4	1	MeOH	25	0.05	50	1000	50	91.6 (*S*)	91b
1093	3-MeO-4-AcO-C_6H_3	H	Me	[Rh(COD)**97c**]BF_4	1	MeOH	25	0.15	50	333	50	95.2 (*S*)	91b

Table 27.5 (continued)

Entry	Substrate			Catalyst	Conditions				TON	TOF [h^{-1}]	Conv. [%]	ee [%]	Reference(s)
	R^1	R^2	R^3		$P[H_2]$ [bar]	Solvent	Temp. [°C]	Time [h]					
1094	3,4-$(MeO)_2$-C_6H_3	H	Ph	[Rh(COD)**97a**]BF_4	1	MeOH	25	0.08	50	625	50	95.1 (*S*)	91b
1095	3,4-$(MeO)_2$-C_6H_3	H	Ph	[Rh(COD)**97c**]BF_4	1	MeOH	25	0.07	50	714	50	92.0 (*S*)	91b
1096	3-MeO-4-HO-C_6H_3	H	Ph	[Rh(COD)**97a**]BF_4	1	MeOH	25	0.07	50	714	50	96.9 (*S*)	91b
1097	3-MeO-4-HO-C_6H_3	H	Ph	[Rh(COD)**97c**]BF_4	1	MeOH	25	0.07	50	714	50	94.1 (*S*)	91b
1098	3,4-$(MeO)_2$-C_6H_3	Et	Me	[Rh(COD)**97a**]BF_4	1	MeOH	25	0.13	50	385	50	90.6 (*S*)	91b
1099	3,4-$(MeO)_2$-C_6H_3	Et	Me	[Rh(COD)**97c**]BF_4	1	MeOH	25	0.08	50	625	50	95.2 (*S*)	91b
1100	3,4-$(MeO)_2$-C_6H_3	Et	Ph	[Rh(COD)**97a**]BF_4	1	MeOH	25	0.17	50	294	50	88.9 (*S*)	91b
1101	3,4-$(MeO)_2$-C_6H_3	Et	Ph	[Rh(COD)**97c**]BF_4	1	MeOH	25	0.13	50	385	50	90.5 (*S*)	91b
1102	3-MeO-4-AcO-C_6H_3	Et	Ph	[Rh(COD)**97a**]BF_4	1	MeOH	25	0.08	50	625	50	87.2 (*S*)	91b
1103	3-MeO-4-AcO-C_6H_3	Et	Ph	[Rh(COD)**97c**]BF_4	1	MeOH	25	0.08	50	625	50	90.5 (*S*)	91b
1104	3,4-$(MeO)_2$-C_6H_3	*i*-Pr	Me	[Rh(COD)**97a**]BF_4	1	MeOH	25	0.37	50	135	50	91.3 (*S*)	91b
1105	3,4-$(MeO)_2$-C_6H_3	*i*-Pr	Me	[Rh(COD)**97c**]BF_4	1	MeOH	25	0.18	50	278	50	94.7 (*S*)	91b
1106	3,4-$(MeO)_2$-C_6H_3	*i*-Pr	Ph	[Rh(COD)**97a**]BF_4	1	MeOH	25	0.33	50	152	50	89.1 (*S*)	91b
1107	3,4-$(MeO)_2$-C_6H_3	*i*-Pr	Ph	[Rh(COD)**97c**]BF_4	1	MeOH	25	0.23	50	217	50	92.7 (*S*)	91b
1108	3,4-$(MeO)_2$-C_6H_3	Me	Me	[Rh(COD)**97a**]BF_4	1	MeOH	25	0.37	50	135	50	92.4 (*S*)	91b
1109	3,4-$(MeO)_2$-C_6H_3	Me	Me	[Rh(COD)**97c**]BF_4	1	MeOH	25	0.2	50	250	50	95.7 (*S*)	91b
1110	3-MeO-4-AcO-C_6H_3	Me	Me	[Rh(COD)**97a**]BF_4	1	MeOH	25	0.08	50	625	50	92.4 (*S*)	91 b
1111	3-MeO-4-AcO-C_6H_3	Me	Me	[Rh(COD)**97c**]BF_4	1	MeOH	25	0.05	50	1000	50	95.6 (*S*)	91 b
1112	3-M3O-4AcO-C_6H_3	Me	Me	[Rh(COD)**97a**]BF_4	1	MeOH	25	0.2	50	250	50	91.7 (*S*)	91 b
1113	3-M3O-4AcO-C_6H_3	Me	Me	[Rh(COD)**97c**]BF_4	1	MeOH	25	0.08	50	625	50	95.0 (*S*)	91 b
1114	3,4-$(MeO)_2C_6H_3$	Me	Ph	[Rh(COD)**97a**]BF_4	1	MeOH	25	0.13	50	385	50	87.7 (*S*)	91 b
1115	3,4-$(MeO)_2C_6H_3$	Me	Ph	[Rh(COD)**97c**]BF_4	1	MeOH	25	0.12	50	417	50	91.2 (*S*)	91b

1116	3-MeO-4-AcO-C_6H_3	Me	Ph	[Rh(COD)**97a**]BF_4	1	MeOH	25	0.08	50	625	50	87.2 (*S*)	91b
1117	3-MeO-4-AcO-C_6H_3	Me	Ph	[Rh(COD)**97c**]BF_4	1	MeOH	25	0.07	50	714	50	91.3 (*S*)	91b
1118	3-MeO-4-HO-C_6H_3	Me	Ph	[Rh(COD)**97a**]BF_4	1	MeOH	25	0.18	50	278	50	89.0 (*S*)	91b
1119	3-MeO-4-HO-C_6H_3	Me	Ph	[Rh(COD)**97c**]BF_4	1	MeOH	25	0.1	50	500	50	92.1 (*S*)	91b

a) Optical yield.
b) Estimated by proton NMR spectra.
c) Substrate : catalyst ratio.
d) Crude reaction yields were determined by ^{1}H-NMR and found to be quantitative.
f) t/2 for half-life time.
g) Surfactants.
h) Reaction time in the second cycle using recovered aqueous phase containing the catalyst.
i) Value obtained from the second cycle.
j) Determined as its methyl ester.
k) Catalysis carried out with preformed catalyst.
l) Ligand : metal ratio = 2

Table 27.6 Enantiomeric hydrogenation of tetrasubstituted substrates using bisphosphinite ligands.

Entry	Substrate				Catalyst	Conditions				TON	TOF $[h^{-1}]$	Conv. [%]	ee [%]	Reference
	R^1	R^2	R^3	R^4		$P(H_2)$ [bar]	Solvent	Temp [°C]	Time [h]					
1	Me	Me	CO_2H	NHAc	[Rh(COD)**98a**]BF_4	1	MeOH	25	16.7[a]	50	3	50	26.3 (*S*)	91f
2	Me	Me	CO_2H	NHAc	[Rh(COD)**98a**]BF_4	100	MeOH	25	1.2[a]	50	42.9	50	21 (*S*)	91f
3	Me	Me	CO_2H	NHAc	[Rh(COD)**98b**]SbF_6	2–2.8	THF	r.t.	2–3	–	–	–	15.5	95b
4	Me	Me	CO_2Me	NHAc	[Rh(COD)**98b**]SbF_6	2–2.8	Propylene carbonate	r.t.	2–3	–	–	–	28.4 (S)	95b
5	Me	Me	CO_2Me	NHAc	[[Rh(COD)**98d**]SbF_6	2–2.8	THF	r.t.	2–3	–	–	–	7.8 (R)	95b
6	Me	Me	CO_2Me	NHAc	[Rh(COD)**103a**]SbF_6	2–2.8	THF	r.t.	2–3	–	–	–	10.1 (R)	95b

a) t/2 for half-life time.

The enantioselectivities were found to be relatively independent of the solvent used. In 1998, Zhang reported a bisphosphinite based on a rigid bis-cyclopentyl ring system (BICPO, **95**, **96**) which induced 45.7 to 95% ee in the hydrogenation of α-dehydroamino acid derivatives [85].

Carbohydrate-based ligands represent an interesting area in the field of asymmetric catalysis (Tables 27.5 and 27.6). Apart from their unique biological properties, carbohydrates are highly functionalized inexpensive chiral-scaffolds. Various ligands derived from sugars, including glucose, galactose, mannitol, xylose, and trehalose, were synthesized and their effectiveness in asymmetric hydrogenation was differentiated by modulation of the steric and electronic properties. Claver and Diéguez summarized the application of carbohydrates in asymmetric catalysis in a recent review [86]. Other reviews relevant to this field also provided excellent information on their characterization and application [87]. Among these ligands, bidentate phosphorus donors were widely used in the form of phosphines, phosphinites, phosphites, or other mixed donor ligands.

Cullen [88], Thompson [89], Descotes [90] and Selke [91] were the early contributors to the use of a carbohydrate backbone in the Rh-catalyzed asymmetric hydrogenation of α-dehydroamino acid derivatives. A wide variety of 2,3-diphenylphosphinite pyranoside ligands (Fig. 27.7) were synthesized in order to probe the enantiodiscrimination from the stereocenters of the backbone. Among these, the best system was found by Selke to be based on β-glucopyranoside 2,3-diphosphinite ligand (i.e., **98a**), which provided up to 96% ee in the hydrogenation of 2-acetamidocinnamic acid [26, 91c]. The company VEB-ISIS produced L-DOPA in the former German Democratic Republic for many years based on an asymmetric olefin hydrogenation step using Selke's Ph -GLUP ligand [92].

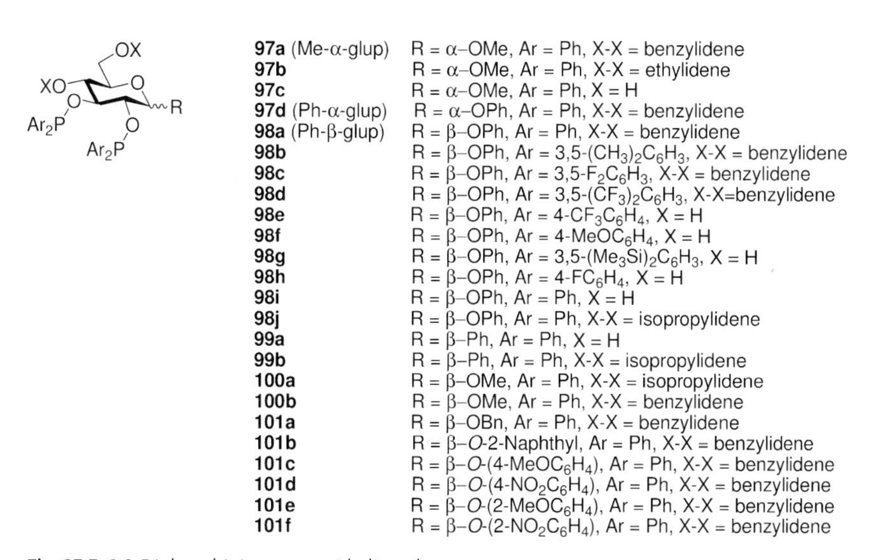

Fig. 27.7 2,3-Diphosphinite pyranoside ligands.

Šunjić [93] and Snatzke [94] systematically designed a series of pyranoside ligands **107–110** (see Fig. 27.10) for use in Rh-catalyzed hydrogenation. Ligand **108** proposed by Šunjić gave the highest ee-value (up to 90.4%). Thompson found poor results in the hydrogenation of (*Z*)-methyl *α*-acetamido-cinnamate (20–46% ee) using *β*-galactoside-based **113b** (see Fig. 27.10) [89]. Selke showed that *α*-galactose-based ligand **113a** induced higher enantioselectivity (86% ee) [91e].

In 1994, RajanBabu carried out systematic studies on the electronic and steric properties of the diphosphinite ligands (**102–106**) [95]. It was determined that, in the Rh-catalyzed hydrogenation of a wide variety of dehydroamino acid derivatives, high enantioselectivities (ee-values up to 99% in *S*-configuration) were obtained with **98b** and **98g** bearing electron-rich substituents, whereas poor selectivity was obtained using the electron-deficient ligands. These results raised the question of the preparation of products with the *R*-configuration. Preparing the other enantiomer of **98** from L-glucose would be prohibitively expensive. Nonetheless, RajanBabu developed *pseudo*-enantiomeric diphosphinite ligands based on the relationship of the 2,3-diphenylphosphinite and its corresponding 3,4-diphosphinite ligands (**98** and **103**; Fig. 27.8). Again, electron-rich phosphinites provided up to 99% ee of the products (dehydroamino acids) with the *R*-configuration. This might be the most convenient way to synthesize both enantiomers of aromatic and heteroaromatic alanines when using sugar-based diphosphinite ligands.

Two-phase catalysis has been established as a new field of study, and has achieved industrial-scale importance in olefin hydroformylation [96]. A significant advantage is the ease of separation of catalyst and product, which may have economic and environmental impact. Thus, removal of the 4,6-*O*-protecting group in 2,3-diphosphinite ligand **98** easily generated a water-soluble catalyst. The effectiveness of using Rh–complexes of diphosphinite **98a** in an aqueous system was proved successfully. Oehme reported that use of Ph-*β*-glup with free hydroxy groups (i.e., **98i** in Fig. 27.7) resulted in 84% ee with 100% conversion in the hydrogenation of (*Z*)-methyl *α*-acetamidocinnamate in water (95% ee in MeOH) [97]. The enantioselectivity was further improved (up to 97% ee) using a surfactant such as sodium dodecylsulfate (SDS) or Triton X-100. Similar results were obtained in the hydrogenation of methyl *α*-acetamidoacrylate. Selke reported more experimental results with **98i**. The enantioselectivities were similar to those obtained with protected 2,3-disphosphinite ligand **98** using metha-

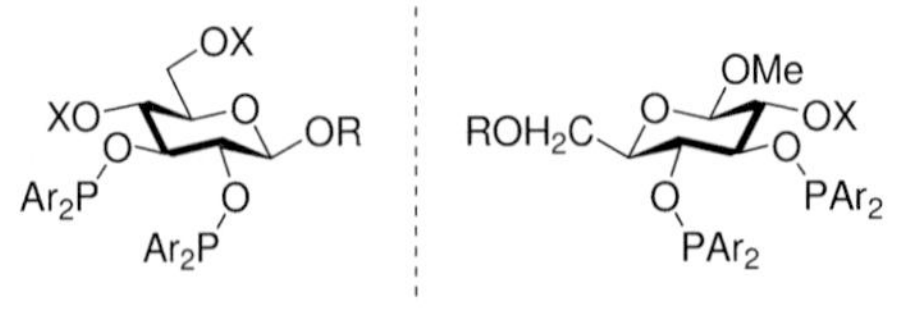

2,3-diphosphinite **98** 3,4-diphosphinite **103**, **104**, **112**

Fig. 27.8 *Pseudo*-enantiomeric diphosphinite pyranoside ligands.

102a R^1 = β–OMe, R^2 = NHAc, R^3 = OTBDMS, Ar = 3,5-$(CH_3)_2C_6H_3$
102b R^1 = β–OMe, R^2 = NHAc, R^3 = OTBDMS, Ar = Ph
102c R^1 = β–OMe, R^2 = NHAc, R^3 = OTBDMS, Ar = 3,5-$F_2C_6H_3$
102d R^1 = β–OMe, R^2 = NHAc, R^3 = OTBDMS, Ar = 3,5-$(CF_3)_2C_6H_3$
102e R^1 = β–OMe, R^2 = NHAc, R^3 = OTBDMS, Ar = 4-$CF_3C_6H_4$
102f R^1 = β–OMe, R^2 = NHAc, R^3 = OTBDMS, Ar = 4-$MeOC_6H_4$
103a R^1 = α–OMe, R^2 = R^3 = OBz, Ar = 3,5-$(CH_3)_2C_6H_3$
103b R^1 = α–OMe, R^2 = R^3 = OBz, Ar = Ph
103c R^1 = α–OMe, R^2 = R^3 = OBz, Ar = 3,5-$F_2C_6H_3$
103d R^1 = α–OMe, R^2 = R^3 = OBz, Ar = 3,5-$(CF_3)_2C_6H_3$
103e R^1 = α–OMe, R^2 = R^3 = OBz, Ar = 4-$CF_3C_6H_4$
103f R^1 = α–OMe, R^2 = R^3 = OBz, Ar = 4-$MeOC_6H_4$
103g R^1 = α–OMe, R^2 = R^3 = OBz, Ar = 3,5-$(Me_3Si)_2C_6H_3$
103h R^1 = α–OMe, R^2 = R^3 = OBz, Ar = 4-FC_6H_4
104a R^1 = α–OMe, R^2 = R^3 = OPiv, Ar = 3,5-$(CH_3)_2C_6H_3$
104b R^1 = α–OMe, R^2 = R^3 = OPiv, Ar = Ph
104c R^1 = α–OMe, R^2 = R^3 = OPiv, Ar = 3,5-$F_2C_6H_3$
104d R^1 = α–OMe, R^2 = R^3 = OPiv, Ar = 3,5-$(CF_3)_2C_6H_3$
104e R^1 = α–OMe, R^2 = R^3 = OPiv, Ar = 4-$CF_3C_6H_4$
104f R^1 = α–OMe, R^2 = R^3 = OPiv, Ar = 4-$MeOC_6H_4$
104g R^1 = α–OMe, R^2 = R^3 = OPiv, Ar = 3,5-$(Me_3Si)_2C_6H_3$
104h R^1 = α–OMe, R^2 = R^3 = OPiv, Ar = 4-FC_6H_4
105a R^1 = α–OMe, R^2 = H, R^3 = OTBDMS, Ar = 3,5-$(CH_3)_2C_6H_3$
105b R^1 = α–OMe, R^2 = H, R^3 = OTBDMS, Ar = Ph
105c R^1 = α–OMe, R^2 = H, R^3 = OTBDMS, Ar = 3,5-$F_2C_6H_3$
105d R^1 = α–OMe, R^2 = H, R^3 = OTBDMS, Ar = 3,5-$(CF_3)_2C_6H_3$
105e R^1 = α–OMe, R^2 = H, R^3 = OTBDMS, Ar = 4-$CF_3C_6H_4$
105f R^1 = α–OMe, R^2 = H, R^3 = OTBDMS, Ar = 4-$MeOC_6H_4$
106a R^1 = R^2 = H, R^3 = OTr, Ar = 3,5-$(CH_3)_2C_6H_3$
106b R^1 = R^2 = H, R^3 = OTr, Ar = Ph
106c R^1 = R^2 = H, R^3 = OTr, Ar = 3,5-$F_2C_6H_3$
106d R^1 = R^2 = H, R^3 = OTr, Ar = 3,5-$(CF_3)_2C_6H_3$
106e R^1 = R^2 = H, R^3 = OTr, Ar = 4-$CF_3C_6H_4$
106f R^1 = R^2 = H, R^3 = OTr, Ar = 4-$MeOC_6H_4$

Fig. 27.9 Bisphosphinite–3,4-diphosphinite pyranoside ligands.

nol as solvent [91 a, b]. Attempts also were made by RajanBabu using modified D-salicin with pendant quaternary ammonium groups; however, the result in water (61% ee with **115g** in the hydrogenation of methyl *α*-acetamidoacrylate) [98] was inferior to that obtained in organic solvents (up to 96% ee). Attempts were also made using mannoside-based 3,4-diphosphinite **112**, but only with moderate (72.2%) ee. Glucosamine-based 3,4-diphosphinite **102a**, on the other hand, induced very high enantioselectivity (95–98.4%) in the Rh-catalyzed hydrogenation of various dehydroamino acids.

Recently, Miethchen modified diphosphinite **97d** with a crown-ether linker in the 1,4-positions in order to study the effect on enantioselectivity in Rh-catalyzed asymmetric hydrogenation reactions [99]. Introduction of the crown ether in the 1,4-position of the carbohydrate allows the enantioselectivity to be tuned, based on a strong effect of the formation of cryptate species with alkali ions.

Unfortunately, the application of this new ligand **114** (Fig. 27.10) in the hydrogenation of various dehydroamino acid derivatives gave poorer results in comparison to the parent ligand **97d**.

In 1998, Uemura developed novel disaccharide diphosphinite ligands **119a** and **116a** (Fig. 27.10) from *α,α*-trehalose. Rh-catalyzed asymmetric hydrogenation of *α*-acetamidoacrylic and cinnamic acid derivatives afforded amino acids with up to 84% ee (*S*) (with ligand **119a**) and 72% ee (*R*) (with ligand **116a**), respectively [100]. The deprotected-hydroxyl diphosphinite ligand **119e** also enabled hydroge-

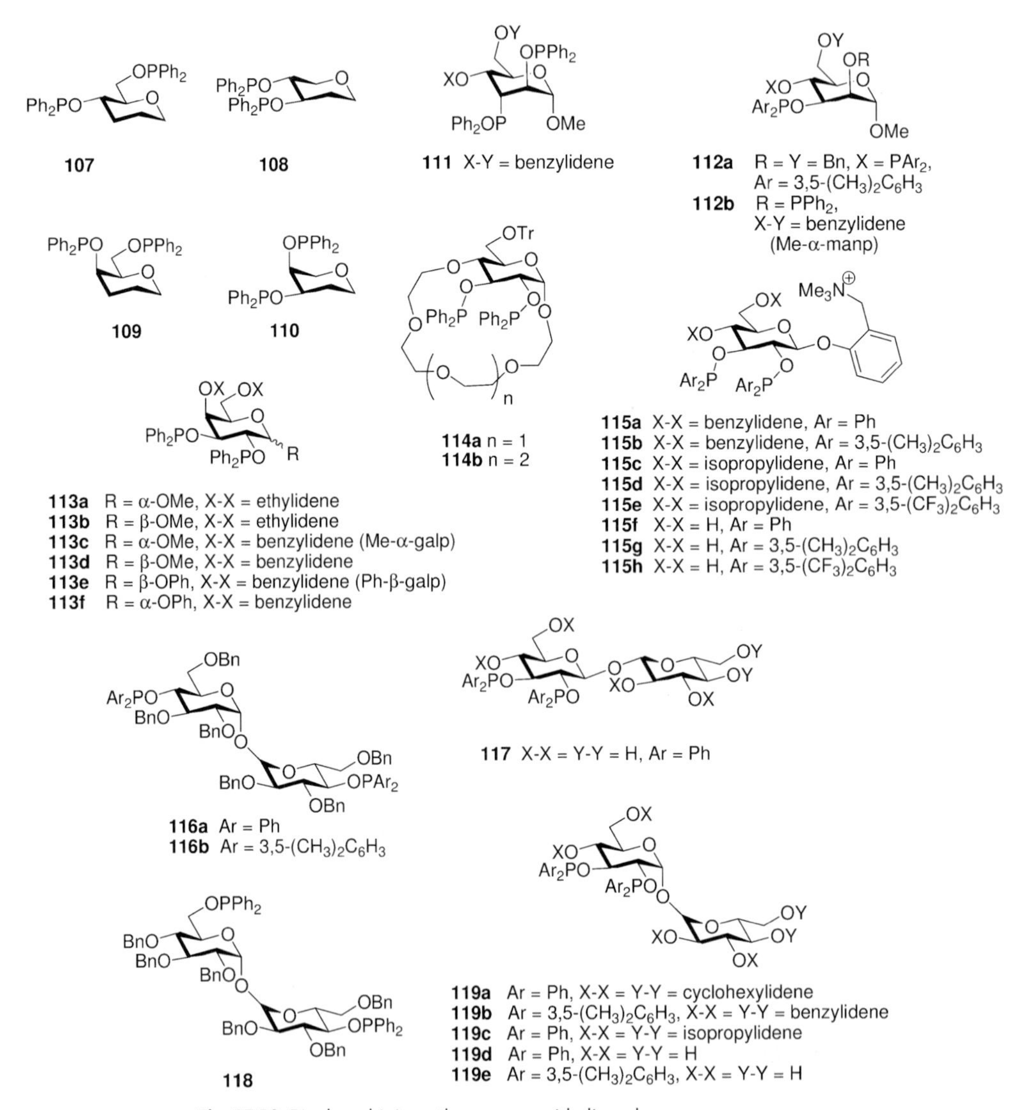

Fig. 27.10 Bisphosphinite–others pyranoside ligands.

nation of enamides and itaconic acid in aqueous solution with enhanced enantioselectivities (ee-values up to 99%) [101]. Similar reports by RajanBabu showed its application with moderate to good enantioselectivity [102].

In the light of the fruitful results obtained with the pyranoside-based bisphosphonites, RajanBabu also used a series of 3,4-diphosphinite ligands with a fructofuranoside backbone (i.e., **123**; Fig. 27.11) in the Rh-catalyzed hydrogenation of α-dehydroamino acids. However, the results were unsatisfactory (with only 49–57% ee) [95 b]. Similar results were found by Johnson with ligands based on α-D-glucofuranose (**124**) and α-L-idofuranose (**125**) with highest enantioselectivity (54% ee) obtained in the hydrogenation of α-methylcinnamic acid [103]. Diéguez and Ruiz described a facile synthesis of 3,4-diphosphinites **126** and **127** from D-(+)-xylose [104]. Application of these ligands in asymmetric hydrogenation showed that the enantioselectivity was strongly dependent on the absolute configuration of the C-3 stereocenter and the metal source. When ligand **126** was used in the rhodium-catalyzed hydrogenation of 2-acetamidoacrylic acid, the product was obtained with 76% ee. On the other hand, 78% ee was obtained using the Ir–**127** complex.

Díaz and Castillón reported new modular C_2 symmetric ligands prepared from D-glucosamine, D-glucitol and tartaric acid [105]. Ligand **122e** was found to induce the highest ee-value (93%), with full conversion in hydrogenation of methyl 2-acetamidoacrylate. In comparison to **ent-120**, the enantioselectivities of *N*-acetyl-L-alanine methyl ester induced by the catalysts based on **122** and **121** were strongly influenced by the stereocenters at positions 2 and 5 of the tetrahydrofuran ring and steric effect of the R groups. The configuration of the hydrogenation product (methyl 2-acetamidoacrylate and acetamidocinnamic acid ester) was influenced by the stereocenters at C-3 and C-4.

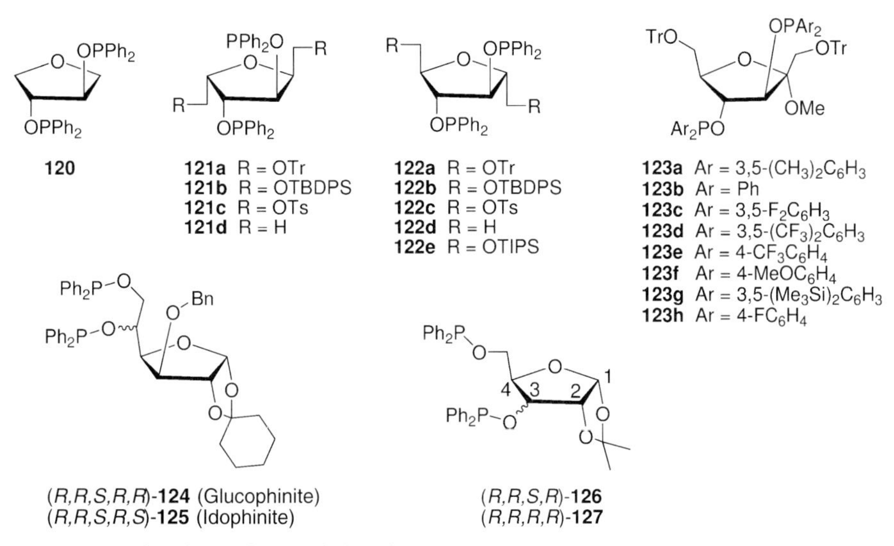

Fig. 27.11 Bisphosphinite–furanoside ligands.

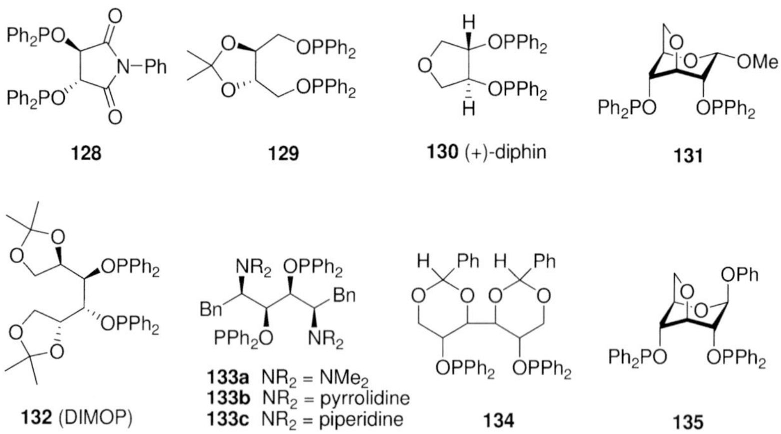

128 **129** **130** (+)-diphin **131**

132 (DIMOP) **133a** NR_2 = NMe_2 **133b** NR_2 = pyrrolidine **133c** NR_2 = piperidine **134** **135**

Fig. 27.12 Bisphosphinite–other carbohydrate-derived ligands.

Increasing the ligand rigidity provides one possibility of increasing enantioselectivity. Jackson and Lovel reported a ligand [(+)-Diphin **130**; Fig. 27.12] derived from natural L-tartaric acid [68 b], but use of this ligand containing a rigid tetrahydrofuran ring in the rhodium-catalyzed hydrogenation of α-acetamidocinnamic acid led to poor results (2% ee); in contrast, DIOP **129** induced 88% ee in the same reaction [68]. Bourson and Oliveros also developed a bisphosphonite ligand based on the *N*-phenylimide of natural L-tartaric acid **128** [106]. Unfortunately, the Rh-catalyzed hydrogenation of prochiral olefins gave unsatisfactory results with this ligand (1 to 44% ee). In 1999, we developed a new C_2 ligand (DIMOP **132**) from inexpensive D-mannitol, and found it to be highly effective in the Rh-catalyzed asymmetric hydrogenation of α-amidoacrylic acid and its derivatives [107]. For example, in the hydrogenation of 2-acetamidoacrylic acid, the product was obtained with full conversion in 15 min and 96.7% ee (SCR = 100). In all cases the desired products were found to have ee-values in excess of 90%. Lu and Jiang introduced three analogues based on D-mannitol and D-glucose, (**131**, **134**, and **135**). All ligands led to highly active catalysts with rhodium, but these were less enantioselective in the hydrogenation of α-acetamidocinnamic acid and its methyl ester (24.6–46.2% ee) [108]. Through structural modification of D-mannitol, Jiang and Zhang synthesized three bulky analogues (**133 a–133 c**), each of which induced moderate to excellent ee-values in the hydrogenation of dehydroamino acid derivatives (41–97% ee) [48].

27.6
Bisphosphonite Ligands (Two P–O Bonds)

In recent years, there is no doubt that BINOL is one of the most extensively studied motifs. Incorporating a chiral binol unit into the chiral or achiral backbone constitutes a straightforward way in which to generate new chiral ligands [109].

Both ligands **138** (ferrocene backbone; Fig. 27.13) and **136** (ethylene backbone) performed very well in the hydrogenation of itaconic acid dimethyl ester (97–99.5% ee) and 2-acetamido methyl acrylate (90–99.5% ee). Pringle and Orpen reported poor results in the hydrogenation of methyl 2-acetamido acrylate with the new modified ligand **140**, although the monodentate analogues performed surprisingly well [110]. An enhancement of enantioselectivity may be achieved by combining a chiral backbone with binol in a matching sense. Switching from an achiral backbone to chiral paracyclophane was successful, as reported by Zanotti-Gerosa [111]. Ligands **141 b** and **141 c** displayed a very strong matching/mismatching effect in the Rh-catalyzed hydrogenations of methyl 2-acetamido acrylate, inducing 99% ee and 0% ee, respectively, with the stereochemistry of the product being mainly controlled by the chirality of the backbone. Rh-**141 a** was a faster catalyst (TOF 2500 h^{-1}) than Rh-**141 b** (TOF 833 h^{-1}), albeit at the expense of a few percent lower ee. Bakos used (*S,S*)-pentane-2,3-diol as the chiral backbone leading to ligands **137 a** and **b** that induced moderate to good ee-values in the hydrogenation of dimethyl 2-methylsuccinate (59.7–88.5% ee) [112]. Vogt developed a new bisphosphonite based on 9,9-dimethylxanthene (**139**) and, by applying it to

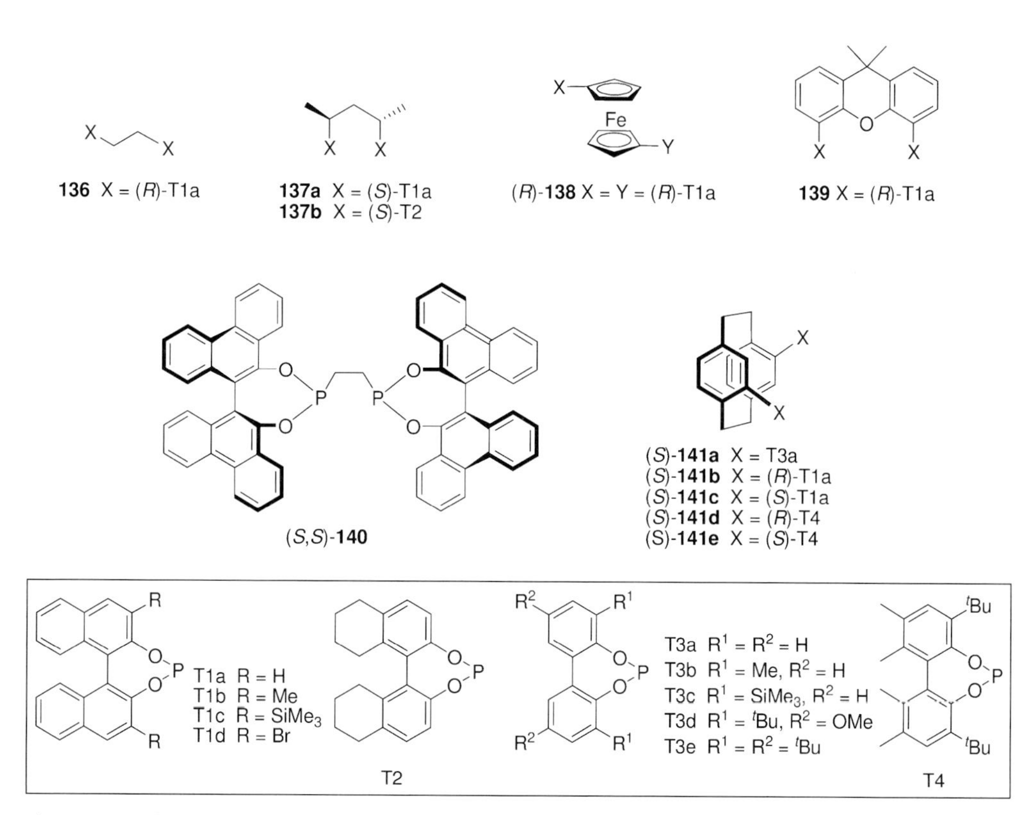

Fig. 27.13 Bisphosphonite ligands (two P–O bonds).

the rhodium-catalyzed hydrogenation of methyl (*Z*)-2-acetamidocinnamate, obtained 54% ee with full conversion [113].

27.7
Bisphosphite Ligands (Three P–O Bonds)

Wink reported the use of bisphosphite ligands in the asymmetric hydrogenation of enamides (2–10% ee) [114]. In 1998, Selke synthesized a series of analogues based on **98a**. Of these compounds, **147** (Fig. 27.14) was selected as ligand for the Rh-catalyzed hydrogenation of methyl (*Z*)-2-acetamidocinnamate, though it induced only low enantioselectivity (13% ee) [115].

In 1999, Reetz established a class of bidentate bisphosphite ligands **149** (Fig. 27.14) based on C_2-symmetric 1,4:3,6-dianhydro-D-mannite [116]. These ligands induced high enantioselectivity in the hydrogenation of dimethyl itaconate (98.2% ee) and methyl *N*-2-acetamidoacrylate (88.8% ee). The results also indicated a cooperative effect between the stereogenic centers of the ligand backbone and the axial chiral binaphthyl phosphite moieties, although the sense of enantioselectivity was predominantly controlled by the binaphthyl moieties (**149e** versus **149b** and **149c**). The use of biphenyl phosphite moieties led to ligands with better performance than those carrying binaphthyls, in spite of their easy epimerization.

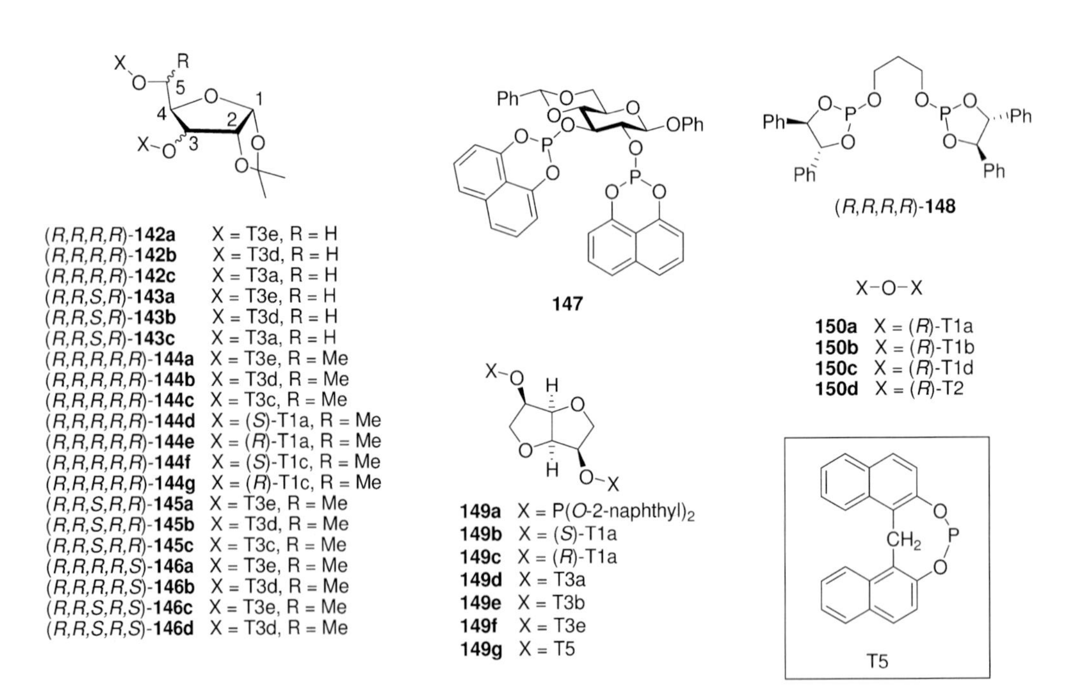

Fig. 27.14 Bisphosphite ligands (three P–O bonds).

Recently, Claver and co-workers developed a series of highly effective modular C_1 diphosphite ligands **142–146** (Fig. 27.14) with a furanoside backbone [117]. Excellent enantioselectivities (ee-values up to >99%) and good activities were achieved in the Rh-catalyzed hydrogenation of dimethyl itaconate, methyl (*Z*)-2-acetamidocinnamate and methyl (*Z*)-2-acetamidoacrylate [117b]. Systematic variation of the stereocenters C-3 and C-5 at the ligand backbone showed that the enantiomeric excesses depended strongly on the absolute configuration of C-3 and only slightly on that of the stereocenter carbon C-5. Similar to Reetz's observation, the axially chiral binaphthyl substituent predominantly controlled the sense of the enantiodiscrimination. Bulky substituents at the *ortho*-positions of the achiral biaryl diphosphite moieties have a positive effect on enantioselectivity, especially with *o*-trimethylsilyl substituents in the biphenyl moieties of **144c**.

Börner reported the synthesis of pyrophosphites **149** with chiral binaphthyl substituents [118]. The results showed that the H_8-binaphthyl unit was the best for the Rh-catalyzed hydrogenation of methyl (*Z*)-2-acetamidocinnamate (48% ee) and dimethyl itaconate (70% ee).

27.8
Other Mixed-Donor Bidentate Ligands

In 1982, Yamashita reported the application of L-talopyranoside-based phosphine-phosphinite ligand **165** (Fig. 27.15), and found that it induced low enantioselectivity (4.7–13% ee) in the hydrogenation of *α*-acetamidocinnamic acid [119]. Reetz introduced the phosphine-phosphonite ligand (**151–153**), which led to moderate enantioselectivity (52–88% ee) in the Rh-catalyzed hydrogenation of dimethyl itaconate [120]. The binaphthyl unit remained an essential element in the system.

Claver and Ruiz reported excellent enantioselectivity (>99% ee) and good activities (TOF >1200 h^{-1}) in the hydrogenation of methyl *N*-acetamidoacrylate and methyl *N*-acetamidocinnamate using phosphine–phosphite ligand **167** [121]. Again, ligands based on the biphenyl unit (especially with bulky *tert*-butyl groups in the *ortho* and *para positions*) showed a strong enantioinduction. Interestingly, **167** induced a higher activity and enantioselectivity than its corresponding diphosphine [122].

van Leeuwen and Claver designed a new class of chiral phosphine–phosphite ligands **159** and **160** with a stereogenic phosphine for the hydrogenation of methyl *N*-2-acetamidoacrylate and methyl *N*-2-acetamidocinnamate [123]. Up to 99% ee was achieved after systematically tuning the steric and electronic properties of the biaryl phosphite unit.

Pizzano and Suárez described a convenient preparation of a series of new chiral phosphine–phosphites based on the easy demethylation of *o*-anisyl phosphines [124]. Rh–**156a** complex was found to be the most effective catalyst for the hydrogenation of dimethyl itaconate (99.6% ee), whereas **155b** and **156a** induced >99% ee in the hydrogenation of methyl *N*-2-acetamidocinnamate. Reetz

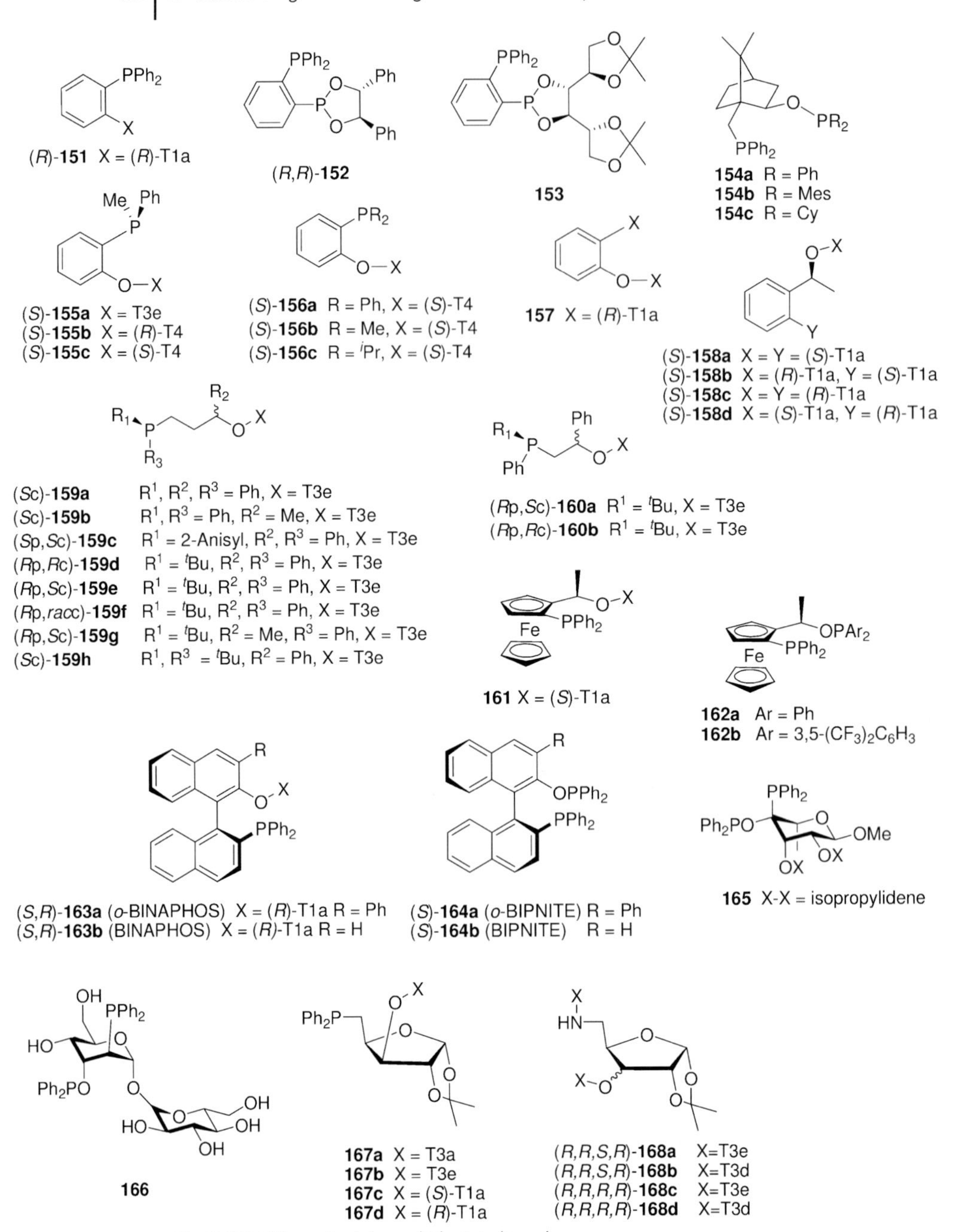

Fig. 27.15 Other mixed-donor bidentate ligands.

used (*S*)-1-(2-bromophenyl)ethanol together with binol to make ligands **158a**–**158d**. Use of these ligands in the Rh-catalyzed hydrogenation of itaconic acid dimethyl ester gave up to 79% ee [125].

The use of phosphite–phosphoramidite ligands **168a** and **b** provided up to 98% ee in the hydrogenation of methyl (*Z*)-*N*-2-acetylaminocinnamate, but the activities were rather low when compared to **167** or to the corresponding diphosphine ligand [126].

In contrast to the extensive studies on phosphine–phosphites, the corresponding phosphine–phosphinites are rarely exploited. Laschat introduced this design with a bicyclic chiral skeleton derived from (1*S*)-(+)-camphorsulfonic acid [127]. The Rh–complex based on dimesitylphosphinite **154b** was found to be the most reactive catalyst, and was used to produce methyl *N*-2-acetamidocinnamate, with 89% ee.

In 2004, we introduced new phosphine–phosphite ligands with a ferrocenyl scaffold derived from Ugi's amine [61]. Ligand **161** was found to exhibit good enantioselectivity in the hydrogenation of methyl *N*-2-acetamidocinnamate (85–89% ee). Ligand **162b** was also found to be highly effective in the hydrogenation of methyl *N*-2-acetamidocinnamate (95.3–99.6% ee) and *N*-acetyl-*α*-arylenamides (83–91% ee).

Zhang reported two new (*S*)-BINOL based ligands: phosphine–phosphite (*S*,*R*)-*o*-BINAPHOS **163** and phosphine–phosphinite (*S*)-*o*-BIPNITE **164** [128]. Applications of these ligands in the Rh-catalyzed hydrogenation of methyl *N*-2-acetamidocinnamate and methyl *N*-2-acetamidoacrylate induced very high enantioselectivities (>99% ee), and with a wide range of substrates.

Uemura developed a water-soluble phosphine–phosphinite ligand (derived from *α*,*α*-trehalose) (**166**) for the Rh-catalyzed hydrogenation of enamide derivatives; this induced only moderate enantioselectivity [129].

27.9 Ligands Containing Neutral S-Donors

Ligands containing thioethers are stereochemically very interesting, because upon coordination, the sulfur atom becomes a stereogenic center. In the absence of any stereocontrol, the *S*-center can be either (*R*)- or (*S*)-configured. However, if one imposes an efficient stereochemical control through judicious selection of the backbone chirality, it is possible to stabilize the configuration of the sulfur atom and thereby confer chiral information to the metal center. During the past few years, a number of reports have been disclosed describing attempts to harness this special property of thioethers in the asymmetric hydrogenation of a variety of prochiral olefins.

A number of dithioethers **169–173** (Fig. 27.16) based on the chiral skeleton of some well-known phosphines such as DIOP, Deguphos and BINAP, have been reported. The use of 1,4-dithioether ligands which lack contiguous chiral centers such as (+)-DiopsR_2 **169** [130], BINASR_2 **172** [131] and **173** [132] in the Ir- or

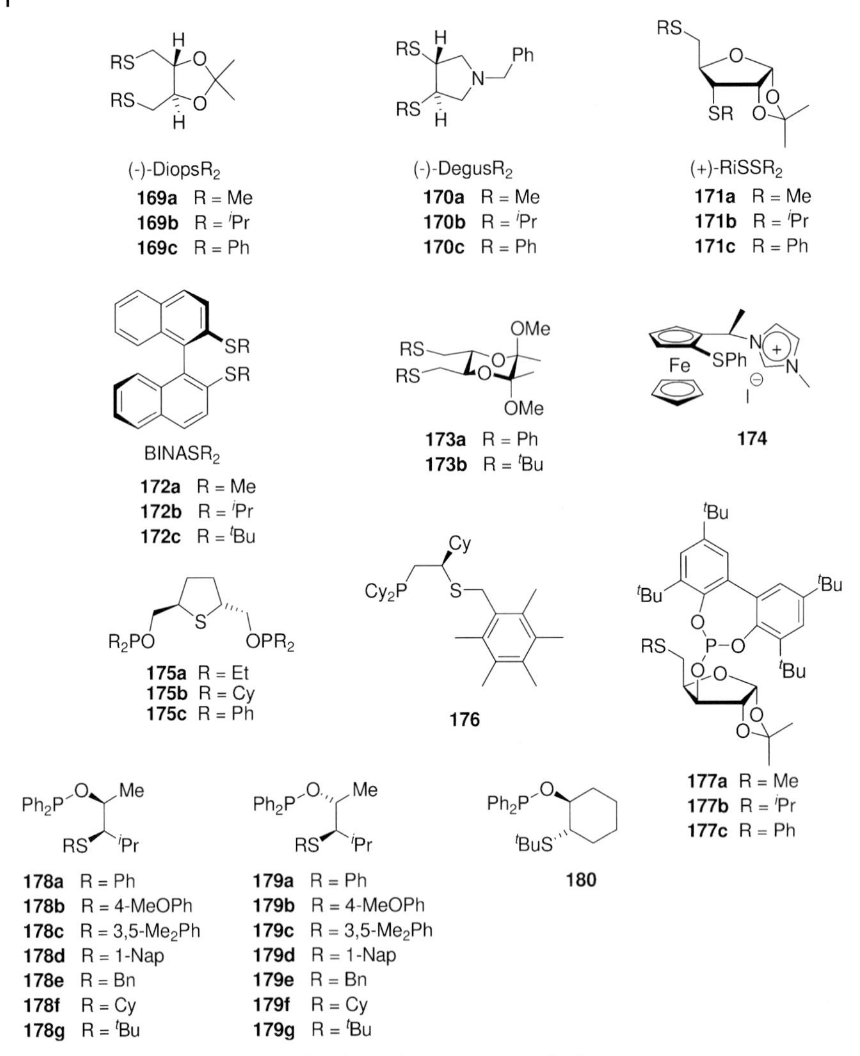

Fig. 27.16 Thioether-containing chiral ligands in asymmetric hydrogenation.

Rh-catalyzed asymmetric hydrogenation of itaconic acid and its derivatives, dehydroamino acid derivatives and enamides led to extremely poor to moderate enantioselectivities (Table 27.7). Although NMR spectroscopic studies of the iridium(I) cyclooctadiene complexes of **169** and **172** suggested that they possessed well-defined C_2-symmetry, implying that both sulfur atoms have the same configuration, their corresponding *cis*-dihydridoiridium(III) adducts appeared in the NMR spectrum as either a mixture of diastereomers or C_1-symmetric complexes, suggestive of the configurational lability of the ligated sulfur atom under remote chiral control in the octahedral complex, thus explaining the observed

Table 27.7 Enantiomeric hydrogenation using ligands containing a neutral S-donor.

R^1–CH=C(R^2)(R^3) → R^1–CH$_2$–C*H(R^2)(R^3)

Entry	Substrate			Catalyst	Conditions				TON	TOF [h^{-1}]	Conv. [%]	ee [%]	Reference
	R^1	R^2	R^3		P(H_2) [bar]	Solvent	Temp [°C]	Time [h]					
1	H	Ph	NHAc	**179c**+[Rh(COD)$_2$]SbF$_6$	35.5	THF	r.t.	18	100	5.6	100	95	139
2	Me(*E*/*Z*)	Ph	NHAc	**173a**+[Rh(NBD)$_2$]SbF$_6$	3.1	CH_3OH	r.t.	24	100	4.0	95	21	132
3	Me(*E*/*Z*)	Ph	NHAc	**173b**+[Rh(NBD)$_2$]SbF$_6$	3.1	CH_3OH	r.t.	24	100	1.5	37	18	132
4	H	CO_2H	NHAc	**169c**+[Ir(COD)$_2$]BF$_4$	1	CH_2Cl_2	20	12	40	3.3	100	10	130
5	Me	CO_2Me	NHAc	**179c** vs. **180**+[Rh(COD)$_2$]SbF$_6$	7.9	THF	r.t.	18	100	5.6	100	97 vs. 98	139
6	Et	CO_2Me	NHAc	**179c** vs. **180**+[Rh(COD)$_2$]SbF$_6$	7.9	THF	r.t.	18	100	5.6	100	94 vs. 94	139
7	*i*Pr	CO_2Me	NHAc	**179c** vs. **180**+[Rh(COD)$_2$]SbF$_6$	7.9	THF	r.t.	18	100	5.6	100	89 vs. 36	139
8	Ph	CO_2H	NHAc	**169b**+[Ir(COD)$_2$]BF$_4$	1	CH_2Cl_2	20	16	40	2.4	96	37	130
9	Ph	CO_2H	NHAc	**170c**+[Ir(COD)$_2$]BF$_4$	1	CH_2Cl_2	20	2	40	20	100	27	133
10	Ph	CO_2Me	NHAc	**169c**+[Ir(COD)$_2$]BF$_4$	1	CH_2Cl_2	20	48	40	0.4	50	13	130
11	Ph	CO_2Me	NHAc	**172a–172c** +[Ir(COD)$_2$]BF$_4$	1	CH_3OH	25	0.5	100	–	–	–	131
12	Ph	CO_2Me	NHAc	**175b**+[Rh(COD)$_2$]OTf	4.1	CH_3OH	r.t.	O/N	50	–	100	55	134
13	Ph	CO_2Me	NHAc	**176**+[Rh(COD)$_2$]OTf	5.5	CH_3OH	r.t.	16	50	3.1	100	39	137
14	Ph	CO_2Me	NHAc	**178a**+[Rh(COD)$_2$]SbF$_6$	7.9	THF	r.t.	18	100	5.6	100	84	139
15	Ph	CO_2Me	NHAc	**179a**+[Rh(COD)$_2$]SbF6	7.9	THF	r.t.	18	100	5.6	100	95	139
16	Ph	CO_2Me	NHAc	**178c**+[Rh(COD)$_2$]SbF$_6$	7.9	THF	r.t.	18	100	5.6	100	81	139
17	Ph	CO_2Me	NHAc	**179c**+[Rh(COD)$_2$]SbF$_6$	7.9	THF	r.t.	18	100	5.6	100	97	139
18	Ph	CO_2Me	NHAc	**178f**+[Rh(COD)$_2$]SbF$_6$	7.9	THF	r.t.	18	100	5.6	100	84	139
19	Ph	CO_2Me	NHAc	**179f**+[Rh(COD)$_2$]SbF$_6$	7.9	THF	r.t.	18	100	5.6	NR		139
20	Ph	CO_2Me	NHAc	**178g**+[Rh(COD)2]SbF$_6$	7.9	THF	r.t.	18	100	5.6	100	82	139

Table 27.7 (continued)

Entry	Substrate			Catalyst	$P(H_2)$ [bar]	Conditions			TON	TOF [h^{-1}]	Conv. [%]	ee [%]	Reference
	R^1	R^2	R^3			Solvent	Temp [°C]	Time [h]					
21	Ph	CO_2Me	NHAc	**179g**+$[Rh(COD)_2]SbF_6$	7.9	THF	r.t.	18	100	5.6	20	68	139
22	Ph	CO_2Me	NHAc	**179c** vs. **180**+$[Rh(COD)_2]SbF_6$	7.9	THF	r.t.	18	100	5.6	100	97 vs. 97	139
23	3-Br-Ph	CO_2Me	NHAc	**179c** vs. **180**+$[Rh(COD)_2]SbF_6$	7.9	THF	r.t.	18	100	5.6	100	94 vs. 95	139
24	4-F,3-NO_2Ph	CO_2Me	NHAc	**179c** vs. **180**+$[Rh(COD)_2]SbF_6$	7.9	THF	r.t.	18	100	5.6	100	92 vs. 94	139
25	4-MeO-Ph	CO_2Me	NHAc	**179c** vs. **180**+$[Rh(COD)_2]SbF_6$	7.9	THF	r.t.	18	100	5.6	100	96 vs. 98	139
26	2-thienyl	CO_2Me	NHAc	**176**+$[Rh(COD)_2]OTf$	5.5	CH_3OH	r.t.	16	50	3.1	100	19	137
27	4-F,3-NO_2Ph	CO_2Me	NHAc	**176**+$[Rh(COD)_2]OTf$	5.5	CH_3OH	r.t.	16	50	3.1	100	51	137
28	H	CO_2H	CH_2CO_2H	**169b**+$[Ir(COD)_2]BF_4$	1	CH_2Cl_2	20	6	40	6.1	91	47	130
29	H	CO_2H	CH_2CO_2H	**170c**+$[Ir(COD_2]BF_4$	1	CH_2Cl_2	20	12	40	3.3	100	68	133
30	H	CO_2H	CH_2CO_2H	**171b**+$[Ir(COD_2]BF_4$	1	CH_2Cl_2	20	12	100	8.3	100	62	134
31	H	CO_2H	CH_2CO_2H	**172a172c**+$[Ir(COD_2]BF_4$	1	CH_3OH	25	0.5	100		[b]	[c]	131
32	H	CO_2H	CH_2CO_2H	**177b**+$[Ir(COD_2]BF_4$	1	N/A	40	12	50	4.2	100	51	138
33	H	CO_2Me	CH_2CO_2Me	**172a172c**+$[Ir(COD)_2]BF_4$	1	CH_3OH	25	0.5	100		[b]	[c]	131
34	H	CO_2Me	CH_2CO_2Me	**174**+$[Rh(COD_2]BF_4$	10.1	N/A	50	12	N/A		44	18	133

poor enantioselectivity [135, 136]. Slight improvements resulted when neighboring stereocenters were introduced, as in (–)-DegusR_2 **170** [133] and **171** [134] with S-substituents larger than a methyl group (Table 27.7, entries 29 and 30).

An unusual carbene-thioether hybrid ligand **174** was synthesized and applied in the rhodium-catalyzed asymmetric hydrogenation of dimethyl itaconate by Chung and co-workers; however, the selectivity and activity were low (Table 27.7, entry 34) [135].

Another major class of ligands containing a thioether functionality is the phosphorus–sulfur (P/S) mixed donor family. To date, only a few ligands of this type have been tested. The tridentate tetrahydrothiophene **175** flanked by two *trans-O*-methylene phosphinites was among the first P/S-ligands examined by Hauptman and co-workers, but only mediocre enantioselectivity was recorded in the hydrogenation of methyl α-acetamidoacrylate (Table 27.7, entry 12) [136]. Whilst the mode of coordination of **175** in the actual operating Rh-catalyst was unknown, the bidentate phosphine–thioethers **176**, prepared by the same team, also showed unsatisfactory results [137]. The xylofuranose-based phosphite–thioether **177** was also found to be inefficient (Table 27.7, entry 32) [138]. A breakthrough was unveiled by the Evans team [139], when Rh(I) complexes based on phosphinite–thioethers **178** and **179** were found to be highly efficient catalysts in the hydrogenation of a variety of enamide substrates. A side-by-side comparison revealed that skeleton **179** was generally more efficient than **178**, and sterically more encumbered *thio*-aryl substituents were generally superior than the less bulky ones or *thio*-alkyls. Remarkably, the meta-dialkyl effect, which was commonly noted in the phosphorus counterparts [140], also appeared to be operative here as **179** was found to be the optimal ligand (Table 27.7 entry 15). Moreover, the latter was found also to be effective in the enantioselective hydrogenation of β,β-disubstituted dehydroamino acids (Table 27.8, entries 1

Table 27.8 Enantiomeric hydrogenation of β,β-disubstituted dehydroamino acids and enamides.

S1: CO_2Me, NHAc; S2: CO_2Me, NHAc; S3: NHAc

Entry	Substrate			Catalyst]	Conditions				TON	TOF $[h^{-1}]$	Conv. [%]	ee [%]	Reference
	R^1	R^2	R^3		$P(H_2$ [bar]	Solvent	Temp [°C]	Time [h]					
1		S1		**179c**+[Rh(COD$_2$]SbF_6	7.9	THF	r.t.	18	100	5.6	100	93	139
2		S2		**179c**+[Rh(COD$_2$]SbF_6	1	THF	r.t.	18	100	5.6	100	95	139
3		S3		**179c**+[Rh(COD$_2$]SbF_6	7.9	THF	r.t.	18	100	5.6	100	92	139

and 2) and enamides (Table 27.7, entry 1; Table 27.8, entry 3). The more rigid ligand **180** also proved to be comparable to **179c**. In contrast to **178g** and **179g**, the *S*-*t*Bu group in **180** exerted a positive effect in the stereodifferentiating process and induced much better reactivity. The elegant investigations of Evans and co-workers recapitulated the fact that meticulous screening of the modifiable units – the *S*-substituents in this case – was the key to finding effective ligands [141, 142].

Acknowledgments

The authors thank the University Grants Committee Areas of Excellence Scheme in Hong Kong (AoE P/10-01) and the Hong Kong Polytechnic University Area of Strategic Development Fund for financial support of this study.

Abbreviations

AMPP	aminophosphine–phosphinite
DCE	dichloroethane
DCM	dichloromethane
IPA	isopropyl alcohol
r.t.	room temperature
$scCO_2$	supercritical CO_2
SCR	substrate:catalyst ratio
SDS	sodium dodecylsulfate
THF	tetrahydrofuran
TOF	turnover frequency
TON	turnover number

References

1 Agbossou, F., Suisse, I., *Coord. Chem. Rev.* **2003**, *242*, 145; Agbossou, F., Carpentier, J.-F., Hapiot, F., Suisse, I., *Coord. Chem. Rev.* **1998**, *178–180*, 1615.

1 (a) Cesarotti, E., Chiesa, A., D'Alfonso, G., *Tetrahedron Lett.*, **1982**, *23*, 2995; (b) Cesarotti, E., Chiesa, A., *J. Organomet. Chem.* **1983**, *251*, 79.

3 Pracejus, G., Pracejus, H., GDR Patent Appl. WPC07F/240486 (1982).

4 Döbler, C., Kreuzfeld, H.-J., Krause, H.W., Michalik, M., *Tetrahedron: Asymm.* **1993**, *4*, 1833.

5 Döbler, C., Kreuzfeld, H.-J., Michalik, M., Krause, H.W., *Tetrahedron: Asymm.* **1996**, *7*, 117.

6 Döbler, C., Kreuzfeld, H.-J., Krause, H.W. German Patent Appl. DE44344293 A1 (1996).

7 Kreuzfeld, H.-J., Döbler, C., Schmidt, U., Krause, H.W., *Chirality* **1998**, *10*, 535.

8 Krause, H.W., Schmidt, U., Taudien, S., Costisella, B., Michalik, M., *J. Mol. Cat. A: Chemical* **1995**, *104*, 147.

9 Kreuzfeld, H.-J., Schmidt, U., Döbler, C., Krause, H.W., *Tetrahedron: Asymm.* **1996**, *7*, 1011.

10 Kreuzfeld, H.-J., Döbler, Ch., Krause, H.W., Facklam, C., *Tetrahedron: Asymm.* **1993**, *4*, 2047.

11 Taudien, S., Schinkowski, K., *Tetrahedron: Asymm.* **1993**, *4*, 73.

12 Schmidt, U., Fisher, C., Grassert, I., Kempe, R., Fröhlich, R., Drauz, K., Oehme, G., *Angew. Chem. Int. Ed.* **1998**, *37*, 2851.

13 Heller, D., Kadyrov, R., *Tetrahedron: Asymm.* **1996**, *7*, 3025.

14 Krause, H.W., Oehme, G., Michalik, M., Fisher, C., *Chirality* **1998**, *10*, 564.

15 Krause, H.W., Foken, H., Pracejus, H., *New. J. Chem.* **1989**, *13*, 615.

16 Oehme, G., Dwars, T., Schmidt, U., Fisher, C., Krause, H.W., Drauz, K. German Patent Appl. DE19801952 C1 (1999).

17 Kreuzfeld, H.-J., Döbler, C., *J. Mol. Cat. A: Chemical* **1998**, *136*, 105.

18 Lou, R.L., Mi, A.Q., Jiang, Y.H., Qin, Y., Li, Z., Fu, F.M., Chan, A.S.C. *Tetrahedron* **2000**, *56*, 5857.

19 Mi, A.Q., Lou, R.L., Jiang, Y.H., Deng, J.G., Qin, Y., Fu, F.M., Li, Z., Hu, W.H., Chan, A.S.C., *Synlett* **1998**, 847.

20 Burk, M., *J. Am. Chem. Soc.* **1991**, *113*, 8518.

21 Knowles, W.S., *J. Chem. Edu.* **1986**, *63*, 222.

22 Sawamura, M., Kuwano, R., Ito, Y., *J. Am. Chem. Soc.* **1995**, *117*, 8602.

23 Kagan, H.B., Dang, T.P., *J. Am. Chem. Soc.* **1972**, *94*, 6429.

24 Ojima, I., Yoda, N., *Tetrahedron Lett.* **1980**, *21*, 1051.

25 Hayashi, T., Kumada, M., *Acc. Chem. Res.* **1982**, *15*, 395.

26 Selke, R., Pracejus, H., *J. Mol. Cat.* **1986**, *37*, 213.

27 Chan, A.S.C., Jiang, Y.Z., Hu, W.H., Mi, A.Q., Yan, M., Pai, C.C., Sun, J., Lau, C.P., Lou, R.L., Deng, J.G., *J. Am. Chem. Soc.* **1997**, *119*, 9570.

28 Knowles, W.S., *Acc. Chem. Res.* **1983**, *16*, 106.

29 Spindler, F. Pittelkow, U., Blaser, H.U., *Chirality* **1991**, *3*, 370.

30 Xie, Y.O., Lou, R.L., Li, Z., Mi, A.G., Jiang, Y.Z., *Tetrahedron: Asymm.* **2000**, *11*, 1487.

31 Moulin, D., Darcel, C., Jugé S., *Tetrahedron: Asymm.* **1999**, *10*, 4729.

32 Karim, A., Mortreux, A., Petit, F., *J. Organomet. Chem.* **1986**, *317*, 93.

33 Yasuda, A., Toriumi, K., Ito, T., Souchi, T., Noyori, R., *J. Am. Chem. Soc.* **1980**, *102*, 7932; Miyashita, A., Takaya, H., Souchi, T., Noyori, R., *Tetrahedron* **1984**, *40*, 1245.

34 Broger, E.A., Burkart, W., Henning, M., Scalone, M., Schmid, R., *Tetrahedron: Asymm.* **1998**, *9*, 4043.

35 Meyer, H., *Pure Appl. Chem.* **1979**, *51*, 300.

36 Pracejus, G., Pracejus, H., *J. Mol. Cat.* **1984**, *24*, 227.

37 Döbler, C., Kreuzfeld, H.-J; Pracejus, H., *J. Organomet. Chem.* **1988**, *344*, 89.

38 Arias, L.A., Adkins, S., Nagel, C.J., Bach, R.D., *J. Org. Chem.*, **1983**, *48*, 888.

39 Li, X.S., Lou, R.L., Yeung, C.H., Chan, A.S.C., Wong, W.K., *Tetrahedron: Asymm.* **2000**, *11*, 2077.

40 Döbler, C., Schmidt, U., Krause, H.W., Kreuzfeld, H.-J; Michalik, M., *Tetrahedron: Asymm.* **1995**, *6*, 385.

41 Dubrovina, N.V., Tararov, V.I., Kadyrova, Z., Monsees, A., Börner, A., *Synthesis* **2004**, 2047.

42 Krause, H.W., Kreuzfeld, H.-J; Döbler, C., *Tetrahedron: Asymm.* **1992**, *3*, 555.

43 Fiorini, M., Giongo, G.M., Marcati, F., Marconi, W., *J. Mol. Catal.* **1975/76**, *1*, 451.

44 Pracejus, G., Pracejus, H., *Tetrahedron Lett.* **1977**, *28*, 3497.

45 (a) Fiorini, M., Marcati, F., Giongo, G.M., *J. Mol. Catal.* **1978**, *4*, 125; (b) Fiorini, M., Giongo, G.M., *J. Mol. Catal.* **1979**, *5*, 303; (c) Fiorini, M., Giongo, G.M., *J. Mol. Catal.* **1980**, *7*, 411.

46 Onuma, K.-I., Ito, T., Nakamura, A., *Tetrahedron Lett.* **1979**, *30*, 3163.

47 (a) Valentini, C., Cernia, E., Fiorini, M., Giongo, G.M., *J. Mol. Catal.* **1984**, *23*, 81; (b) Ait Ali, M., Allaoud, S., Karim, A., Roucoux, A., Mortreux, A., *Tetrahedron: Asymm.* **1995**, *6*, 369.

48 Zhang, A., Jiang, B., *Tetrahedron Lett.* **2001**, *42*, 1761.

49 Brenchley, G., Fedouloff, M., Merrifield, E., Wills, M., *Tetrahedron: Asymm.* **1996**, *7*, 2809.

50 Miyano, S., Nawa, M., Hashimoto, H., *Chem. Lett.* **1980**, 729.

51 (a) Zhang, F.-Y., Pai, C.-C., Chan, A. S. C., *J. Am. Chem. Soc.* **1998**, *120*, 5808; (b) Chan, A. S. C., Zhang, F.-Y., US Patent Appl. US5919981 (1999).

52 Zhang, F.-Y., Kwok, W. H., Chan, A. S. C., *Tetrahedron: Asymm.* **2001**, *12*, 2337.

53 Guo, R., Li, X., Wu, J., Kwok, W. H., Chen, J., Choi, M. C. K., Chan, A. S. C., *Tetrahedron Lett.* **2002**, *43*, 6803.

54 Guo, R., Ph. D. Thesis, The Hong Kong Polytechnic University, **2003**.

55 Lin, C. W., Lin, C.-C., Lam, L. F.-L., Au-Yeung, T. T.-L., Chan, A. S. C., *Tetrahedron Lett.* **2004**, *45*, 7379.

56 Chen, Y.-X., Li, Y.-M., Lam, K.-H., Chan, A. S. C., *Chin. J. Chem.* **2003**, *21*, 66.

57 Boaz, N. W., Debenham, S. D., Mackenzie, E. B., Large, S. E., *Org. Lett.* **2002**, *4*, 2421.

58 Boaz, N. W., Patent Appl. WO 0226750 (2002).

59 Maligres, P. E., Krska, S. W., Humphrey, G. R., *Org. Lett.* **2004**, *6*, 3147.

60 PPFA: (*R*)-1-[(*S*)-2-(diphenylphosphino)-ferrocenyl]-*N,N*-Dimethylethylamine; Ref.: Hayashi, T., Mise, T., Fukushima, M., Kagotani, M., Nagashima, N., Hamada, Y., Matsumoto, A., Kawakami, S., Konishi, M., Yamamoto, K., Kumada, M., *Bull. Chem. Soc. Jpn.* **1980**, *53*, 1138.

61 Jia, X., Li, X., Lam, W. S., Kok, S. H. L., Xu, L., Lu, G., Yeung, C. H., Chan, A.S.C., *Tetrahedron: Asymm.* **2004**, *15*, 2273.

62 Hu, X. -P., Zheng, Z., *Org. Lett.* **2004**, *6*, 3585.

63 Li, X., Jia, X., Xu, L., Kok, S. H. L., Yip, C. W., Chan, A. S. C., *Adv. Synth. Catal.* **2005**, *347*, 1904.

64 Gokel, G. W., Ugi, I. K., *J. Chem. Ed.* **1972**, *49*, 294.

65 Lam, W. S., Kok, S. H. L., Au-Yeung, T. T.-L., Wu, J., Cheung, H. Y., Lam, F.-L., Yeung, C. H., Chan, A. S. C., *Adv. Synth. Catal.* **2006**, *348*, 370.

66 Franci, G., Faraone, F., Leitner, W., *Angew. Chem. Int. Ed.* **2000**, *39*, 1428.

67 Tanaka, M., Ogata, I., *J. S. C. Chem. Commun.* **1975**, 735.

68 (a) Hayashi, T., Tanaka, M., Ogata, I., *Tetrahedron Lett.* **1977**, *3*, 295; (b) Jackson, W. R., Lovel, C. G., *Aust. J. Chem.* **1982**, *35*, 2069.

69 Lange, S., Brinkmann, A., Trautner, P., Woelk, K., Bargon, J., Leitner, W., *Chirality* **2000**, *12*, 450.

70 Ohta, T., Takaya, H., Noyori, R., *Inorg. Chem.* **1988**, *27*, 566.

71 Grubbs, R. H., DeVries, R. A., *Tetrahedron Lett.* **1977**, *18*, 1879.

72 (a) Zhang, X., Taketomi, T., Yoshizumi, T., Kumobayashi, H., Akutagawa, S., Mashima, K., Takaya, H., *J. Am. Chem. Soc.* **1993**, *115*, 3318; (b) Zhang, X., Uemura, T., Matsumura, K., Kumobayashi, H., Sayo, N., Takaya, H., *Synlett* **1994**, *1*, 501; (c) Uemura, T., Zhang, X., Matsumura, K., Sayo, N., Kumobayashi, H., Ohta, T., Nozaki, K., Takaya, H., *J. Org. Chem.* **1996**, *61*, 5510.

73 (a) Zhang, F.-Y., Chan, A. S. C., *Tetrahedron: Asymm.* **1997**, *8*, 3651; (b) Chan, A.S.C., Zhang, F.-Y., Yip, C.-W., *J. Am. Chem. Soc.* **1997**, *119*, 4080.

74 Liu, G.-B., Tsukinoki, T., Kanda, T., Mitoma, Y., Tashiro, M., *Tetrahedron Lett.* **1998**, *39*, 5991.

75 Zhang, F.-Y., Chan, A. S. C., *Tetrahedron: Asymm.* **1998**, *9*, 1179.

76 Wang, Y., Guo, H., Ding, K., *Tetrahedron: Asymm.* **2000**, *11*, 4153.

77 Au-Yeung, T. T.-L., Chan, S.S., Chan, A. S. C., *Adv. Synth. Catal.* **2003**, *345*, 537.

78 Gergely, I., Hegedüs, C., Szöllösy, Á., Monsees, A., Riermeier, T., Bakos, J., *Tetrahedron Lett.* **2003**, *44*, 9025.

79 Zhou, Y.-G; Zhang, X., *Chem. Commun.* **2002**, *10*, 1124.

80 Zhou, Y.-G., Tang, W., Wang, W.-B., Li, W., Zhang, X., *J. Am. Chem. Soc.* **2002**, *124*, 4952.

81 Johnson, T. H., Pretzer, D. K., Thomen, S., Chaffin, V. J. K., Rangarajan, G., *J. Org. Chem.* **1979**, *44*, 1878.

82 Fuerte, A., Igesias, M., Sánchez, F., *J. Organomet. Chem.* **1999**, *588*, 186.

83 Chan, A. S. C., Hu, W., Pai, C.-C., Lau, C.P., Jiang, Y., Mi, A., Yan, M., Sun, J., Lou, R. L., Deng, J., *J. Am. Chem. Soc.* **1998**, *120*, 9975.

84 Hu, W., Yan, M., Lau, C.-P., Yang, S. M., Chan, A. S. C., *Tetrahedron Lett.* **1999**, *40*, 973.

85 Zhu, G., Zhang, X., *J. Org. Chem.* **1998**, *63*, 3133.

86 (a) Diéguez, M., Pámies, O., Claver, C., *Chem. Rev.* **2004**, *104*, 3189; (b) Diéguez, M., Pámies, O., Ruiz, A., Díaz, Y., Castillón, S., Claver, C., *Coord. Chem. Rev.* **2004**, *248*, 2165.

87 (a) RajanBabu, T. V., *Chem. Rev.* **2003**, *103*, 2645; (b) Ohe, K., Yonehara, K., Uemura, S., *Yuki Gosei Kagaku Kyokaishi* **2001**, *59*, 185; (c) Liu, X., Wang, Y., Miao, Q., Jin, Z., *Youji Huaxue* **2001**, *21*, 191; (d) Gyurcsik, B., Nagy, L., *Coord. Chem. Rev.* **2000**, *203*, 81; (e) Steinborn, D., Junicke, H., *Chem. Rev.* **2000**, *100*, 4283; (f) Chen, M., Lu, S., *Fenzi Cuihua* **2000**, *14*, 441; (g) Ayers, T. A., RajanBabu, T. V., in: Gadamasetti, K. G. (Ed.), *Process Chemistry in the Pharmaceutical Industry.* Dekker: New York, 1999; pp. 327–345; (h) RajanBabu, T. V., Casalnuovo, A. L., *Pure Appl. Chem.* **1994**, *66*, 1535; (i) Blaser, H.-U., *Chem. Rev.* **1992**, *92*, 935.

88 Cullen, W. R., Sugi, Y., *Tetrahedron Lett.* **1978**, *19*, 1635.

89 Jackson, R., Thompson, D. J., *J. Organomet. Chem.* **1978**, *159*, C29–C31.

90 Sinou, D., Descotes, G., *React. Kinet. Catal. Lett.* **1980**, *14*, 463.

91 (a) Selke, R., Ohff, M., Riepe, A., *Tetrahedron* **1996**, *52*, 15079; (b) Selke, R., Facklam, C., Foken, H., Heller, D., *Tetrahedron: Asymm.* **1993**, *4*, 369; (c) Selke, R., Schwarze, M., Baudisch, H., Grassert, I., Michalik, M., Oehme, G., Stoll, N., Costisella, B., *J. Mol. Catal.* **1993**, 84, 223; (d) Selke, R., *J. Organomet. Chem.* **1989**, *370*, 241; (e) Selke, R., *J. Prakt. Chem.* **1987**, *329*, 717; (f) Selke, R., *React. Kinet. Catal. Lett.* **1979**, *10*, 135.

92 Vocke, W., Hänel, R., Flöther, F.-U., *Chem. Techn.*, **1987**, *39*, 123.

93 Habůs, I., Raza, Z., Šunjić, V., *J. Mol. Catal.* **1987**, *42*, 173.

94 Snatzke, G., Raza, Z., Habůs, I., Šunjić, V., *J. Mol. Catal.* **1988**, *182*, 179.

95 (a) RajanBabu, T. V., Ayers, T. A., Cassalnuovo, A. L., *J. Am. Chem. Soc.* **1994**, *116*, 4101; (b) RajanBabu, T. V., Ayers, T. A., Halliday, G. A., You, K. K., Calabrese, J. C., *J. Org. Chem.* **1997**, *62*, 6012.

96 (a) Haggin, J., *Chem. Eng. News* **1994**, *72*, 28; (b) Cornils, B., *Nachr. Chem. Tech. Lab.* **1994**, *42*, 1136; (c) Cornils, B., *Angew. Chem.* **1995**, *107*, 1709; *Angew. Chem. Int. Ed. Engl.* **1995**, *34*, 1575; (d) Wiebus, E., Cornils, B., *Chem. Ing. Tech.* **1994**, *66*, 916; (e) Cornils, B., Wiebus, E., *Chemtech* **1995**, 25; (f) Trzeciak, A. M., Ziolkowski, J. J., *Coord. Chem. Rev.* **1999**, 883; (g) Lindner, E., Schneller, T., Auer, F., Mayer, H. A., *Angew. Chem. Int. Ed.* **1999**, *38*, 2154; (h) Herrmann, W. A., Elison, M., Fischer, J., Koecher, C., German Patent Appl. DE4447067 (1995); (i) Hermann, W., Elison, M., Fischer, J., Koecher, C., Oefele, K., German Patent Appl. DE4447066 (1995).

97 Oehme, G., Paetzold, E., Selke, R., *J. Mol. Catal.* **1992**, *71*, L1–L5.

98 Yan, Y. Y., RajanBabu, T. V., *J. Org. Chem.* **2001**, *66*, 3277.

99 Faltin, F., Fehring, V., Kadyrov, R., Arrieta, A., Schareina, T., Selke, R., Miethchen, R., *Synthesis* **2001**, 638.

100 Yonehara, K., Hashizume, T., Ohe, K., Uemura, S., *Bull. Chem. Soc. Jpn.* **1998**, *71*, 1967.

101 Yonehara, K., Hashizume, T., Mori, K., Ohe, K., Uemura, S., *J. Org. Chem.* **1999**, *64*, 5593.

102 Shin, S., RajanBabu, T. V., *Org. Lett.* **1999**, *1*, 1229.

103 Johnson, T. H., Rangarajan, G., *J. Org. Chem.* **1980**, *45*, 62.

104 Guimet, E., Diéguez, M., Ruiz, A., Claver, C., *Tetrahedron: Asymm.* **2004**, *15*, 2247.

105 Aghmiz, M., Aghmiz, A., Díaz, Y., Masdeu-Bultó, A., Claver, C., Castillón, S., *J. Org. Chem.* **2004**, *69*, 7502.

106 Bourson, J., Oliveros, L., *J. Organomet. Chem.* **1982**, *229*, 77.

107 Chen, Y., Li, X., Tong, S.-K., Choi, M.C.K., Chan, A.S.C., *Tetrahedron Lett.* **1999**, *40*, 957.

108 Jiang, P., Lu, S.J., *Chin. Chem. Lett.* **2000**, *11*, 587.

109 Reetz, M.T., Gosberg, A., Goddard, R., Kyung, S.-H., *Chem. Commun.* **1998**, 2077.

110 Claver, C., Fernandez, E., Gillon, A., Heslop, K., Hyett, D.J., Martorell, A., Orpen, A.G., Pringle, P.G., *Chem. Commun.* **2000**, 961.

111 Zanotti-Gerosa, A., Malan, C., Herzberg, D., *Org. Lett.* **2001**, *3*, 3687.

112 Gergely, I., Hegedüs, C., Gulyás, H., Szöllösy, Á., Monsees, A., Riermeier, T., Bakos, J., *Tetrahedron: Asymm.* **2003**, *14*, 1087.

113 Vlugt, J.I., Paulusse, J.M.J., Zijp, E.J., Tijmensen, J.A., Mills, A.M., Spek, A.L., Claver, C., Vogt, D., *Eur. J. Inorg. Chem.* **2004**, *21*, 4193.

114 Wink, D.J., Kwok, T.J., Yee, A., *Inorg. Chem.* **1990**, *29*, 5006.

115 Kadyrov, R., Heller, D., Selke, R., *Tetrahedron: Asymm.* **1998**, *9*, 329.

116 Reetz, M.T., Neugebauer, T., *Angew. Chem. Int. Ed.* **1999**, *38*, 179.

117 (a) Diéguez, M., Ruiz, A., Claver, C., *Dalton Trans.* **2003**, 2957; (b) Diéguez, M., Ruiz, A., Claver, C., *J. Org. Chem.* **2002**, *67*, 3796; (c) Pámies, O., Net, G., Ruiz, A., Claver, C., *Tetrahedron: Asymm.* **2000**, *11*, 1097; (d) Pámies, O., Net, G., Ruiz, A., Claver, C., *Eur. J. Inorg. Chem.* **2000**, 1287.

118 Korostylev, A., Selent, D., Monsees, A., Borgmann, C., Börner, A., *Tetrahedron: Asymm.* **2003**, *14*, 1905.

119 Yamashita, M., Hiramatsu, K., Yamada, M., Suzuki, N., Inokawa, S., *Bull. Chem. Soc. Jpn.* **1982**, *55*, 2917.

120 Reetz, M.T., Gosberg, A., *Tetrahedron: Asymm.* **1999**, *10*, 2129.

121 (a) Pámies, O., Diéguez, M., Net, G., Ruiz, A., Claver, C., *Chem. Comm.* **2000**, 2383; (b) Pámies, O., Diéguez, M., Net, G., Ruiz, A., Claver, C., *J. Org. Chem.* **2001**, *66*, 8364.

122 (a) Pámies, O., Net, G., Ruiz, A., Claver, C., *Eur. J. Inorg. Chem.* **2000**, 2011; (b) Diéguez, M., Pámies, O., Ruiz, A., Castillón, S., Claver, C., *Tetrahedron: Asymm.* **2000**, *11*, 4701.

123 Deerenberg, S., Pámies, O., Diéguez, M., Claver, C., Kamar, P.C.J., van Leeuwen, P.W.N.M., *J. Org. Chem.* **2001**, *66*, 7626.

124 (a) Suárez, A., Méndez-Rojas, M.A., Pizzano, A., *Organometallics* **2002**, *21*, 4611; (b) Suárez, A., Pizzano, A., *Tetrahedron: Asymm.* **2001**, *12*, 2501.

125 Reetz, M.T., Maiwald, P., *C. R. Chimie* **2002**, *5*, 341.

126 Diéguez, M., Ruiz, A., Claver, C., *Chem. Commun.* **2001**, 2702.

127 Monsees, A., Laschat, S., *Synlett* **2002**, *6*, 1011.

128 Yan, Y., Chi, Y., Zhang, X., *Tetrahedron: Asymm.* **2004**, *15*, 2173.

129 Ohe, K., Morioka, K., Yonehara, K., Uemura, S., *Tetrahedron: Asymm.* **2002**, *13*, 2155.

130 Diéguez, M., Orejón, A., Masdeu-Bultó, A.M., Echarri, R., Castillón, S., Claver, C., Ruiz, A., *J. Chem. Soc., Dalton Trans.* **1997**, 4611.

131 Diéguez, M., Ruiz, A., Claver, C., Doro, F., Sanna, M.G., Gladiali, S., *Inorg. Chim. Acta* **2004**, *357*, 2957.

132 Li, W., Waldkirch, J.P., Zhang, X., *J. Org. Chem.* **2002**, *67*, 7618.

133 Diéguez, M., Ruiz, A., Claver, C., Pereira, M.M., Rocha Gonsalves, A.M.d'A., *J. Chem. Soc., Dalton Trans.* **1998**, 3517.

134 Pámies, O., Diéguez, M., Net, G., Ruiz, A., Claver, C., *J. Chem. Soc., Dalton Trans.* **1999**, 3439.

135 Seo, H., Park, H.-J., Kim, B.Y., Lee, J.H., Son, S.U., Chung, Y.K., *Organometallics* **2003**, *22*, 618.

136 Hauptman, E., Shapiro, R., Marshall, W., *Organometallics* **1998**, *17*, 4976.

137 Hauptman, E., Fagan, P.J., Marshall, W., *Organometallics* **1999**, *18*, 2061.

138 Pámies, O., Diéguez, M., Net, G., Ruiz, A., Claver, C., *Organometallics* **2000**, *19*, 1488.

139 Evans, D.A., Michael, F.E., Tedrow, J.S., Campos, K.R., *J. Am. Chem. Soc.* **2003**, *125*, 3534.

140 For examples, see Guo, R., Au-Yeung, T. T.-L., Wu, J., Choi, M. C. K., Chan, A. S. C., *Tetrahedron: Asymm.* **2002**, *13*, 2519 and references therein.

141 Kawabata, Y., Tanaka, M., Ogata, I., *Chem. Lett.* **1976**, 1213.

142 Berens, U., Fischer, C., Selke, R., *Tetrahedron: Asymm.* **1995**, *6*, 1105.

28
Enantioselective Alkene Hydrogenation: Monodentate Ligands

Michel van den Berg, Ben L. Feringa, and Adriaan J. Minnaard

28.1
Introduction

In 1968, Knowles et al. [1] and Horner et al. [2] independently reported the use of a chiral, enantiomerically enriched, monodentate phosphine ligand in the rhodium-catalyzed homogeneous hydrogenation of a prochiral alkene (Scheme 28.1). Although enantioselectivities were low, this demonstrated the transformation of Wilkinson's catalyst, $Rh(PPh_3)_3Cl$ [3] into an enantioselective homogeneous hydrogenation catalyst [4].

In order to enhance enantioselective induction by preventing rotation around the rhodium–phosphorus bond, Dang and Kagan developed a chelating bidentate phosphine; DIOP [5]. By using tartaric acid as a starting material from the chiral pool, and by situating the chirality in the backbone, and not on phosphorus, synthesis of the ligand was simplified. In addition, it was the first example of a C_2 symmetric ligand, designed in this way to minimize the number of diastereomeric rhodium–ligand–substrate complexes. This strategy proved to be very effective, being confirmed several years later by Knowles et al. in the dimerization of PAMP to DIPAMP, which raised the enantioselectivity in the hydrogenation of methyl 2-acetamido-cinnamate from 55% to 95% [6].

The trend to develop chiral ligands devoid of chirality on phosphorus simplified the synthesis and led to the preparation of literally hundreds of chiral bi-

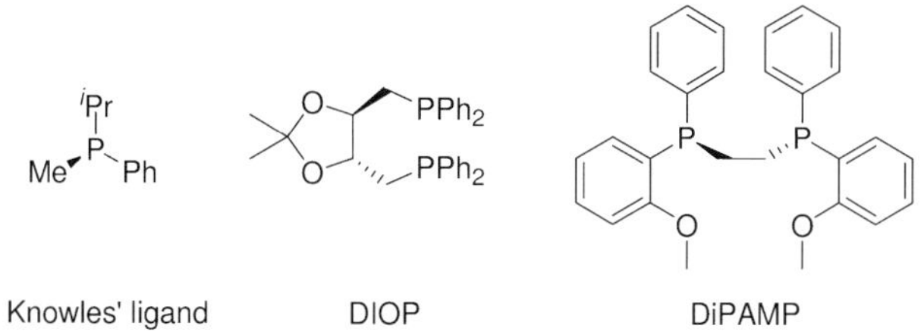

Knowles' ligand DIOP DiPAMP

Scheme 28.1 Some of the first monodentate and bidentate ligands in enantioselective hydrogenation.

The Handbook of Homogeneous Hydrogenation.
Edited by J. G. de Vries and C. J. Elsevier
Copyright © 2007 WILEY-VCH Verlag GmbH & Co. KGaA, Weinheim
ISBN: 978-3-527-31161-3

sphosphines [7]. Together with the application of DIPAMP and Ph-β-Glup in commercial processes for L-DOPA [8], this established the use of bidentate phosphorus ligands as a *conditio sine qua non* for high ee-values in asymmetric hydrogenation. This was apparently underscored by the development of the very successful ligands BINAP, especially versatile with ruthenium, and DuPhos.

Knowles et al. had shown that the use of the P-chiral monodentate CAMP gave rise to an *e.e.* of 88% [9] in the formation of *N*-acyl-phenylalanine. However, due to the superior results obtained using bidentate ligands and the difficult preparation of P-chiral phosphines, this route was rarely followed for a long time [10, 11].

It thus came as a surprise that in the year 2000, three groups independently reported the use of three new classes of monodentate ligands (Scheme 28.2) [12]. The ligands induced remarkably high enantioselectivities, comparable to those obtained using the best bidentate phosphines, in the rhodium-catalyzed enantioselective alkene hydrogenation. All three being based on a BINOL backbone, and devoid of chirality on phosphorus, these monophosphonites [13], monophosphites [14] and monophosphoramidites [15] are very easy to prepare and are equipped with a variable alkyl, alkoxy, or amine functionality, respectively.

These reports announced the rapid development of a large variety of monodentate ligands for rhodium-catalyzed enantioselective hydrogenation. It was shown that the substrate scope for catalysts based on monodentate ligands is most probably at least as big as for their bidentate counterparts. Also, initial doubts about the activity and stability of the monodentate ligand-catalysts have been taken away. Several reports show that substrate:catalyst ratios (SCRs) of 10^3 or higher, essential for industrial application, are possible. In addition, reaction rates are in the studied cases comparable to those reached by catalysts based on state-of-the-art bidentate ligands [16].

The mechanism of the rhodium-catalyzed enantioselective hydrogenation has been thoroughly studied, and a wealth of information is now available. Logically, these studies have been performed using bidentate ligands. It will be very interesting to see whether catalytic cycles that have been proposed will also hold for catalysts equipped with monodentate ligands. Although a mechanistic study is still lacking [17], Zhou et al. performed a kinetic study of hydrogenations using the monodentate phosphoramidite SIPHOS [18]. As noticed earlier for MonoPhos,

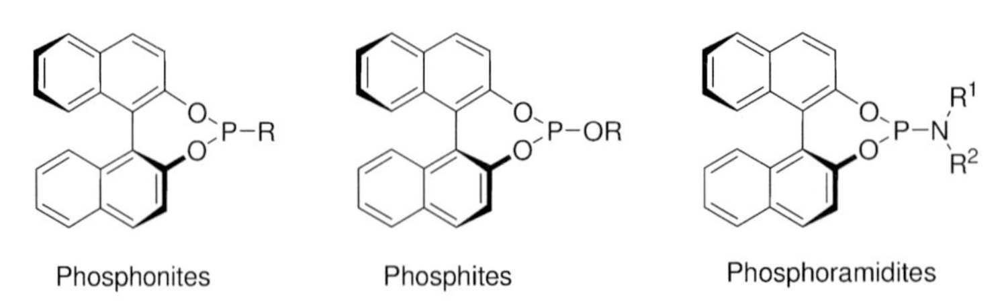

Scheme 28.2 New classes of monodentate ligands used in asymmetric hydrogenation. R = alkyl or aryl.

the enantioselectivity of the reactions was shown to be independent of the hydrogen pressure (e.g., hydrogen concentration) between 1 and 50 bar. This seems to be more general for monodentate ligands. In addition, the enantioselectivity decreases slightly with increasing temperature, and *vice versa*. Both observations disagree with the "major/minor" diastereomer part of the Halpern mechanism.

Both for MonoPhos and SIPHOS, a positive non-linear effect was observed with respect to the ee of the ligand. The observation that for several ligands a ligand:rhodium ratio of 1:1 gives a faster reaction than a L:Rh ratio of 2:1, with preservation of ee, tempted Zhou et al. to propose a mechanism with only one ligand on rhodium in the enantiodiscriminating step. This seems to contradict the recent results obtained using mixtures of ligands, a synergy that logically can only arise from a catalyst containing two different ligands. The application of these mixtures of monodentate ligands in catalysis, first shown by the group of Reetz and discussed in Chapter 36, in a number of cases affords higher ee-values than the corresponding pure ligands [19]. Very recent reports show that also the combination of chiral and achiral ligands can lead to unprecedented ee-values in enantioselective hydrogenation [20]. Combined with the modular construction of most monodentate ligands, and therefore the easy variation of their structure, this offers a tremendous opportunity for high throughput catalyst screening, as discussed in Chapter 36.

The present chapter provides a comprehensive overview of the literature relating to monodentate ligands in enantioselective hydrogenation until the end of 2004. Patent literature has not been covered. As the large majority of the ligands is available in both enantiomeric forms, the absolute configuration of the products has not been indicated. As most authors focus on the enantioselectivity of their catalysts, this will be reflected in this chapter. Whenever possible, attention will be given to turnover frequencies (TOF) and turnover numbers (TON). Parts of this chapter have been covered recently by a review of Jerphagnon, Renaud and Bruneau [21] and by De Vries and Ager [22].

28.2 Monodentate Phosphines

Although, in the past, most attention was paid to the use of bidentate phosphines, a number of monodentate phosphines has also been developed and applied in the rhodium-catalyzed hydrogenation of alkenes. These earlier-developed ligands are chiral on phosphorus (**1**) [9] and usually equipped with a phenyl and a methyl moiety (Scheme 28.3). The third substituent varies in size in order to maximize the chiral induction in the hydrogenation. The enantioselectivity in the hydrogenation of substrates such as the precursors of L-DOPA varies over a broad range, from 1% to 90% *e.e.* using ligands **1m** and **1l** (CAMP), respectively. The results of the other ligands fall between these values.

Ligands **2a–2d**, which are also chiral on phosphorus, and **3**, were used in the chemo- and enantioselective hydrogenation of (*E*)-3,7-dimethyl-2,6-dienoic acid

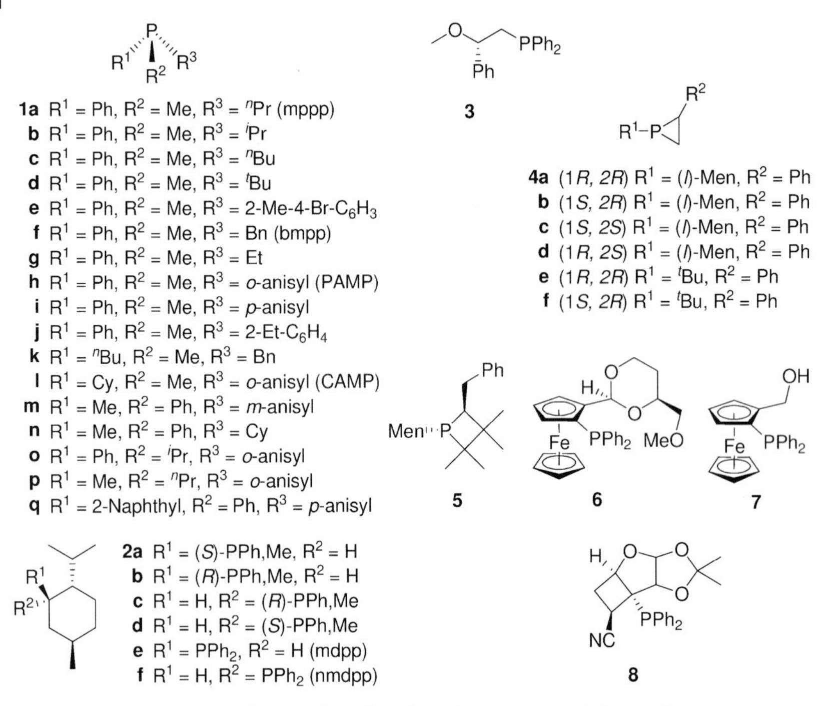

Scheme 28.3 Chiral monodentate phosphine ligands (men = menthyl, see **2**).

[23]. The *e.e.*-values were moderate, an improved result of 79% *e.e.* being obtained with nmdpp (**2f**) which is not chiral on phosphorus. The use of the other, non P-chiral, ligand mdpp (**2e**) gave low enantioselectivity. Ligand **2f** was also successfully used in the hydrogenation of 2-methylcinnamic and (*E*)-3-methylcinnamic acid [24].

Both ligands **4** [25] and **5** [26], in which phosphorus is part of a ring, were tested in the hydrogenation of *α*-acetamidocinnamic acid. Of these phosphirane ligands, only **4b** possessing a *trans*-configuration was able to induce reasonable enantioselectivities. Ferrocenyl-based monodentate phosphine **6**, used in a 4:1 ratio with $[Rh(COD)Cl]_2$, afforded an ee of 87% in this reaction, albeit with incomplete conversion [27]. Using ligand **7** under identical conditions, full conversion was reached, though with an *e.e.*-value of only 30%. Ligand **8**, derived from a carbohydrate, has also been applied with reasonable success [28].

Cyclic, C_2-symmetric monodentate phosphines with the phosphorus atom in a four-, five-, six-, or seven-membered ring have frequently been used in enantioselective hydrogenation (Scheme 28.4). The use of the six-membered oxaphosphiranes **9**, demonstrates that with these secondary phosphines high ee-values can be obtained in the hydrogenation of dehydroamino esters and methyl itaconate [29]. The atropisomeric ligands **10a–o** show a large effect of the size of the substituent on the enantioselectivity of the hydrogenation. Low ee-values in the hydrogenation of methyl *N*-acylcinnamate are obtained using ligand **10c** which

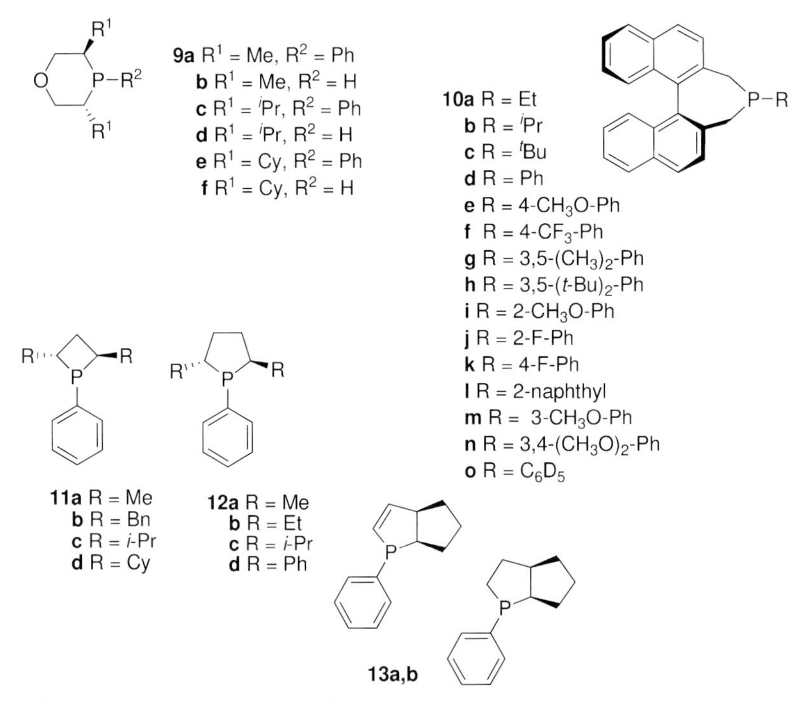

Scheme 28.4 Cyclic monodentate phosphine ligands.

contains a bulky *t*-Bu group [30]. When this group is replaced by a phenyl moiety, the *e.e.* obtained is 90%. After optimization by varying the substituent, excellent ee-values could be obtained [31]. Using these ligands, the first highly enantioselective ruthenium-catalyzed hydrogenation of *β*-ketoesters with monodentate ligands was also achieved [32].

As for ligands **11**, containing a four-membered ring [33, 34], ligands **12** which also contain a five-membered ring afford good enantioselectivities [35], especially **12d**. One could consider these ligands as monodentate analogues of DuPhos. The group of Fiaud [36] reported the existence of **12d** a year before the publication of the BINOL-based monodentate phosphonite, phosphite and phosphoramidite ligands.

Recently, two new P- and C-chiral monodentate phosphines **13** were reported. The ligands were applied in a number of transition metal-catalyzed reactions, though ee-values in the rhodium-catalyzed hydrogenation of *N*-acyl dehydrophenylalanine were only moderate [37].

28.3 Monodentate Phosphonites

Rhodium-catalyzed enantioselective hydrogenation using monodentate phosphonite ligands was first reported by the group of Pringle [13], followed by the group of Reetz (Scheme 28.5) [38]. The reported ligands are easily synthesized from an alkyl- or arylphosphorus dichloride and the appropriate BINOL or 9,9′-bisphenanthrol. The ligands are easily hydrolyzed in the presence of moisture, but are considerably more stable as their rhodium complexes [39].

Enantioselectivities obtained in the hydrogenation of methyl 2-acetamido-cinnamate, methyl 2-acetamido-acrylate and dimethyl itaconate are surprisingly high (up to 94% *e.e.*). The TOFs of the hydrogenation reactions using these monodentate phosphonites is fairly high, with most of the reactions with a SCR of 500 reaching TOFs of 250–300 mol $mol^{-1} \cdot h$ and full conversion at 1.5 bar. In addition, ligand **15d** has been studied in the rhodium- and iridium-catalyzed hydrogenation of a benzyl imine, but no chiral induction was observed [40].

The first – and until now only – case of ruthenium-catalyzed enantioselective ketone hydrogenation using monodentate ligands concerns phosphonite ligands [41]. In several cases, but especially with **15f**, excellent ee-values are obtained. The simple synthesis of these phosphonites makes them an interesting class of ligands for the synthesis of a ligand library for high-throughput experimentation (HTE). In addition, mixtures of ligands can be used (see Chapter 36).

28.4 Monodentate Phosphites

The chiral monodentate phosphites presented in Scheme 28.6 are easily prepared from a diol, phosphorus trichloride, and an alcohol. Usually, the diol is converted into the corresponding phosphoro chloridite, followed by reaction

15a R^1 = H, R^2 = Me
b R^1 = H, R^2 = Et
c R^1 = H, R^2 = ^{t}Bu
d R^1 = H, R^2 = Ph
e R^1 = H, R^2 = *o*-biphenyl
f R^1 = H, R^2 = *o*-BrPh
g R^1 = H, R^2 = *o*-Tol
h R^1 = H, R^2 = *o*-Anisyl
i R^1 = H, R^2 = *c*-C_6H_{11}
j R^1 = H, R^2 = Cl
k R^1 = C_6F_{13}, R^2 = Ph

16a R = Me
b R = ^{t}Bu
c R = Ph

Scheme 28.5 Monodentate phosphonite ligands.

with the appropriate alcohol. The reversed approach – for example, reaction of a phosphoric dichloride with a diol – has also been used.

The application of monodentate phosphites as ligands in the rhodium-catalyzed enantioselective hydrogenation was first reported by the group of Reetz [14]. Initially, bidentate phosphites based on dianhydro-D-mannitol and two BINOL moieties were used, but it transpired that by substituting one of the BINOL moieties for methanol, leading to **17bh** and **17bi**, enantioselectivities in the rhodium-catalyzed hydrogenation were surprisingly high.

Based on a comparison of matched and mismatched ligands, it was shown that the BINOL moiety had the largest influence on the enantioselectivity of the reaction. Elaborating on this finding, a number of simple BINOL-based monodentate phosphite ligands was synthesized. The use of these ligands in the rhodium-catalyzed hydrogenation revealed their excellent behavior, resulting in high *e.e.*-values in the products. The group of Xiao reported monodentate phosphite ligands based on BINOL and L-menthol, **17bc** and **17bd** [42], while more recently Bakos et al. reported ligands derived from octahydro-BINOL, **18a** and **18b** [43] and the groups of Börner [44] and Helmchen [45] reported substituted BINOL-based phosphites. Large, helicene-like phosphites **19** have also been reported recently [46].

The initial report of Reetz describes the use of a Rh:L ratio of 1:1, although more recent experiments were conducted using a ratio of 1:2. Within this range, the enantioselectivities are unaffected. The combination of rhodium with ligands **17** used in the hydrogenation of methyl 2-acetamido cinnamate afforded enantioselectivities ranging from 2% to 99%. However, the majority of the results ranged from 75% to 99%. The use of ligands **17am** and **17an** gave the highest ee-values. Particularly striking was the influence of the BINOL moiety, which completely dominate the configuration of the product. The chiral alcohol present does not seem to have any influence. Similar to the use of monodentate phosphonites, the hydrogenations using monodentate phosphites are best performed in non-protic solvents. The rate of the reactions is high; even at a hydrogen pressure of 1.3 bar rates of 300 mol $mol^{-1} \cdot h$ were obtained. At an elevated pressure of 20 bar, TOFs up to 120000 mol $mol^{-1} \cdot h$ were obtained in the hydrogenation of dimethyl itaconate with **18a**. This increase in pressure had only a marginal, if any, effect on the enantioselectivity. Recently, Reetz et al. reported the use of the "parent" phosphite ligand **17be** (a phosphoric acid diester) which led to *e.e.*-values of up to 85%. This was only slightly lower than the results obtained using ligand **17aa** [47]. The related phosphite **20**, based on TADDOL, was tested in an iridium-catalyzed imine hydrogenation but produced disappointing results [48].

Besides the phosphite ligands based on BINOL, phosphite ligands based on bisphenol are also used in rhodium-catalyzed hydrogenation. These ligands are shown in Scheme 28.7 and consist of a bisphenol with different substituents on the 3,3′,5,5′, and 6,6′-positions. The ligands without substituents on the 6,6′-positions are only fluxionally chiral. The use of readily available chiral alcohols (**21aa–21aj**) such as menthol in combination with bisphenol was thought to induce one of the bisphenol conformations in preponderant amounts [49]. The

18a 18b

17aa R^1 = H, R^2 = Me
ab R^1 = H, R^2 = Et
ac R^1 = H, R^2 = Ph
ad R^1 = H, R^2 = ^{i}Pr
ae R^1 = H, R^2 = *c*-C_5H_9
af R^1 = H, R^2 = *c*-C_6H_{11}
ag R^1 = H, R^2 = Bn
ah R^1 = H, R^2 = ^{t}Bu
ai R^1 = H, R^2 = ^{i}Bu [(*R*)-BINOL]
aj R^1 = H, R^2 = 9-Fluorenyl
ak R^1 = H, R^2 = $CH_2C(CH_3)_3$ [(*R*)-BINOL]
al R^1 = H, R^2 = $CH_2CH(C_2H_5)_2$ [(*R*)-BINOL]
am R^1 = H, R^2 = ^{i}Bu
an R^1 = H, R^2 = (*S*)-^{i}Bu [(*R*)-BINOL]
ao R^1 = H, R^2 = (*S*)-$CH_2CH(CH_3)C_2H_5$
ap R^1 = H, R^2 = (*S*)-$CH_2CH(CH_3)C_2H_5$ [(*R*)-BINOL]
aq R^1 = H, R^2 = (*R*)-$CH(CH_3)Ph$
ar R^1 = H, R^2 = (*S*)-$CH(CH_3)Ph$
as R^1 = H, R^2 = (*rac*)-$CH(CH_3)Ph$
at R^1 = H, R^2 = (*S*)-$CH(OCH_3)Ph$
au R^1 = H, R^2 = $CH_2CH_2N[CH(CH_3)_2]_2$
av R^1 = H, R^2 = $CH_2CH_2OCH_3$

17aw R^1 = H, R^2 = $CH_2CH_2OCH_3$ [(*R*)-BINOL]
ax R^1 = H, R^2 = CH_2CH_2Cl [(*R*)-BINOL]
ay R^1 = H, R^2 = CH_2CCl_3 [(*R*)-BINOL]
az R^1 = H, R^2 = 2-Br-Ph
ba R^1 = H, R^2 = 2,6-(CH_3)-Ph
bb R^1 = H, R^2 = 2,6-$(Ph)_2$-Ph
bc R^1 = H, R^2 = *L*-Menthyl
bd R^1 = H, R^2 = *L*-Menthyl [(*R*)-BINOL]
be R^1 = R^2 = H
bf R^1 = H, R^2 = $C(O)CH_3$ [(*R*)-BINOL]
bg R^1 = H, R^2 = C(O)Ph [(*R*)-BINOL]
bh R^1 = H, R^2 = see below [(*R*)-BINOL]
bi R^1 = H, R^2 = see below
bj R^1 = Br, R^2 = C(O)Ph [(*R*)-BINOL]
bk R^1 = Me, R^2 = C(O)Ph [(*R*)-BINOL]
bl R^1 = Ph, R^2 = C(O)Ph [(*R*)-BINOL]
bm R^1 = $SiMe_3$, R^2 = C(O)Ph [(*R*)-BINOL]

OR^2 in **17bh**, **17bi** =

19a (*M,M,S,L*), R = *L*-Menthyl
b (*M,M,S,S*), R = 1-(*S*)-Phenylethyl
c (*M,M,S,R*), R = 1-(*R*)-Phenylethyl
d (*M,M,R,S*), R = 1-(*S*)-Phenylethyl
e (*M,M,R,R*), R = 1-(*R*)-Phenylethyl

20

Scheme 28.6 Monodentate phosphite ligands derived from BINOL or related diols.

2:1 complexation of **21ac** with rhodium resulted in a 5:1 diastereomeric mixture of [Rh(**21ac**)$_2$COD]BF_4. Within the two complexes, the bisphenol part of the ligands has the same conformation, and no complexes were found in which the two bisphenol parts have different conformations.

The axially chiral ligands **21ak–21ba** were recently reported by Ojima et al. [50]. In addition, the group of Driessen-Hölscher prepared a series of monodentate phosphites based on 5-Cl-6-MeO-bisphenol (**21bb–21bf**) [51]. Both series of ligands are successful. The use of phosphite ligand **22**, which has a chiral diol

21aa-ad	$R^1 = R^2 = R^3 = H$, R^4 = see below
ae-ah	$R^1 = H$, $R^2 = R^3 = {}^t$Bu, R^4 = see below
ai	$R^1 = R^2 = R^3 = H$, R^4 = *L*-Menthyl
aj	$R^1 = H$, $R^2 = R^3 = {}^t$Bu, R^4 = *L*-Menthyl
ak	$R^1 = R^2$ = Me, $R^3 = H$, R^4 = Ph
al	$R^1 = R^2$ = Me, $R^3 = {}^t$Bu, R^4 = Ph
am	$R^1 = R^2$ = Me, $R^3 = H$, R^4 = 2-naphthyl
an	$R^1 = R^2$ = Me, $R^3 = {}^t$Bu, R^4 = 2-naphthyl
ao	$R^1 = R^2$ = Me, $R^3 = H$, R^4 = see below
ap	$R^1 = R^2$ = Me, $R^3 = {}^t$Bu, R^4 = see below
aq	$R^1 = R^2$ = Me, $R^3 = H$, R^4 = see below
ar	$R^1 = R^2 = R^3$ = Me, R^4 = see below
as	$R^1 = R^2$ = Me, R^3 = Br, R^4 = see below
at	$R^1 = R^2$ = Me, $R^3 = {}^t$Bu, R^4 = see below
au	$R^1 = R^2$ = Me, R^3 = Ph, R^4 = see below
av	$R^1 = R^2$ = Me, $R^3 = H$, R^4 = *L*-Menthyl
aw	$R^1 = R^2$ = Me, $R^3 = {}^t$Bu, R^4 = *L*-Menthyl
ax	$R^1 = R^2$ = Me, $R^3 = {}^t$Bu, R^4 = see below
ay	$R^1 = R^2$ = Me, $R^3 = H$, R^4 = *L*-Menthyl [(*R*)-biphenyl]
az	$R^1 = R^2$ = Me, $R^3 = {}^t$Bu, R^4 = *L*-Menthyl [(*R*)-biphenyl]
ba	$R^1 = R^2$ = Me, $R^3 = {}^t$Bu, R^4 = see below [(*R*)-biphenyl]
bb	R^1 = OMe, R^2 = Cl, $R^3 = H$, $R^4 = {}^i$Pr
bc	R^1 = OMe, R^2 = Cl, $R^3 = H$, R^4 = Cy
bd	R^1 = OMe, R^2 = Cl, $R^3 = H$, R^4 = (*R*)-Phenethyl
be	R^1 = OMe, R^2 = Cl, $R^3 = H$, R^4 = Ph
bf	R^1 = OMe, R^2 = Cl, $R^3 = H$, R^4 = 2,6-$(CH_3)_2$-Ph

22

OR^4 in **21** =

aa, ae **ab, af** **ac, ag** **ad, ah** **ao, ap** **aq - au** **ax, ba**

Scheme 28.7 Monodentate phosphite ligands derived from bisphenol.

bridging the 6,6′ position of the bisphenol backbone locking its conformation, was reported recently [52]. The use of **22** resulted in an excellent *e.e.* of 96% in the hydrogenation of dimethyl itaconic acid.

In general, the results obtained with monodentate ligands based on bisphenol are comparable to those obtained using ligands based on BINOL. Large substituents on the 3,3′-positions of the bisphenol results in lower ee-values. In some cases even the absolute configuration of the products is reversed. This is unfortunate, as bulky substituents on the 3,3′-positions increase the stability of the ligands towards hydrolysis, though the rate of the hydrogenation is not greatly influenced. The result of hydrogenations using these ligands is very solvent-dependent. The preferred solvents are dichloromethane and 1,2-dichloroethane, but when other solvents such as tetrahydrofuran, methanol, ethyl acetate or chloroform are used, no enantioselectivity is observed. The factor determining the configuration of the product is, as in the case using BINOL-based ligands, the configuration of the biaryl moiety.

Rhodium-catalyzed hydrogenation of enamides has been successfully performed using monodentate phosphites **17**, with enantioselectivities of up to 95% being obtained [53]. The rate of hydrogenation is low; in order to reach full conversion with a SCR of 500, hydrogenation is performed at a pressure of 60 bar for 20 h. The use of ligand **17am** in the rhodium-catalyzed hydrogenation of aromatic enamides resulted in ee-values of up to 95%.

Monodentate phosphites have been used very successfully in the hydrogenation of enol-esters by the group of Reetz [54]. The use of ligands **17** which consist of a BINOL moiety and a simple alcohol gave only moderate results, from 21% to 65% *e.e.* Monodentate phosphite ligands derived from carbohydrates (**23** and **24**), however, afforded considerably higher enantioselectivities in the hydrogenation of enol esters derived from aliphatic alkynes (Scheme 28.8). Especially using an enol ester based on 2-furanoic acid, 90% ee was obtained. Performing the same reaction at –20 °C resulted in an ee of 94%. Thus far, the highest ee-value was obtained using (bidentate) Ru/PennPhos in the hydrogenation of the enol acetate based on 2-hexanone (75%) [55]. As for the *N*-acyl enamides, enol esters are hydrogenated at low rates. To reach full conversion, similar conditions were needed, with a SCR of 200 and a hydrogen pressure of 60 bar for 20 h.

The use of monodentate phosphite ligands in the hydrogenation of β-acylamino acrylates, affording derivatives of β-amino acids, has been demonstrated by Bruneau et al. [56]. Ligand **17bc** is clearly more effective in the hydrogenation of substrates with an *E*-configuration. In contrast, ligand **17bd**, a diastereomer of **17bc**, affords better results in the hydrogenation of substrates with the *Z*-configuration.

The carbohydrate ligands **23** and **24** were also applied in the hydrogenation of itaconate and enamides [57]. Also here, the configuration of the products is predominantly determined by the configuration of the BINOL moiety in the ligand. An extensive study, including the hydrogenation of *N*-acyl β-dehydroamino esters, using carbohydrate-derived monophosphites (also **25** and **26**) was recently reported by Zheng et al. [58].

Very recently, the group of Reetz published details of a monodentate phosphite ligand **27** (together with a large number of comparable phosphoramidite ligands) in which the BINOL unit bears a single *ortho*-substituent. This creates an additional stereocenter at phosphorus, which leads to mixtures of diastereomers. The ligand was found to be very successful in the hydrogenation of *N*-acyl dehydroalanine methyl ester [59].

As described for monodentate phosphonite ligands, monodentate phosphite ligands have also been used in a monodentate ligand combination approach.

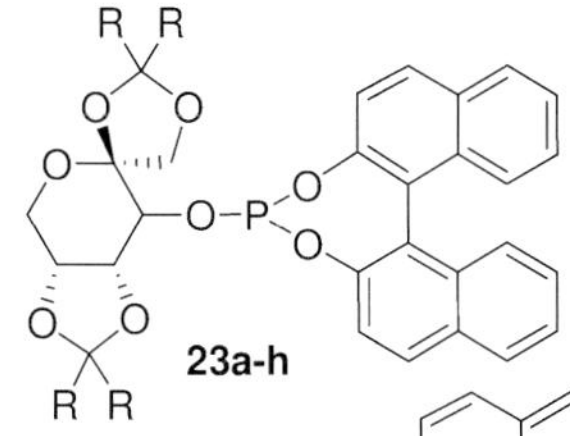

23a-h

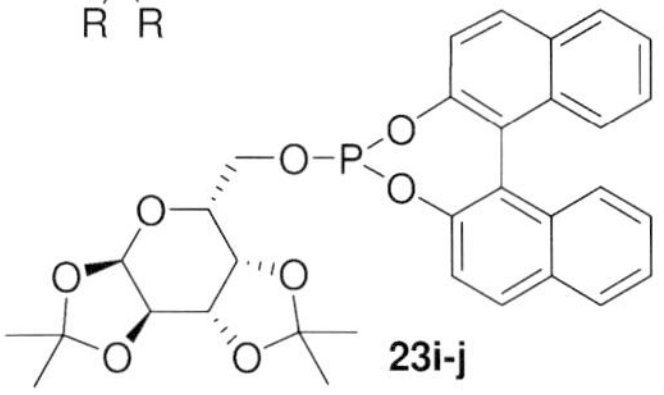

23i-j

23a 4-(*S*), R = CH_3, (*R*)-BINOL
b 4-(*S*), R = CH_3, (*S*)-BINOL
c 4-(*S*), R = -$(CH_2)_5$-, (*R*)-BINOL
d 4-(*S*), R = -$(CH_2)_5$-, (*S*)-BINOL
e 4-(*R*), R = CH_3, (*R*)-BINOL
f 4-(*R*), R = CH_3, (*S*)-BINOL
g 4-(*R*), R = -$(CH_2)_5$-, (*R*)-BINOL
h 4-(*R*), R = -$(CH_2)_5$-, (*S*)-BINOL
i (*R*)-BINOL
j (*S*)-BINOL

24a-h

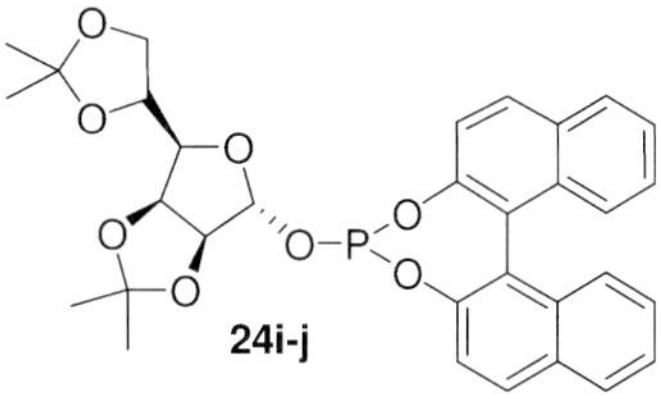

24i-j

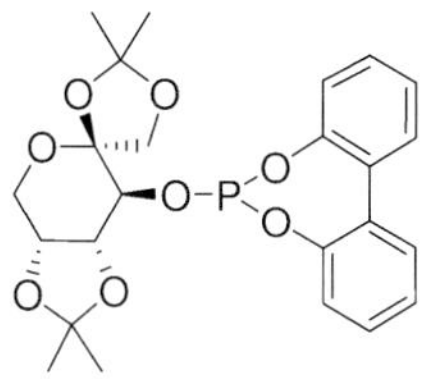

24a 3-(*S*), R = CH_3, (*R*)-BINOL
b 3-(*S*), R = CH_3, (*S*)-BINOL
c 3-(*S*), R = -$(CH_2)_5$-, (*R*)-BINOL
d 3-(*S*), R = -$(CH_2)_5$-, (*S*)-BINOL
e 3-(*R*), R = CH_3, (*R*)-BINOL
f 3-(*R*), R = CH_3, (*S*)-BINOL
g 3-(*R*), R = -$(CH_2)_5$-, (*R*)-BINOL
h 3-(*R*), R = -$(CH_2)_5$-, (*S*)-BINOL
i (*R*)-BINOL
j (*S*)-BINOL

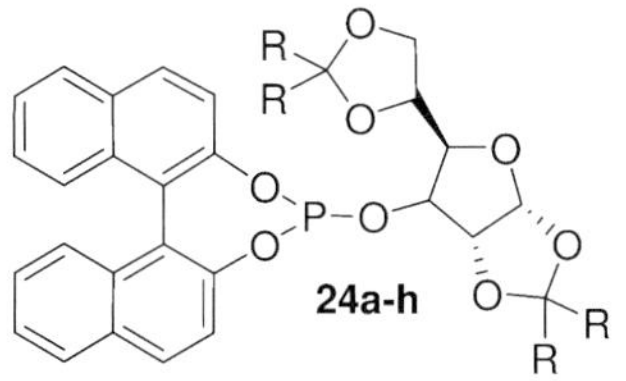

25 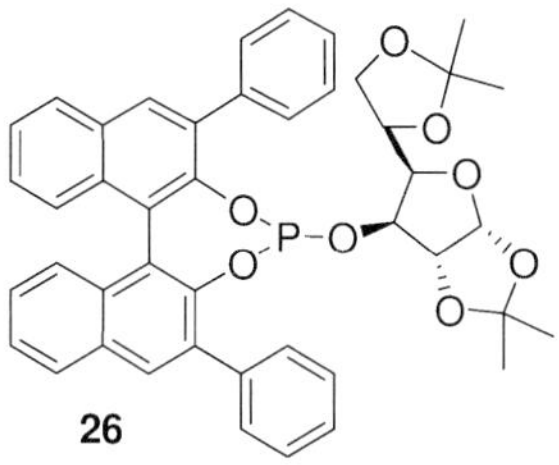 26 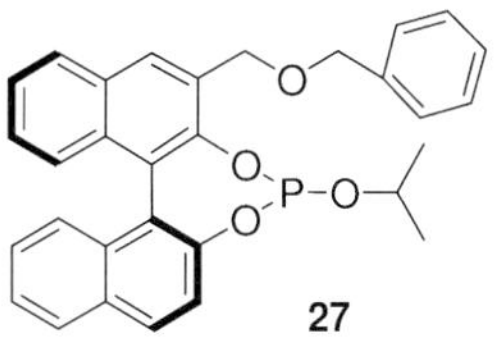 27

Scheme 28.8 Monodentate phosphite ligands based on carbohydrates.

28.5
Monodentate Phosphoramidites

The use of monodentate phosphoramidites in enantioselective hydrogenation was first reported in 2000, together with reports on the use of phosphites and phosphonites [15]. Phosphoramidites are prepared in a variety of ways, but the most common route is the treatment of a diol with PCl_3, followed by addition of an amine [60, 61]. MonoPhos (**29a**), the first reported phosphoramidite used as a ligand, is prepared from BINOL and HMPT in toluene [62]. Phosphoramidites, especially those based on BINOL, have the distinct advantage of being resistant to water and oxygen (Scheme 28.9). Although sensitive to acidic conditions, this is hardly a handicap as their rhodium complexes are considerably less sensitive. This is revealed in the successful hydrogenation of dehydroamino acids. Together with their ease of preparation – mostly in one or two steps – this feature makes them very versatile, and has been employed successfully in HTE using ligands of phosphoramidites and in the use of ligand-mixtures (see Chapter 36).

The majority of the reported phosphoramidite ligands consist of BINOL and a diversity of readily available amines. Excellent enantioselectivities in the hydrogenation of α- and β-dehydroamino acids, itaconates and enamides [63, 64] have been reported. In a recent full report, the group of Minnaard, De Vries and Feringa noted that especially the BINOL-derived ligands containing a piperidine or

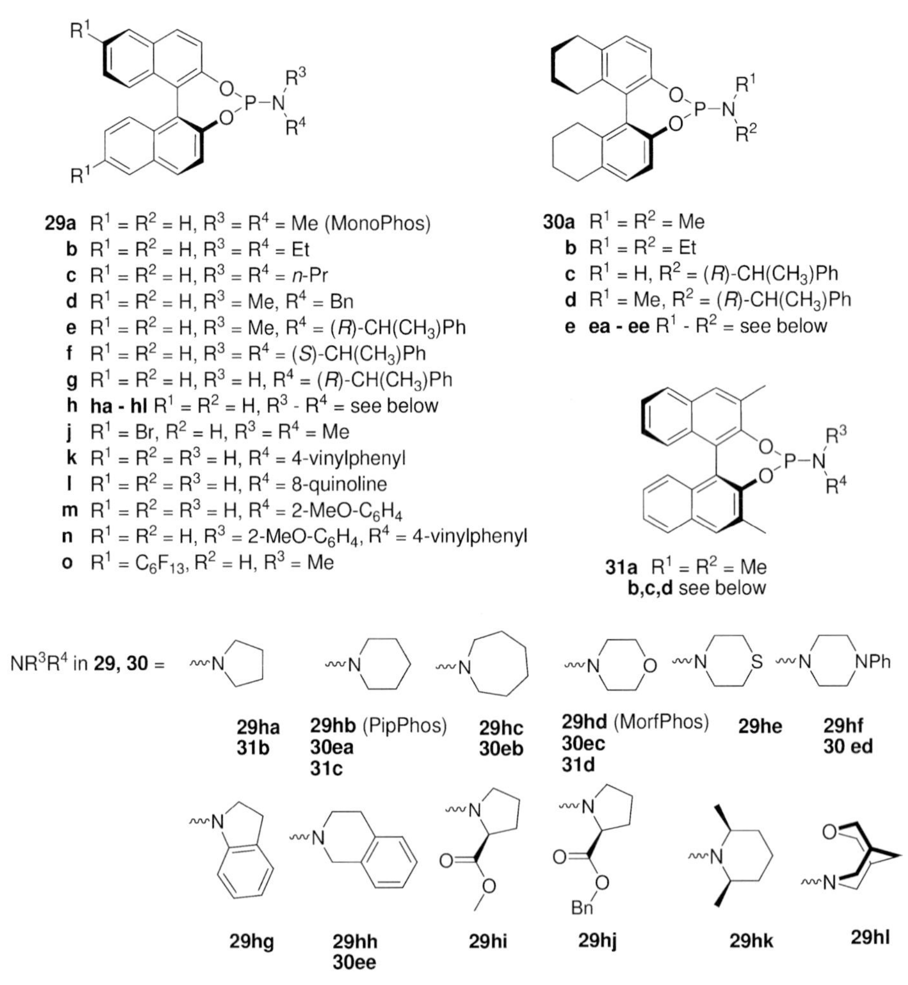

Scheme 28.9 Monodentate phosphoramidites based on BINOL.

a morpholine substituent (PipPhos **29hb** and MorfPhos **29hd**, respectively) are the most privileged ligands [65]. Variations on this theme comprise the use of substituted BINOLs and octahydro-BINOL (H_8-BINOL) [17, 66]. In some cases these ligands afford higher ee-values. Very recently, it was shown that enol acetates and enol carbamates can also be hydrogenated, with excellent ee-values to the corresponding alcohol derivatives using PipPhos **29hb** [67]. Phosphoramidite ligand **29hl**, based on a combination of BINOL and oxa-bispidine, has been reported by the group of Waldmann [68].

Very recently, Reetz, Ma and Goddard reported phosphoramidites based on BINOL bearing a single *ortho*-substituent (Scheme 28.10) [69]. These ligands are also chiral on phosphorus, such that the synthesis results mostly in diastereomers which have to be separated. In several cases, however, one of the diastereomers was formed exclusively. Some of the ligands afford high ee-values in the hydrogenation of methyl *N*-acyl dehydroalanine and dimethyl itaconate.

Zhou et al. have reported extensively on the use of a spiro-biindanediol as the backbone in the ligands **35a–f** (Scheme 28.11, SIPHOS) [70]. Excellent results are obtained for a variety of substrates, and recently a full report has appeared on the use of these ligands [71]. Synthesis of the diol backbone requires a number of steps, including a resolution [72]. An additional and successful spiro-diol-derived phosphoramidite **39** has recently been disclosed by the group of Zhang [73].

Phosphoramidite ligands based on TADDOL (**36**) and on D-mannitol (**37**) [74] have also been used (Scheme 28.11). However, the enantioselectivities reported for the hydrogenation of α-dehydroamino acids and itaconates were generally lower compared to the ligands based on BINOL. A different strategy is the use of ligands **38a–g** based on the achiral diol catechol, and chiral amines [75].

The rate of hydrogenation of dehydroamino acids using [Rh(MonoPhos)$_2$-COD]BF_4 is not very high at 1 bar of hydrogen, though this can be overcome by applying higher pressure. Reactions performed with 5 bar reach TOFs of 200 to 600 mol $mol^{-1} \cdot h$. This increase in rate also allows for a reduction in the amount of catalyst needed to about 0.02–0.1 mol%. The increase in hydrogen pressure, up to 100 bar, does not affect the enantioselectivity; this is in contrast to the de-

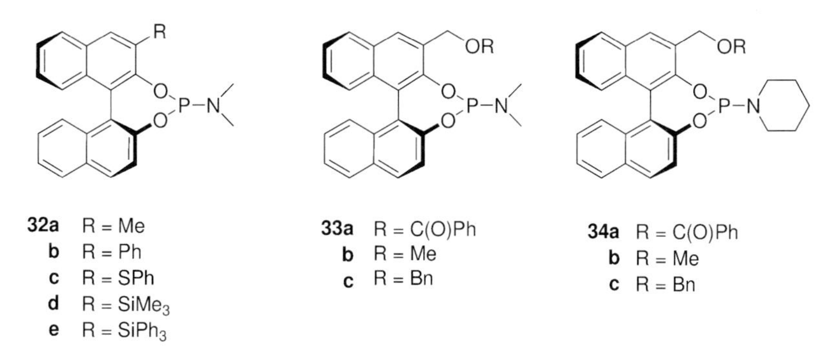

Scheme 28.10 Monodentate phosphoramidites based on monosubstituted BINOL.

35a R^1 = H, R^2 = Me (SIPHOS)
b R^1 = H, R^2 = Et
c R^1 = H, R^2 = ^{i}Pr
d R^1 = H, R^2 = -$(CH_2)_4$-
e R^1 = Br, R^2 = Me
f R^1 = Ph, R^2 = Me
g R^1 = OMe, R^2 = Me

36a R = ^{i}Pr
b R = Bn

37a R = H
b R = Me
c R = Et
d R = ^{i}Bu
e R = Ph
f R = Me (*S,S,S,S*)
g R = Ph (*S,S,S,S*)

NR^1R^2 in **38g** =

38a R^1 = (*S*)-$CH(CH_3)Ph$, R^2 = H
b R^1 = (*S*)-$CH(CH_3)Ph$, R^2 = Me
c R^1 = (*S*)-$CH(CH_3)Ph$, R^2 = Et
d R^1 = (*R*)-$CH(CH_3)Ph$, R^2 = ^{i}Pr
e R^1 = (*S*)-$CH(CH_3)Ph$, R^2 = Bn
f R^1 = (*S*)-$CH(CH_3)Ph$, R^2 = (*S*)-$CH(CH_3)Ph$
g R^1 - R^2 = see above

39

Scheme 28.11 Phosphoramidite ligands based on alternative backbones.

crease in *e.e.* which is seen upon increasing hydrogen pressure when using most bidentate ligands, and is a distinct advantage.

Initially, it was not clear whether monodentate ligands could, apart from enantioselectivity, compete with the established bidentate ligands. Activity and stability of the catalyst are also important parameters. Using MonoPhos **29a** and related phosphoramidites, it has been shown that monodentate ligands can (at least in a number of cases) keep up with the best bidentate ligands. Itaconic acid has been hydrogenated using $[Rh(MonoPhos)_2COD]BF_4$ at a SCR of 10 000 on 100-g scale. This reaction gave full conversion in 3 h, with 97.5% *e.e.* [76]. In the hydrogenation of *β*-dehydroamino esters, the reaction rates of MonoPhos **29a** and **29g** were compared with the bidentate ligands DuPhos, JosiPhos and PhanePhos; subsequently, the reactions with ligand **29g**, together with DuPhos, proved to be the fastest [16].

Monodentate phosphoramidite ligands were also employed in the rhodium-catalyzed hydrogenation of enamides. These hydrogenations generally require longer reaction times and higher hydrogen pressures (e.g., 1 to 20 h at 10 to 20 bar with 0.1–2 mol% catalyst). Good to excellent enantioselectivities are obtained using a variety of phosphoramidites. Very high enantioselectivities were reached using PipPhos **29hb**, MorfPhos **29hd** and members of the SIPHOS family.

Phosphoramidite ligands have also been very successful in the asymmetric hydrogenation of *β*-dehydroamino esters. As with bidentate ligands, there is a large difference in behavior during hydrogenation of the *E*- and *Z*-substrates. The *E*-isomer is generally hydrogenated more easily, and with a higher ee-value. However, by varying the ligand's structure and solvent, both *E*- and *Z*-*β*-dehydroamino esters could be hydrogenated with excellent ee-values using **29d** and **29g**, thereby surpassing – at that time – the bidentate ligands [77]. Recently, the SIPHOS ligands have also been shown as successful in the hydrogenation of *E*- and *Z*-*β*-dehydroamino esters.

Since monodentate phosphoramidites are so successful in asymmetric hydrogenation – both because of their performance and their ease of preparation – a logical extension is their application in recyclable systems. Doherty et al. were the first to prepare polymer-supported phosphoramidites by using the monomers **40** and **41** (Scheme 28.12); these led to high ee-values which fell somewhat upon polymerization [78]. The catalyst was shown to be capable of being recycled at least four times.

One highly successful approach was demonstrated by Ding et al., using self-supporting heterogeneous catalysts consisting of ligands such as **42** to create a polymer-type catalyst [79]. Both "dents" in the ligand coordinate to different rhodium ions. The results obtained using this self-supporting catalyst were comparable or better than those obtained using the analogous monomers. The reusability of the catalyst was at least seven cycles.

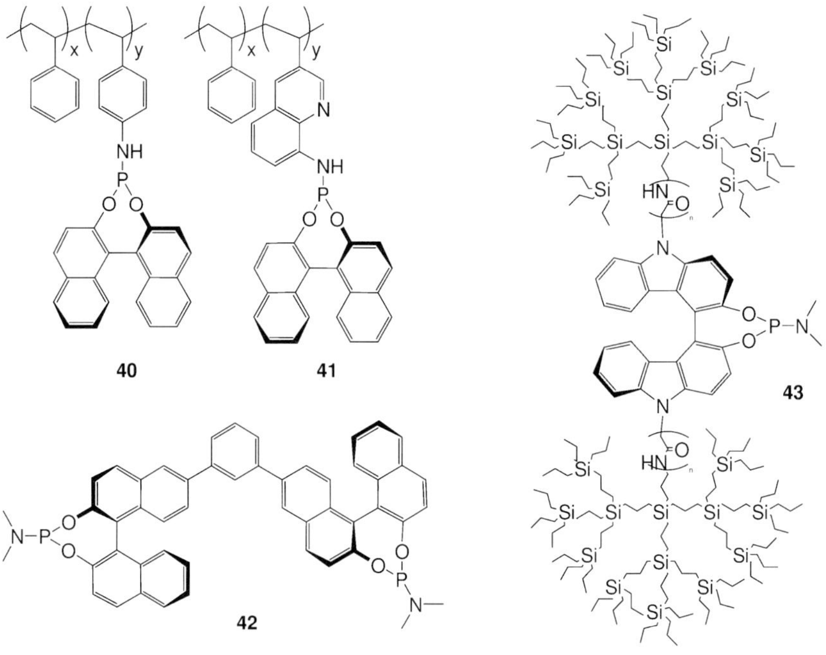

Scheme 28.12 Immobilized phosphoramidite ligands.

The immobilization of catalysts on a solid support is a well-known approach to render a system recycleable, and this has been performed recently by the immobilization of rhodium-MonoPhos **29a** on aluminosilicate AlTUD-1. The resultant system showed high efficiency in water, and could be recycled [80].

Another way of retaining the catalyst is to create dendrimer-supported ligands, thereby allowing separation of the product and catalyst by membranes. Based on the readily modified BICOL backbone, two dendrimer-ligands **43** were prepared that had performance comparable to that of MonoPhos **29a** in the hydrogenation of methyl *N*-acyl dehydrophenylalanine [81].

28.6
Monodentate Phosphinites, Aminophosphinites, Diazaphospholidines and Secondary Phosphine Oxides

In the quest for effective ligands in enantioselective hydrogenation, a number of groups have varied the atoms surrounding the phosphorus atom in order to develop new and hopefully successful ligands. An additional argument for ligand development is to circumvent existing patent literature. Next to phosphines, phosphonites, phosphites and phosphoramidites, attention has been paid to monodentate phosphinites. Surprisingly, as early as 1986 an excellent ee was reported in the hydrogenation of dimethyl itaconate using phosphinite **45** (Scheme 28.13) [82]. Most likely because ee-values were determined somewhat

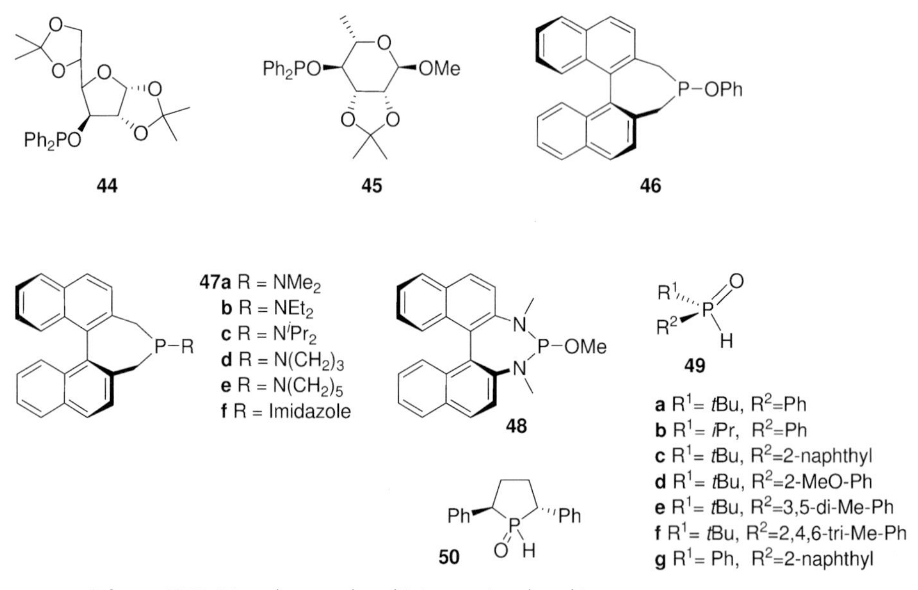

Scheme 28.13 Monodentate phosphinites, aminophosphinites, diazaphospholidines and secondary phosphine oxides.

inaccurately by measuring optical rotations, this result was generally overlooked. Nevertheless, it must be considered as an early example of a successful monodentate ligand. Other monodentate phosphinites, **44** and the recently reported **46** [83], were much less successful. Although phosphinites are rather stable in air, rapid hydrolysis takes place in the presence of moisture.

Monodentate aminophosphines **47** have been developed recently in analogy to the other BINOL-based ligands and related bidentate aminophosphines. Although the scope of these ligands has not been studied in depth, good ee-values can be obtained in rhodium-catalyzed enantioselective hydrogenation [84]. One recent study reports the preparation and use of diazaphospholidine **48** as a logical extension of the BINOL-based phosphites and phosphoramidites [85]. This ligand has not yet been studied in depth, mainly because the synthesis is rather laborious and the ligand is sensitive to hydrolysis.

Secondary phosphine oxides are known to be excellent ligands in palladium-catalyzed coupling reactions and platinum-catalyzed nitrile hydrolysis. A series of chiral enantiopure secondary phosphine oxides **49** and **50** has been prepared and studied in the iridium-catalyzed enantioselective hydrogenation of imines [48] and in the rhodium- and iridium-catalyzed hydrogenation functionalized olefins [86]. Especially in benzyl substituted imine-hydrogenation, **49a** ranks among the best ligands available in terms of *e.e.*

28.7 Hydrogenation of N-Acyl-α-Dehydroamino Acids and Esters

The hydrogenation of methyl *N*-acyl-dehydrophenylalanine **52a** and methyl *N*-acyl-dehydroalanine **53a** to their corresponding amino acid derivatives **54a** and **55a** are the benchmark reactions for rhodium-catalyzed enantioselective hydrogenation. Most newly developed ligands are tested in the hydrogenation of these substrates, and good enantioselectivities are often obtained. As the number of reports is overwhelming, a selection of the results is presented in Table 28.1. Only ligands that afford ee-values of 95% and higher have been included.

It transpires that most classes of monodentate ligands include members that are able to induce high enantioselectivity in the hydrogenation of the two benchmark substrates **52a** and **53a**. It is not clear whether their corresponding acids **52b** and **53b** have been studied or, alternatively, if the authors decided not to include (disappointing) ee-values. For phosphoramidite MonoPhos (**29a**), however, the ee-values are invariably excellent. Overall, the TOFs range from 50 to 170 h^{-1}, but have not been optimized in most cases. Unfortunately, with one exception [87], the hydrogenation of dehydroamino esters in which R^1 is a (functionalized) alkyl substituent has not been studied, probably because of their difficult accessibility.

As the hydrogenation of substituted dehydrophenylalanines is important from an industrial point of view, and the substrates are easily accessible, some phosphoramidites have been screened against a series of these substrates. According

Table 28.1 Enantioselective hydrogenation of *N*-acyl-dehydrophenylalanine and *N*-acyl-dehydroalanine.

R^1–CH=C(NHAc)–CO_2R^2 —[Rh]-cat / Ligand, solvent, H_2→ R^1–CH$_2$–CH(NHAc)–CO_2R^2

52a R^1 = Ph, R^2 = Me
52b R^1 = Ph, R^2 = H
53a R^1 = H, R^2 = Me
53b R^1 = H, R^2 = H

54a R^1 = Ph, R^2 = Me
54b R^1 = Ph, R^2 = H
55a R^1 = H, R^2 = Me
55b R^1 = H, R^2 = H

Entry	Ligand	54a (54b) ee [%]	55a (55b) ee [%]
1	**10d** [h)]	95	67
2	**10h** [h)]	95	94
3	**15b** [a)]	89	94
4	**17ac** [b)]		95
5	**17ae** [b)]		97
6	**17af** [b)]		95
7	**17ah** [b)]		95
8	**17am** [b)]		96
9	**17an** [b)]		96
10	**17aq** [b)]		96
11	**17at** [b)]		96
12	**29a** [c)]	97 (97)	>99 (>99)
13	**29b** [d)]	98	97
14	**29hb** [g)]	>99	>99
15	**29hc** [g)]	97	97
16	**29hd** [g)]	98	99
17	**29hf** [g)]	97	96
18	**29hh** [g)]	99	99
19	**30a** [e)]	94	>99
20	**30ea** [g)]	>99	97
21	**30eb** [g)]	96	
22	**30ec** [g)]	99	95
23	**30ed** [g)]	99	96
24	**30ee** [g)]	98	96
25	**35a** [f)]	98	97
26	**35d** [f)]	98	
27	**35e** [f)]	98	
28	**35f** [f)]	97	
29	**33a** [i)]		>98
30	**33b** [i)]		>98
31	**39** [j)]	98	98

a) Reactions carried out with SCR 500, in CH_2Cl_2, $p(H_2) = 1.5$ bar, 25 °C, 3 h.
b) Reactions carried out with SCR 1000, in CH_2Cl_2 or $ClCH_2CH_2Cl$, $p(H_2) = 1.3$ bar, 25 °C, 20 h.
c) Reactions carried out with SCR 100 or 1000, in CH_2Cl_2 or EtOAc, $p(H_2) = 1$ bar, 25 °C, 1–3 h.
d) Reactions carried out in THF.
e) Reactions carried out in acetone.
f) Reactions carried out in toluene.
g) Reactions carried out with SCR 50, in CH_2Cl_2, $p(H_2) = 5$ bar, 25 °C, 3 h.
h) Reactions carried out with SCR 100, in toluene + SDS, $p(H_2) = 1$ bar, 25 °C, $t/2 = 0.5$–3 h.
i) Reactions carried out with pure diastereomers, SCR 200, in CH_2Cl_2, $p(H_2) = 1.3$ bar, 20 h.
j) Reactions carried out with SCR 100, in CH_2Cl_2, $p(H_2) = 1.3$ bar, r.t., 12 h.

to the data in Table 28.2, it is safe to assume that a large variety of sterically and electronically different substituents are tolerated without repercussions on the ee-value. This also leads to the assumption, for example, that the recently developed PipPhos **29hb** and MorfPhos **29hd**, being the second generation of MonoPhos **29a**, will perform very well. On average, TOFs using MonoPhos **29a** are around 500 h^{-1} at 5 bar, increasing to 1700 h^{-1} at 60 bar. In the case of a cyano substituent, a strongly decreased rate was observed, probably because of coordination of this group to rhodium. For SIPHOS **35** and **39**, the TOFs are about 50 to 100 h^{-1} at 1 bar.

Table 28.2 Enantiomeric hydrogenation of substituted methyl *N*-acyl-dehydrophenylalanine.

R^1-CH=C(NHAc)-CO_2R^2 —[Rh]-cat / Ligand, H_2→ R^1-CH_2-CH(NHAc)-CO_2R^2

Entry	Substituent	29a	29b	30a	35a	39
1	R^1=3-MeO-Ph, R^2=Me	97				
2	R^1=4-MeO-Ph, R^2=Me	94	99	94	96	
3	R^1=3-MeO-4-AcO-Ph, R^2=Me	96		96		
4	R^1=4-F-Ph, R^2=Me	96	>99			>99
5	R^1=4-F-Ph, R^2 = H	93				
6	R^1=3-F-Ph, R^2=Me	95				
7	R^1=3-F-Ph, R^2=H	96				
8	R^1=2-F-Ph, R^2=Me	95				
9	R^1=4-Cl-Ph, R^2 = Me	94	99	98	99	99
10	R^1=4-Cl-Ph, R^2 = H			83		
11	R^1=3,4-Cl_2-Ph, R^2=H	97				
12	R^1=3,4-Cl_2-Ph, R^2=Me	99				
13	R^1=3-NO_2-Ph, R^2=Me	95			99	
14	R^1=4-NO_2-Ph, R^2=Me	95	>99	96	99	
15	R^1=3-NO_2-4-F-Ph, R^2=Me	95				
16	R^1=4-biphenyl, R^2=Me	95				
17	R^1=3-F-4-biphenyl, R^2=Me	93				
18	R^1=4-Ac-Ph, R^2=Me	99	99			
19	R^1=4-Bz-Ph, R^2=Me	94				
20	R^1=4-CN-Ph, R^2=Me	92 [a]				
21	R^1=1-naphthyl, R^2=Me	93				
22	R^1=4-Br-Ph, R^2=Me		99	91		
23	R^1=2-Cl-Ph, R^2=Me		99	93	97	99
24	R^1=3-Cl-Ph, R^2=Me		99			
25	R^1=3-Cl-Ph, R^2=H			74		
26	R^1=4-Me-Ph, R^2=Me		98		98	
27	R^1=3-Br-Ph, R^2=Me					>99
28	R^1=2-naphthyl, R^2=Me					>99

a) Very slow reaction was observed.

1 mol% $[Ir(COD)Cl]_2$, **49a**

CH_2Cl_2, 10 bar H_2

85% ee

Scheme 28.14 Enantioselective hydrogenation of methyl *N*-acyl dehydrocyclohexylglycine.

Studies have been limited to substrates containing the *N*-acyl or *N*-benzoyl stereodirecting groups. On occasion, for further synthetic applications, a carbamate protecting group is preferred [88]. Substrates possessing two substituents at the β-position have also been ignored, with one exception (Scheme 28.14) [84]. In that report, secondary phosphine oxide **49a** induced 85% ee in the iridium-catalyzed hydrogenation of methyl *N*-acyl dehydrocyclohexylglycine, with a low TOF (1 h^{-1}). This (sub)class of substrates clearly deserves further investigation, as the number of bidentate ligands that induces excellent enantioselectivity is also limited.

28.8
Hydrogenation of Unsaturated Acids and Esters

Next to the hydrogenation of α-dehydroamino acids and esters, the hydrogenation of itaconic acid **56** and its corresponding dimethyl ester **57** is considered to be a benchmark reaction. In addition, the substrates are cheap and the products are valuable intermediates in natural product synthesis. A large number of monodentate ligands has been reported to give good and often excellent results in the hydrogenation of itaconic acid and its corresponding dimethyl ester; hence, only a selection is provided here.

All classes of ligands have members that perform well, although excellent ee-values are rare when phosphines are used, with the exception of ligand **9f** developed by the group of Helmchen. A large number of phosphite ligands has been explored, and both ligands based on BINOL and bisphenols give excellent ee-values. Only **57** has been used as a substrate, and not itaconic acid **56**. Phosphoramidites also perform extremely well, especially PipPhos **29hb**, MorfPhos **29hd**, and related ligands. One remarkable finding was the 100-g scale hydrogenation of itaconic acid **56** using a SCR of 10 000 (TOF 5000 h^{-1}) with MonoPhos **29a** at high pressure, giving quantitative yield and 97% ee [89] (Table 28.3). An even more impressive result was the hydrogenation of **57** with a S/C of 10 000 (TOF 40 000 at 20 bar, 98% ee) when applying phosphite **18a**. As mentioned previously, phosphinite **45** is an early example of a successful monodentate ligand in the hydrogenation of dimethyl itaconate. On average, TOFs range from 20 to 50 h^{-1} at 1 bar, and to 1300 h^{-1} at 10 bar.

Although some attention has been paid to the hydrogenation of β-substituted itaconates, that can be prepared by Stobbe condensation, this class of com-

Table 28.3 Enantioselective hydrogenation of itaconic acid and its dimethyl ester.

RO_2C–C(=CH_2)–CH_2–CO_2R → ([RhL_2(COD)]BF_4; H_2, CH_2Cl_2) → RO_2C–CH(CH_3)–CH_2–CO_2R

56 R = H
57 R = Me

58 R = H
59 R = Me

Entry	Ligand	58 [%]	59 [%]	Entry	Ligand	58	59
1	**9f** [a]	96		25	**21bd** [d]		99
2	**17ac** [b]		98	26	**22** [d]		96
3	**17ad** [b]		99	27	**23e** [e]		>99
4	**17ae** [b]		99	28	**23f** [e]		99
5	**17af** [b]		97	29	**23g** [e]		99
6	**17ai** [b]		99	30	**24b** [e]		99
7	**17aj** [b]		96	31	**24d** [e]		97
8	**17am** [b]		99	32	**29a** [f]	97	94
9	**17an** [b]		99	33	**29hb** [g]		99
10	**17ap** [b]		97	34	**30ea** [g]		99
11	**17aq** [b]		99	35	**29hd** [g]		98
12	**17ar** [b]		99	36	**30ec** [g]		98
13	**17as** [b]		99	37	**30ed** [g]		99
14	**17aw** [b]		95	38	**29hh** [g]		97
15	**17bc** [b]		95	39	**30ee** [g]		95
16	**17bh** [b]		95	40	**33a** [h]		96
17	**17bi** [b]		98	41	**33b** [h]		95
18	**18a** [b]		99	42	**33c** [h]		96
19	**18b** [b]		98	43	**34a** [h]		96
20	**21ak** [c]		97	44	**34b** [h]		96
21	**21am** [c]		96	45	**34c** [h]		97
22	**21as** [c]		97	46	**45**		99
23	**21au** [c]		98				
24	**21bb** [d]		97				

a) Reactions carried out with SCR 500, in CH_2Cl_2, $p(H_2) = 1.1$ bar, 20 °C, 24 h.
b) Reactions carried out with SCR 1000, in CH_2Cl_2, $p(H_2) = 1.3$ bar, 20 °C, 20 h.
c) Reactions carried out with SCR 200, in CH_2Cl_2 or CH_2ClCH_2Cl, $p(H_2) = 7$ bar, 25 °C, 20 h. In some cases $[Rh(COD)_2]SbF_6$ was used instead of $[Rh(COD)_2]BF_4$ at 50 °C.
d) Reactions carried out with SCR 1000, in CH_2Cl_2, $p(H_2) = 3$ bar, 25 °C, 20 h. $[Rh(COD)_2]SbF_6$ was used instead of $[Rh(COD)_2]BF_4$ at 50 °C.
e) Reactions carried out with SCR 100, in CH_2Cl_2, $p(H_2) = 10$ bar, 25 °C, 12 h.
f) The reaction with itaconic acid was carried out on 100-g scale with SCR 10 000, in CH_2Cl_2, $p(H_2) = 100$ bar, 25 °C (at the start of the reaction), 2 h.
g) Reactions carried out with SCR 50, in CH_2Cl_2, $p(H_2) = 5$ bar, 25 °C, 4 h. For **29hb**, reaction carried out with SCR 1000, in CH_2Cl_2, $p(H_2) = 10$ bar, 25 °C, TOF 1300 h^{-1}.
h) Reactions carried out with SCR 200, in CH_2Cl_2, $p(H_2) = 1.3$ bar, 25 °C, 20 h.

pounds seems to have escaped attention in the hydrogenation using monodentate ligands, until now.

In general, unsaturated esters and acids have hardly been studied in rhodium-catalyzed hydrogenation. This is not surprising, as a carbonyl group at a suitable position is generally thought to be essential for obtaining high ee-values [90]. Using monodentate ligands, some studies were performed during the early years of asymmetric hydrogenation, with most providing low ee-values. An exception was the hydrogenation of (*E*)-3,7-dimethyl-2,6-dienoic acid that afforded the product in 79% ee using monophosphine **2f**. All the more surprising, therefore, was a recent study in which tiglic acid and a series of substituted cinnamic acids were hydrogenated using a combination of a monodentate phosphoramidite and a monodentate phosphine [91]. The rates were very high and excellent ee-values were obtained (details of this study are provided in Chapter 36).

28.9
Hydrogenation of *N*-Acyl Enamides, Enol Esters and Enol Carbamates

Rhodium-catalyzed enantioselective hydrogenation of *N*-acyl enamides provides access to enantioenriched amides which can be hydrolyzed to the free amines. The synthesis of the substrates is considerably less straightforward than that of *N*-acyl dehydroamino acids, which explains the smaller number of reports devoted to *N*-acyl enamides.

Nevertheless, a number of monodentate ligands have shown good performance in this hydrogenation. A selection of results for the hydrogenation of acyclic terminal enamides is listed in Table 28.4. Only the most successful ligands in terms of ee-value are reported, though both phosphites and phosphoramidites perform very well. The phosphite ligands are based on BINOL, and in particular **23c**, which contains a carbohydrate unit, provides excellent ee-values. As phosphoramidites, PipPhos **29hb**, SIPHOS **35a** and the phosphoramidite ligand based on catechol **38g** are excellent ligands. On average, the TOFs are approximately 25 h^{-1} at pressures of 10 to 25 bar, though PipPhos **29hb** is especially impressive, with a TOF of 250 h^{-1} at 25 bar.

N-Acyl enamides substituted at C2 and cyclic enamides have received considerably less attention. A selection of the results is listed in Table 28.5. Phosphite ligand **23c** appears to be the only ligand that gives excellent ee-values for both **62*Z*** and **62*E*** (the mixture was used). PipPhos **29hb** is the monodentate ligand of choice in the hydrogenation of cyclic enamides, and ranks between the best bidentate ligands that can handle these substrates. Clearly, **63** – and especially **67** – are substrates that escape selective hydrogenation with these ligands, until now. The TOFs generally range from 10 to 25 h^{-1}.

The rhodium-catalyzed enantioselective hydrogenation of enol esters is an alternative to the asymmetric reduction of ketones. Although enol esters are accessible both from ketones and alkynes, the number of studies reporting successful asymmetric hydrogenation has been limited. It appears that, compared

Table 28.4 Enantioselective hydrogenation of acyclic terminal N-acyl enamides.

NHAc NHAc $\xrightarrow[\text{Ligand, } H_2]{[Rh(COD)_2]BF_4}$ NHAc NHAc

60a R = H
b R = Cl
c R = MeO
d R = CF_3
e 2-naphthyl

60f X = S
g R = O

61a R = H
b R = Cl
c R = MeO
d R = CF_3
e R = 2-naphthyl

61f X = S
g R = O

Entry	Ligand	61 a [%]	61 b [%]	61 c [%]	61 d [%]	61 e [%]	61 f [%]	61 g [%]
1	**17 ag** [a)]	95				94		
2	**17 am** [a)]	95						
3	**23 c** [f)]	95	98	96	98	97		
4	**29 b** [b)]	97	90	98	99			
5	**29 hb** [c)]	99	99	99				
6	**29 hd** [c)]	99	99	99				
7	**29 hf** [c)]	99	99	98				
8	**30 a** [b)]	96	86	92	99			81
9	**30 ea** [c)]	98	98	99				
10	**30 ec** [c)]	97	97	98				
11	**35 a** [d)]	98	99		99		96	99
12	**35 d** [d)]	97	99					
13	**35 e** [d)]	98	98					
14	**35 f** [d)]	95	94					
15	**38 g** [e)]	97	94	97				

a) Reactions carried out with SCR 500, in CH_2Cl_2, $p(H_2) = 60$ bar, 30 °C, 20 h.
b) Reactions carried out with SCR 200, in THF, $p(H_2) = 20$ bar, 5 °C, 8 h. More substrates were tested than shown in the table.
c) Reactions carried out with SCR 50, in CH_2Cl_2, $p(H_2) = 25$ bar, 25 °C, 20 h. One substrate was hydrogenated using **29 hb** at SCR 1000 with an overall TOF of 250 h^{-1} giving the same ee and full conversion.
d) Reactions carried out with SCR 100, in toluene, $p(H_2) = 50$ bar, 5 °C, 12 h. More substrates were tested than shown in the table.
e) Reactions carried out with SCR 100, in EtOAc, $p(H_2) = 25$ bar, 25 °C, 16 h. More substrates were tested than shown in the table.
f) Reactions carried out with SCR 100, in CH_2Cl_2, $p(H_2) = 10$ bar, 25 °C, 12 h.

Table 28.5 Enantioselective hydrogenation of cyclic and substituted *N*-acyl enamides.

Entry	Ligand	68 [62*Z*]	68 [62*E*]	69 [%]	70 [%]	71 [%]	72a [%]	72b [%]	73 [%]
1	**17aq**[a)]	97	76						
2	**23c**[b)]	97	97						
3	**29hb**[c)]	96	3	–17	82*	98*	98	99	21
4	**30ea**[c)]	97	5	–1	44	82		99	28
5	**29hd**[c)]	98	23		27	97*	97*	99	13
6	**30ec**[c)]	99	26		13	89	88	99	15
7	**29hf**[c)]	98	17		21	87	82	99	8
8	**35a**[d)]					94			
9	**38g**[e)]	99	88		70		35		9

a) Reactions carried out with SCR 500, in CH_2Cl_2, $p(H_2)=60$ bar, 30 °C, 20 h.
b) Reactions carried out with SCR 100, in CH_2Cl_2, $p(H_2)=10$ bar, 25 °C, 12 h.
c) Reactions carried out with SCR 50, in CH_2Cl_2, $p(H_2)=25$ bar, 25 °C, 20 h.
d) Reactions carried out at $p(H_2)=100$ bar, 0 °C. The 5-bromo- and 6-methoxy-substituted compounds were obtained in 88% and 95% ee, respectively.
e) Reactions carried out with SCR 100, in CH_2Cl_2, $p(H_2)=25$ bar, 25 °C, 16 h.
* Reactions carried out at –20 °C.

to the corresponding *N*-acyl enamides, the enantioselective hydrogenation of enol esters is considerably more difficult in terms of TOF and *e.e.*. An exception is the hydrogenation of enol esters derived from *α*-keto esters. Nevertheless, a limited number of bidentate ligands have been reported that afford *e.e.*-values >90% in aryl-, vinyl- or trifluoromethyl-substituted enol esters. For alkyl-substituted enol esters, the *e.e.*-values have only been moderate [92].

Reetz and Goossen et al. reported recently the asymmetric hydrogenation of a series of enol esters using monodentate phosphite ligands **17** and **24** based on a combination of BINOL and carbohydrates or simple alcohols; the results of these studies are shown in Table 28.6.

Unprecedented *e.e.*-values were obtained using ligand **24b** in the hydrogenation of aliphatic enol esters. A furyl substituent on the carboxylate is apparently

Table 28.6 Enantioselective hydrogenation of aliphatic enol esters.

$[Rh\mathbf{24}_2(COD)]BF_4$, H_2, CH_2Cl_2

74a R^1 = ^{n}Bu, R^2 = Ph
b R^1 = ^{n}Bu, R^2 = Me
c R^1 = ^{n}Bu, R^2 = Et
d R^1 = ^{n}Bu, R^2 = ^{t}Bu
e R^1 = ^{n}Bu, R^2 = 2-Furyl
f R^1 = Et, R^2 = Ph
g R^1 = Et, R^2 = 2-*N*-Me-pyrrolyl
h R^1 = Et, R^2 = 2-Furyl

75a-h

Entry	Ligand	Product							
		75a	75b	75c	75d	75e	75f	75g	75h
1	24b[a)]	86	74	74	42	90*	80	72	84
2	24f[a), b)]	13	32	6	10	22	11	5	34

a) Reactions carried out with SCR 200, in CH_2Cl_2, $p(H_2) = 60$ bar, 30 °C, 20 h (TOF = 5 h^{-1}).
b) Using ligand **24f** the conversion was between 76% and 100%.
* At –20 °C, 94% ee was obtained.

1 mol% $[Rh(PipPhos)_2(COD)]BF_4$, 5 bar H_2, CH_2Cl_2, 4 - 16 h

74i R^1 = Ph, R^2 = Me
j R^1 = Ph, R^2 = NEt_2
k R^1 = 4-Cl-Ph, R^2 = Me
l R^1 = 4-NO_2-Ph, R^2 = Me
m R^1 = 4-NO_2-Ph, R^2 = NEt_2
n R^1 = ^{n}Bu, R^2 = NEt_2
o R^1 = Bn, R^2 = NEt_2
p R^1 = Me_3Si, R^2 = NEt_2
q R^1 = 1*E*-heptenyl, R^2 = NEt_2
r R^1 = styryl, R^2 = NEt_2

ee%
75i 90
j 96
k 90
l 98
m 98
n 63
o 73
p 43
q 97
r 76

Scheme 28.15 Enantioselective hydrogenation of enol acetates and enol carbamates.

beneficial for the enantioselectivity. The reactions are most likely not very fast, which is also the case using bidentate ligands.

In order to mimic the electronic properties of the corresponding *N*-acyl enamide, enol carbamate **74j** (Scheme 28.15) has been introduced in the enantioselective hydrogenation using rhodium and a series of secondary phosphinoxide ligands **49**. The use of 2 mol% of Rh/L and 1 bar of hydrogen gave full conversion and 81% *e.e.* in a slow reaction. Unfortunately, the *e.e.*-values fell upon increasing the hydrogen pressure.

Very recently, however, the use of Rh/PipPhos **29hb** was reported as an excellent catalyst for the hydrogenation of both enol acetates and enol carbamates (Scheme 28.15). The carbamate group induced higher enantioselectivities compared to the corresponding acyl group, and the hydrogenations were faster (TOFs up to 25 h^{-1} at 5 bar). Remarkably, dienol carbamates **74q** and **74r** were hydrogenated to the corresponding allylic carbamate, leaving the additional double bond intact.

28.10
Hydrogenation of *N*-Acyl-*β*-Dehydroamino Acid Esters

Enantiopure *β*-amino acids can efficiently be obtained using rhodium-catalyzed asymmetric hydrogenation. The substrates are synthesized by reacting the *β*-keto esters with NH_4OAc and subsequent acylation with acetic anhydride. This reaction generally results in a mixture of double bond isomers [93]. Compared to the corresponding *α*-dehydroamino acids and esters, their *β*-analogues are considerably more challenging. There is a large difference in behavior of the hydrogenation of the *E*- and *Z*-stereoisomers. The *E*-isomer is generally hydrogenated at higher rate and with considerably higher ee than the *Z*-isomer. A few monodentate ligands have been studied for the rhodium-catalyzed hydrogenation of this class of compounds. Phosphites **17bc** and **24c** induce high ee-values but require a high catalyst loading and long reaction times. In addition, for **24c** the conversion is incomplete. Better results have been obtained using phosphoramidites; for example, it has been shown that using BINOL-based phosphoramidites, different ligands and different solvents were necessary to hydrogenate the different double bond isomers. Excellent ee-values were obtained, however. In particular for the Z-isomers, **29g** was the best ligand available at the time, also taking into account the bidentate ligands. SIPHOS ligand **35a** also provides high ee-values, with the advantage that mixtures of *E* and *Z* substrates can be used (Table 28.7). The reactions are slow (TOF 1 h^{-1} at 100 bar), however. In general, the TOFs vary considerable among the ligands, ranging from 3 h^{-1} at 15 bar to 200 h^{-1} at 10 bar.

Table 28.7 Enantioselective hydrogenation of *N*-Acyl-β-dehydroamino acid esters.

[Rh(COD)$_2$]BF$_4$
Ligand, H$_2$ (10 bar)

78a R^1 = Me, R^2 = Me
b R^1 = Me, R^2 = Et
d R^1 = Ph, R^2 = Et
e R^1 = Et, R^2 = Me
f R^1 = ^{i}Pr, R^2 = Et
g R^1 = *p*-F-Ph, R^2 = Me
h R^1 = *o*-Br-Ph, R^2 = Me
i R^1 = *m*-Br-Ph, R^2 = Me
j R^1 = *p*-Br-Ph, R^2 = Me
k R^1 = *p*-Cl-Ph, R^2 = Me
l R^1 = *p*-Me-Ph, R^2 = Me
m R^1 = *p*-MeO-Ph, R^2 = Me

79a-m

	Ligand	79a *E/Z*	79b *E/Z*	79d *Z*	79e *E/Z*	79f *E/Z*	79g *Z*	79h	79i	79j	79k	79l	79m
1	**17bc** a)	91/55	94/38	52									
2	**24c** b)	96/–	96/–	93	98/–								
3	**29a** c)	91/–											
4	**29d** c)	99/–	98/–		99/–	99/–							
5	**29g** d)	–/95	–/94	92	–/94	–/92	94						
6	**35** e)	89	87	90				91	92	94	91	91	93

a) Reactions carried out with SCR 100, in CH_2Cl_2, $p(H_2) = 15$ bar, 30 h.
b) Reactions carried out with SCR 50, in CH_2Cl_2, $p(H_2) = 30$ bar, 12–48 h. Conversions were incomplete.
c) Reactions carried out with SCR 50, in CH_2Cl_2, $p(H_2) = 10$ bar, 4 h or SCR of 200, in CH_2Cl_2, $p(H_2) = 25$ bar, 6 h.
d) Reactions carried out with SCR 50, in *i*-PrOH, $p(H_2) = 10$ bar, 0.3 h or with SCR 200, in *i*-PrOH, $p(H_2) = 10$ bar, 1 h.
e) Reactions carried out with SCR 50, in CH_2Cl_2, $p(H_2) = 100$ bar, 48 h. Mixtures of *Z* and *E* were used.

28.11
Hydrogenation of Ketones and Imines

Although this chapter is devoted to the hydrogenation of alkenes, it is interesting to include the studies that have appeared on the hydrogenation of imines and ketones. Surprisingly, the enantioselective imine hydrogenation using monodentate ligands has been reported in only a few studies. BINOL-based monodentate phosphonite ligand **15d** has been studied in the rhodium- and iridium-catalyzed hydrogenation of benzyl imine **80** (Scheme 28.16) [40]. No chiral

80 $R_1 = R_2 = H$
81 R_1 = OMe, R_2 = H
82 R_1 = H, R_2 = OMe
83 R_1 = H, R_2 = Cl
84 R_1 = Cl, R_2 = H

85 **86** **87**

Scheme 28.16 Imines for enantioselective hydrogenation.

2.5 mol% $[Ir(COD)Cl]_2$
10 mol% **49a**, pyridine
25 bar H_2, toluene, rt, 24h
MeO
100% conv.
80% ee

Scheme 28.17 Enantioselective hydrogenation of benzyl imines using iridium/secondary phosphinoxide ligands.

induction was observed. In a thorough study, secondary phosphine oxide ligands **49** and **50** were used in the iridium-catalyzed hydrogenation of a series of imines [48]. Enantioselectivities up to 80% and full conversion were reached with Ir/**49a** in the hydrogenation of benzyl imine **81** in toluene, adding pyridine as a co-ligand (Scheme 28.17). This places the catalyst among the best catalysts

Table 28.8 Enantioselective hydrogenation of β-keto esters using monodentate phosphine ligands.

1% $[Ru\mathbf{10}_2Br_2]$
H_2, MeOH

88a R^1 = Me, R^2 = Me
88b R^1 = Et, R^2 = Me
88c R^1 = CH_2Cl, R^2 = Et
88d R^1 = Ph, R^2 = Et

89a R^1 = Me, R^2 = Me
89b R^1 = Et, R^2 = Me
89c R^1 = CH_2Cl, R^2 = Et
89d R^1 = Ph, R^2 = Et

Entry	Ligand	Substrate	Yield [%]	ee [%]
1	**10d** [a]	**88a**	95	84
2	**10e** [a]	**88a**	97	92
3	**10e** [a]	**88b**	99	94
4	**10e** [a]	**88c**	77	38
5	**10e** [a]	**88d**	99	95
6	**10o** [a]	**88a**	98	64

a) Reactions carried out with SCR 1000, in MeOH, $p(H_2)$ = 40–80 bar, at 50 °C, 16 h.

0.05 mol% [Ru**15f**$_2$(S,S-DPEN)Cl$_2$]
0.5 mol% KO*t*-Bu
50 bar H_2, 2-propanol, 0 °C, 4h
95% conv.
93% ee

Scheme 28.18 Enantioselective hydrogenation of aryl-methyl ketones.

known for benzyl imine hydrogenation in terms of enantioselectivity, although the reactions are slow. Low *e.e.*-values were obtained with *N*-diphenylphosphinoyl ketimine **85**, benzhydryl imine **86** and aryl imine **87**. A chiral phosphoric acid diester **20**, based on TADDOL was also tested, but gave very low *e.e.*-values.

At the same time, however, the iridium-catalyzed hydrogenation of **80** was reported using chiral phosphoric acid diester **17be** based on BINOL [47a]. Full conversion and a maximum *e.e.* of 50% was observed, again in a slow reaction. Interestingly, a catalyst based on palladium and **17be** afforded 39% *e.e.* and full conversion in the hydrogenation of aryl imine **87**.

In an early report, the *β*-keto ester methyl acetylacetate was hydrogenated with 71% ee using Rh/CAMP [94]. In a thorough study, the group of Beller reported excellent results in the ruthenium-catalyzed hydrogenation of *β*-keto esters using monodentate phosphines based on the binaphthyl skeleton. The phosphoramidite MonoPhos **29a** and a related phosphonite gave only low ee-values in this reaction. A selection of the results is presented in Table 28.8. One remarkable point was the difference between the non-deuterated ligand **10d** and its deuterated analogue **10o**.

Recently, the first report was made on the ruthenium-catalyzed enantioselective hydrogenation of aryl-methyl ketones using monodentate phosphonites (Scheme 28.18). In particular, ligand **15f** induced excellent ee-values. One very early report on rhodium-catalyzed hydrogenation of ketones using the monophosphine bmpp **1f** met with a low *e.e.* [95].

28.12 Conclusions

It is safe to state that monodentate ligands have rapidly found their place in rhodium-catalyzed enantioselective hydrogenation, and the high speed at which new ligands appear continuously is the best illustration of their versatility. On the one hand, the straightforward preparation and the use of ligand libraries (see Chapter 36) makes the rapid development of tailor-made ligands possible. On the other hand, it clear from the information provided in this chapter that certain ligands, such as PipPhos **29hb**, SIPHOS **35a**, and some BINOL and biphenol-based phosphites, are so-called privileged ligands with a large scope.

Nevertheless, there remains a plethora of substrates that have not yet been studied using the monodentate ligand approach. In the application of asymmetric hydrogenation, it is very important to go beyond the benchmark substrates, and several studies have already shown that the scope of enantioselective hydrogenation might be much broader than was originally assumed.

One especially underexposed aspect in most reports is that of the TOF and TON of the catalyst. Not only from a scientific point of view, but also because of the costs of the precious metals used, this must be an important characteristic of catalytic systems.

Abbreviations

HTE high-throughput experimentation
TOF turnover frequency
TON turnover number

References

1 (a) W. S. Knowles, M. J. Sabacky, *J. Chem. Soc. Chem. Commun.* **1968**, 1445; (b) W. S. Knowles, *Acc. Chem. Res.* **1983**, *16*, 106.
2 L. Horner, H. Siegel, H. Büthe, *Angew. Chem. Int. Ed. Engl.* **1968**, *7*, 942.
3 J. A. Osborn, F. H. Jardine, J. F. Young, G. Wilkinson, *J. Chem. Soc. A* **1966**, 1711.
4 For an excellent overview of the field of enantioselective hydrogenation, see: J. M. Brown, in: E. N. Jacobsen, A. Pfaltz, H. Yamamoto (Eds.), *Comprehensive Asymmetric Catalysis*, Springer, Berlin, **1999**, Vol. 1, Chapter 5.1.
5 (a) T. P. Dang, H. B. Kagan, *J. Chem. Soc. Chem. Commun.* **1971**, 481; (b) H. B. Kagan, T. P. Dang, *J. Am. Chem. Soc.* **1972**, *94*, 6429.
6 B. D. Vineyard, W. S. Knowles, M. J. Sabacky, G. L. Bachman, D. J. Weinkauff, *J. Am. Chem. Soc.* **1977**, *99*, 5946.
7 (a) H. Brunner, W. Zettlmeier, *Handbook of Enantioselective Catalysis*, VCH, Weinheim, **1993**; (b) R. Noyori, *Asymmetric Catalysis in Organic Synthesis*, Wiley, New York, **1993**; (c) H. B. Kagan, in: J. D. Morrison (Ed.), *Asymmetric Synthesis*, Academic Press, Inc., Orlando, **1985**, Volume 5; (d) H. Brunner, *Top. Stereochem.* **1988**, *18*, 129; (e) H.-U. Blaser, C. Malan, B. Pugin, F. Spindler, H. Steiner, M. Studer, *Adv. Synth. Catal.* **2003**, *345*, 103.
8 H. U. Blaser, E. Schmidt (Eds.), *Asymmetric Catalysis on Industrial Scale: Challenges, Approaches and Solutions*, Wiley-VCH, **2004**.
9 (a) W. S. Knowles, M. J. Sabacky, B. D. Vineyard, *J. Chem. Soc. Chem. Commun.* **1972**, 10; (b) J. Solodar, *J. Org. Chem.* **1978**, *43*, 1787.
10 For a discussion on this topic, see: X. Zhang, *Enantiomer* **1999**, *4*, 541.
11 For a review on the use of chiral monodentate phosphines in asymmetric catalysis, see: F. Lagasse, H. B. Kagan, *Chem. Pharm. Bull.* **2000**, *48*, 315.
12 For a perspective on the use of chiral monodentate phosphorus ligands in enantioselective olefin hydrogenations see: I. V. Komarov, A. Börner, *Angew. Chem. Int. Ed.* **2001**, *40*, 1197. As a forerunner, the development of a new, though less enantioselective, monodentate phosphine ligand was reported; F. Guillen, J.-C. Fiaud, *Tetrahedron Lett.* **1999**, *40*, 2939.
13 C. Claver, E. Fernandez, A. Gillon, K. Heslop, D. J. Hyett, A. Martorell, A. G. Orpen, P. G. Pringle, *Chem. Commun.* **2000**, 961.
14 M. T. Reetz, G. Mehler, *Angew. Chem. Int. Ed.* **2000**, *39*, 3889.
15 M. van den Berg, A. J. Minnaard, E. P. Schudde, J. van Esch, A. H. M. de Vries, J. G. de Vries, B. L. Feringa, *J. Am. Chem. Soc.* **2000**, *122*, 11539.
16 D. Peña, A. J. Minnaard, A. H. M. de Vries, J. G. de Vries, B. L. Feringa, *Org. Lett.* **2003**, *5*, 475.
17 For an initial study using the phosphoramidite MonoPhos, see: M. van den Berg, A. J. Minnaard, R. M. Haak, M. Leeman, E. P. Schudde, A. Meetsma, B. L. Feringa,

A. H. M. de Vries, C. E. P. Maljaars, C. E. Willans, D. Hyett, J. A. F. Boogers, H. J. W. Henderickx, J. G. de Vries, *Adv. Synth. Catal.* **2003**, *345*, 308.

18 Y. Fu, X. X. Guo, S. F. Zhu, A. G. Hu, J. H. Xie, Q. L. Zhou, *J. Org. Chem.* **2004**, *69*, 4648.

19 M.T. Reetz, T. Sell, A. Meiswinkel, G. Mehler, *Angew. Chem. Int. Ed.* **2003**, *42*, 790.

20 (a) M.T. Reetz, X. Li, *Angew. Chem. Int. Ed.* **2005**, *44*, 2959; (b) see Ref. 91.

21 T. Jerphagnon, J.-L. Renaud, C. Bruneau, *Tetrahedron; Asymm.* **2004**, *15*, 2101.

22 J. G. de Vries, in: D. Ager (Ed.), *Handbook of Chiral Chemicals*, M. Dekker, **2004**.

23 D. Valentine, Jr., K. K. Johnson, W. Priester, R. C. Sun, K. Toth, G. Saucy, *J. Org. Chem.* **1980**, *45*, 3698.

24 (a) J. D. Morrison, R. E. Burnett, A. M. Aguiar, C. J. Morrow, C. Phillips, *J. Am. Chem. Soc.* **1971**, *93*, 1301; (b) J. D. Morrison, W. F. Masler, *J. Org. Chem.* **1974**, *39*, 270.

25 A. Marinetti, F. Mathey, L. Ricard, *Organometallics* **1993**, *12*, 1207.

26 A. Marinetti, L. Ricard, *Organometallics* **1994**, *13*, 3956.

27 O. Riant, O. Samuel, T. Flessner, S. Taudien, H. B. Kagan, *J. Org. Chem.* **1997**, *62*, 6733.

28 S. Saito, Y. Nakamura, Y. Morita, *Chem. Pharm. Bull.* **1985**, *33*, 5284.

29 M. Ostermeier, J. Priess, G. Helmchen, *Angew. Chem. Int. Ed.* **2002**, *41*, 612.

30 K. Junge, G. Oehme, A. Monsees, T. Riermeier, U. Dingerdissen, M. Beller, *Tetrahedron Lett.* **2002**, 43, 4977.

31 K. Junge, B. Hagemann, S. Enthaler, A. Spannenberg, M. Michalik, G. Oehme, A. Monsees, T. Riermeier, M. Beller, *Tetrahedron; Asymm.* **2004**, *15*, 2621.

32 K. Junge, B. Hagemann, S. Enthaler, G. Oehme, M. Michalik, A. Monsees, T. Riermeier, U. Dingerdissen, M. Beller, *Angew. Chem. Int. Ed.* **2004**, *43*, 5066.

33 A. Marinetti, J.-P. Gent, *C. R. Chimie* **2003**, *6*, 507.

34 A. Marinetti, S. Jus, F. Labrue, A. Lemarchand, J.-P. Gent, L. Ricard, *Synthesis*, **2001**, 2091.

35 M. J. Burk, J. E. Feaster, R. L. Harlow, *Tetrahedron; Asymm.* **1991**, *2*, 569. Ligands **12b** and **12c** have not been tested in asymmetric hydrogenation.

36 (a) F. Guillen, M. Rivard, M. Toffano, J. Y. Legros, J. C. Daran, J. C. Fiaud, *Tetrahedron* **2002**, *58*, 5895; (b) see Ref. 12; (c) C. Dobrota, M. Toffano, J. C. Fiaud, *Tetrahedron Lett.* **2004**, *45*, 8153.

37 Z. Pakulski, O. M. Demchuk, J. Frelek, R. Luboradzki, K. M. Pietrusiewicz, *Eur. J. Org. Chem.* **2004**, 3913.

38 M.T. Reetz, T. Sell, *Tetrahedron Lett.* **2000**, *41*, 6333.

39 S. Trinkhaus, R. Kadyrov, R. Selke, J. Holz, L. Götze, A. Borner, *J. Mol. Cat. A. Chem.* **1999**, *144*, 15.

40 A. Martorell, C. Claver, E. Fernandez, *Inorg. Chem. Comm.* **2000**, *3*, 132.

41 Y. Xu, N. W. Alcock, G. J. Clarkson, G. Docherty, G. Woodward, M. Wills, *Org. Lett.* **2004**, *6*, 4105.

42 W. Chen, J. Xiao, *Tetrahedron Lett.* **2001**, *42*, 2897.

43 I. Gergely, C. Hegedüs, H. Gulyás, Á. Szöllösy, A. Monsees, T. Riermeier, J. Bakos, *Tetrahedron: Asymmetry* **2003**, *14*, 1087.

44 A. Korostylev, A. Monsees, C. Fischer, A. Börner, *Tetrahedron: Asymmetry* **2004**, *15*, 1001.

45 M. Ostermeier, B. Brunner, C. Korff, G. Helmchen, *Eur. J. Org. Chem.* **2003**, 3453.

46 D. Nakano, M. Yamaguchi, *Tetrahedron Lett.* **2003**, *44*, 4969.

47 (a) M.T. Reetz, T. Sell, R. Goddard, *Chimia* **2003**, *57*, 290; (b) M.T. Reetz, *Russ. J. Org. Chem.* **2003**, *39*, 392.

48 X.-B. Jiang, A. J. Minnaard, B. Hessen, B. L. Feringa, A. L.L. Duchateau, J. G. O. Andrien, J. A. F. Boogers, J. G. de Vries, *Org. Lett.* **2003**, *5*, 1503.

49 W. Chen, J. Xiao, *Tetrahedron Lett.* **2001**, *42*, 8737.

50 Z. Hua, V. C. Vassar, I. Ojima, *Org. Lett.* **2003**, *5*, 3831.

51 B. Meseguer, T. Prinz, U. Scholz, H.-C. Militzer, F. Agel, B. Driessen-Hölscher EP 1 298 136, 2003, to Bayer AG.

52 P. Hannen, H.-C. Millitzer, E. M. Vogl, F. A. Rampf, *Chem. Commun.* **2003**, 2210.

53 M.T. Reetz, G. Mehler, A. Meiswinkel, T. Sell, *Tetrahedron Lett.* **2002**, *43*, 7941.
54 M.T. Reetz, L.J. Goossen, A. Meiswinkel, J. Paetzold, J. Feldthusen Jensen, *Org. Lett.* **2003**, *5*, 3099.
55 Q. Jiang. D. Xiao, P. Cao, X. Zhang, *Angew. Chem. Int. Ed.* **1998**, *37*, 1100.
56 T. Jerphagnon, J.-L. Renaud, P. Demonchaux, A. Ferreira, C. Bruneau, *Adv. Synth. Catal.* **2004**, *346*, 33.
57 H. Huang, Z. Zheng, H. Luo, C. Bai, X. Hu, H. Chen, *Org. Lett.* **2003**, *5*, 4137.
58 (a) H. Huang, X. Liu, S. Chen, H. Chen, Z. Zheng *Tetrahedron: Asymm.* **2004**, *15*, 2011; (b) Note added in proof: a new class of monodentate phosphites based on a combination of D-mannitol and BINOL or bisphenols affords excellent ee s in the hydrogenation of *α*-and *β*-*α*-acyl-dehydroamino esters, *N*-acyl enamides and itaconates, see: H. Huang, Z. Zheng, H. Luo, C. Bai, X. Hu, H. Chen, *J. Org. Chem.* **2004**, *69*, 2355.
59 M.T. Reetz, J.-A. Ma, R. Goddard, *Angew. Chem. Int. Ed.* **2005**, *44*, 412.
60 (a) K. Nozaki, N. Sakai, T. Nanno, T. Higashijima, S. Mano, T. Horiuchi, H. Takaya, *J. Am. Chem. Soc.* **1997**, *119*, 4413; (b) A. Duursma, J.-G. Boiteau, L. Lefort, J.A.F. Boogers, A.H.M. de Vries, J.G. de Vries, A.J. Minnaard, B.L. Feringa, *J. Org. Chem.* **2004**, *69*, 8045.
61 For alternative routes, see: (a) Ref. 17; (b) A. van Rooy, D. Burgers, P.C.J. Kamer, P.W.N.M. van Leeuwen, *Recl. Trav. Chim. Pays-Bas* **1996**, *115*, 492.
62 R. Hulst, N.K. de Vries, B.L. Feringa, *Tetrahedron: Asymmetry* **1994**, *5*, 699.
63 M. Van den Berg, R.M. Haak, A.J. Minnaard, A.H.M. de Vries, J.G. de Vries, B.L. Feringa, *Adv. Synth. Catal.* **2002**, *344*, 1003.
64 (a) X. Jia, X. Li, L. Xu, Q. Chi, X. Yao, A.S.C. Chan, *J. Org. Chem.* **2003**, *68*, 4539; (b) X. Jia, R. Guo, X. Li, X. Yao, A.S.C. Chan, *Tetrahedron Lett.* **2002**, *43*, 5541.
65 H. Bernsmann, M. van den Berg, R. Hoen, A.J. Minnaard, G. Mehler, M.T. Reetz, J.G. de Vries, B.L. Feringa, *J. Org. Chem.* **2005**, *70*, 943.
66 (a) Q. Zeng, H. Liu, X. Cui, A. Mi, Y. Jiang, X. Li, M.C.K. Choi, A.S.C. Chan, *Tetrahedron: Asymmetry* **2002**, *13*, 115; (b) Q. Zeng, H. Liu, A. Mi, Y. Jiang, X. Li, M.C.K. Choi, A.S.C. Chan, *Tetrahedron* **2002**, *58*, 8799; (c) X. Li, X. Jia, G. Lu, T.T.-L. Au-Yeung, K.-H. Lam, T.W.H. Lo, A.S.C. Chan, *Tetrahedron: Asymmetry* **2003**, *14*, 2687.
67 L. Panella, B.L. Feringa, J.G. de Vries, A.J. Minnaard, *Org. Lett.* **2005**, *7*, 4177.
68 O. Huttenloch, J. Spieler, H. Waldmann, *Chem. Eur. J.* **2000**, *6*, 671.
69 M.T. Reetz, J.-A. Ma, R. Goddard, *Angew. Chem. Int. Ed.* **2005**, *44*, 412.
70 (a) A.-G. Hu, Y. Fu, J.-H. Xie, H. Zhou, L.-X. Wang, Q.-L. Zhou, *Angew. Chem. Int. Ed.* **2002**, *41*, 2348; (b) Y. Fu, J.-H. Xie, A.-G. Hu, H. Zhou, L.-X. Wang, Q.-L. Zhou, *Chem. Commun.* **2002**, 480; (c) S.-F. Zhu, Y. Fu, J.-H. Xie, B. Liu, L. Xing, Q.-L. Zhou, *Tetrahedron: Asymmetry* **2003**, *14*, 3219.
71 Y. Fu, X.-X. Guo, S.-F. Zhu, A.-G. Hu, J.-H. Xie, Q.-L. Zhou, *J. Org. Chem.* **2004**, *69*, 4648.
72 V.B. Birman, A.L. Rheingold, K.-C. Lam, *Tetrahedron: Asymmetry* **1999**, *10*, 125.
73 S. Wu, W. Zhang, Z. Zhang, X. Zhang, *Org. Lett.* **2004**, *6*, 3565.
74 A. Bayer, P. Murszat, U. Thewalt, B. Rieger, *Eur. J. Inorg. Chem.* **2002**, 2614.
75 R. Hoen, M. van den Berg, H. Bernsmann, A.J. Minnaard, J.G. de Vries, B.L. Feringa, *Org. Lett.* **2004**, *6*, 1433.
76 M. van den Berg, PhD. Thesis **2005**, University of Groningen, The Netherlands.
77 D. Peña, A.J. Minnaard, J.G. de Vries, B.L. Feringa, *J. Am. Chem. Soc.* **2002**, *124*, 14552.
78 S. Doherty, E.G. Robins, I. Pál, C.R. Newman, C. Hardacre, D. Rooney, D.A. Mooney, *Tetrahedron: Asymmetry* **2003**, *14*, 1517. Of course, once polymerized the ligands are no longer monodentate.
79 X. Wang, K. Ding, *J. Am. Chem. Soc.* **2004**, *126*, 10524.
80 C. Simons, U. Hanefeld, I.W.C.E. Arends, A.J. Minnaard, T. Maschmeyer, R.A. Sheldon, *Chem. Comm.* **2004**, 2830.
81 P.N.M. Botman, A. Amore, R. van Heerbeek, J.W. Back, H. Hiemstra, J.N.H. Reek, J.H. van Maarseveen *Tetrahedron Lett.* **2004**, *45*, 5999.

82 M. Yamashita, M. Kobayashi, M. Sugiura, K. Tsunekawa, T. Oshikawa, S. Inokawa, H. Yamamoto, *Bull. Chem. Soc. Jpn.* **1986**, *59*, 175.

83 Y. Chi, X. Zhang, *Tetrahedron Lett.* **2002**, *43*, 4849.

84 K. Junge, G. Oehme, A. Monsees, T. Riermeier, U. Dingerdissen, M. Beller, *J. Organomet. Chem.* **2003**, *675*, 91.

85 M.T. Reetz, H. Oka, R. Goddard, *Synthesis* **2003**, 1809.

86 X.-B., Jiang, M. van den Berg, A.J. Minnaard, B.L. Feringa, J.G. de Vries, *Tetrahedron: Asymmetry* **2004**, *15*, 2223.

87 The methyl-substituted substrate has been hydrogenated with SIPHOS in excellent ee; see: Ref. [68b].

88 MonoPhos **29a** has shown to give the same ee-values for *N*-acyl, *N*-BOC and *N*-Z protection (unpublished results).

89 M. van den Berg, E.P. Schudde, A.J. Minnaard, B.L. Feringa, J. G. de Vries (unpublished results).

90 (a) E. Farrington, M.C. Franchini, J.M. Brown, *Chem. Comm.* **1998**, 277; (b) J.M. Brown, *Angew. Chem. Int. Ed. Engl.* **1987**, *26*, 191.

91 R. Hoen, J.A.F. Boogers, H. Bernsmann, A.J. Minnaard, A. Meetsma, T.D. Tiemersma-Wegman, A.H.M. de Vries, J.G. de Vries, B.L. Feringa, *Angew. Chem. Int. Ed.* **2005**, *44*, 4209.

92 N.W. Boaz, *Tetrahedron Lett.* **1998**, *39*, 5505.

93 See, however J. You, H.-J. Drexler, S. Zhang, C. Fischer, D. Heller, *Angew. Chem. Int. Ed.* **2003**, *42*, 913 for an in-depth study of the synthesis of these stereoisomers.

94 (a) J. Solodar, *Chemtech* **1975**, 421. For the enantioselective hydrogenation of cyclopenta-1,3,4,-trione using Rh/CAMP, see (b) C.J. Sih, J.B. Heather, G.P. Peruzzotti, P. Price, R. Sood, L.-F.H. Lee, *J. Am. Chem. Soc.* **1973**, *95*, 1676.

95 P. Bonvicini, A. Levi, G. Modena, G. Scorrano, *J. Chem. Soc., Chem. Comm.* **1972**, 1188.

29
P,N and Non-Phosphorus Ligands

Andreas Pfaltz and Sharon Bell

29.1
Introduction

The first homogeneous enantioselective hydrogenation catalysts were developed during the 1960s [1, 2]. Since then the range and scope of chiral hydrogenation catalysts has expanded to the point where many functionalized substrates can be hydrogenated in good enantiomeric excess. This breakthrough is largely due to the C_2-symmetric diphosphine catalysts such as DIOP [3], the first such diphosphine, and DIPAMP [4], used in the first industrial-scale enantioselective homogeneous hydrogenation to produce L-DOPA. Since that time, numerous other phosphorus ligands have been introduced, which have considerably expanded the scope of enantioselective hydrogenation.

More recently, there has been increasing interest in the synthesis of hydrogenation catalysts with ligands other than the diphosphines. The initial success of titanocene catalysts in the hydrogenation of unfunctionalized alkenes (see Section 29.5) led to a great deal of research into metallocene catalysts. Iridium complexes with ligands bearing a coordinating phosphorus (P) and nitrogen (N) atoms have shown considerable success in recent years.

P,N and non-phosphorus ligands have been most successful in the enantiomeric iridium-catalyzed hydrogenation of unfunctionalized alkenes [5], and for this reason this chapter necessarily overlaps with Chapter 30. Here, the emphasis is on ligand synthesis and structure, whereas Chapter 30 expands on substrates, reaction conditions and reaction optimization. However, a number of specific substrates are mentioned in the comparison of catalysts, and their structures are illustrated in Figure 29.1.

The Handbook of Homogeneous Hydrogenation.
Edited by J. G. de Vries and C. J. Elsevier
Copyright © 2007 WILEY-VCH Verlag GmbH & Co. KGaA, Weinheim
ISBN: 978-3-527-31161-3

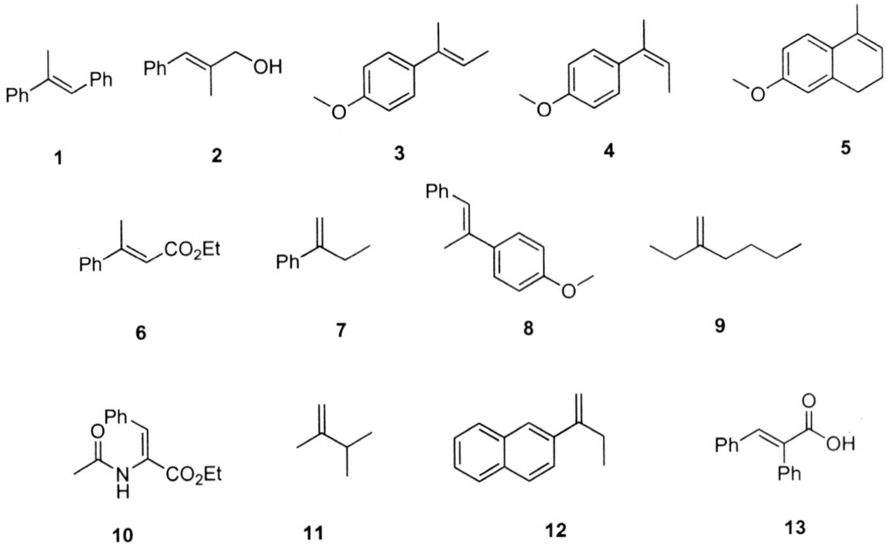

Fig. 29.1 Substrates discussed in this chapter.

29.2
Oxazoline-Derived P,N Ligands

The Crabtree catalyst ($[Ir(PCy_3)(py)(COD)]PF_6$) [6] shows remarkable activity in the hydrogenation of alkenes, particularly sterically hindered tri- and even tetra-substituted alkenes. Its structure has inspired a great deal of research into chiral P,N ligands for enantioselective hydrogenation, producing a variety of useful catalysts. The largest and most successful group of chiral analogues of the Crabtree catalyst are iridium complexes with oxazoline-derived P,N ligands.

The oxazoline-derived P,N ligands can be classified into four groups according to structure: phosphino-oxazolines; phosphite- and phosphinite-oxazolines; catalysts containing a P–N bond; and structurally related non-oxazoline catalysts.

29.2.1
Phosphino-oxazolines

The most extensively studied of these systems are the phosphino-oxazoline (PHOX) catalysts **14** (Fig. 29.2). Good enantioselectivity has been achieved with these catalysts over a broad range of substrates [7].

The highest enantioselectivity in the hydrogenation of unfunctionalized tri-substituted alkenes has been achieved with catalyst **14a**. The same catalyst was also used to hydrogenate α,β-unsaturated phosphonates with enantiomeric excesses (ee) of 70 to 94% [8].

The PHOX ligands have a modular structure (Scheme 29.1). They are synthesized from chiral amino alcohols and benzonitrile or bromobenzonitrile: the

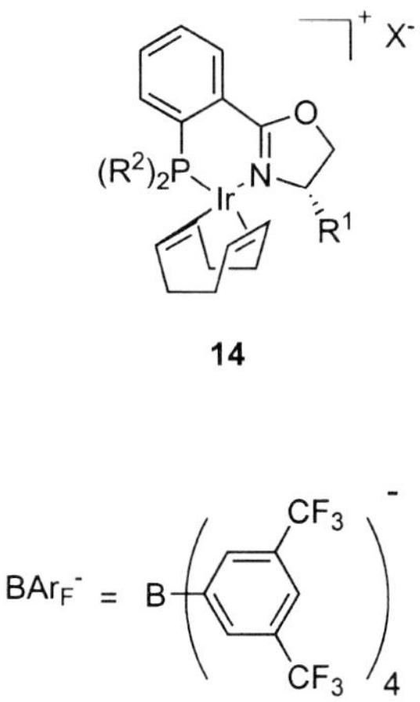

	R^1	R^2	X
a)	*t*Bu	*o*-tol	BAr_F^-
b)	*t*Bu	*o*-tol	PF_6^-
c)	CH_2*t*Bu	Ph	BAr_F^-
d)	*t*Bu	Cy	BAr_F^-
e)	*i*Pr	Ph	PF_6^-
f)	*i*Pr	Ph	BAr_F^-
g)	*t*Bu	Ph	PF_6^-
h)	*i*Pr	*o*-tol	PF_6^-

Fig. 29.2 PHOX catalyst **14**.

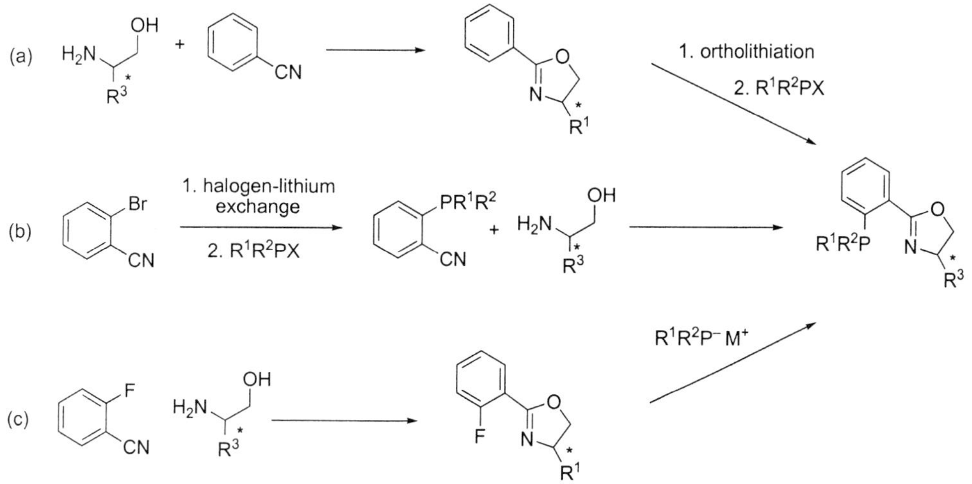

Scheme 29.1 Synthesis routes to the PHOX ligand.

phosphine is introduced *via* ortholithiation [9, 10] or lithium–halogen exchange as the first step [10–12] in ligand synthesis. Alternatively, the phosphine moiety can be introduced by nucleophilic substitution using an *ortho*-fluorophenyloxazoline as precursor (Scheme 29.1, route c) [13].

The stability of the PHOX catalysts containing the BAr_F counterion was found to be higher than that of those containing a PF_6^- counterion. The catalysts were more stable to air and moisture and, during hydrogenation, were deactivated more slowly than the corresponding PF_6^- catalysts, giving full conversion at >0.02 mol% catalyst loading [10].

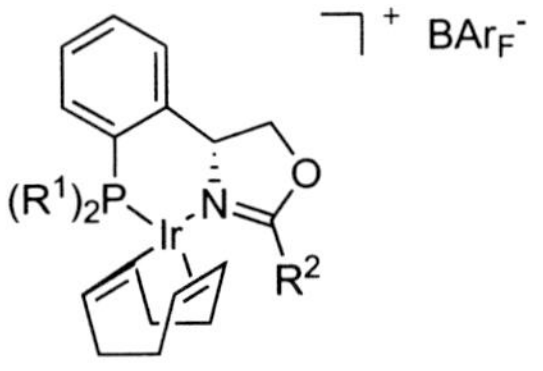

	R^1	R^2
a)	Ph	*t*Bu
b)	Ph	$CHPh_2$
c)	Ph	3,5-*t*$Bu_2C_6H_3$
d)	Ph	adamantyl
e)	Cy	*t*Bu

Fig. 29.3 PHOX catalysts **15**.

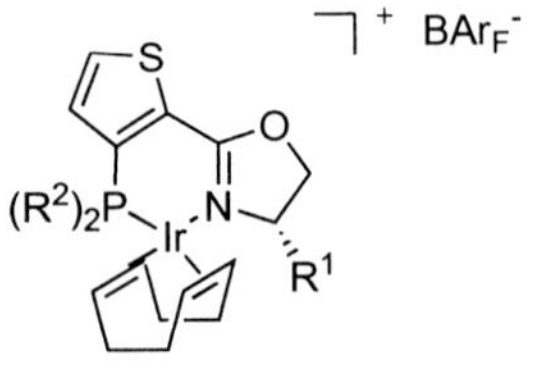

26a-d

	R^1	R^2
a)	*i*Pr	Ph
b)	*t*Bu	Ph
c)	*t*Bu	*o*-tol
d)	*t*Bu	Cy

Fig. 29.4 HetPHOX catalyst **6**.

A series of analogues of the original PHOX catalyst, in which the phenyl bridge is attached to C(4) instead of C(2) of the oxazoline ring (**15**, Fig. 29.3), were recently synthesized [14]. These catalysts were used to hydrogenate a number of substrates, including a range of 1-phenylbutenoic acids, with >90% ee.

Another PHOX analogue has the aryl ring of the PHOX catalyst replaced by a thiophene unit **16** (Fig. 29.4) [15]. The synthesis is similar to that of the PHOX catalysts, starting with *ortho*-metallation of the thiophene. The catalysts showed similar selectivity to PHOX, and were used to hydrogenate substrates **1** and **2** with maximum enantioselectivities of 99% and 94%, respectively.

JM-Phos, a PHOX analogue with an alkyl backbone (**17**, Fig. 29.5) has been used to hydrogenate a number of substrates with moderate to good enantioselectivity [16].

Another alkyl-bridged PHOX (**18**, Fig. 29.6) was recently synthesized [17], and used to hydrogenate a series of substituted methylstilbenes in 75–95% ee, and *β*-methylcinnamic esters in 80–99% ee. The hydrogenation results suggest that the selectivity of these catalysts is mainly derived from the substitution at the stereogenic center on the oxazoline ring, with the other stereocenter having a relatively minor effect on the ee-value.

The ligand synthesis is straightforward, using amino alcohols as the source of chirality in the oxazoline ring, whereas the stereochemistry in the phospholane ring is controlled by an enantioselective deprotonation using sparteine (Scheme 29.2).

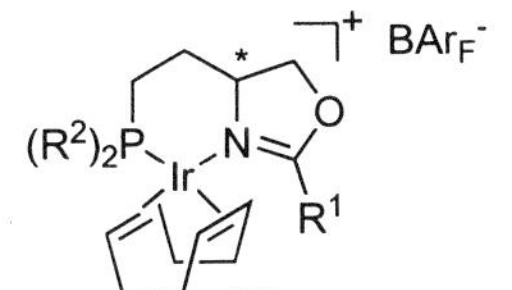

17

1-Ad = 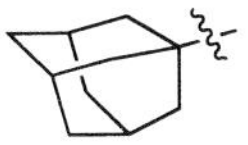

	R^1	R^2		R^1	R^2
a)	Me	Ph	f)	$CH(Ph)_2$	Ph
b)	*t*Bu	Ph	g)	$CH(Ph)_2$	*o*-tol
c)	*t*Bu	*o*-tol	h)	$C(Ph)_2Me$	*o*-tol
d)	1-Ad	Ph	i)	Ph	Ph
e)	$C(Ph)_3$	Ph	j)	3,5-$t$$Bu_2C_6H_3$	Ph

Fig. 29.5 JM-PHOS catalyst **17**.

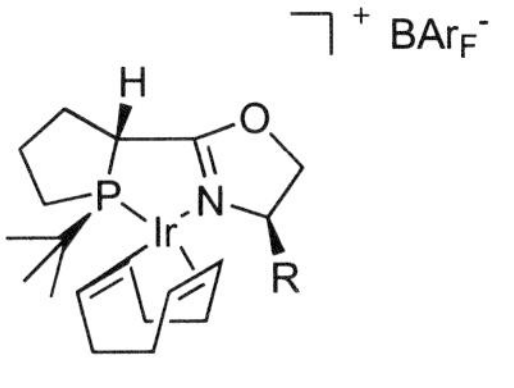

18

	R
a)	*i*Pr
b)	*t*Bu
c)	Ph
d)	CH_2Ph
e)	*i*Bu

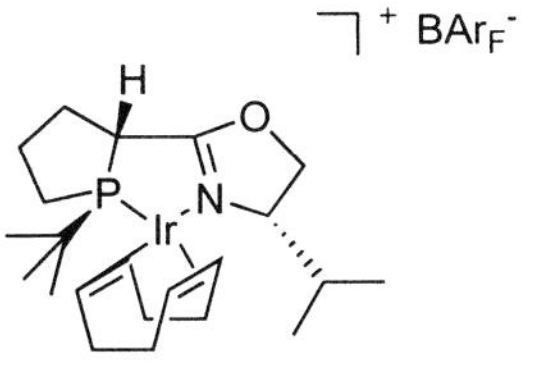

f)

Fig. 29.6 Phospholane-oxazoline catalyst **18**.

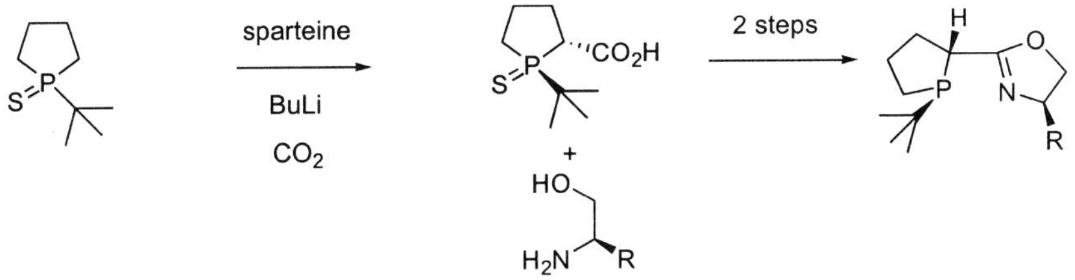
sparteine
BuLi
CO_2
CO_2H
+
HO
H_2N
R
2 steps
H
O
P
N
R

Scheme 29.2 Synthesis of ligands for catalyst **18**.

The ferrocene-oxazoline catalyst **19** (Fig. 29.7) has recently been used to hydrogenate substituted quinolines [18]. The ligand synthesis is again similar to that of the original PHOX ligand, with introduction of phosphorus *via* orthometallation.

29.2.2
Phosphite and Phosphinite Oxazolines

A PHOX analogue containing a P–O bond, the TADDOL-derived phosphite oxazoline catalyst **20a** (Fig. 29.8) has been used in the hydrogenation of a number of substituted styrenes, as well as in asymmetric allylic alkylation [19]. However,

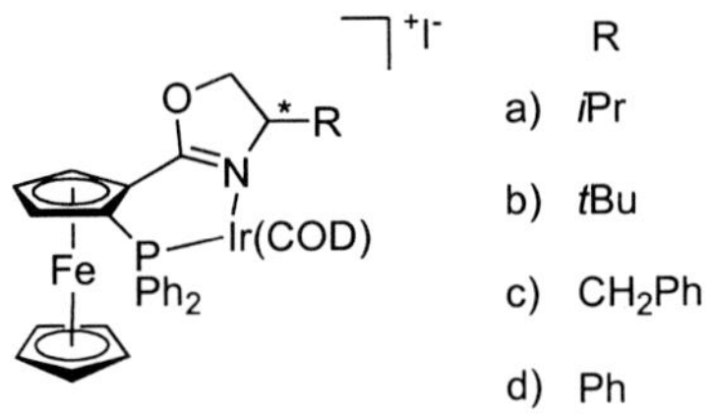

R
a) *i*Pr
b) *t*Bu
c) CH_2Ph
d) Ph

19

Fig. 29.7 Ferrocene-oxazoline catalyst **19**.

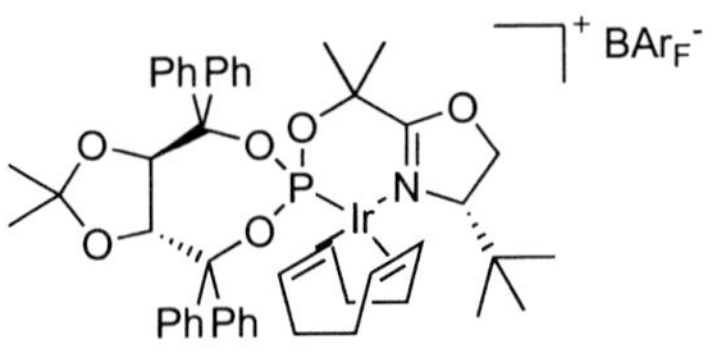

20

Fig. 29.8 Catalyst **20**.

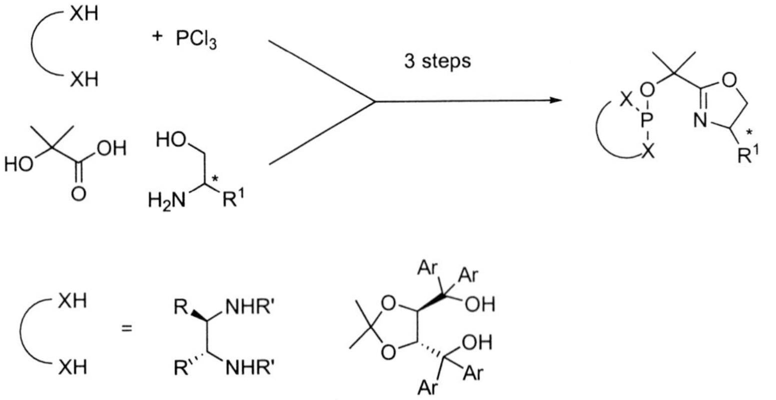

3 steps

21a, 21b

20b

Scheme 29.3 Synthesis of aminophosphine and phosphinite catalysts **20b**, **21a** and **21b**.

the enantioselectivities were only moderate and catalytic activity was low, requiring 4 mol% catalyst.

The synthesis strategy is shown in Scheme 29.3. The incorporation of a diol or diamine gives ligand **20b** or ligands **21a** and **21b**, respectively.

More recent developments are the SerPHOX catalyst **22** [20] and the ThrePHOX catalyst **23** [21] (Fig. 29.9), derived from serine or threonine, respectively (Scheme 29.4).

In many cases, these catalysts proved to be superior to the PHOX complexes **14**. Complexes **23a** and **23g** are among the most efficient catalysts (and are

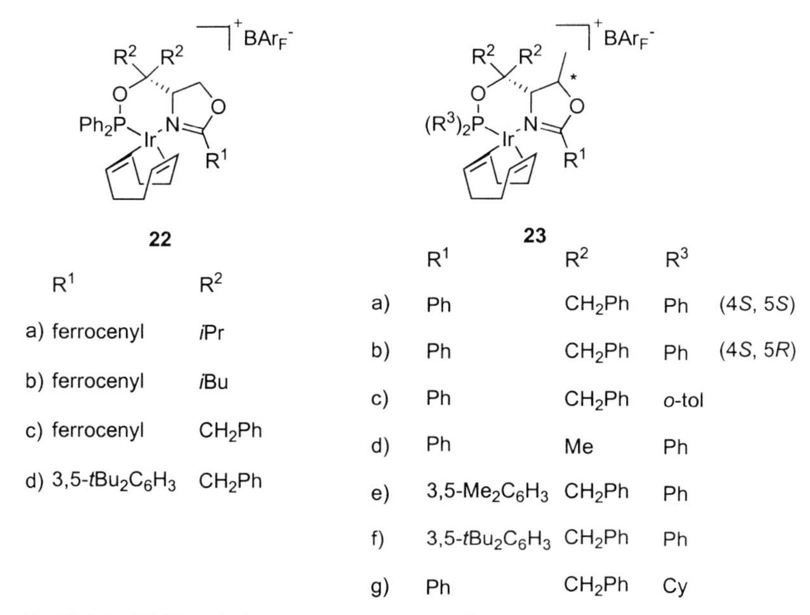

Fig. 29.9 SerPHOX and ThrePHOX catalysts **22** and **23**.

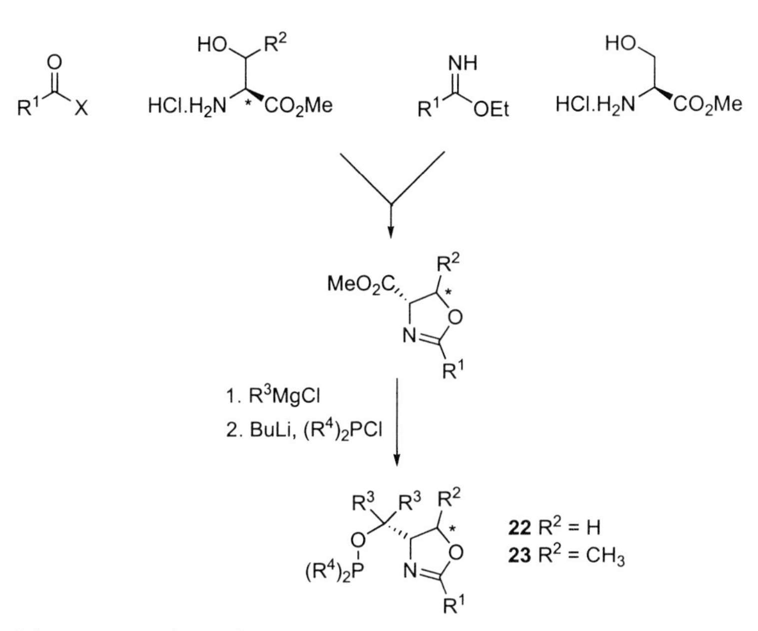

Scheme 29.4 Synthesis of SerPHOX (**22**) and ThrePHOX (**23**) catalysts.

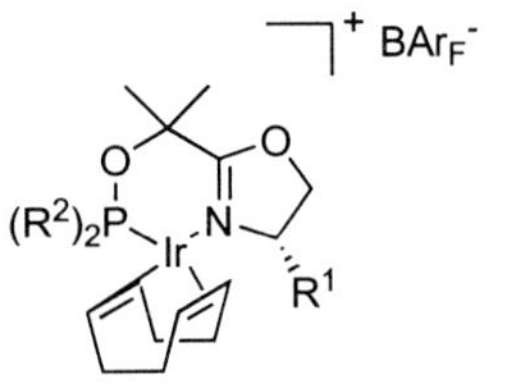

	R^1	R^2
a)	*t*Bu	Ph
b)	*t*Bu	*o*-tol
c)	*i*Pr	Ph
d)	*i*Pr	*o*-tol

Fig. 29.10 SimplePHOX catalyst **24**.

Scheme 29.5 Synthesis of SimplePHOX catalysts.

now commercially available [22]), giving high enantioselectivities and turnover numbers (TONs) for a range of trisubstituted and 1,1-disubstituted alkenes. The additional methyl group on the oxazoline ring of ThrePHOX ligands led to greater selectivity in the hydrogenation of substrates **1**, **3**, and **4** [21]. The synthesis is again modular (Scheme 29.4).

Another efficient, readily available catalyst is the SimplePHOX complex (**24**, Fig. 29.10) [23].

The ligand synthesis requires only two steps from simple starting materials. As with the PHOX type catalysts, chirality is built in through the use of a chiral amino alcohol (Scheme 29.5).

Complex **24a** proved to be a particularly efficient catalyst for the hydrogenation of the cyclic substrate **5**, affording 95% ee, which currently is the highest ee-value obtained with any catalyst for this substrate. High enantioselectivities were also obtained with α,β-unsaturated substrate **6**. Catalyst **24a** also gave higher selectivity than the ThrePHOX catalysts **23** in the hydrogenation of substrates **2** and **4**.

29.2.3 Oxazoline-Derived Ligands Containing a P–N Bond

The first oxazoline-derived catalysts containing P–N bonds to be synthesized were the diaminophosphines **21** (Fig. 29.11) (details of the synthesis are shown in Scheme 29.3). This class of catalysts gave good enantiomeric excesses in the hydrogenations of some unfunctionalized olefins, albeit with low conversions and high catalyst loading (4 mol%) [19, 24]. Analogous complexes with N-aryl-

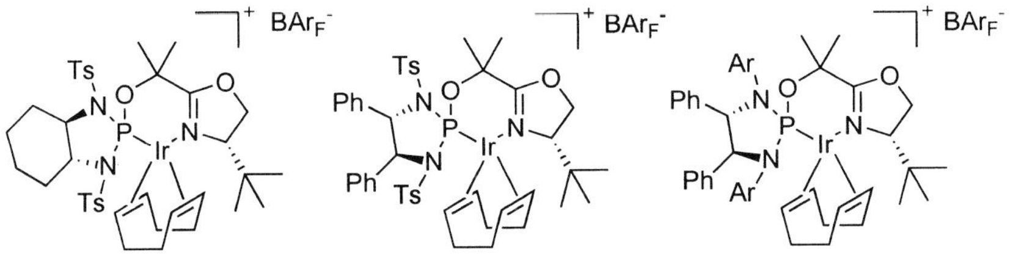

21a 21b 21c

Fig. 29.11 Diaminophosphine catalysts **21**.

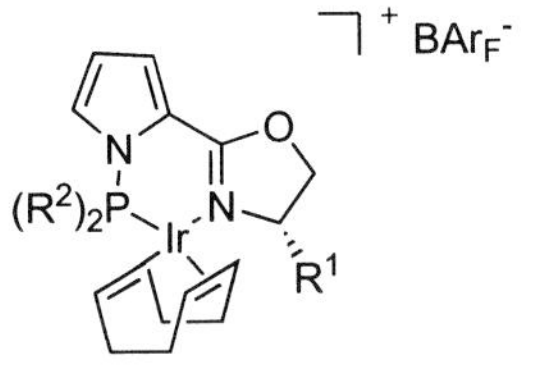

25

	R^1	R^2
a)	*i*Pr	Ph
b)	*t*Bu	Ph
c)	*t*Bu	*o*-tol
d)	*t*Bu	Cy

Fig. 29.12 PyrPHOX catalyst **25**.

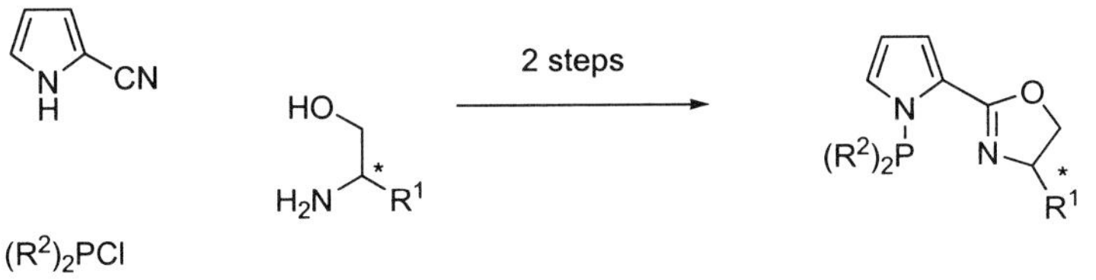

2 steps

$(R^2)_2PCl$

Scheme 29.6 Synthesis of PyrPHOX catalysts.

substituted ligands **21 c** demonstrate much higher reactivity, and have been shown to be particularly efficient catalysts for the hydrogenation of α,β-unsaturated carboxylic esters [5].

PyrPHOX catalysts **25** (Fig. 29.12) [25], with a pyrrole instead of a phenyl bridge, showed selectivity which was comparable to – and in some cases better than – that of the PHOX catalysts **14**.

The synthesis is similar to that of the PHOX catalysts (Scheme 29.6), starting with oxazoline formation from the pyrrole nitrile and an amino alcohol, followed by introduction of the phosphine group.

The pyrrolidine-oxazoline catalysts **26** (Fig. 29.13) can be conveniently synthesized from proline in five steps, and have been used to hydrogenate the methylstilbenes **1** and **8** in 92% ee and 94% ee, respectively [26].

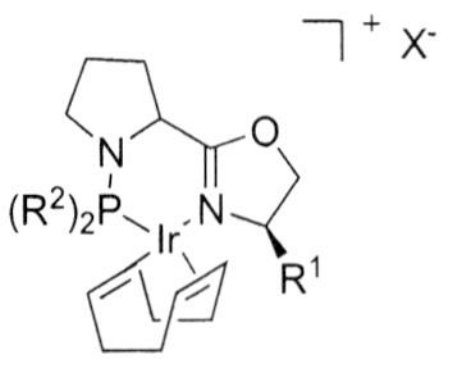

	R^1	R^2	X^-
a)	*i*Pr	Ph	PF_6^-
b)	*i*Pr	Ph	BAr_F^-
c)	*t*Bu	Ph	BAr_F^-
d)	*t*Bu	*o*-tol	PF_6^-
e)	*t*Bu	*o*-tol	BAr_F^-
f)	*t*Bu	2-Ethylphenyl	BAr_F^-
g)	*t*Bu	2,3-$Me_2C_6H_3$	BAr_F^-
h)	*t*Bu	2,4-$Me_2C_6H_3$	BAr_F^-

Fig. 29.13 Pyrrolidine-oxazoline catalyst **26**.

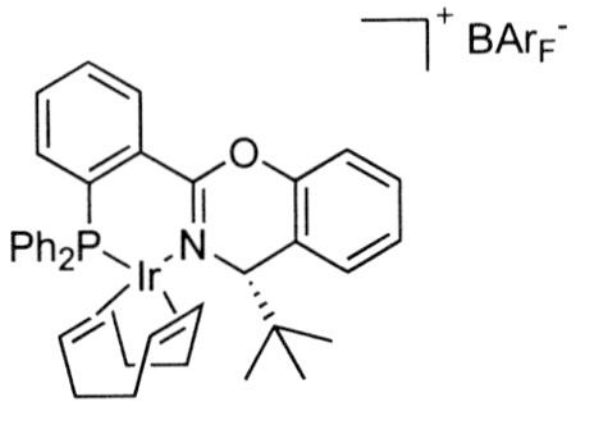

Fig. 29.14 Phosphino-benzoxazine catalyst **27**.

29.2.4
Structurally Related Ligands

A number of ligands have been synthesized which have structural similarities to the oxazoline-derived ligands.

The first of these was the phosphino-benzoxazine catalyst **27** (Fig. 29.14) [27]. This catalyst contains a six-membered benzoxazine ring in place of the five-membered oxazoline ring in the PHOX catalysts. It was hoped that this change would bring the chiral center on the ring into closer contact with the metal, resulting in higher enantioselectivities. However, enantiomeric excesses were only modest with the substrates chosen.

The phosphino-imidazoline catalyst **28** (Fig. 29.15) is a close analogue of the original PHOX ligand.

Phosphino-imidazoline ligands of this type were originally synthesized by Busacca and coworkers and used in an enantioselective Heck reaction [28].

The additional nitrogen atom in the ring gives another site for substitution (Scheme 29.7), and thus for steric and electronic tuning of the ligand structure. A range of catalysts were prepared and used in the hydrogenation of unfunctionalized alkenes. These outperformed the original PHOX catalyst in the hydrogenation of substrates **3**, **4**, and **5** [29]. They have also been successfully applied to the hydrogenation of terpenoid dienes and trienes [30].

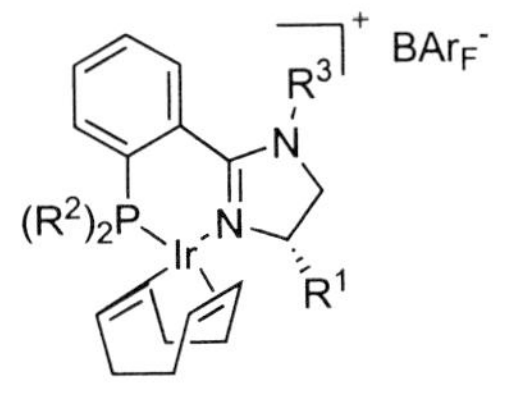

28

	R^1	R^2	R^3
a)	*i*Pr	Ph	*i*Pr
b)	*i*Pr	Ph	Cy
c)	*i*Pr	Ph	CH_2Ph
d)	*i*Pr	Ph	Ph
e)	*t*Bu	Ph	Ph
f)	*t*Bu	Ph	*p*-$MeOC_6H_4$
g)	*t*Bu	Ph	*p*-$CF_3C_6H_4$
h)	*t*Bu	Ph	*p*-tol
i)	*t*Bu	*o*-tol	Ph
j)	*t*Bu	*o*-tol	CH_2Ph

Fig. 29.15 Phosphino-imidazoline catalysts **28**.

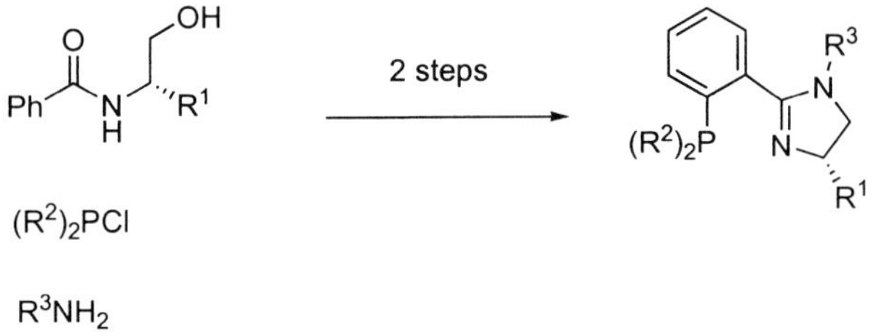

2 steps

$(R^2)_2PCl$

R^3NH_2

Scheme 29.7 Synthesis of phosphino-imidazoline ligands.

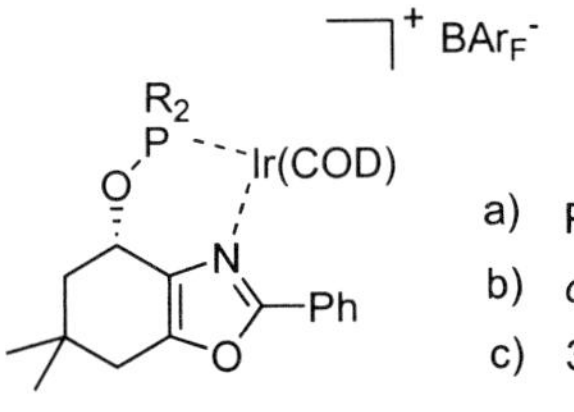

29

	R
a)	Ph
b)	*o*-tol
c)	3,5-$Me_2C_6H_3$

Fig. 29.16 Phosphinite-oxazole catalyst **29**.

Finally, the phosphinite-oxazole catalyst **29** (Fig. 29.16) was recently reported and used to hydrogenate a series of functionalized and unfunctionalized alkenes [31]. It was anticipated that the planar oxazole unit and the fused ring system would improve the enantioselectivity compared to the PHOX catalyst by increasing rigidity in the six-membered chelating ring [32]. Indeed, these catalysts

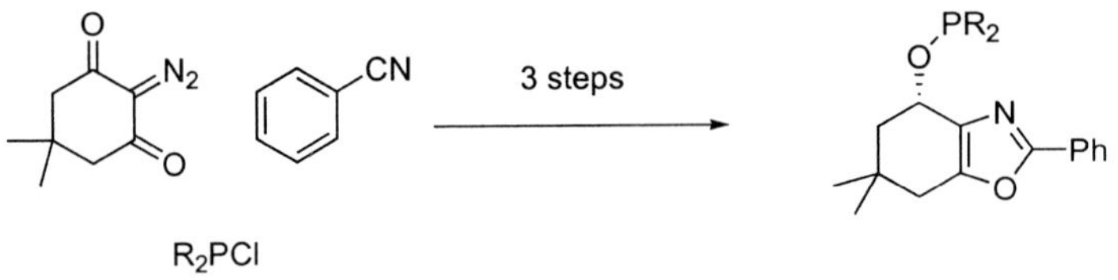

N_2 CN 3 steps PR_2 N Ph O

R_2PCl

Scheme 29.8 Synthesis of phosphinite-oxazole catalysts.

proved to be highly selective, rivaling the most selective oxazoline-based catalysts in the hydrogenation of a range of substrates.

The ligands are synthesized from achiral starting materials using a ketocarbene-nitrile cycloaddition as the key step (Scheme 29.8). The stereogenic center is introduced by enantioselective reduction of the carbonyl group in the cycloaddition product.

29.3 Pyridine and Quinoline-Derived P,N Ligands

Knochel and coworkers synthesized a series of camphor-derived pyridine and quinoline P,N ligands. The catalysts **30** (Fig. 29.17) were used to hydrogenate substrates **1** and **2** in up to 95% and 96% ee, respectively [33]. The selectivities were moderate for other unfunctionalized alkenes; however, a high enantioselectivity was reported for the hydrogenation of ethyl acetamidocinnamate **10** [34].

The ligands derive their chirality from the ring bearing the phosphorus atom, synthesized from (+)-camphor (R=H) or (+)-nopinone (R=CH_3) (Scheme 29.9).

Recently, a series of pyridine- and quinoline-derived catalysts **31** (Fig. 29.18) have been developed. Ligands **31a–k** were obtained by reduction of pyridyl ke-

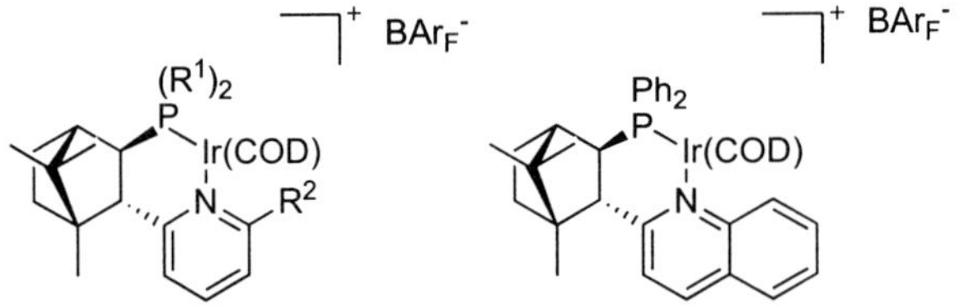

BAr_F^- $(R^1)_2$ P Ir(COD) N R^2 **30a-c**

BAr_F^- Ph_2 P Ir(COD) N **30d**

	R^1	R^2
a)	Ph	H
b)	Ph	Ph
c)	Cy	H

Fig. 29.17 Catalyst **30**.

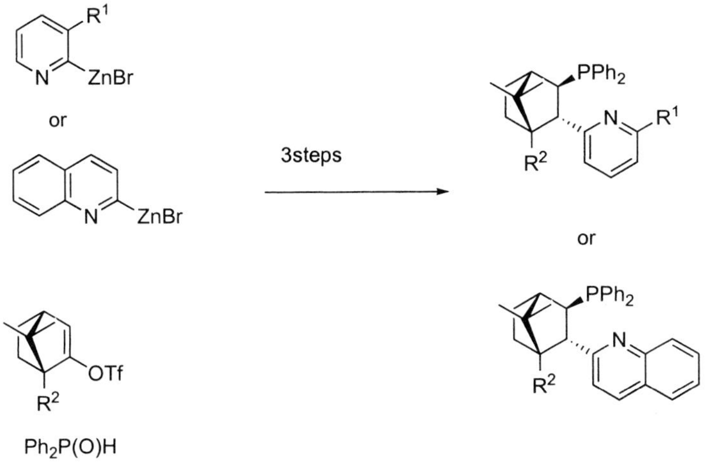

Scheme 29.9 Synthesis of pyridine-derived ligands for catalysts **30**.

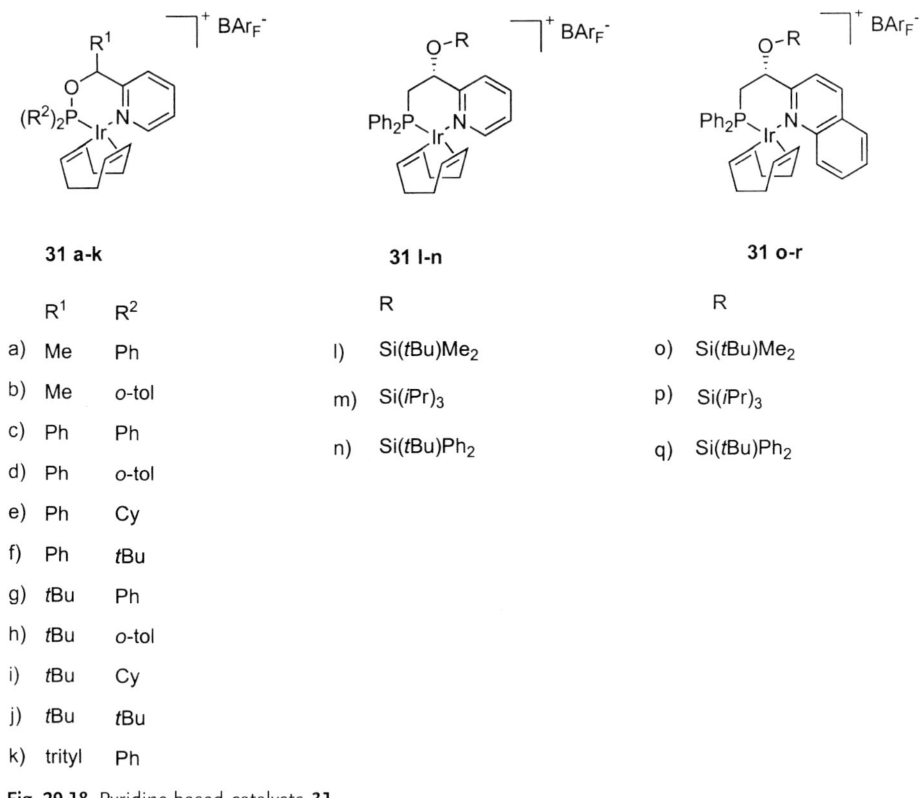

Fig. 29.18 Pyridine-based catalysts **31**.

tones and resolution of the resulting alcohols. Ligands **31 l–n** were synthesized *via* enantioselective reduction of pyridyl ketones and protection as silyl ethers, and ligands **31 o–q** were accessed through Sharpless dihydroxylation of a quinolyl alkene, followed by tosylation and introduction of the phosphine group.

In the hydrogenation of **1**, catalysts **31 a–j** gave 87–97% ee, whereas catalyst **31 k**, with the very bulky trityl substituent on the alkyl backbone, only gave 38% ee. The silyl substituent was found to have a significant effect on selectivity: catalyst **31 l** gave 88% ee, while catalyst **31 n** gave only 4% ee with substrate **1** [35]. This remarkable effect was rationalized based on X-ray structural data.

29.4 Carbenoid Imidazolylidene Ligands

The first application of a heterocyclic carbenoid achiral ligand for hydrogenation of alkenes was reported in 2001 by Nolan and coworkers. Both ruthenium [36] and iridium [37] complexes proved to be active catalysts. Turnover frequency (TOF) values of up to 24 000 h^{-1} (at 373 K) were measured for a ruthenium catalyst in the hydrogenation of 1-hexene.

Another series of achiral iridium catalysts containing phosphine and heterocyclic carbenes have also been tested in the hydrogenation of unfunctionalized alkenes [38]. These showed similar activity to the Crabtree catalyst, with one analogue giving improved conversion in the hydrogenation of **11**.

Subsequently, Burgess and coworkers developed the chiral carbenoid ligands **32** (Scheme 29.10), which were structurally related to the JM-PHOS ligand [39, 40].

The most selective catalyst in this series, complex **32 c**, with very bulky substituents at the oxazoline and the imidazolylidene moiety, was used to hydrogenate a range of trisubstituted alkenes with enantiomeric excesses of greater than

32

	R^1	R^2
a)	Ph	2,6-(iPr)$_2$C$_6$H$_3$
b)	tBu	2,6-(iPr)$_2$C$_6$H$_3$
c)	1-Ad	2,6-(iPr)$_2$C$_6$H$_3$
d)	CHPh$_2$	2,6-(iPr)$_2$C$_6$H$_3$
e)	1-Ad	2,5-(Et)$_2$C$_6$H$_3$

Scheme 29.10 Carbenoid imidazolylidene ligands based on the JM-PHOS ligand (**32**).

	R^1	R^2	R^3
a)	Me	Me	Me
b)	H	*t*Bu	Me
c)	Me	Me	mesityl

	R	X
a)	Ph	BAr_F^-
b)	mesityl	BAr_F^-
c)	2,6-*i*Pr_2Ph	BAr_F^-
d)	mesityl	PF_6^-

Fig. 29.19 Paracyclophane catalysts.

90%. It was found that the R^1 substituent on the carbenoid ring had a greater effect on selectivity than the R^2 substituent on the oxazoline. All the other catalysts **32a**, **32b**, **32d** and **32e** gave distinctly lower enantioselectivities.

Recently, carbene-oxazoline catalysts **33** and carbene-phosphine catalysts **34** (Fig. 29.19) with a chiral paracyclophane backbone have been synthesized and used to hydrogenate a variety of alkenes, with modest selectivity [41].

29.5 Metallocenes

The metallocenes are described in greater detail in Chapter 30.

Catalyst **35b** was used by Buchwald and coworkers with $[PhMe_2NH]^+$-$[(BC_6F_5)_4]^-$ to hydrogenate tetrasubstituted unfunctionalized cyclic olefins with up to 97% ee [42].

The enantiopure catalyst **35b** was synthesized from **35c** [43] *via* the binaphtholate **35a** [44, 45].

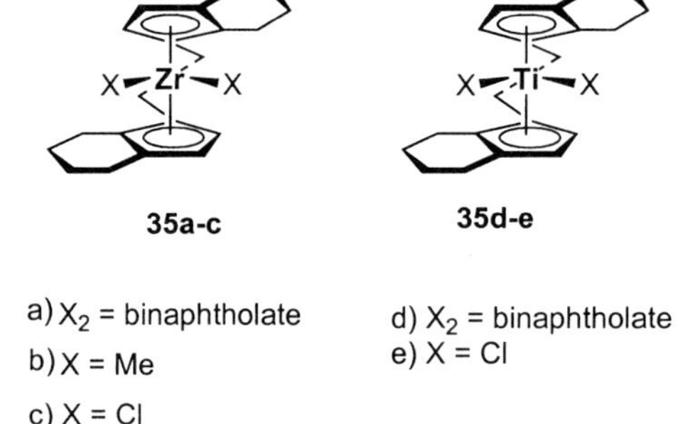

Fig. 29.20 EBTHI catalysts **35**.

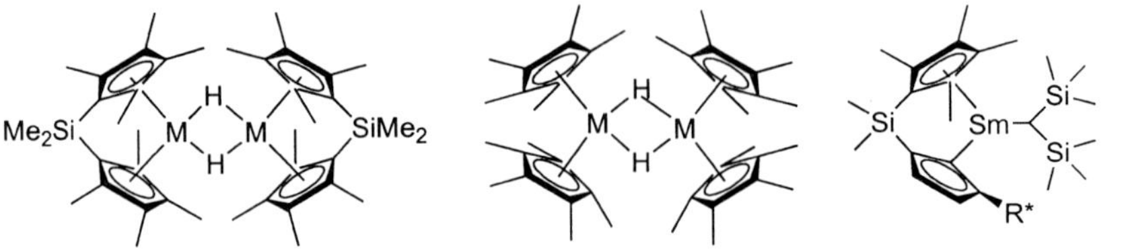

Fig. 29.21 Cyclopentadienyl lanthanide catalysts **36**.

The most selective – and also most general – titanocene catalyst is complex **35d**, also studied by Buchwald and coworkers. This catalyst was used to hydrogenate a variety of functionalized and unfunctionalized cyclic and acyclic alkenes with excellent ee-values in most cases [46]. Enamines could also be hydrogenated with enantiomeric excesses of 80–90% [47]. However, high catalyst loadings (5–8 mol%) and long reaction times were required to drive the reactions to completion.

Marks and coworkers developed a series of cyclopentadienyl-lanthanide complexes. In the initial investigations on achiral catalysts **36a** and **36b** (Fig. 29.21), TOFs greater than 100 000 h^{-1} were observed in the hydrogenation of 1,2-disubstituted unfunctionalized alkenes [48].

Chiral analogue **36c** was then synthesized [49]. Catalyst **36c** (R=menthyl) hydrogenated 2-phenyl-1-butene **7** with 96% ee at 195 K, with enantioselectivity being found to be highly temperature-dependent.

29.6 Other Ligands

The cobalt complex **37** was used in combination with quinine as a chiral coordinating base to hydrogenate 1,2-diphenyl-2-propene-1-one in 49% ee (Fig. 29.22) [50]. However, no further studies of this type of catalyst were reported.

Corma and coworkers tested a number of rhodium and other transition metal complexes with ligands based on proline (Fig. 29.23). These authors reported ee-values of 54–90% for the hydrogenation of dehydroamino acid derivatives with a catalyst prepared from ligand **38** [51]. With ligand **39**, an ee-value of 34% was recorded for the hydrogenation of ethyl acetamidocinnamate **10** [52].

Brauer and coworkers also prepared a rhodium catalyst, using an amino-phospholane ligand, catalyst **40** (Fig. 29.24). This was used to hydrogenate ethyl acetamidocinnamate **10** in 60% ee [53].

Catalyst **41** (Fig. 29.29) is based on a sulfinyl imine, with the stereogenic center at the sulfinamide group. The best catalyst in this series, **41a**, gave 94% ee in the hydrogenation of substrate **1** [54]. The catalyst has low activity, with 5 mol% catalyst being required.

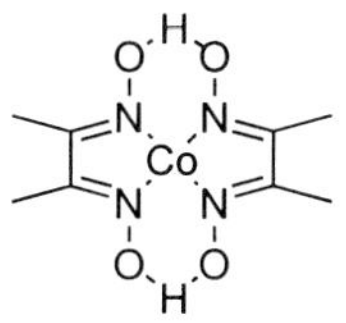

Fig. 29.22 Cobalt catalyst **37**.

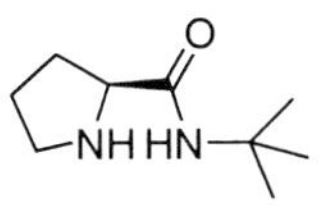

Fig. 29.23 Ligands **38** and 39.

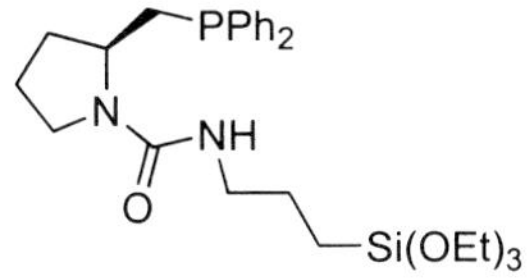

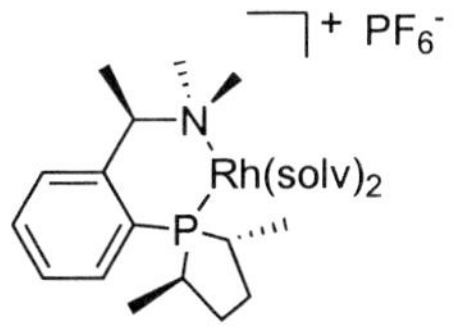

Fig. 29.24 Catalyst **40**.

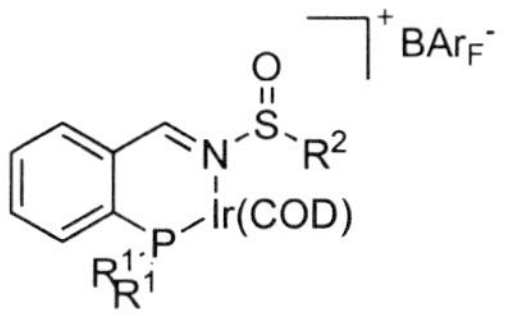

	R^1	R^2
a)	*o*-tol	*t*Bu
b)	*o*-tol	1-Ad
c)	*o*-tol	3-ethylpentyl
d)	*o*-tol	*p*-tol
e)	*o*-tol	mesityl
f)	3,5-$Me_2C_6H_3$	*t*Bu
g)	Ph	*t*Bu

Fig. 29.25 Sulfinyl imine catalyst **41**.

Finally, another group of rhodium catalysts based on ligands **42** to **45** (Fig. 29.26) has recently been synthesized and tested on the enantiomeric hydrogenation of α-phenylcinnamic acid **13**, giving 59 to 84% ee [55, 56]. The same complexes, when immobilized on mesoporous silica, were found to be even more enantioselective. The results obtained with rhodium complexes based on ligands such as **38**, **43**, **44**, or **45** are in contrast to the general experience that catalysts containing only nitrogen ligands are catalytically inactive and show a tendency to decompose under hydrogen pressure to give metallic rhodium.

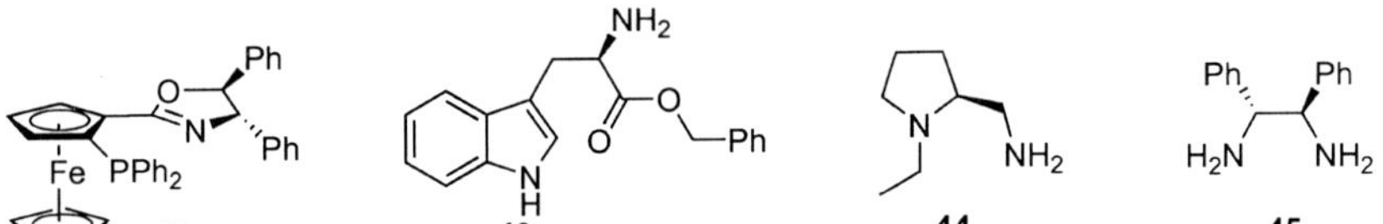

42-45 $[Rh(COD)L]^+CF_3SO_3^-$

Fig. 29.26 Catalysts **42** to **45**.

Further studies will be necessary to confirm these results and to establish the potential of catalysts of this type.

29.7
Conclusions

The development of chiral P,N and non-phosphorus ligands for the asymmetric hydrogenation of C=C bonds remains a young and rapidly developing field of research. The application range of these ligands is largely complementary to chiral phosphines. While chiral phosphines form efficient catalysts in combination with Rh and Ru, the primary domain of P,N ligands is Ir-catalyzed hydrogenation (see Chapter 30). Iridium complexes with chiral P,N ligands have considerably expanded the application range of asymmetric hydrogenation to unfunctionalized alkenes as well as various functionalized olefins, which give poor results with Rh or Ru catalysts. The most efficient P,N ligands developed to date have been oxazoline derivatives such as ThrePHOX 23 or SimplePHOX 24, and the bicyclic oxazole derivative **29**. Promising results have also been obtained with the carbenoid P,C ligand **32**. Metallocene catalysts have also been successfully applied for asymmetric hydrogenation of unfunctionalized alkenes, though as yet this class of catalysts has found only limited use.

Abbreviations

ee	enantiomeric excess
PHOX	phosphino-oxazoline
TOF	turnover frequency
TON	turnover number

References

1 L. Horner, H. Siegel, H. Büthe, *Angew. Chem.* **1968**, 1034.

2 W.S. Knowles, M.J. Sabacky, *J. Chem. Soc. Chem. Commun.* **1968**, 1445.

3 T.P. Dang, H.B. Kagan, *J. Chem. Soc. Chem. Commun.* **1971**, 481.

4 B.D. Vineyard, W.S. Knowles, M.J. Sabacky, G.L. Bachman, D.J. Weinkauff, *J. Am. Chem. Soc.* **1977**, *99*, 5946.

5 A. Pfaltz, J. Blankenstein, R. Hilgraf, E. Hörmann, S. McIntyre, F. Menges, M. Schönleber, S.P. Smidt, B. Wüstenberg, N. Zimmermann, *Adv. Synth. Catal.* **2003**, *345*, 33.

6 R. Crabtree, *Acc. Chem. Res.* **1979**, *12*, 331.

7 A. Lightfoot, P. Schnider, A. Pfaltz, *Angew. Chem. Int. Ed.* **1998**, *37*, 2897.

8 N.S. Goulioukina, M. Dolgina, G.N. Bondarenko, I.P. Beletskaya, M.M. Ilyin, V.A. Davankov, A. Pfaltz, *Tetrahedron Asymm.* **2003**, *14*, 1397.

9 D.G. Blackmond, A. Lightfoot, A. Pfaltz, T. Rosner, P. Schnider, N. Zimmermann, *Chirality* **2000**, *12*, 442.

10 G. Koch, G.C. Lloyd-Jones, O. Loiseleur, A. Pfaltz, R. Pretot, S. Schaffner, P. Schnider, P. von Matt, *Rec. Trav. Chim. Pays-Bas* **1995**, *114*, 206.

11 P. Von Matt, A. Pfaltz, *Angew. Chem. Int. Ed. Engl.* **1993**, *32*, 556.

12 J.V. Allen, G.J. Dawson, C.G. Frost, J.M.J. Williams, *Tetrahedron* **1994**, *50*, 799.

13 M. Peer, J.C. de Jong, M. Kiefer, T. Langer, H. Rieck, H. Schell, P. Sennhenn, J. Sprinz, H. Steinhagen, B. Wiese, G. Helmchen, *Tetrahedron* **1996**, *52*, 7547.

14 D. Liu, W. Tang, X. Zhang, *Org. Lett.* **2004**, *6*, 513.

15 P.G. Cozzi, F. Menges, S. Kaiser, *Synlett* **2003**, *6*, 829.

16 D.R. Hou, J.H. Reibenspies, T.J. Colacot, K. Burgess, *Chem. Eur. J.* **2001**, *7*, 5391.

17 W. Tang, W. Wang, X. Zhang, *Angew. Chem. Int. Ed.* **2003**, *42*, 943.

18 S. Lu, X. Han, Y. Zhou, *Adv. Synth. Catal.* **2004**, *346*, 909.

19 R. Hilgraf, A. Pfaltz, *Synlett* **1999**, *11*, 1814.

20 J. Blankenstein, A. Pfaltz, *Angew. Chem. Int. Ed.* **2001**, *40*, 4445.

21 F. Menges, A. Pfaltz, *Adv. Synth. Catal.* **2002**, *344*, 40.

22 Available from STREM: complex **23a** catalog number 77-5020, complex **23g** catalog number 77-5010.

23 S.P. Smidt, F. Menges, A. Pfaltz, *Org. Lett.* **2004**, *6*, 2023.

24 R. Hilgraf, A. Pfaltz, *Adv. Synth. Catal.* **2005**, *347*, 61.

25 P.G. Cozzi, N. Zimmermann, R. Hilgraf, S. Schaffner, A. Pfaltz, *Adv. Synth. Catal.* **2001**, *343*, 450.

26 G. Xu, S.R. Gilbertson, *Tetrahedron Lett.* **2003**, *44*, 953.

27 G.H. Bernardinelli, E.P. Kündig, P. Meier, A. Pfaltz, K. Radkowski, N. Zimmermann, M. Neuburger-Zehnder, *Helv. Chim. Acta* **2001**, *84*, 3233.

28 C. Busacca, WO2001018012, **2001**.

29 F. Menges, M. Neuburger, A. Pfaltz, *Org. Lett.* **2002**, *4*, 4173.

30 Patent application: B. Wüstenberg, F. Menges, T. Netscher, A. Pfaltz, (DSM), 22. 12. 2004.

31 K. Källstrom, C. Hedberg, P. Brandt, A. Bayer, P.G. Andersson, *J. Am. Chem. Soc.* **2004**, *126*, 14308.

32 P. Brandt, C. Hedberg, P.G. Andersson, *Chem. Eur. J.* **2003**, *9*, 339.

33 T. Bunlaksananusorn, K. Polborn, P. Knochel, *Angew. Chem.* **2003**, *42*, 3941.

34 T. Bunlaksananusorn, P. Knochel, *J. Org. Chem.* **2004**, *69*, 4595.

35 W.J. Drury, N. Zimmermann, M. Keenan, M. Hayashi, S, Kaiser, R. Goddard, A. Pfaltz, *Angew. Chem. Int. Ed.* **2004**, *43*, 70.

36 H.M. Lee, D.C. Smith, Z. He, E.D. Stevens, C.S. Yi, S.P. Nolan, *Organometallics* **2001**, *20*, 794.

37 H.M. Lee, T. Jiang, E.D. Stevens, S.P. Nolan, *Organometallics* **2001**, *20*, 1255.

38 L.D. Vázquez-Serrano, B.T. Owens, J.M. Buriak, *J. Chem. Soc. Chem. Commun.* **2002**, 2518.

39 M.T. Powell, D.R. Hou, M.C. Perry, X. Cui, K. Burgess, *J. Am. Chem. Soc.* **2001**, *123*, 8878.

40 M.C. Perry, X. Cui, M.T. Powell, D.R. Hou, J.H. Riebenspies, K. Burgess, *J. Am. Chem. Soc.* **2003**, *125*, 113.

41 C. Bolm, T. Focken, G. Raabe, *Tetrahedron Asymm.* **2003**, *14*, 1733.

42 M.V. Troutman, D.H. Appella, S.L. Buchwald, *J. Am. Chem. Soc.* **1999**, *121*, 4916.

43 F.R.W.P. Wild, L. Zsolnai, G. Huttner, H.H. Brintzinger, *J. Organomet. Chem.* **1985**, *232*, 233.

44 R.B. Grossman, W.M. Davis, S.L. Buchwald, *J. Am. Chem. Soc.* **1991**, *113*, 2321.

45 R.M. Waymouth, F. Bangerter, P. Pino, *Inorg. Chem.* **1988**, *27*, 758.

46 R.D. Broene, S.L. Buchwald, *J. Am. Chem. Soc.* **1993**, *115*, 12569.

47 N.E. Lee, S.L. Buchwald, *J. Am. Chem. Soc.* **1994**, *116*, 5985.

48 G. Jeske, H. Lauke, H. Mauermann, H. Schumann, T.J. Marks, *J. Am. Chem. Soc.* **1985**, *107*, 8111.

49 V.P. Conticello, L. Brard, M.A. Giardello, Y. Tsuji, M. Sabat, C.L Stern, T.J. Marks, *J. Am. Chem. Soc.* **1992**, *114*, 2761.

50 Y. Ohgo, S. Takeuchi, Y. Natori, J. Yoshimura, *Bull. Chem. Soc. Jpn.* **1981**, *54*, 2124.

51 A. Corma, M. Iglesias, C. Del Pino, F. Sánchez, *J. Chem. Soc. Chem. Commun.* **1991**, *1252*, 1253.

52 A. Carmona, A. Corma, M. Iglesias, A. San José, F. Sánchez, *J. Organomet. Chem.* **1995**, *492*, 11.

53 D.J. Brauer, K.W. Kottsieper, S. Rossenbach, O. Stelzer, *Eur. J. Inorg. Chem.* **2003**, 1748.

54 L.B. Schenkel, J.A. Ellman, *J. Org. Chem.* **2003**, *69*, 1800.

55 J. Rouzaud, M.D. Jones, R. Raja, B.F.G. Johnson, J.M. Thomas, M.J. Duer, *Helv. Chim. Acta* **2003**, *86*, 1753.

56 M.D. Jones, R. Raja, J.M. Thomas, B.F.G. Johnson, D.W. Lewis, J. Rouzaud, K.D.M. Harris, *Angew. Chem. Int. Ed.* **2003**, *42*, 4326.